ATLAS
of the
BRITISH FLORA

INCLUDING

DISTRIBUTION MAPS

OF

1700 FLOWERING PLANTS AND FERNS

ATLAS
of the
BRITISH FLORA

Edited by
F. H. PERRING and S. M. WALTERS

Published for the
BOTANICAL SOCIETY OF THE BRITISH ISLES
by
EP PUBLISHING LIMITED
1976

This edition first published 1976 by EP Publishing Ltd
East Ardsley, Wakefield, England

First edition originally published by Thomas Nelson & Sons Ltd

Second Impression 1963
Second Edition 1976

ISBN 0 7158 1199 1

TEXT PRINTED IN BEMBO

MADE AND PRINTED IN GREAT BRITAIN BY THE SCOLAR PRESS LIMITED, ILKLEY

Contents

Foreword

A FEATURE OF MANY FLORAS, including some which have long been in general use in Britain, is a reference under the description of each species to its geographical distribution. Such references can only be in general terms, since it is not the purpose of an ordinary descriptive Flora to provide detailed distributional data about every species. Nevertheless, the need for factual information about the distribution of plants in Britain was recognised long ago by H. C. Watson, who was the first to undertake a scientific study of the subject—a study which occupied his attention for more than forty years, and culminated in the production of two major works, *Cybele Britannica* and *Topographical Botany*, a second edition of which appeared in 1883, two years after the author's death. The vice-comital system which Watson finally adopted as a method for recording the distribution of species in this country is familiar to British botanists and has long been in general use.

Watson did not deal with Ireland, but an essentially similar scheme of vice-county divisions for that country was developed by Praeger, who gives in *Irish Topographical Botany* a carefully prepared and reliable record of the distribution of plants in Ireland.

These topographical statistics both for Britain and Ireland are conveniently summarised in *The Comital Flora of the British Isles* by Druce, and allowing for corrections and additions, they have frequently served as a basis for the construction of maps to illustrate species distribution as well as for more general studies in the phytogeography of the British Isles.

Mapping on a vice-county basis, however, may and frequently does obscure important distributional features which become apparent only when localised records are used; and at a conference on the distribution of British plants held by the Botanical Society of the British Isles in 1950, it became evident that an advance on the Watsonian vice-county system was not only desirable but long overdue. To ensure a greater measure of detail and greater scientific accuracy in the production of maps of British plants, it was suggested by Professor Clapham that they should be based on the 10 km. square of the national grid as the unit area, a unit very much smaller than the average vice-county.

This proposal was welcomed and it was decided to refer the matter to a special "Maps Committee" which would consider ways and means. After making detailed investigations requiring several years to complete, an agreed scheme was officially launched in April 1954, and it requires no words of mine to emphasise the magnitude of the task to be undertaken—a task now successfully accomplished. Within the space of five full summers records have been obtained of the plants occurring in approximately 3,500 squares covering the British Isles. Tribute must be paid to the officials for their sustained efforts and enthusiasm in carrying out the plan and to the large body of workers in the field for their active co-operation. For the first time we have before us a comprehensive series of maps illustrating in detail the distribution of plants in Great Britain and Ireland—maps which I, personally, would have greatly welcomed when engaged in my own studies in British phytogeography.

The Atlas of the British Flora takes its place alongside Hulten's *Atlas över Kärlväxterna i Norden* published in 1950, and will be invaluable to botanists, professional and amateur alike, who are in any way interested in the study of our native flora. It will make a special appeal to the ecologist as well as to the taxonomist and plant-geographer, and in view of the possible significance of polyploidy in phytogeography, the cytogeneticist may turn over its pages not without interest. Examination of the maps reveals a surprising number of distribution patterns, some of which recall Watson's "Types of Distribution", but the interpretation of the range of any particular species requires an investigation of all the factors which play a part in determining its range. Marked discontinuities in distribution are exhibited by some species, the explanation of which may be found by taking into account what is known of their history in late-glacial and post-glacial times, as detailed by Godwin in *The History of the British Flora*. Present and past distributions can thus be compared, but whatever the interpretation of different distribution patterns, it is hoped that the maps themselves, whether of common or rare species, may serve as an impetus to further autecological studies of our native plants. Be that as it may, the publication of the Atlas is an important landmark in the scientific progress of British Botany.

Dinnet, Aberdeenshire
March 1961

J. R. Matthews

FOREWORD TO THE SECOND EDITION

THE FIRST EDITION of this *Atlas* has now been out of print for many years and a new generation of botanists has grown up to whom it is no longer easily available. For their sakes the Botanical Society of the British Isles was glad to accept the present publisher's offer to reprint this work.

As the offer coincided with the virtual completion of a manuscript for a *Red Data Book of British Vascular Plants* which was the result of 7 years' survey of the detailed distribution of about 350 of our rarest species, it was decided to update the maps of these species, to increase the accuracy of the *Atlas* and to have available in print the data on which the *Red Data Book* was based. In addition any errors in the first edition brought to our attention were removed and any other observed blemishes corrected. Any maps which have been altered carry a cross (+) to the right of the species number in the inset.

The maps of rare species which have been altered are virtually all those natives or well-established introductions which have been recorded in 15 or fewer 10 km squares in Great Britain (excluding the Channel Isles) since 1930. Many of these rare species maps show significant increases in the number of records, notably amongst the more difficult groups like grasses and sedges, where the changes in *Eriophorum gracile* and *Agropyron donianum* are outstanding examples. The *Atlas* also shows that other species have declined. *Carex depauperata* may now be only in one locality in England; *Otanthus maritimus* is probably extinct, though still in one Irish station, and *Spiranthes aestivalis* and *Bromus interruptus* can no longer be found in the British Isles. This increase in information is a tribute to the many members of the BSBI whose enthusiasm for field work has never ceased. They have helped to provide a knowledge of the distribution of rare species which is essential to a sound basis for conservation and the fact that 35 species previously thought to be rare have been removed from that category as a result of their work means that efforts can now be concentrated on those species which are truly endangered.

Change is almost certainly also taking place amongst the less rare species. Had time and finance permitted we would like to have prepared a completely revised edition incorporating all the additional records since the first edition went to press nearly 15 years ago. Had this been undertaken, however, the complete remaking of the book from new original maps would have meant a published price beyond the private purses of most botanists, so that the work would still have been inaccessible to them.

For those still only requiring a general picture of the distribution of species in Great Britain and Ireland the reprinted maps are largely adequate, and the Biological Records Centre can provide more up-to-date maps for those specialist taxonomists and ecologists who require them. In the course of time revised maps will be published in special volumes covering particular groups in more detail than the present work: a volume on the Pteridophyta is in preparation.

Cost has also prevented the inclusion in this

edition of all the 12 transparent overlays of physical and other features which were a feature of the first. Only two are available: those for altitude and chalk and limestone outcrops. Each of these does, however, show one 100 km square divided into 100 10 km squares so the user can refer any record to the correct square. The only essential information omitted is the overlay showing under-recorded squares. These are now given as a list on page 408.

This revision would not have been possible without the combined support of the BSBI membership, particularly its County Recorders, and to each and every one who has contributed I am deeply grateful. But their efforts would have been pointless without the dedicated concentration of the two ladies who have successively been at the receiving end of the data —first Miss M. N. Hamilton and more recently Miss L. Farrell. I hope this printed thanks will partly atone for my irascible moments born of my impatience to see so much wonderful co-operation come to fruition as rapidly as possible.

F. H. Perring

Biological Records Centre
Institute of Terrestrial Ecology
Monks Wood,
Huntingdon.
10 November 1975

Introduction

HISTORY OF THE MAPPING OF PLANT DISTRIBUTION

EARLY HISTORY

As long ago as 1836, Hewett Cottrell Watson, an English amateur botanist who pioneered the study of the distribution of British vascular plants, had seen the possibility of preparing a series of plant distribution maps. In an article on the construction of maps for illustrating the distribution of plants he wrote as follows:

To represent the distribution of individual forms or species let us first imagine a common geographical map, in outline, of such dimensions as would render it possible to mark every locality for any given species, by some sign, or spot of colour, covering a corresponding space on the map. This would give an exact picture of topographical distribution; but, as it would require to be made on the scale of at least a yard to the mile, it is obviously out of the question. With less precision . . . we might greatly reduce the scale by indicating all localities within certain distances of each other as single ones.

From this passage it is clear that Watson would have been a wholehearted supporter of the work which has gone on during the last eight years to produce this Atlas, and that he might have attempted such a work himself had he lived long enough and had the circumstances been more favourable. As it was, during his lifetime he gradually divided Great Britain into smaller and smaller recording units: from the six of *Outlines of the Distribution of British Plants* (1832) to the 112 vice-counties of *Topographical Botany* (1873). In 1843, in the third edition of *The Geographical Distribution of British Plants* we find what might be called "provincial distribution maps" for thirty-nine native British species, together with the data on which the maps are based. At this period Watson had divided Britain into eighteen provinces and for each species he has a diagram in which "by omitting the figures from those squares which correspond to districts within which the species has not been ascertained to grow, a tolerably exact notion of its topographical area may be instantly conveyed to the eye of the reader". He goes on to say that botanists who wished might colour the maps following the data given in the text, and that had not expense ruled it out he would have produced a printed map for each of the 1,200 species listed.

Finally, to appreciate how completely Watson had grasped the essential points to be borne in mind in recording and mapping plant distributions, one should turn to the introduction to *Cybele Britannica* (1847). Here he distinguishes two "circumstances of distribution—first the extent of geographical surface over which the plant is spread", for which he suggests the term *Area* should be used, "and secondly the greater or less frequency of the species within that space", which he suggests might be called *Census*. He sees clearly that, by making the recording units sufficiently small, some sort of measure of the second attribute of distribution can be obtained. Thus:

Tormentilla officinalis [i.e. *Potentilla erecta*] and *Hypericum pulchrum* are found in every province; and it is not improbable that they would equally be found in every county, if looked for. But if we could subdivide all the counties into sections of a square mile each, the *Tormentilla* would assuredly be found in many more of these square mile sections than would the *Hypericum*.

Now, well over 100 years later, we can test the validity of Watson's hypothesis, if for "county" we read "vice-county", and for "square mile" we read "a 10-kilometre square", by comparing the maps of the *Potentilla* and the *Hypericum* on pages 122 and 59 respectively. He might equally well have referred to the two species of nettle, *Urtica dioica* and *U. urens*; both are recorded from every county in the British Isles and so have similar "Areas", but a glance at page 182 will show that their "Census" is very different.

Before he died in 1881 Watson had therefore prepared the way for the study of the distribution of British plants on a unit area basis. Unfortunately, he had no successor with the same interest and ability, and as late as the early 1950's British botanists were still using the vice-county system, as for example in the accounts in the "Biological Flora" series published in the *Journal of Ecology*.

Development of techniques of mapping plant distribution in the late nineteenth and the first half of the twentieth century nearly all took place on the continent of Europe. One of the most outstanding contributions was made by Hermann Hoffmann, who between 1860 and 1880 published a series of papers which contain not only much the earliest published examples of the use of dot distribution maps, both on a local and on a European scale, but also an example of the use of the grid square method about half a century earlier than any other that has been traced. How far Hoffmann was influenced by Watson's earlier works we may never know: there is no indication in his writings that he had ever heard of Watson. However, Watson's work was widely known to Continental plant-geographers by the middle of the nineteenth century and there is nothing improbable in Hoffmann having read, for example, the introduction to *Cybele* or *Remarks on the Distribution of British Plants* which was translated into German in 1837. The first of Hoffmann's published works of interest was a paper in 1860 in which he deals with two calcicolous species *Prunella grandiflora* and *Dianthus carthusianorum*. Maps of these two species which accompany this paper have a good claim to be the earliest published examples of dot distribution maps. Two further papers (1867 and 1869) provide many examples of Hoffmann's local mapping, and also, quite unexpectedly, what must be the first European dot-map, that of *Gentiana verna*, based on data from De Candolle's *Prodromus*. Finally, Hoffmann invented a simple grid method of recording and used it throughout his *Flora des Mittelrheingebietes* (1879).

Hoffmann's example fell on more fertile ground than had

Watson's, and from his time dot-mapping in Europe became a more or less continuously expanding activity. Already, by 1879, Ihne had published a dot-map using several symbols to distinguish records made in different years and therefore to indicate change of range. In Germany others began to follow the established methods and the influence gradually spread to Scandinavia and Finland—but not to the British Isles. There appear to be no published distribution maps of any British plant in the nineteenth century after Watson's originals, except the classic of Miller Christy on the oxlip (*Primula elatior*), first published in 1884. It was not until the beginning of this century that interest revived, when we find that the "pioneer" was Praeger (1902 and 1909) for the Irish flora only. It was, in fact, 1917 before a distribution map for the British Isles as a whole appeared, when several were published by Moss in his *Cambridge British Flora*. Further maps were published by Stapf (1917), by Sir Edward Salisbury in his deservedly famous study of the East Anglian flora (1932), and by Professor J. R. Matthews in his important paper on the *Geographical relationships of the British flora* (1937), though none of these used dot-maps.

THE PRODUCTION OF DOT-MAPS IN BRITAIN

The credit for the publication of the first dot-map of a British plant must go to Good who included one in his study of the Lizard Orchid, *Himantoglossum hircinum* (1936). But it is not until after the war of 1939–45 and the advent of the "Biological Flora" and the "New Naturalist" series that dot-distribution maps may be said to have become a standard requirement of British ecologists and taxonomists.

Could this need be met from the data available in 1950? For the rare species the answer was probably "yes". At a Conference of the Botanical Society of the British Isles that year on *Aims and Methods in the Study of the Distribution of British Plants* (published in 1951) a number of speakers produced evidence to show that data for rare species were available and how much more informative were dot-maps than those based on vice-comital distributions. It was also made clear that for common species the answer would have to be "No"; for many parts of Scotland, Ireland and Wales no other data were available than that a species does or does not occur in a particular vice-county. During this conference the difficulties and possibilities of mapping the British flora were fully discussed. Notice was taken of recent work on the Continent, notably that of Hultén, whose magnificent atlas of maps showing the distribution of vascular plants in north-west Europe, *Atlas över Kärlväxterna i Norden* (1950), was published shortly after the conference finished. The work of the Instituut voor det Vegetatie-Onderzoek in Nederland (I.V.O.N.), which was carrying out a survey of the distribution of vascular plants in Holland using a basic square of approximately 5×4 kilometres, was also carefully considered.

The final paper of the conference was read by Professor A. R. Clapham. In the course of it he suggested that the Botanical Society of the British Isles should take steps to ensure that before long we had a set of distribution maps of British species; that the maps, when produced, should be comprehensive and accurate; that they should also be of small scale (1 in 10 million) so that they could be printed four on each page of a single volume; that the unit area should be the 10-kilometre grid square; that overlays should show ground over 1,000 and 2,000 feet and also basic substrata; and that certain further information about infraspecific units and certain historical data should also be supplied if possible. After these suggestions had been discussed the following proposal was made:

> That the conference expresses its wish that the Council of the B.S.B.I. should discuss the possibility of preparing and producing a series of maps of the British Flora and that if they consider it practicable they should enlist the co-operation of such other bodies as may be desirable. . . .

This proposal was carried with acclamation.

THE LAUNCHING OF THE MAPS SCHEME

The Council of the B.S.B.I. met in May 1950 and discussed the proposal. They appointed a committee to consider the part the Society might play in the project. This Maps Committee consisted of:

Professor A. R. Clapham
Mr J. E. Lousley
Mr E. Milne-Redhead
Professor T. G. Tutin
Mr E. C. Wallace
Dr E. F. Warburg

and at its first meeting Mr Lousley was elected Chairman and Professor Clapham, Secretary. After much careful investigation of the practical difficulties the Committee felt, early in 1953, that, given adequate financial support, the project to prepare and publish an atlas of British plants was a practical one, and accordingly an approach was made to the Nuffield Foundation for a grant for a five years' project. This approach was successful and the offer of a grant of £10,000 for the scheme was gratefully accepted by the Council in December 1953. Valuable financial assistance was also given by the Nature Conservancy who made a grant of £1,000 per annum for four years (commencing in April 1955) and also agreed to meet the cost of the Powers-Samas punched-card recording system adopted by the Committee for the incorporation of the data and the mechanised map production, on the understanding that, at the end of the Scheme, the Conservancy would take over the machinery and the punched cards as the basis of a permanent recording system. Through the kind co-operation of the Professor of Botany, Professor G. E. Briggs, and the General Board of the University, accommodation was made available to the organisation in the Botany School, Cambridge. The following societies have made grants to the Scheme since it commenced: The Botanical Society of the British Isles, The British Ecological Society, and the Royal Irish Academy.

Now that the future of the project was assured the Maps Committee was reformed and enlarged, with powers to co-opt representatives of other bodies, and the following also served on the committee or its sub-committees during the course of the scheme: R. W. David, Dr J. G. Dony, J. C. Gardiner, Professor

H. Godwin, R. A. Graham, Professor J. Heslop Harrison, R. D. Meikle, Dr F. H. Perring, Dr C. D. Pigott, Professor M. E. D. Poore, J. G. Skellam and other representatives of the Nature Conservancy, Dr W. T. Stearn, E. L. Swann, Dr S. M. Walters, Professor D. A. Webb. To run the scheme the following appointments were made:

**Director:*	S. M. Walters from 6 April 1954
Senior Worker:	F. H. Perring from 1 October 1954
Secretary:	Miss A. Matthews from 6 April 1954
Punched-card Operator:	Mrs S. Fincham from 1 October 1954

* On 31 March 1959 Dr Walters' term of office came to an end and the scheme has been directed by Dr Perring since that date.

The "Maps Office" thus came into being just before the scheme was officially launched at the Conference of the Botanical Society on 9 April 1954, exactly four years after its creation had been first suggested.

In addition to the central organisation in Cambridge regional offices were set up in Scotland and Ireland, and part-time paid assistants organised the enrolment of volunteers and the collection and refereeing of data before passing the information on to headquarters. Mrs M. E. D. Poore served as the Scottish Regional officer from 1954–6, and was followed by Miss E. Beattie who carried on until the office was closed in 1959. Accommodation for the office was kindly made available by the Scottish Nature Conservancy. In Ireland, Professor D. A. Webb set up a small regional office to assemble the data for Eire, whilst Professor J. Heslop Harrison co-ordinated workers in Northern Ireland, with the able assistance of Miss M. P. H. Kertland. The National Museum of Wales in Cardiff has provided advice and assistance with Welsh records throughout the Scheme.

OUTLINE OF THE METHOD USED IN PREPARING THE MAPS

THE RECORDING UNIT

The basis of the scheme is to indicate by means of a symbol the presence of each species of vascular plant in every 10-kilometre square of the Ordnance Survey National Grid in which it occurs, thus producing a distribution map of each. The National Grid covers Great Britain and the eastern half of Ireland, and is represented on all recent British Ordnance Survey Maps of a scale between 1:1,000,000 and 1:25,000. The maps in the 1:25,000 or $2\frac{1}{2}$ inches to 1 mile scale are issued in sheets which exactly correspond to the 10-kilometre square, and these were recommended for field workers, being available for almost the whole of Great Britain. In Ireland not only was the National Grid missing from half the country, but no large-scale maps were available even for the gridded half. The problem of producing gridded maps for Ireland was presented to Professor Webb, who solved it elegantly and published the skeleton of his agony in a short paper in 1955. He prepared originals on the $\frac{1}{2}$ inch to 1 mile Irish Ordnance Survey Maps and copies were made for use in Cambridge and Northern Ireland. Volunteer recorders in Ireland have had their own maps marked with the grid for the area they have been visiting, or have been requested to limit their recording of one list of plants to within a mile radius of any one locality, the list being conventionally gridded when sent in. The grid system has been strictly adhered to with a few exceptions:

(i) Some coastal squares contain such a small area of land that a round dot on the final map would appear to fall in the sea. In these cases the records for the square concerned have been incorporated with those of an adjacent square.

(ii) The records for the Orkneys and Shetlands have been conventionally gridded so that they are placed in an inset. This is more convenient for publication and also avoids the difficulty of duplicated grid references: N30, 31, 40, 41 and 42 become 57, 58, 67, 68 and 69.

(iii) The records for the Channel Isles have also been conventionally gridded so that they appear in an inset: 90/15 Guernsey, 90/25 Herm, 90/35 Sark, 90/42 Jersey, 90/48 Alderney. All records for those islands are placed in these squares.

N.B. Records for N.E. Caithness (v.-c. 109), S. Orkneys (v.-c. 111) and Alderney in the Channel Isles appear twice on the map, once in the true position and once in the conventional one.

THE RECORD CARDS

Regional

The first task of the scheme was to produce record cards which would enable data to be collected quickly and accurately in the field, whilst at the same time enabling the processing of the data to be carried out efficiently at headquarters. To this end seven Regional Record Cards were designed, each an 8×5-inch card printed with six columns (five in Ireland) on each side, carrying up to 900 species which were perhaps the most likely to be recorded in each region (Fig. 1). The scientific names of the species are conventionally abbreviated to five letters of the genus and three of the species (with few exceptions where this would be ambiguous). Each species has a code number—the species number—used to identify the species in the punched card system. In addition space is provided at the head of the card for grid reference, locality, habitat, date, altitude, vice-county and the recorder's personal code number. At the bottom of the reverse side space is left for "Other Species". Such cards had been pioneered by I.V.O.N. in Holland, and in England by the Cambridge Natural History Society. The seven regions were:

(1) South-west (v.-cs 1–14 and 22).
(2) South-east (v.-cs 15–21, 24–32).
(3) Midlands (v.-cs 23, 33, 34, 36–40, 53–65).
(4) North (v.-cs 66–71 and Scotland south of the Clyde-Forth Line).
(5) Scotland (north of the Clyde-Forth line).
(6) Wales.
(7) Ireland.

The nomenclature used was that of the *Flora of the British Isles* by Clapham, Tutin and Warburg (1952), though the Irish card differs in some respects. In addition a card was produced based on the nomenclature used by Bentham and Hooker (ed. vii, 1924) for the benefit of those who had not adopted the new names. The grasses and sedges were largely excluded from this card. A ninth card was called the "Common Species Card" and carried the English and scientific names of 120 widespread species, which were selected as being readily recognisable and having no taxonomic complications, so that it might be used by school children and other beginners. When the scheme started we were uncertain of the response which would come from volunteers and we felt these two additional cards might be important. In practice, however, their use has been very limited in the field, and because their design differed from the regional cards they proved to be very inconvenient in the office when data had to be transferred from one type of card to another.

Individual

In addition field recorders were supplied with individual record cards (Fig. 2). These are $4\frac{3}{4} \times 2$ inches with spaces reserved for the following information: species, locality, habitat, grid reference, date, vice-county, altitude, status, collector's name, determiner's name (if other than the collector), and source (whether field record, or from literature or herbarium). There were three colours of card to be used for the three main sources:

(i) Buff—field records
(ii) Yellow—herbarium records
(iii) Pink—literature records.

The use of these cards is described on page xviii. The buff cards are the same as those used to incorporate the field records; they are dual-purpose cards written on and punched as individual records, punched only as field records.

COLLECTION OF THE FIELD RECORDS

One of the essential requirements of a scheme of this kind is that it should cover the country as evenly as possible. Our main objective therefore has been to try to acquire lists of species from each of the approximately 3,500 10-kilometre squares which cover the British Isles. To this end appeals for help were made in 1954 to all members of the Botanical Society, the British Ecological Society, local Natural History Societies, Field Study Centres, University Botany Departments, and to the public in general through articles and letters to national and local newspapers and scientific periodicals. As far as possible particular squares were allocated to individuals or groups in an attempt to avoid duplication. Cards and instructions were sent to all who volunteered with a request for them to send in their data at the end of each season. Further stimulus was given from time to time by publishing, or circulating to selected centres of activity, situation maps which showed squares from which records had already been received or were expected. Gradually the cover of records became more complete. Botanical Society field meetings were organised mainly to visit underworked areas, and in this work the Council for the Study of the Scottish Flora joined with enthusiasm. During the last two or three years of the scheme, when some information had been received from almost every square, volunteers were asked to make a copy of our master card (see page xviii) for a square, and send only additions to us at the end of each year. New people were asked to concentrate on underworked squares whereas they had been directed to unworked squares in the past.

Enormous credit is due to British botanists for their achievement in the five effective full seasons which were available for them to complete the task. Over 3,000 volunteered, though only about 1,500 actually sent in returns and the great bulk of the records were contributed by about 250. Nevertheless during that period they visited all but seven of the 3,500 squares, and failed to record more than 150 species in only 137 squares. The total number of records made was about $1\frac{1}{2}$ million—an average of about 400 records per square (Fig. 3). The whole effort was almost entirely voluntary; people gave freely of their time and went to considerable expense to pay their own travelling and hotel bills, and without this service the Atlas could never have been produced.

When the scheme started some data were already available in the form of published lists and others in private notebooks. It was our intention to concentrate on lists made from 1950 onwards and to distinguish between these and earlier records by the use of symbols on the published maps. However, as the scheme developed it became clear that 1930 should be used as the basic date-line, for a number of reasons:

(i) The botanical survey of Dorset made by Professor Good was started in 1931. As Professor Good's survey had been intensive and thorough, there seemed no point in repeating his work when he had generously agreed to make his, largely unpublished, basic data available to us.
(ii) The study of the flora of the Inner and Outer Hebrides by Professor J. W. Heslop Harrison *et al.* had begun about 1935 and had continued ever since. The work of A. J. Wilmott *et al.* in the Outer Hebrides also began in the 1930s. There seemed no justification for repeating this work in one of the most inaccessible parts of Britain, and one where the flora changes least, when so many other areas had never been visited.
(iii) The second supplement to *Topographical Botany* was published in 1929–1930 and the *Comital Flora* by G. C. Druce in 1932. These combined to provide an effective summary of knowledge up to about 1930, and by extracting the reports of the *Botanical Exchange Club* and the *Botanical Society of the British Isles* from 1930 onwards we hoped to be able to locate most vice-county records.

It was also fortunate that at the time the scheme commenced a number of botanists were engaged in writing Floras of particular counties, and all generously agreed to make their invaluable data available to us—in most cases producing cards for each square in the county themselves and in others lending us the data in a readily assimilable form. Others have subsequently begun to work on their Floras and have contributed likewise. We wish to acknowledge with sincere thanks the contribution of all those who have supplied more or less complete county lists (see Appendix III, page 419).

In many cases we know that the county organisers did not work

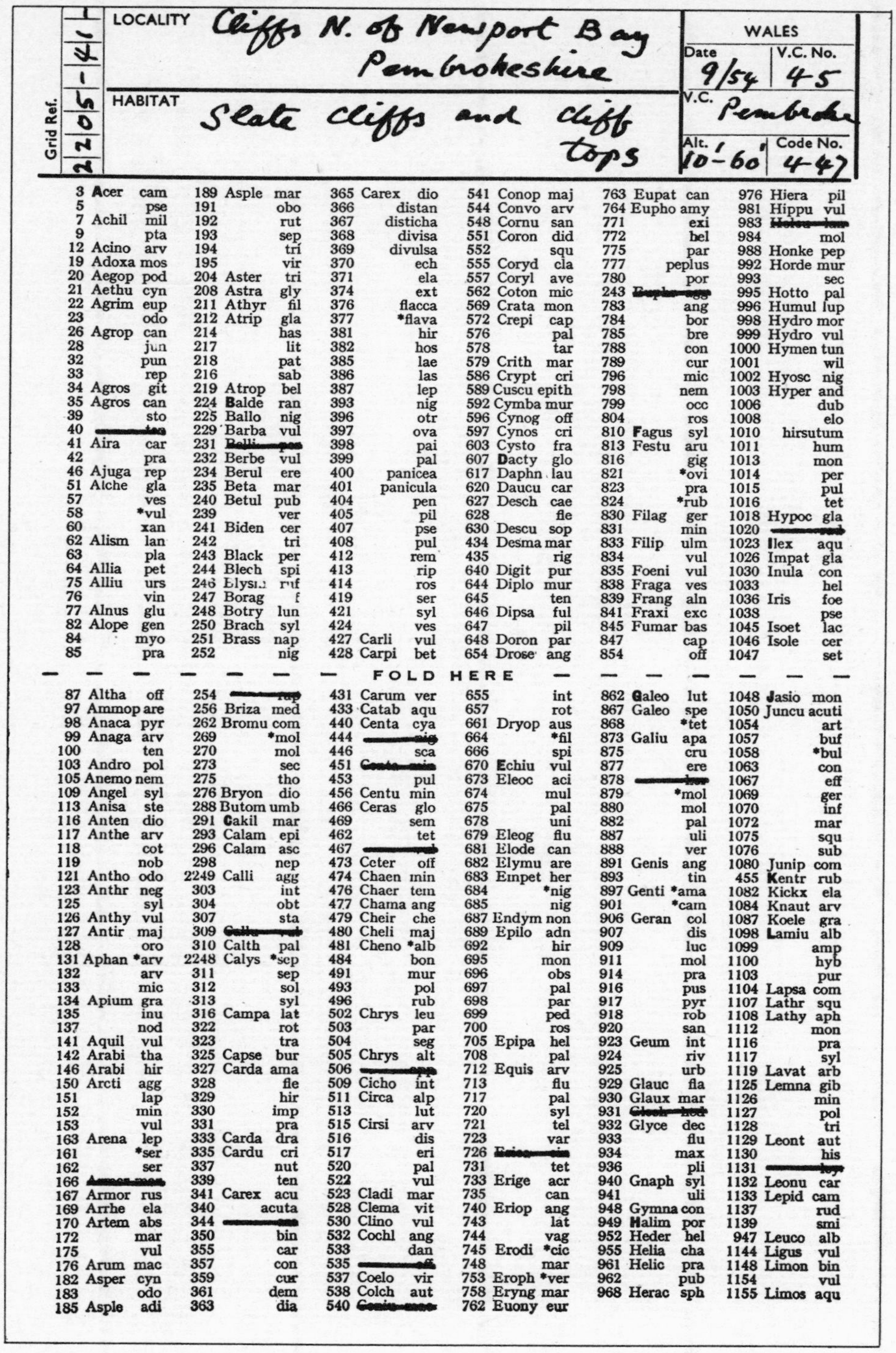

Grid Ref.	LOCALITY	WALES
2205-41	Cliffs N. of Newport Bay Pembrokeshire	Date 9/54 · V.C. No. 45
	HABITAT Slate cliffs and cliff tops	V.C. Pembroke
		Alt. 10′-60′ · Code No. 4-47

3	Acer	cam	189	Asple	mar	365	Carex	dio	541	Conop	maj	763	Eupat	can	976	Hiera	pil
5		pse	191		obo	366		distan	544	Convo	arv	764	Eupho	amy	981	Hippu	vul
7	Achil	mil	192		rut	367		disticha	548	Cornu	san	771		exi	983	~~Holcu~~	~~lan~~
9		pta	193		sep	368		divisa	551	Coron	did	772		bel	984		mol
12	Acino	arv	194		tri	369		divulsa	552		squ	775		par	988	Honke	pep
19	Adoxa	mos	195		vir	370		ech	555	Coryd	cla	777		peplus	992	Horde	mur
20	Aegop	pod	204	Aster	tri	371		ela	557	Coryl	ave	780		por	993		sec
21	Aethu	cyn	208	Astra	gly	374		ext	562	Coton	mic	243	~~Eupho~~	~~agg~~	995	Hotto	pal
22	Agrim	eup	211	Athyr	fil	376		flacca	569	Crata	mon	783		ang	996	Humul	lup
23		odo	212	Atrip	gla	377		*flava	572	Crepi	cap	784		bor	998	Hydro	mor
26	Agrop	can	214		has	381		hir	576		pal	785		bre	999	Hydro	vul
28		jun	217		lit	382		hos	578		tar	788		con	1000	Hymen	tun
32		pun	218		pat	385		lae	579	Crith	mar	789		cur	1001		wil
33		rep	216		sab	386		las	586	Crypt	cri	796		mic	1002	Hyosc	nig
34	Agros	git	219	Atrop	bel	387		lep	589	Cuscu	epith	798		nem	1003	Hyper	and
35	Agros	can	224	Balde	ran	393		nig	592	Cymba	mur	799		occ	1006		dub
39		sto	225	Ballo	nig	396		otr	596	Cynog	off	804		ros	1008		elo
40		~~[illegible]~~	229	Barba	vul	397		ova	597	Cynos	cri	810	Fagus	syl	1010		hirsutum
41	Aira	car	231	~~Belli~~	~~per~~	398		pai	603	Cysto	fra	813	Festu	aru	1011		hum
42		pra	232	Berbe	vul	399		pal	607	Dacty	glo	816		gig	1013		mon
46	Ajuga	rep	234	Berul	ere	400		panicea	617	Daphn	lau	821		*ovi	1014		per
51	Alche	gla	235	Beta	mar	401		panicula	620	Daucu	car	823		pra	1015		pul
57		ves	240	Betul	pub	404		pen	627	Desch	cae	824		*rub	1016		tet
58		*vul	239		ver	405		pil	628		fle	830	Filag	ger	1018	Hypoc	gla
60		xan	241	Biden	cer	407		pse	630	Descu	sop	831		min	1020		~~[illegible]~~
62	Alism	lan	242		tri	408		pul	434	Desma	mar	833	Filip	ulm	1023	Ilex	aqu
63		pla	243	Black	per	412		rem	435		rig	834		vul	1026	Impat	gla
64	Allia	pet	244	Blech	spi	413		rip	640	Digit	pur	835	Foeni	vul	1030	Inula	con
75	Alliu	urs	246	Blysm	ruf	414		ros	644	Diplo	mur	838	Fraga	ves	1033		hel
76		vin	247	Borag	f	419		ser	645		ten	839	Frang	aln	1036	Iris	foe
77	Alnus	glu	248	Botry	lun	421		syl	646	Dipsa	ful	841	Fraxi	exc	1038		pse
82	Alope	gen	250	Brach	syl	424		ves	647		pil	845	Fumar	bas	1045	Isoet	lac
84		myo	251	Brass	nap	427	Carli	vul	648	Doron	par	847		cap	1046	Isole	cer
85		pra	252		nig	428	Carpi	bet	654	Drose	ang	854		off	1047		set

— — — — — — — FOLD HERE — — — — — — —

87	Altha	off	254		~~[illegible]~~	431	Carum	ver	655		int	862	Galeo	lut	1048	Jasio	mon
97	Ammop	are	256	Briza	med	433	Catab	aqu	657		rot	867	Galeo	spe	1050	Juncu	acuti
98	Anaca	pyr	262	Bromu	com	440	Centa	cya	661	Dryop	aus	868		*tet	1054		art
99	Anaga	arv	269		*mol	444		~~[illegible]~~	664		*fil	873	Galiu	apa	1057		buf
100		ten	270		mol	446		sca	666		spi	875		cru	1058		*bul
103	Andro	pol	273		sec	451	~~Centa~~	~~min~~	670	Echiu	vul	877		ere	1063		con
105	Anemo	nem	275		tho	453		pul	673	Eleoc	aci	878		~~[illegible]~~	1067		eff
109	Angel	syl	276	Bryon	dio	456	Centu	min	674		mul	879		*mol	1069		ger
113	Anisa	ste	288	Butom	umb	466	Ceras	glo	675		pal	880		mol	1070		inf
116	Anten	dio	291	Cakil	mar	469		sem	678		uni	882		pal	1072		mar
117	Anthe	arv	293	Calam	epi	462		tet	679	Eleog	flu	887		uli	1075		squ
118		cot	296	Calam	asc	467		~~[illegible]~~	681	Elode	can	888		ver	1076		sub
119		nob	298		nep	473	Ceter	off	682	Elymu	are	891	Genis	ang	1080	Junip	com
121	Antho	odo	2249	Calli	agg	474	Chaen	min	683	Empet	her	893		tin	455	Kentr	rub
123	Anthr	neg	303		int	476	Chaer	tem	684		*nig	897	Genti	*ama	1082	Kickx	ela
125		syl	304		obt	477	Chama	ang	685		nig	901		*cam	1084	Knaut	arv
126	Anthy	vul	307		sta	479	Cheir	che	687	Endym	non	906	Geran	col	1087	Koele	gra
127	Antir	maj	309	~~[illegible]~~	~~[illegible]~~	480	Cheli	maj	689	Epilo	adn	907		dis	1098	Lamiu	alb
128		oro	310	Calth	pal	481	Cheno	*alb	692		hir	909		luc	1099		amp
131	Aphan	*arv	2248	Calys	*sep	484		bon	695		mon	911		mol	1100		hyb
132		arv	311		sep	491		mur	696		obs	914		pra	1103		pur
133		mic	312		sol	493		pol	697		pal	916		pus	1104	Lapsa	com
134	Apium	gra	313		syl	496		rub	698		par	917		pyr	1107	Lathr	squ
135		inu	316	Campa	lat	502	Chrys	leu	699		ped	918		rob	1108	Lathy	aph
137		nod	322		rot	503		par	700		ros	920		san	1112		mon
141	Aquil	vul	323		tra	504		seg	705	Epipa	hel	923	Geum	int	1116		pra
142	Arabi	tha	325	Capse	bur	505	Chrys	alt	708		pal	924		riv	1117		syl
146	Arabi	hir	327	Carda	ama	506		~~[illegible]~~	712	Equis	arv	925		urb	1119	Lavat	arb
150	Arcti	agg	328		fle	509	Cicho	int	713		flu	929	Glauc	fla	1125	Lemna	gib
151		lap	329		hir	511	Circa	alp	717		pal	930	Glaux	mar	1126		min
152		min	330		imp	513		lut	720		syl	931	~~[illegible]~~	~~hed~~	1127		pol
153		vul	331		pra	515	Cirsi	arv	721		tel	932	Glyce	dec	1128		tri
163	Arena	lep	333	Carda	dra	516		dis	723		var	933		flu	1129	Leont	aut
161		*ser	335	Cardu	cri	517		eri	726	~~[illegible]~~	~~[illegible]~~	934		max	1130		his
162		ser	337		nut	520		pal	731		tet	936		pli	1131		~~[illegible]~~
166	~~[illegible]~~	~~[illegible]~~	339		ten	522		vul	733	Erige	acr	940	Gnaph	syl	1132	Leonu	car
167	Armor	rus	341	Carex	acu	523	Cladi	mar	735		can	941		uli	1133	Lepid	cam
169	Arrhe	ela	340		acuta	528	Clema	vit	740	Eriop	ang	948	Gymna	con	1137		rud
170	Artem	abs	344		~~[illegible]~~	530	Clino	vul	743		lat	949	Halim	por	1139		smi
172		mar	350		bin	532	Cochl	ang	744		vag	952	Heder	hel	947	Leuco	alb
175		vul	355		car	533		dan	745	Erodi	*cic	955	Helia	cha	1144	Ligus	vul
176	Arum	mac	357		con	535		~~off~~	748		mar	961	Helic	pra	1148	Limon	bin
182	Asper	cyn	359		cur	537	Coelo	vir	753	Eroph	*ver	962		pub	1154		vul
183		odo	361		dem	538	Colch	aut	758	Eryng	mar	968	Herac	sph	1155	Limos	aqu
185	Asple	adi	363		dia	540	~~[illegible]~~	~~[illegible]~~	762	Euony	eur						

Fig. 1. REGIONAL RECORD CARD

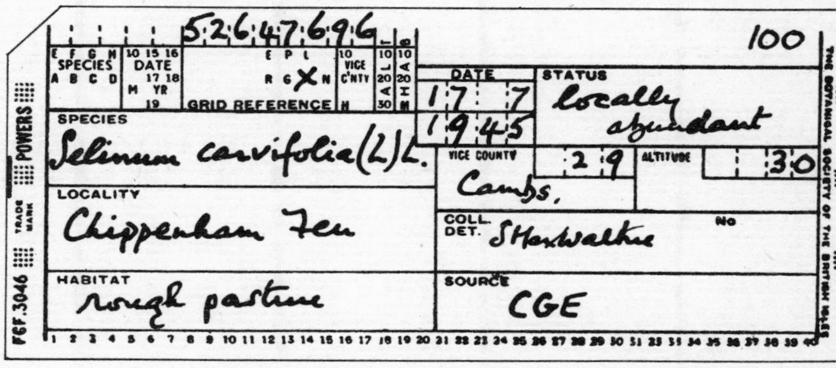

5:2:6:4:7:6:9:6 100

SPECIES E F G H A B C D · DATE 10 15 16 17 18 M YR 19 · GRID REFERENCE E P L R G X N H · 10 VICE C'NTY · 10 ALT 20 30 · 10 HAB 20 30

DATE 17 7 1945 · STATUS locally abundant

SPECIES Selinum carvifolia (L) L.

VICE COUNTY Cambs. 29 · ALTITUDE 30

LOCALITY Chippenham Fen

COLL. DET. S Max Walters · NO

HABITAT rough pasture

SOURCE CGE

POWERS TRADE MARK F6F.3046

THE BOTANICAL SOCIETY OF THE BRITISH ISLES

1 2 3 4 5 6 7 8 9 10 11 12 13 14 15 16 17 18 19 20 21 22 23 24 25 26 27 28 29 30 31 32 33 34 35 36 37 38 39 40

Fig. 2. INDIVIDUAL RECORD CARD

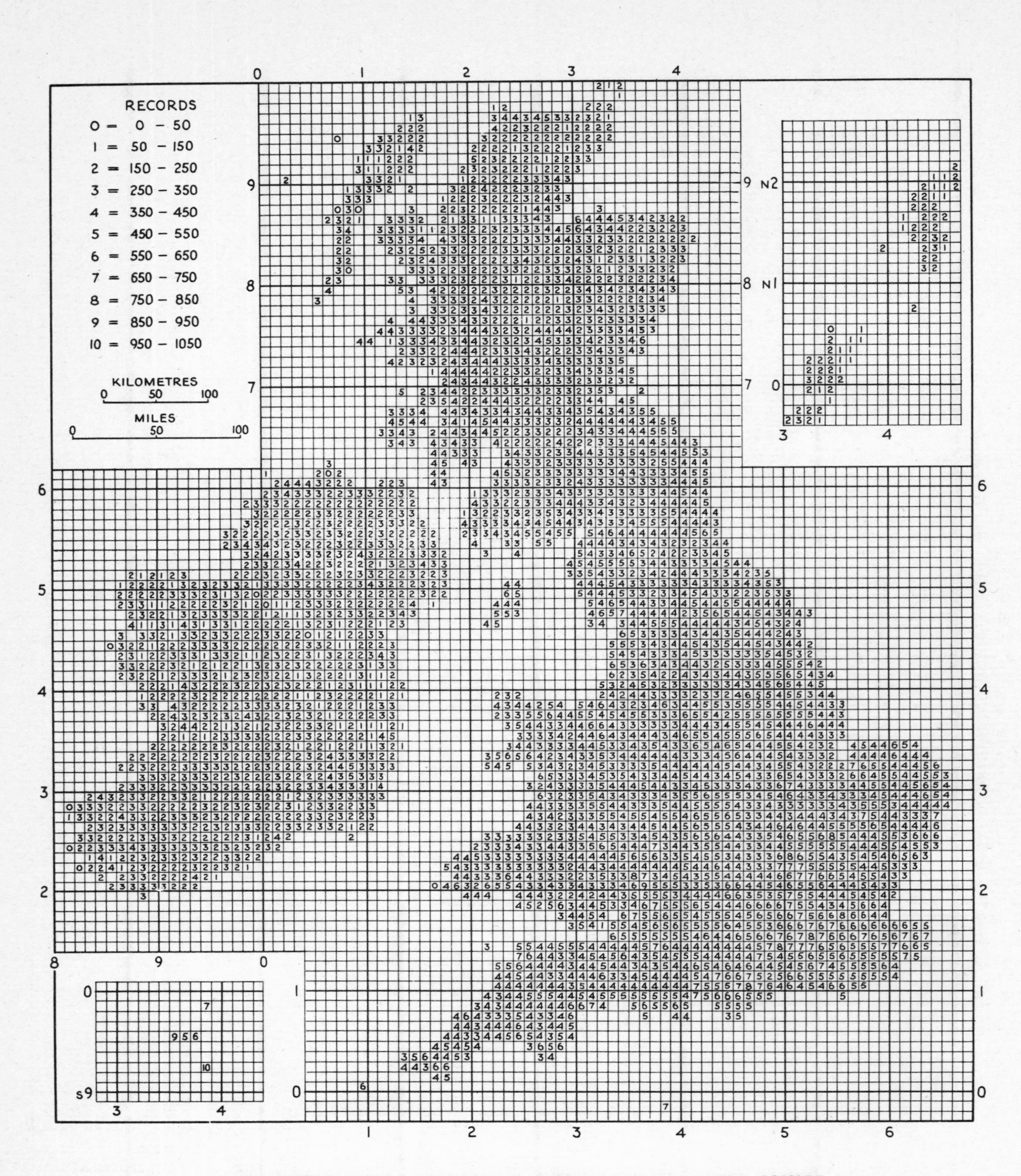

Fig. 3. RECORDS RECEIVED AND INCORPORATED FOR EACH 10 KM. SQUARE, INCLUDING PRE-1930 RECORDS BUT EXCLUDING INDIVIDUAL RECORDS

alone. That we never corresponded with their assistants directly may mean we are not aware of their help and cannot acknowledge their contribution personally. We must apologise if their names do not appear in the list of acknowledgements but we are not unmindful of the importance of their work. Many others have made contributions at least equal in quantity to those listed, but because their loyalties were not so restricted to particular counties it is less easy to acknowledge their work. We hope they will forgive this omission; it does not mean we are any the less grateful.

Frequency data

The list of species from a 10-kilometre square has been used only qualitatively in the production of the maps. At the beginning of the scheme it was realised that with a great many recorders over a short period it would be senseless to ask for frequency data; nevertheless we thought that it might be possible to get an estimate of rarity on the following basis:

> A species known or reasonably suspected to be restricted to one particular locality in the 10-kilometre square, the "locality" not exceeding 1-kilometre square, should be marked as rare.

This request was only complied with by those submitting complete county lists or by those working intensively particular squares; it was found to be impractical over large areas of the country where squares were only visited once by a party for a day or less, so this distinction between the records has not been used in compiling the maps.

THE EVENNESS OF THE SURVEY

Despite the vigorous attempts made, the survey has not been as even as we would have liked: some parts of the British Isles have been less well recorded than others. An impression of this uneven coverage can be obtained by studying the map (Fig. 3) showing by hundreds the records received from each square. Thus, when analysing a distribution map, apparent areas of rarity or absence for which no plant geographical reason appears likely should be regarded with caution, and the possibility of a recording omission should be considered. This is particularly aggravating for the common species, because local floras often give no localities for such species. Other inadequately recorded areas which are not clear from the map (Fig. 3) are listed below. Some visual impression of these areas is given by overlay *6a* (see page xxv).

Great Britain

(1) V.-cs 24 and 32, N. Buckinghamshire, S. Northamptonshire. Squares 42/62–5, 72–5, 82, 83 were under-recorded, particularly for late-flowering species. The area was mainly covered in May 1956, by a B.S.B.I. meeting, but has not been re-visited later in the year.

(2) V.-c. 35, Monmouth, near Newport. Squares 31/28, 29, 38 have not been visited recently. Records used date from *circa* 1928 and were made by A. E. Wade. The absence of post-1930 records from them is not significant.

Ireland

(1) Because of the scarcity of botanists living in the country and the difficulty of persuading great numbers of volunteers to make a visit, much greater use has had to be made of old (pre-1930) records in Ireland than elsewhere. That only an old record is shown for any particular square is *not* significant. However, these old records are reasonably evenly distributed and the absence of any recent (post-1930) records within an area showing old records may be significant, particularly if the change in range suggested corresponds with a similar trend in Great Britain.

(2) An additional complication in Ireland is that, in the south-west particularly, the flowering season starts earlier than elsewhere in the British Isles. Moreover, the south-west is so far removed from the main centres of botanical activity that visitors tend to arrive rather late in the season. Therefore some species, e.g. *Anemone nemorosa* (page 19) and *Ranunculus ficaria* (page 24) may appear to be rarer than they really are in south-west Ireland because they had died down by the time the observers arrived.

Lists of old (pre-1930) records

Besides the post-1950 records collected in the field since the scheme began (and post-1930 in the exceptions cited), a considerable amount of data has been assembled from work done before 1930 which has either been published in Floras and journals, or preserved in manuscripts and notebooks. An above-average concentration of old records may be expected for the following vice-counties:

7 & 8 Wiltshire	70 Cumberland
21 Middlesex	71 Isle of Man
30 Bedfordshire	S Sarnia
35 Monmouthshire	78 Peeblesshire
36 Herefordshire	80 Roxburghshire
40 Shropshire	81 Berwickshire
41 Glamorgan	109 Caithness
42 Breconshire	H. 20 Wicklow
60 West Lancashire	H. 34 & 35 Donegal

In addition to the "county" lists there are two other fairly large sources of data:

(1) Scotland. G. West, *Flora of Scottish Lakes* (1904, 1910).

(2) Ireland. Lists of data collected by Praeger and others when preparing *Irish Topographical Botany* (1901) and the *Botanist in Ireland* (1934).

Some other lists were acquired, but only those west of the major grid line 4 were incorporated. There has not been time to deal with old records east of this line—but as this is the most heavily populated part of the British Isles and the most heavily botanised the omission will, if anything, redress the balance rather than disturb it.

Additional records from herbaria and literature

Besides the 1½ million or so field records at least 150,000 individual record cards have been accumulated. Many of the buff cards were duplicates sent in to amplify records submitted on regional cards—this information was asked for if the record was of special interest for reasons of taxonomy, status, rarity, etc. Most were collected deliberately and represent an important additional source of grid square records, particularly for the rare species.

Information has been extracted from the following herbaria:

Royal Botanic Gardens, Kew
British Museum (Natural History)
Royal Botanic Garden, Edinburgh
National Museum of Wales
National Museum of Ireland
Cambridge University
Oxford University
Trinity College, Dublin
Queen's University, Belfast
Belfast Museum
Stratford Museum, London E.11
Somerset County Museum, Taunton
Warwick County Museum
Horsham School, Horsham, Sussex
Sexey's School, Bruton, Somerset
T. J. and E. Foggitt, Huddersfield, W. Yorks
R. A. Graham
J. E. Lousley
N. D. Simpson
E. C. Wallace

The following publications have been widely used:

Botanical Exchange Club Reports (1930–47).
Watsonia and *Proceedings of the Botanical Society* (1947–).
County Floras, etc. of which an almost complete set was made available on loan to the scheme through the generosity of the University Library, the Botany School Library and the University Botanic Garden Library, Cambridge, and many members of the Botanical Society of the British Isles who gave or lent copies of Floras or papers.

INCORPORATION OF THE RECORDS

Field records

From the field record cards for a square, of which many may have been received from different recorders, a single master card is compiled; thus we have collected 3,500 master cards, one for each square. It is from these cards that the data are incorporated into the "system" by punching a single Powers-Samas forty-column card for each species recorded in each square. All the information can be coded into numerical form: each species has a number with suffixes for sub-species, etc., and the date (usually 1950, which was used conventionally for all records made during the period 1950–60), grid reference of the 10-kilometre square and the vice-county,* are already numerical. Using an Automatic Key Punch (A.K.P.) it is only necessary to punch separately the code number of each species which has been recorded from the square. The other information which is common for all the species is punched into the pack of cards thus produced by setting up the A.K.P. in the "repeat" position and passing the pack through a second time.

The punched cards are checked by feeding them into a Powers 3 Tabulator which prints a list of the data punched into the cards. First the "common" information is checked, then the variable information; the species numbers are compared with the master card by two members of the office staff calling over the numbers and checking the tabulation visually.

After checking, the cards are sorted according to the species number, so as to bring together all the cards for each species, which are then filed in numerical order to await preparation for mapping.

* Sarnia was numbered 113 but for all other numbers Watson and Praeger were followed.

Individual records

The individual record cards are edited to ensure that all the essential information is complete and has been altered into a suitable numerical code for punching, and then passed to the punched-card operator who punches each card separately on the A.K.P. This machine is ideal for this purpose as the face of the card is visible whilst the keys are being set up. The whole card is then punched in one action. These are then sorted and stored in numerical order. Every individual card is punched in a certain position so that all the individual cards for a species can be retrieved from the pack for the whole species in order that the extra information they give can be available for scrutiny.

The information on these cards (including date to the nearest month, an eight-figure grid reference, status, source, vice-county, altitude and habitat) can also be produced in numerical form on the tabulator.

CONTENTS OF THE ATLAS

The decision as to which species were to be included in the Atlas was left to a sub-committee of the Maps Committee. The resulting list has been used as a basis for the contents of the Atlas, but at a late stage it was realised that it was not practical to include all the micro-species listed in the *List of British Vascular Plants* by Dandy (1958), and that though data on some groups were adequate it would be better to leave all microspecies and hybrids until a later volume for two reasons:

(1) Future users of the Atlas would find it more convenient for reference if there were two distinct volumes with a clear difference of contents.
(2) The preparation of maps of some microspecies for the main volume would hold up its publication and leave so few maps for a second volume that it would perhaps be too small to produce at an economic price.

This volume therefore contains maps of all generally accepted native British species (excluding critical segregates) and most well-established introductions. The choice of the latter group has necessarily been arbitrary, and opinions will differ as to the wisdom of the final selection. In certain cases it has only been possible to publish aggregate maps; the reasons for these are various but are outlined on page xviii.

PREPARATION OF THE MAPS

The list of species approved by the sub-committee was divided into three classes depending upon how much editing a species would probably require before an adequate map could be produced. These classes were:

A. Rare species—recorded from not more than twenty vice-counties in the *Comital Flora*.
B. Medium species—recorded from 21–100 vice-counties.
C. Common species—recorded from more than 100 vice-counties.

In the Atlas each species is labelled with its category, indicating the amount of editing which it has received. The vice-county distribution has only been used as a guide, and some species have received more attention than their vice-comital distribution alone demanded. Special attention was given to species formerly more common whose distribution has become much more restricted during the last few centuries. For example, *Lycopodium selago* (page 1) is recorded in the *Comital Flora* from 129 vice-counties, but to have obscured its almost complete extinction in southern England would clearly have been misleading, so it has been treated as a *B* instead of a *C* species. The maps were prepared in taxonomic order and final revision and despatch to the printers began in December 1960. Records have continued to come in and many have been incorporated; thus the later maps should be more complete than the earlier ones. However, as it is the grasses and sedges which have benefited from this, and as they always tend to be less well recorded, this fact may have preserved the balance between the beginning and end of the Atlas.

A. Rare species

The distribution of these species has been investigated as fully as possible. In each case the following procedure was adopted. To the field records were added data from the herbarium of Cambridge University, the relevant county Floras, papers and monographs and individual records from other sources. After editing, all the cards were punched and a number of tabulations prepared on to which the localities were typed. The tabulations were circulated to the following:

J. E. Lousley
A. E. Wade, National Museum of Wales
E. C. Wallace
Dr E. F. Warburg, The Herbarium, University of Oxford
Mrs B. Welch, British Museum (Natural History)
P. S. Green and (later) Dr J. Milne, The Herbarium, Royal Botanic Garden, Edinburgh (only if species occurred in Scotland)
Professor D. A. Webb, Trinity College, Dublin (only if species occurred in Ireland)

These authorities checked their herbaria for additional records and added considerably from their own knowledge of the species in the field or of overlooked sources in the literature. They also made recommendations about taxonomy, status and validity. Further individual cards were prepared and others altered before the data were finally assembled for map-making. On these maps the old and recent records are distinguished by symbols except in a few cases where taxonomic difficulty has resulted in under-recording, so that the distinction would be meaningless, or where the species is now so rare that to disclose its present distribution might be dangerous.

Another device has also been used occasionally to protect very rare species. The dot for a locality, instead of being placed in the correct grid square, has been placed in one of the eight adjacent grid squares.

B. Medium species

The distribution of these species has been checked at the vice-county level and records have been added from the readily available literature for all those vice-counties from which no *recent* (post-1930) record has been fed into the system.

In most cases records from about 10 per cent of the counties listed in the *Comital Flora* still remained untraced at this stage. Thanks to the unselfish co-operation of a number of people, who worked upon lists of desiderata sent to them from time to time, it was possible to reduce the number of vice-county records untraced. Credit for this considerable contribution is due to the following:

A. E. Wade, Wales
Miss J. Gibbons, Lincolnshire
Mr and Mrs R. C. L. Howitt, Nottinghamshire
Miss M. Scannell, Ireland

and to many others whom we have bothered less persistently but who have always replied so readily.

On these maps the old and the new records are distinguished by symbols. It must be remembered that a *complete* search for old records has not been made. They are only included for those counties where no recent records are available and for a few others which have been fed into the system. The maps are not as complete as they could be, but we have tried to include the most important data. What we aim to show is the outline of the present distribution and an indication of those areas where the species is very rare or entirely absent.

C. Common species

The distribution of these species has been checked at the vice-county level and records have been added from the literature from all those vice-counties from which no records have been received. Otherwise the procedure has been identical to that carried out for *B* species. However, on these maps no distinction is made between old and new records.

Despite the efforts made, records from many vice-counties have remained untraced, particularly in Scotland where there are few county Floras and where many species reach the limit of their range and are probably rare or casual only. These missing vice-counties are listed in Appendix II (page 409). Search of H. C. Watson's herbarium at Kew, the Herbarium of the Royal Botanic Garden, Edinburgh, or the *Annals of the Scottish Natural History Society* would probably produce records from a great many of these counties, but in very few cases are these omissions so great as to be misleading. There is, however, one group of plants, the arable weeds, for which some of the maps are unsatisfactory. This is particularly true of those species which have decreased considerably in abundance and frequency since the improvements in the purity of seed at the beginning of this century. *Agrostemma githago* (page 65) for example was so widespread in eastern Scotland in the nineteenth century that it is recorded as "Common" without localities in all the Floras. It is now apparently absent, but very few old localities have been traced and placed on the map, so that the dramatic change which has taken place is under-emphasised unless reference is also made to the list of "missing vice-counties".

Special help

Though most of the editing has been carried out at head-quarters with the guidance of the various people mentioned above, we have also been fortunate in the number of those who have provided data for particular species in which they were interested. We wish to acknowledge this help, which means that the maps of many species are much more complete than would otherwise have been possible. See Appendix III (p. 419).

Special mention must also be made of the work of A. Slack and A. McG. Stirling who unearthed many additional records from literature and herbaria for the less rare Scottish mountain plants.

VERIFICATION OF RECORDS

The records received by the scheme have been checked in a number of ways and elimination of challenged records has taken place at various stages.

County referees

The field cards, grouped into counties or larger areas, were sent to the referees who had agreed to check the cards and who had a special knowledge of the area concerned. Upon the basis of their observations records were challenged or withdrawn.

We wish to acknowledge the assistance of all those who undertook this work. A list of their names appears in Appendix III (page 420).

Preparation of Master Cards

All the field records were transferred to Master Cards, and this process gave the staff an opportunity of considering every record. Possible errors overlooked by the county referees could often be reconsidered at this stage.

Preparation of data for mapping

When the cards for a species were arranged in vice-county and grid reference order, apparent new county records or outlying squares could be noted and rejected unless confirmation was found in Floras published since the *Comital Flora*, or in the revision of the Irish Flora by Webb (1952), or in the form of reliable individual record cards. There are also certain counties, e.g. E. Ross and E. Sutherland (v.-cs 106 and 107), which were very much under-recorded in the past, particularly in the lowlands, and from which a great many new county records have been made in recent years, often by several independent recorders; it became clear that these should be accepted. There are also a number of species about which no taxonomic confusion seems likely (e.g. *Cymbalaria muralis*), which have been spreading continuously over the last thirty years and for which the vice-county distribution in the *Comital Flora* is now out of date.

Inspection of the map

The map itself when produced may show certain isolated records and the origin and validity of these can be checked. A final inspection of all the maps was made by the editors with the assistance of C. D. Pigott.

By these methods we hope to have eliminated any flagrant errors, so that the general distribution pattern is not misleading. Nevertheless errors within the distribution range may still remain. These are unfortunate but not serious.

Despite the use of these various devices the maps of some species still suggest difficulties in identification reflected in a high percentage of apparent new county records or a great disjunction of distribution between one recorder's area and another's. Various methods have been used to reduce the seriousness of these difficulties.

Reliance on experts

For two difficult genera, *Fumaria* and *Potamogeton*, we have been fortunate in having the assistance of acknowledged experts who have generously made available to us the results of their work over many years. In these cases we have used symbols to distinguish between those records which have been confirmed by the experts and those which have not. For most species of these genera we have not included any records from a county from which the experts have seen no material, except in Ireland where the amount of herbarium material was limited and to follow this rule would have been more misleading than to ignore it. In one or two species of a very critical nature only the records passed by the experts have been accepted.

Mapping of aggregates

This device has been used in a number of cases where two or more closely related species have been confused or seriously under-recorded. In some of these cases the specific distinctions are not clear; in other cases it has been apparent that some recorders have found more difficulty than others.

Such aggregate maps are clearly labelled and a complete list will be found in Appendix I (page 407).

Aggregate maps have also been included for a number of familiar critical genera. The segregates will in most cases be included in the critical supplement, but it was thought worth while to include maps of the following in this volume: *Rubus fruticosus*, *Alchemilla vulgaris*, *Sorbus aria*, *Euphrasia officinalis*.

Provisional maps

Where treatment as an aggregate would have concealed some valuable information, and yet doubt still remained about the accuracy or completeness of the records, it has been decided to include maps in the Atlas but label them provisional. These maps are clearly marked with a "P" in the top right-hand corner of the legend. A list of provisional maps with explanation is given in Appendix I (page 407).

STATUS OF RECORDS

There is no difficulty with two classes of species when maps are being prepared. One class consists of species certainly native throughout their range, the other of alien species entirely introduced. These aliens are marked as such with an asterisk before the species name in the legend. In a few cases aliens have persisted for many years in some stations whilst being only casual in others; here a symbol to indicate this has been used (e.g. *Crocus purpureus*, page 330).

Some indication of the history of the spread of the alien species has been given in five cases where adequate information in the form of dated records existed. These species are: *Buddleja davidii*,

Veronica filiformis, *Galinsoga parviflora*, *Galinsoga ciliata* and *Senecio squalidus*. In addition the first record has been indicated for eleven species where the spread has been rapid.

The maps of a few conspicuous aliens, particularly trees, e.g. *Laburnum anagyroides* and *Aesculus hippocastanum*, are inadequate because some recorders ignored them.

Other species vary widely in the relative proportion of native and introduced records and the ease with which their status in different areas can be ascertained. Each species has been treated on its merits, but three main types of treatment have been used.

(i) Rare natives, widely introduced.

Native, or possibly native records alone, have been mapped (e.g. *Pinus sylvestris*, *Tilia platyphyllos*, *Buxus sempervirens*).

(ii) Fairly frequent natives, rarely introduced.

This is the largest group. An attempt has been made based on statements in local Floras, the *Comital Flora* and field observations to distinguish between native and introduced localities (e.g. *Helleborus* spp., *Tilia cordata*, *Viburnum lantana*).

(iii) Common natives, commonly introduced.

This group consists mainly of our common trees which have been so widely planted as to make determination of native status impossible. In these cases we have made a statement on the possible native range based on the judgement of experts (e.g. *Carpinus betulus*, *Fagus sylvatica*, *Salix alba*.)

Recorders were asked to note on their field cards whether or not a species was in their opinion only an introduction in the square. Though the request was not always complied with, the distribution of these observations is often a useful guide. These records have been distinguished by using the formula "recorded introduction" (one symbol) and "all other records" (another symbol). The maps of *Ilex aquifolium*, *Fraxinus excelsior* and *Sambucus nigra*, for example, have been prepared in this way, and the information on introduction, though incomplete, seems helpful.

MAP PRODUCTION

At a late stage in the planning of the Maps Scheme, when it had already been agreed that the use of punched-card machinery was essential for the preparation of the data for mapping, the idea arose that it might be possible to use the punched cards themselves to produce the maps mechanically. The problem was presented to L. R. Smith of Powers-Samas (now part of International Computers and Tabulators) and within days he had solved it and was able to demonstrate the technique when the scheme was launched in April 1954. So thorough was Mr Smith's first analysis of the problem that the technique has been altered little since it was first developed.

The forty-column cards are prepared for mapping by sorting all those for one species on the first figure of the grid reference. The cards are divided into four groups, each for a vertical strip of the British Isles. The four groups proceeding from west to east across the map have the first figure of the grid reference as follows:

8 and 9   0 and 1   2 and 3   4, 5 and 6

Each of these groups is twenty 10-kilometre squares wide except the last which is twenty-five—the maximum number of print-units available on the tabulator used.

Within each strip the cards are arranged in vertical order: 100 master cards are interspersed between them, one for each of the 100 possible vertical positions. (By insetting the Channel Islands, Orkneys and Shetlands, the British Isles can be shown on a map as less than 1,000 kilometres from north to south.)

The cards are transferred to a standard tabulator modified only so that the vertical and horizontal throws are equal. An outline map is set in a fixed position in the tabulator, and the 100 master cards operate the mechanism so that the map turns through 100 positions, one for each 10-kilometre square, during the course of a run. When a species card punched with a grid reference is fed into the tabulator the map remains stationary, the card is sensed and the print-unit is operated to bring up a symbol in the correct position.

By this method a map of 3,500 dots can be produced in less than an hour, and an average map of about 1,000 dots in 20 minutes.

During the scheme the tabulator has been used to place the solid dots on the maps. Circles and other symbols have been added by hand.

We would like to take this opportunity of expressing our gratitude to Mr Smith for his help with this particular problem, and also for the many other ways in which he ensured that the mechanisation equipment gave such reliable service. To his name we would like to add that of W. Wadeson whose administration on behalf of Powers-Samas meant that our association with that firm was always a happy one.

INTERPRETATION OF THE MAPS

It was never intended that the Atlas should include a phytogeographical essay based upon the distribution maps. It is a factual document which, we hope, details the distribution of the British flora with sufficient accuracy to make it a valuable tool for biologists and all others whom it may happen to interest.

So far the introduction has explained exactly how the maps have been prepared, a necessary preliminary to any interpretation of the published maps. What follows is an attempt to interpret to the reader the maps in relation to the work done in preparing them. This we feel is of greater importance than settling down to a detailed analysis of types of distribution. Such an analysis may well be considered worth while in the future, but if so it can be published elsewhere. Nevertheless the reader may wish to consider for himself the distribution patterns in relation to possible limiting factors, and as an aid to this interpretation we have included in a pocket six transparent overlays each consisting of a pair of maps on the same scale as those in the text. The six pairs are:

1*a* Vice-counties   *b* Rivers
2*a* and *b* Altitude
3*a* Geology   *b* Temperature
4*a* and *b* Temperature
5*a* Rainfall   *b* Humidity
6*a* Underworked squares   *b* National Grid

VICE-COUNTIES

1a This shows the sub-divisions of the British Isles laid down for Britain by H. C. Watson and for Ireland by R. Ll Praeger. The map should be useful in at least two ways. It can be used in conjunction with the data on missing vice-county records to extend, in the mind's eye, the limits of distribution; it may also be useful to local botanists who wish to know which species occur in their county. We are grateful to J. E. Dandy for his advice in the preparation of this map. The lists of vice-counties and of vice-county records omitted from the maps are given in Appendix II (page 409).

RIVERS

1b The main rivers have been included both as a topographical guide and for their occasional value in interpreting distribution patterns.

ALTITUDE

2a For mountain plants. This overlay shows the distribution of high ground by means of three symbols. The circle indicates those 10-kilometre squares which have some land over 1,000 feet but not exceeding 2,500 feet. The black dot indicates squares with some land between 2,500 feet and 3,500 feet, and the dot surrounded by a circle shows squares where the altitude exceeds 3,500 feet.

The main value of this map will be found in studying the tolerances of mountain plants. It is possible to build up a series of increasing tolerance to low altitudes from those like *Gnaphalium norvegicum* (page 278) confined to the highest mountains, through intermediate species like *Saxifraga aizoides* (page 139) which indicates so clearly the "Highland Line" in Scotland, to widespread mountain plants like *Saxifraga stellaris* (page 136) and *Oxyria digyna* (page 178) which are found as low down as 1,000 feet or so throughout their range. Some species occur at even lower altitudes in the north and west where many of our mountain plants, apparently intolerant of high summer temperatures, come down to sea-level. Examples of this type are *Carex capillaris* (page 356) and *Sedum rosea* (page 133).

2b For lowland plants. This overlay shows by dots all those squares which contain no land below 500 feet. This is intended to be useful in interpreting the distribution of some of our strictly lowland species, particularly the arable weeds which are scarce or absent from most of north and west Britain. Here again a series can be produced from the more exacting species like *Papaver rhoeas* (page 28), through more tolerant species like *Potentilla reptans* (page 123), to species which only fail to reach the highest areas like *Stellaria media* (page 70). The three common species of *Sonchus* (pages 297–8) also form an interesting and informative series.

GEOLOGY

3a It would have been impossible on this small scale to produce a map showing all the main geological divisions for an area as complex as the British Isles. Instead we have included a simple map which shows by dots those 10-kilometre squares in which there is an outcrop of limestone which exceeds 5 per cent of the area of the square. We would have liked also to include calcarcous shell sands, but adequate data were not available for the whole area. These are found on the west and north coasts of Britain, and are absent from the east and south coasts from Durham to Cornwall, where calcareous species only occur on the coast if there is a limestone outcrop. The value of this map in distinguishing between calcicole and calcifuge species can perhaps be best appreciated by comparing a widespread calcicole, *Anthyllis vulneraria* (page 109) with a widespread calcifuge, *Erica cinerea* (page 195). These illustrate the basic patterns of the two types. Less widespread species reflect their preferences over a part of the British Isles, but are limited by climate. Thus we have southern calcicoles like *Phyteuma tenerum* (page 258) and *Thesium humifusum* (page 152), intermediate species like *Actaea spicata* (page 19) and *Sesleria caerulea* (page 384), and northern ones like *Veronica fruticans* (page 230).

This overlay is of help in interpreting local distribution patterns in many of the maps. The only other significant group of species for which another geological overlay might have been valuable is the plants of the heavy clays. *Pimpinella major* (page 161) and *Sanguisorba officinalis* (page 126) are good examples of this type of distribution, and may be used as reference maps in discovering other species with similar preferences.

CLIMATE

There is an infinite number of ways in which the climate, the average weather of a period of years, can be presented in graphic or cartographic form. The effect of climate upon different species varies widely, and it may take years of research to determine what is the critical factor which has limited a particular species to its present range. Moreover, the limiting factor is not the same in different parts of the range and in many cases it will be a combination of factors which will be of importance. Thus it must be clear that the selection of climatic maps which we have chosen to include is a subjective one, and is only to be used as a general guide to the type of factors which may be of prime importance for a particular species. However, our experience suggests that the maps we have chosen are helpful with a wide range of species.

3b Average mean of daily minimum temperature, February (1901–30). The lowest mean temperatures occur in February. The most obvious features are that the minimum values occur in eastern Scotland and England (34° F., 1.1° C.), and the lowest values in Ireland (36° F., 2.2° C.) are found in the north. In contrast high values (over 40° F., 4.4° C.) are confined to south and west Ireland and to south-west England.

The distribution of two groups of species seems to be closely correlated with winter temperature. Firstly there are those which are mainly northern and eastern: *Linnaea borealis* (page 265) occurs only in eastern Scotland from Sutherland southwards, though it was once known from Northumberland (v.-c. 67) and north-east Yorkshire* (v.-c. 62), and appears to be limited by a February mean minimum temperature of over 34° F. A more widespread species with a similar distribution pattern but which probably tolerates slightly higher winter temperatures is *Trientalis europaea* (page 204). This almost reaches the west coast of Scotland, but it is still essentially an eastern species, reaching its southern limit in Britain near the east coast in Suffolk (v.-c. 25). Both these species

* There is a possibility that *Linnaea* was an introduction with pine trees here.

occur widely in the coniferous forests of northern Europe. The climate of north-east Scotland has affinities with that of northern continental Europe. It is possible that the Scots Pine, *Pinus sylvestris* (page 16), is truly native only in this area.

A more widespread species which may nevertheless belong to this group is *Cardamine amara* (page 44). Though it occurs from Inverness to the south coast of England it reaches the west coast only between Lancashire (v.-c. 60) and Argyll (v.-c. 98) and it may be limited by the 36° F. isotherm. It is therefore significant to find that this species also occurs in Ireland, but only in the north. *Geranium pratense* and *Geum rivale* are rather common species which might be included in this group. The former (page 90) is also confined in Ireland to the north, and it becomes much rarer near the south coast of England where it is often only of garden origin. The latter (p. 124) does occur more widely in Ireland, but it is more frequent in the north and more or less confined to the mountains in the south.

It is possible that in some of these cases at least a low winter temperature is necessary for the production of viable seed. Matthews (1942) found evidence to suggest that low temperatures increase the germination of *Trientalis europaea*, and Kinzel (in Skene, 1924, page 425) showed that the seeds of *Menyanthes trifoliata*, *Gentiana nivalis* and *Adoxa moschatellina* require freezing before germination takes place in nature.

There is thus a fairly large group of species with a mainly eastern tendency in Britain which only occur in the north of Ireland, where they are often restricted to a single locality (e.g. *Helianthemum chamaecistus*, page 60, and *Adoxa moschatellina*, page 266). These isolated occurrences are not concentrated in any one area; they suggest a relict distribution from a period when winter temperatures were lower in the northof Ireland than they are today.

A second group of species whose distribution appears to be controlled in part at least by winter temperatures are those which are intolerant of low values. The most exacting species are those which occur in south and west Ireland and in south-west England. Mean February minimum temperatures have almost the same values in these two areas, 41° F. (5.0° C.) being the highest temperature on the mainland. The species in this group are clearly relics of a time when a warmer winter prevailed, and their exact localities now show no significant grouping into particular areas. Thus it is possible to arrange a series of species which are mainly to be found in the Mediterranean or south-west Europe, which occur in different parts of the total area with mean February minima above 38° F. *Erica mediterranea* (page 195) is concentrated in West Mayo (v.-c. H.27). *Daboecia cantabrica* (page 193) is confined to a small area centred on West Galway (v.-c. H.16). *Pinguicula grandiflora* (page 239) is almost restricted to Cork and Kerry (v.-cs H.1–5) but has become naturalised when introduced outside the main area in Clare, Wicklow, Devon and Cornwall. *Erica vagans* (page 195) has generally been regarded as native only on the Lizard Peninsula of West Cornwall (v.-c. 1), but there are old records from Tramore, Waterford (v.-c. H.6) and a colony has recently been discovered some distance from human habitation in Fermanagh (v.-c. H.33); both these localities are within the total range of this group of south-western species. A last example in the series could be *Erica ciliaris* (page 194) which has not been found in Ireland but is limited to two main areas with temperatures between about 38 and 40° F.

Other species showing similar tendencies but which are more widespread include *Cicendia filiformis* (page 207), *Carex punctata* (page 354), and *Ranunculus lenormandii* (page 23). It is interesting to note that the last two occur in south-west Scotland where there are relatively high values along the northern edge of the Irish Sea. Likewise all three species have been recorded from East Anglia mainly on or near the coast, and here again the maritime influence maintains higher winter temperatures near the coast than inland. The end-point of this series might be seen in species like *Phyllitis scolopendrium* (page 7) and *Epipactis helleborine* (page 332) which occur throughout Great Britain except in north-east Scotland, and are thus almost complete opposites of the northern-continental *Linnaea* and *Trientalis*.

It is probable that this intolerance of low winter temperatures involves at least two different factors limiting plant distribution. Firstly low temperatures affect those species which are truly frost-sensitive. Amongst these one would include *Arbutus unedo*, the Strawberry Tree. Many of our aliens which have found normally suitable growing conditions in the south and east nevertheless suffer occasional catastrophes if caught by a rare frost, when, for example, whole populations of *Carpobrotus edulis* (page 80), and *Eucalyptus* spp. are destroyed. The map of *Fuchsia magellanica* (page 149), which indicates the areas where it is naturalised, is probably one of the best to indicate the effectiveness of frost. It would be interesting to study this in relation to a map showing the mean of the lowest annual temperature recorded for the last thirty years.

The other effect of low winter temperatures is to limit those species which use the winter and spring for growth and become dormant during the heat of high summer. These species are not necessarily frost-sensitive but probably fail to compete successfully after a series of cold winters. Some of the most widespread examples include *Umbilicus rupestris* (page 135), and *Ceterach officinarum* (page 9).

4*a Average mean of daily mean temperature, January* (1901–30). This map may perhaps give a better correlation with the group of species mentioned in the last paragraph, where it is the cumulative effect of winter temperatures above a certain figure which is important for full development.

4*b Average mean of daily mean temperature, July* (1901–30). The main feature of this map is that the isotherms run almost horizontally across our islands, with a slight dip in the west. Highest values occur in the Channel Islands, the lowest in the Shetlands, with a difference between the two of almost 9° F. (5° C.). The highest values in Ireland occur in the south-east.

A series can be made of species showing a decreasing demand for high summer temperatures. The northern limits of these species are almost horizontal. There are about forty species recorded from the Channel Islands which do not occur as natives in the rest of the British Isles, where the summer temperatures are presumably not high enough. There is a fairly significant group of species which are confined to England south of the Thames, e.g. *Phyteuma tenerum* (page 258), which appear to be limited by the 62° F. isotherm. Other species, less exacting, reach their northern limit in the south Midlands, e.g. *Polygala calcarea* (page 56), *Cerastium pumilum* (page 68), and *Lathyrus aphaca* as a native (page 116). None of these species occurs in Ireland. However, *Campanula trachelium* (page 256) is more widespread in England reaching north to Lincolnshire (v.-c. 54) and south Yorkshire (v.-c. 63), a region which lies between the 60° F. and 61° F. isotherm. It is perhaps significant that this species *does* occur in Ireland, and in the south-east, the only part of that country with mean July temperatures over 60° F. Another species of very similar type is *Euphorbia amygdaloides* (page 173), very local in southern Ireland though not in the same area as the *Campanula*. As with the apparently relict species of the north of Ireland, which have a scattered, almost random, distribution, there is a similar group of species of southern distribution in Britain which occur somewhere in south-east Ireland, but are not concentrated in one particular area.

The series can be extended northwards with species of increasing tolerance. *Blackstonia perfoliata* (page 208) has a northern limit in Berwickshire, whilst *Rosa arvensis* (page 127) occurs throughout Ireland and just reaches the south of Scotland as a native.

The converse of the northern limit is the southern limit. This is most clearly seen in coastal plants where the distribution is not confused by altitude. Here again a series can be recognised. *Blysmus rufus* (page 351) reaches as far south as Kerry (v.-c. H.1) and North Lincolnshire (v.-c. 54) and is probably limited by temperatures over 60° F. (15.6° C.); *Mertensia maritima* (page 216) was once reported as far south as Kerry and West Norfolk (v.-c. 28), but now has its southern limit in Down (v.-c. H.38) and West Lancashire (v.-c. 60), and appears to be limited by slightly lower temperatures than *Blysmus*. *Ligusticum scoticum* (page 166) is restricted even further—to the coasts of Scotland and the north of Ireland. These species which appear to be intolerant of high summer temperatures are mainly found in northern Europe, some even having a circumpolar distribution, whereas the species with horizontal northern limits are found in continental and southern Europe.

5a Rainfall. This map may be useful in understanding the limits of hydrophytes on the one hand and xerophytes on the other, but it is unsatisfactory in that it does not include the factor of evaporation, which varies considerably throughout the British Isles and is of equal importance with rainfall in determining the water regime of the soil. However, what this map shows clearly is the very low rainfall experienced on the east side of Britain as far north as Caithness (v.-c. 109). This may account for the rarity or absence from such areas of species like *Chrysosplenium oppositifolium* (page 139) and *Nardus stricta* (page 405), and the presence of many of our arable weeds and plants of waste places in a narrow coastal strip, particularly concentrated at the head of the Moray Firth in v.-cs 95, 96, 106 and 107.

5b Humidity. Precipitation/Saturation deficit ratio. This map is based upon the method described by Meyer (1926). The Precipitation/Saturation deficit ratio (P/SD) is calculated by dividing the annual precipitation in millimetres by the absolute saturation deficit of the air expressed in millimetres of mercury. This ratio can be determined for any meteorological station recording precipitation, temperature and vapour pressure. In the British Isles these data are available for forty-five stations, and by interpolating from a map of mean annual vapour pressure (Meteorological Office, 1938), data from another sixty stations can be used. This is rather too few for detailed work, but this map should give a general indication of the trend of humidity distribution in these islands.

The most outstanding features are the very high values experienced all along the west coast of Ireland and Scotland; the low values in the counties of Essex and Cambridgeshire (v.-cs 18, 19 and 29); and the strong influence of the coast in maintaining high values, there being a very rapid gradient between the south coast of England and the inland areas to the north.

Here again we can recognise two main groups of species, those which are confined to the driest area and those which are absent from it. Of those which occur within it, the main concentration is in the "Breckland" on the borders of West Norfolk, West Suffolk and Cambridgeshire (v.-cs 26, 28 and 29). This is an area of very porous sands which, combined with the low humidity of the climate, make it probably the driest ecologically in the British Isles. Examples of this group include *Silene otites* (page 63), *Veronica praecox* and *Veronica verna* (pages 230 and 231). Other species with different edaphic requirements are also confined to central-eastern England; *Trifolium ochroleucon* (page 105) in hedgerow banks on the clay; *Primula elatior* (page 201) mainly in boulder-clay woods, and *Melampyrum cristatum* (page 234) on their margins. *Calamintha nepeta* (page 244) is locally abundant in Essex and Suffolk. Many of these species of eastern England are truly continental and some have their centre of distribution as far east as the Russian steppes.

A species of wider tolerance is *Pulsatilla vulgaris* (page 19) which occurs almost entirely in areas with P/SD values below 150. The absence of this species from the chalk of the North and South Downs may be for purely climatic reasons. It does not occur immediately on the other side of the Channel in France, but appears again near Rouen where the P/SD value is *c.* 125, about the same as Cambridge.

Other species have more widespread but parallel types of distribution, and a remarkably large group have a northern and western limit along a line joining the Humber to the Severn Estuary. This line is probably the resultant of a number of factors; nearly all our lowland limestones lie to the south and east of it, and the higher ground to the north and west begins to affect the species most sensitive to altitude; nevertheless humidity must also be of great importance. Good examples are *Chenopodium hybridum* (page 82), *Lactuca serriola* (page 296) and *Medicago polymorpha* (page 103) as a native. It is noteworthy that in a number of these cases the species become rare near the coast, particularly in Sussex (v.-cs 13 and 14) and in north-east Norfolk (v.-c.27). More tolerant species occur widely over south, east and central England but become limited to the coast in the west, and may be almost entirely coastal in Ireland, e.g. *Cerastium semidecandrum* (page 69), or only occur inland in the south-east corner of that country, e.g. *Erodium cicutarium* (page 95).

Humidity alone can hardly be sufficient explanation for all species having distribution patterns of this type; some of them, for example, occur in seasonally wet soils, e.g. *Pulicaria vulgaris* (page 276) and *Lythrum hyssopifolia* (page 143). These species have declined with improved agricultural drainage. The probable explanation of their occurrence in "continental" Britain is that in this area alone is there a combination of a high winter water-table followed by a dry spring and early summer during which they can develop and reach maturity in the absence of competition.

In contrast to the species mainly found in the drier parts of the British Isles are those which are confined to or are most abundant in the north and west. Although the uneven distribution of acid and basic rocks, and the high winter temperatures of the west compared with the east, make interpretation difficult, it is hard to believe that in a group like the ferns humidity is not one of the most significant limiting factors. This is emphasised by the fact that the area which lacks many of these apparently high-humidity demanding species is almost the same as that in which the drought-resistant species are centred, i.e. central East Anglia. In Cambridgeshire (v.-c. 29) ferns are generally rare and only thirteen of the forty-seven species in the British flora occur at the present day, whereas Sussex (v.-cs 13 and 14) on the same longitude, despite the fact that it is 100 miles further south, has twenty-six species. This must surely be due to the coastal position and higher humidity values of the latter county.

UNDERWORKED SQUARES

6a Certain variations in the evenness with which the British Isles have been surveyed during the scheme have already been referred to (page xvii). A map (fig. 3) gives an account of the actual numbers of species recorded from each 10-kilometre square. However, this does not readily give a visual impression of the areas which have been somewhat neglected. This overlay is therefore provided which indicates by dots those squares which we believe have been underworked. It attempts to take account of the total flora likely to be found in an area before deciding whether the list we have received is adequate or not. "Adequate" might be taken as meaning that over 60 per cent of the possible flora has been recorded.

NATIONAL GRID

6b This overlay makes it possible to find the four-figure co-ordinates of any dot in the Atlas. Those interested in having further details on any of the 10-kilometre square records in this work can obtain them by writing to the headquarters of the scheme and quoting species number and grid reference.

Besides those to whom we have paid tribute in the text, and those many others whose help in the collection of the records we have acknowledged in Appendix III, we would like to thank most sincerely the office staff who have been responsible for dealing with all the correspondence and the incorporation of the records during the eight years it has taken to produce this Atlas. Most particularly we would like to mention Mrs S. Fincham, the punched-card operator, who has been with us throughout, and whose cheerful enthusiasm has never wavered, and Miss A. Matthews, who was our secretary from 1954–9, and whose able direction of the office administration enabled us to deal smoothly with some almost overwhelming situations. We would also like to thank Miss J. Nicholson, who was secretary from July 1959 to July 1961, and Miss A. Hughes, who has been helping us for the last few months. We are very grateful to Mrs R. Marriott and Miss J. Daltry, who gave general assistance in the office between 1957 and 1959.

This list of acknowledgements would not be complete without special reference to those members of the Society who have served on the Maps Committee and have given freely of their time to advise and encourage. We are particularly grateful to J. E. Lousley and A. R. Clapham, respectively the permanent chairman and secretary of that Committee, and to the two treasurers of the Society, E. L. Swann and J. C. Gardiner, who served during the course of the scheme and who undertook the not inconsiderable additional business without demur.

We also wish to record our sincere thanks to the Nuffield Foundation for their grant for the first five years of the scheme, and to the Nature Conservancy for their continued financial assistance which has enabled us to produce a more complete work than was at first thought to be possible. To the General Board of the University of Cambridge we are grateful for making accommodation available throughout the scheme and, for the last few years, providing it without charge; and to our hosts, the Professor of Botany (1954–5) and the Director of the University Botanic Garden (since 1955) for tolerating our presence in their departments and making us feel so much at home.

Finally, we wish to record our appreciation of the help given us by W. T. Stearn in preparing this Introduction, by C. D. Pigott in editing the maps, by Mrs M. Pigott in producing the base-map, and by Miss C. A. Lambert in drawing the overlays.

University Botanic Garden,
Cambridge, March 1962

F. H. Perring
S. M. Walters

The Distribution Maps

EXPLANATION OF THE LEGEND

Amount of editing

A Search made for all records: probably almost complete. Old and recent records generally distinguished.

B Search made for records from vice-counties from which no recent (post-1930) records had been received. Old and recent records distinguished.

C As for B but old and recent records not distinguished on the map.

For a fuller account of these categories see the Introduction (page xix).

Provisional maps

P These species are discussed in the Introduction (page xx), and a complete list with explanation appears in Appendix 1 (page 407).

The species number

These are taken from Dandy (1958). The maps follow the order of that work except in cases where it has been convenient to place two species on one map. This has generally only been done with species of the same genus and, with large genera, only with species of the same section.

Nomenclature

Dandy (1958) has been followed with one or two minor exceptions. Synonyms have been given where the nomenclature used by Clapham, Tutin and Warburg (1952) differs from that used by Dandy.

Status

* An asterisk before a species name indicates that the species is almost certainly not native anywhere in the British Isles including the Channel Islands.

Symbols

The meaning of the symbols is given on each map so that care should be taken, when comparing maps with identical symbols, to ensure that they are being used in the same way. In preparing the maps primary consideration has always been given to making the symbols easily visible. No symbol has any quantitative significance: each one indicates only that one or more specimens of a species have been recorded in a 10-kilometre square.

Dates

onwards

A date followed by "onwards" refers to records made in 10-kilometre squares in that or subsequent years (recent records).

Before

A date preceded by "Before" refers to records made in 10-kilometre squares from which no more recent record has been traced (old records). Recent records which are now known or thought to be extinct are also mapped as though they were old. The *spread* of a few interesting aliens has been shown in a special way by indicating the first record and giving its date or by showing all the 10-kilometre squares from which a species was known before a certain date. In all other cases the old records only indicate the possible *decline* of a species.

Scale

The maps in this Atlas are reproduced at a scale of approximately 1 to 8,000,000.

The maps published in this Atlas disclose the distribution of species to the nearest 10-kilometre square, and this may give rise to misgivings that the publication of this information will endanger the survival of some of our rarest species. To avoid possible risks we have, in certain cases, conventionally disguised the 10-kilometre squares where the species are known to occur today (see Introduction, page xix): in all cases the more exact localities for rare species which are held in our files are treated as confidential and are revealed, if at all, only to bona fide students and research workers.

The Atlas also shows how local many quite common species have become in certain parts of the British Isles in recent times, and stresses the importance of conserving locally not only the classical rarities but some of the more widespread British species. This publication should guide naturalists in two ways to a more effective conservation of our flora; firstly by acting as a warning about areas where species should on no account be collected, and secondly by pointing to areas where active conservation is necessary and giving factual evidence to support arguments against the destruction of threatened sites.

For both rare and local species we believe that education and information will serve the ends of conservation better than secrecy and ignorance. We are confident that in the hands of British botanists, without whose enthusiasm this volume could not have been produced, the information published will never be misused.

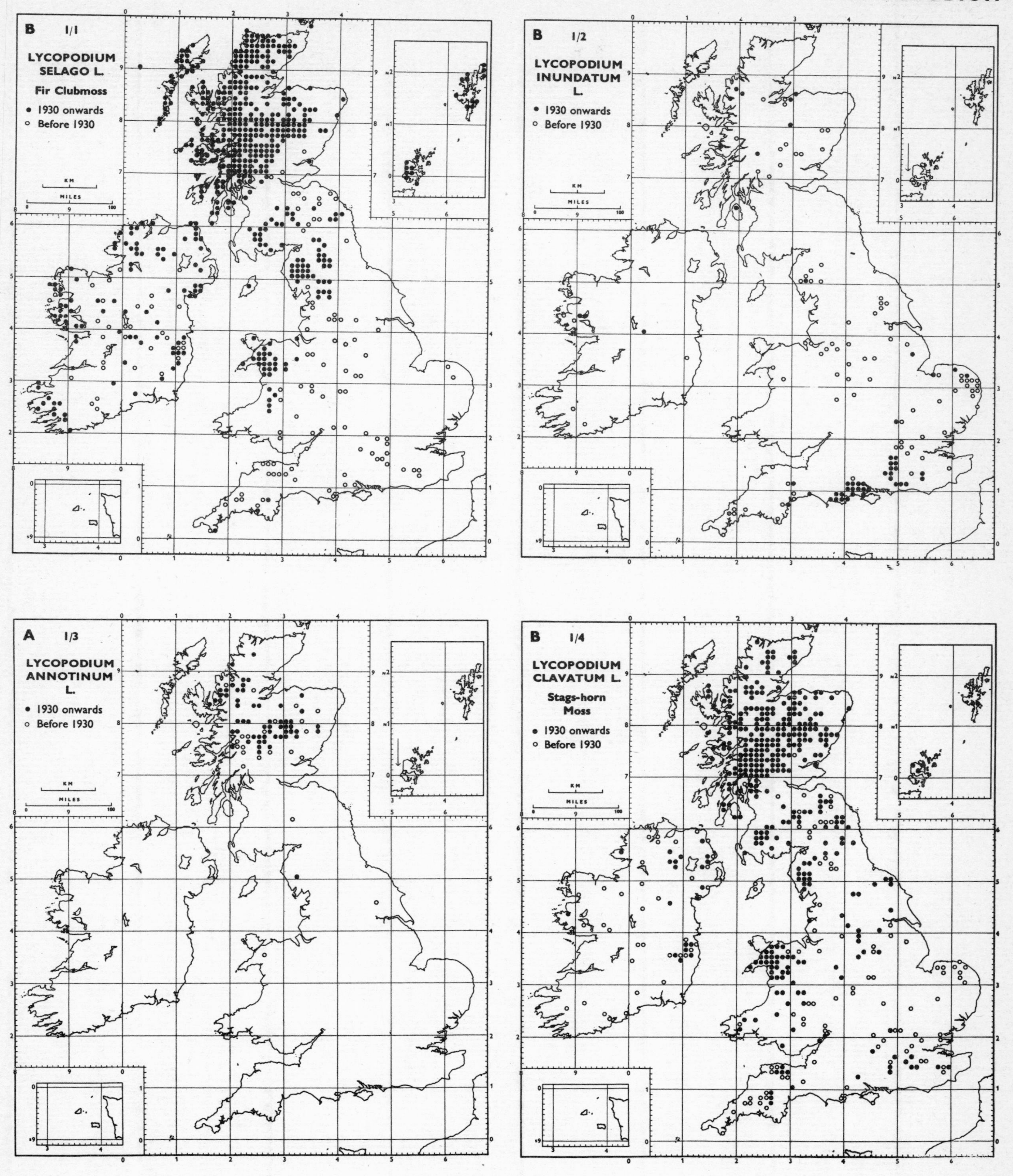
B 1/1
LYCOPODIUM SELAGO L.
Fir Clubmoss
• 1930 onwards
○ Before 1930
KM
MILES
B 1/2
LYCOPODIUM INUNDATUM L.
• 1930 onwards
○ Before 1930
A 1/3
LYCOPODIUM ANNOTINUM L.
• 1930 onwards
○ Before 1930
B 1/4
LYCOPODIUM CLAVATUM L.
Stags-horn Moss
• 1930 onwards
○ Before 1930

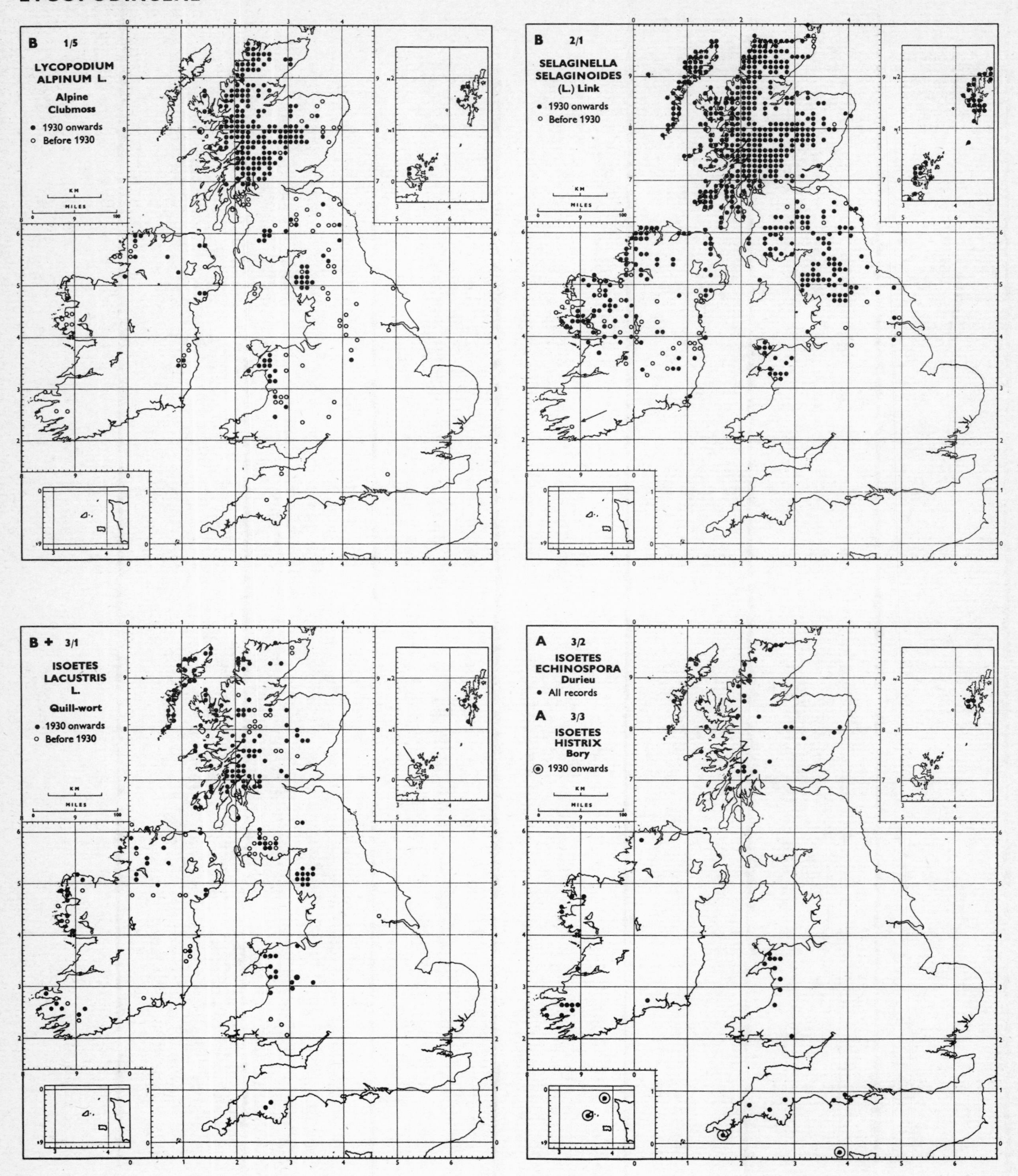
B 1/5
LYCOPODIUM ALPINUM L.
Alpine Clubmoss
• 1930 onwards
○ Before 1930
KM
MILES
B 2/1
SELAGINELLA SELAGINOIDES (L.) Link
• 1930 onwards
○ Before 1930
B + 3/1
ISOETES LACUSTRIS L.
Quill-wort
• 1930 onwards
○ Before 1930
A 3/2
ISOETES ECHINOSPORA Durieu
• All records
A 3/3
ISOETES HISTRIX Bory
⊙ 1930 onwards

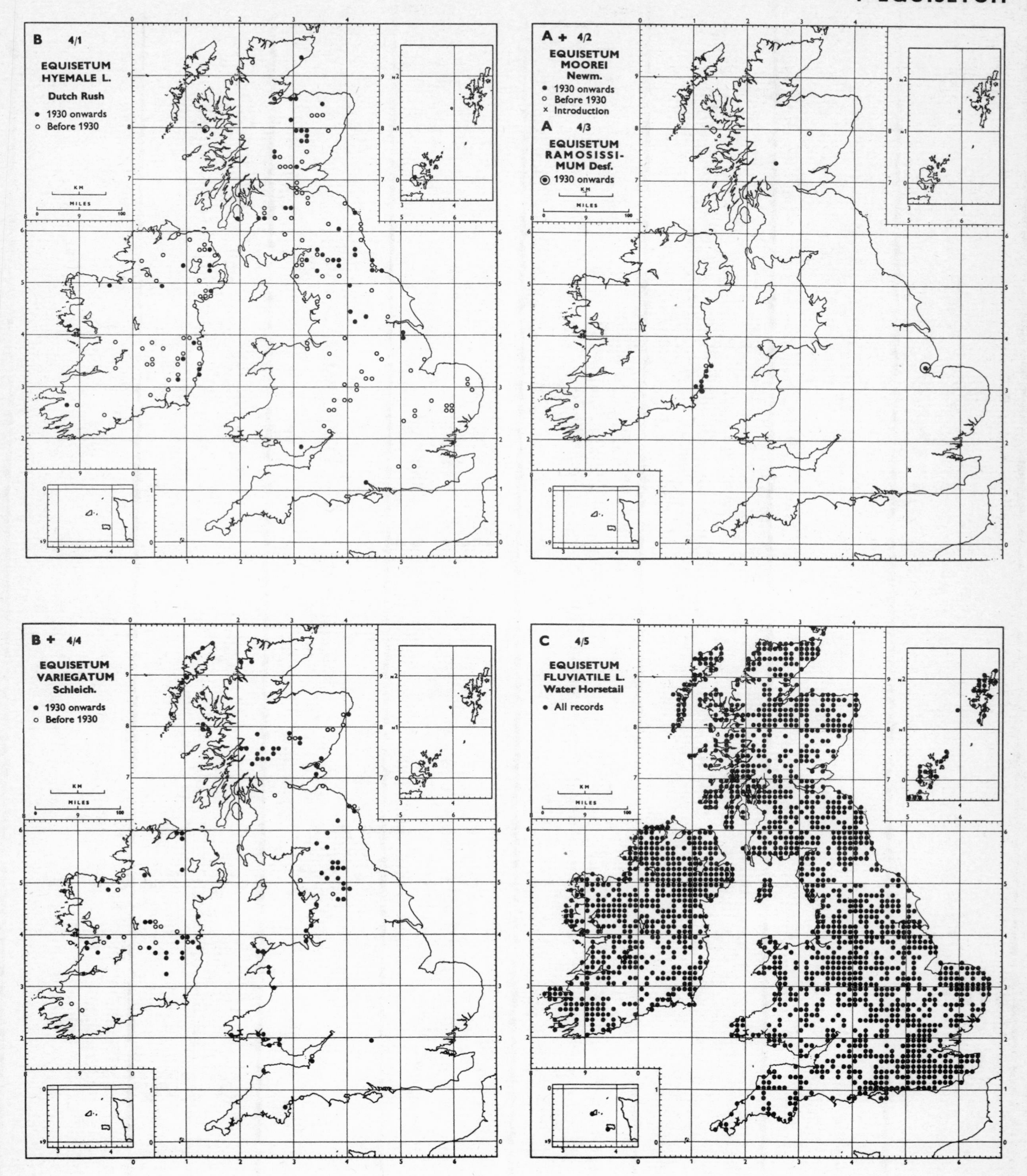
B 4/1
EQUISETUM HYEMALE L.
Dutch Rush
1930 onwards
Before 1930
A + 4/2
EQUISETUM MOOREI Newm.
1930 onwards
Before 1930
Introduction
A 4/3
EQUISETUM RAMOSISSIMUM Desf.
1930 onwards
B + 4/4
EQUISETUM VARIEGATUM Schleich.
1930 onwards
Before 1930
C 4/5
EQUISETUM FLUVIATILE L.
Water Horsetail
All records

EQUISETACEAE

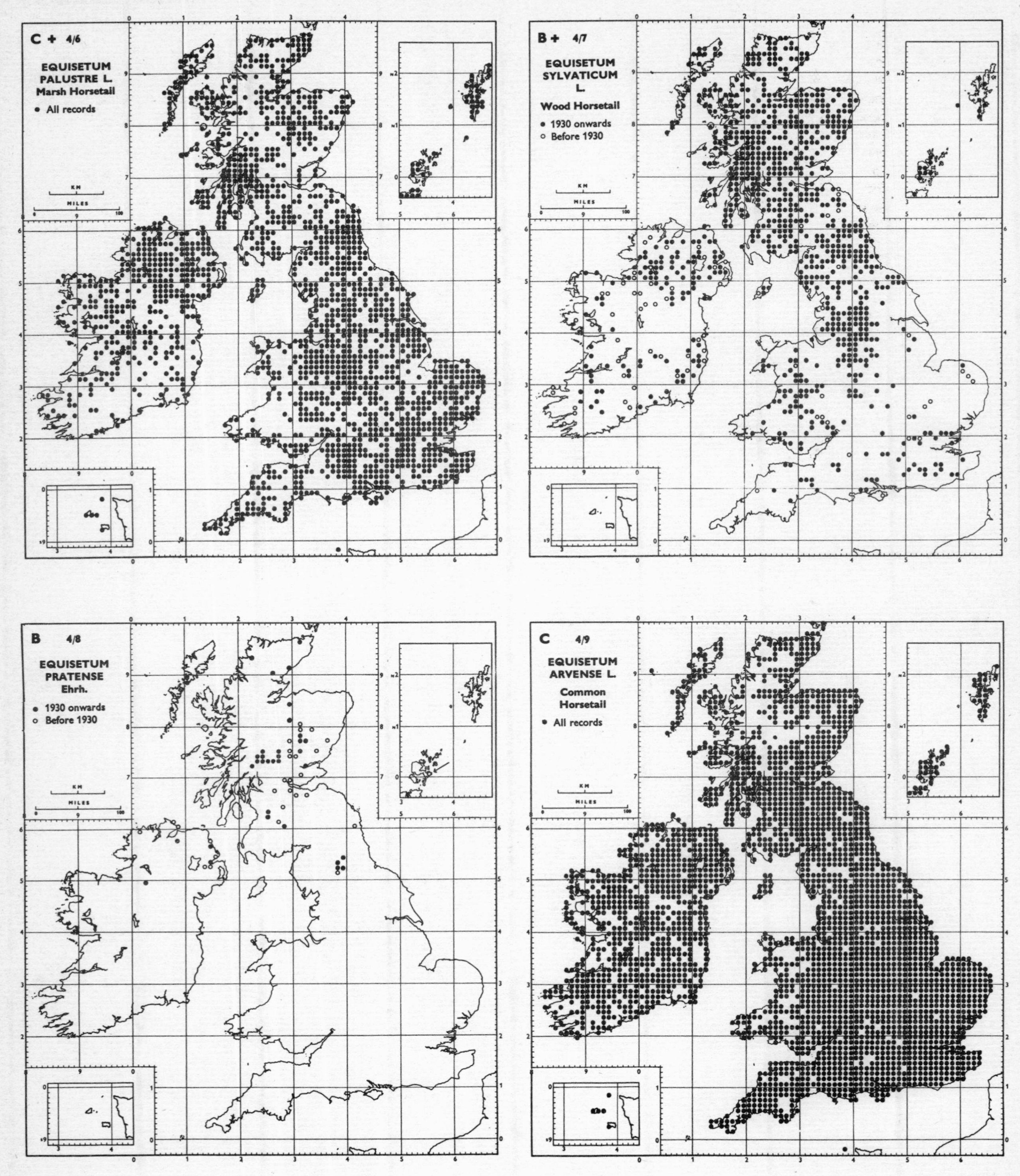

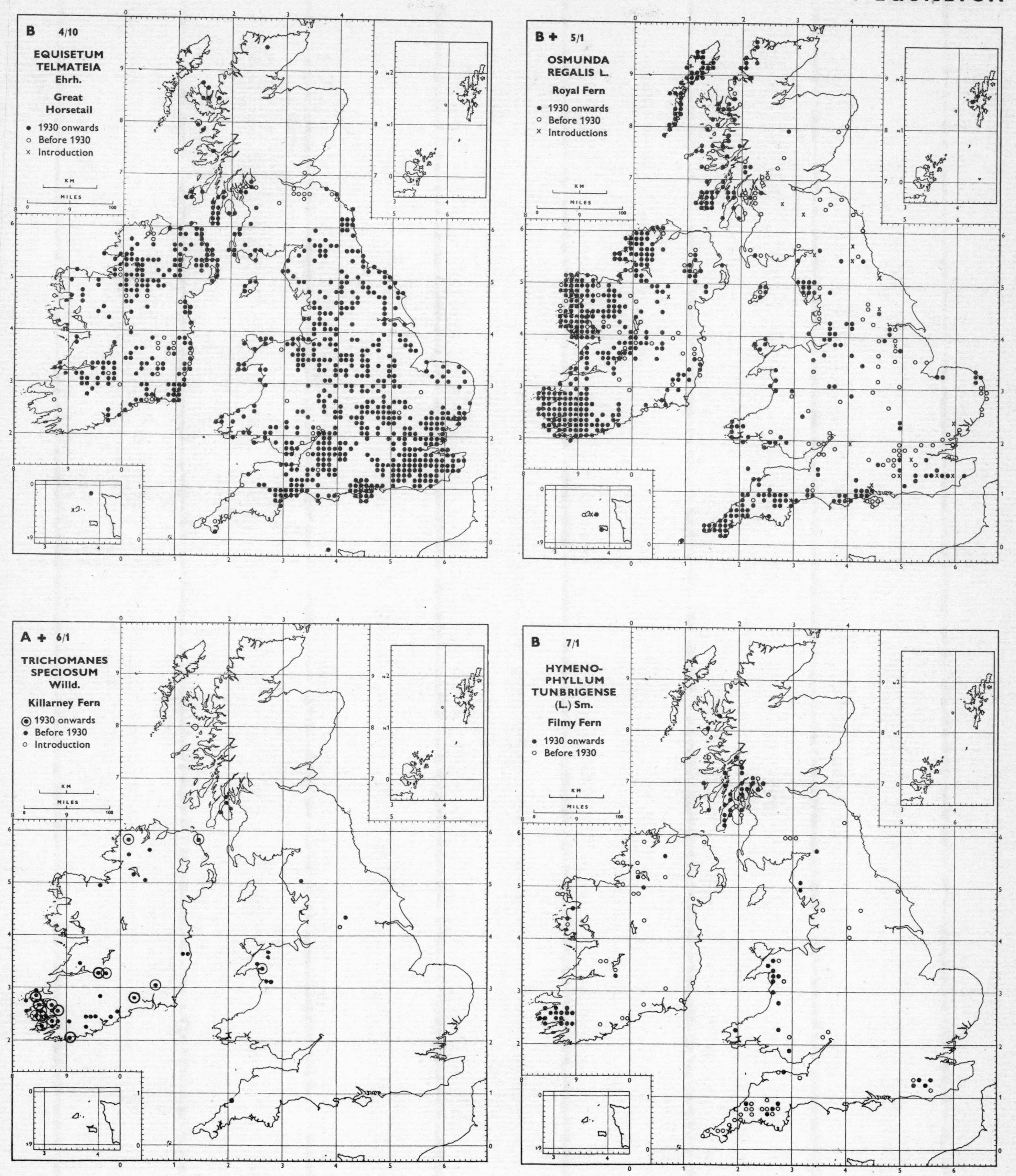
B 4/10
EQUISETUM TELMATEIA Ehrh.
Great Horsetail
1930 onwards
Before 1930
Introduction
B + 5/1
OSMUNDA REGALIS L.
Royal Fern
1930 onwards
Before 1930
Introductions
A + 6/1
TRICHOMANES SPECIOSUM Willd.
Killarney Fern
1930 onwards
Before 1930
Introduction
B 7/1
HYMENOPHYLLUM TUNBRIGENSE (L.) Sm.
Filmy Fern
1930 onwards
Before 1930
KM
MILES

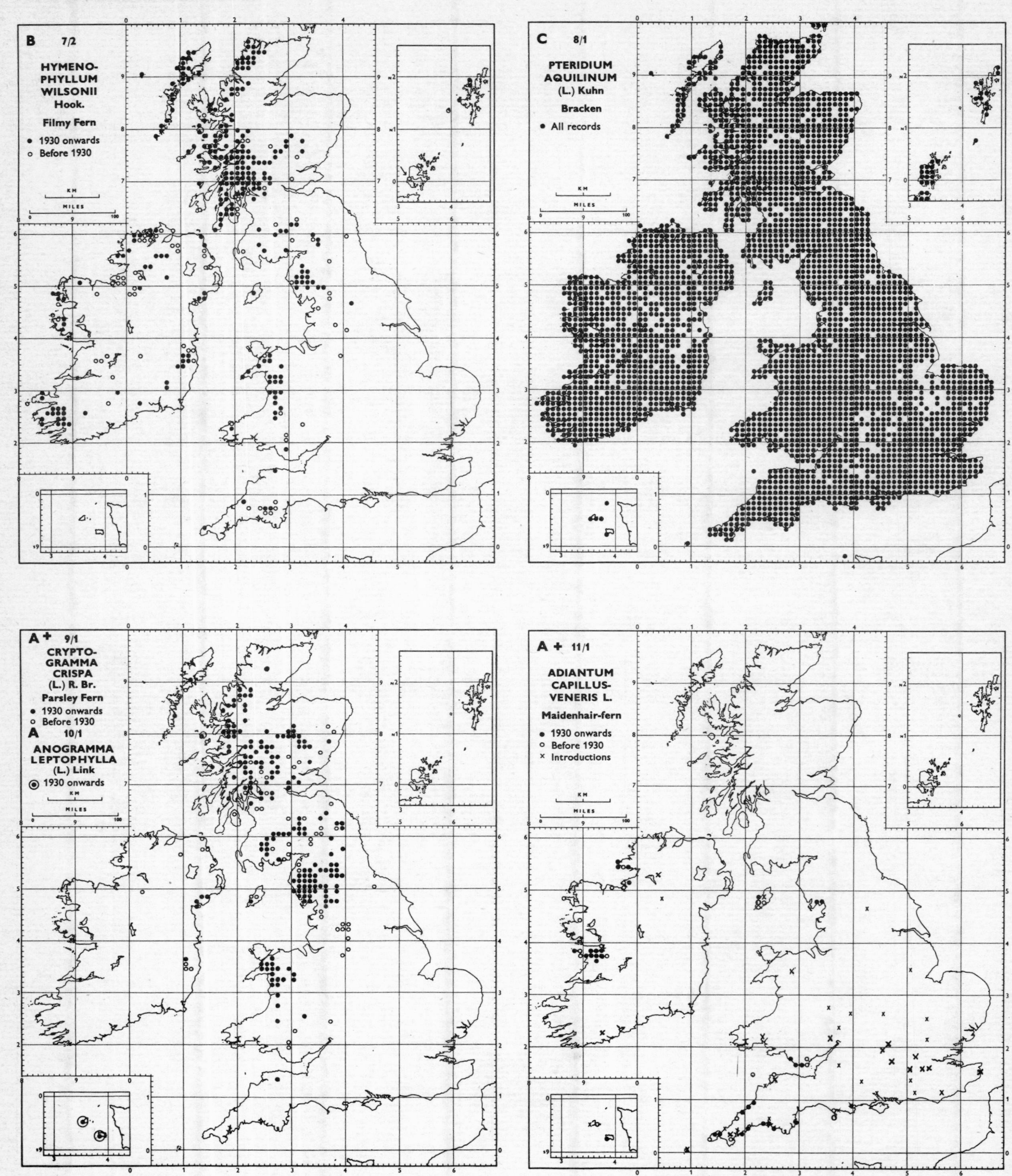
B 7/2
HYMENOPHYLLUM WILSONII Hook.
Filmy Fern
• 1930 onwards
○ Before 1930
C 8/1
PTERIDIUM AQUILINUM (L.) Kuhn
Bracken
• All records
A+ 9/1
CRYPTOGRAMMA CRISPA (L.) R. Br.
Parsley Fern
• 1930 onwards
○ Before 1930
A 10/1
ANOGRAMMA LEPTOPHYLLA (L.) Link
◉ 1930 onwards
A + 11/1
ADIANTUM CAPILLUS-VENERIS L.
Maidenhair-fern
• 1930 onwards
○ Before 1930
× Introductions
KM
MILES

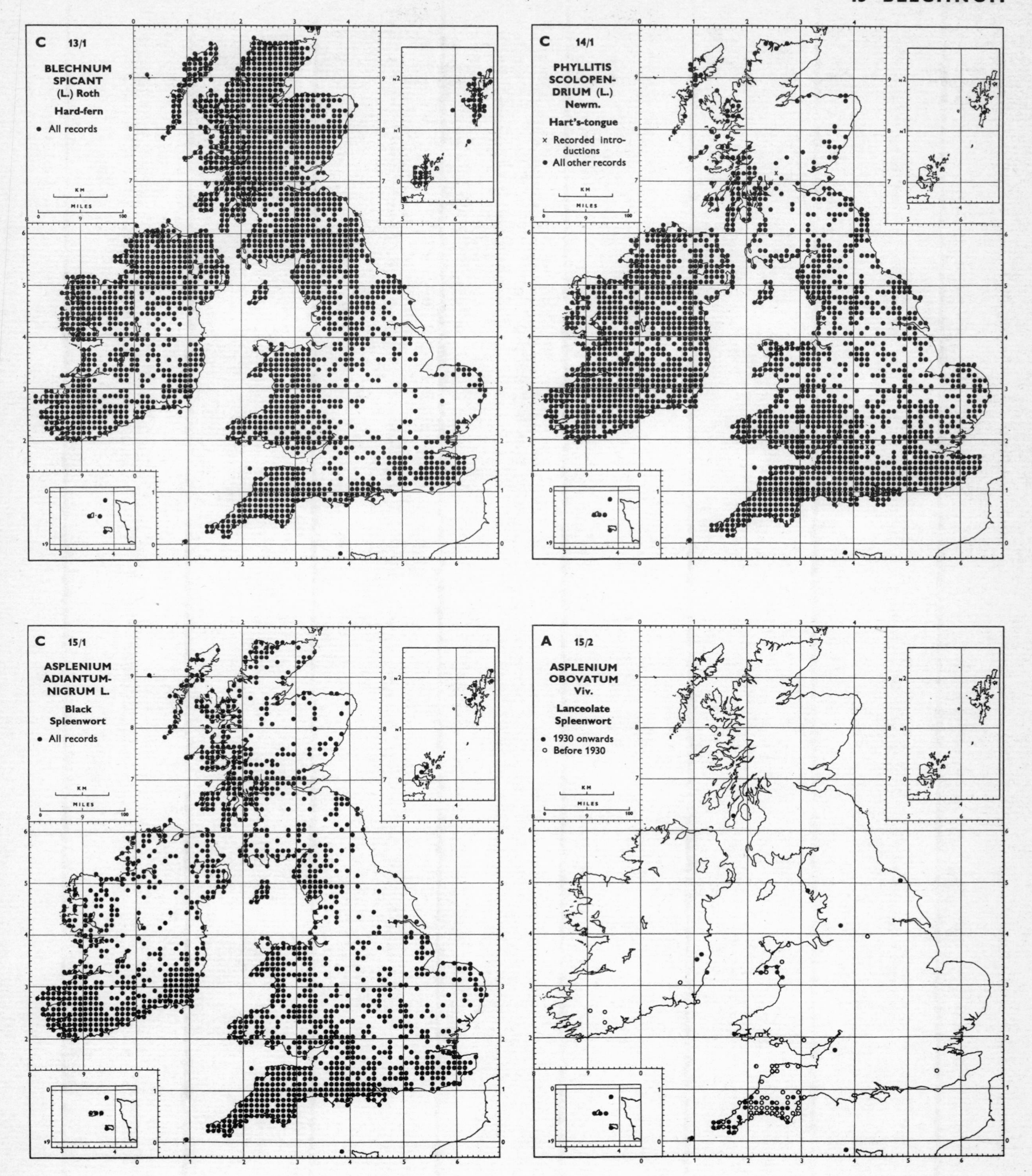
C 13/1
BLECHNUM SPICANT (L.) Roth
Hard-fern
• All records
C 14/1
PHYLLITIS SCOLOPENDRIUM (L.) Newm.
Hart's-tongue
× Recorded introductions
• All other records
C 15/1
ASPLENIUM ADIANTUM-NIGRUM L.
Black Spleenwort
• All records
A 15/2
ASPLENIUM OBOVATUM Viv.
Lanceolate Spleenwort
• 1930 onwards
○ Before 1930

ASPLENIACEAE

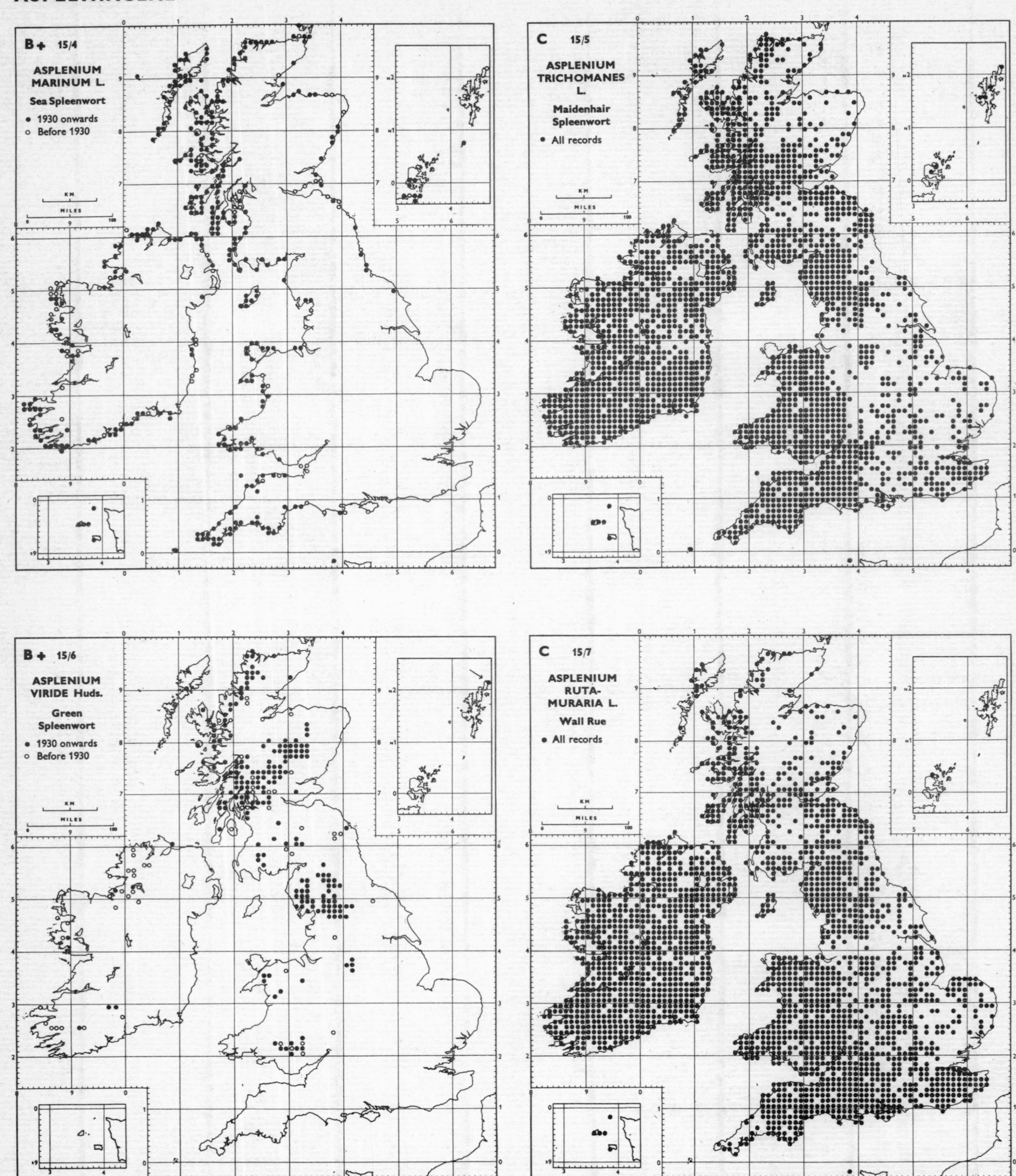

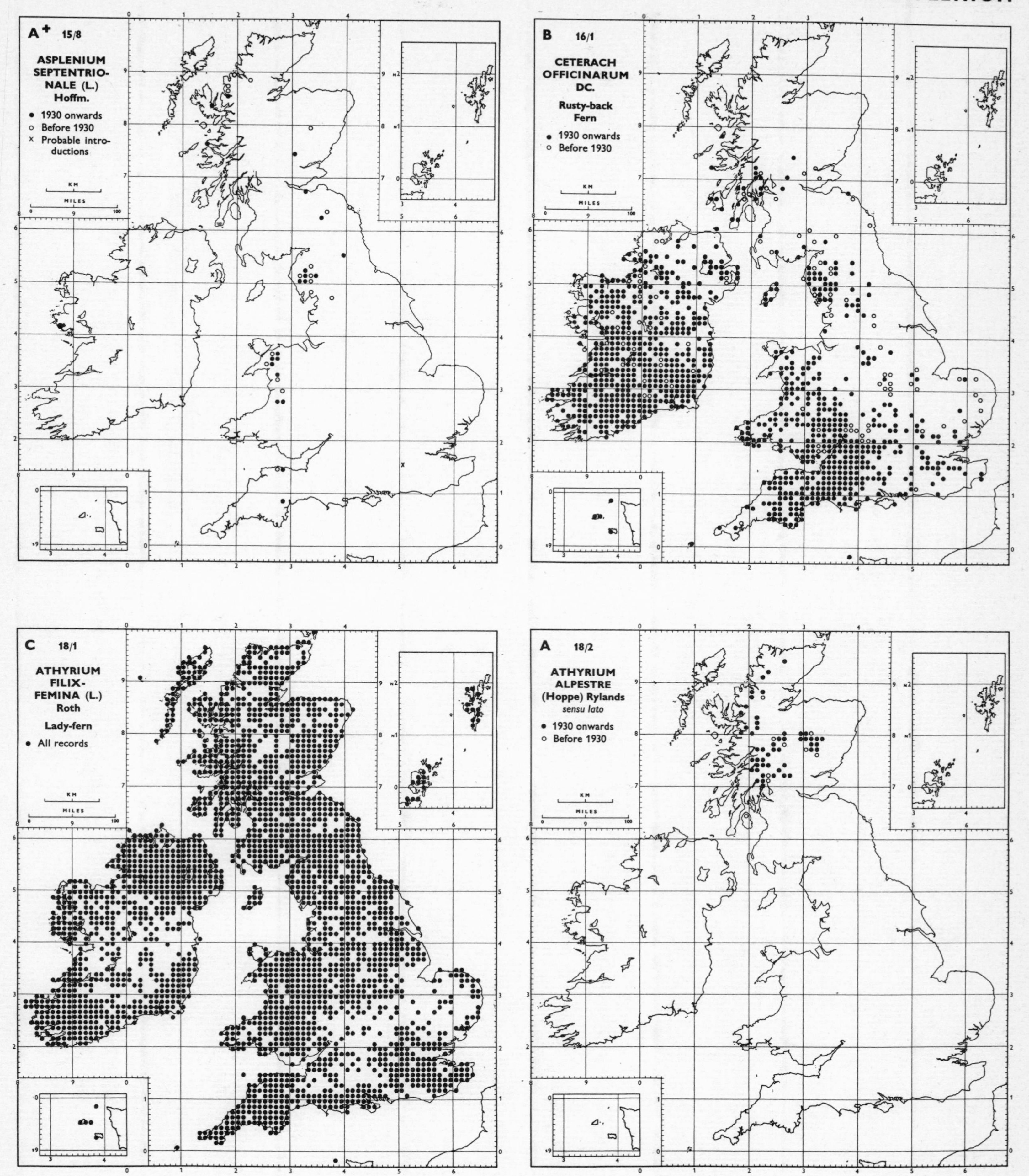
A+ 15/8
ASPLENIUM SEPTENTRIONALE (L.) Hoffm.
• 1930 onwards
○ Before 1930
× Probable introductions
B 16/1
CETERACH OFFICINARUM DC.
Rusty-back Fern
• 1930 onwards
○ Before 1930
C 18/1
ATHYRIUM FILIX-FEMINA (L.) Roth
Lady-fern
• All records
A 18/2
ATHYRIUM ALPESTRE (Hoppe) Rylands
sensu lato
• 1930 onwards
○ Before 1930
KM
MILES

ATHYRIACEAE

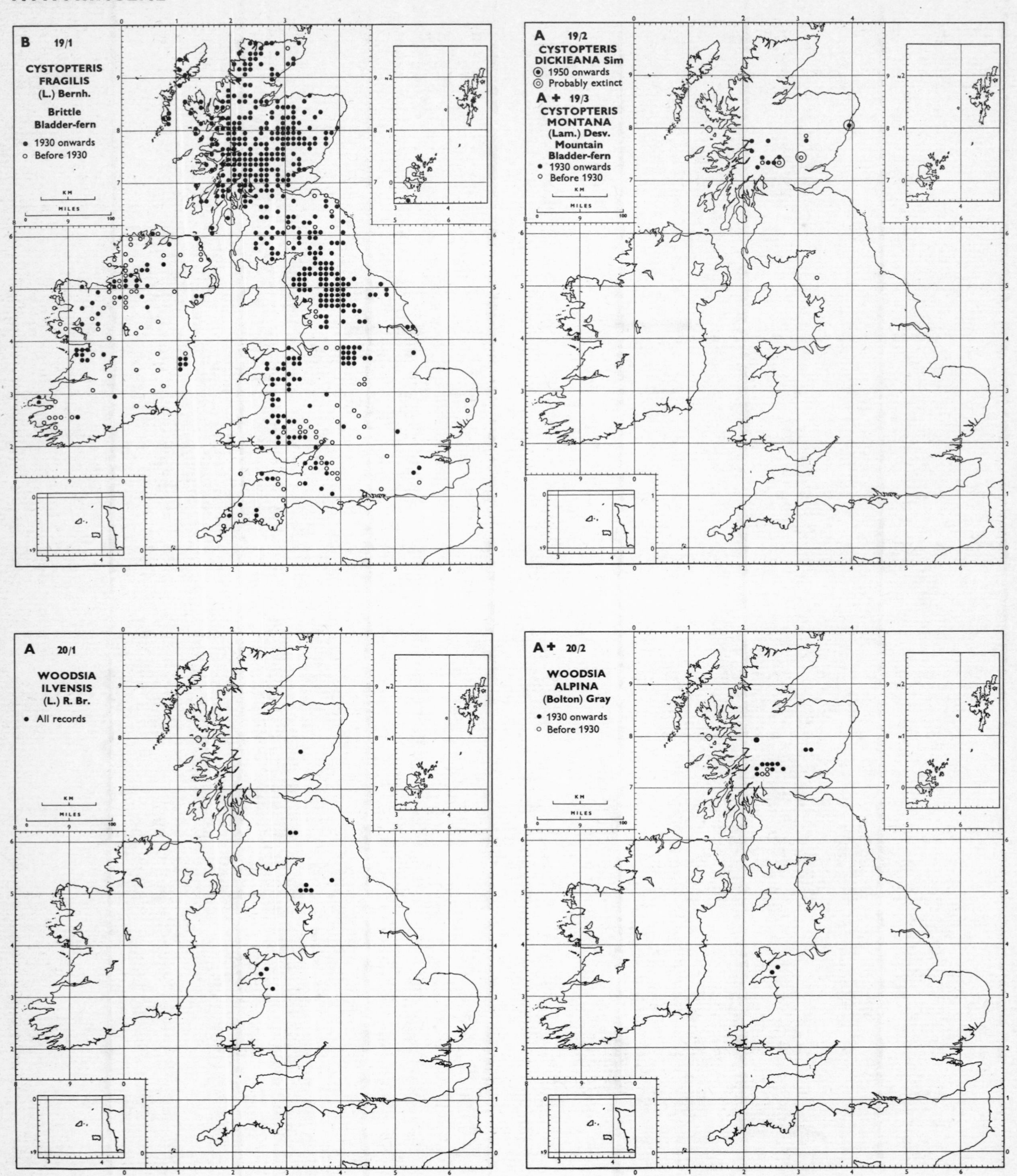

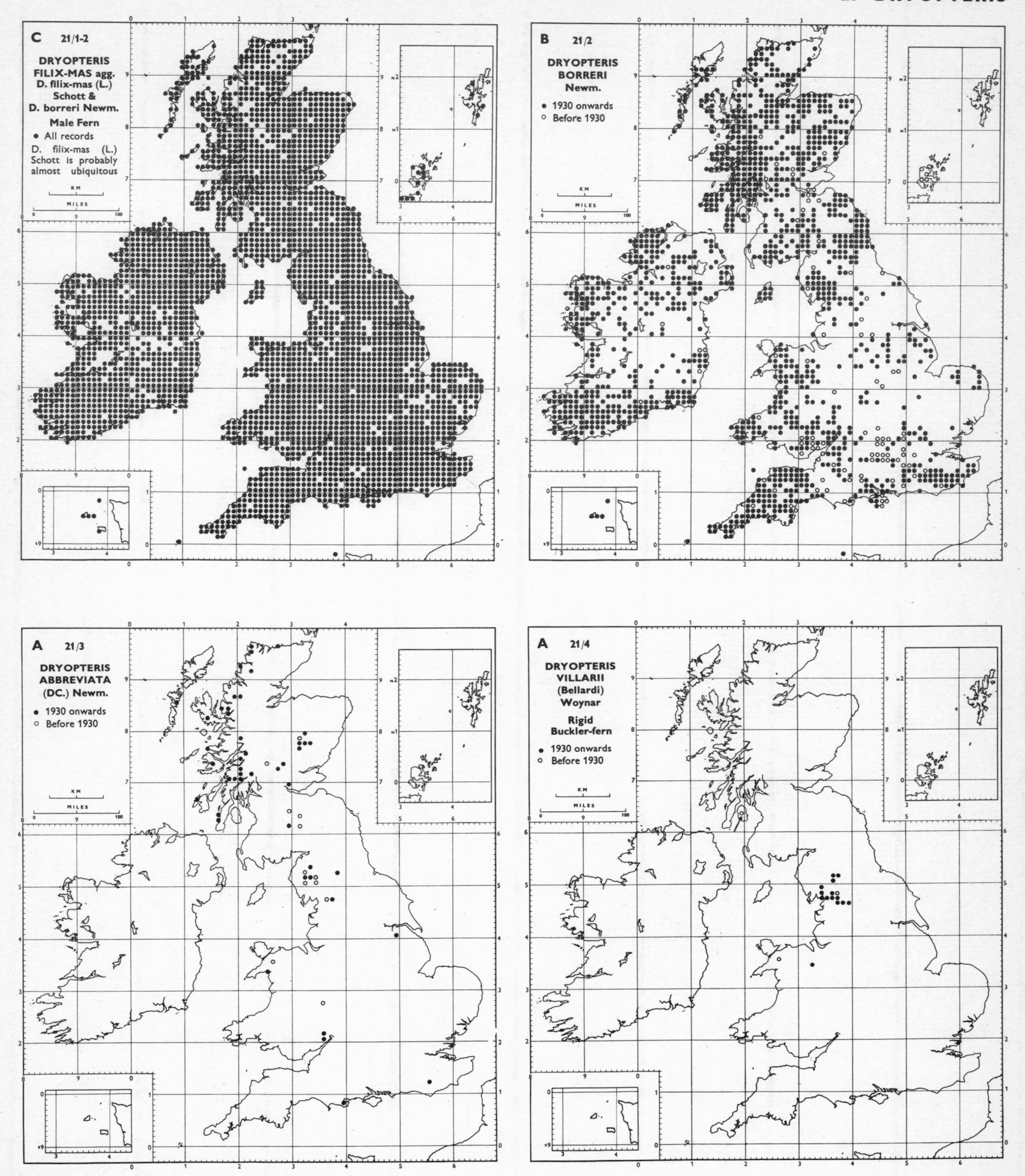
C 21/1-2
DRYOPTERIS FILIX-MAS agg. D. filix-mas (L.) Schott & D. borreri Newm.
Male Fern
• All records
D. filix-mas (L.) Schott is probably almost ubiquitous
B 21/2
DRYOPTERIS BORRERI Newm.
• 1930 onwards
○ Before 1930
A 21/3
DRYOPTERIS ABBREVIATA (DC.) Newm.
• 1930 onwards
○ Before 1930
A 21/4
DRYOPTERIS VILLARII (Bellardi) Woynar
Rigid Buckler-fern
• 1930 onwards
○ Before 1930

ASPIDIACEAE

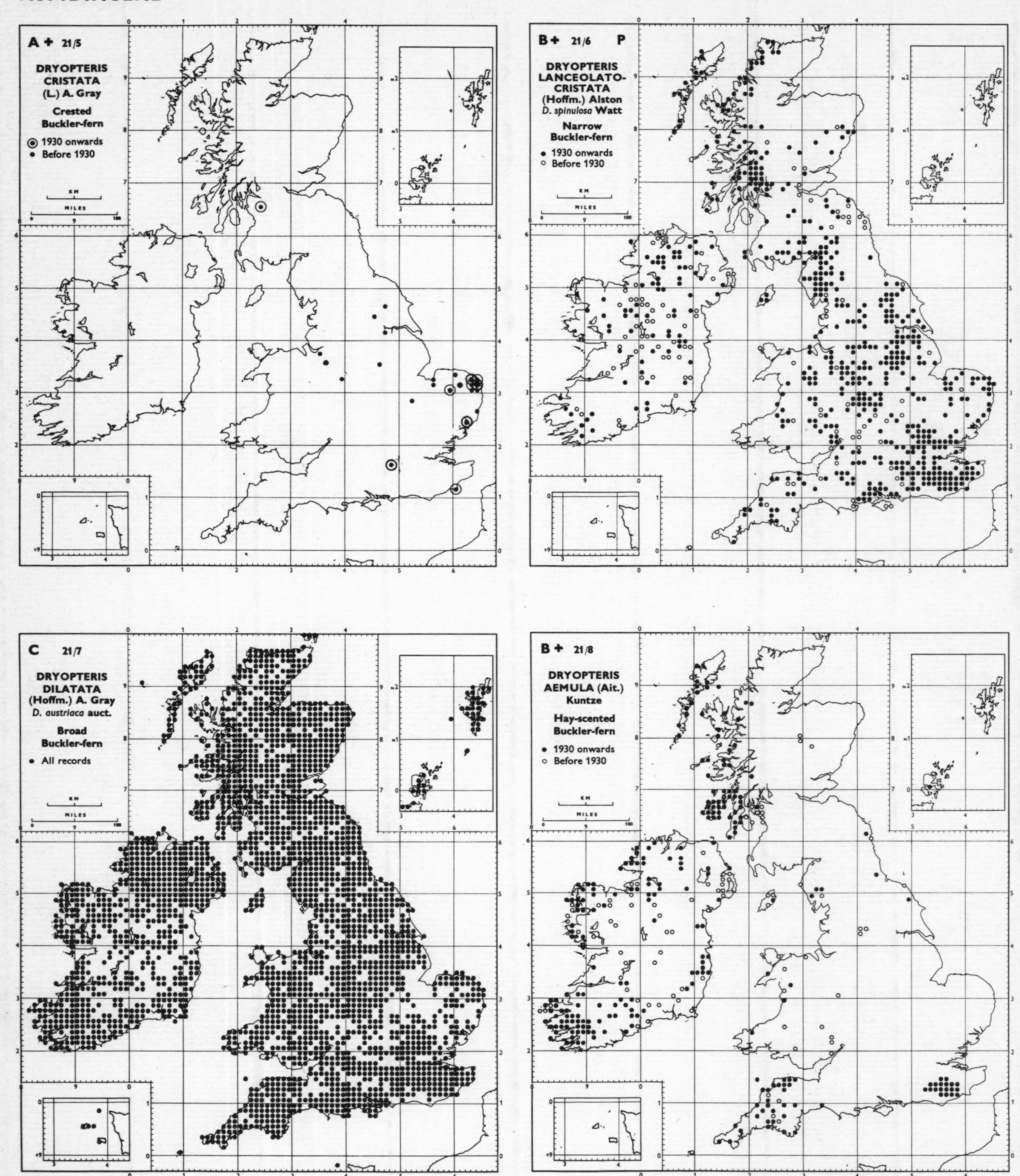

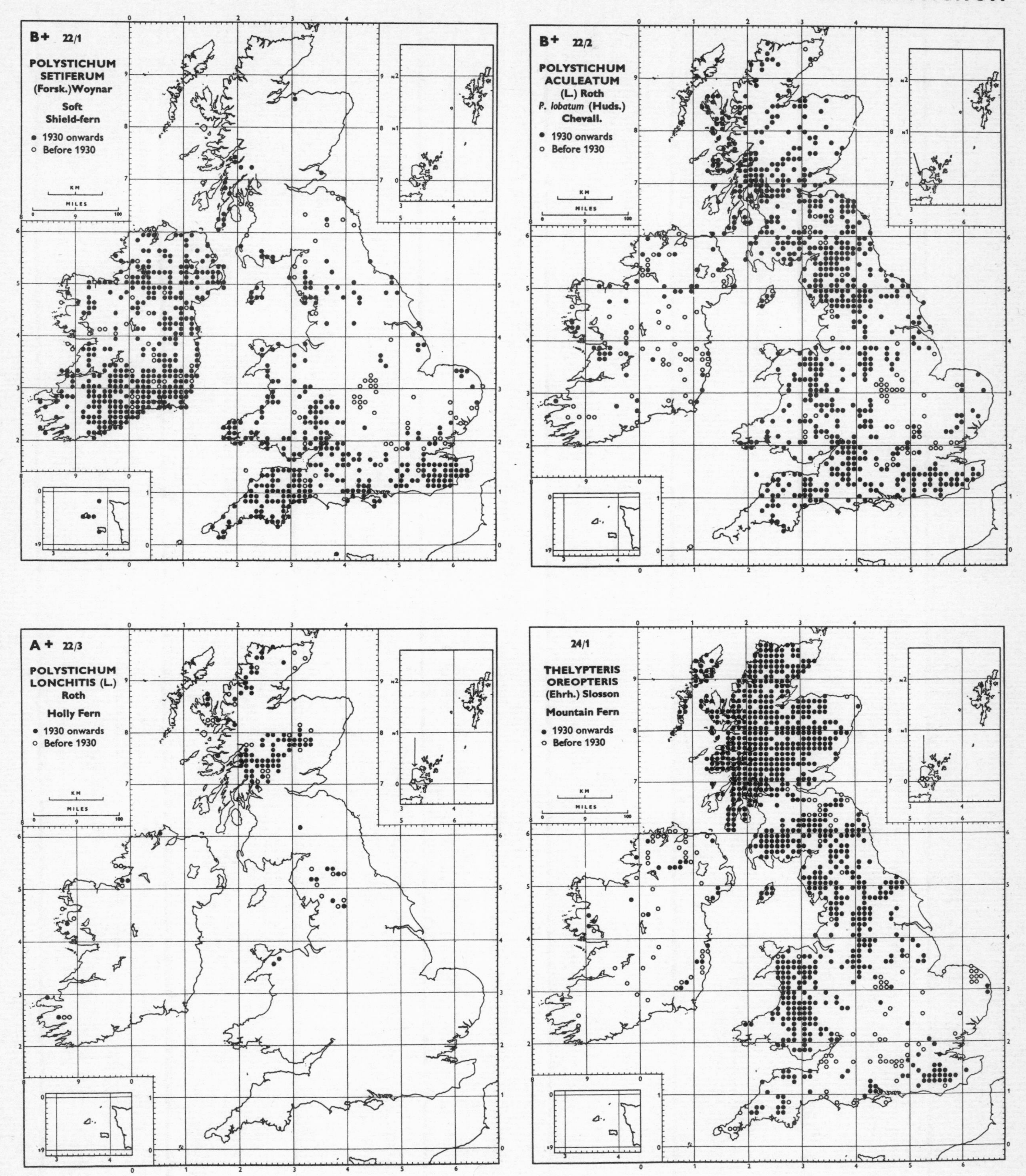

B+ 22/1
POLYSTICHUM SETIFERUM (Forsk.) Woynar
Soft Shield-fern
● 1930 onwards
○ Before 1930
B+ 22/2
POLYSTICHUM ACULEATUM (L.) Roth
P. lobatum (Huds.) Chevall.
● 1930 onwards
○ Before 1930
A+ 22/3
POLYSTICHUM LONCHITIS (L.) Roth
Holly Fern
● 1930 onwards
○ Before 1930
24/1
THELYPTERIS OREOPTERIS (Ehrh.) Slosson
Mountain Fern
● 1930 onwards
○ Before 1930

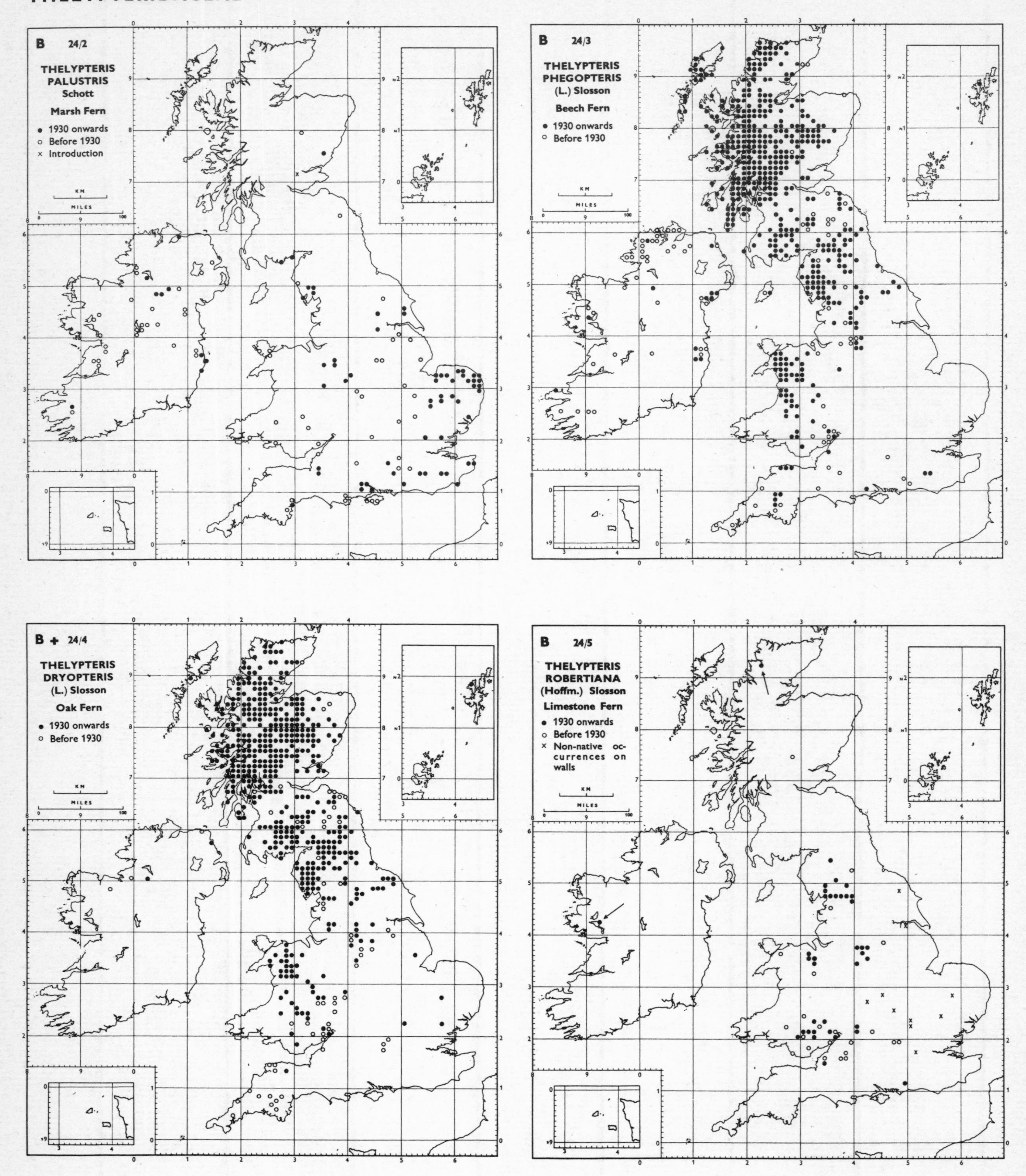
B 24/2
THELYPTERIS PALUSTRIS Schott
Marsh Fern
1930 onwards
Before 1930
Introduction
B 24/3
THELYPTERIS PHEGOPTERIS (L.) Slosson
Beech Fern
1930 onwards
Before 1930
B + 24/4
THELYPTERIS DRYOPTERIS (L.) Slosson
Oak Fern
1930 onwards
Before 1930
B 24/5
THELYPTERIS ROBERTIANA (Hoffm.) Slosson
Limestone Fern
1930 onwards
Before 1930
Non-native occurrences on walls
KM
MILES

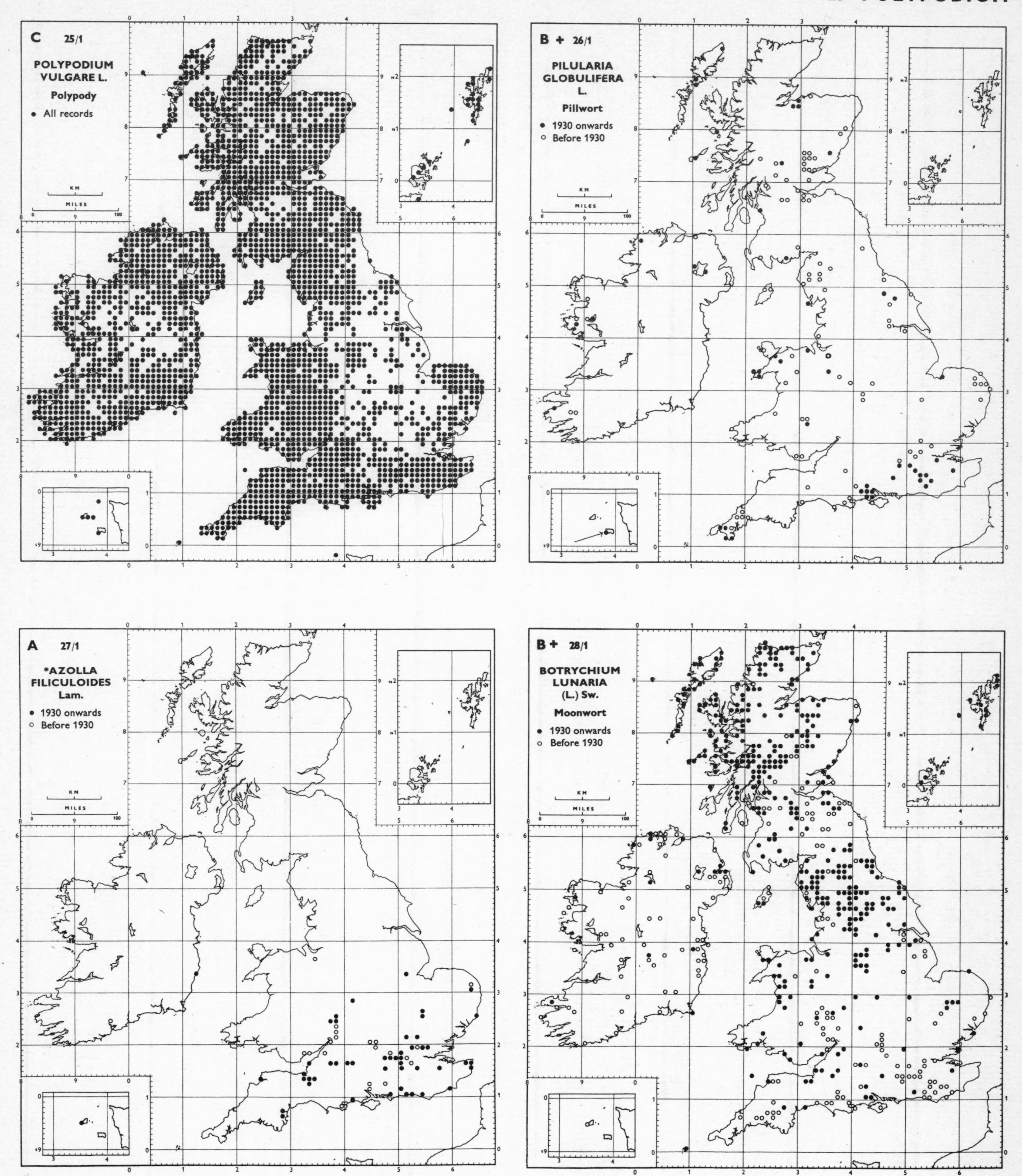
C 25/1
POLYPODIUM VULGARE L.
Polypody
● All records
B + 26/1
PILULARIA GLOBULIFERA L.
Pillwort
● 1930 onwards
○ Before 1930
A 27/1
*AZOLLA FILICULOIDES Lam.
● 1930 onwards
○ Before 1930
B + 28/1
BOTRYCHIUM LUNARIA (L.) Sw.
Moonwort
● 1930 onwards
○ Before 1930

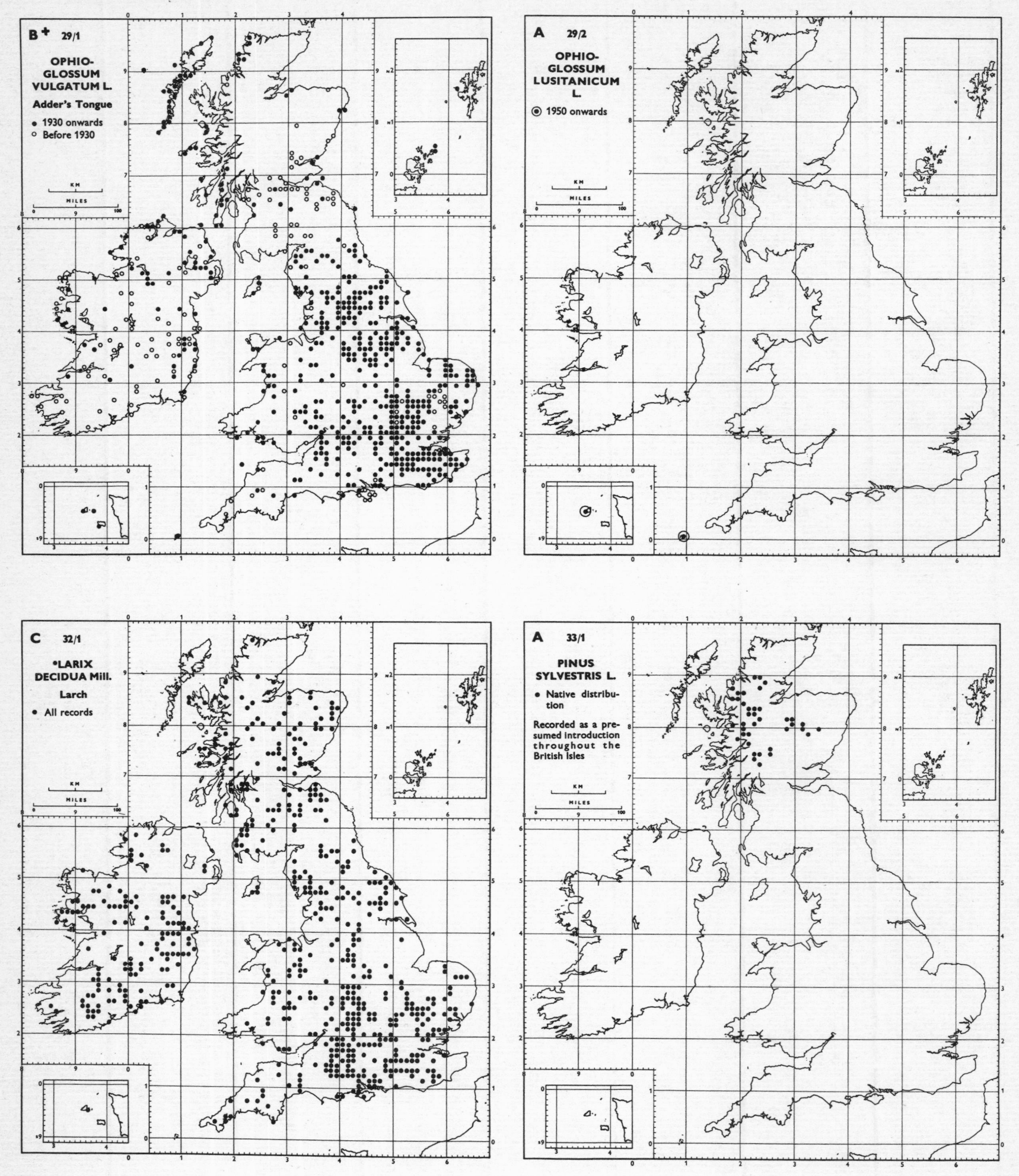

B+ 29/1
OPHIOGLOSSUM VULGATUM L.
Adder's Tongue
1930 onwards
Before 1930
A 29/2
OPHIOGLOSSUM LUSITANICUM L.
1950 onwards
C 32/1
*LARIX DECIDUA Mill.
Larch
All records
A 33/1
PINUS SYLVESTRIS L.
Native distribution
Recorded as a presumed introduction throughout the British Isles

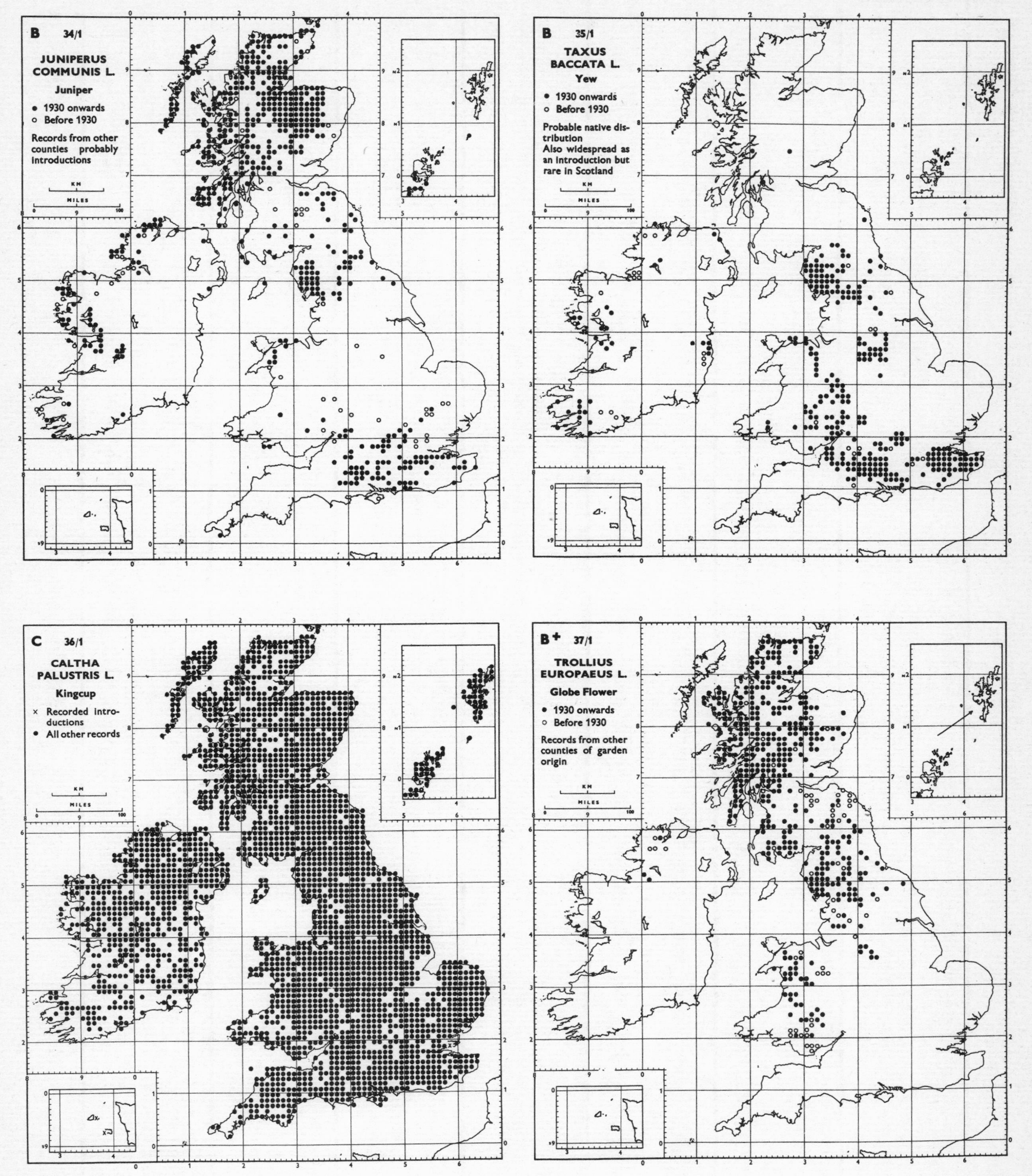
B 34/1
JUNIPERUS COMMUNIS L.
Juniper
• 1930 onwards
○ Before 1930
Records from other counties probably introductions
KM
MILES
B 35/1
TAXUS BACCATA L.
Yew
• 1930 onwards
○ Before 1930
Probable native distribution
Also widespread as an introduction but rare in Scotland
C 36/1
CALTHA PALUSTRIS L.
Kingcup
× Recorded introductions
• All other records
B+ 37/1
TROLLIUS EUROPAEUS L.
Globe Flower
• 1930 onwards
○ Before 1930
Records from other counties of garden origin

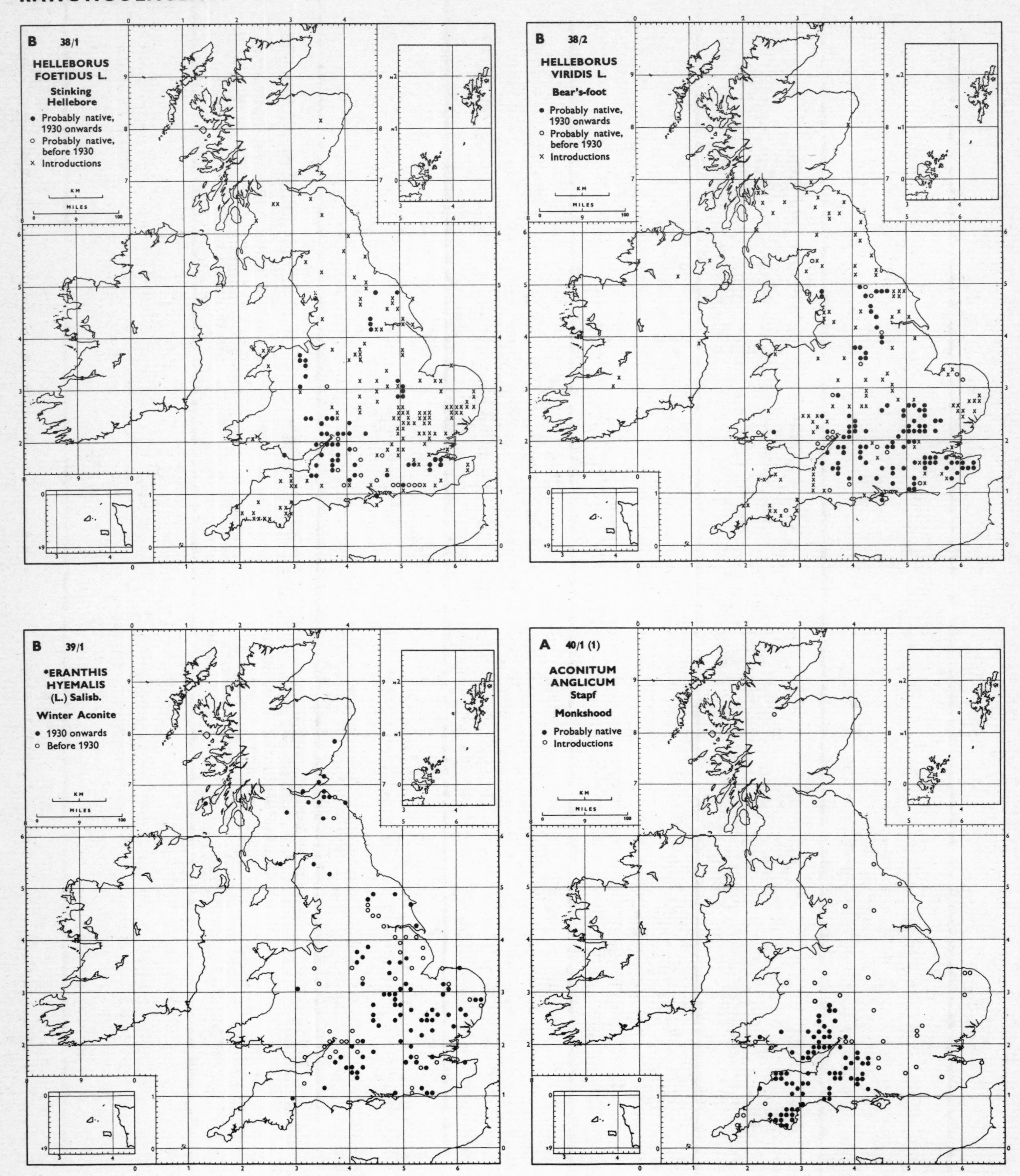
B 38/1
HELLEBORUS FOETIDUS L.
Stinking Hellebore
● Probably native, 1930 onwards
○ Probably native, before 1930
× Introductions
B 38/2
HELLEBORUS VIRIDIS L.
Bear's-foot
● Probably native, 1930 onwards
○ Probably native, before 1930
× Introductions
B 39/1
•ERANTHIS HYEMALIS (L.) Salisb.
Winter Aconite
● 1930 onwards
○ Before 1930
A 40/1 (1)
ACONITUM ANGLICUM Stapf
Monkshood
● Probably native
○ Introductions

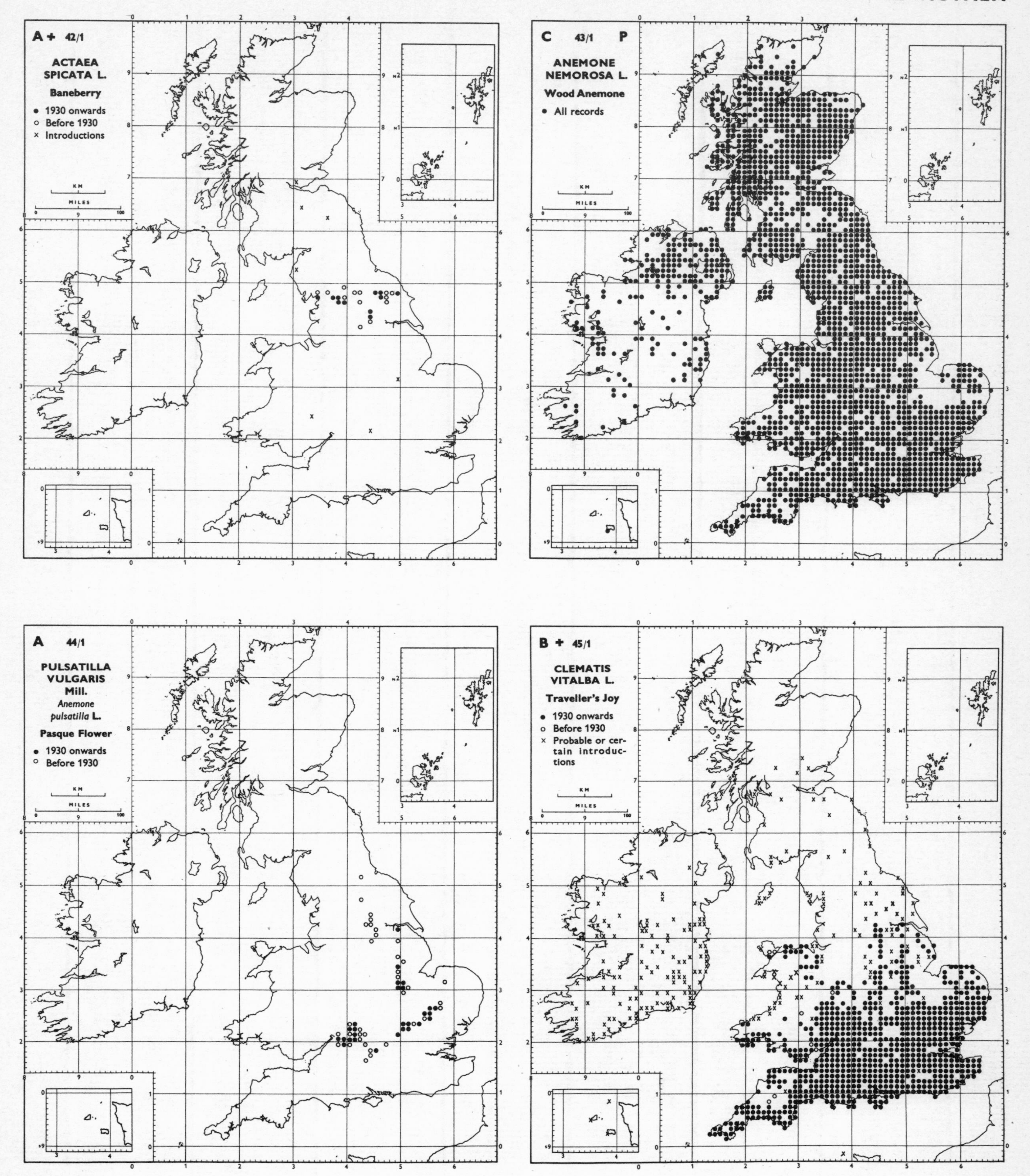
A + 42/1
ACTAEA SPICATA L.
Baneberry
• 1930 onwards
○ Before 1930
× Introductions
C 43/1 P
ANEMONE NEMOROSA L.
Wood Anemone
• All records
A 44/1
PULSATILLA VULGARIS Mill.
Anemone pulsatilla L.
Pasque Flower
• 1930 onwards
○ Before 1930
B + 45/1
CLEMATIS VITALBA L.
Traveller's Joy
• 1930 onwards
○ Before 1930
× Probable or certain introductions

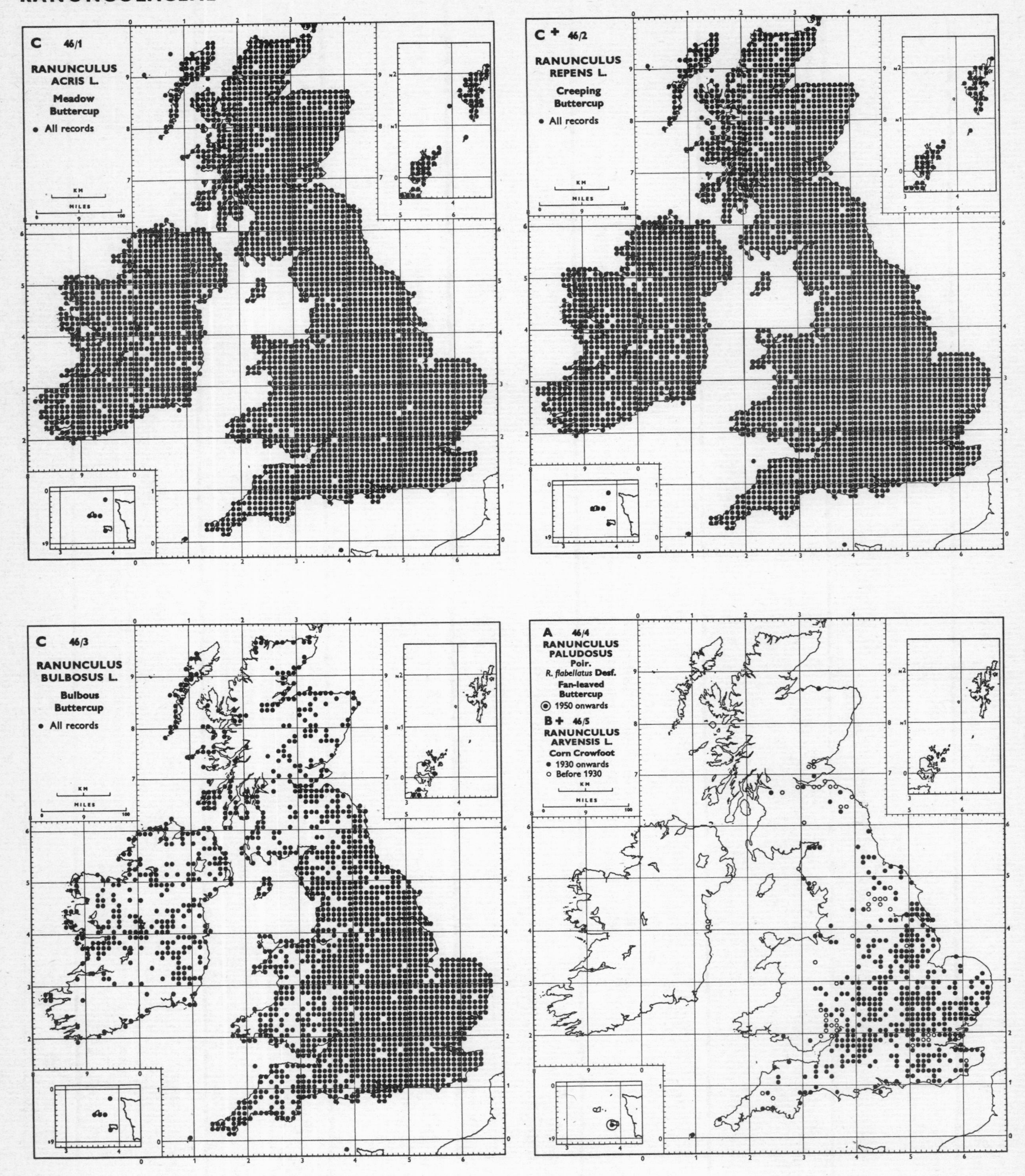
C 46/1
RANUNCULUS ACRIS L.
Meadow Buttercup
• All records
C + 46/2
RANUNCULUS REPENS L.
Creeping Buttercup
• All records
C 46/3
RANUNCULUS BULBOSUS L.
Bulbous Buttercup
• All records
A 46/4
RANUNCULUS PALUDOSUS Poir.
R. flabellatus Desf.
Fan-leaved Buttercup
1950 onwards
B + 46/5
RANUNCULUS ARVENSIS L.
Corn Crowfoot
• 1930 onwards
○ Before 1930
KM
MILES

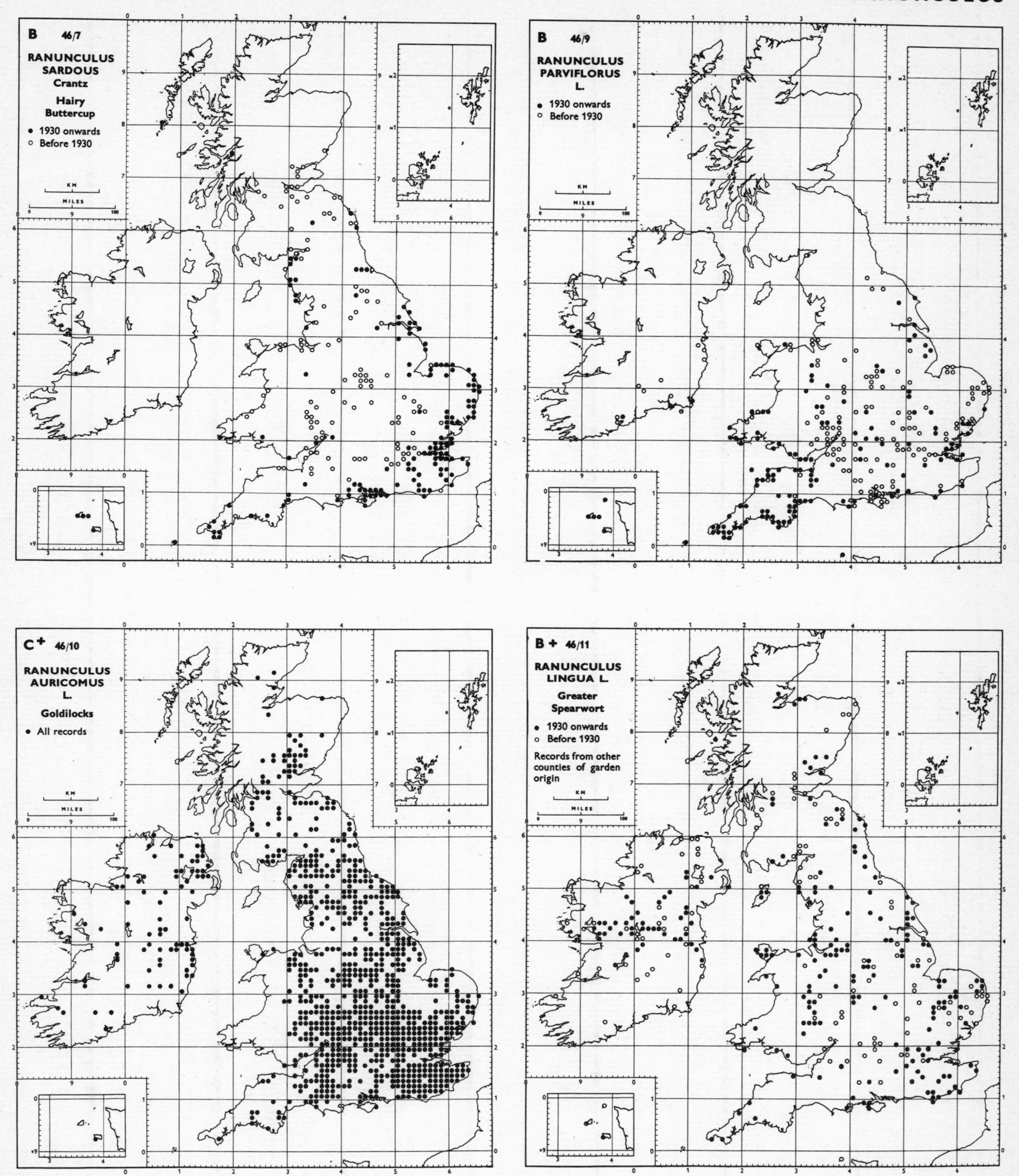

B 46/7
RANUNCULUS SARDOUS Crantz
Hairy Buttercup
• 1930 onwards
○ Before 1930
B 46/9
RANUNCULUS PARVIFLORUS L.
• 1930 onwards
○ Before 1930
C+ 46/10
RANUNCULUS AURICOMUS L.
Goldilocks
• All records
B + 46/11
RANUNCULUS LINGUA L.
Greater Spearwort
• 1930 onwards
○ Before 1930
Records from other counties of garden origin

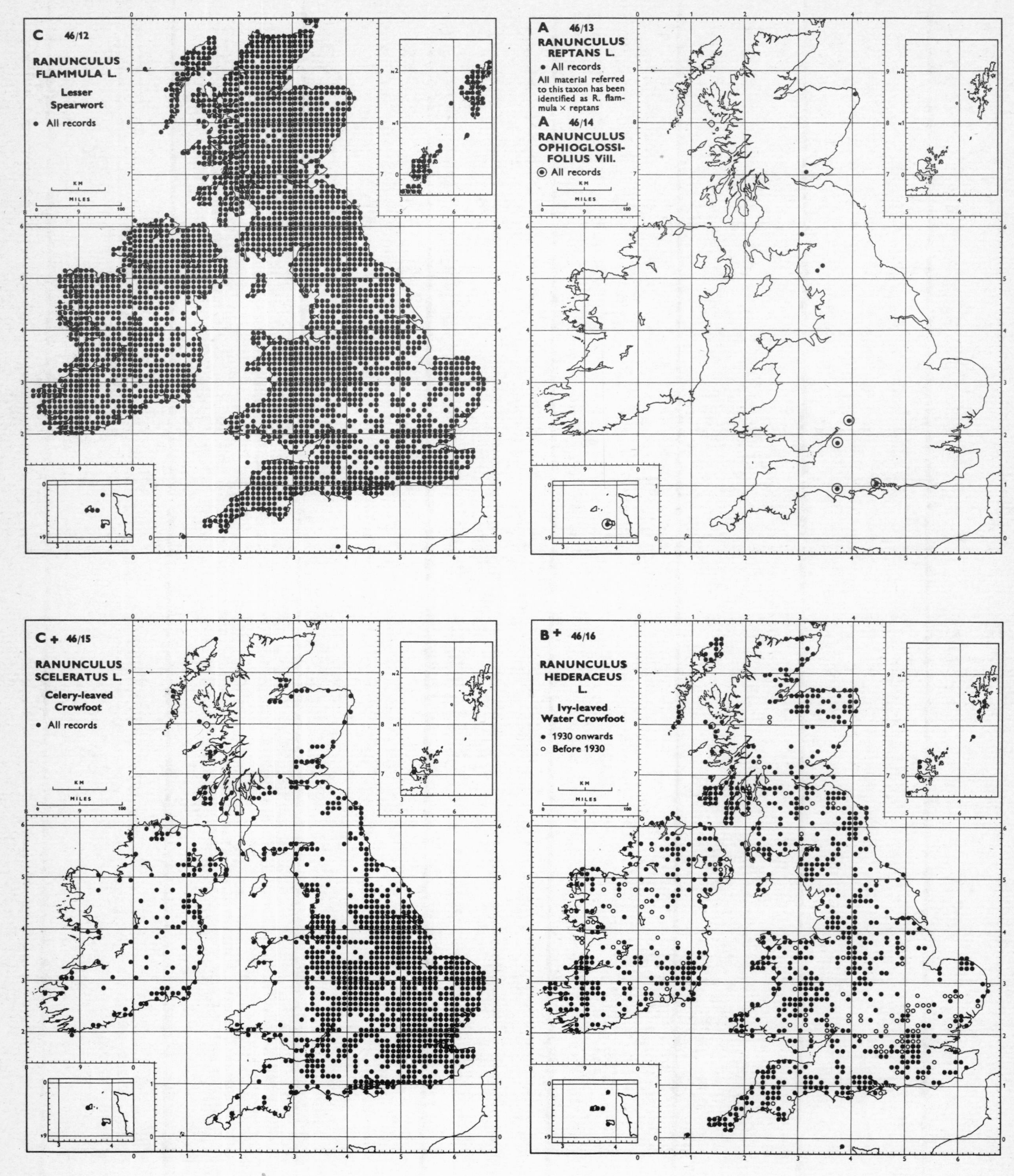
C 46/12
RANUNCULUS FLAMMULA L.
Lesser Spearwort
• All records
A 46/13
RANUNCULUS REPTANS L.
• All records
All material referred to this taxon has been identified as R. flammula × reptans
A 46/14
RANUNCULUS OPHIOGLOSSIFOLIUS Vill.
◉ All records
C + 46/15
RANUNCULUS SCELERATUS L.
Celery-leaved Crowfoot
• All records
B + 46/16
RANUNCULUS HEDERACEUS L.
Ivy-leaved Water Crowfoot
• 1930 onwards
○ Before 1930
KM
MILES

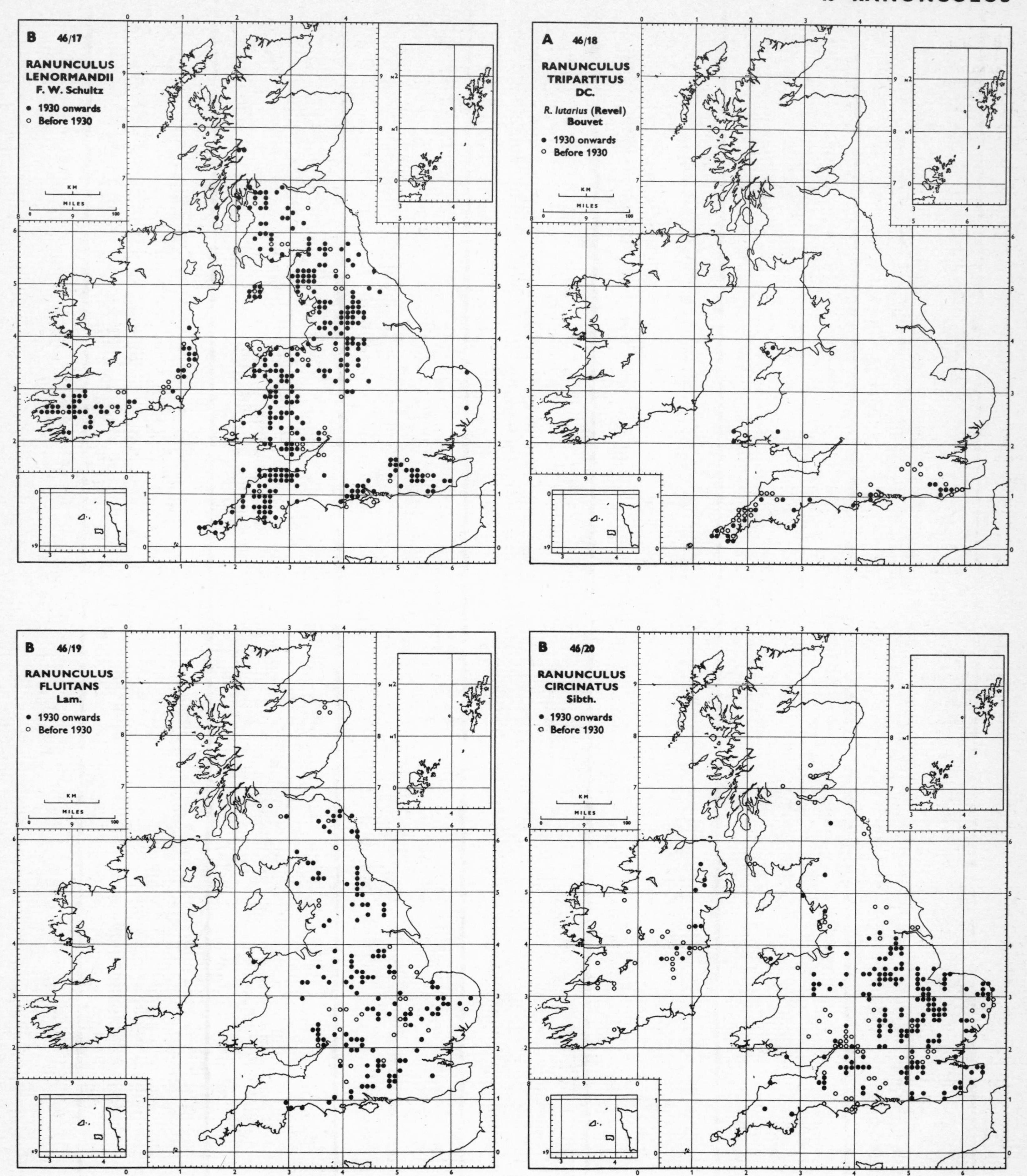
B 46/17
RANUNCULUS
LENORMANDII
F. W. Schultz
• 1930 onwards
○ Before 1930
A 46/18
RANUNCULUS
TRIPARTITUS
DC.
R. lutarius (Revel)
Bouvet
• 1930 onwards
○ Before 1930
B 46/19
RANUNCULUS
FLUITANS
Lam.
• 1930 onwards
○ Before 1930
B 46/20
RANUNCULUS
CIRCINATUS
Sibth.
• 1930 onwards
○ Before 1930
KM
MILES

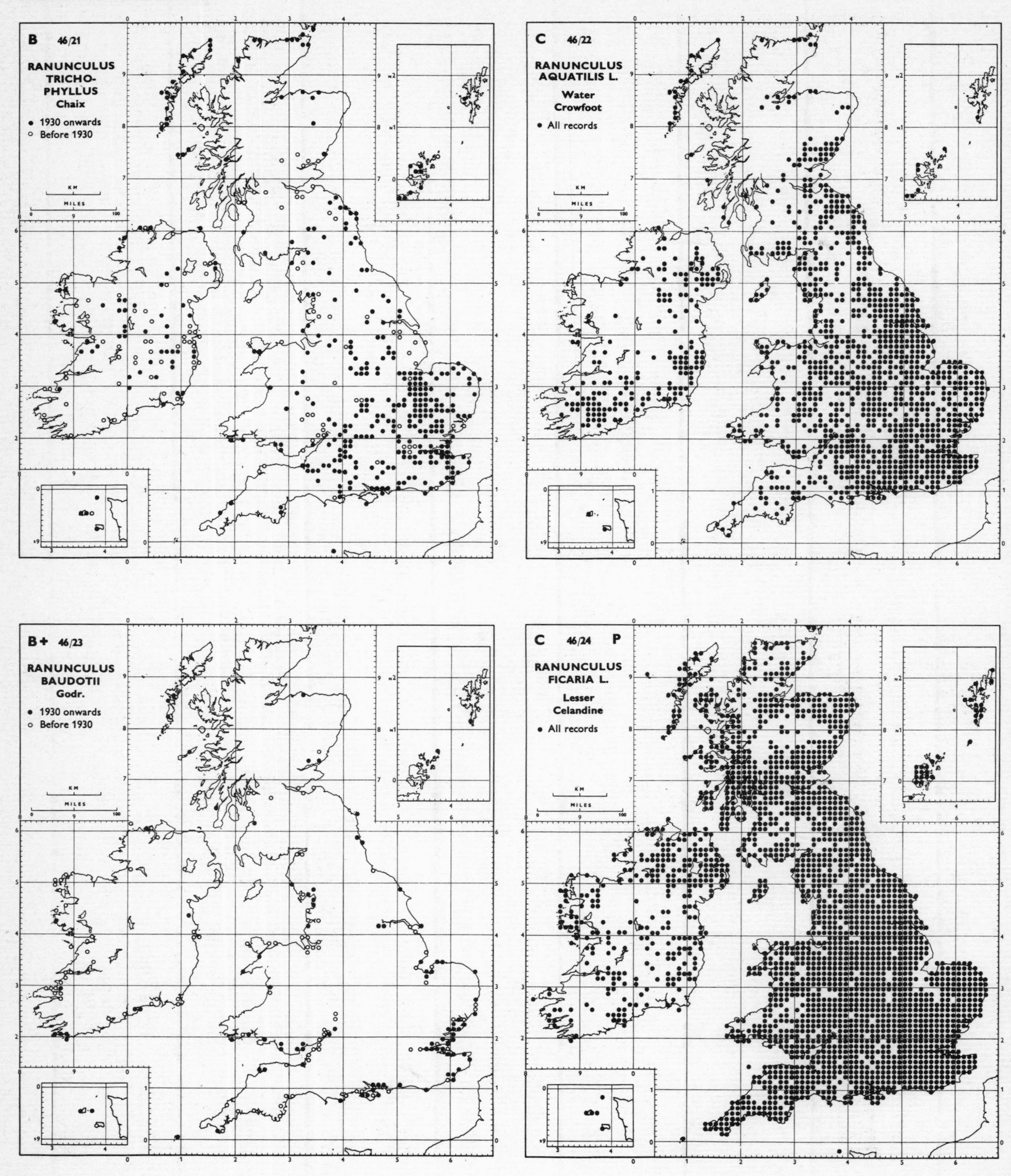
B 46/21
RANUNCULUS TRICHO-PHYLLUS Chaix
• 1930 onwards
○ Before 1930
C 46/22
RANUNCULUS AQUATILIS L.
Water Crowfoot
• All records
B+ 46/23
RANUNCULUS BAUDOTII Godr.
• 1930 onwards
○ Before 1930
C 46/24 P
RANUNCULUS FICARIA L.
Lesser Celandine
• All records

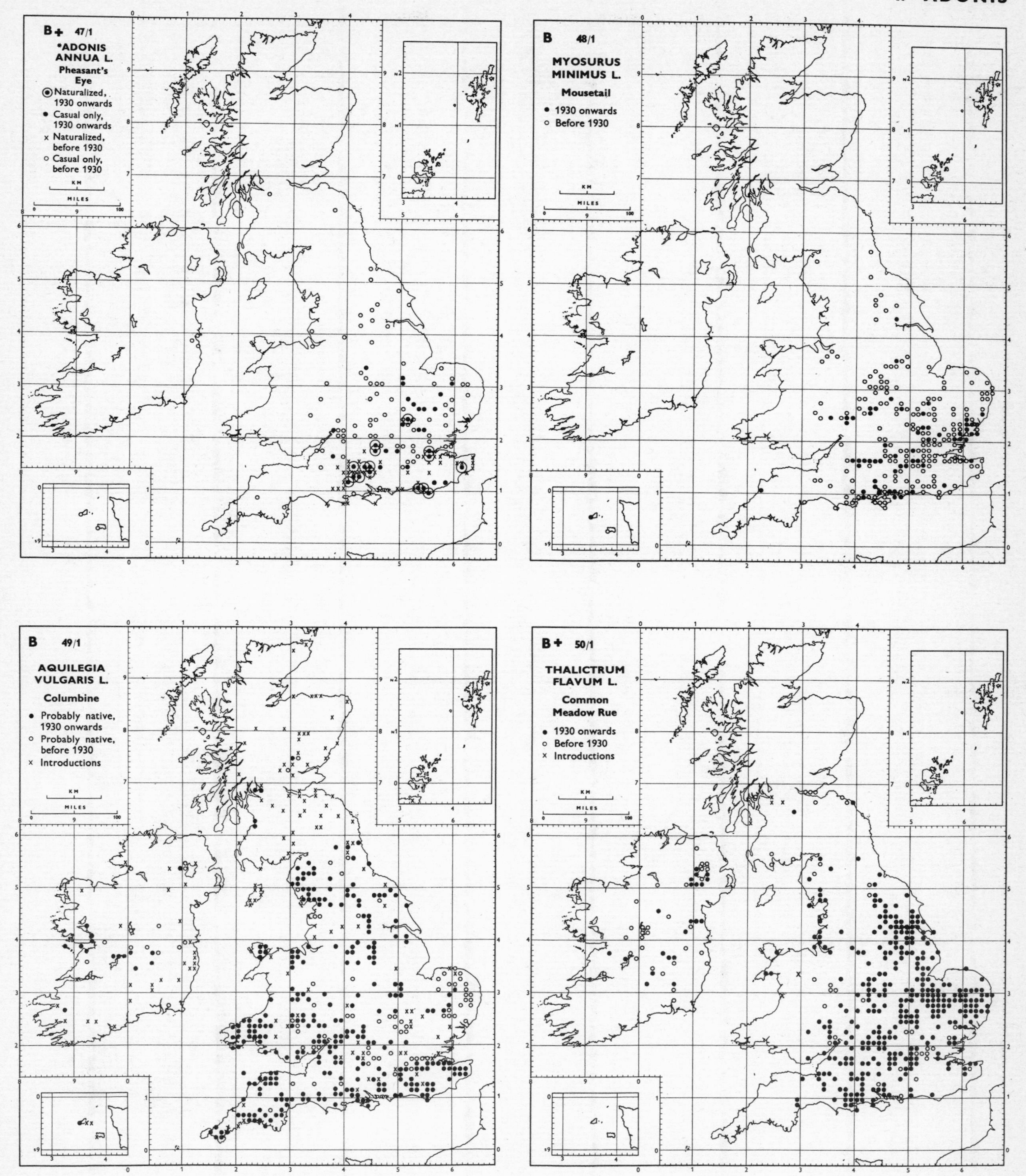
B+ 47/1
*ADONIS ANNUA L.
Pheasant's Eye
Naturalized, 1930 onwards
Casual only, 1930 onwards
Naturalized, before 1930
Casual only, before 1930
KM
MILES
B 48/1
MYOSURUS MINIMUS L.
Mousetail
1930 onwards
Before 1930
B 49/1
AQUILEGIA VULGARIS L.
Columbine
Probably native, 1930 onwards
Probably native, before 1930
Introductions
B+ 50/1
THALICTRUM FLAVUM L.
Common Meadow Rue
1930 onwards
Before 1930
Introductions

RANUNCULACEAE

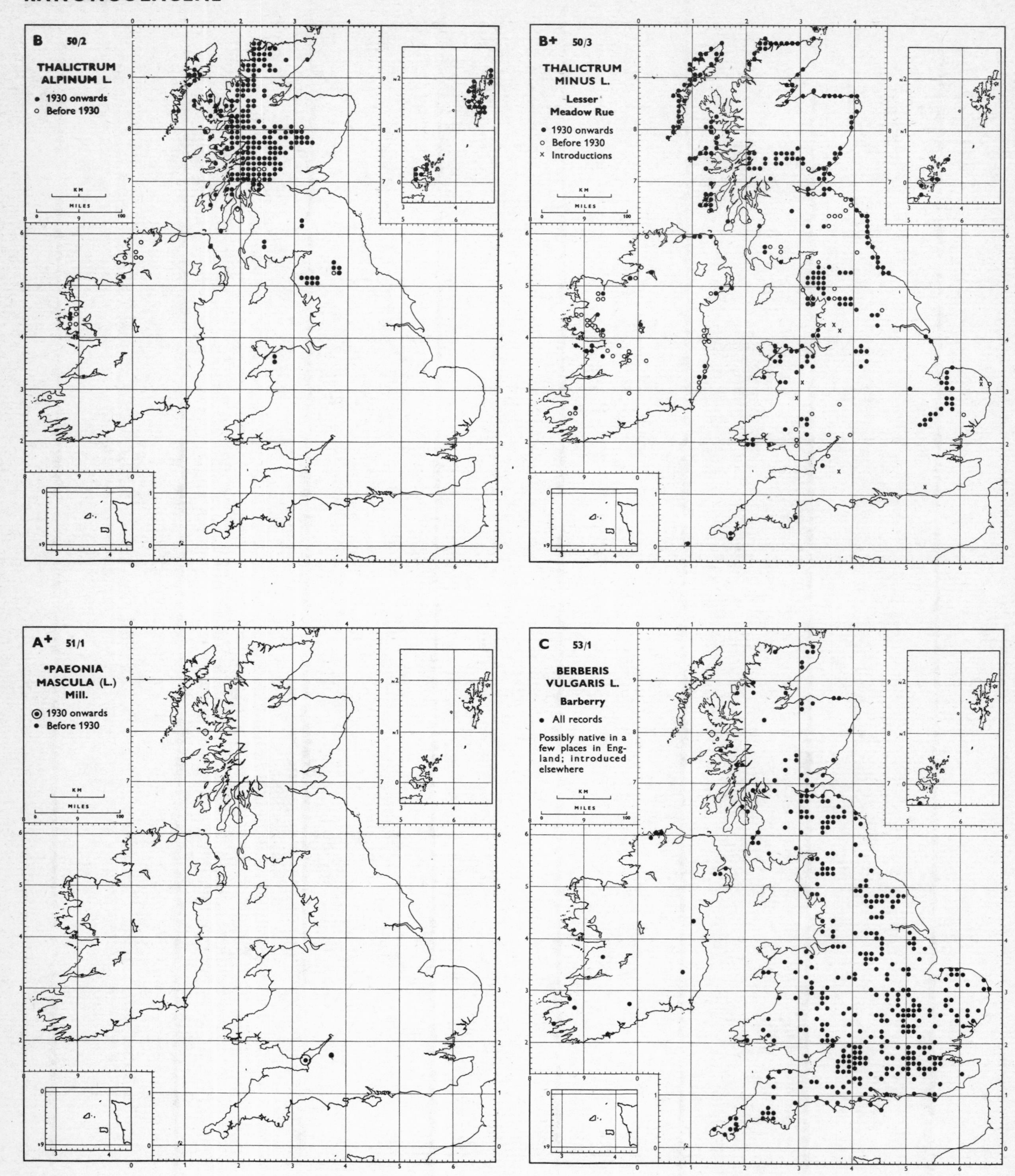

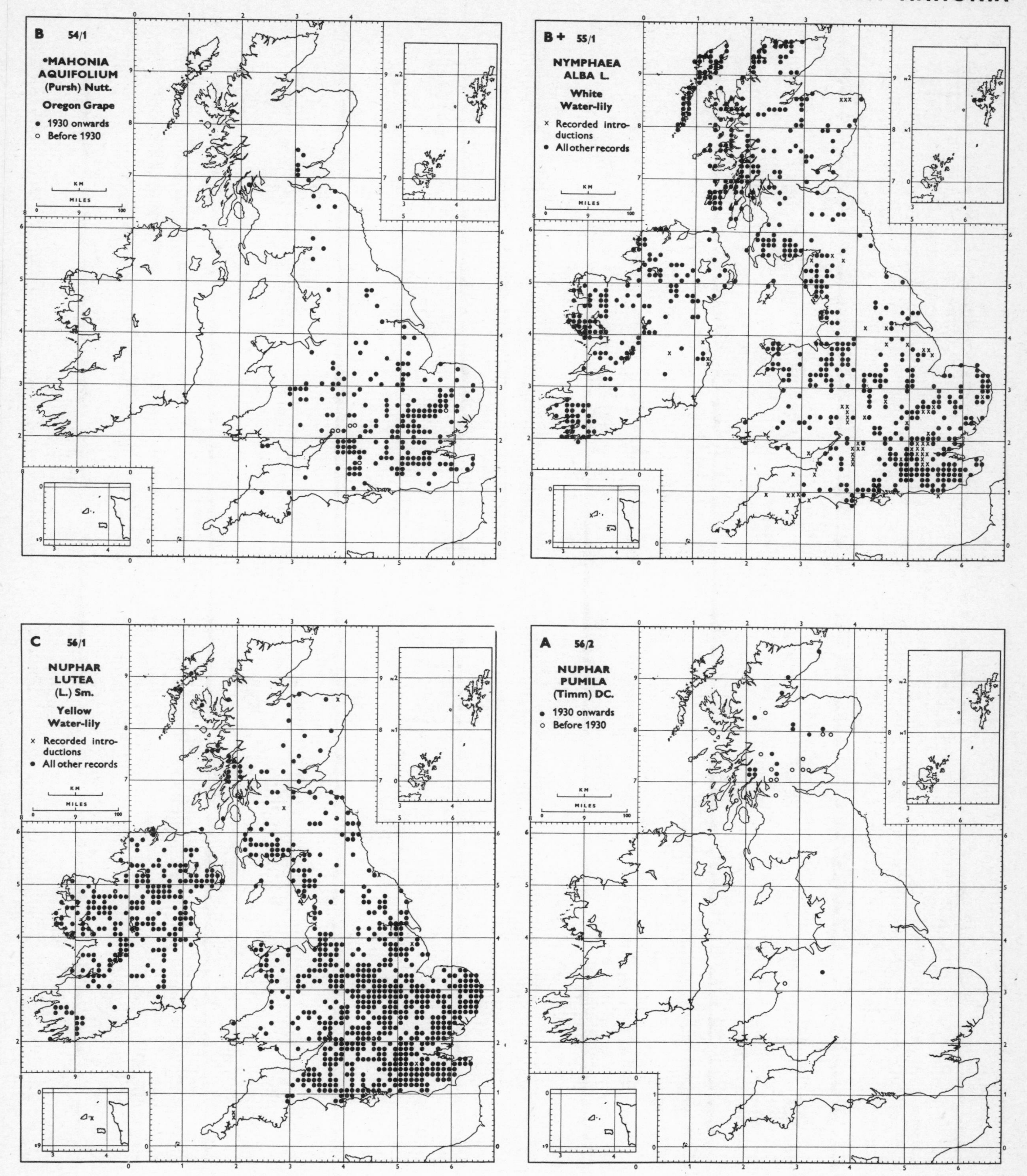
B 54/1
*MAHONIA AQUIFOLIUM (Pursh) Nutt.
Oregon Grape
1930 onwards
Before 1930
B+ 55/1
NYMPHAEA ALBA L.
White Water-lily
Recorded introductions
All other records
C 56/1
NUPHAR LUTEA (L.) Sm.
Yellow Water-lily
Recorded introductions
All other records
A 56/2
NUPHAR PUMILA (Timm) DC.
1930 onwards
Before 1930

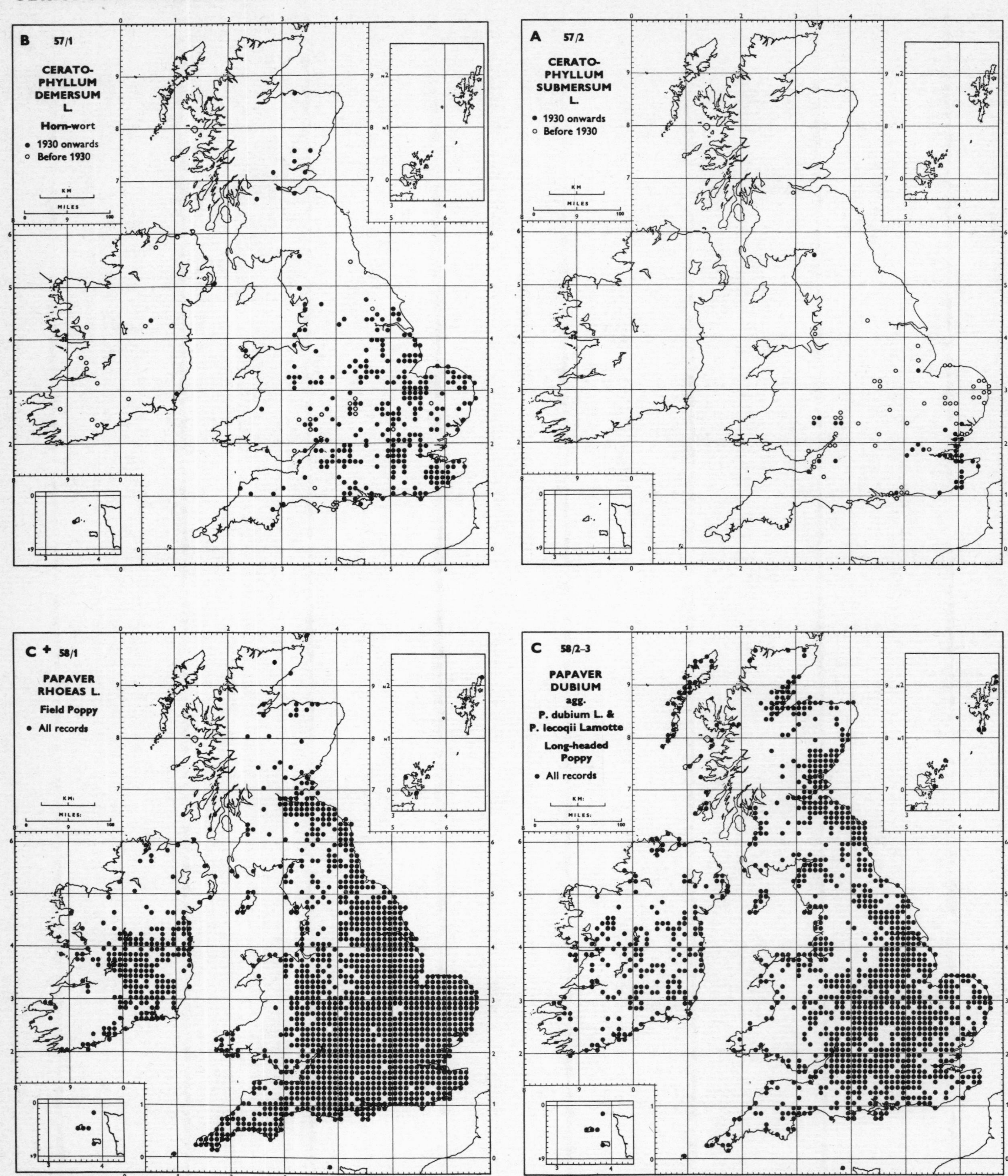
B 57/1
CERATO-PHYLLUM DEMERSUM L.
Horn-wort
1930 onwards
Before 1930
KM
MILES
A 57/2
CERATO-PHYLLUM SUBMERSUM L.
1930 onwards
Before 1930
C + 58/1
PAPAVER RHOEAS L.
Field Poppy
All records
C 58/2-3
PAPAVER DUBIUM agg.
P. dubium L. &
P. lecoqii Lamotte
Long-headed Poppy
All records

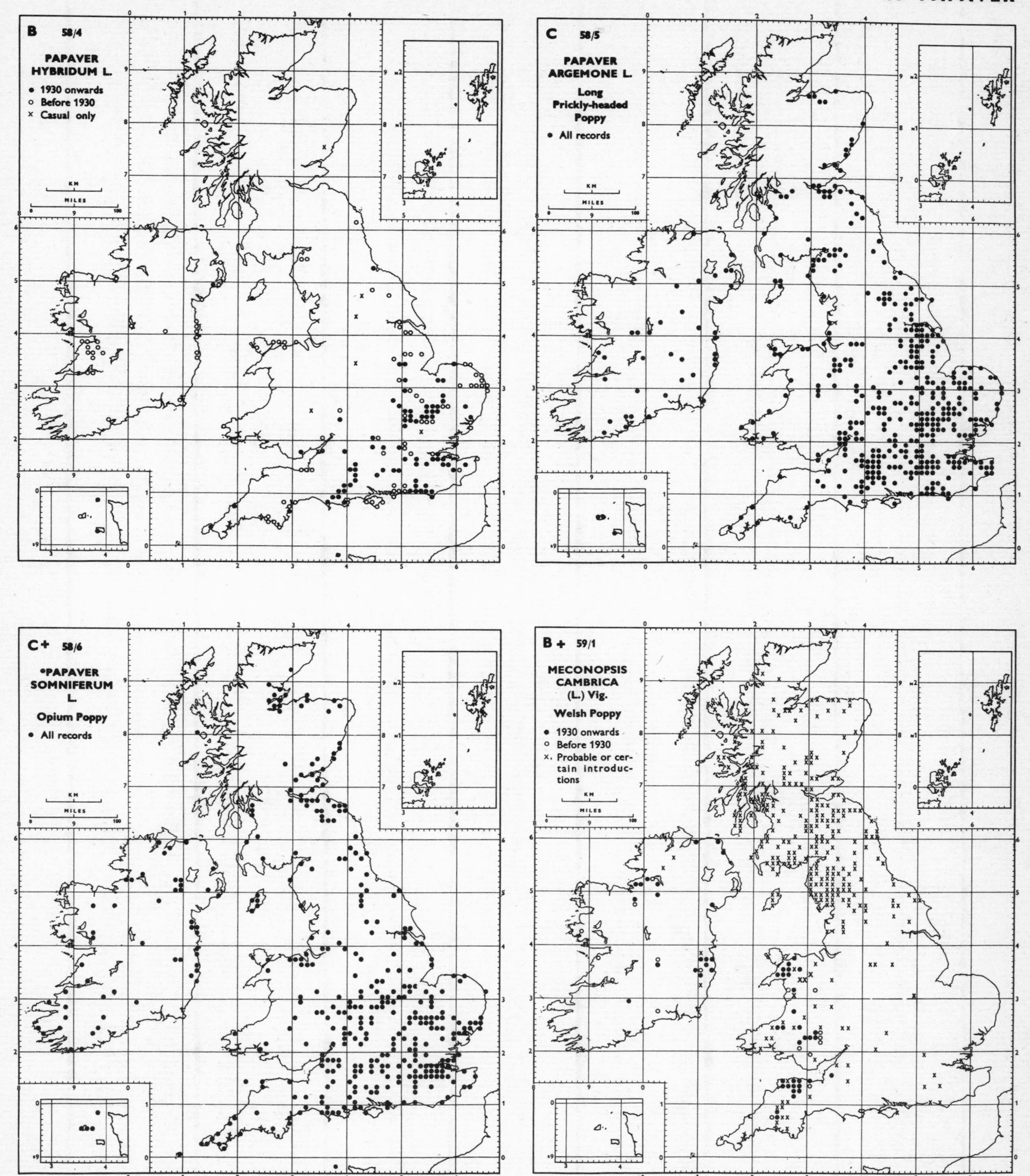
B 58/4
PAPAVER HYBRIDUM L.
• 1930 onwards
○ Before 1930
× Casual only
C 58/5
PAPAVER ARGEMONE L.
Long Prickly-headed Poppy
• All records
C+ 58/6
*PAPAVER SOMNIFERUM L.
Opium Poppy
• All records
B+ 59/1
MECONOPSIS CAMBRICA (L.) Vig.
Welsh Poppy
• 1930 onwards
○ Before 1930
×. Probable or certain introductions
KM
MILES

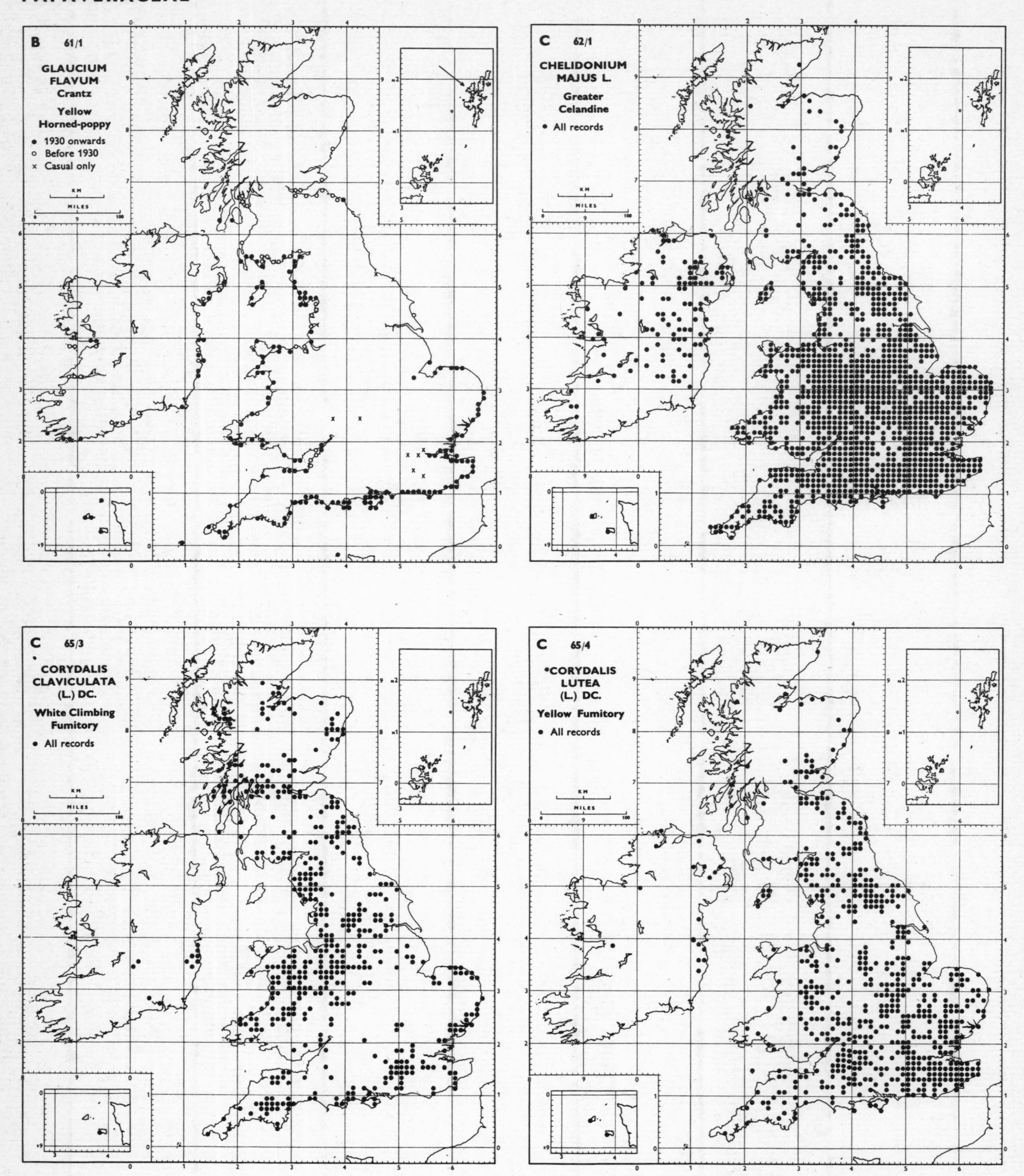
B 61/1
GLAUCIUM FLAVUM Crantz
Yellow Horned-poppy
• 1930 onwards
○ Before 1930
× Casual only
C 62/1
CHELIDONIUM MAJUS L.
Greater Celandine
• All records
C 65/3
CORYDALIS CLAVICULATA (L.) DC.
White Climbing Fumitory
• All records
C 65/4
*CORYDALIS LUTEA (L.) DC.
Yellow Fumitory
• All records

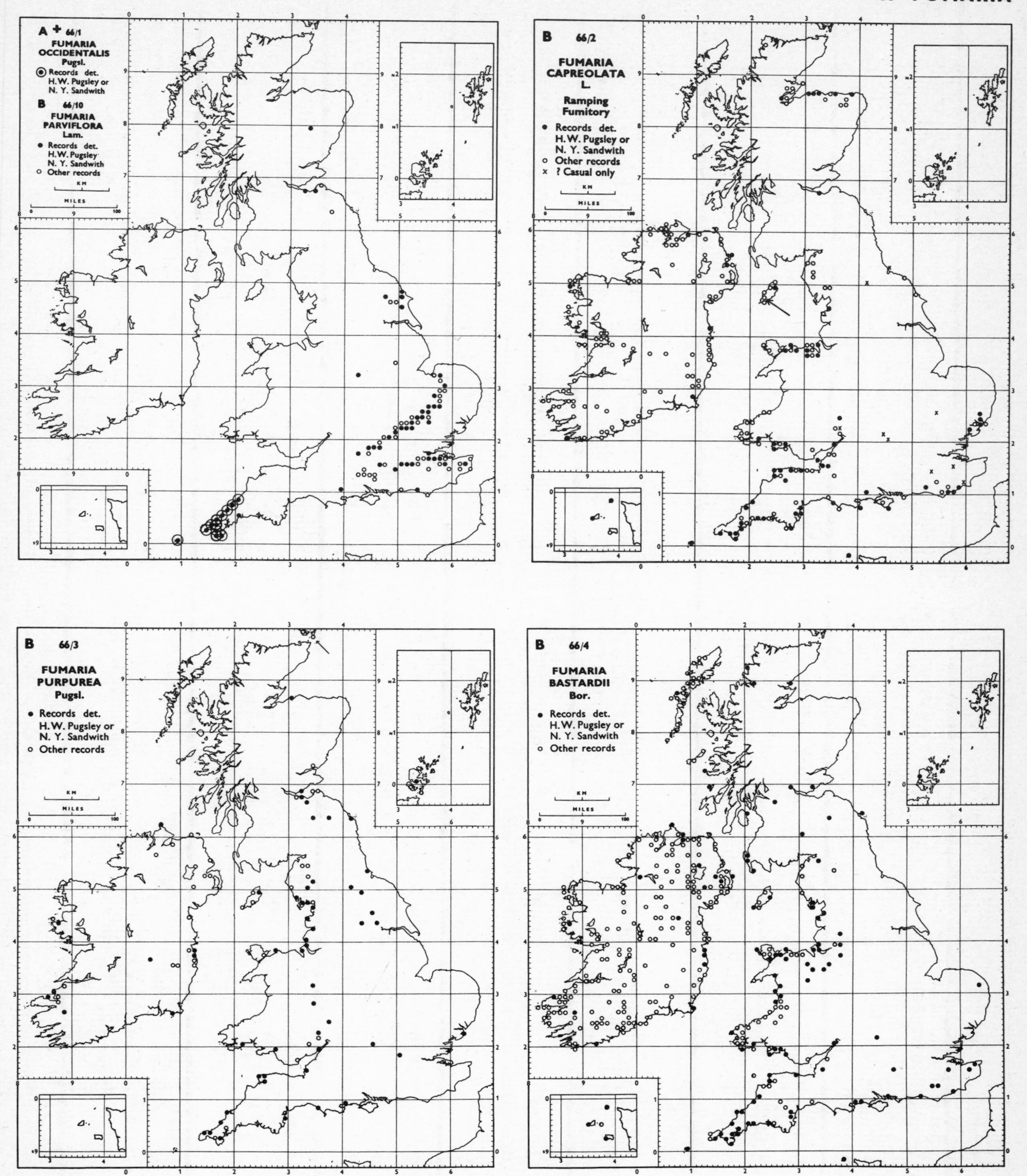
A + 66/1
FUMARIA OCCIDENTALIS Pugsl.
Records det. H.W. Pugsley or N. Y. Sandwith
B 66/10
FUMARIA PARVIFLORA Lam.
Records det. H.W. Pugsley N. Y. Sandwith
Other records
KM
MILES
B 66/2
FUMARIA CAPREOLATA L.
Ramping Fumitory
Records det. H.W. Pugsley or N. Y. Sandwith
Other records
? Casual only
KM
MILES
B 66/3
FUMARIA PURPUREA Pugsl.
Records det. H.W. Pugsley or N. Y. Sandwith
Other records
KM
MILES
B 66/4
FUMARIA BASTARDII Bor.
Records det. H.W. Pugsley or N. Y. Sandwith
Other records
KM
MILES

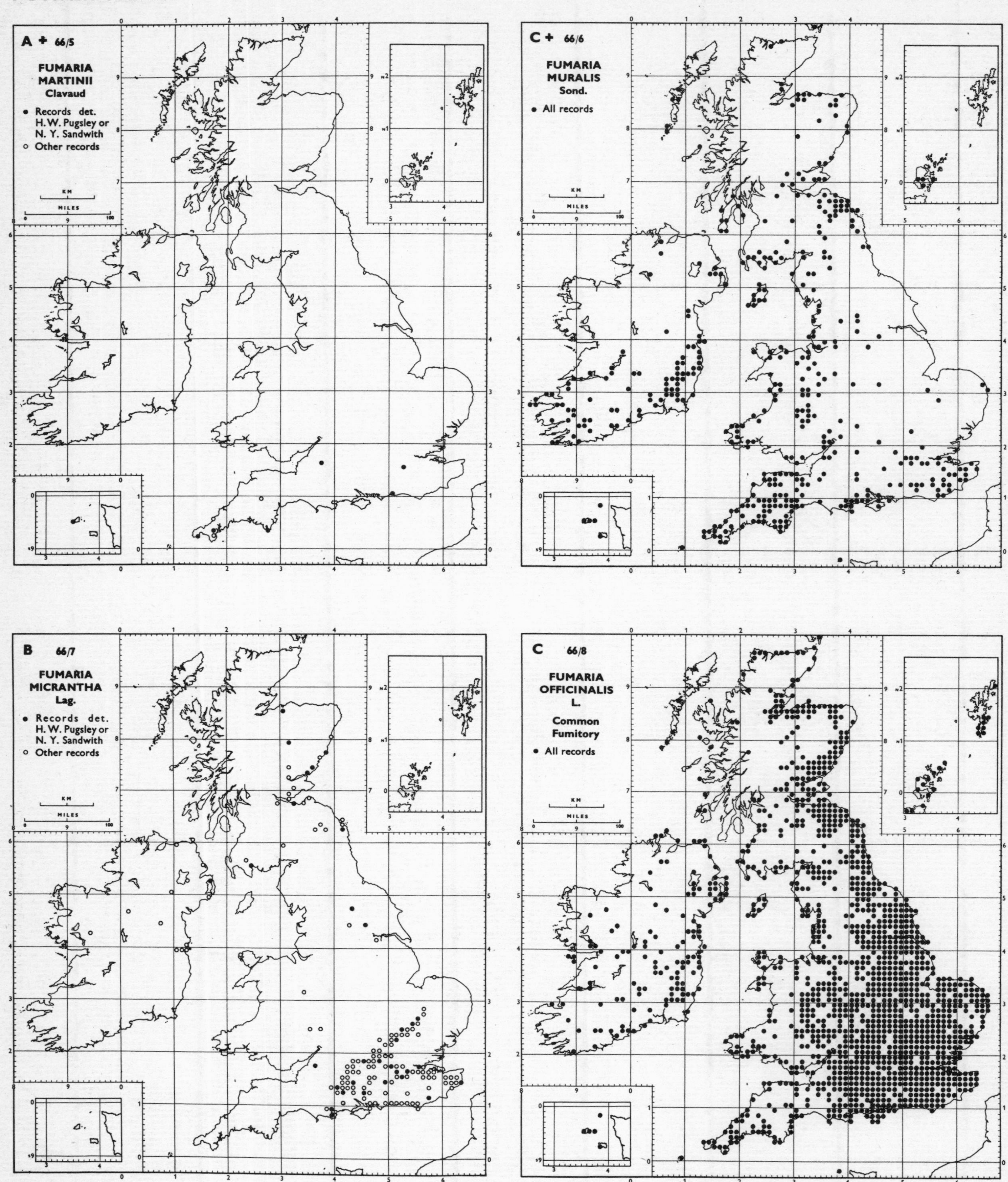
A + 66/5
FUMARIA MARTINII Clavaud
• Records det. H. W. Pugsley or N. Y. Sandwith
○ Other records
C + 66/6
FUMARIA MURALIS Sond.
• All records
B 66/7
FUMARIA MICRANTHA Lag.
• Records det. H. W. Pugsley or N. Y. Sandwith
○ Other records
C 66/8
FUMARIA OFFICINALIS L.
Common Fumitory
• All records
KM
MILES

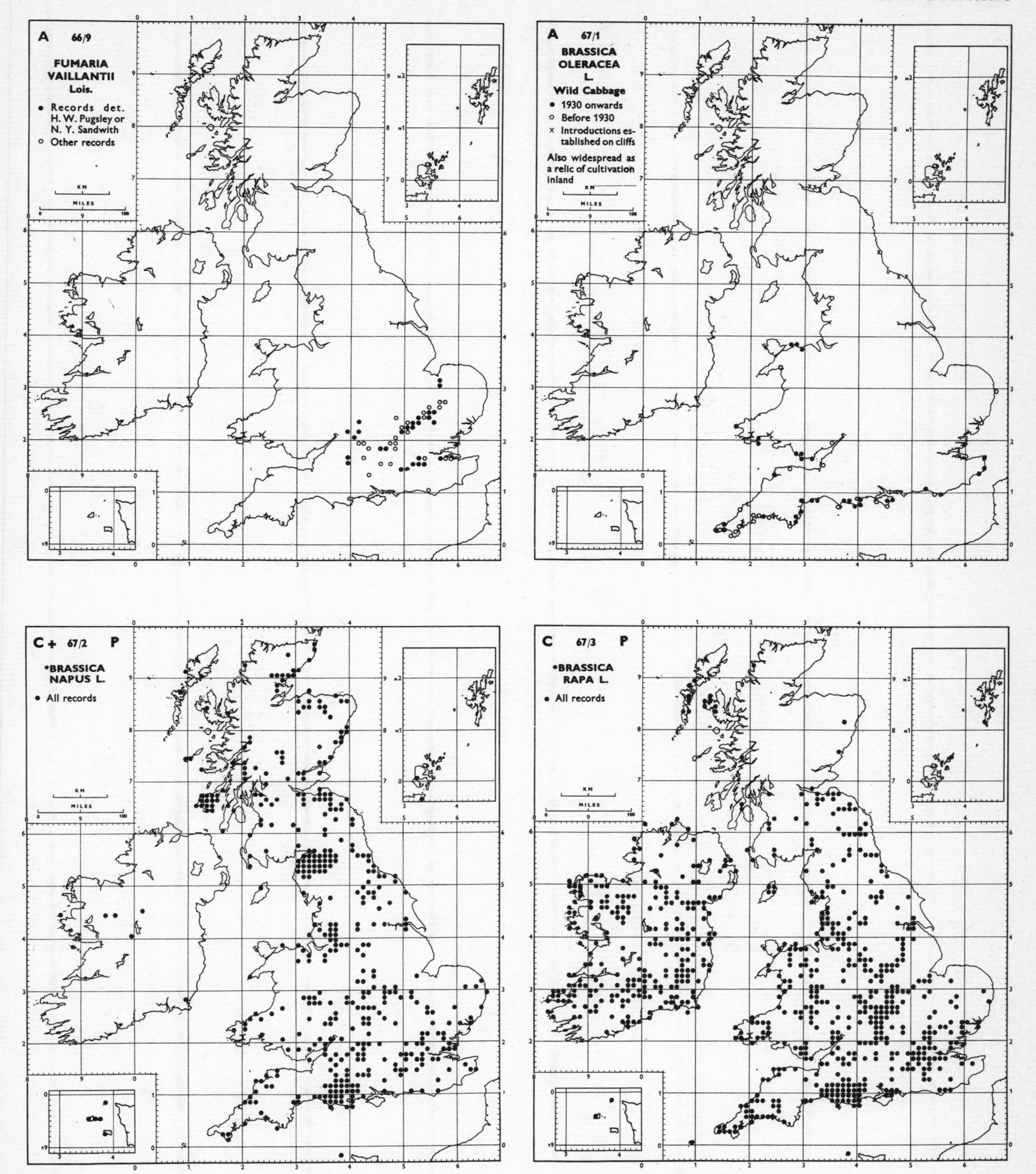

A 66/9
FUMARIA VAILLANTII Lois.
• Records det. H. W. Pugsley or N. Y. Sandwith
○ Other records
A 67/1
BRASSICA OLERACEA L.
Wild Cabbage
• 1930 onwards
○ Before 1930
× Introductions established on cliffs
Also widespread as a relic of cultivation inland
C + 67/2 P
*BRASSICA NAPUS L.
• All records
C 67/3 P
*BRASSICA RAPA L.
• All records
KM
MILES

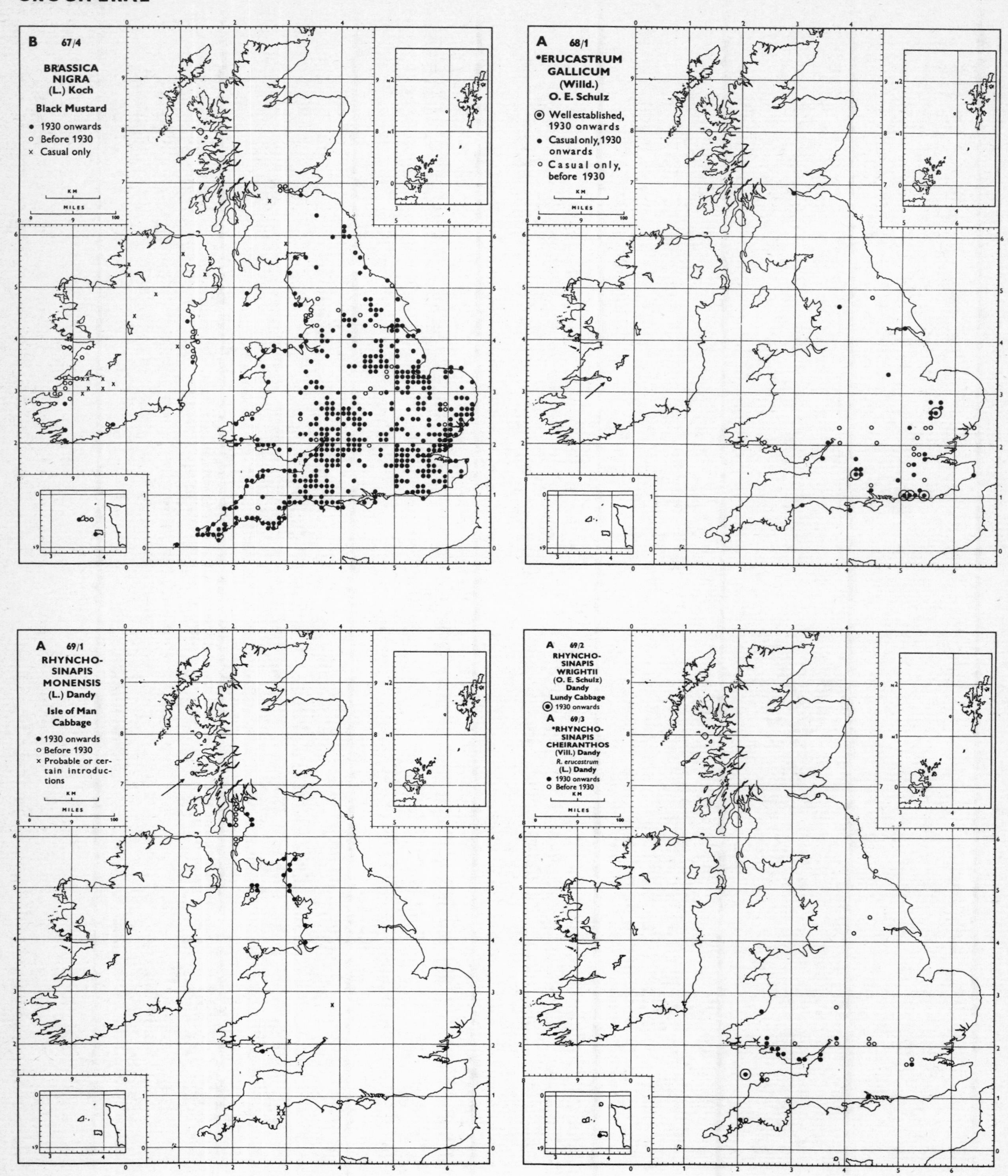
B 67/4
BRASSICA NIGRA (L.) Koch
Black Mustard
• 1930 onwards
○ Before 1930
× Casual only
A 68/1
*ERUCASTRUM GALLICUM (Willd.) O. E. Schulz
⊙ Well established, 1930 onwards
• Casual only, 1930 onwards
○ Casual only, before 1930
A 69/1
RHYNCHO-SINAPIS MONENSIS (L.) Dandy
Isle of Man Cabbage
• 1930 onwards
○ Before 1930
× Probable or certain introductions
A 69/2
RHYNCHO-SINAPIS WRIGHTII (O. E. Schulz) Dandy
Lundy Cabbage
⊙ 1930 onwards
A 69/3
*RHYNCHO-SINAPIS CHEIRANTHOS (Vill.) Dandy
R. erucastrum (L.) Dandy
• 1930 onwards
○ Before 1930
KM
MILES

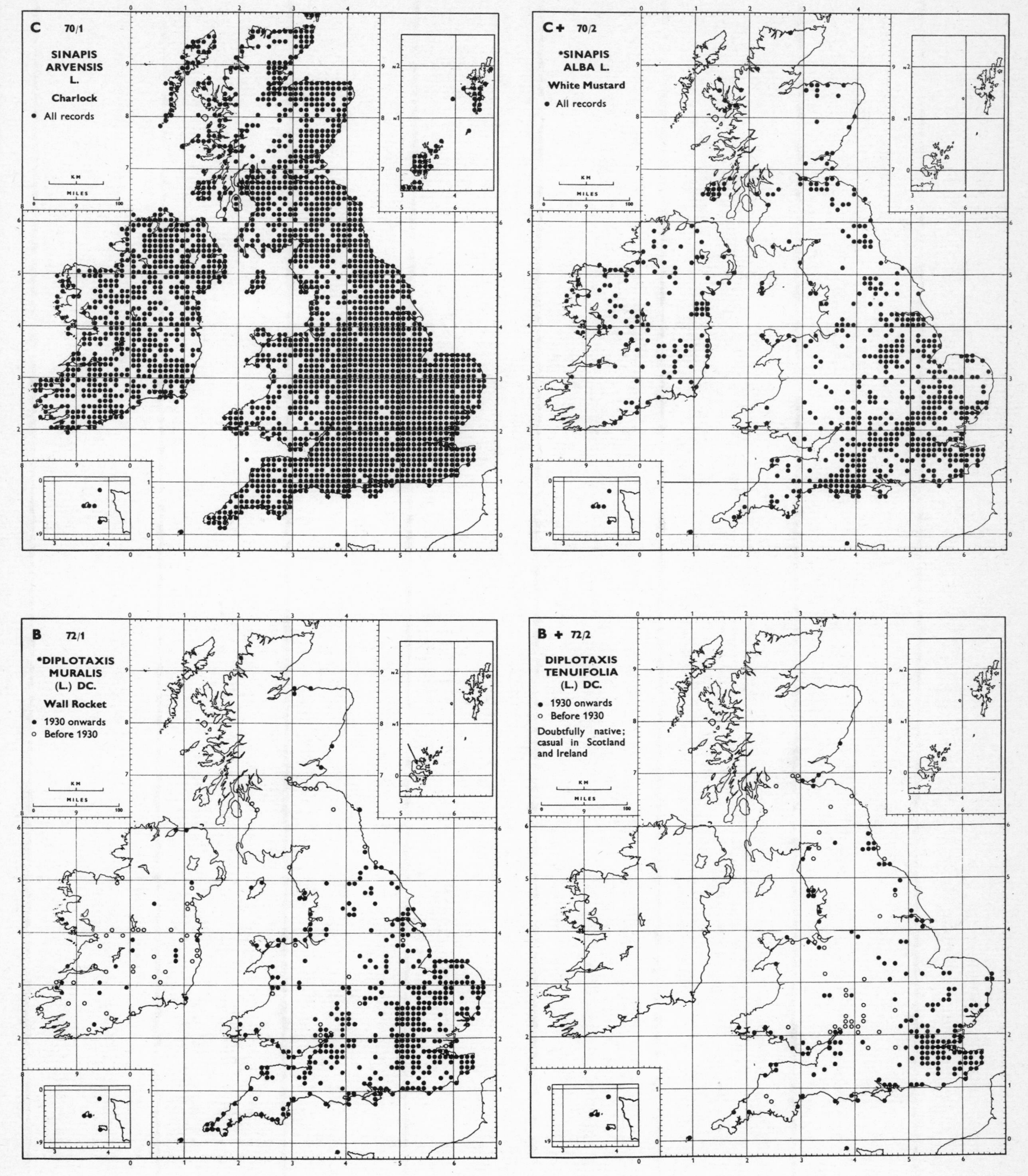
C 70/1
SINAPIS ARVENSIS L.
Charlock
• All records
C+ 70/2
*SINAPIS ALBA L.
White Mustard
• All records
B 72/1
*DIPLOTAXIS MURALIS (L.) DC.
Wall Rocket
• 1930 onwards
○ Before 1930
B + 72/2
DIPLOTAXIS TENUIFOLIA (L.) DC.
• 1930 onwards
○ Before 1930
Doubtfully native; casual in Scotland and Ireland

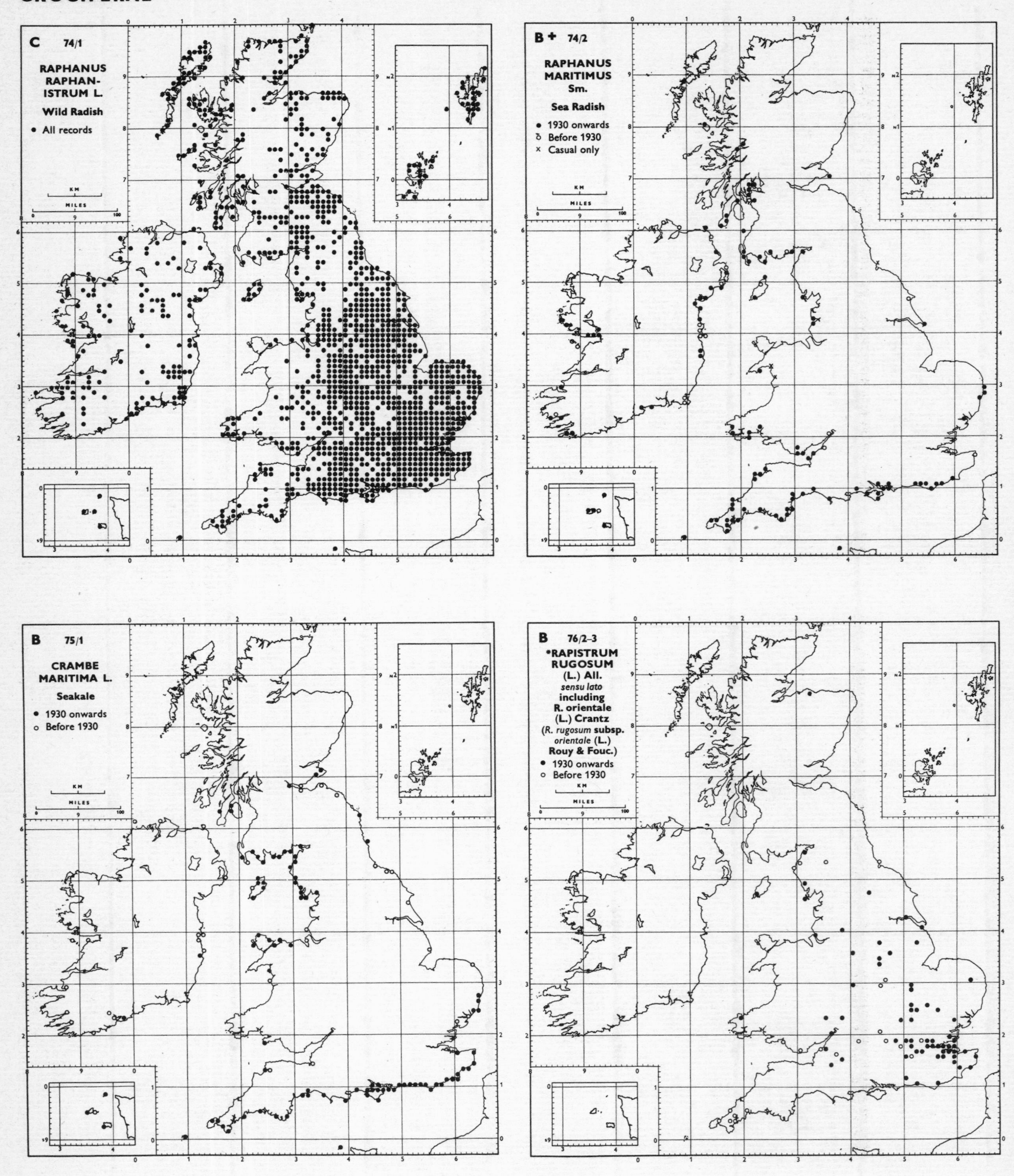
C 74/1
RAPHANUS RAPHAN-ISTRUM L.
Wild Radish
• All records
B + 74/2
RAPHANUS MARITIMUS Sm.
Sea Radish
• 1930 onwards
○ Before 1930
× Casual only
B 75/1
CRAMBE MARITIMA L.
Seakale
• 1930 onwards
○ Before 1930
B 76/2–3
*RAPISTRUM RUGOSUM (L.) All.
sensu lato
including
R. orientale (L.) Crantz
(R. rugosum subsp. orientale (L.) Rouy & Fouc.)
• 1930 onwards
○ Before 1930
KM
MILES

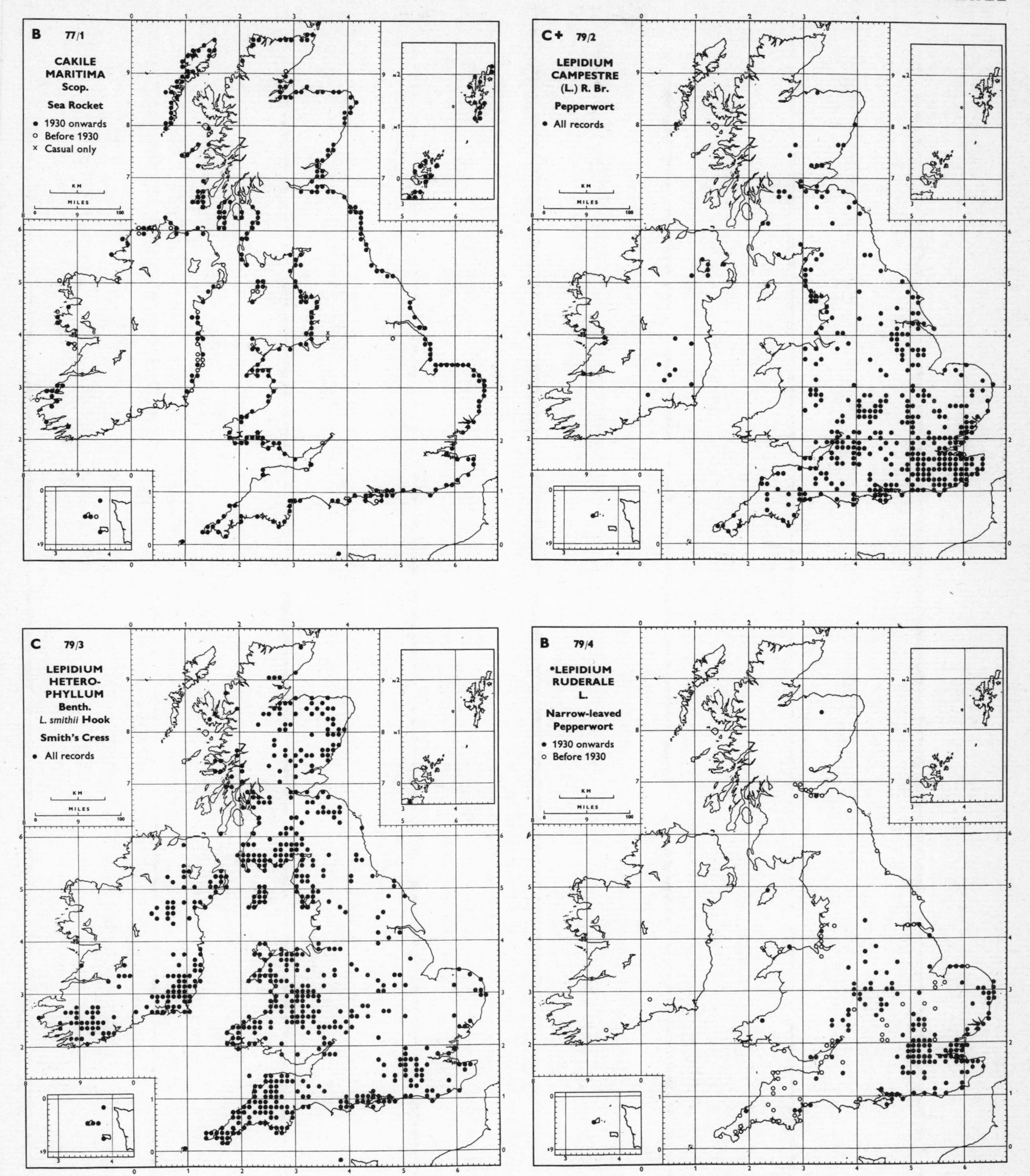
B 77/1
CAKILE MARITIMA Scop.
Sea Rocket
1930 onwards
Before 1930
Casual only
C+ 79/2
LEPIDIUM CAMPESTRE (L.) R. Br.
Pepperwort
All records
C 79/3
LEPIDIUM HETERO-PHYLLUM Benth.
L. smithii Hook
Smith's Cress
All records
B 79/4
*LEPIDIUM RUDERALE L.
Narrow-leaved Pepperwort
1930 onwards
Before 1930
KM
MILES

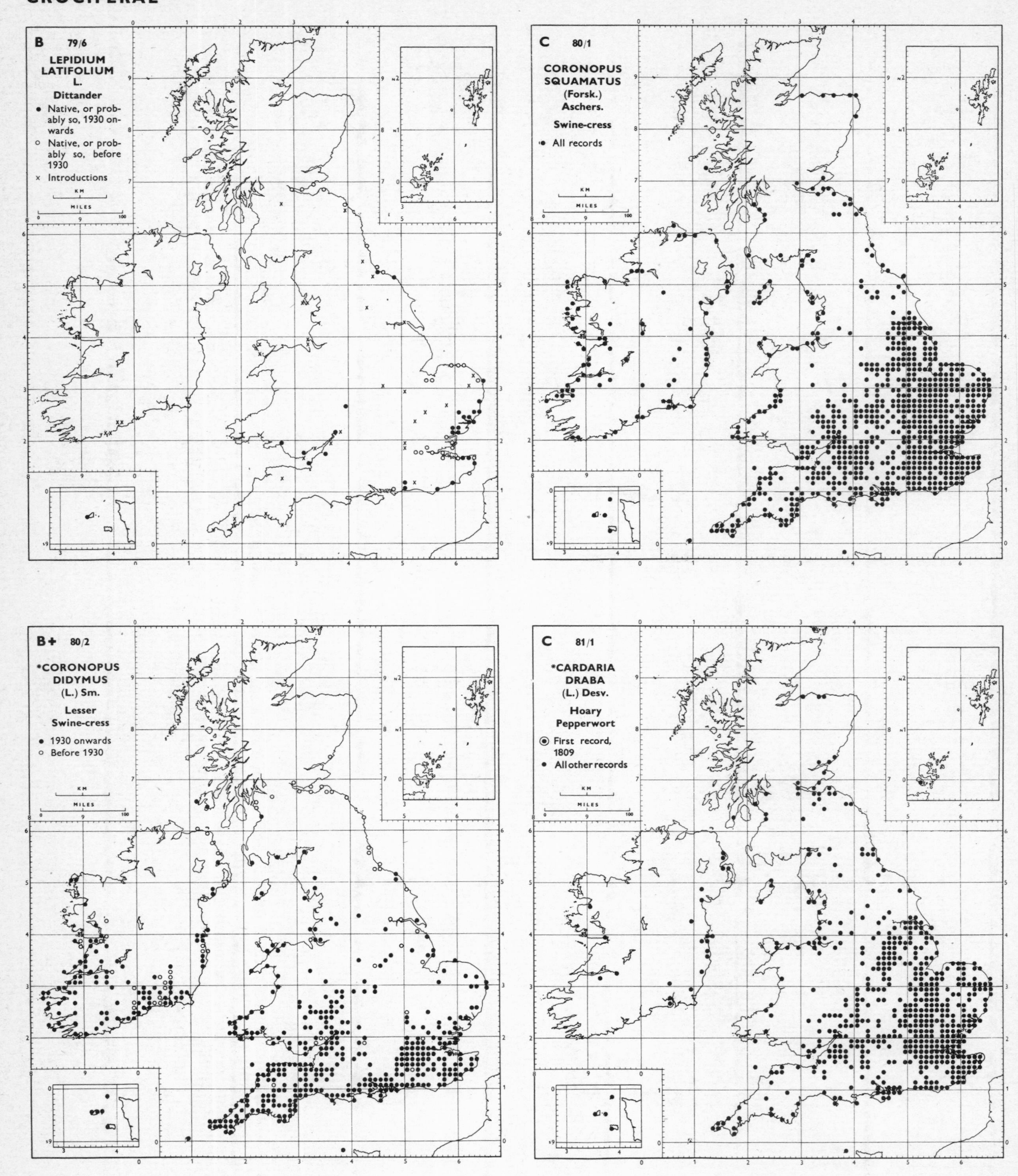
B 79/6
LEPIDIUM LATIFOLIUM L.
Dittander
Native, or probably so, 1930 onwards
Native, or probably so, before 1930
Introductions
C 80/1
CORONOPUS SQUAMATUS (Forsk.) Aschers.
Swine-cress
All records
B+ 80/2
*CORONOPUS DIDYMUS (L.) Sm.
Lesser Swine-cress
1930 onwards
Before 1930
C 81/1
*CARDARIA DRABA (L.) Desv.
Hoary Pepperwort
First record, 1809
All other records

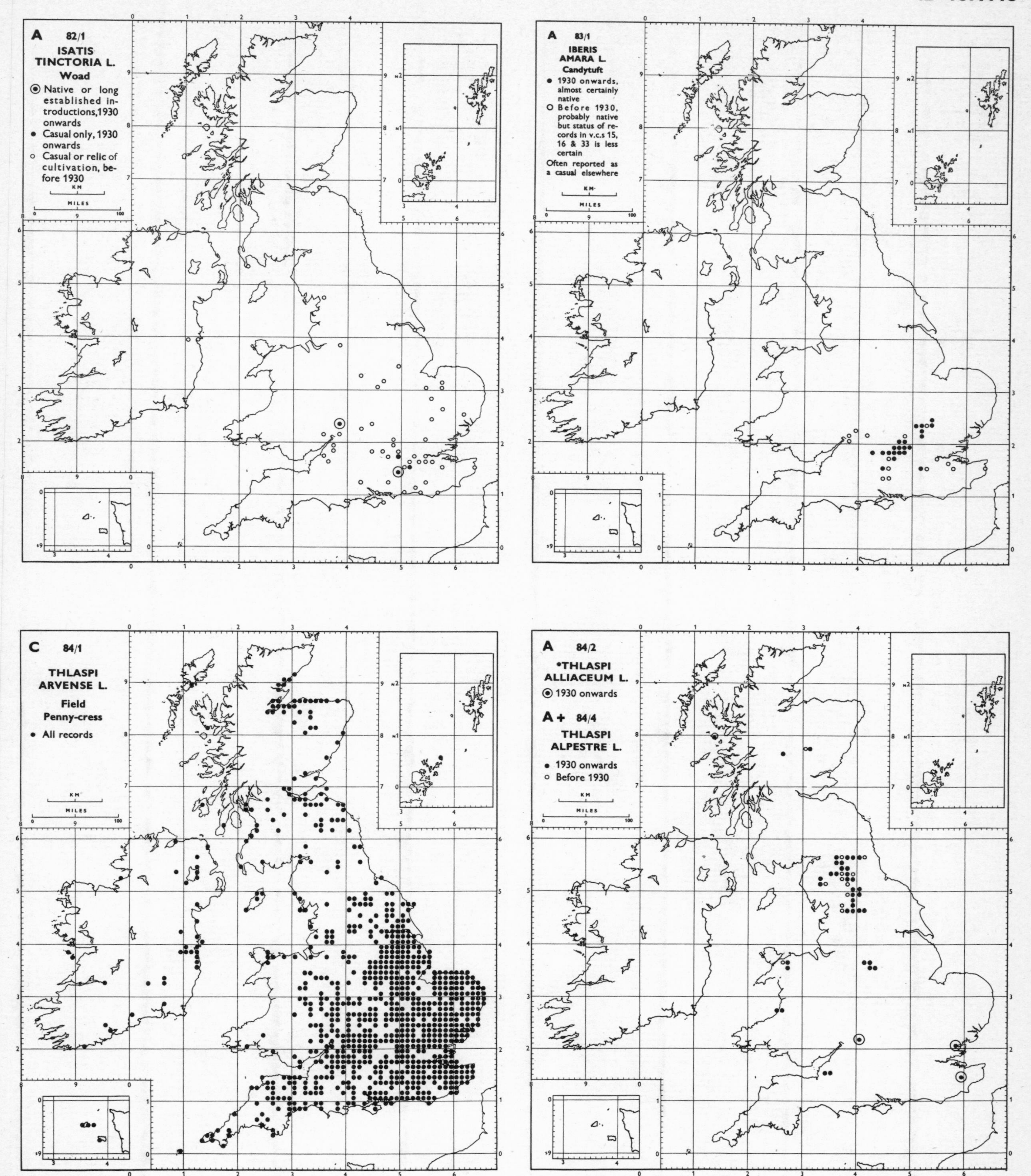
A 82/1
ISATIS TINCTORIA L.
Woad
Native or long established introductions, 1930 onwards
Casual only, 1930 onwards
Casual or relic of cultivation, before 1930
KM
MILES
A 83/1
IBERIS AMARA L.
Candytuft
1930 onwards, almost certainly native
Before 1930, probably native but status of records in v.c.s 15, 16 & 33 is less certain
Often reported as a casual elsewhere
KM
MILES
C 84/1
THLASPI ARVENSE L.
Field Penny-cress
All records
KM
MILES
A 84/2
*THLASPI ALLIACEUM L.
1930 onwards
A+ 84/4
THLASPI ALPESTRE L.
1930 onwards
Before 1930
KM
MILES

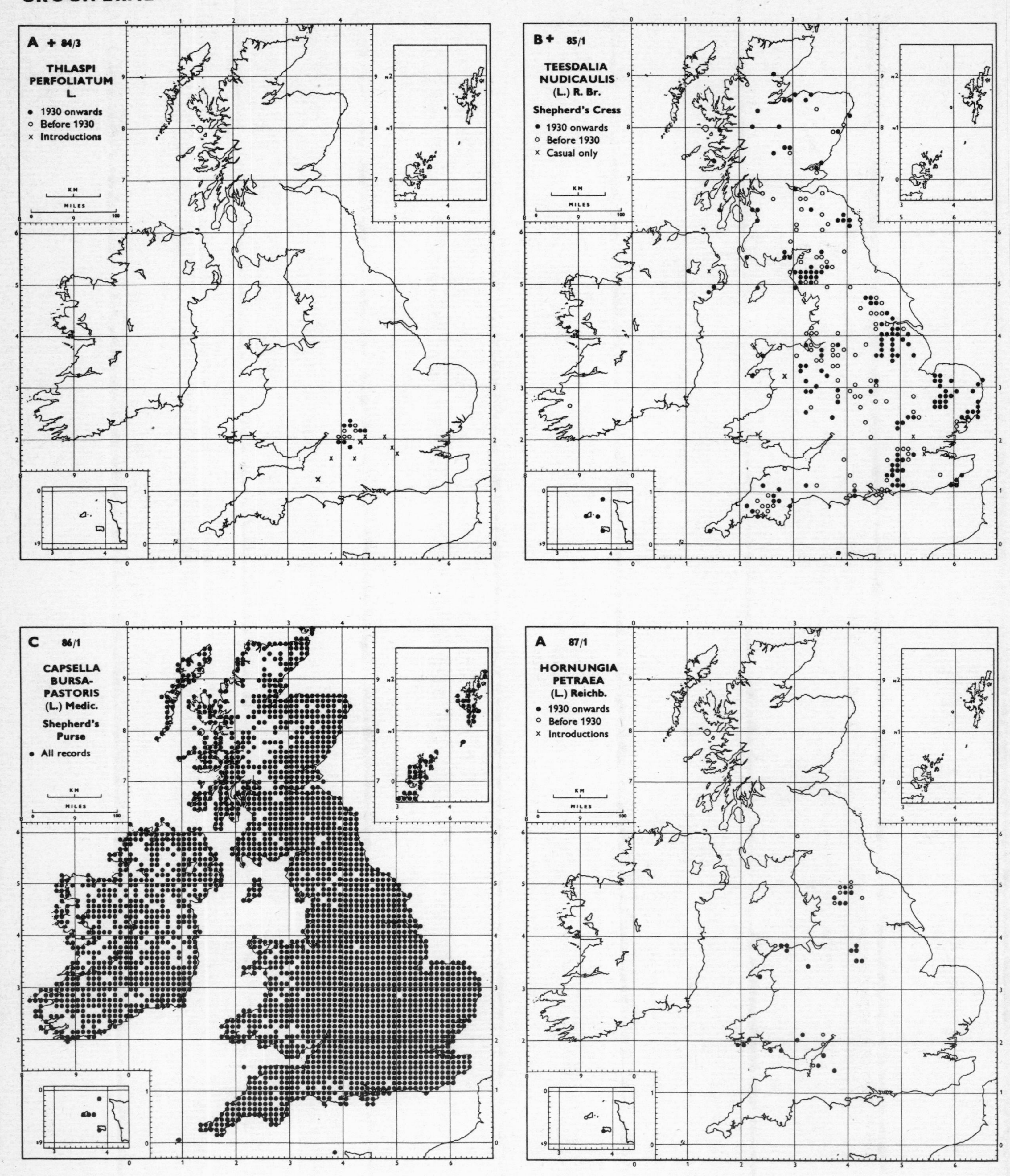
A + 84/3
THLASPI PERFOLIATUM L.
• 1930 onwards
○ Before 1930
× Introductions
B + 85/1
TEESDALIA NUDICAULIS (L.) R. Br.
Shepherd's Cress
• 1930 onwards
○ Before 1930
× Casual only
C 86/1
CAPSELLA BURSA-PASTORIS (L.) Medic.
Shepherd's Purse
• All records
A 87/1
HORNUNGIA PETRAEA (L.) Reichb.
• 1930 onwards
○ Before 1930
× Introductions
KM
MILES

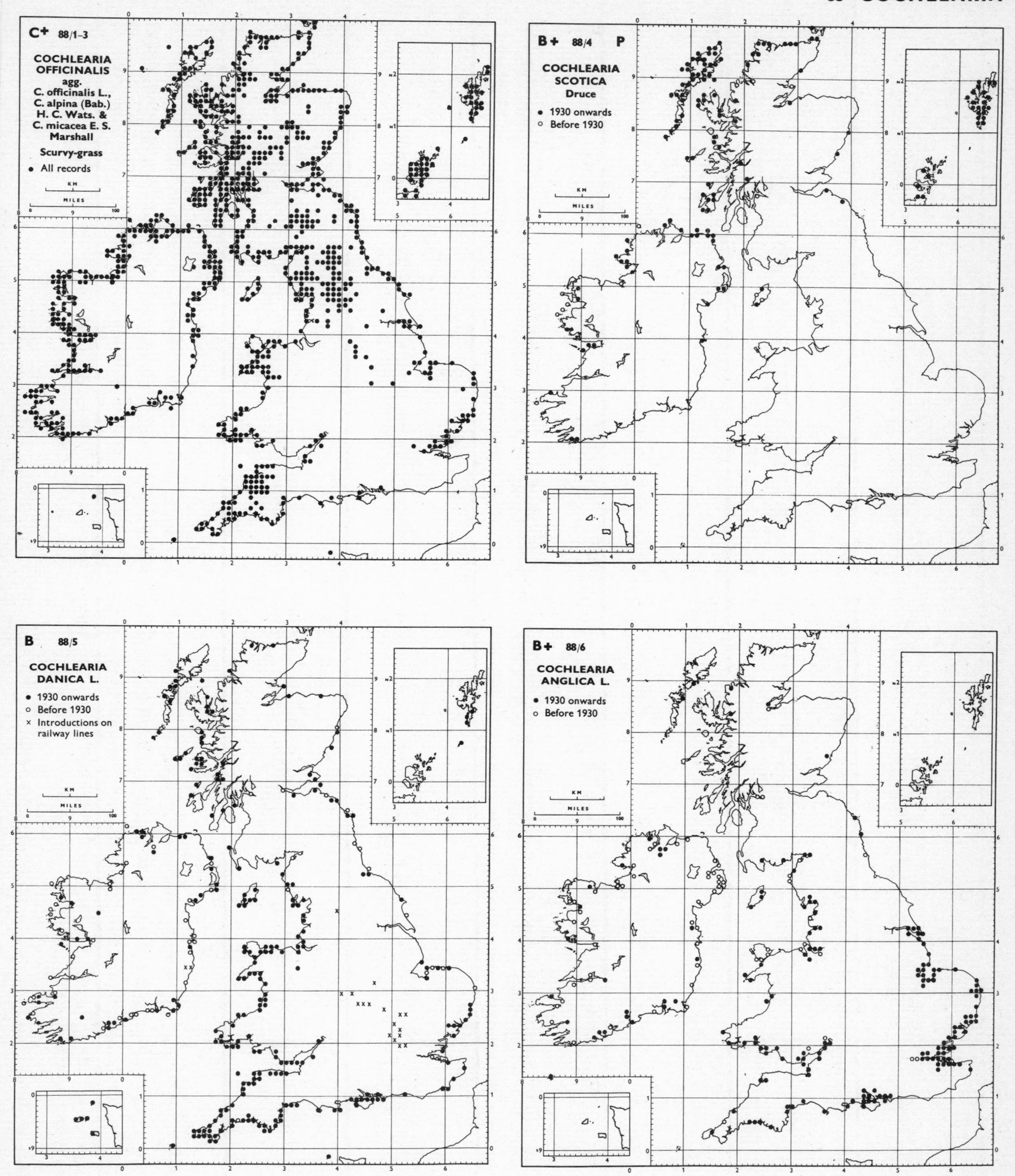
C+ 88/1-3
COCHLEARIA OFFICINALIS agg.
C. officinalis L., C. alpina (Bab.) H. C. Wats. & C. micacea E. S. Marshall
Scurvy-grass
• All records
B+ 88/4 P
COCHLEARIA SCOTICA Druce
• 1930 onwards
○ Before 1930
B 88/5
COCHLEARIA DANICA L.
• 1930 onwards
○ Before 1930
× Introductions on railway lines
B+ 88/6
COCHLEARIA ANGLICA L.
• 1930 onwards
○ Before 1930

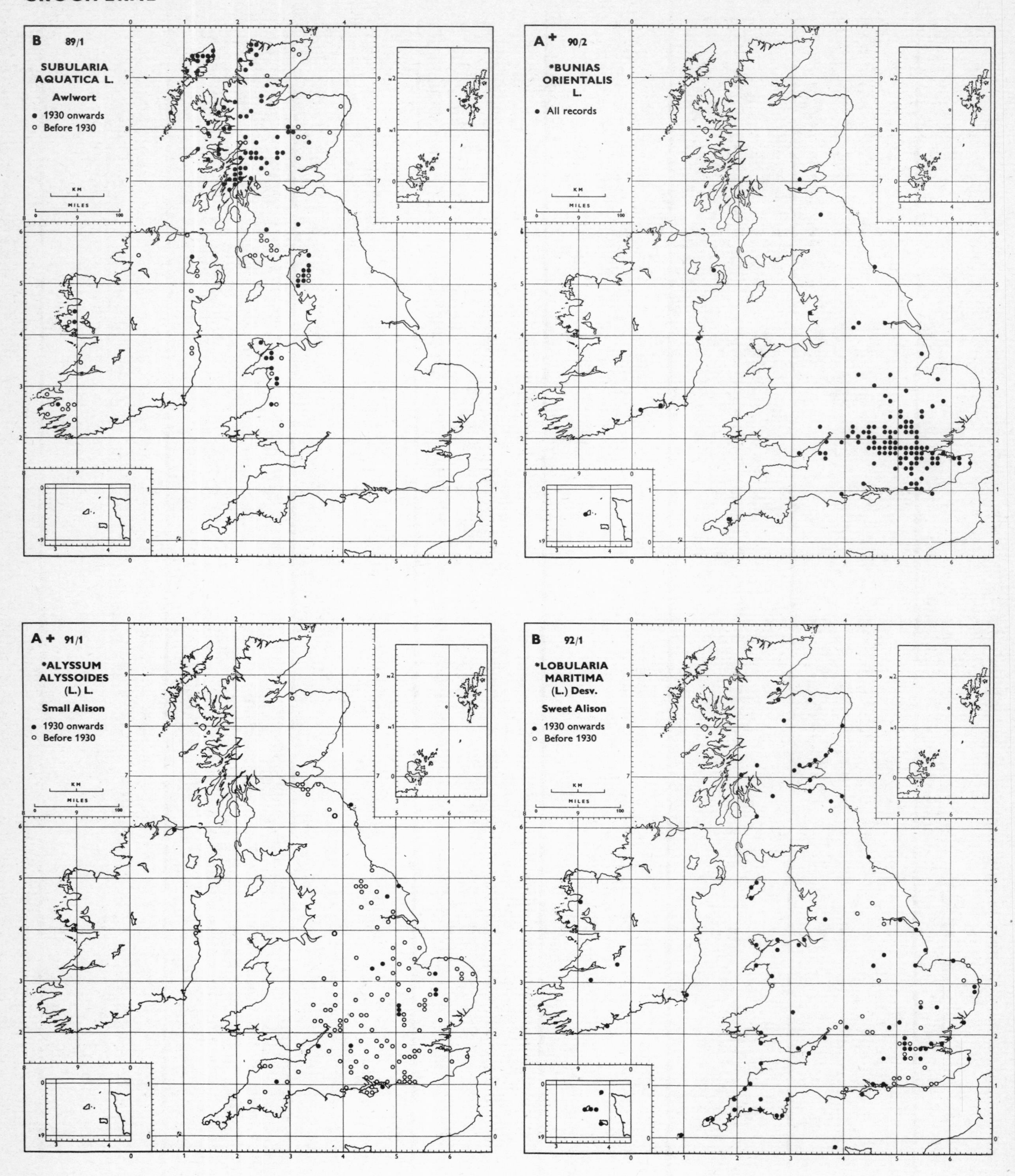
B 89/1
SUBULARIA AQUATICA L.
Awlwort
1930 onwards
Before 1930
A + 90/2
*BUNIAS ORIENTALIS L.
All records
A + 91/1
*ALYSSUM ALYSSOIDES (L.) L.
Small Alison
1930 onwards
Before 1930
B 92/1
*LOBULARIA MARITIMA (L.) Desv.
Sweet Alison
1930 onwards
Before 1930

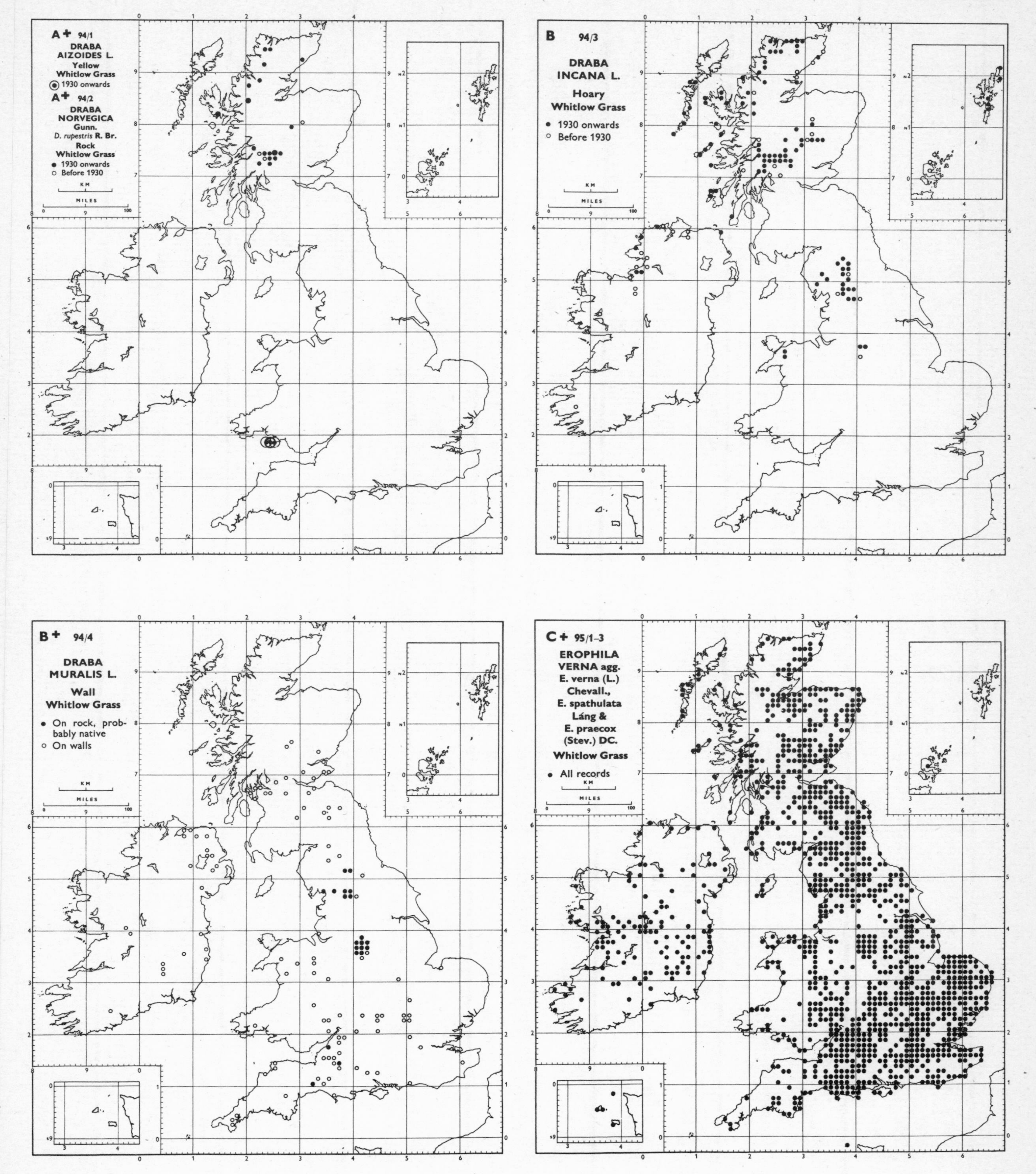

A+ 94/1
DRABA AIZOIDES L.
Yellow Whitlow Grass
1930 onwards
A+ 94/2
DRABA NORVEGICA Gunn.
D. rupestris R. Br.
Rock Whitlow Grass
1930 onwards
Before 1930
B 94/3
DRABA INCANA L.
Hoary Whitlow Grass
1930 onwards
Before 1930
B+ 94/4
DRABA MURALIS L.
Wall Whitlow Grass
On rock, probably native
On walls
C+ 95/1–3
EROPHILA VERNA agg.
E. verna (L.) Chevall.,
E. spathulata Láng &
E. praecox (Stev.) DC.
Whitlow Grass
All records

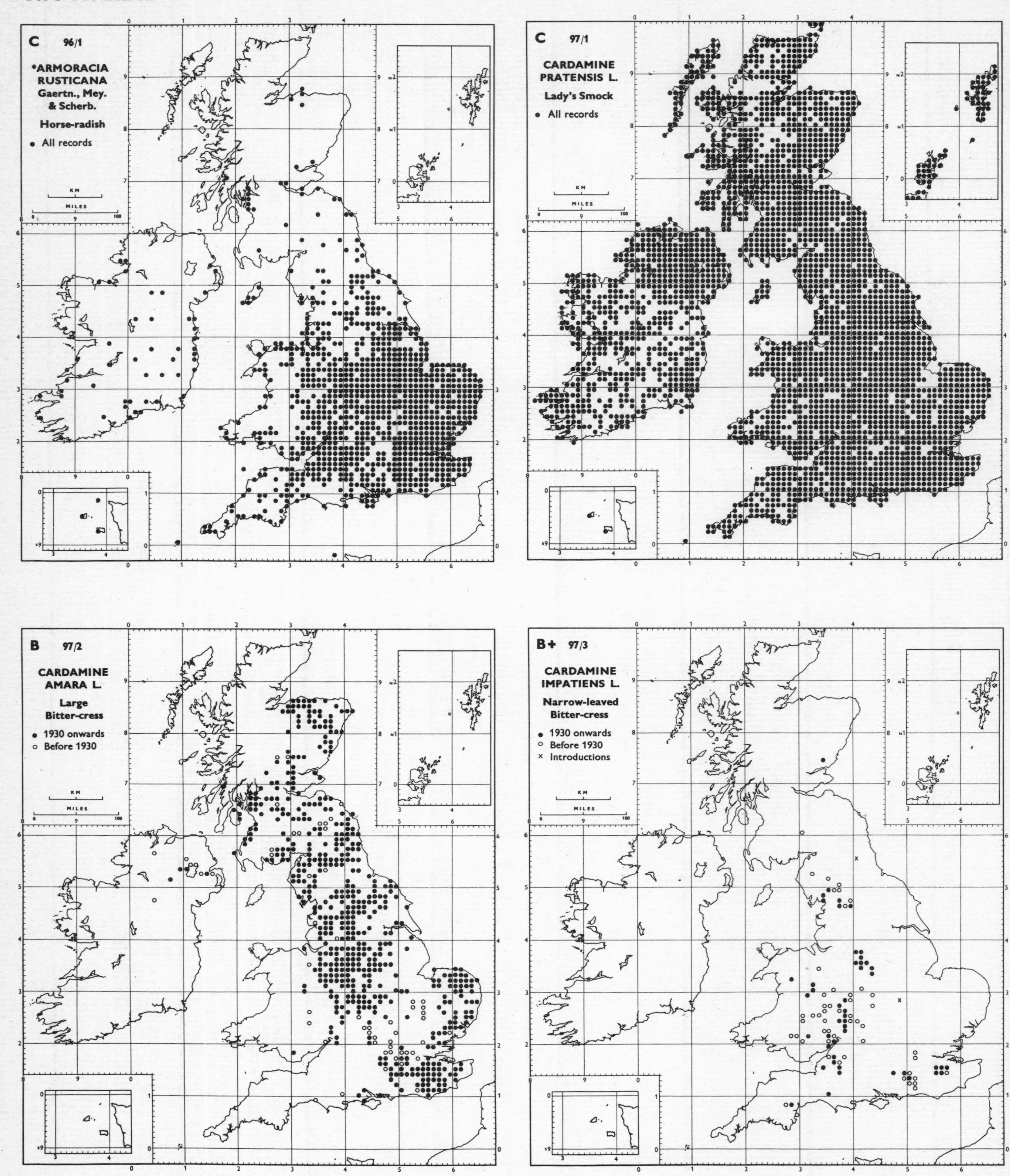
C 96/1
*ARMORACIA RUSTICANA Gaertn., Mey. & Scherb.
Horse-radish
All records
C 97/1
CARDAMINE PRATENSIS L.
Lady's Smock
All records
B 97/2
CARDAMINE AMARA L.
Large Bitter-cress
1930 onwards
Before 1930
B+ 97/3
CARDAMINE IMPATIENS L.
Narrow-leaved Bitter-cress
1930 onwards
Before 1930
Introductions

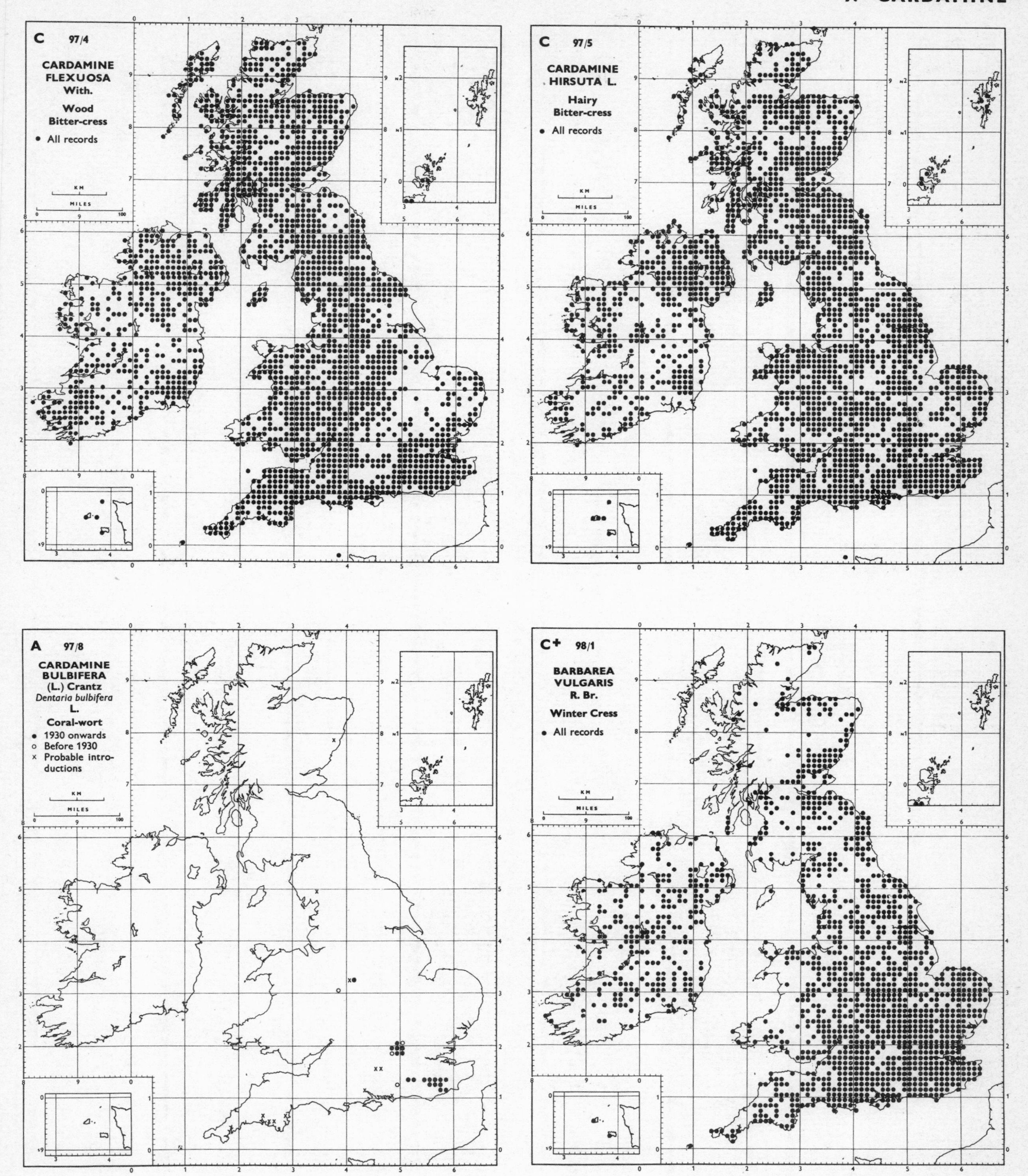
C 97/4
CARDAMINE FLEXUOSA With.
Wood Bitter-cress
● All records
KM
MILES
0 9 100
C 97/5
CARDAMINE HIRSUTA L.
Hairy Bitter-cress
● All records
A 97/8
CARDAMINE BULBIFERA (L.) Crantz
Dentaria bulbifera L.
Coral-wort
● 1930 onwards
○ Before 1930
× Probable introductions
C+ 98/1
BARBAREA VULGARIS R. Br.
Winter Cress
● All records

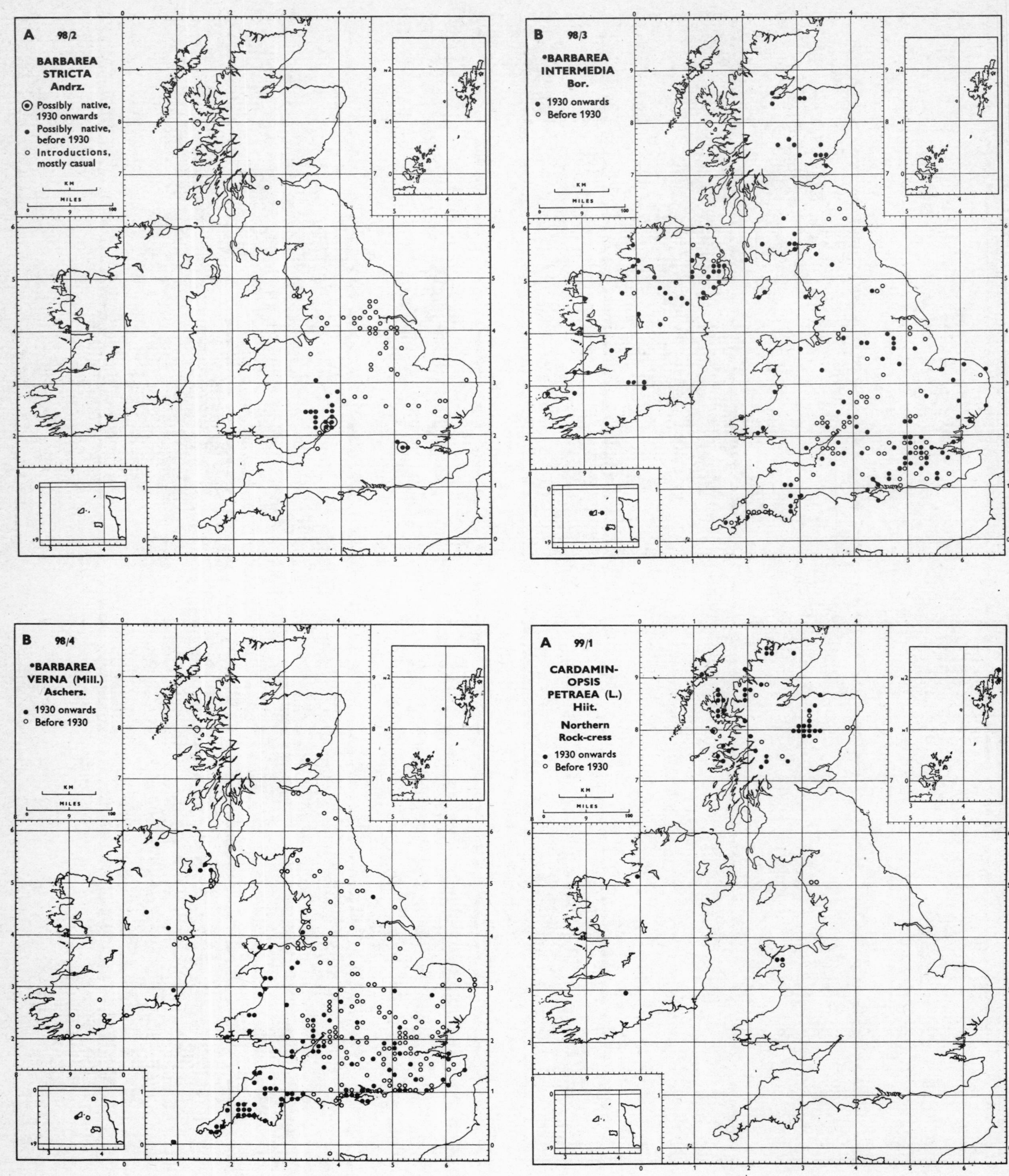
A 98/2
BARBAREA STRICTA Andrz.
Possibly native, 1930 onwards
Possibly native, before 1930
Introductions, mostly casual
B 98/3
*BARBAREA INTERMEDIA Bor.
1930 onwards
Before 1930
B 98/4
*BARBAREA VERNA (Mill.) Aschers.
1930 onwards
Before 1930
A 99/1
CARDAMINOPSIS PETRAEA (L.) Hiit.
Northern Rock-cress
1930 onwards
Before 1930
KM
MILES

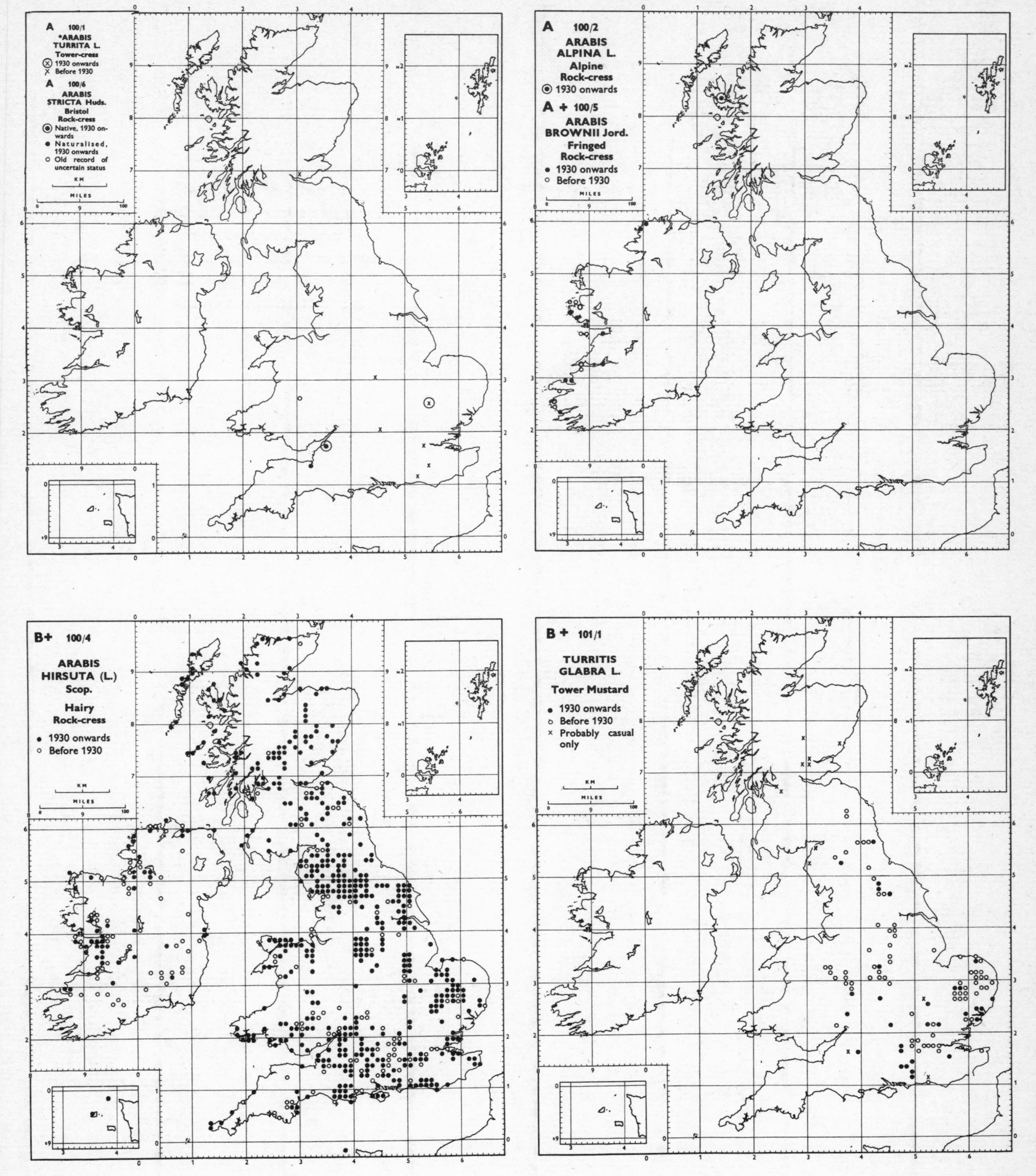
A 100/1
*ARABIS TURRITA L.
Tower-cress
1930 onwards
Before 1930
A 100/6
ARABIS STRICTA Huds.
Bristol Rock-cress
Native, 1930 onwards
Naturalised, 1930 onwards
Old record of uncertain status
KM
MILES
A 100/2
ARABIS ALPINA L.
Alpine Rock-cress
1930 onwards
A + 100/5
ARABIS BROWNII Jord.
Fringed Rock-cress
1930 onwards
Before 1930
MILES
B+ 100/4
ARABIS HIRSUTA (L.) Scop.
Hairy Rock-cress
1930 onwards
Before 1930
KM
MILES
B + 101/1
TURRITIS GLABRA L.
Tower Mustard
1930 onwards
Before 1930
Probably casual only
KM
MILES

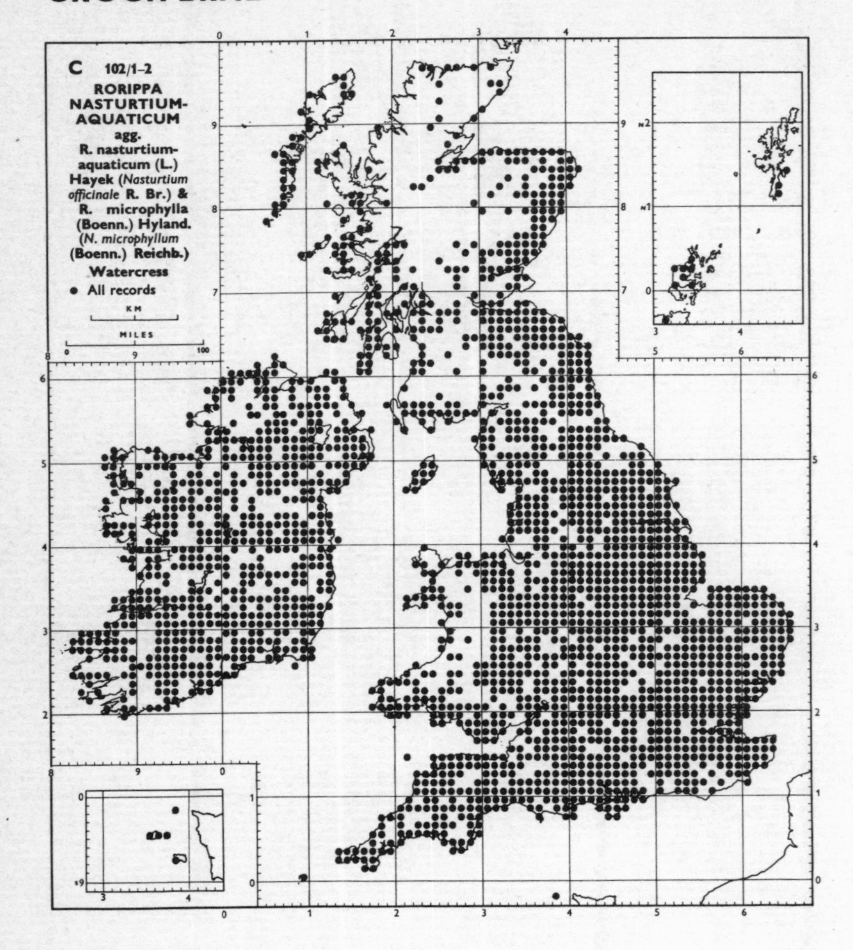

It was intended that *Rorippa nasturtium-aquaticum* and *R. microphylla* should be mapped separately, but the data received proved to be inadequate.

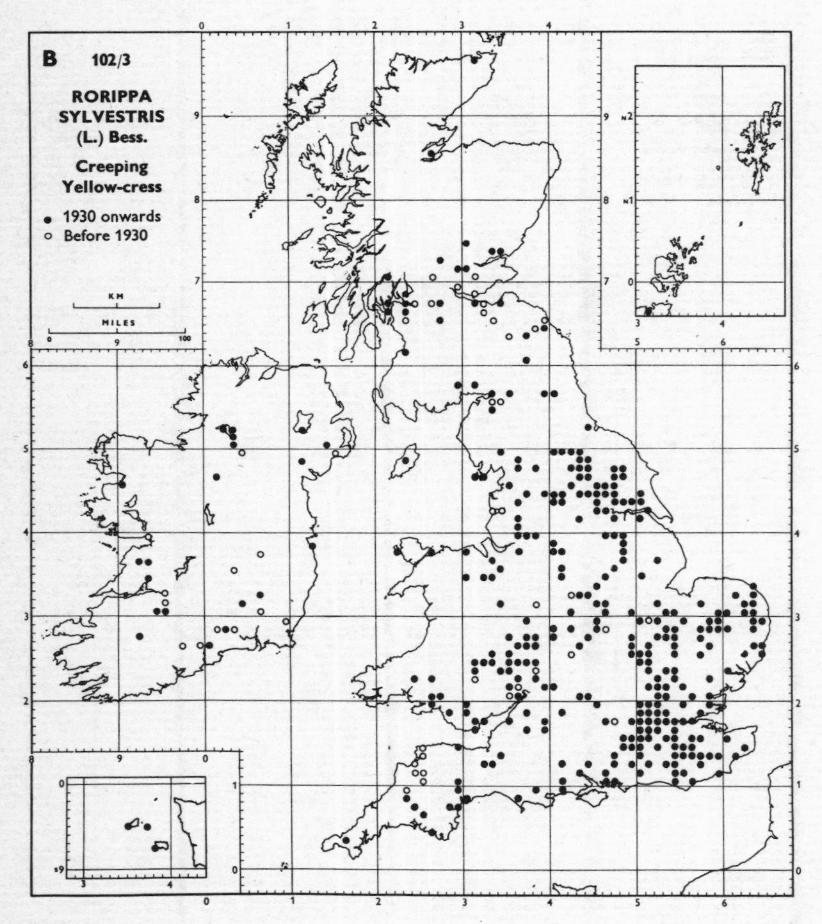

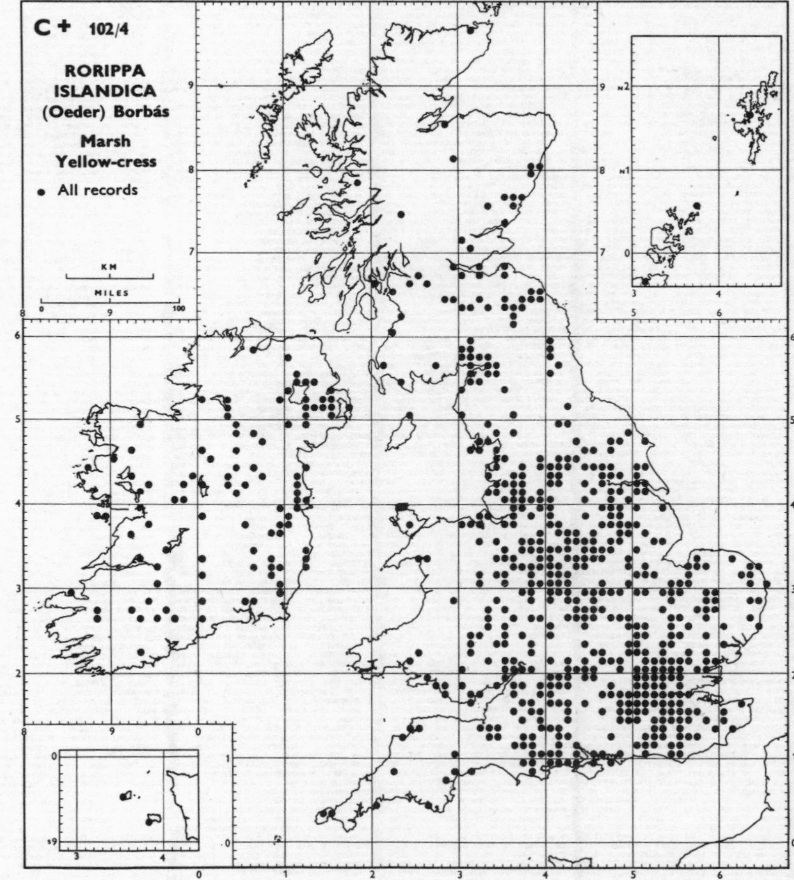

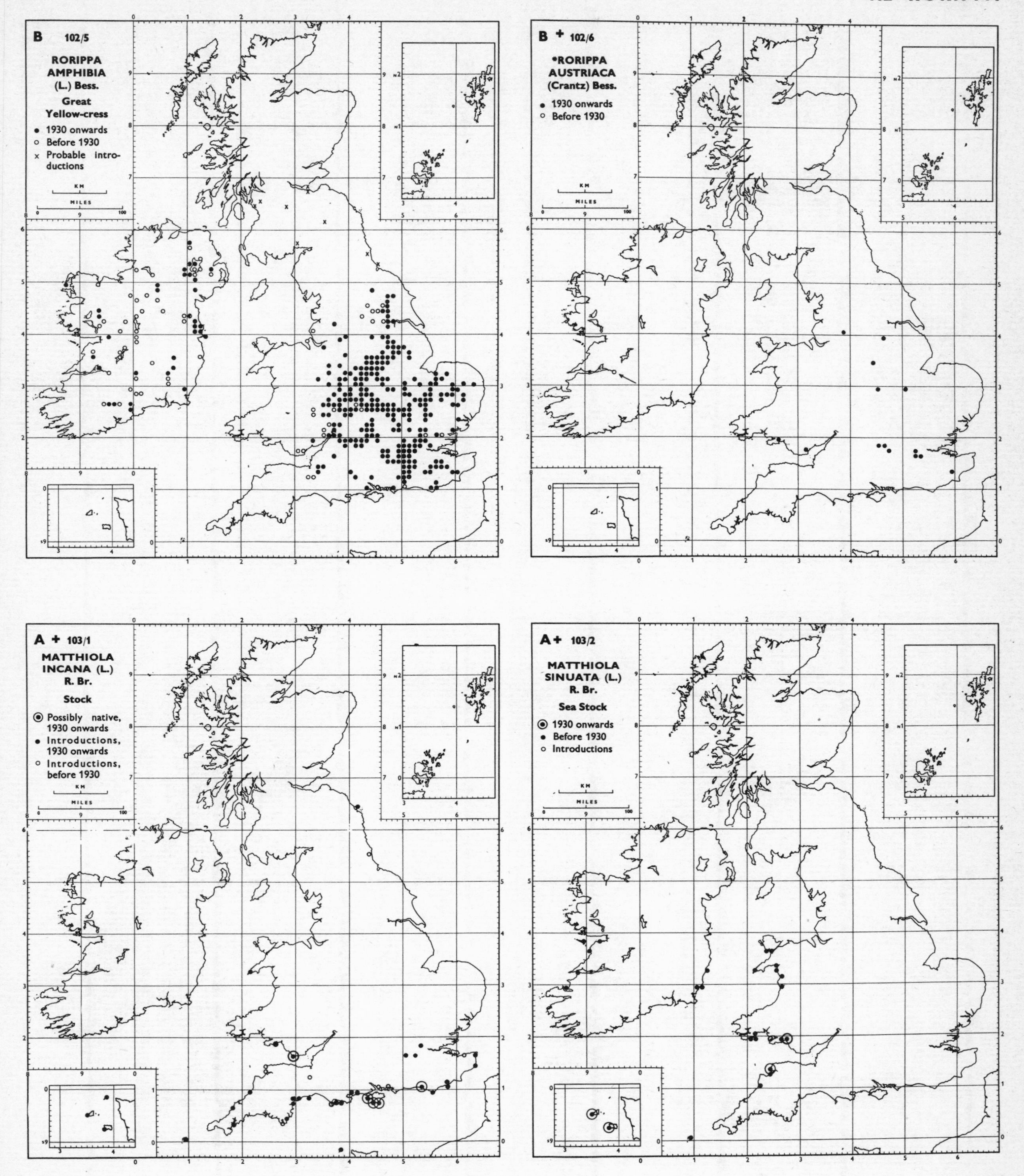
B 102/5
RORIPPA AMPHIBIA (L.) Bess.
Great Yellow-cress
• 1930 onwards
○ Before 1930
x Probable introductions
B + 102/6
*RORIPPA AUSTRIACA (Crantz) Bess.
• 1930 onwards
○ Before 1930
A + 103/1
MATTHIOLA INCANA (L.) R. Br.
Stock
⊙ Possibly native, 1930 onwards
• Introductions, 1930 onwards
○ Introductions, before 1930
A + 103/2
MATTHIOLA SINUATA (L.) R. Br.
Sea Stock
⊙ 1930 onwards
• Before 1930
○ Introductions

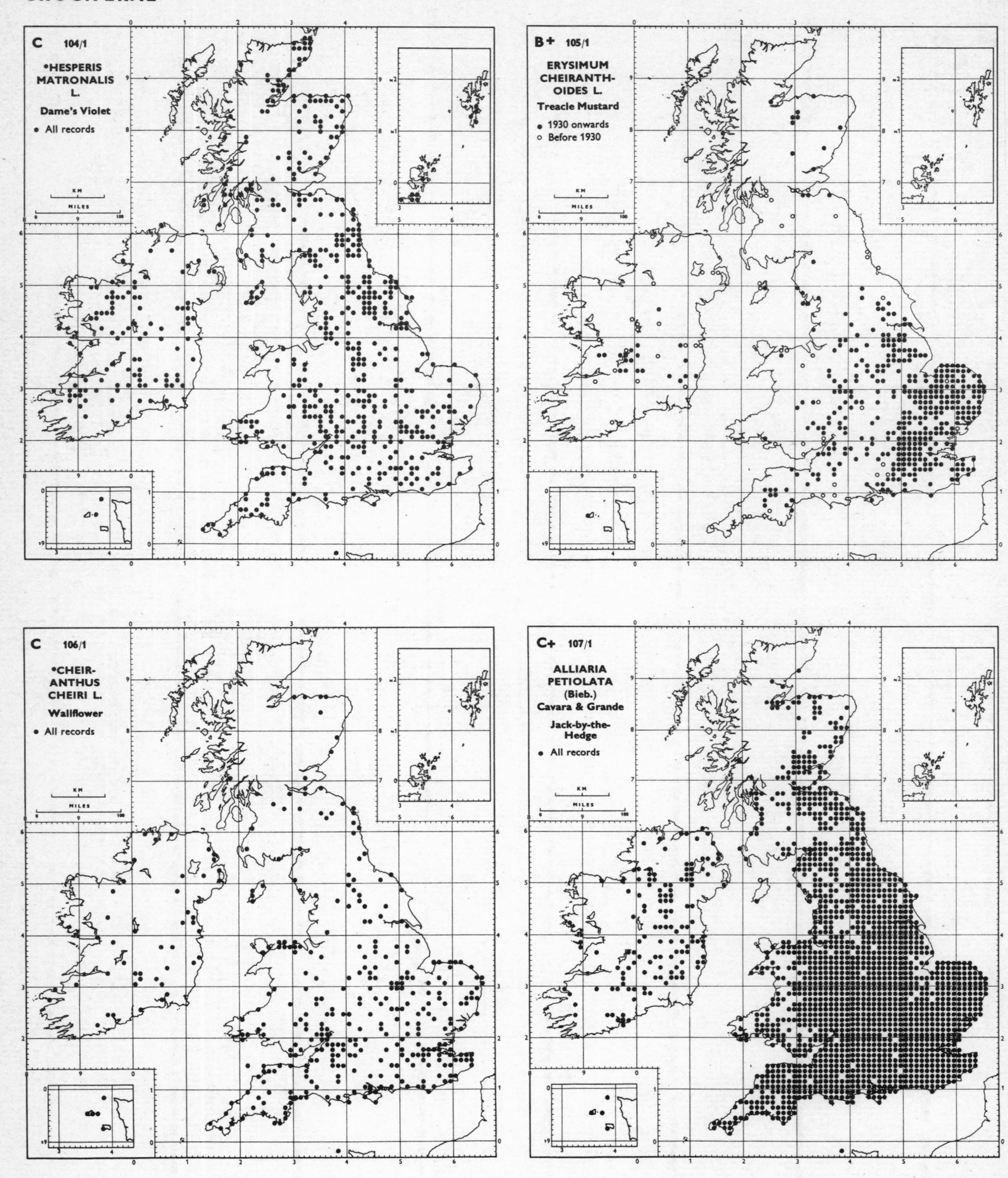
C 104/1
•HESPERIS MATRONALIS L.
Dame's Violet
• All records
B+ 105/1
ERYSIMUM CHEIRANTHOIDES L.
Treacle Mustard
• 1930 onwards
○ Before 1930
C 106/1
•CHEIRANTHUS CHEIRI L.
Wallflower
• All records
C+ 107/1
ALLIARIA PETIOLATA (Bieb.) Cavara & Grande
Jack-by-the-Hedge
• All records

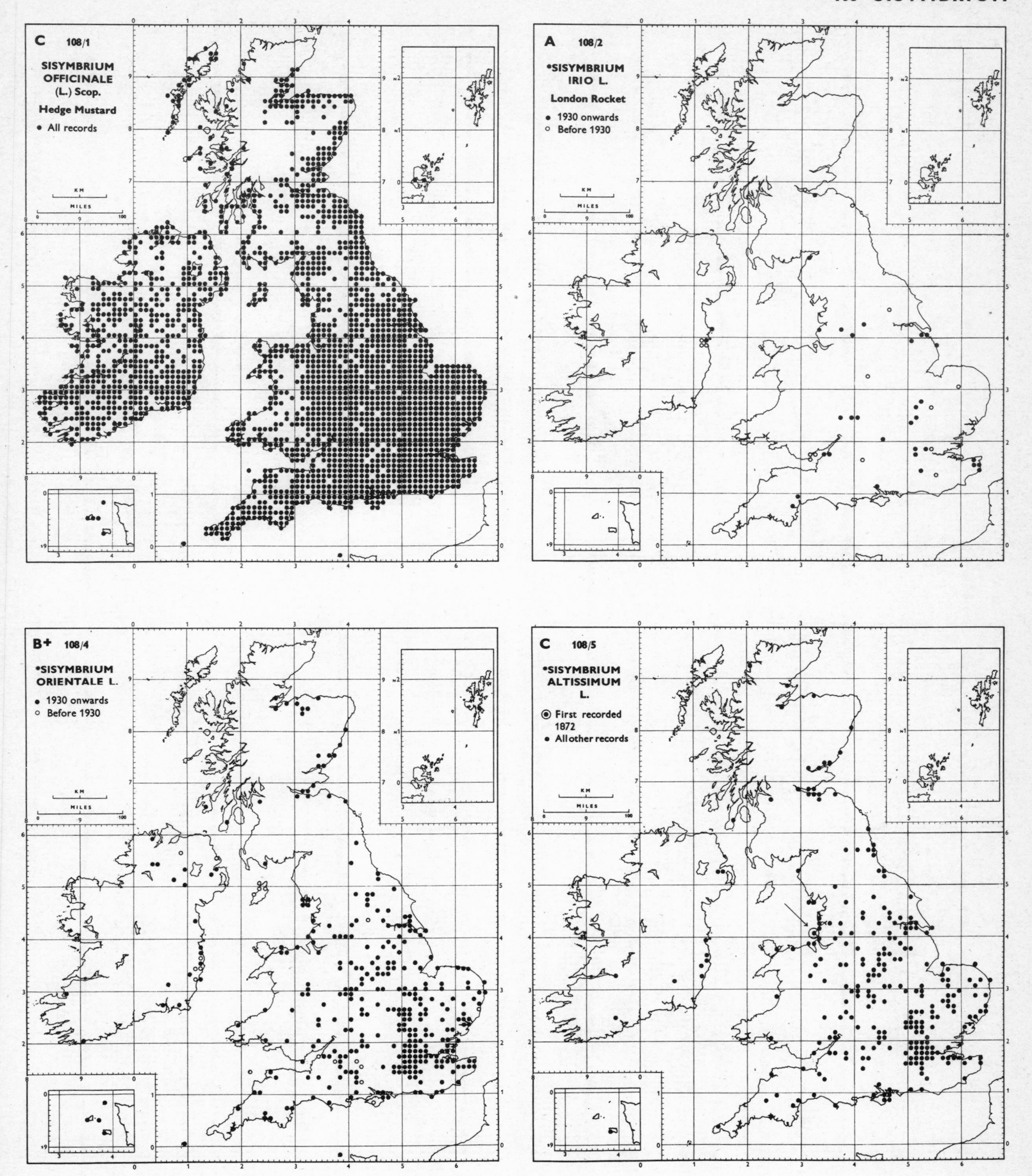
C 108/1
SISYMBRIUM OFFICINALE (L.) Scop.
Hedge Mustard
• All records
A 108/2
*SISYMBRIUM IRIO L.
London Rocket
• 1930 onwards
○ Before 1930
B+ 108/4
*SISYMBRIUM ORIENTALE L.
• 1930 onwards
○ Before 1930
C 108/5
*SISYMBRIUM ALTISSIMUM L.
◉ First recorded 1872
• All other records

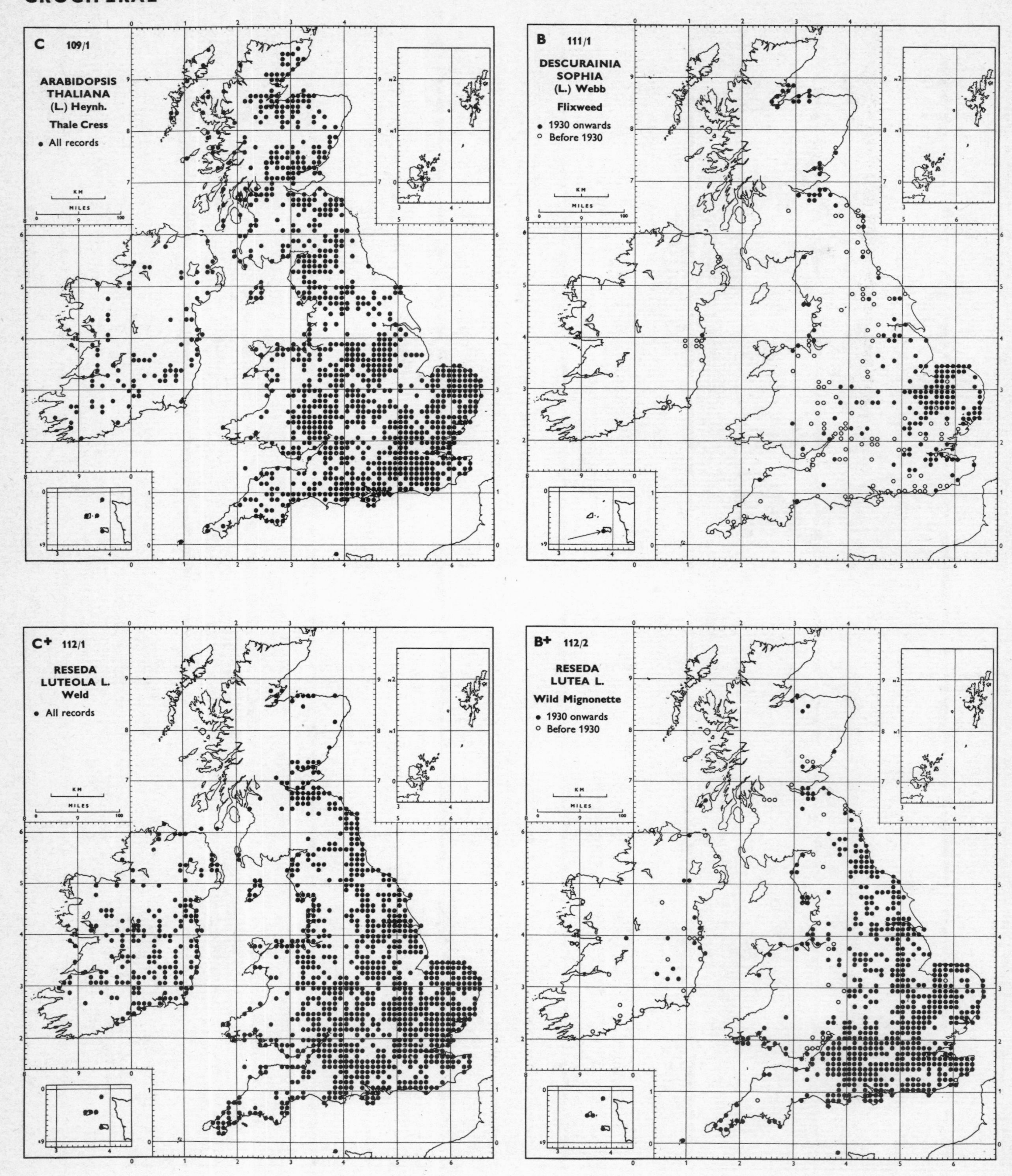
C 109/1
ARABIDOPSIS THALIANA (L.) Heynh.
Thale Cress
• All records
B 111/1
DESCURAINIA SOPHIA (L.) Webb
Flixweed
• 1930 onwards
○ Before 1930
C+ 112/1
RESEDA LUTEOLA L.
Weld
• All records
B+ 112/2
RESEDA LUTEA L.
Wild Mignonette
• 1930 onwards
○ Before 1930
KM
MILES

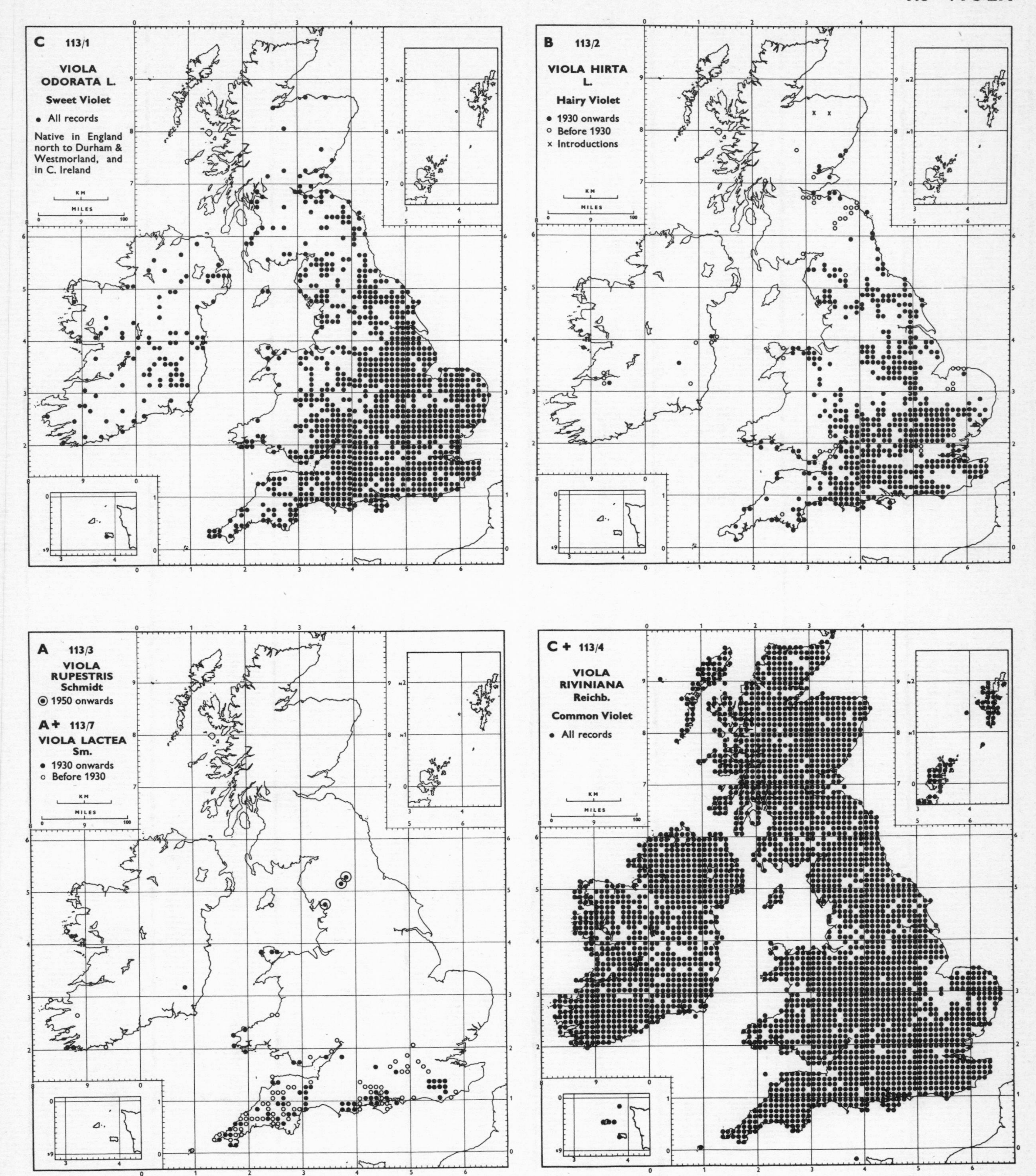
C 113/1
VIOLA ODORATA L.
Sweet Violet
• All records
Native in England north to Durham & Westmorland, and in C. Ireland
B 113/2
VIOLA HIRTA L.
Hairy Violet
• 1930 onwards
○ Before 1930
× Introductions
A 113/3
VIOLA RUPESTRIS Schmidt
◉ 1950 onwards
A+ 113/7
VIOLA LACTEA Sm.
• 1930 onwards
○ Before 1930
C+ 113/4
VIOLA RIVINIANA Reichb.
Common Violet
• All records
KM
MILES

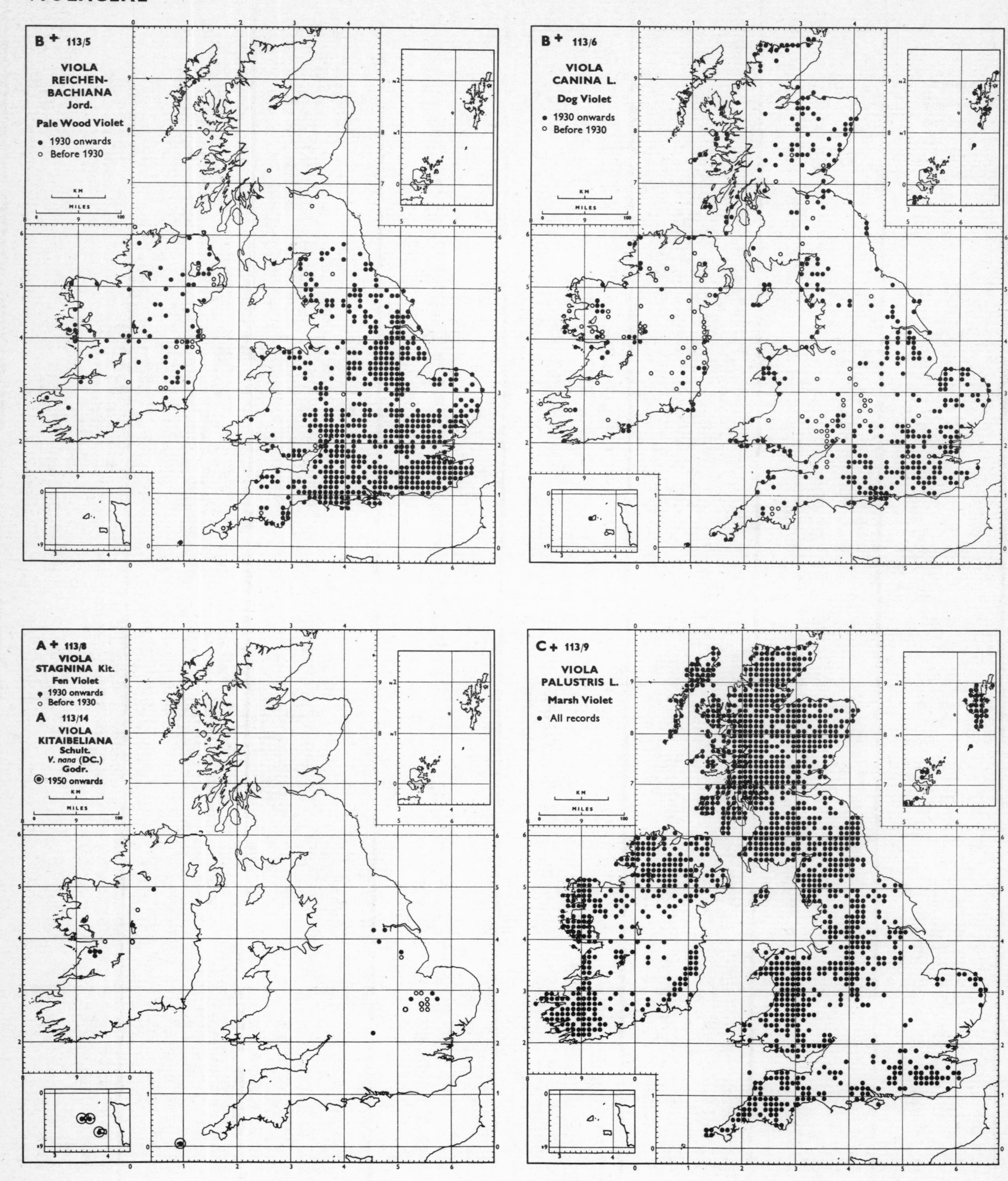
B + 113/5
VIOLA REICHEN-BACHIANA Jord.
Pale Wood Violet
1930 onwards
Before 1930
B + 113/6
VIOLA CANINA L.
Dog Violet
1930 onwards
Before 1930
A + 113/8
VIOLA STAGNINA Kit.
Fen Violet
1930 onwards
Before 1930
A 113/14
VIOLA KITAIBELIANA Schult.
V. nana (DC.) Godr.
1950 onwards
C + 113/9
VIOLA PALUSTRIS L.
Marsh Violet
All records
KM
MILES

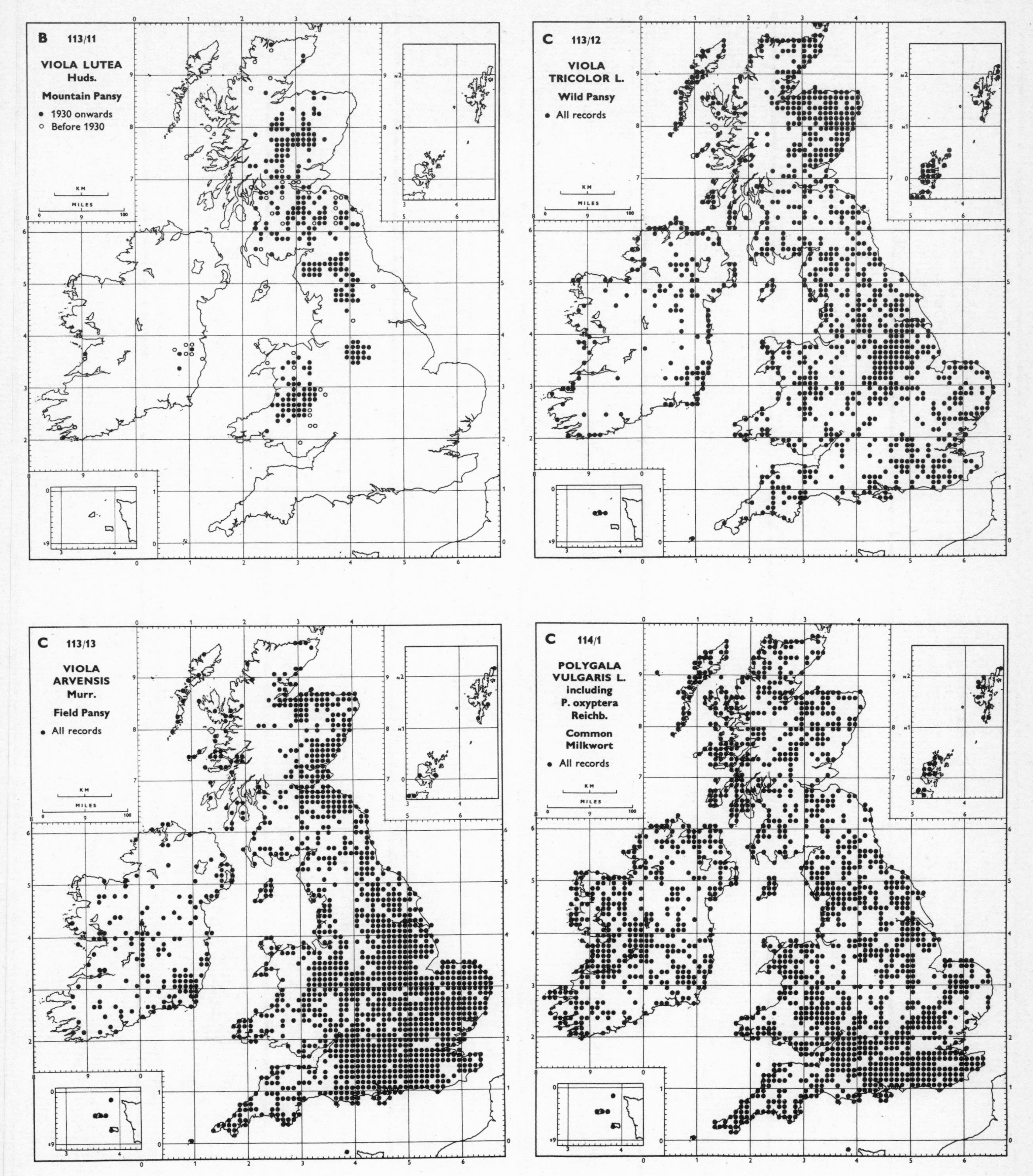
B 113/11
VIOLA LUTEA
Huds.
Mountain Pansy
• 1930 onwards
○ Before 1930
C 113/12
VIOLA
TRICOLOR L.
Wild Pansy
• All records
C 113/13
VIOLA
ARVENSIS
Murr.
Field Pansy
• All records
C 114/1
POLYGALA
VULGARIS L.
including
P. oxyptera
Reichb.
Common
Milkwort
• All records
KM
MILES

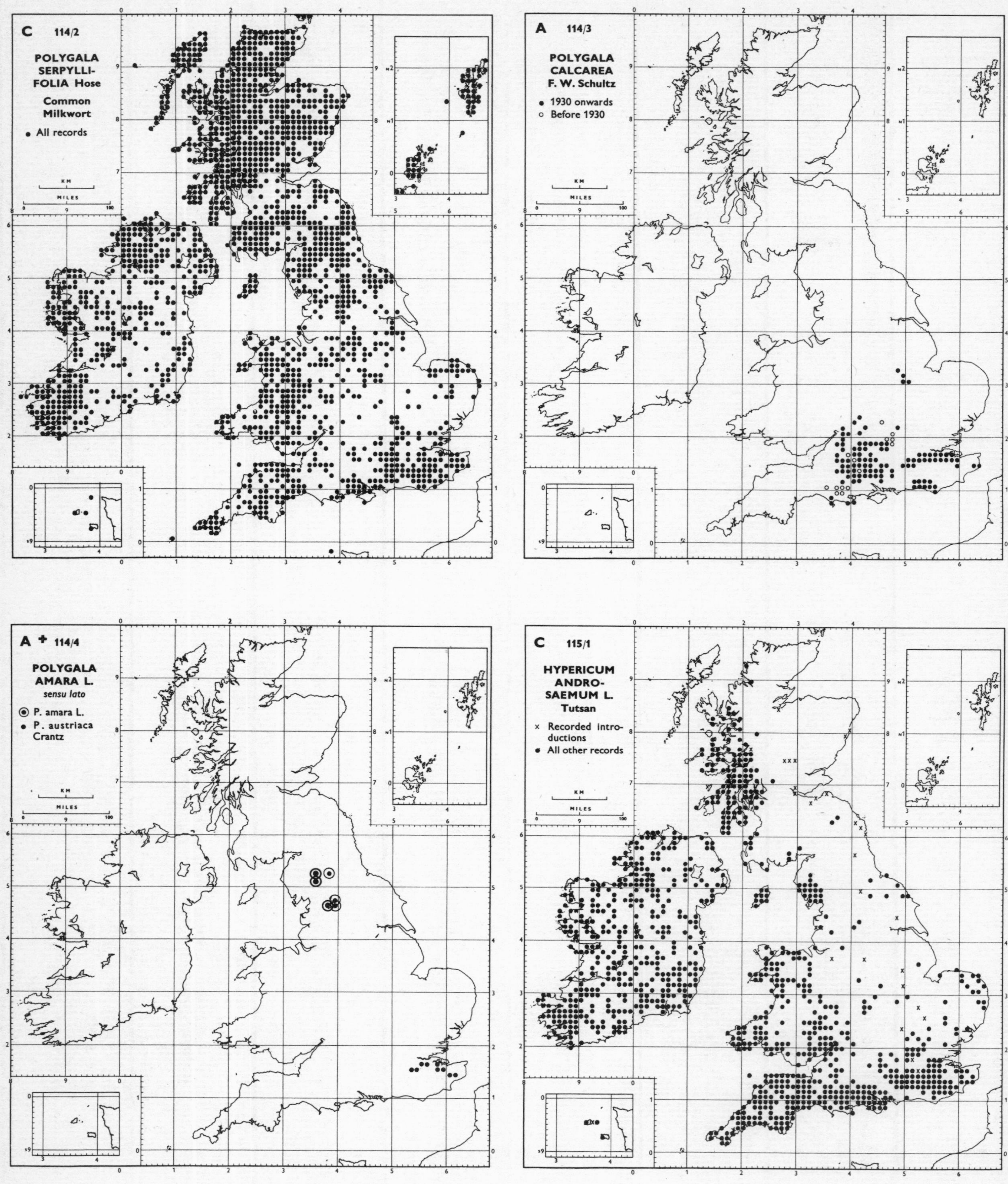
C 114/2
POLYGALA SERPYLLIFOLIA Hose
Common Milkwort
All records
A 114/3
POLYGALA CALCAREA F. W. Schultz
1930 onwards
Before 1930
A + 114/4
POLYGALA AMARA L.
sensu lato
P. amara L.
P. austriaca Crantz
C 115/1
HYPERICUM ANDROSAEMUM L.
Tutsan
Recorded introductions
All other records

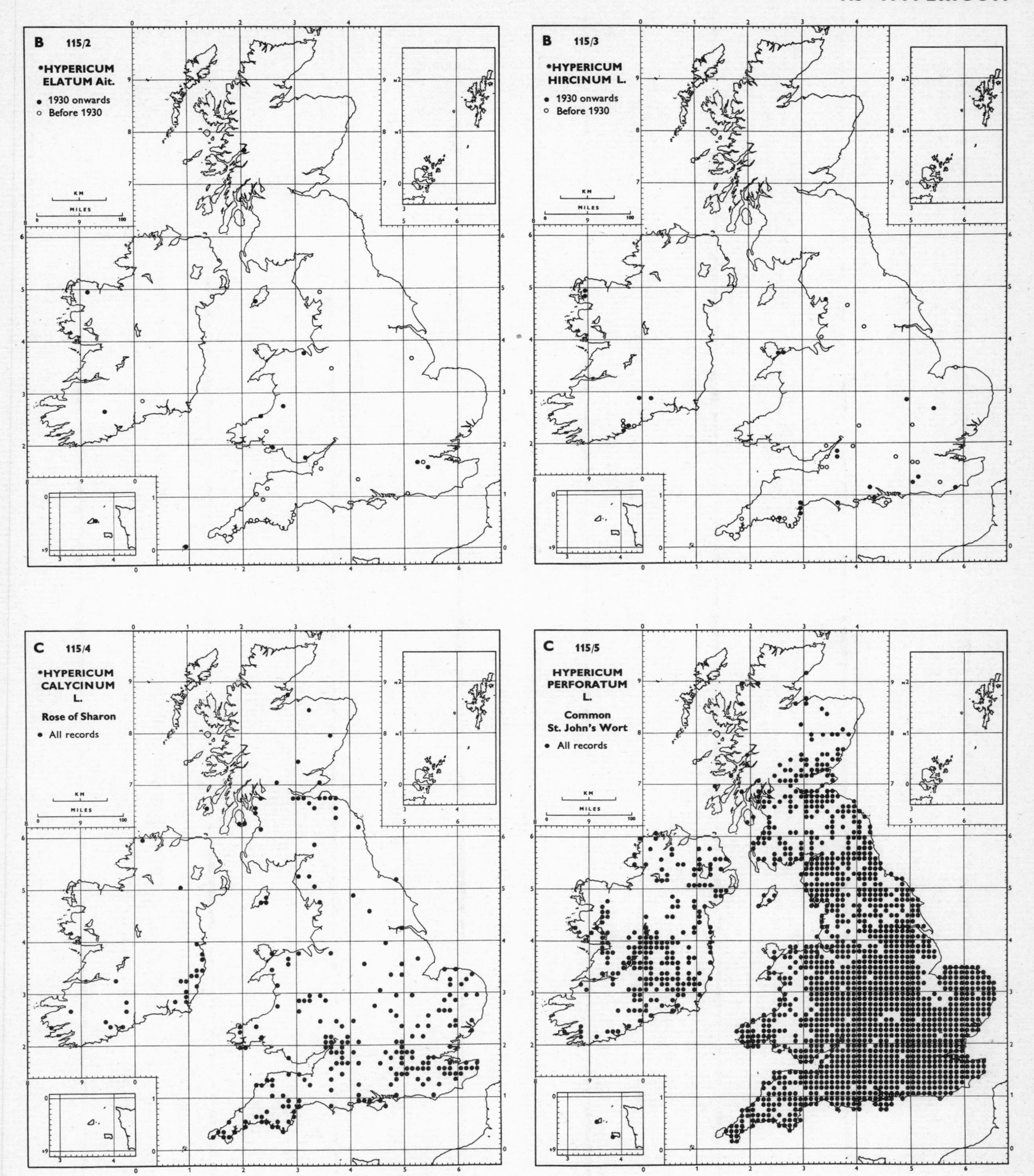
B 115/2
*HYPERICUM ELATUM Ait.
1930 onwards
Before 1930
B 115/3
*HYPERICUM HIRCINUM L.
1930 onwards
Before 1930
C 115/4
*HYPERICUM CALYCINUM L.
Rose of Sharon
All records
C 115/5
HYPERICUM PERFORATUM L.
Common St. John's Wort
All records

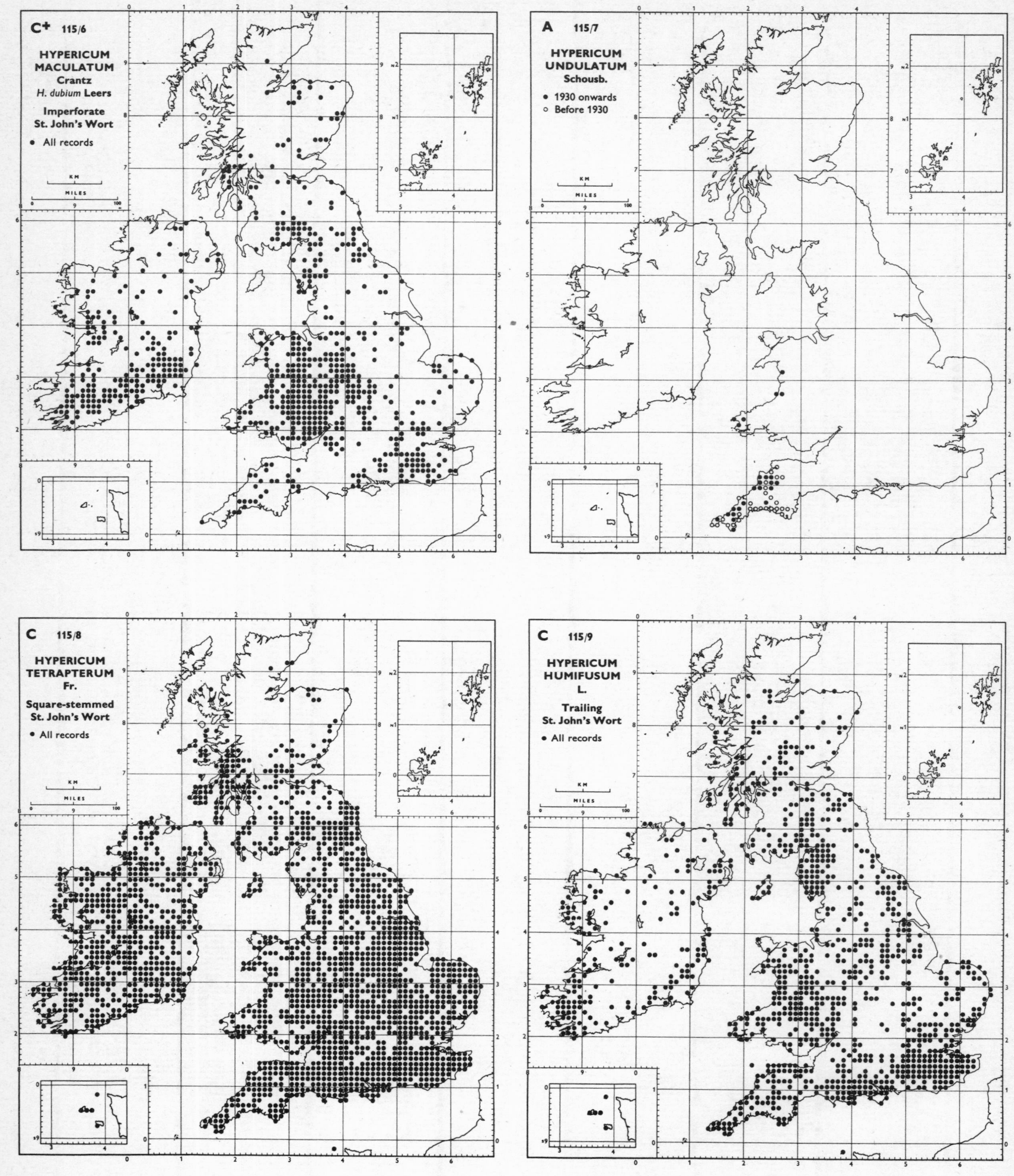
C+ 115/6
HYPERICUM MACULATUM Crantz
H. dubium Leers
Imperforate St. John's Wort
• All records
A 115/7
HYPERICUM UNDULATUM Schousb.
• 1930 onwards
○ Before 1930
C 115/8
HYPERICUM TETRAPTERUM Fr.
Square-stemmed St. John's Wort
• All records
C 115/9
HYPERICUM HUMIFUSUM L.
Trailing St. John's Wort
• All records
KM
MILES

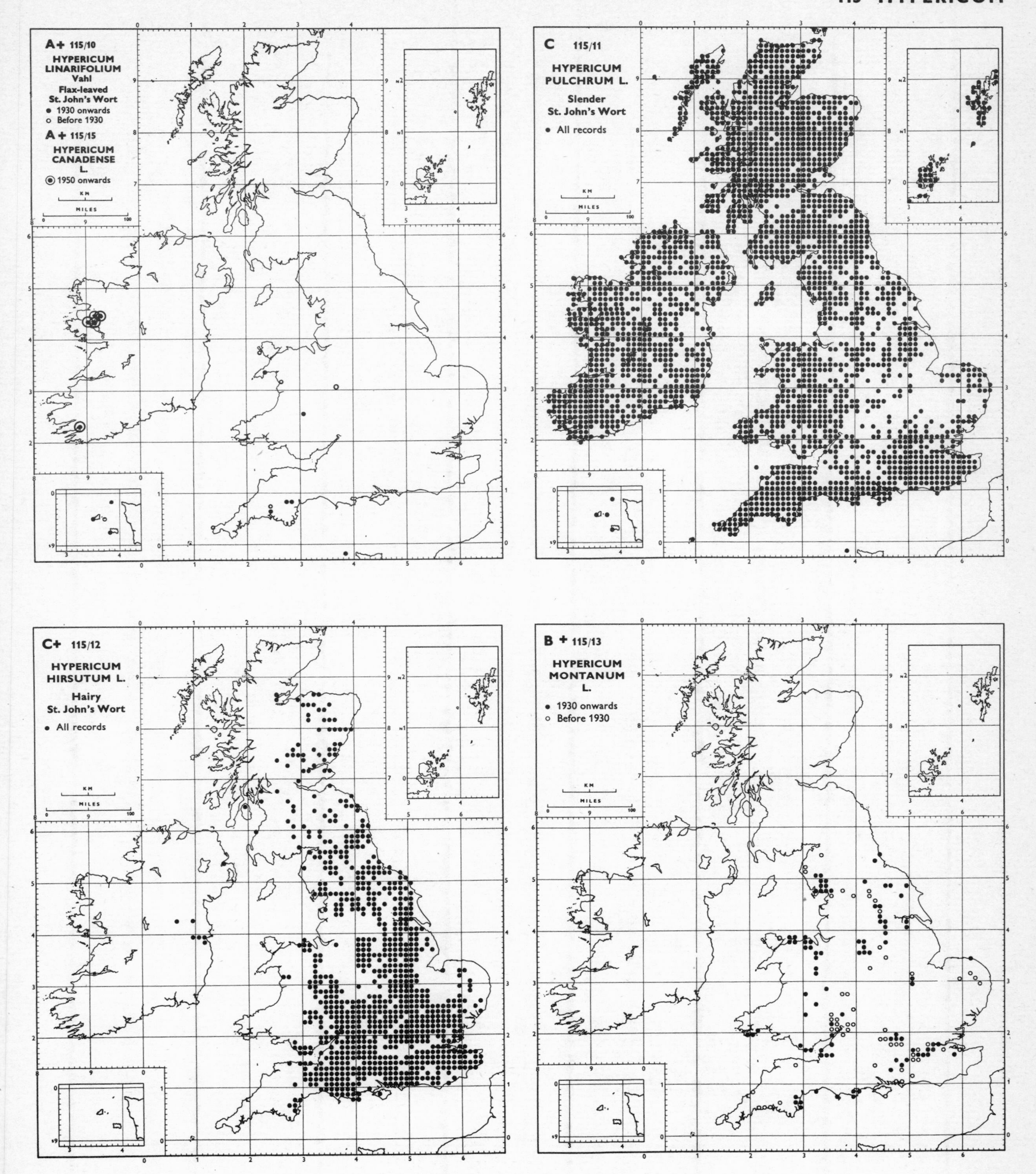
A+ 115/10
HYPERICUM LINARIFOLIUM Vahl
Flax-leaved St. John's Wort
• 1930 onwards
○ Before 1930
A+ 115/15
HYPERICUM CANADENSE L.
◉ 1950 onwards
C 115/11
HYPERICUM PULCHRUM L.
Slender St. John's Wort
• All records
C+ 115/12
HYPERICUM HIRSUTUM L.
Hairy St. John's Wort
• All records
B+ 115/13
HYPERICUM MONTANUM L.
• 1930 onwards
○ Before 1930
KM
MILES

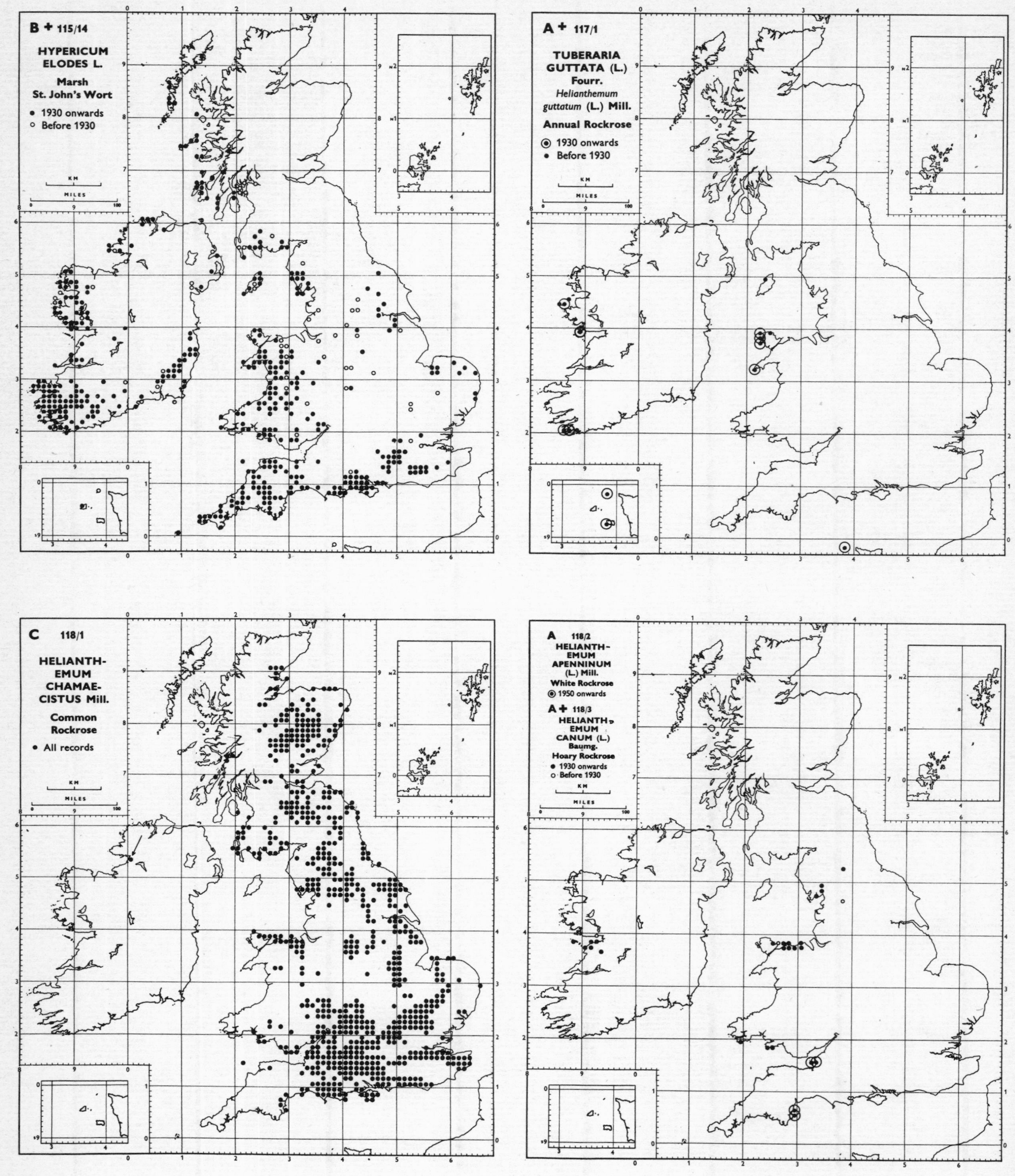
B + 115/14
HYPERICUM ELODES L.
Marsh St. John's Wort
• 1930 onwards
○ Before 1930
A + 117/1
TUBERARIA GUTTATA (L.) Fourr.
Helianthemum guttatum (L.) Mill.
Annual Rockrose
◉ 1930 onwards
• Before 1930
C 118/1
HELIANTHEMUM CHAMAECISTUS Mill.
Common Rockrose
• All records
A 118/2
HELIANTHEMUM APENNINUM (L.) Mill.
White Rockrose
◉ 1950 onwards
A + 118/3
HELIANTHEMUM CANUM (L.) Baumg.
Hoary Rockrose
• 1930 onwards
○ Before 1930
KM
MILES

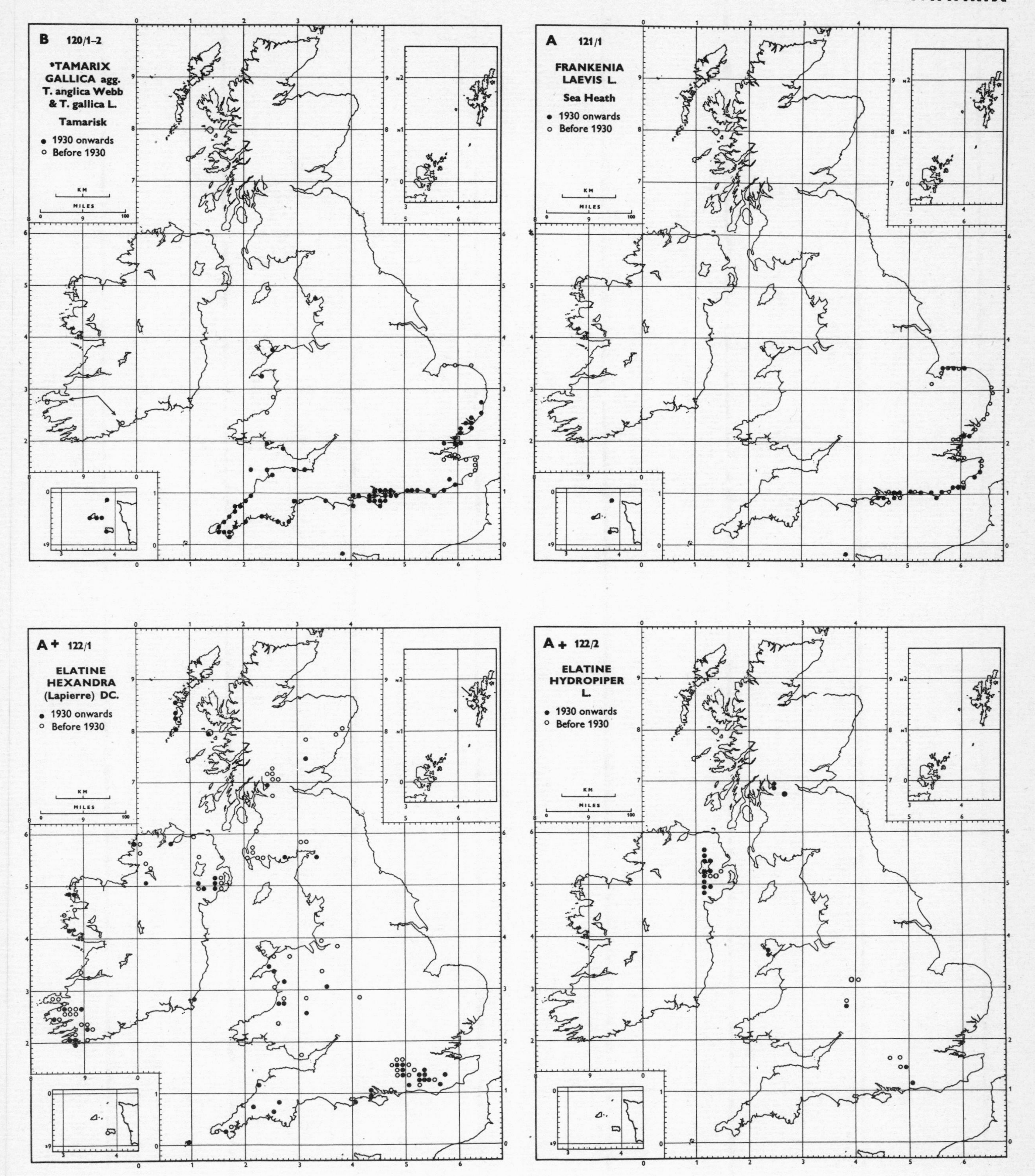
B 120/1–2
*TAMARIX GALLICA agg. T. anglica Webb & T. gallica L.
Tamarisk
1930 onwards
Before 1930
A 121/1
FRANKENIA LAEVIS L.
Sea Heath
1930 onwards
Before 1930
A+ 122/1
ELATINE HEXANDRA (Lapierre) DC.
1930 onwards
Before 1930
A+ 122/2
ELATINE HYDROPIPER L.
1930 onwards
Before 1930
KM
MILES

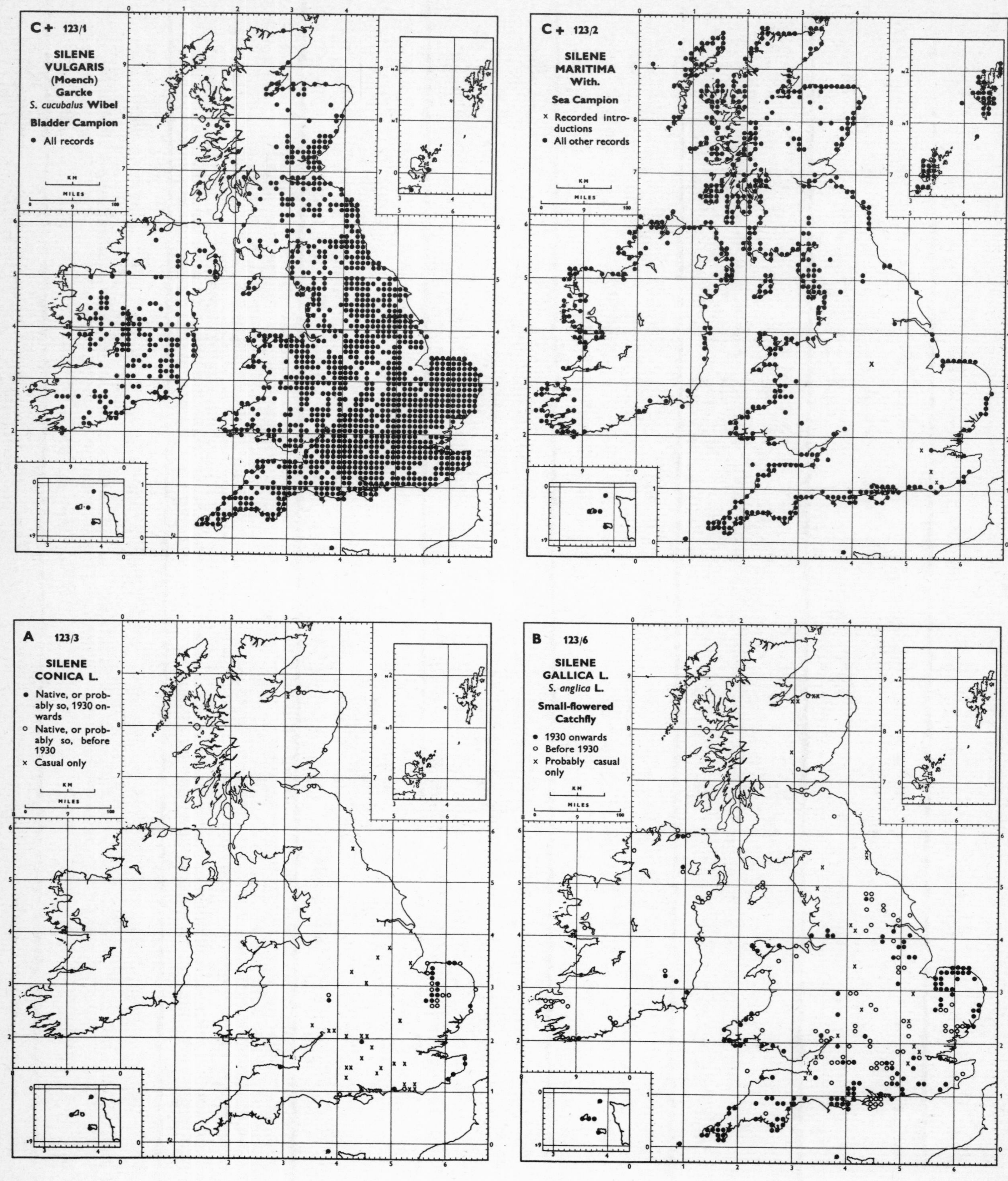
C+ 123/1
SILENE VULGARIS (Moench) Garcke
S. cucubalus Wibel
Bladder Campion
• All records
C+ 123/2
SILENE MARITIMA With.
Sea Campion
× Recorded introductions
• All other records
A 123/3
SILENE CONICA L.
• Native, or probably so, 1930 onwards
○ Native, or probably so, before 1930
× Casual only
B 123/6
SILENE GALLICA L.
S. anglica L.
Small-flowered Catchfly
• 1930 onwards
○ Before 1930
× Probably casual only
KM
MILES

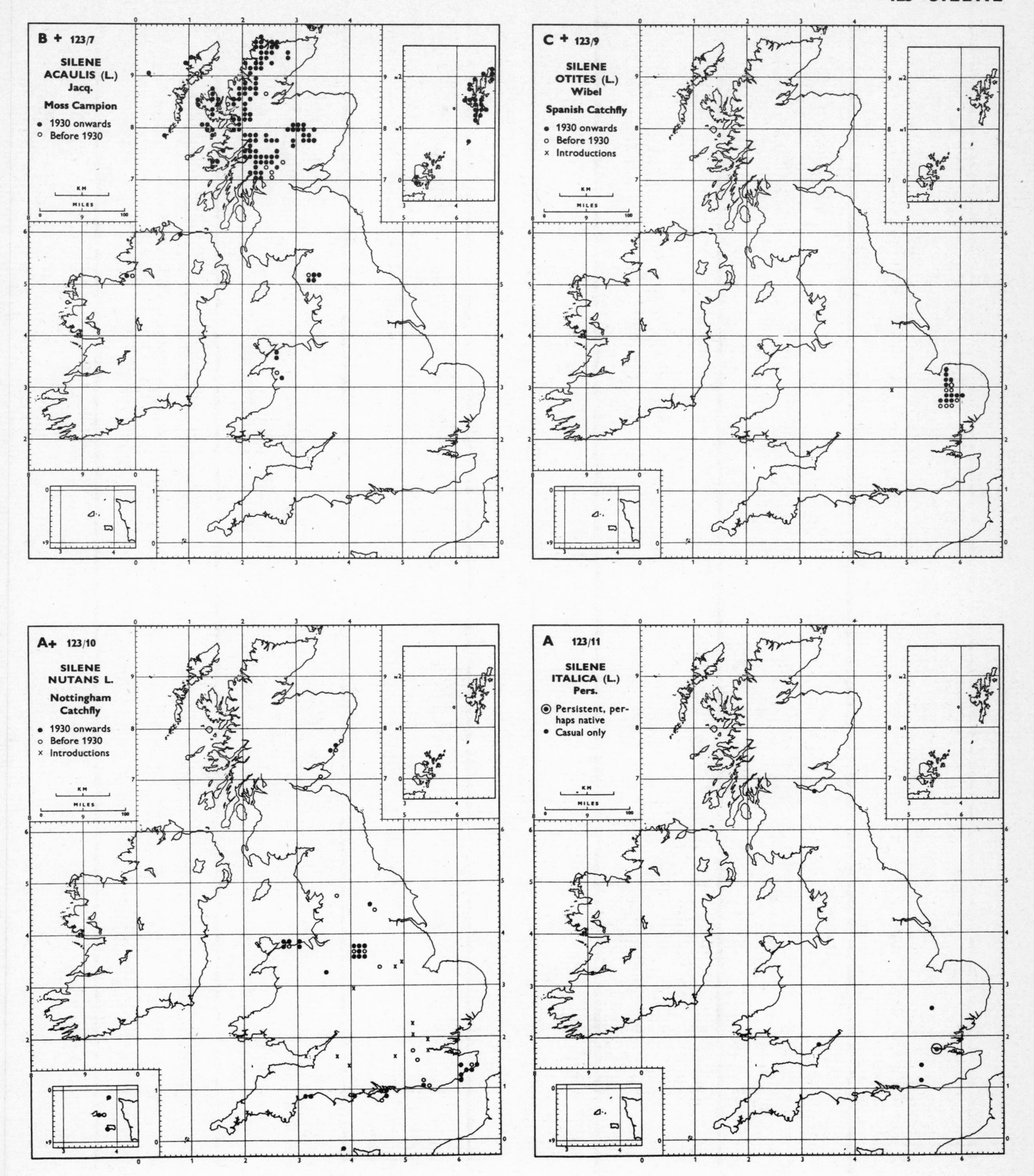
B + 123/7
SILENE
ACAULIS (L.)
Jacq.
Moss Campion
• 1930 onwards
○ Before 1930
C + 123/9
SILENE
OTITES (L.)
Wibel
Spanish Catchfly
• 1930 onwards
○ Before 1930
× Introductions
A+ 123/10
SILENE
NUTANS L.
Nottingham
Catchfly
• 1930 onwards
○ Before 1930
× Introductions
A 123/11
SILENE
ITALICA (L.)
Pers.
◉ Persistent, perhaps native
• Casual only

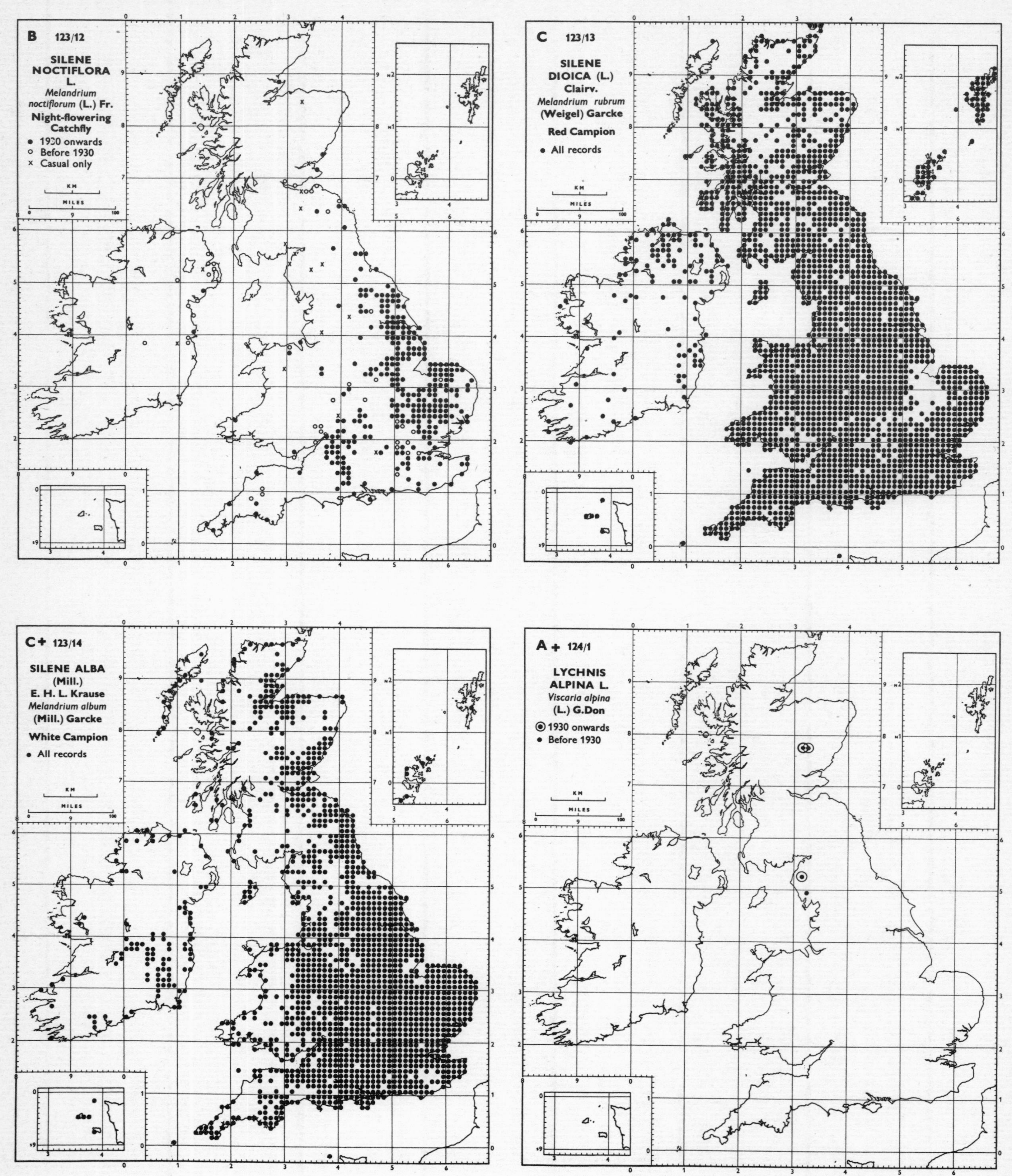
B 123/12
SILENE NOCTIFLORA L.
Melandrium noctiflorum (L.) Fr.
Night-flowering Catchfly
• 1930 onwards
○ Before 1930
× Casual only
C 123/13
SILENE DIOICA (L.) Clairv.
Melandrium rubrum (Weigel) Garcke
Red Campion
• All records
C+ 123/14
SILENE ALBA (Mill.) E. H. L. Krause
Melandrium album (Mill.) Garcke
White Campion
• All records
A+ 124/1
LYCHNIS ALPINA L.
Viscaria alpina (L.) G.Don
⊙ 1930 onwards
• Before 1930

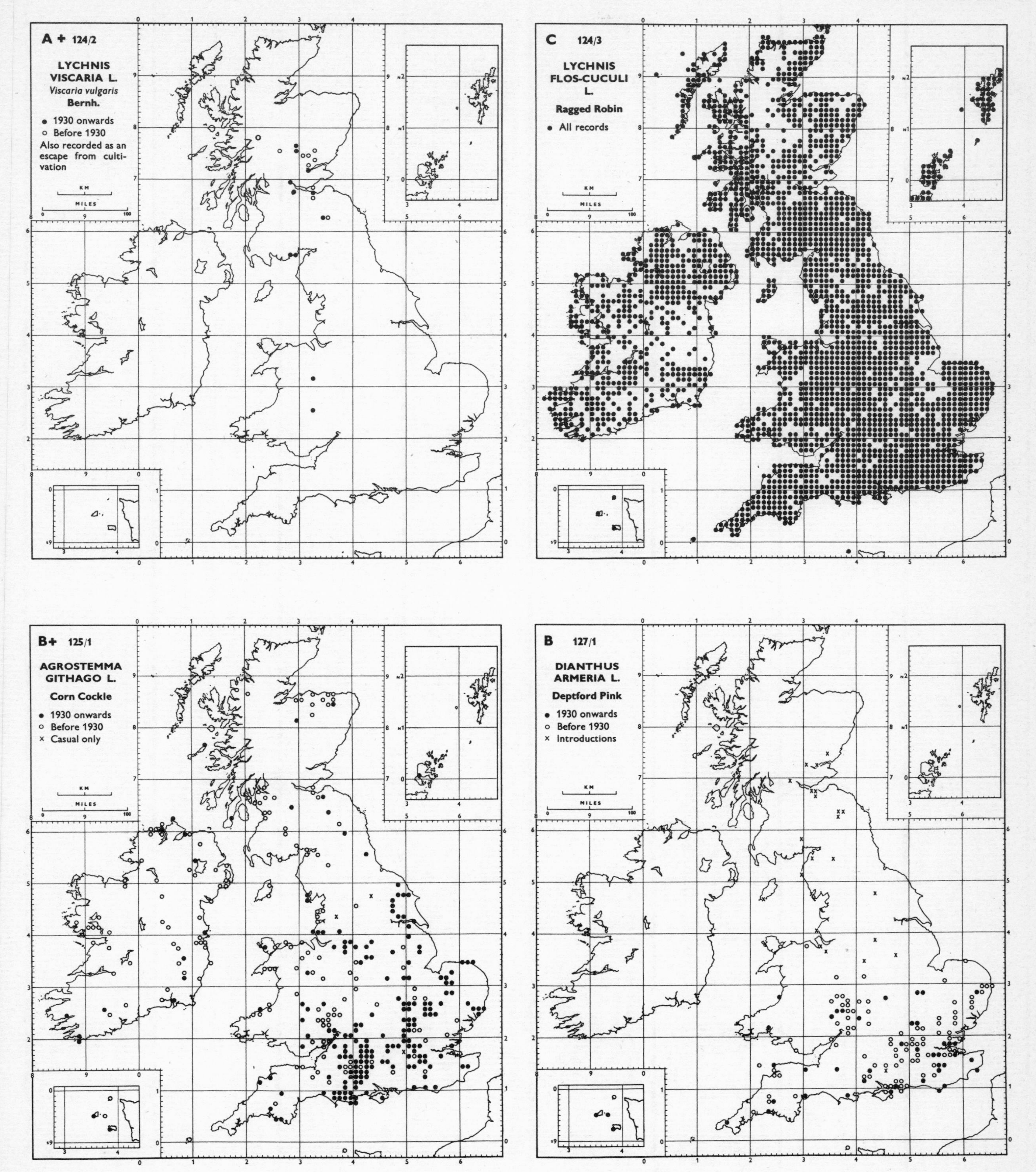
A + 124/2
LYCHNIS VISCARIA L.
Viscaria vulgaris Bernh.
• 1930 onwards
○ Before 1930
Also recorded as an escape from cultivation
C 124/3
LYCHNIS FLOS-CUCULI L.
Ragged Robin
• All records
B+ 125/1
AGROSTEMMA GITHAGO L.
Corn Cockle
• 1930 onwards
○ Before 1930
× Casual only
B 127/1
DIANTHUS ARMERIA L.
Deptford Pink
• 1930 onwards
○ Before 1930
× Introductions
KM
MILES

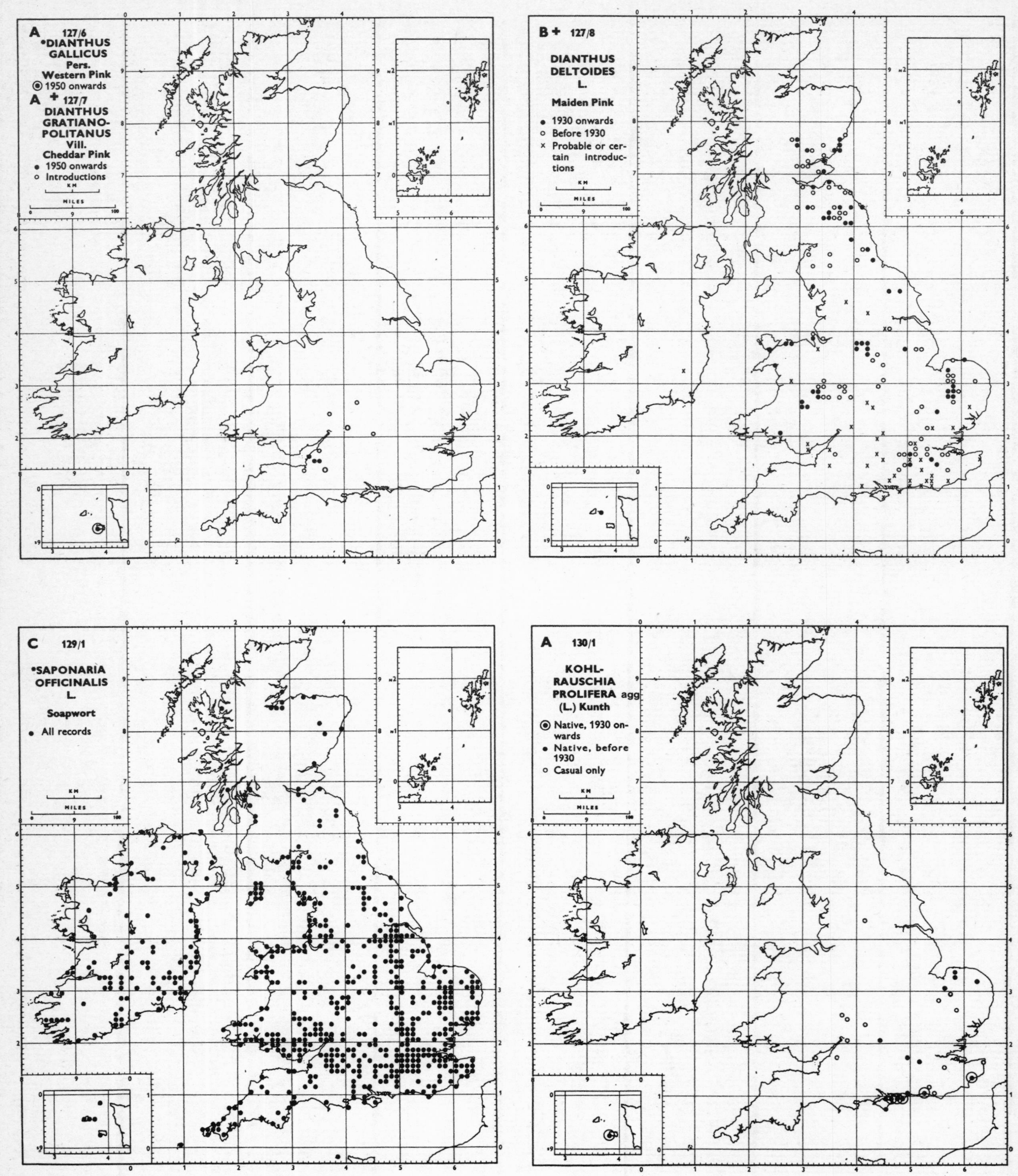
A 127/6
•DIANTHUS GALLICUS Pers.
Western Pink
◉ 1950 onwards
A + 127/7
DIANTHUS GRATIANOPOLITANUS Vill.
Cheddar Pink
• 1950 onwards
○ Introductions
B + 127/8
DIANTHUS DELTOIDES L.
Maiden Pink
• 1930 onwards
○ Before 1930
× Probable or certain introductions
C 129/1
•SAPONARIA OFFICINALIS L.
Soapwort
• All records
A 130/1
KOHLRAUSCHIA PROLIFERA agg (L.) Kunth
◉ Native, 1930 onwards
• Native, before 1930
○ Casual only
KM
MILES

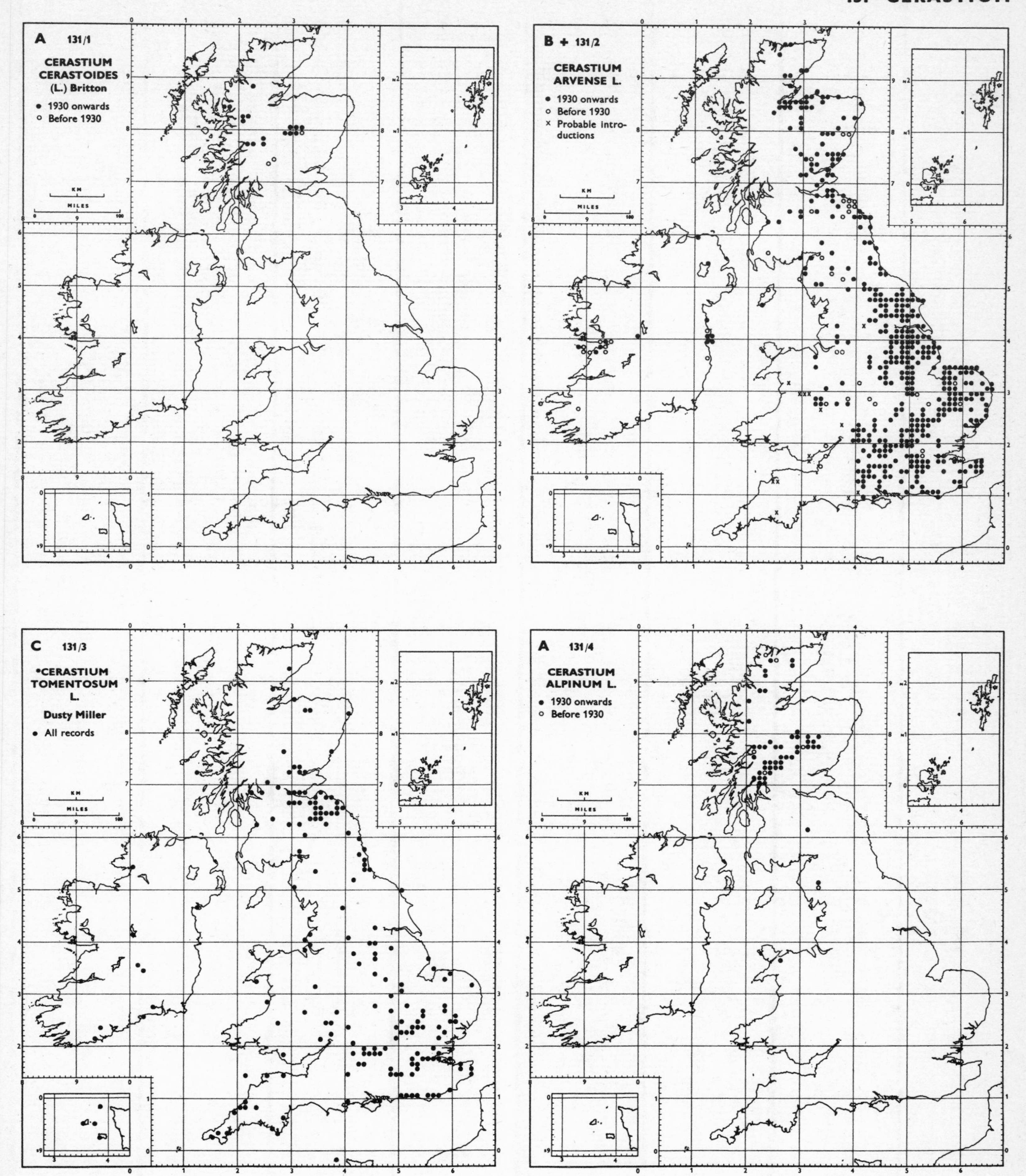
A 131/1
CERASTIUM CERASTOIDES (L.) Britton
• 1930 onwards
○ Before 1930
B + 131/2
CERASTIUM ARVENSE L.
• 1930 onwards
○ Before 1930
× Probable introductions
C 131/3
*CERASTIUM TOMENTOSUM L.
Dusty Miller
• All records
A 131/4
CERASTIUM ALPINUM L.
• 1930 onwards
○ Before 1930
KM
MILES

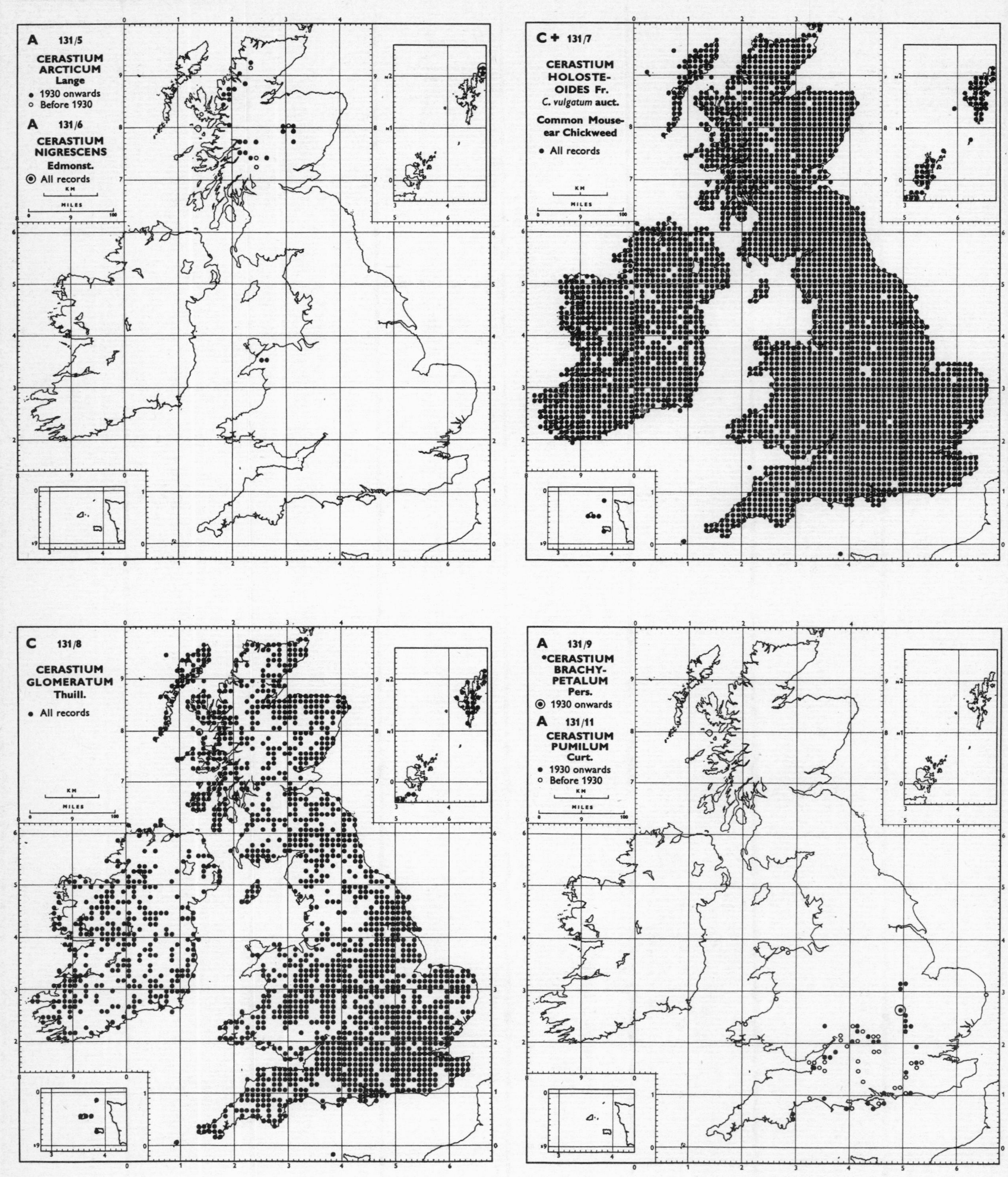
A 131/5
CERASTIUM ARCTICUM Lange
1930 onwards
Before 1930
A 131/6
CERASTIUM NIGRESCENS Edmonst.
All records
C + 131/7
CERASTIUM HOLOSTEOIDES Fr.
C. vulgatum auct.
Common Mouse-ear Chickweed
All records
C 131/8
CERASTIUM GLOMERATUM Thuill.
All records
A 131/9
*CERASTIUM BRACHYPETALUM Pers.
1930 onwards
A 131/11
CERASTIUM PUMILUM Curt.
1930 onwards
Before 1930

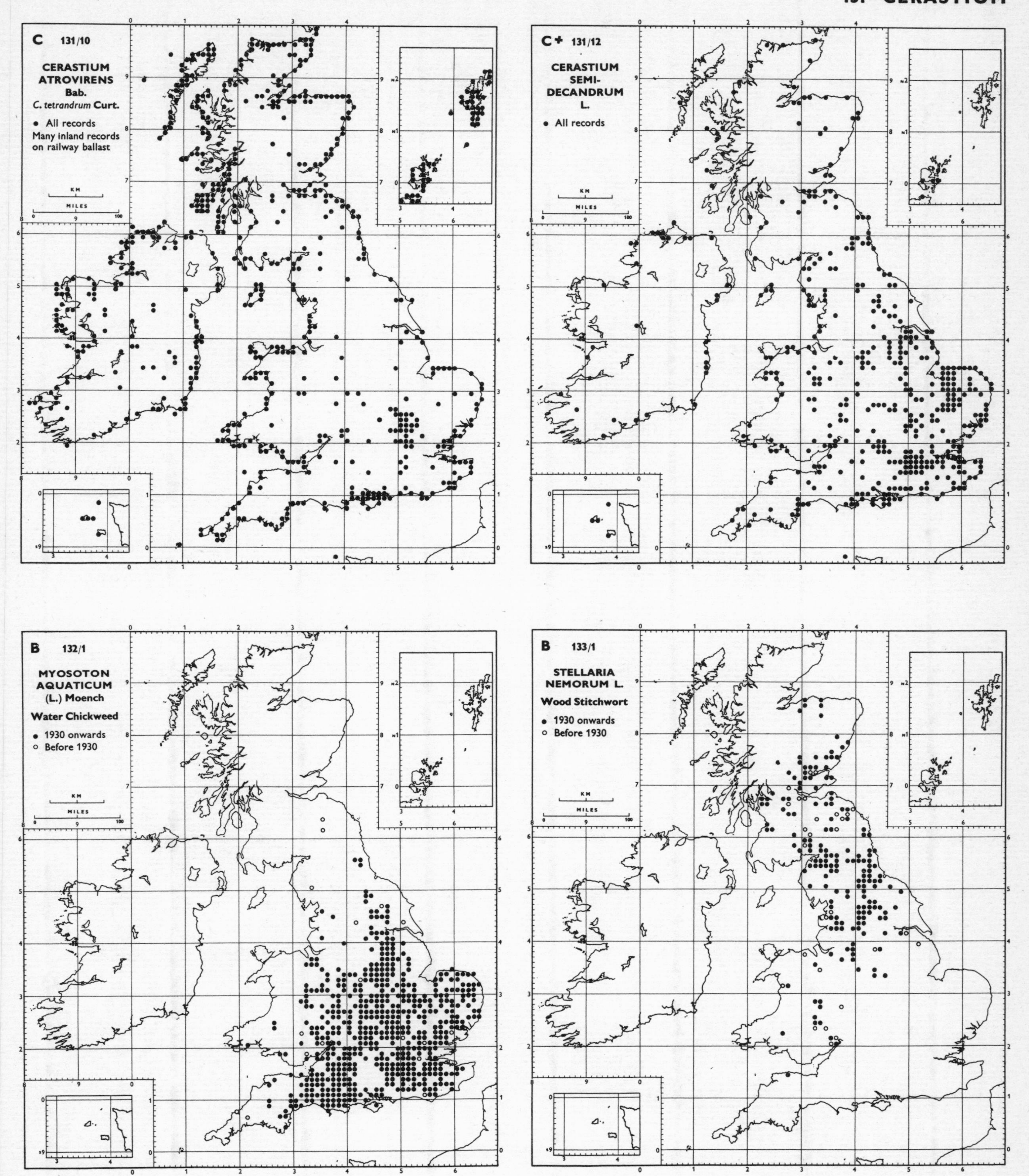
C 131/10
CERASTIUM ATROVIRENS Bab.
C. tetrandrum Curt.
All records
Many inland records on railway ballast
C+ 131/12
CERASTIUM SEMI-DECANDRUM L.
All records
B 132/1
MYOSOTON AQUATICUM (L.) Moench
Water Chickweed
1930 onwards
Before 1930
B 133/1
STELLARIA NEMORUM L.
Wood Stitchwort
1930 onwards
Before 1930
KM
MILES

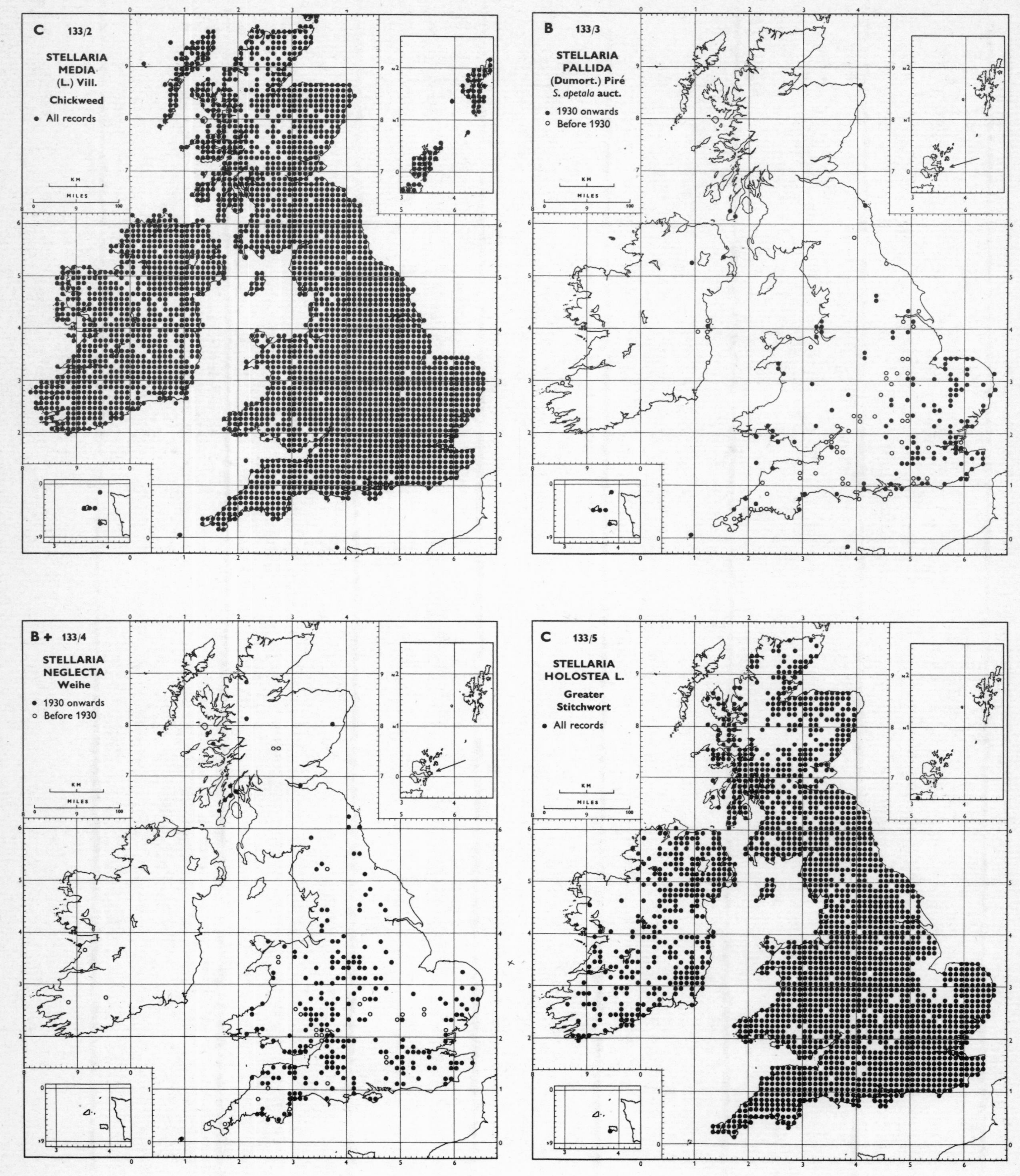
C 133/2
STELLARIA MEDIA (L.) Vill.
Chickweed
• All records
B 133/3
STELLARIA PALLIDA (Dumort.) Piré
S. apetala auct.
• 1930 onwards
○ Before 1930
B + 133/4
STELLARIA NEGLECTA Weihe
• 1930 onwards
○ Before 1930
C 133/5
STELLARIA HOLOSTEA L.
Greater Stitchwort
• All records
KM
MILES

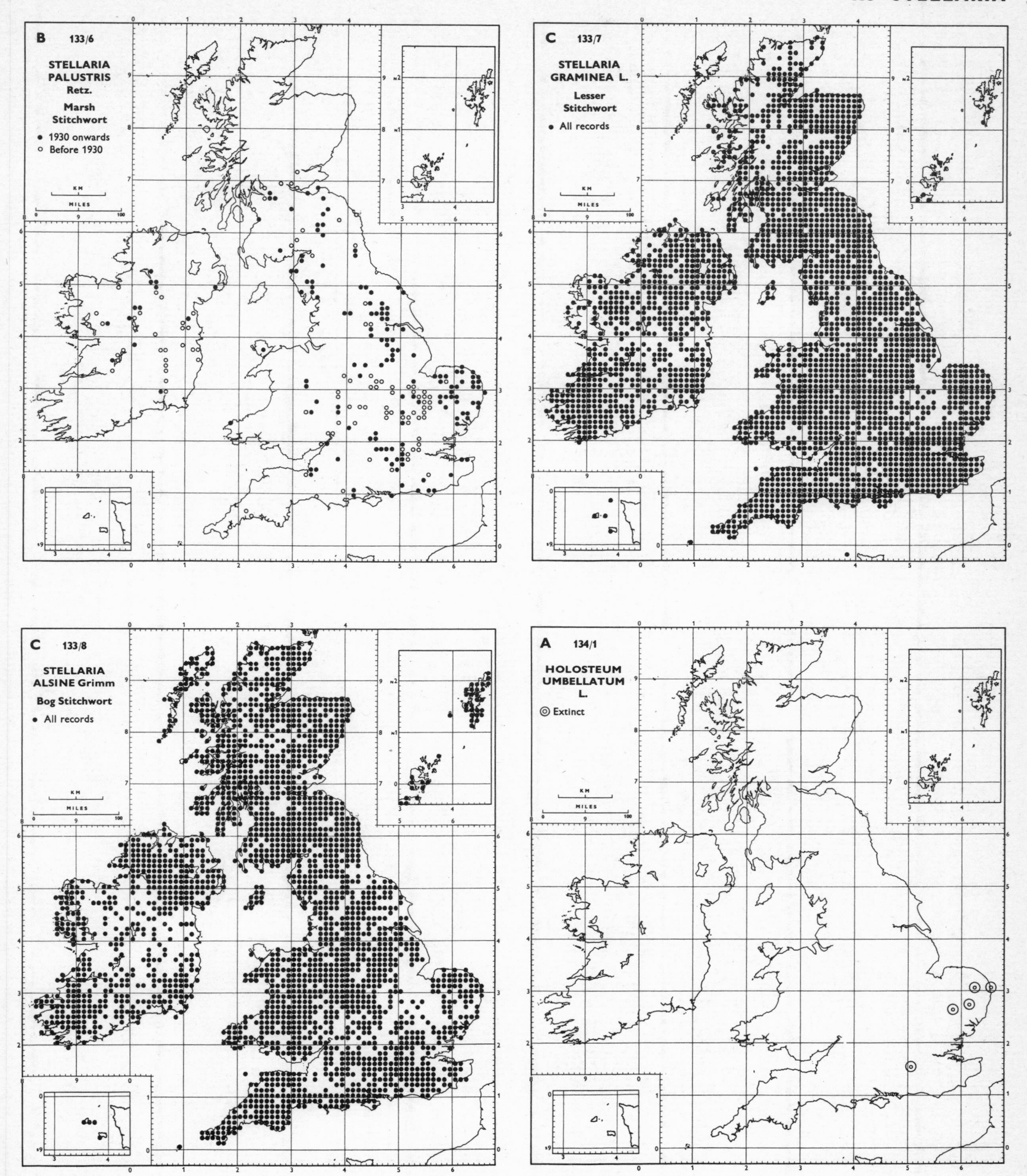
B 133/6
STELLARIA PALUSTRIS Retz.
Marsh Stitchwort
● 1930 onwards
○ Before 1930
C 133/7
STELLARIA GRAMINEA L.
Lesser Stitchwort
● All records
C 133/8
STELLARIA ALSINE Grimm
Bog Stitchwort
● All records
A 134/1
HOLOSTEUM UMBELLATUM L.
◎ Extinct

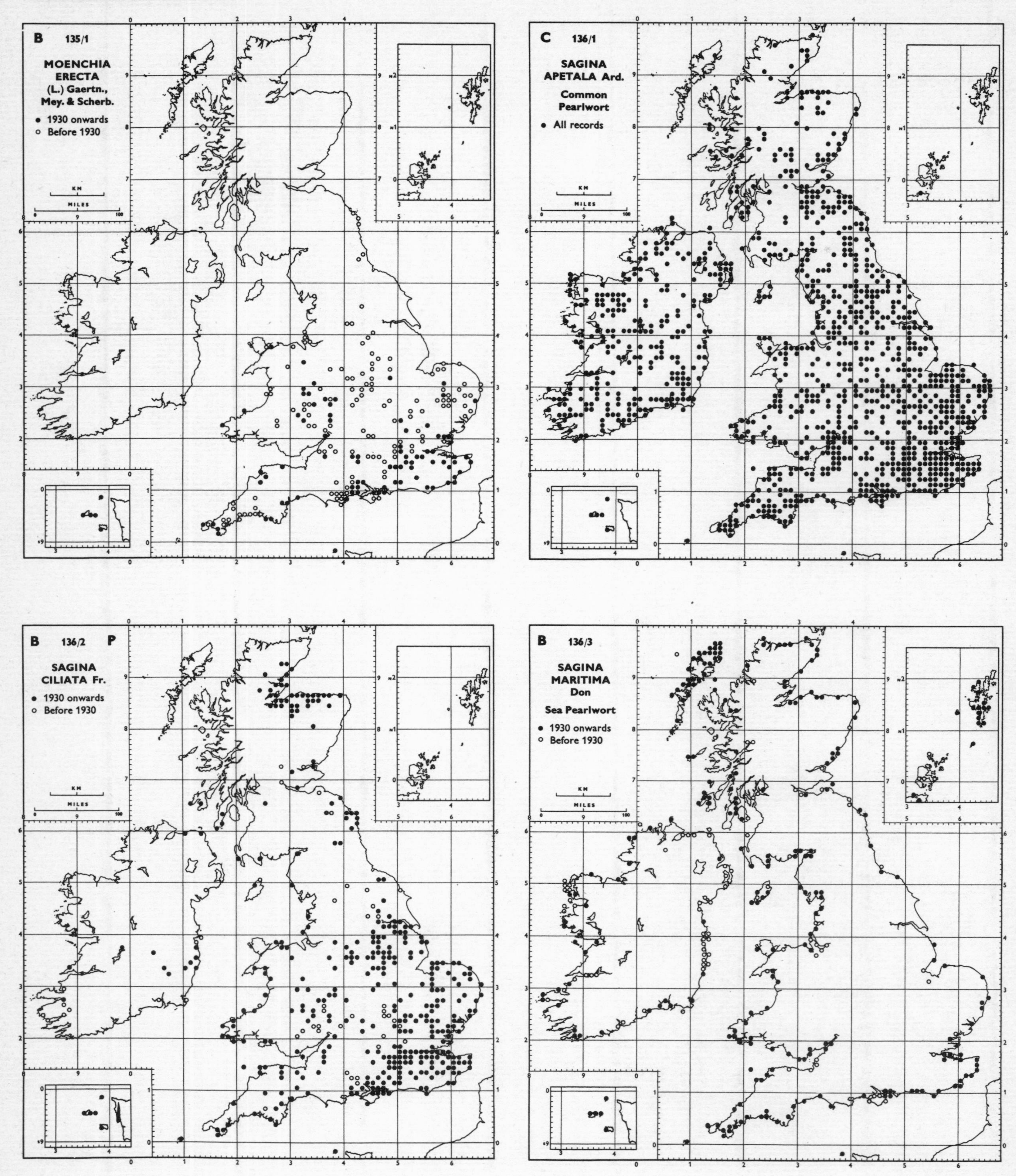
B 135/1
MOENCHIA ERECTA (L.) Gaertn., Mey. & Scherb.
• 1930 onwards
○ Before 1930
C 136/1
SAGINA APETALA Ard.
Common Pearlwort
• All records
B 136/2 P
SAGINA CILIATA Fr.
• 1930 onwards
○ Before 1930
B 136/3
SAGINA MARITIMA Don
Sea Pearlwort
• 1930 onwards
○ Before 1930
KM
MILES

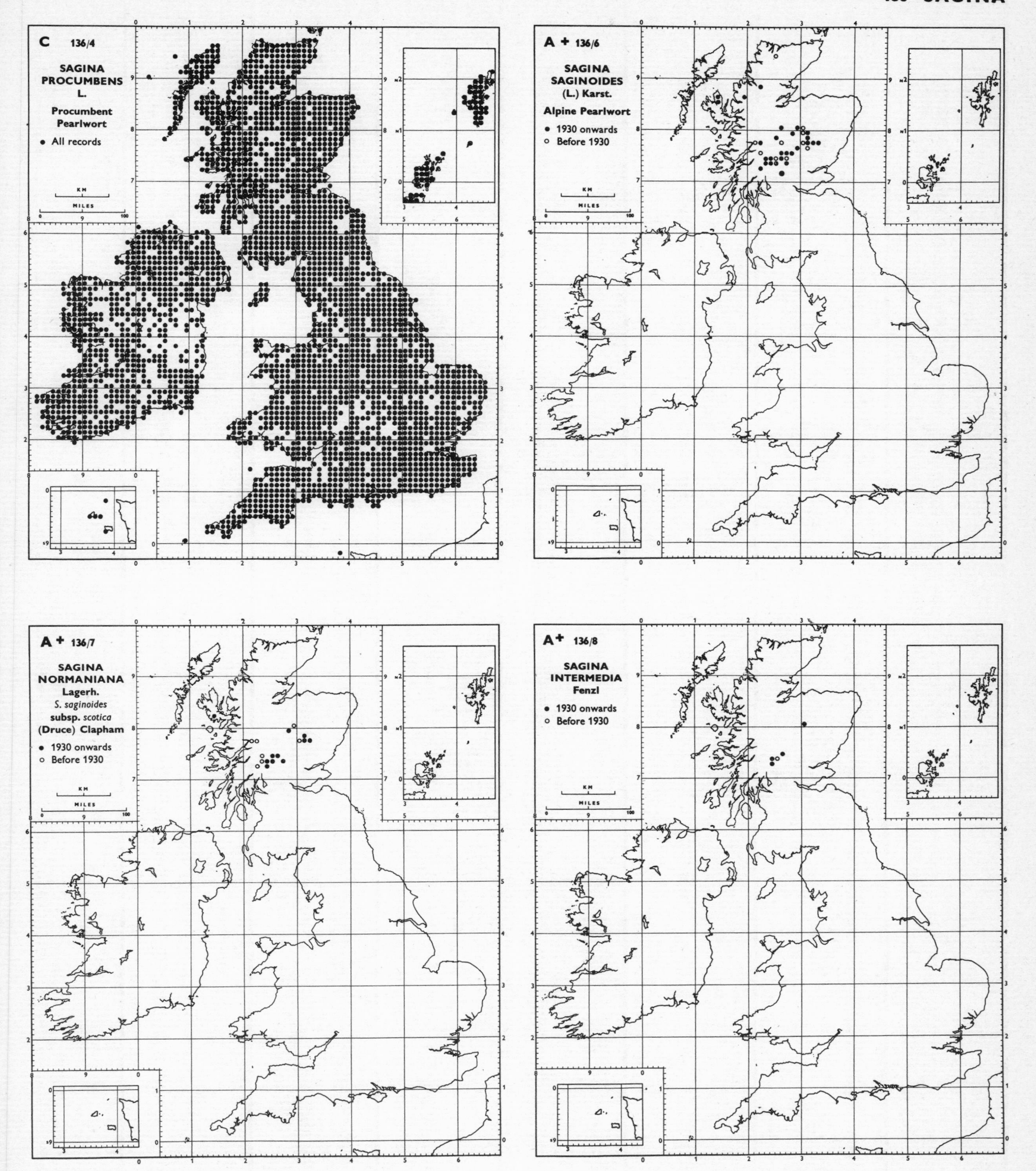
C 136/4
SAGINA PROCUMBENS L.
Procumbent Pearlwort
• All records
A + 136/6
SAGINA SAGINOIDES (L.) Karst.
Alpine Pearlwort
• 1930 onwards
○ Before 1930
A + 136/7
SAGINA NORMANIANA Lagerh.
S. saginoides subsp. scotica (Druce) Clapham
• 1930 onwards
○ Before 1930
A + 136/8
SAGINA INTERMEDIA Fenzl
• 1930 onwards
○ Before 1930

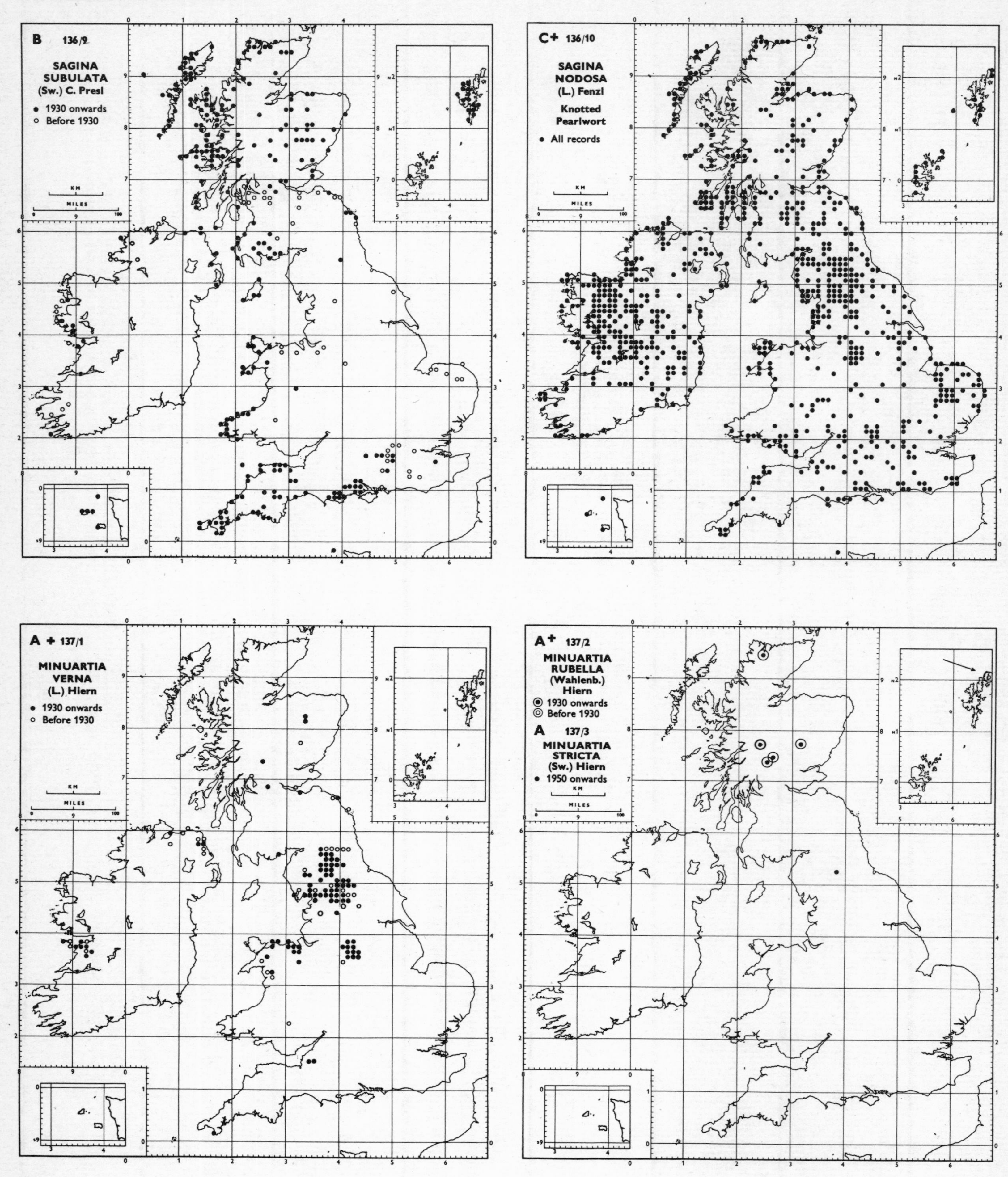
B 136/9
SAGINA SUBULATA (Sw.) C. Presl
• 1930 onwards
○ Before 1930
C+ 136/10
SAGINA NODOSA (L.) Fenzl
Knotted Pearlwort
• All records
A + 137/1
MINUARTIA VERNA (L.) Hiern
• 1930 onwards
○ Before 1930
A+ 137/2
MINUARTIA RUBELLA (Wahlenb.) Hiern
◉ 1930 onwards
◎ Before 1930
A 137/3
MINUARTIA STRICTA (Sw.) Hiern
• 1950 onwards
KM
MILES

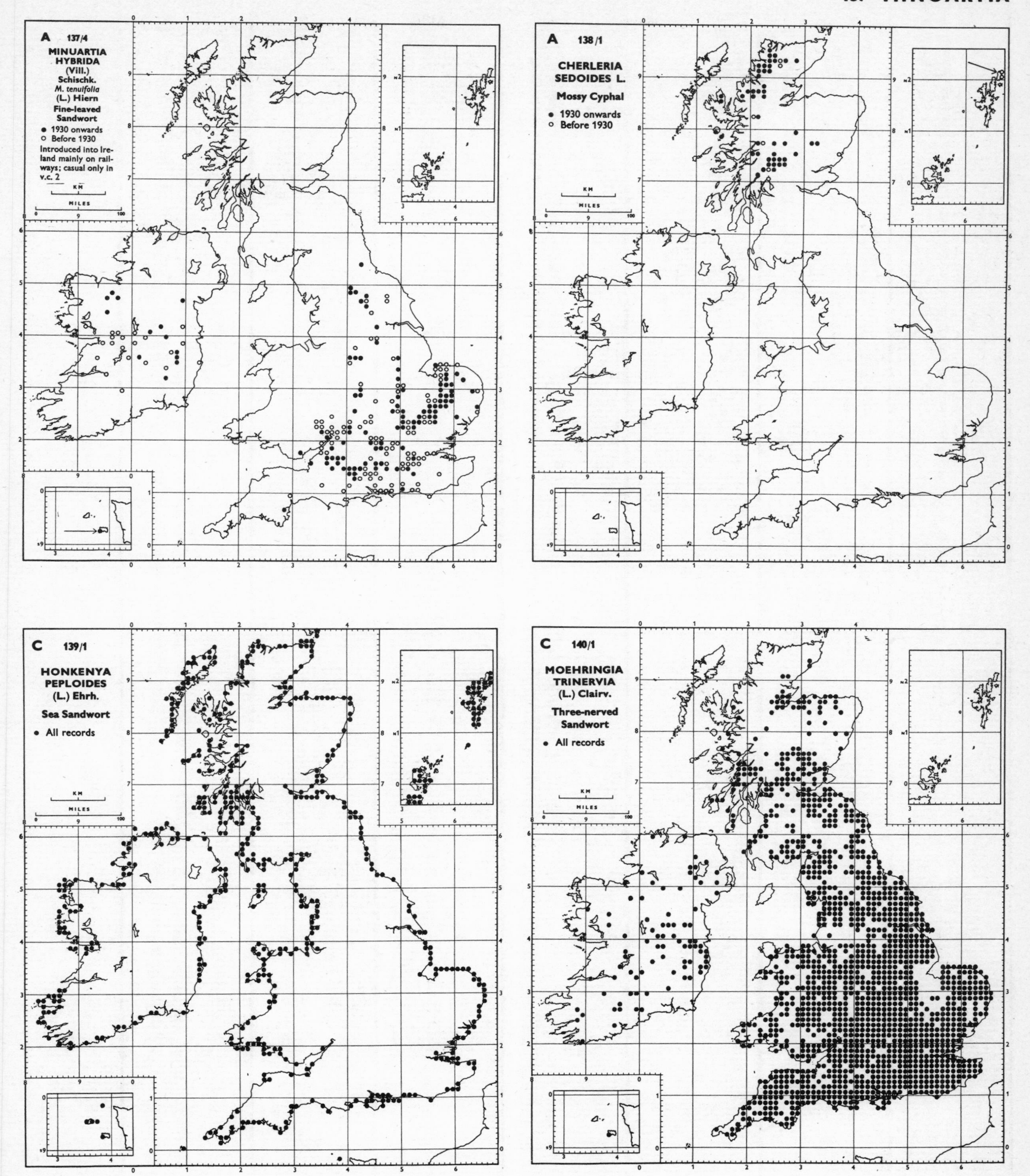
A 137/4
MINUARTIA HYBRIDA (Vill.) Schischk.
M. tenuifolia (L.) Hiern
Fine-leaved Sandwort
● 1930 onwards
○ Before 1930
Introduced into Ireland mainly on railways; casual only in v.c. 2
A 138/1
CHERLERIA SEDOIDES L.
Mossy Cyphal
● 1930 onwards
○ Before 1930
C 139/1
HONKENYA PEPLOIDES (L.) Ehrh.
Sea Sandwort
● All records
C 140/1
MOEHRINGIA TRINERVIA (L.) Clairv.
Three-nerved Sandwort
● All records
KM
MILES

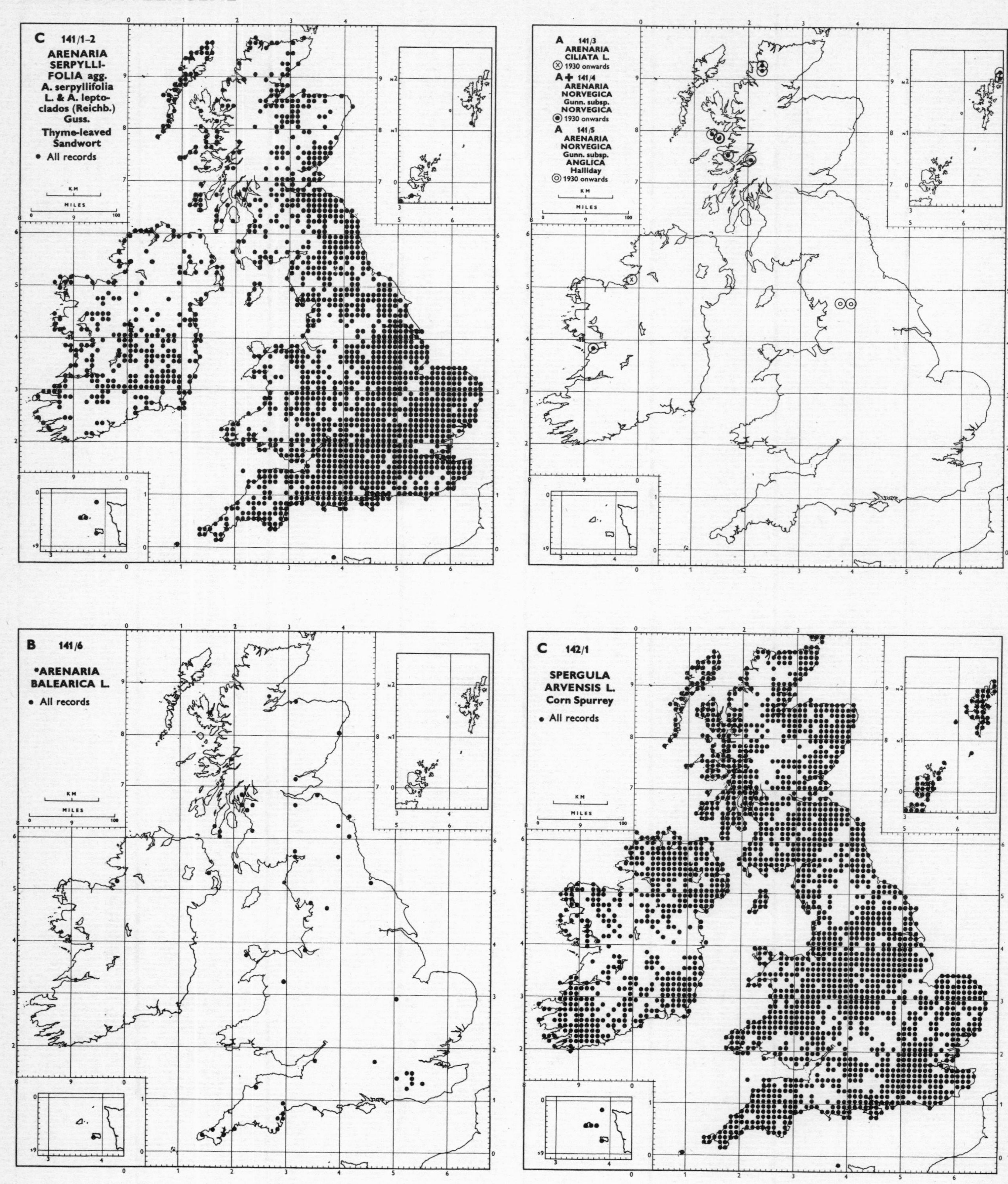
C 141/1–2
ARENARIA SERPYLLIFOLIA agg.
A. serpyllifolia L. & A. leptoclados (Reichb.) Guss.
Thyme-leaved Sandwort
• All records
A 141/3
ARENARIA CILIATA L.
1930 onwards
A 141/4
ARENARIA NORVEGICA Gunn. subsp. NORVEGICA
1930 onwards
A 141/5
ARENARIA NORVEGICA Gunn. subsp. ANGLICA Halliday
1930 onwards
B 141/6
*ARENARIA BALEARICA L.
• All records
C 142/1
SPERGULA ARVENSIS L.
Corn Spurrey
• All records

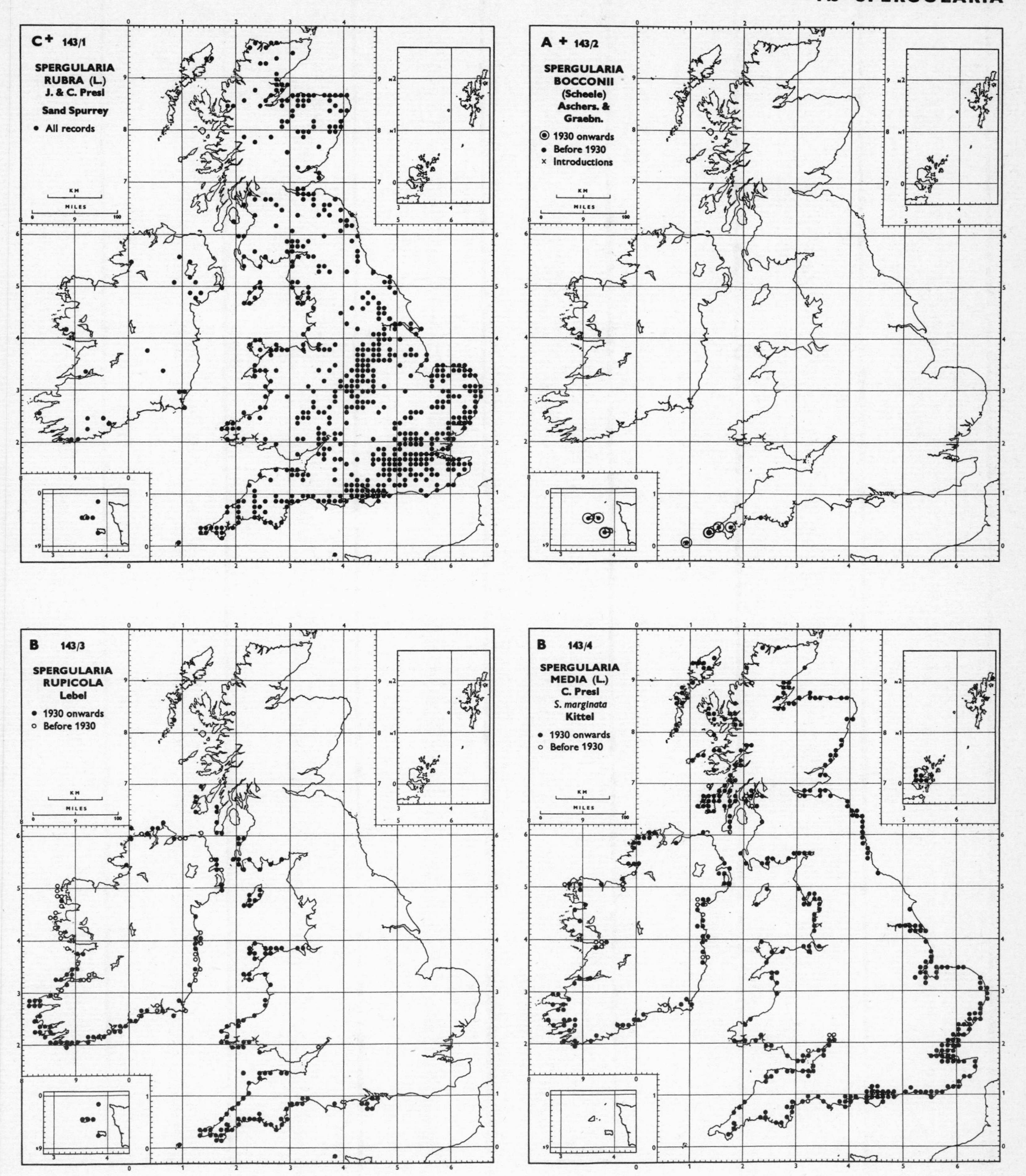
C+ 143/1
SPERGULARIA RUBRA (L.) J. & C. Presl
Sand Spurrey
• All records
A + 143/2
SPERGULARIA BOCCONII (Scheele) Aschers. & Graebn.
◉ 1930 onwards
• Before 1930
× Introductions
B 143/3
SPERGULARIA RUPICOLA Lebel
• 1930 onwards
○ Before 1930
B 143/4
SPERGULARIA MEDIA (L.) C. Presl
S. marginata Kittel
• 1930 onwards
○ Before 1930

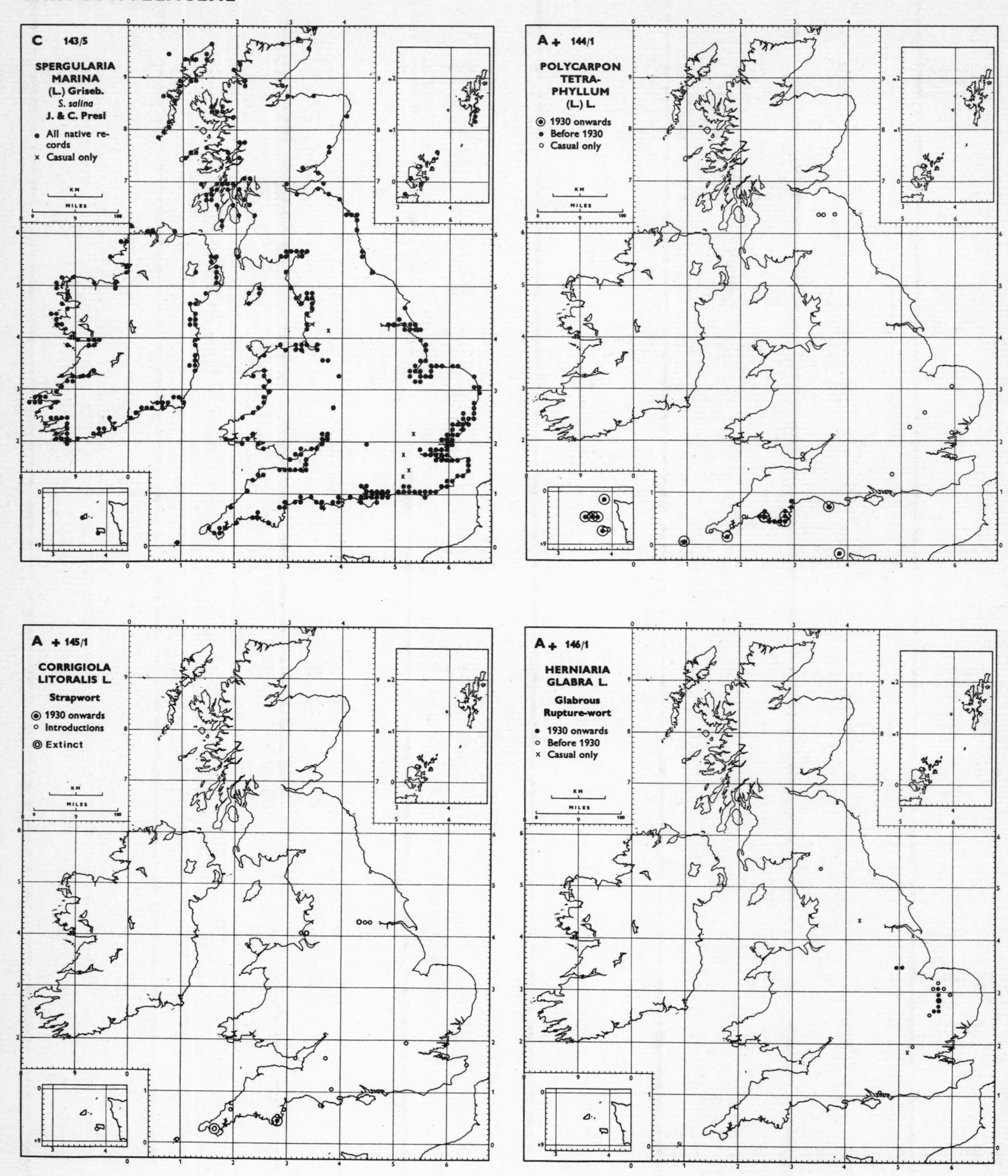
C 143/5
SPERGULARIA MARINA (L.) Griseb.
S. salina J. & C. Presl
• All native records
× Casual only
A + 144/1
POLYCARPON TETRAPHYLLUM (L.) L.
◉ 1930 onwards
• Before 1930
○ Casual only
A + 145/1
CORRIGIOLA LITORALIS L.
Strapwort
◉ 1930 onwards
○ Introductions
◎ Extinct
A + 146/1
HERNIARIA GLABRA L.
Glabrous Rupture-wort
• 1930 onwards
○ Before 1930
× Casual only
KM
MILES

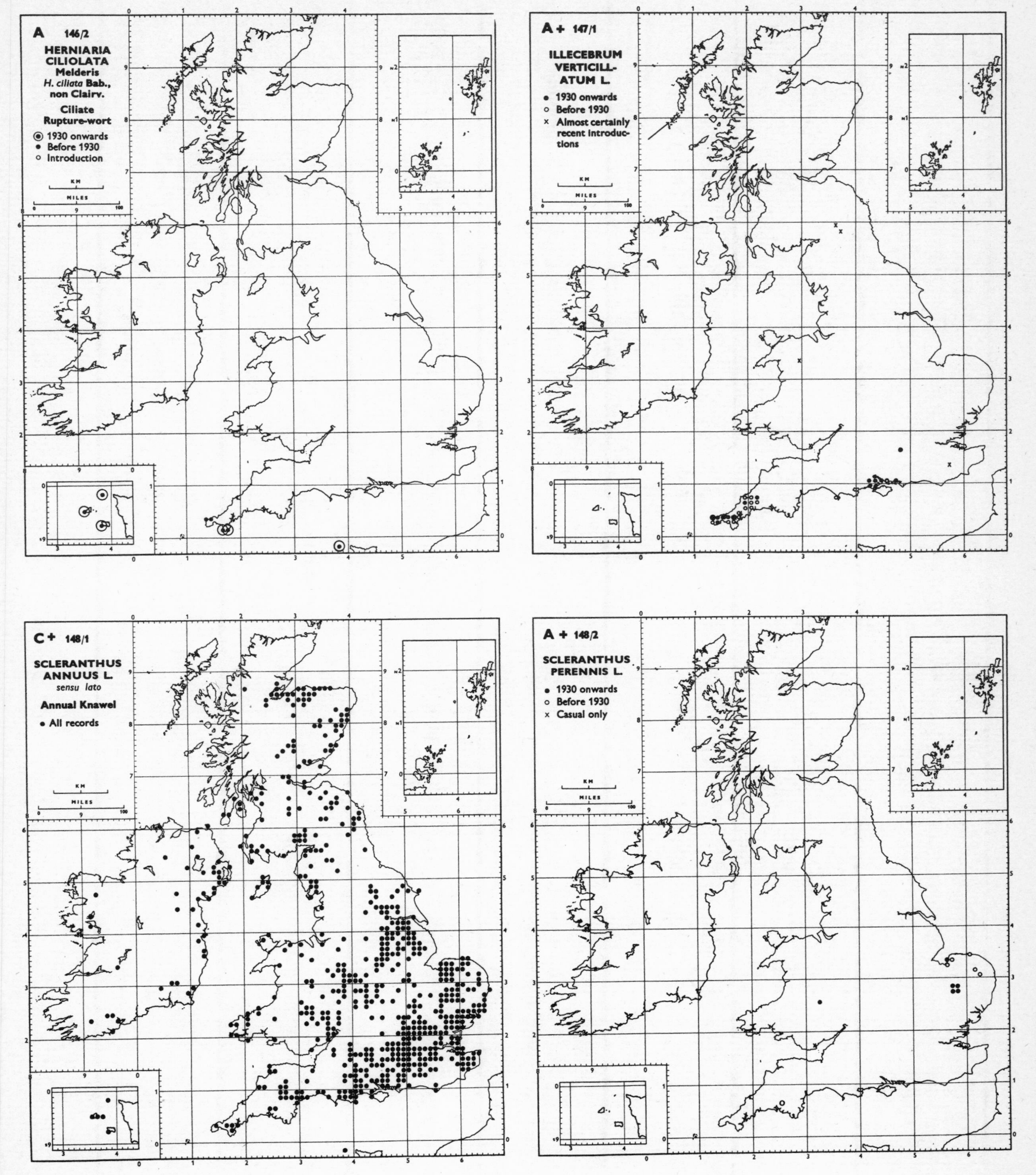
A 146/2
HERNIARIA CILIOLATA Melderis
H. ciliata Bab., non Clairv.
Ciliate Rupture-wort
1930 onwards
Before 1930
Introduction
A + 147/1
ILLECEBRUM VERTICILLATUM L.
1930 onwards
Before 1930
Almost certainly recent introductions
C + 148/1
SCLERANTHUS ANNUUS L. sensu lato
Annual Knawel
All records
A + 148/2
SCLERANTHUS PERENNIS L.
1930 onwards
Before 1930
Casual only

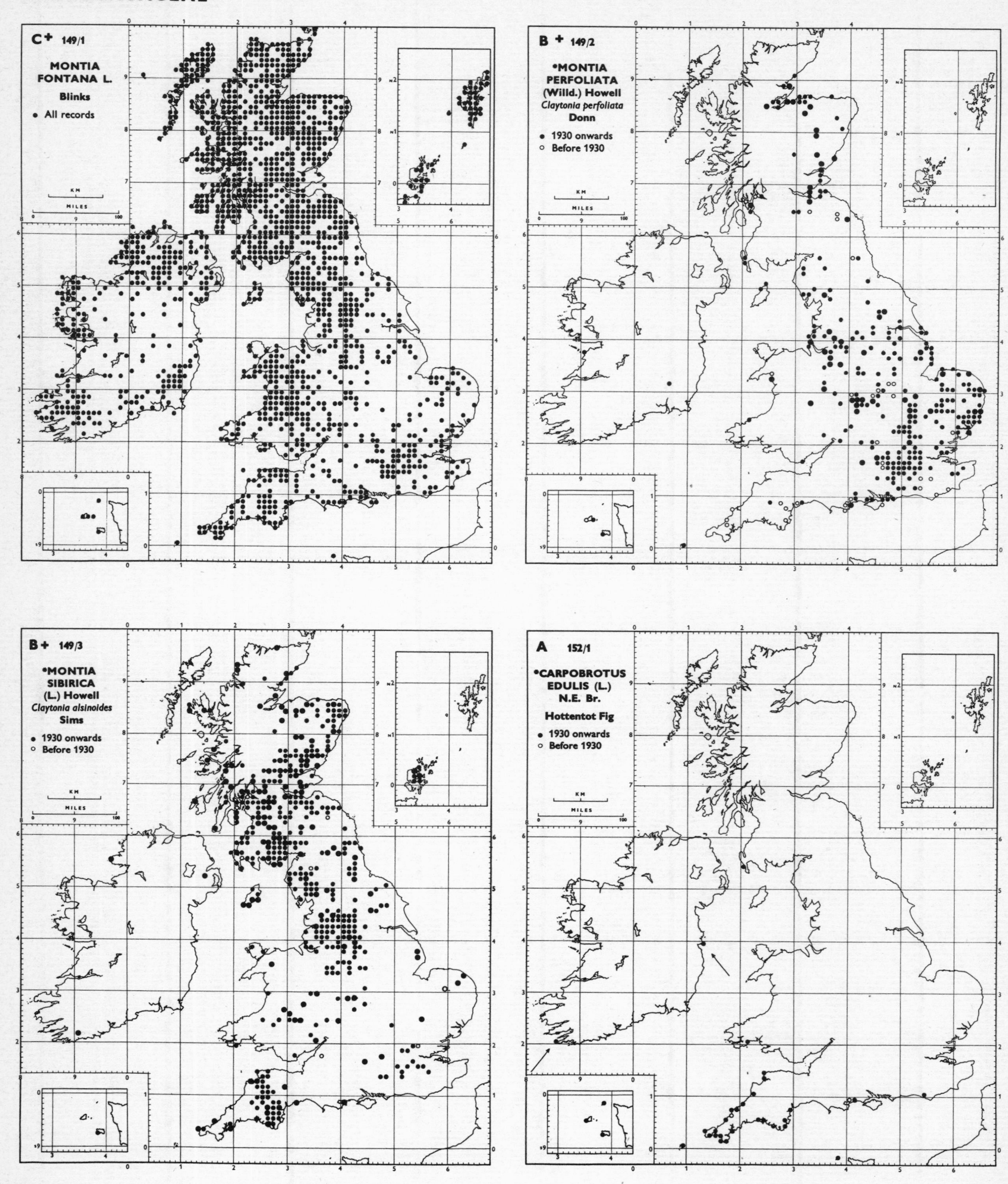

C+ 149/1
MONTIA FONTANA L.
Blinks
• All records
B + 149/2
•MONTIA PERFOLIATA (Willd.) Howell
Claytonia perfoliata Donn
• 1930 onwards
○ Before 1930
B+ 149/3
•MONTIA SIBIRICA (L.) Howell
Claytonia alsinoides Sims
• 1930 onwards
○ Before 1930
A 152/1
•CARPOBROTUS EDULIS (L.) N.E. Br.
Hottentot Fig
• 1930 onwards
○ Before 1930
KM
MILES

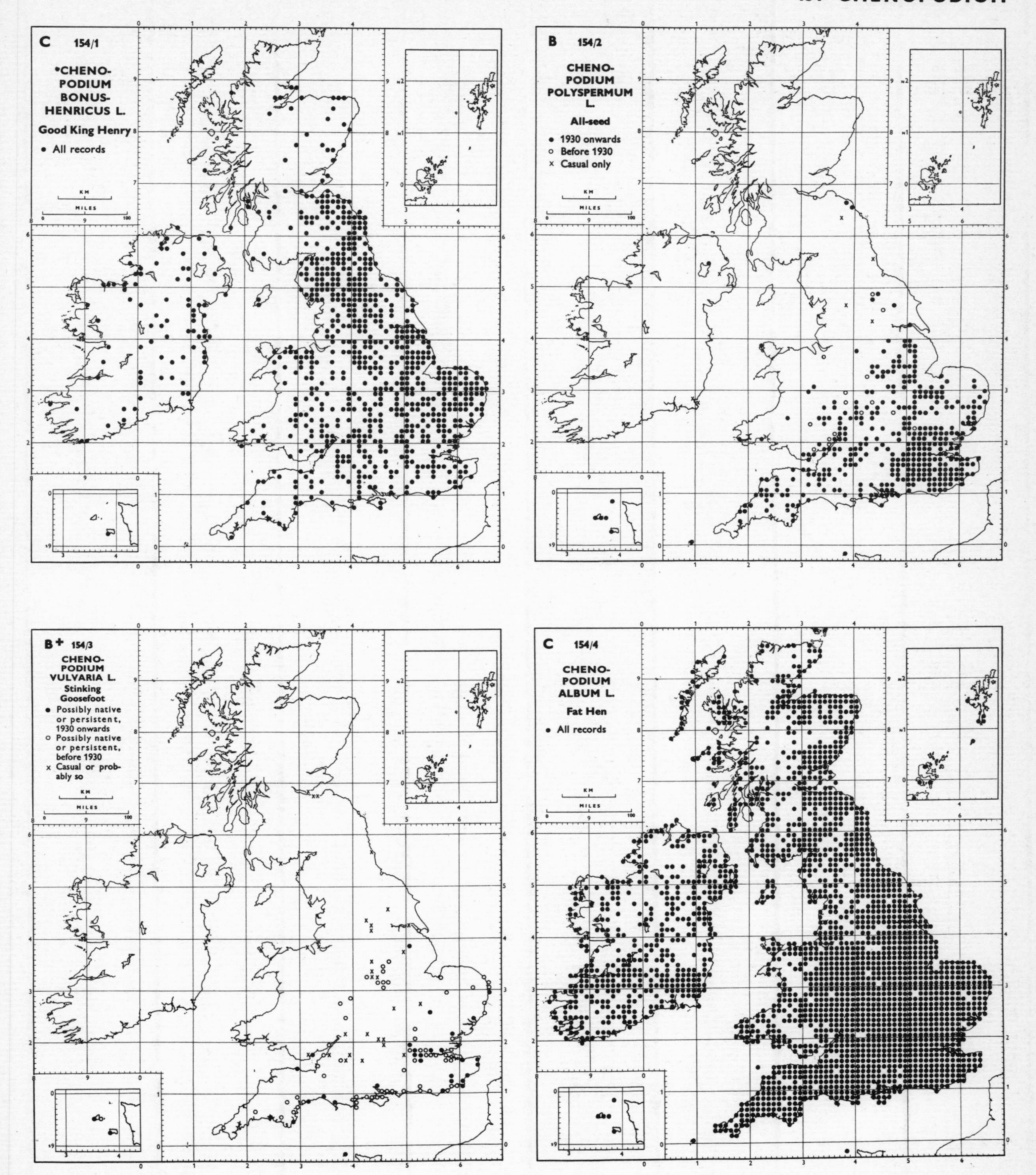
C 154/1
*CHENOPODIUM BONUS-HENRICUS L.
Good King Henry
• All records
B 154/2
CHENOPODIUM POLYSPERMUM L.
All-seed
• 1930 onwards
○ Before 1930
x Casual only
B+ 154/3
CHENOPODIUM VULVARIA L.
Stinking Goosefoot
• Possibly native or persistent, 1930 onwards
○ Possibly native or persistent, before 1930
x Casual or probably so
C 154/4
CHENOPODIUM ALBUM L.
Fat Hen
• All records

CHENOPODIACEAE

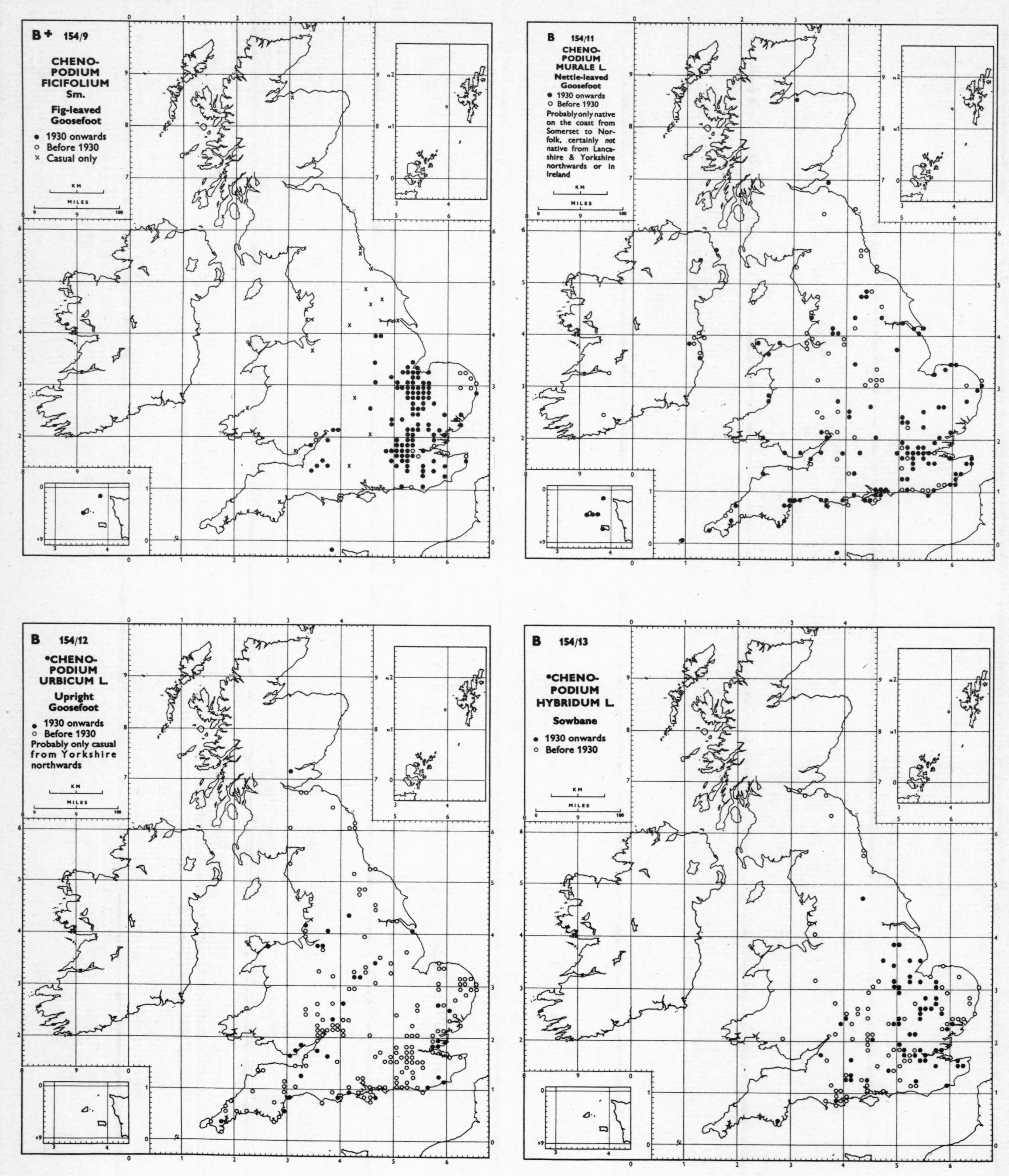

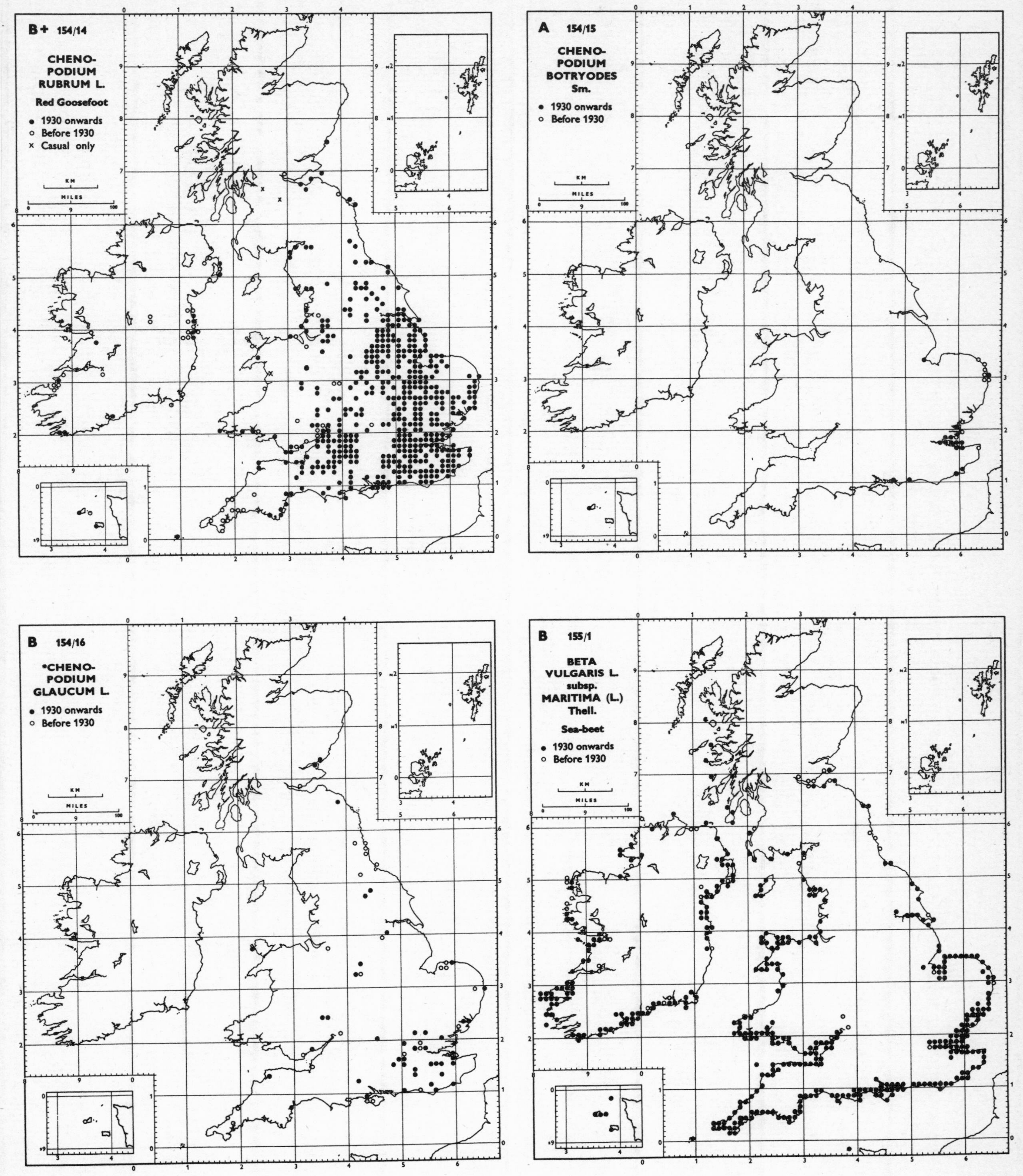
B+ 154/14
CHENO-
PODIUM
RUBRUM L.
Red Goosefoot
• 1930 onwards
○ Before 1930
× Casual only
A 154/15
CHENO-
PODIUM
BOTRYODES
Sm.
• 1930 onwards
○ Before 1930
B 154/16
*CHENO-
PODIUM
GLAUCUM L.
• 1930 onwards
○ Before 1930
B 155/1
BETA
VULGARIS L.
subsp.
MARITIMA (L.)
Thell.
Sea-beet
• 1930 onwards
○ Before 1930

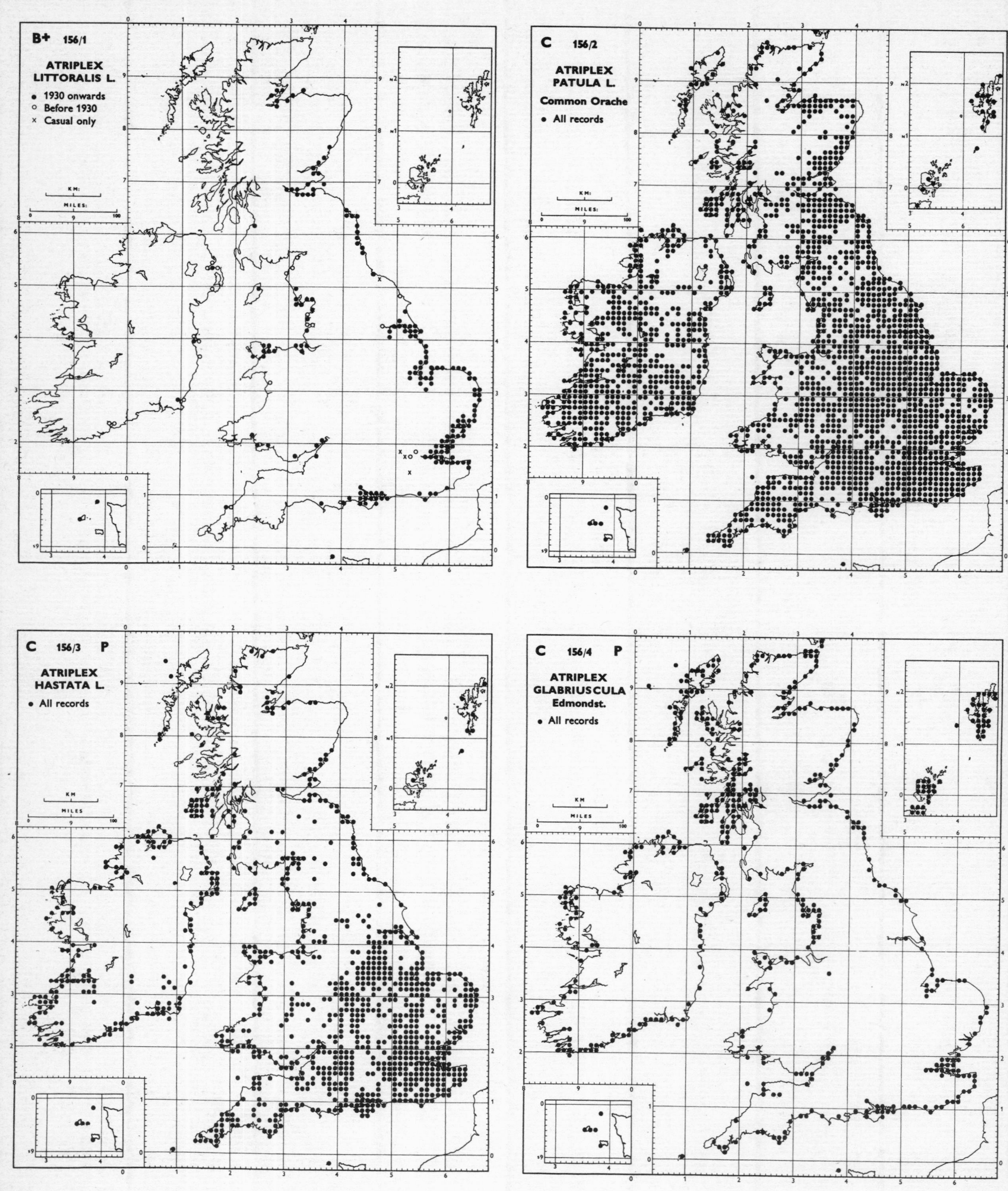
B+ 156/1
ATRIPLEX LITTORALIS L.
• 1930 onwards
○ Before 1930
× Casual only
C 156/2
ATRIPLEX PATULA L.
Common Orache
• All records
C 156/3 P
ATRIPLEX HASTATA L.
• All records
C 156/4 P
ATRIPLEX GLABRIUSCULA Edmondst.
• All records

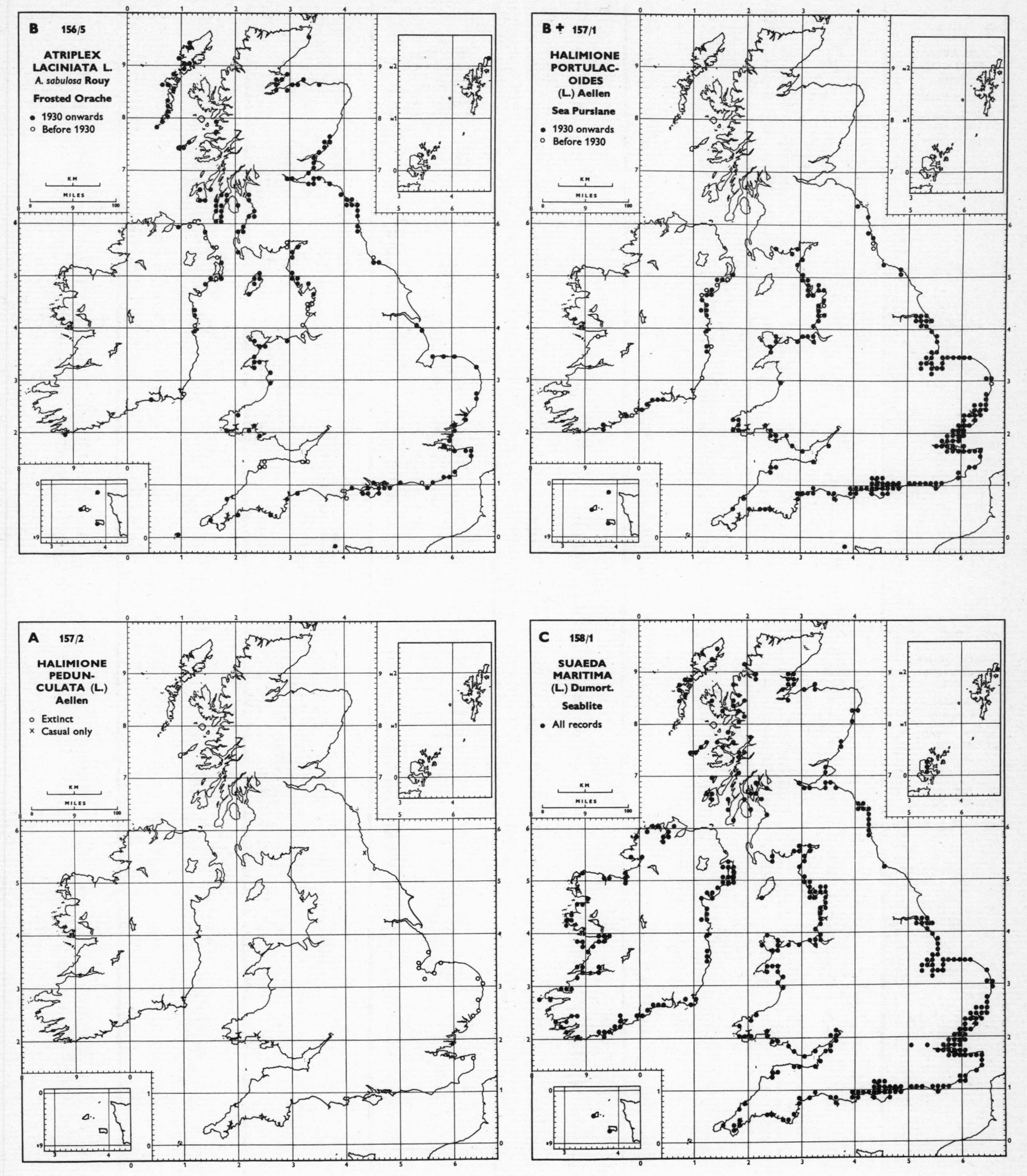
B 156/5
ATRIPLEX LACINIATA L.
A. sabulosa Rouy
Frosted Orache
1930 onwards
Before 1930
B + 157/1
HALIMIONE PORTULACOIDES (L.) Aellen
Sea Purslane
1930 onwards
Before 1930
A 157/2
HALIMIONE PEDUNCULATA (L.) Aellen
Extinct
Casual only
C 158/1
SUAEDA MARITIMA (L.) Dumort.
Seablite
All records
KM
MILES

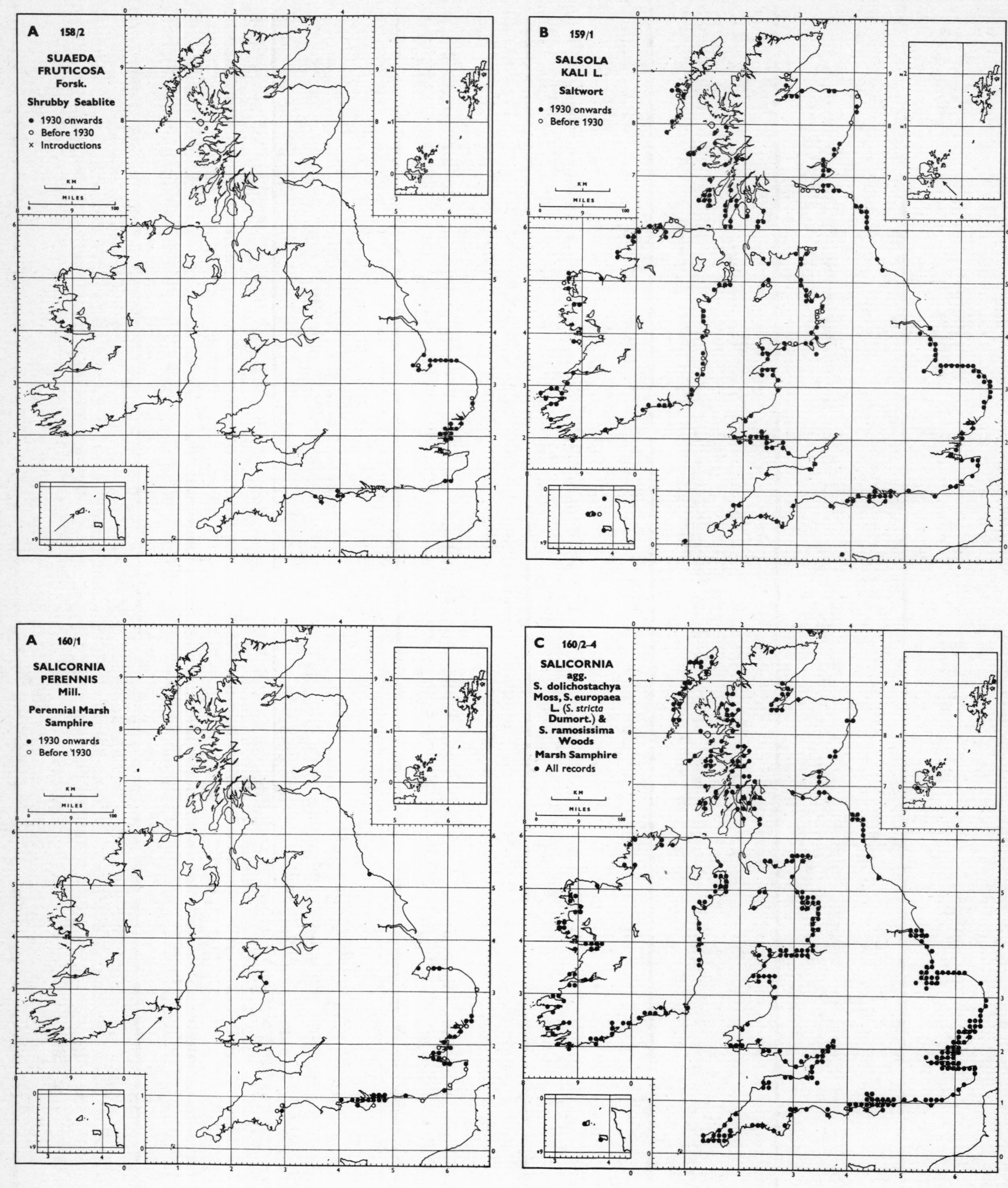
A 158/2
SUAEDA FRUTICOSA Forsk.
Shrubby Seablite
• 1930 onwards
○ Before 1930
× Introductions
KM
MILES
B 159/1
SALSOLA KALI L.
Saltwort
• 1930 onwards
○ Before 1930
A 160/1
SALICORNIA PERENNIS Mill.
Perennial Marsh Samphire
• 1930 onwards
○ Before 1930
C 160/2–4
SALICORNIA agg.
S. dolichostachya Moss, S. europaea L. (S. stricta Dumort.) & S. ramosissima Woods
Marsh Samphire
• All records

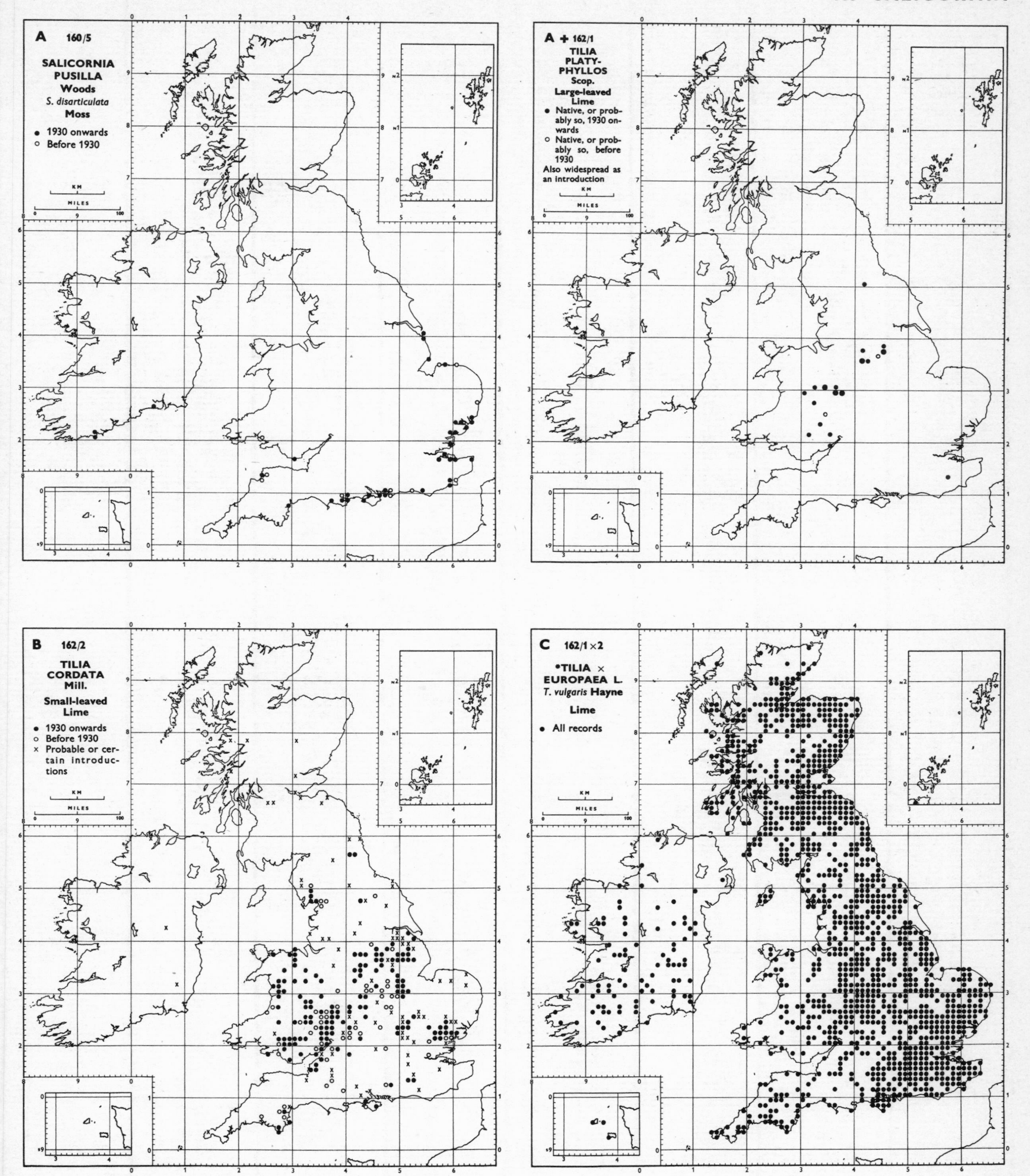
A 160/5
SALICORNIA PUSILLA Woods
S. disarticulata Moss
● 1930 onwards
○ Before 1930
A + 162/1
TILIA PLATYPHYLLOS Scop.
Large-leaved Lime
● Native, or probably so, 1930 onwards
○ Native, or probably so, before 1930
Also widespread as an introduction
B 162/2
TILIA CORDATA Mill.
Small-leaved Lime
● 1930 onwards
○ Before 1930
× Probable or certain introductions
C 162/1×2
*TILIA × EUROPAEA L.
T. vulgaris Hayne
Lime
● All records

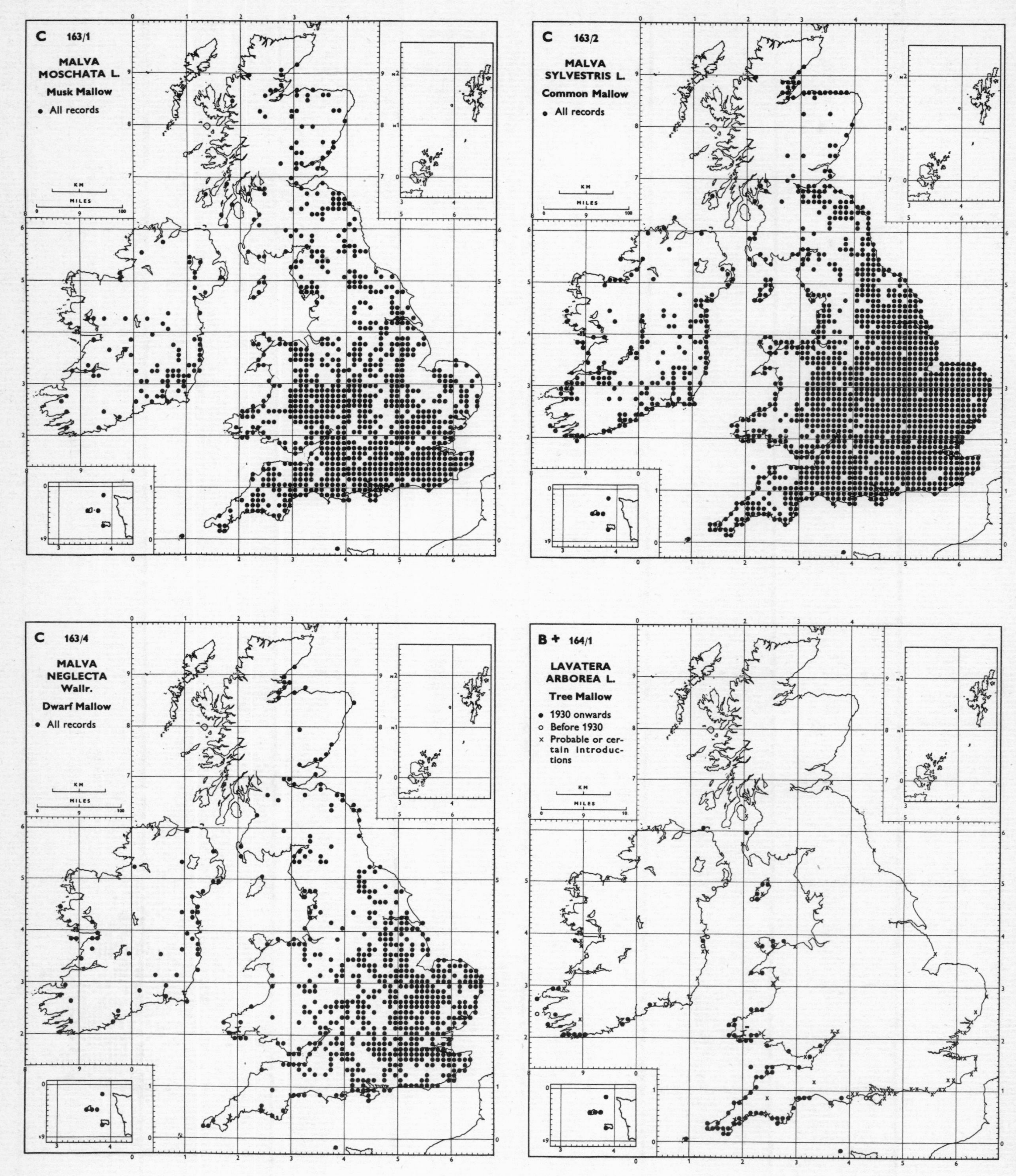
C 163/1
MALVA MOSCHATA L.
Musk Mallow
• All records
C 163/2
MALVA SYLVESTRIS L.
Common Mallow
• All records
C 163/4
MALVA NEGLECTA Wallr.
Dwarf Mallow
• All records
B + 164/1
LAVATERA ARBOREA L.
Tree Mallow
• 1930 onwards
○ Before 1930
× Probable or certain introductions
KM
MILES

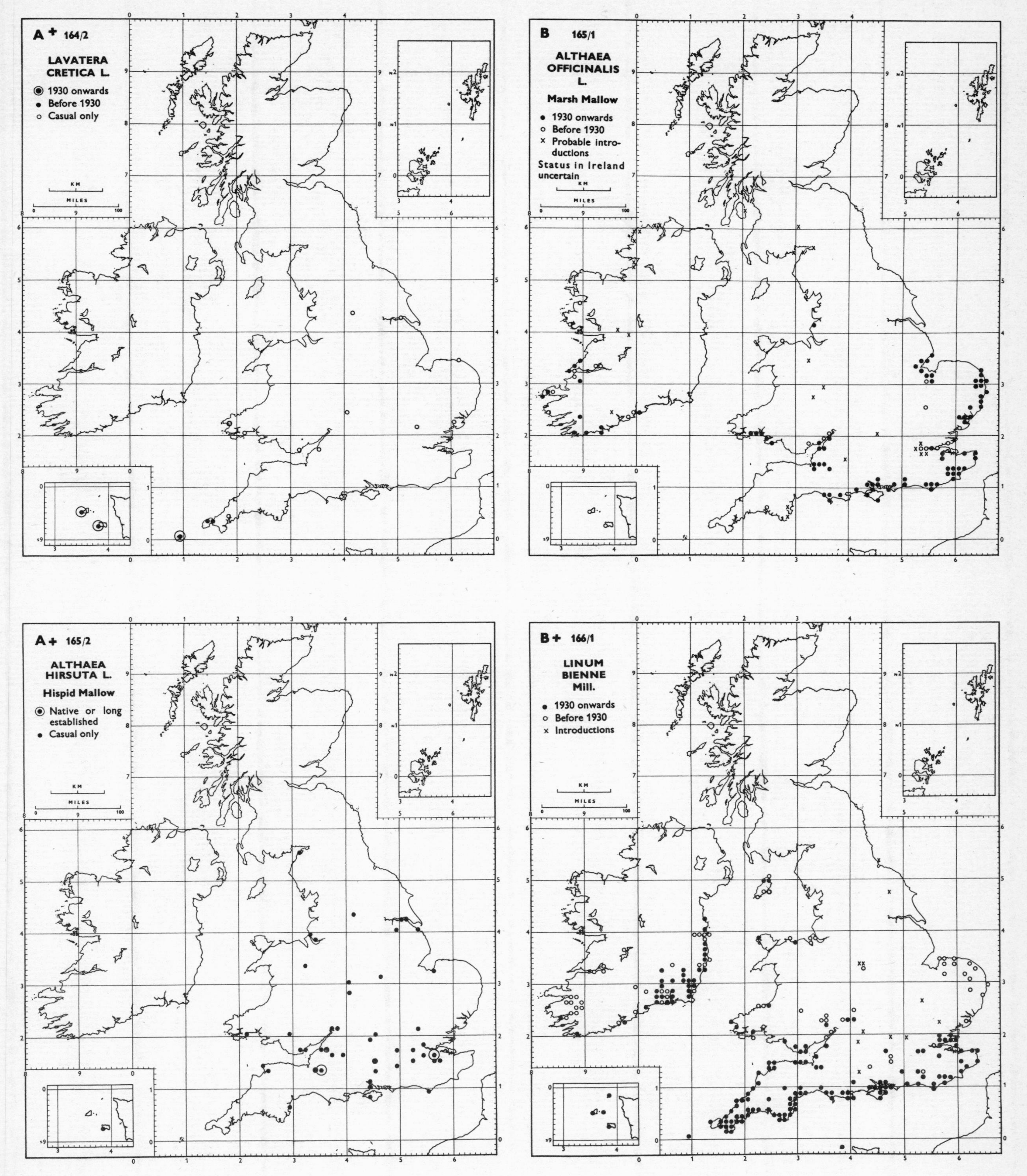
A + 164/2
LAVATERA CRETICA L.
1930 onwards
Before 1930
Casual only
B 165/1
ALTHAEA OFFICINALIS L.
Marsh Mallow
1930 onwards
Before 1930
Probable introductions
Status in Ireland uncertain
A + 165/2
ALTHAEA HIRSUTA L.
Hispid Mallow
Native or long established
Casual only
B + 166/1
LINUM BIENNE Mill.
1930 onwards
Before 1930
Introductions

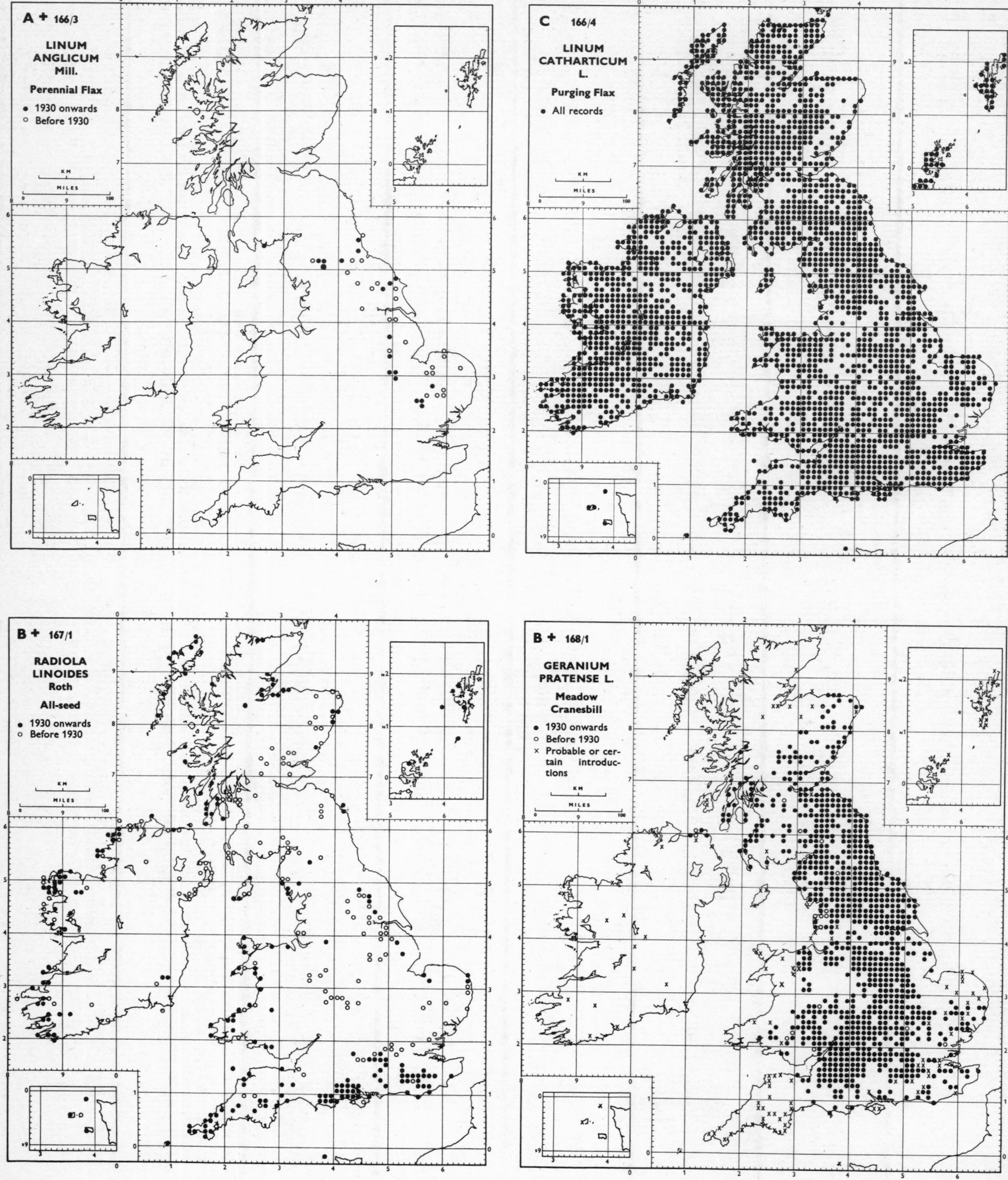
A + 166/3
LINUM ANGLICUM Mill.
Perennial Flax
1930 onwards
Before 1930
C 166/4
LINUM CATHARTICUM L.
Purging Flax
All records
B + 167/1
RADIOLA LINOIDES Roth
All-seed
1930 onwards
Before 1930
B + 168/1
GERANIUM PRATENSE L.
Meadow Cranesbill
1930 onwards
Before 1930
Probable or certain introductions
KM
MILES

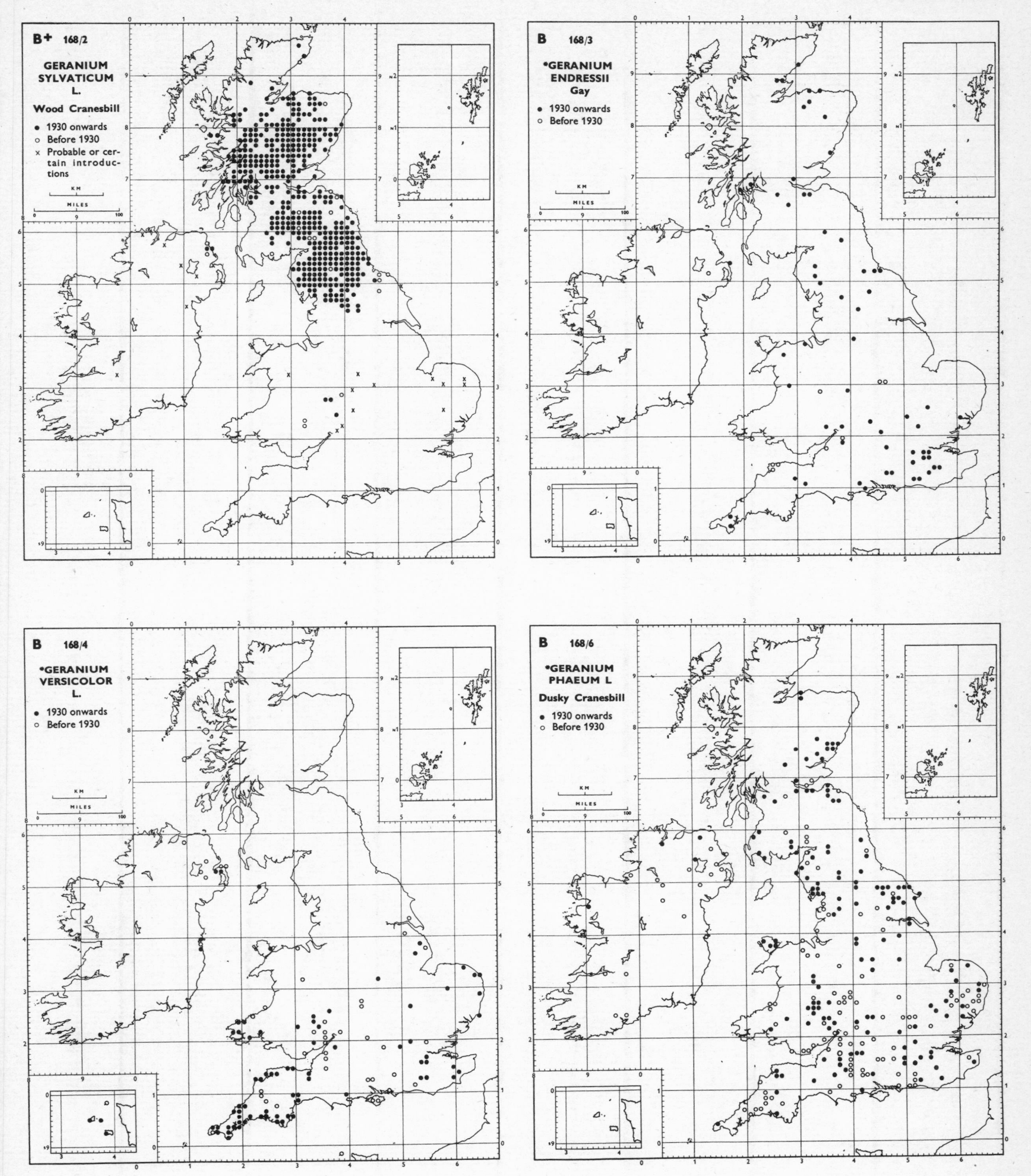
B+ 168/2
GERANIUM SYLVATICUM L.
Wood Cranesbill
1930 onwards
Before 1930
Probable or certain introductions
KM
MILES
B 168/3
*GERANIUM ENDRESSII Gay
1930 onwards
Before 1930
B 168/4
*GERANIUM VERSICOLOR L.
1930 onwards
Before 1930
B 168/6
*GERANIUM PHAEUM L
Dusky Cranesbill
1930 onwards
Before 1930

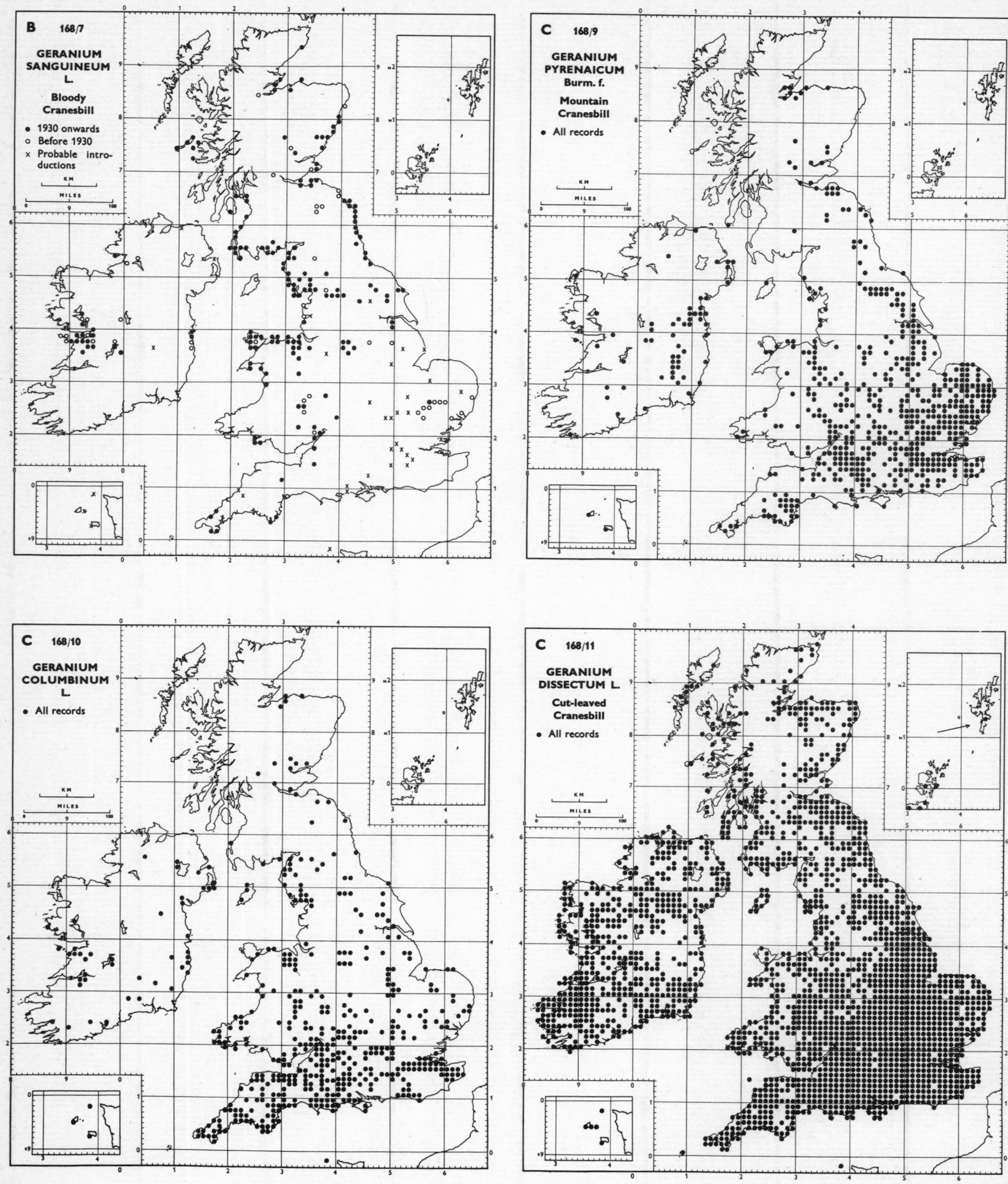
B 168/7
GERANIUM SANGUINEUM L.
Bloody Cranesbill
● 1930 onwards
○ Before 1930
× Probable introductions
KM
MILES
C 168/9
GERANIUM PYRENAICUM Burm. f.
Mountain Cranesbill
● All records
KM
MILES
C 168/10
GERANIUM COLUMBINUM L.
● All records
KM
MILES
C 168/11
GERANIUM DISSECTUM L.
Cut-leaved Cranesbill
● All records
KM
MILES

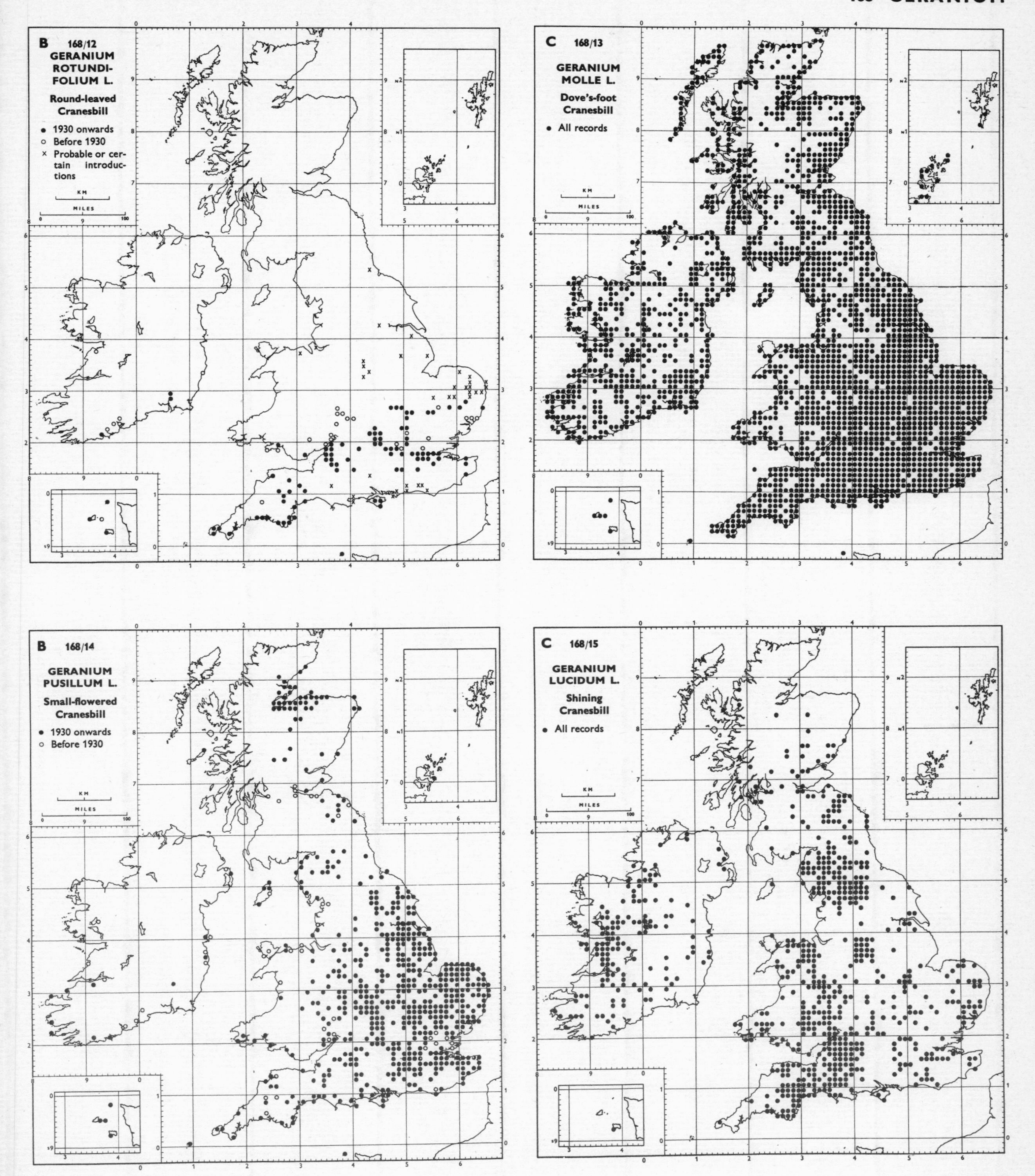
B 168/12
GERANIUM ROTUNDIFOLIUM L.
Round-leaved Cranesbill
• 1930 onwards
○ Before 1930
× Probable or certain introductions
C 168/13
GERANIUM MOLLE L.
Dove's-foot Cranesbill
• All records
B 168/14
GERANIUM PUSILLUM L.
Small-flowered Cranesbill
• 1930 onwards
○ Before 1930
C 168/15
GERANIUM LUCIDUM L.
Shining Cranesbill
• All records
KM
MILES

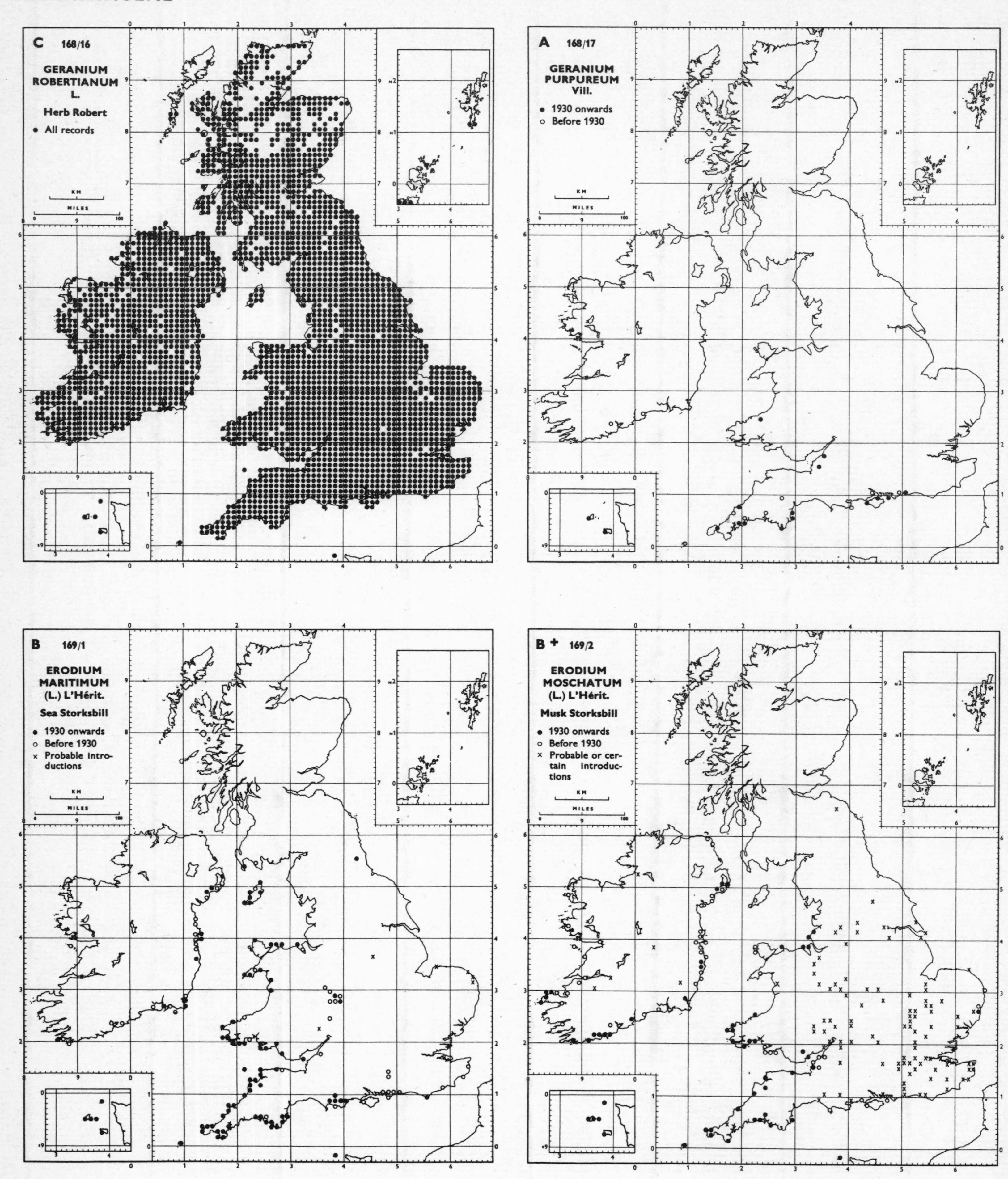

C 168/16
GERANIUM ROBERTIANUM L.
Herb Robert
• All records
A 168/17
GERANIUM PURPUREUM Vill.
• 1930 onwards
○ Before 1930
B 169/1
ERODIUM MARITIMUM (L.) L'Hérit.
Sea Storksbill
• 1930 onwards
○ Before 1930
× Probable introductions
B + 169/2
ERODIUM MOSCHATUM (L.) L'Hérit.
Musk Storksbill
• 1930 onwards
○ Before 1930
× Probable or certain introductions
KM
MILES

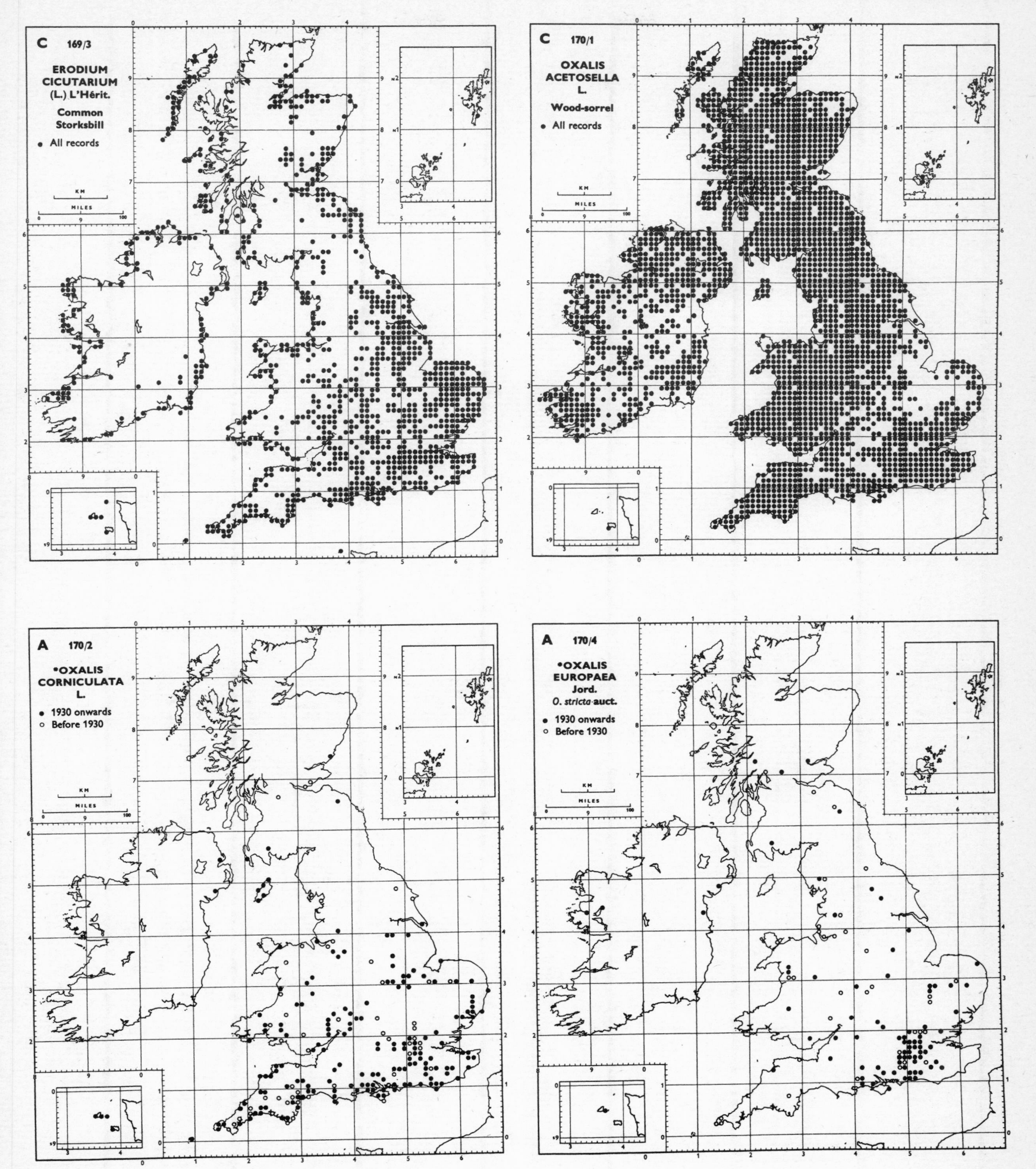
C 169/3
ERODIUM CICUTARIUM (L.) L'Hérit.
Common Storksbill
• All records
C 170/1
OXALIS ACETOSELLA L.
Wood-sorrel
• All records
A 170/2
*OXALIS CORNICULATA L.
• 1930 onwards
○ Before 1930
A 170/4
*OXALIS EUROPAEA Jord.
O. stricta auct.
• 1930 onwards
○ Before 1930

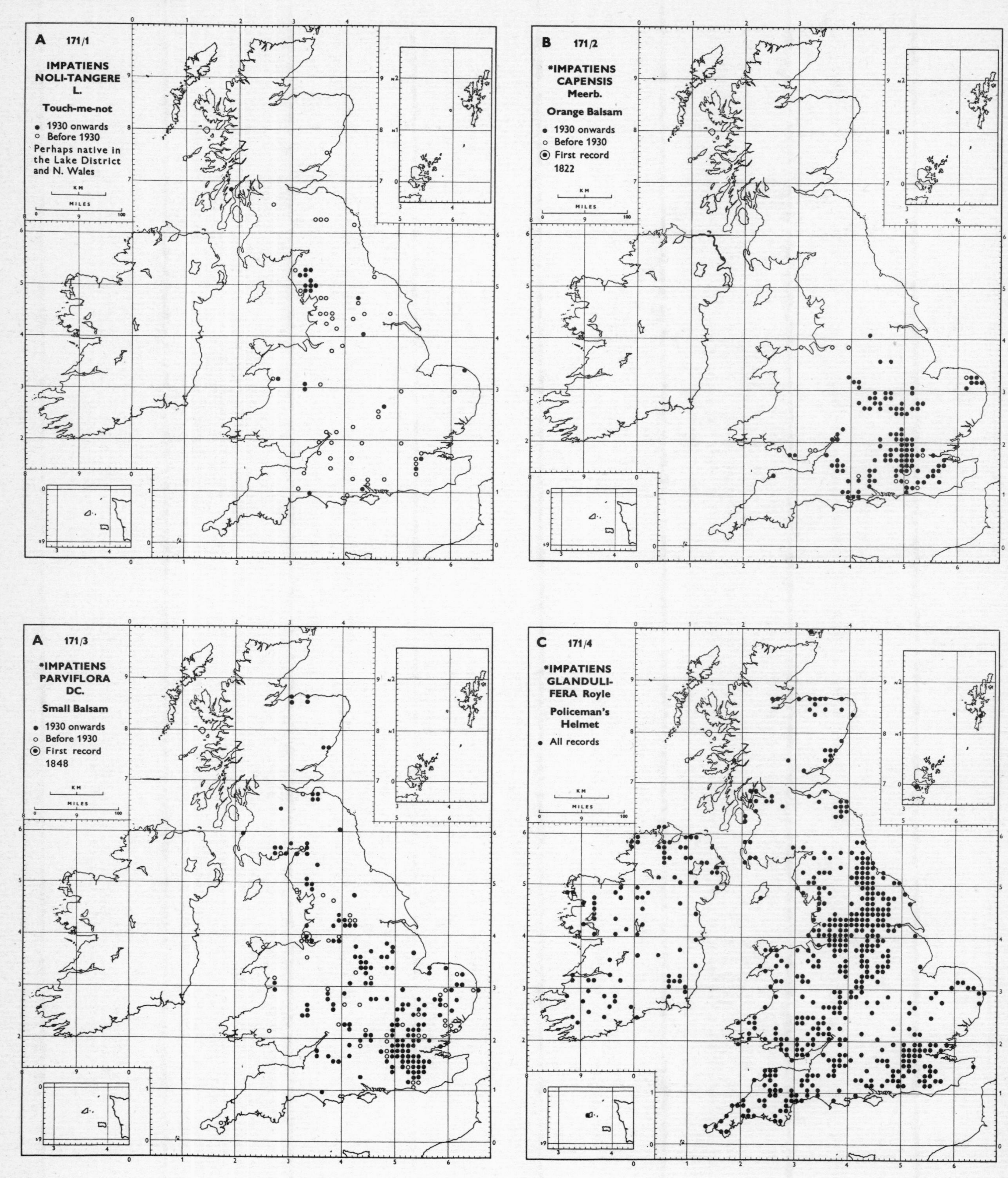
A 171/1
IMPATIENS NOLI-TANGERE L.
Touch-me-not
1930 onwards
Before 1930
Perhaps native in the Lake District and N. Wales
KM
MILES
B 171/2
*IMPATIENS CAPENSIS Meerb.
Orange Balsam
1930 onwards
Before 1930
First record 1822
A 171/3
*IMPATIENS PARVIFLORA DC.
Small Balsam
1930 onwards
Before 1930
First record 1848
C 171/4
*IMPATIENS GLANDULIFERA Royle
Policeman's Helmet
All records

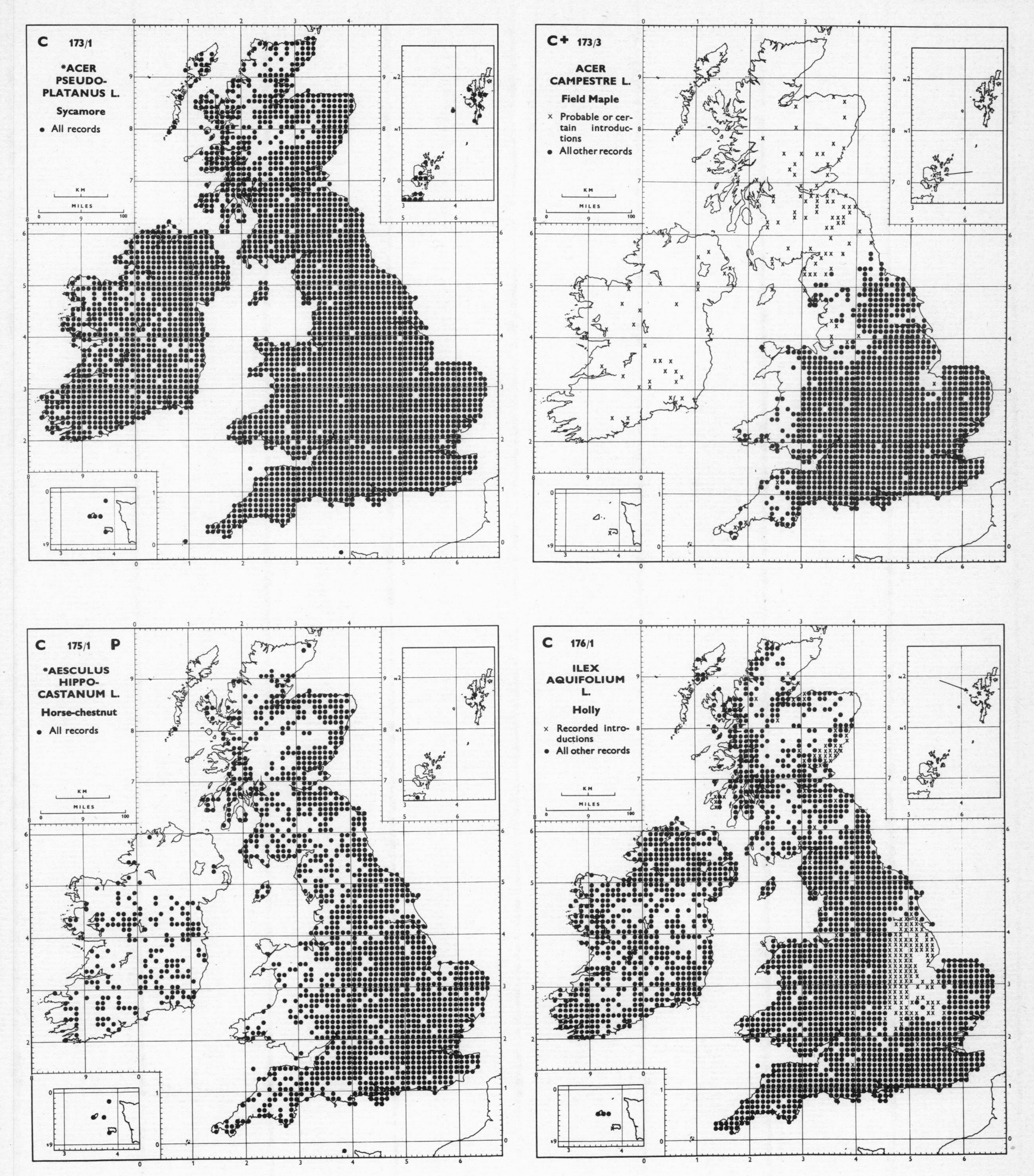
C 173/1
*ACER PSEUDO-PLATANUS L.
Sycamore
● All records
C+ 173/3
ACER CAMPESTRE L.
Field Maple
× Probable or certain introductions
● All other records
C 175/1 P
*AESCULUS HIPPO-CASTANUM L.
Horse-chestnut
● All records
C 176/1
ILEX AQUIFOLIUM L.
Holly
× Recorded introductions
● All other records

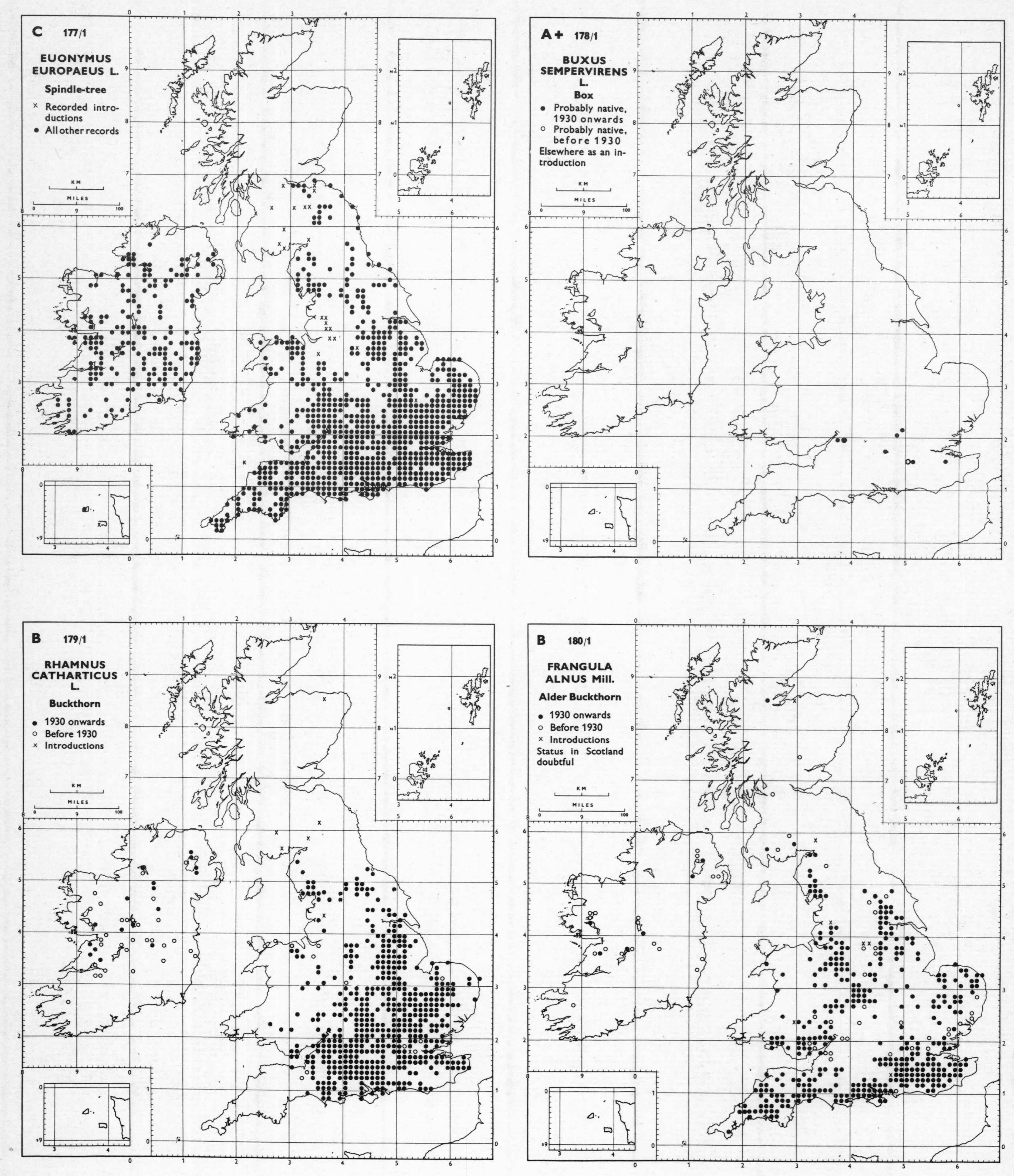
C 177/1
EUONYMUS EUROPAEUS L.
Spindle-tree
× Recorded introductions
● All other records
A+ 178/1
BUXUS SEMPERVIRENS L.
Box
● Probably native, 1930 onwards
○ Probably native, before 1930
Elsewhere as an introduction
B 179/1
RHAMNUS CATHARTICUS L.
Buckthorn
● 1930 onwards
○ Before 1930
× Introductions
B 180/1
FRANGULA ALNUS Mill.
Alder Buckthorn
● 1930 onwards
○ Before 1930
× Introductions
Status in Scotland doubtful

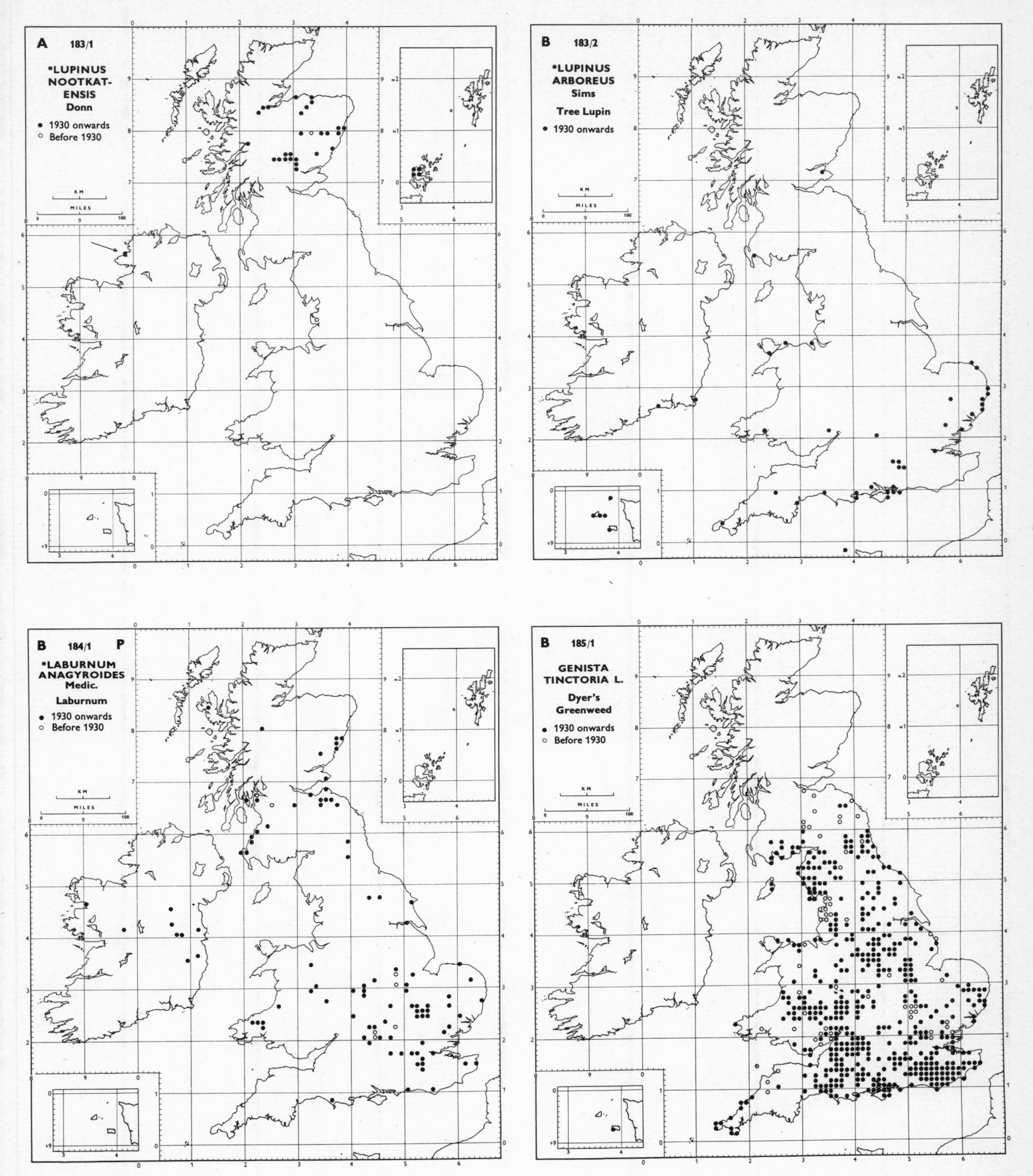
A 183/1
*LUPINUS NOOTKATENSIS Donn
• 1930 onwards
○ Before 1930
B 183/2
*LUPINUS ARBOREUS Sims
Tree Lupin
• 1930 onwards
B 184/1 P
*LABURNUM ANAGYROIDES Medic.
Laburnum
• 1930 onwards
○ Before 1930
B 185/1
GENISTA TINCTORIA L.
Dyer's Greenweed
• 1930 onwards
○ Before 1930
KM
MILES

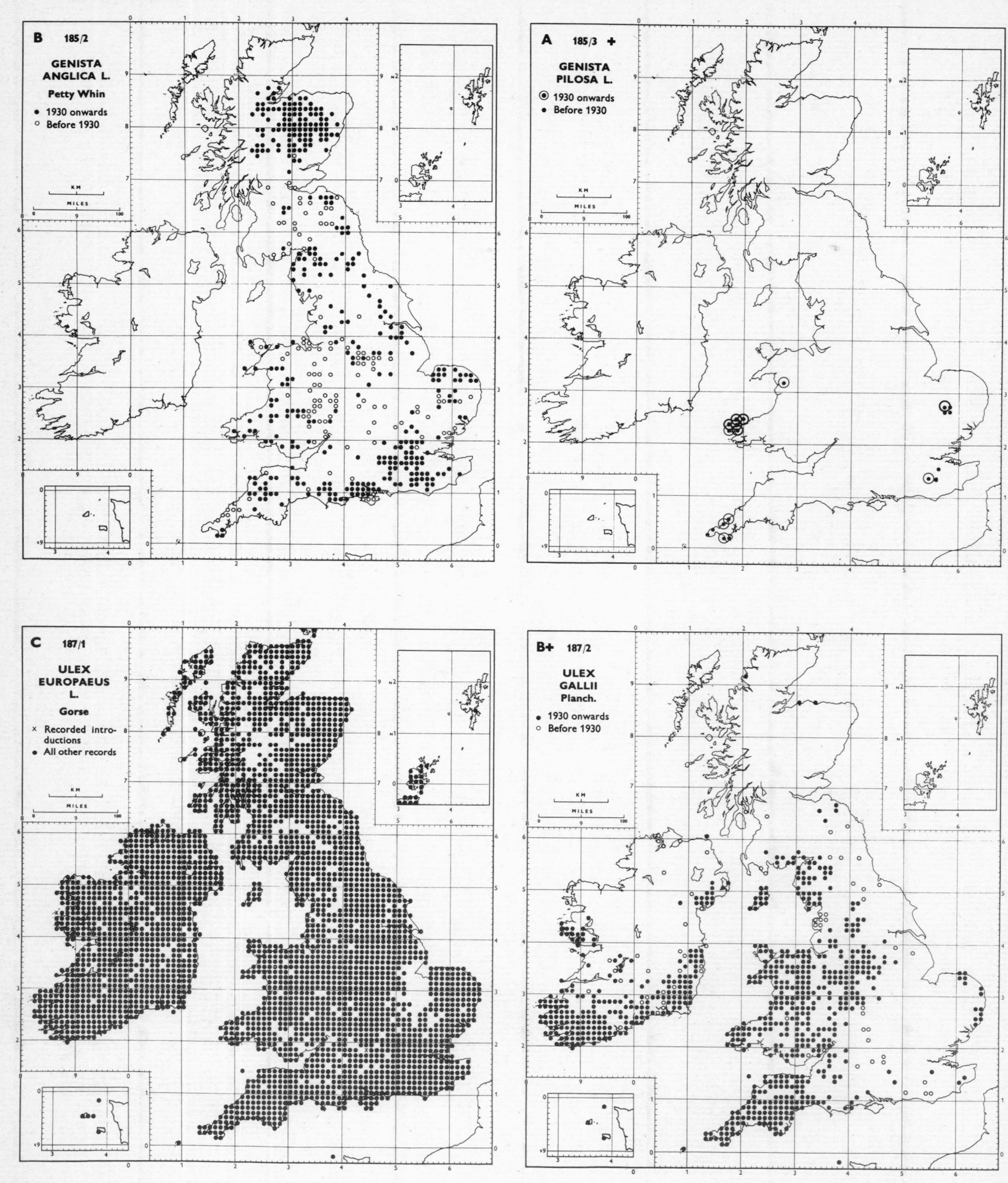
B 185/2
GENISTA ANGLICA L.
Petty Whin
1930 onwards
Before 1930
A 185/3 +
GENISTA PILOSA L.
1930 onwards
Before 1930
C 187/1
ULEX EUROPAEUS L.
Gorse
× Recorded introductions
All other records
B+ 187/2
ULEX GALLII Planch.
1930 onwards
Before 1930
KM
MILES

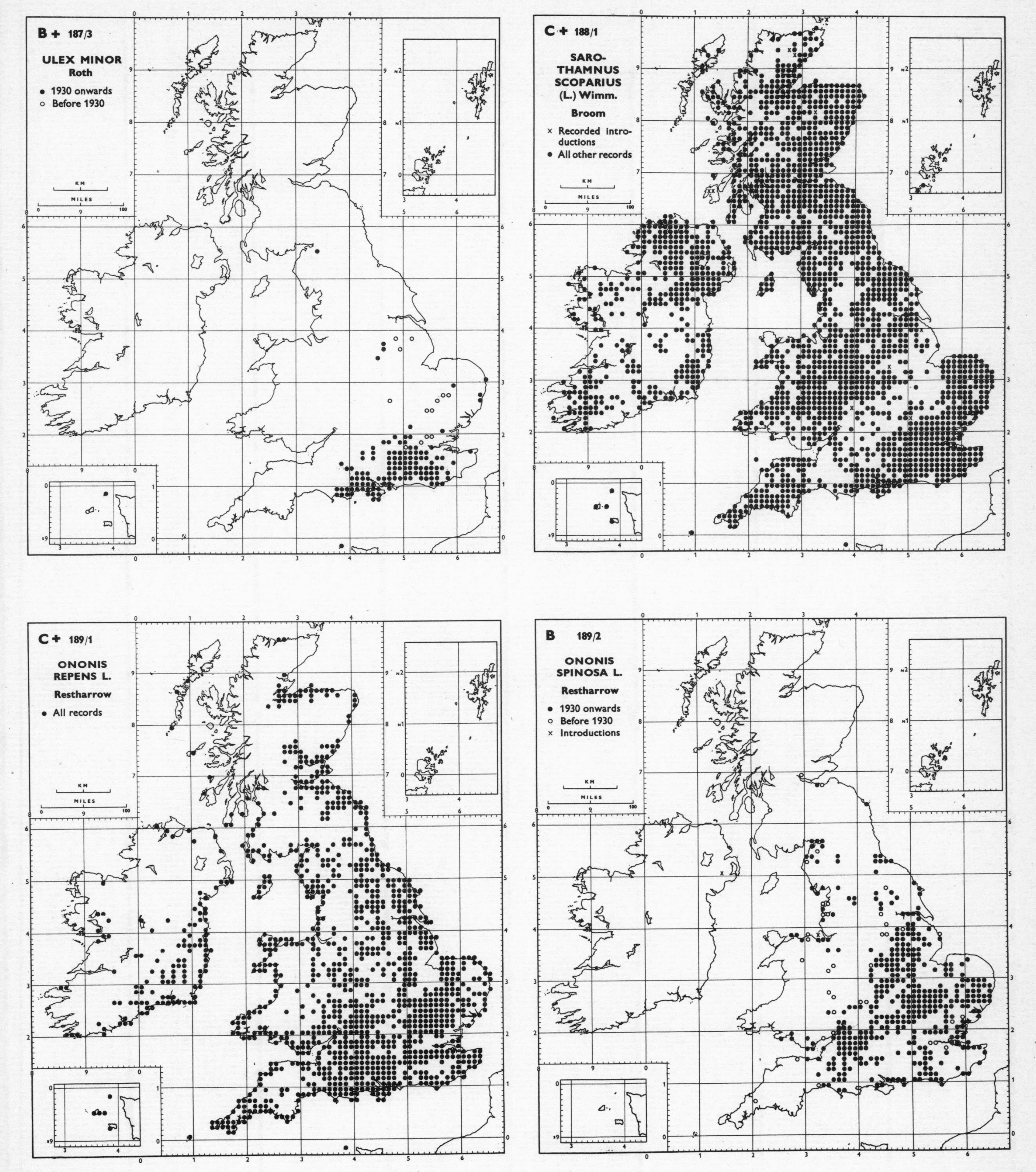
B + 187/3
ULEX MINOR
Roth
1930 onwards
Before 1930
C + 188/1
SARO-
THAMNUS
SCOPARIUS
(L.) Wimm.
Broom
Recorded intro-
ductions
All other records
C + 189/1
ONONIS
REPENS L.
Restharrow
All records
B 189/2
ONONIS
SPINOSA L.
Restharrow
1930 onwards
Before 1930
Introductions

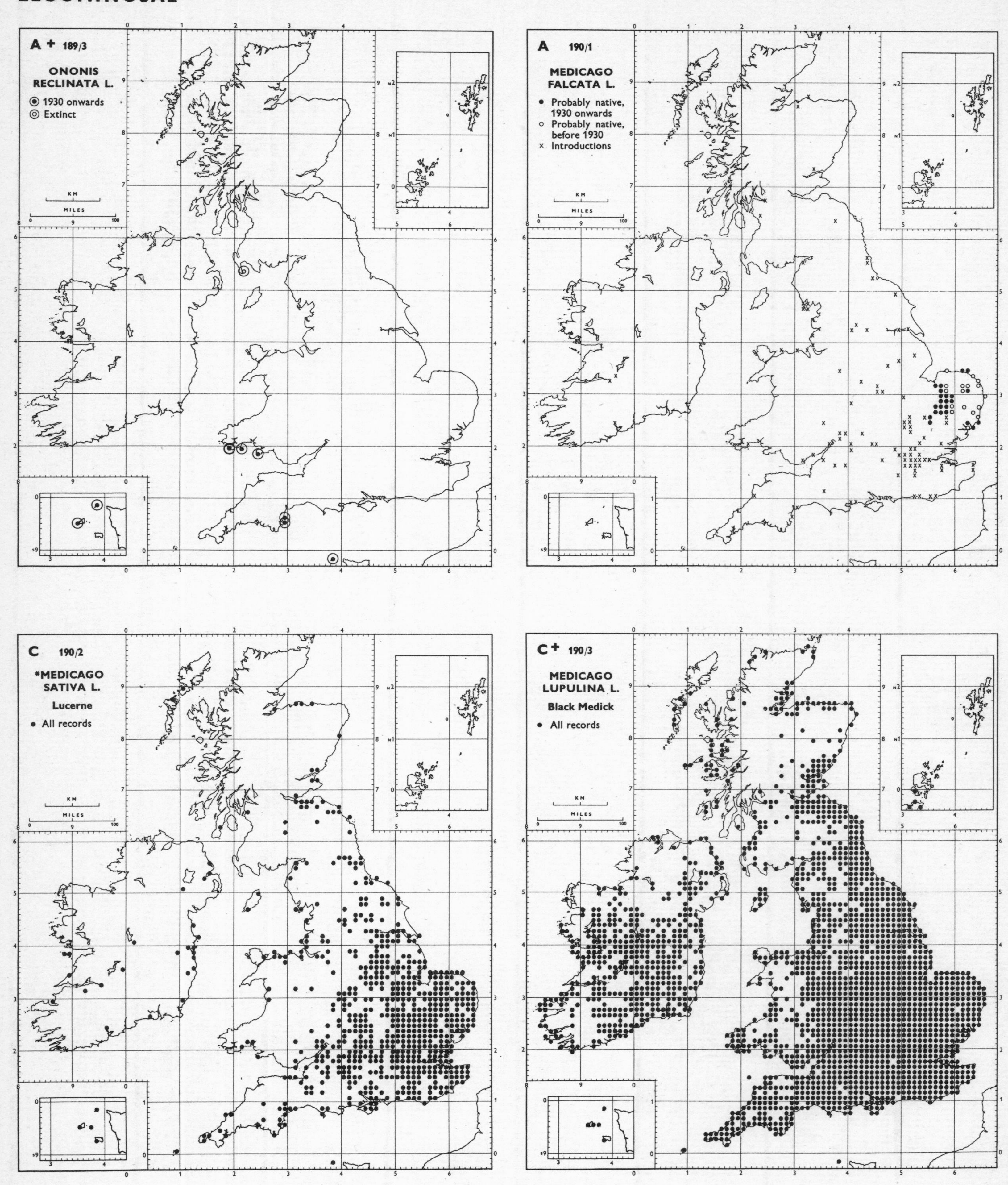
A + 189/3
ONONIS RECLINATA L.
1930 onwards
Extinct
A 190/1
MEDICAGO FALCATA L.
Probably native, 1930 onwards
Probably native, before 1930
Introductions
C 190/2
*MEDICAGO SATIVA L.
Lucerne
All records
C+ 190/3
MEDICAGO LUPULINA L.
Black Medick
All records

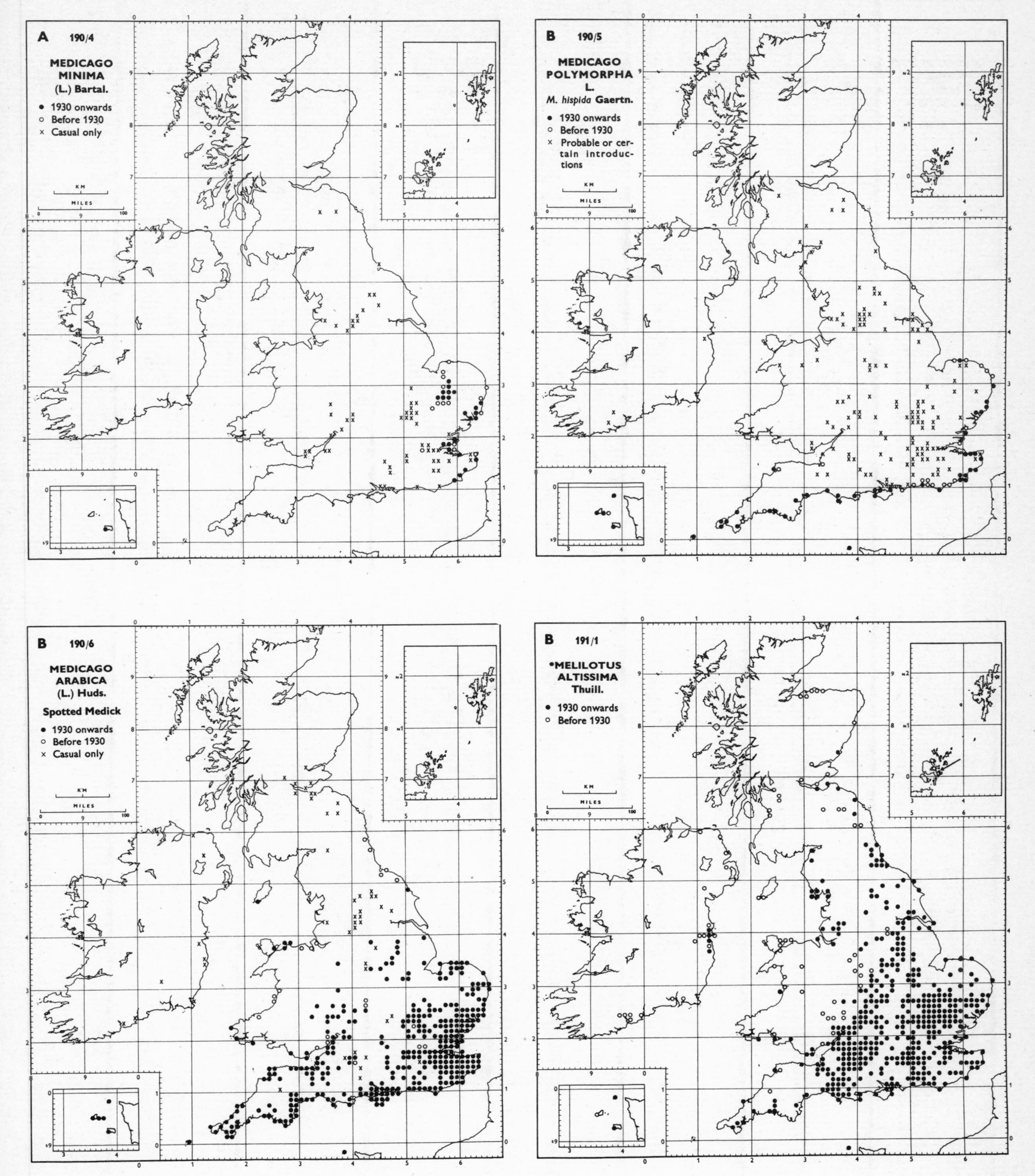
A 190/4
MEDICAGO MINIMA (L.) Bartal.
• 1930 onwards
○ Before 1930
× Casual only
B 190/5
MEDICAGO POLYMORPHA L.
M. hispida Gaertn.
• 1930 onwards
○ Before 1930
× Probable or certain introductions
B 190/6
MEDICAGO ARABICA (L.) Huds.
Spotted Medick
• 1930 onwards
○ Before 1930
× Casual only
B 191/1
*MELILOTUS ALTISSIMA Thuill.
• 1930 onwards
○ Before 1930
KM
MILES

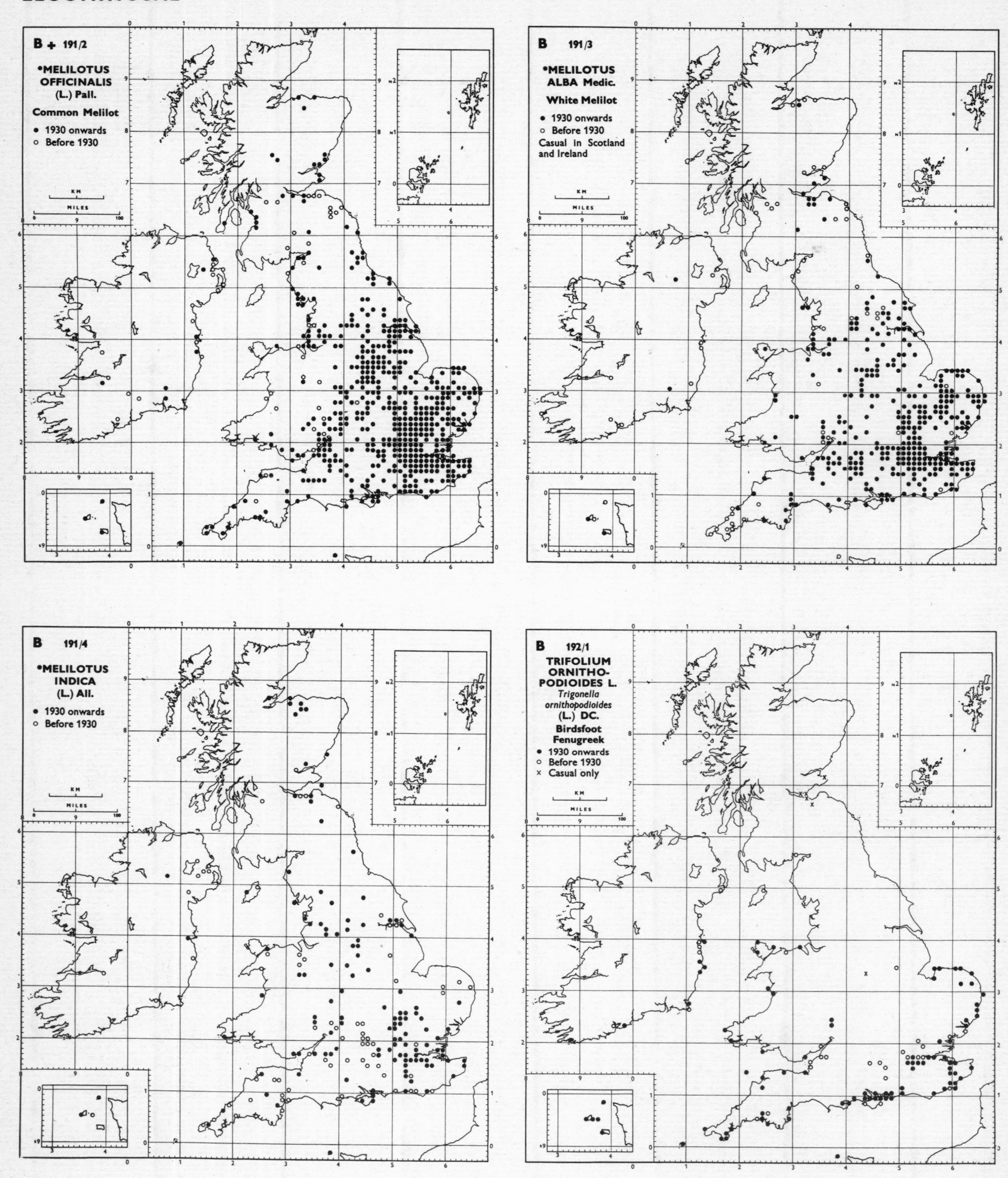
B + 191/2
*MELILOTUS OFFICINALIS (L.) Pall.
Common Melilot
• 1930 onwards
○ Before 1930
B 191/3
*MELILOTUS ALBA Medic.
White Melilot
• 1930 onwards
○ Before 1930
Casual in Scotland and Ireland
B 191/4
*MELILOTUS INDICA (L.) All.
• 1930 onwards
○ Before 1930
B 192/1
TRIFOLIUM ORNITHO-PODIOIDES L.
Trigonella ornithopodioides (L.) DC.
Birdsfoot Fenugreek
• 1930 onwards
○ Before 1930
× Casual only
KM
MILES

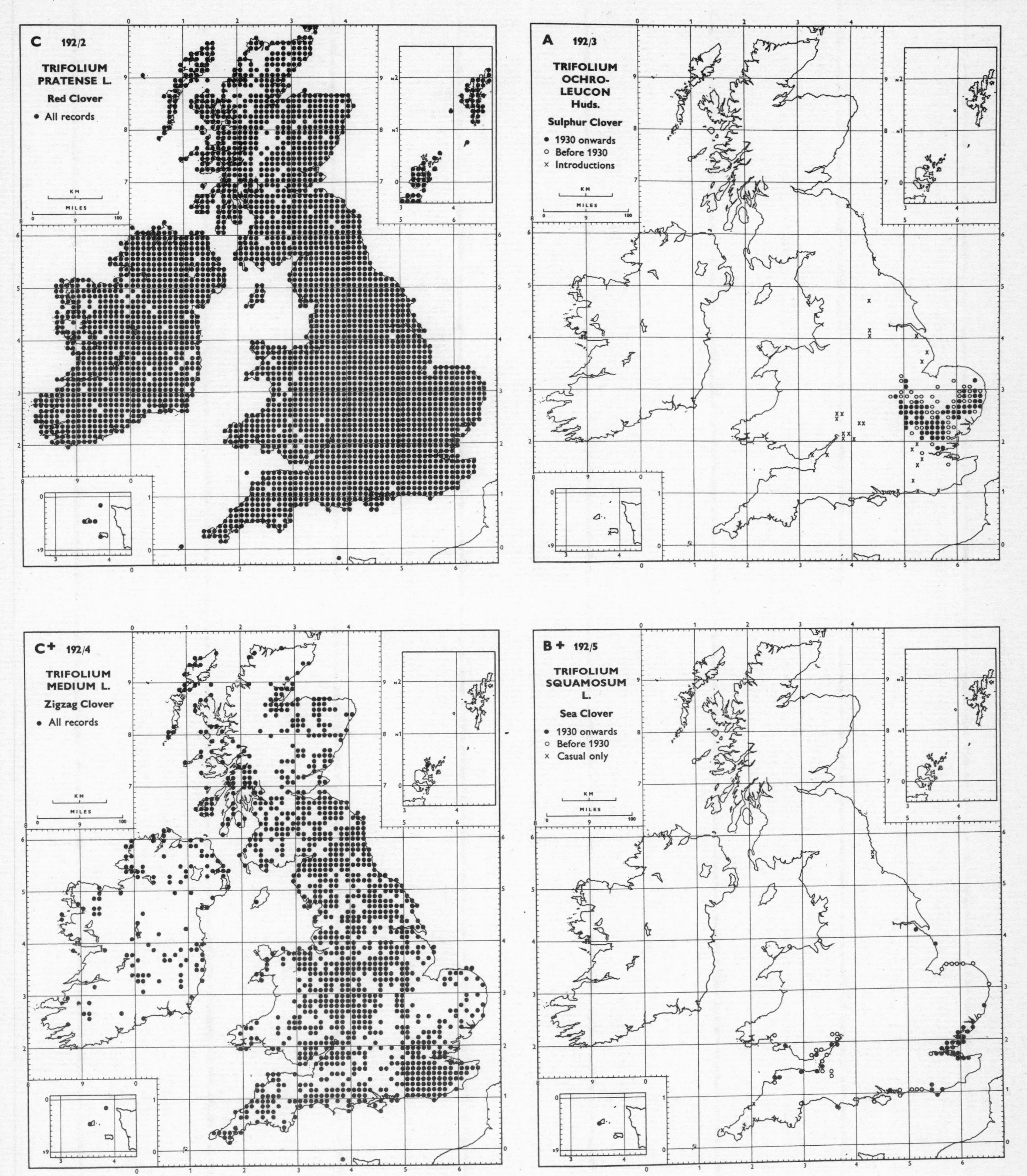
C 192/2
TRIFOLIUM PRATENSE L.
Red Clover
• All records
A 192/3
TRIFOLIUM OCHRO-LEUCON Huds.
Sulphur Clover
• 1930 onwards
○ Before 1930
× Introductions
C+ 192/4
TRIFOLIUM MEDIUM L.
Zigzag Clover
• All records
B+ 192/5
TRIFOLIUM SQUAMOSUM L.
Sea Clover
• 1930 onwards
○ Before 1930
× Casual only

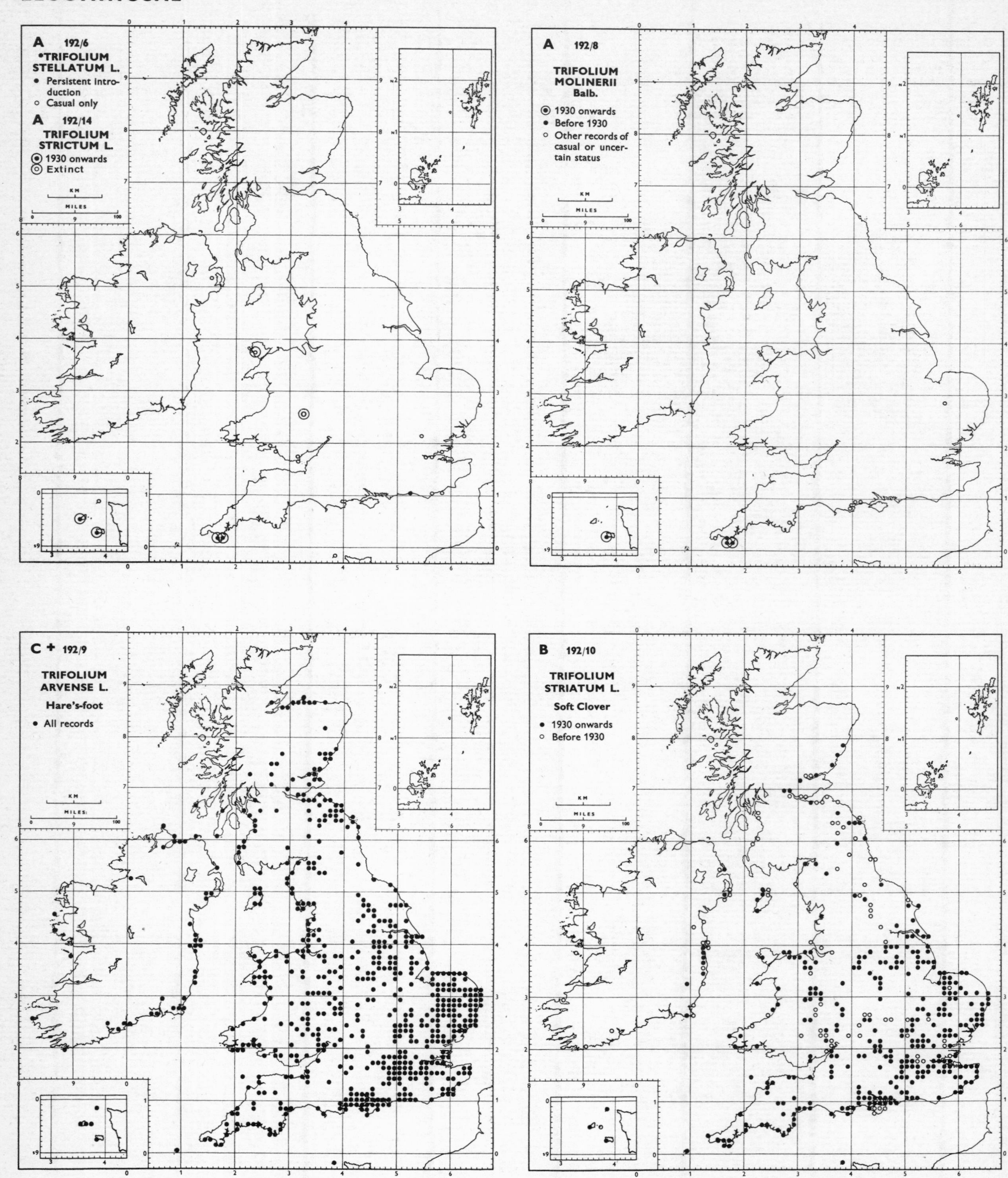
A 192/6
*TRIFOLIUM STELLATUM L.
• Persistent introduction
○ Casual only
A 192/14
TRIFOLIUM STRICTUM L.
⊙ 1930 onwards
◎ Extinct
A 192/8
TRIFOLIUM MOLINERII Balb.
⊙ 1930 onwards
• Before 1930
○ Other records of casual or uncertain status
C + 192/9
TRIFOLIUM ARVENSE L.
Hare's-foot
• All records
B 192/10
TRIFOLIUM STRIATUM L.
Soft Clover
• 1930 onwards
○ Before 1930
KM
MILES

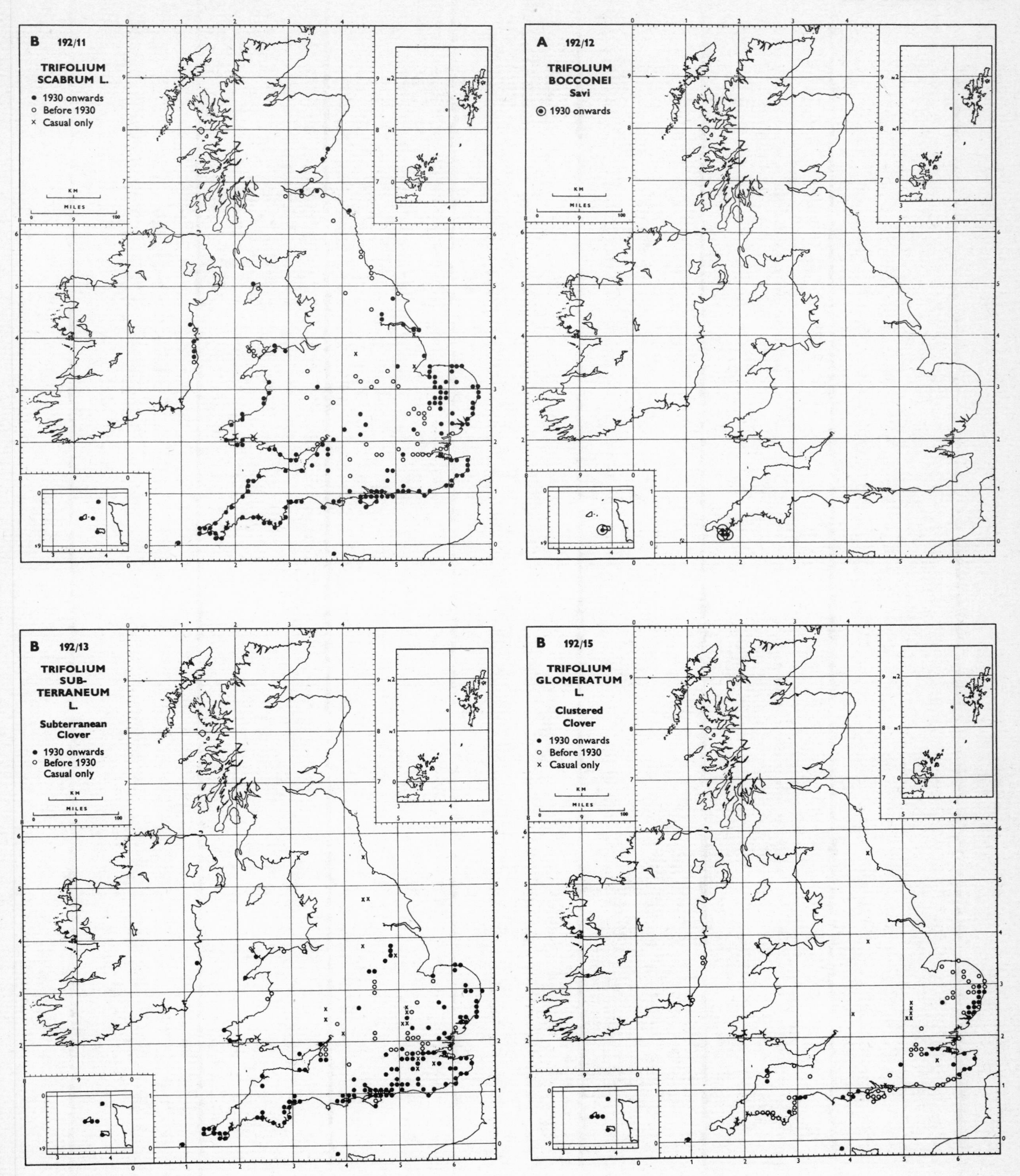
B 192/11
TRIFOLIUM SCABRUM L.
• 1930 onwards
○ Before 1930
× Casual only
KM
MILES
A 192/12
TRIFOLIUM BOCCONEI Savi
⊙ 1930 onwards
B 192/13
TRIFOLIUM SUB-TERRANEUM L.
Subterranean Clover
• 1930 onwards
○ Before 1930
Casual only
B 192/15
TRIFOLIUM GLOMERATUM L.
Clustered Clover
• 1930 onwards
○ Before 1930
× Casual only

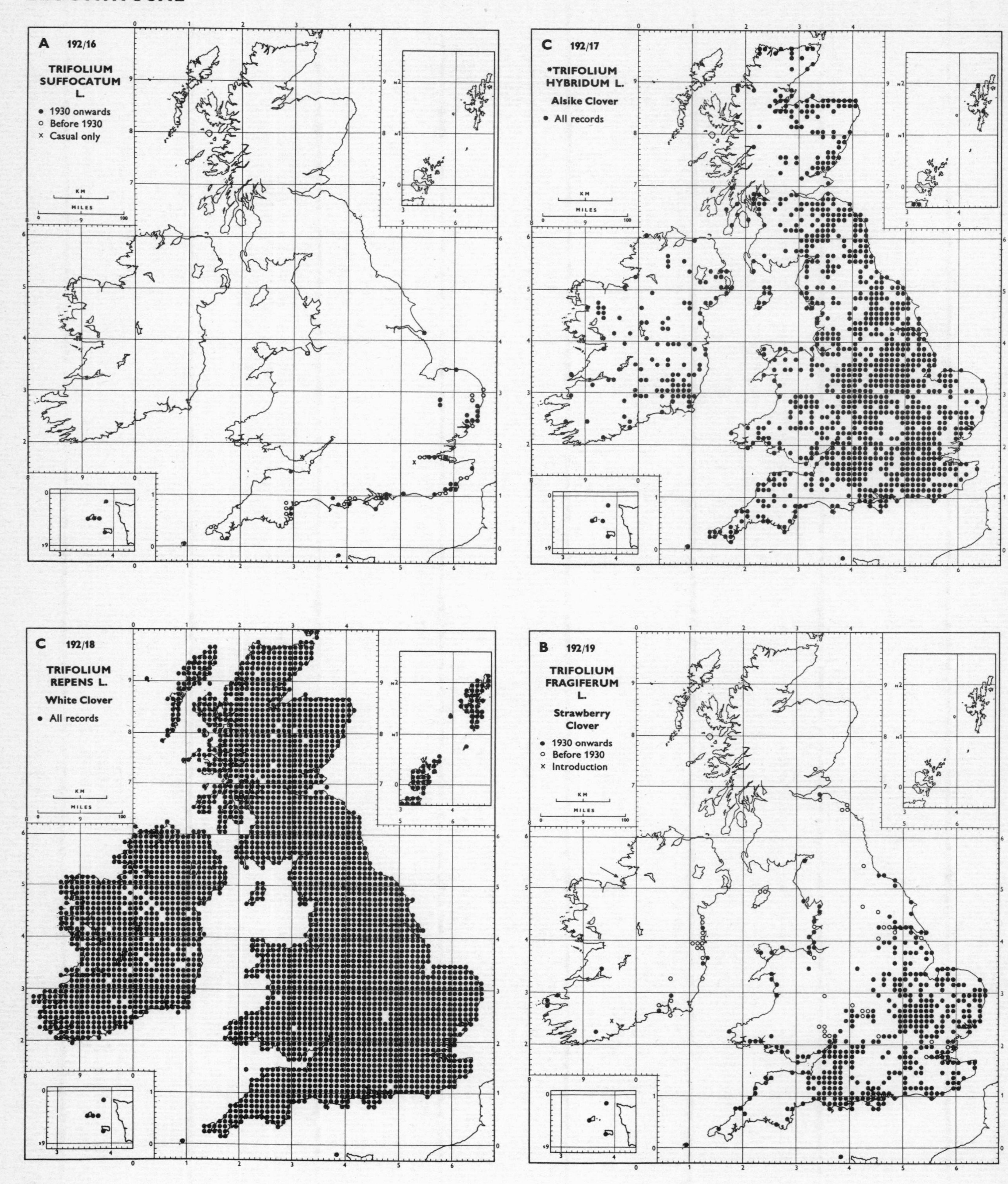
A 192/16
TRIFOLIUM SUFFOCATUM L.
● 1930 onwards
○ Before 1930
× Casual only
C 192/17
*TRIFOLIUM HYBRIDUM L.
Alsike Clover
● All records
C 192/18
TRIFOLIUM REPENS L.
White Clover
● All records
B 192/19
TRIFOLIUM FRAGIFERUM L.
Strawberry Clover
● 1930 onwards
○ Before 1930
× Introduction
KM
MILES

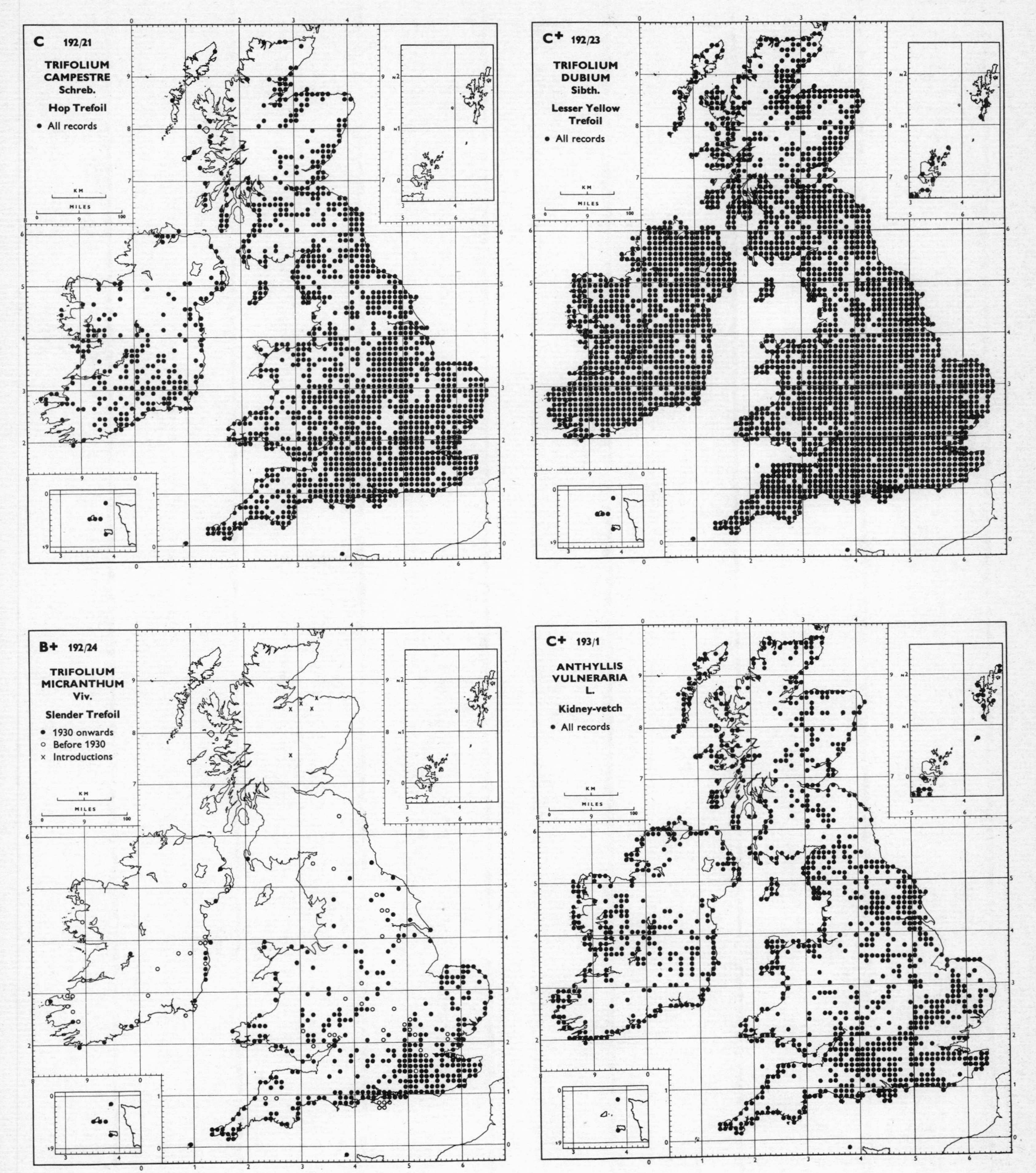
C 192/21
TRIFOLIUM CAMPESTRE Schreb.
Hop Trefoil
• All records
C+ 192/23
TRIFOLIUM DUBIUM Sibth.
Lesser Yellow Trefoil
• All records
B+ 192/24
TRIFOLIUM MICRANTHUM Viv.
Slender Trefoil
• 1930 onwards
○ Before 1930
× Introductions
C+ 193/1
ANTHYLLIS VULNERARIA L.
Kidney-vetch
• All records
KM
MILES

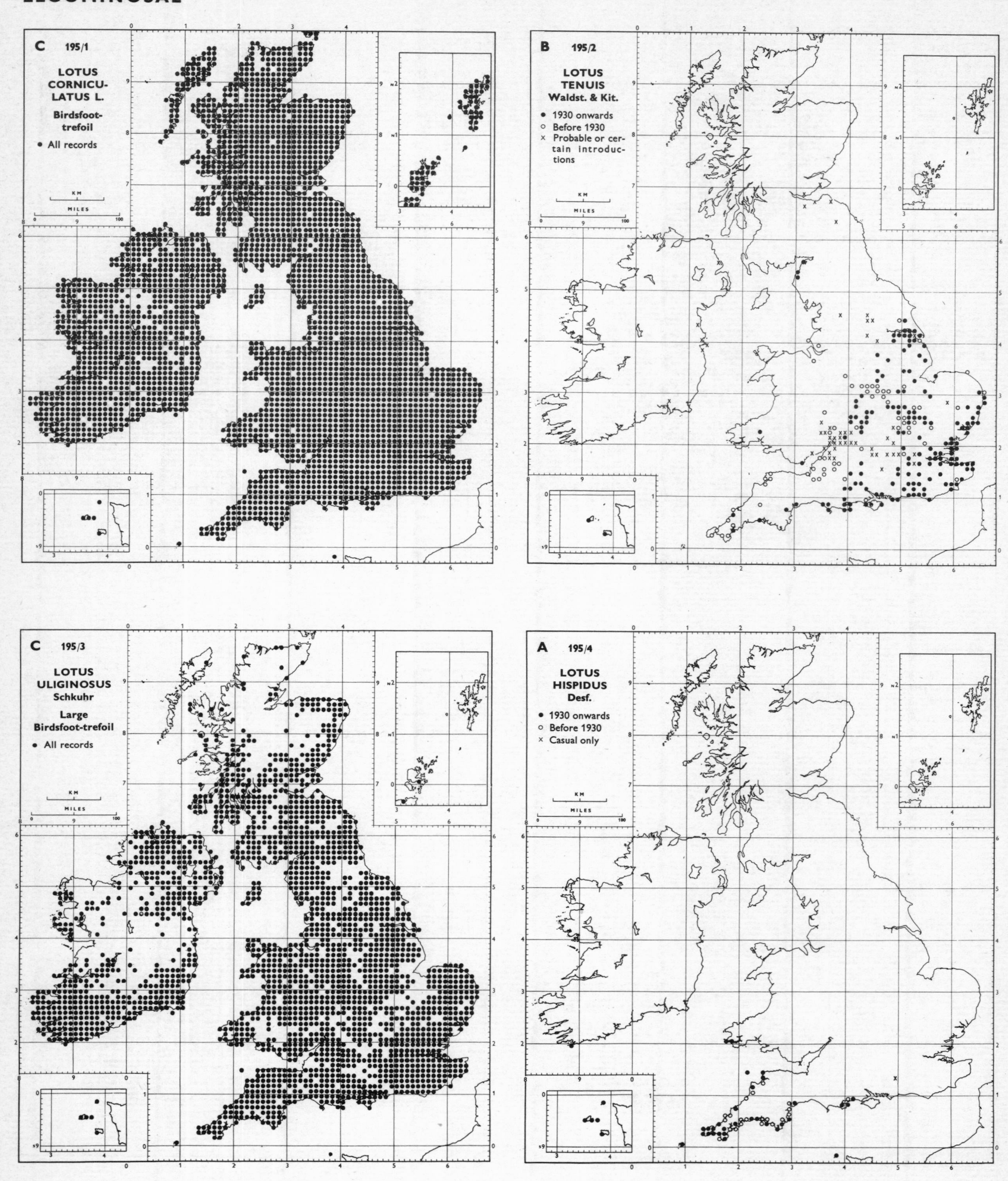
C 195/1
LOTUS CORNICULATUS L.
Birdsfoot-trefoil
• All records
B 195/2
LOTUS TENUIS Waldst. & Kit.
• 1930 onwards
○ Before 1930
× Probable or certain introductions
C 195/3
LOTUS ULIGINOSUS Schkuhr
Large Birdsfoot-trefoil
• All records
A 195/4
LOTUS HISPIDUS Desf.
• 1930 onwards
○ Before 1930
× Casual only
KM
MILES

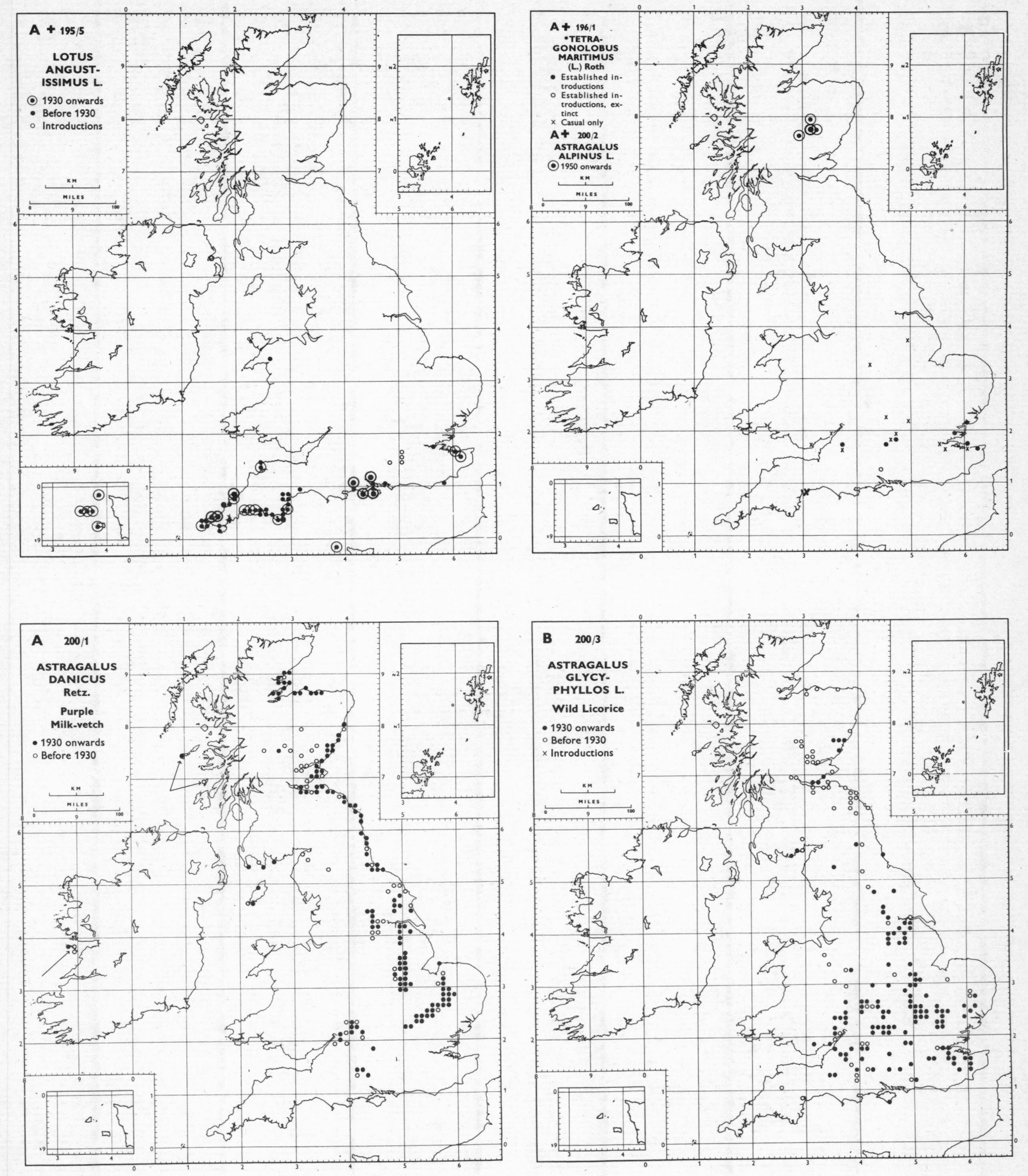
A + 195/5
LOTUS ANGUST-ISSIMUS L.
1930 onwards
Before 1930
Introductions
A + 196/1
*TETRA-GONOLOBUS MARITIMUS (L.) Roth
Established introductions
Established introductions, extinct
Casual only
A + 200/2
ASTRAGALUS ALPINUS L.
1950 onwards
A 200/1
ASTRAGALUS DANICUS Retz.
Purple Milk-vetch
1930 onwards
Before 1930
B 200/3
ASTRAGALUS GLYCY-PHYLLOS L.
Wild Licorice
1930 onwards
Before 1930
Introductions
KM
MILES

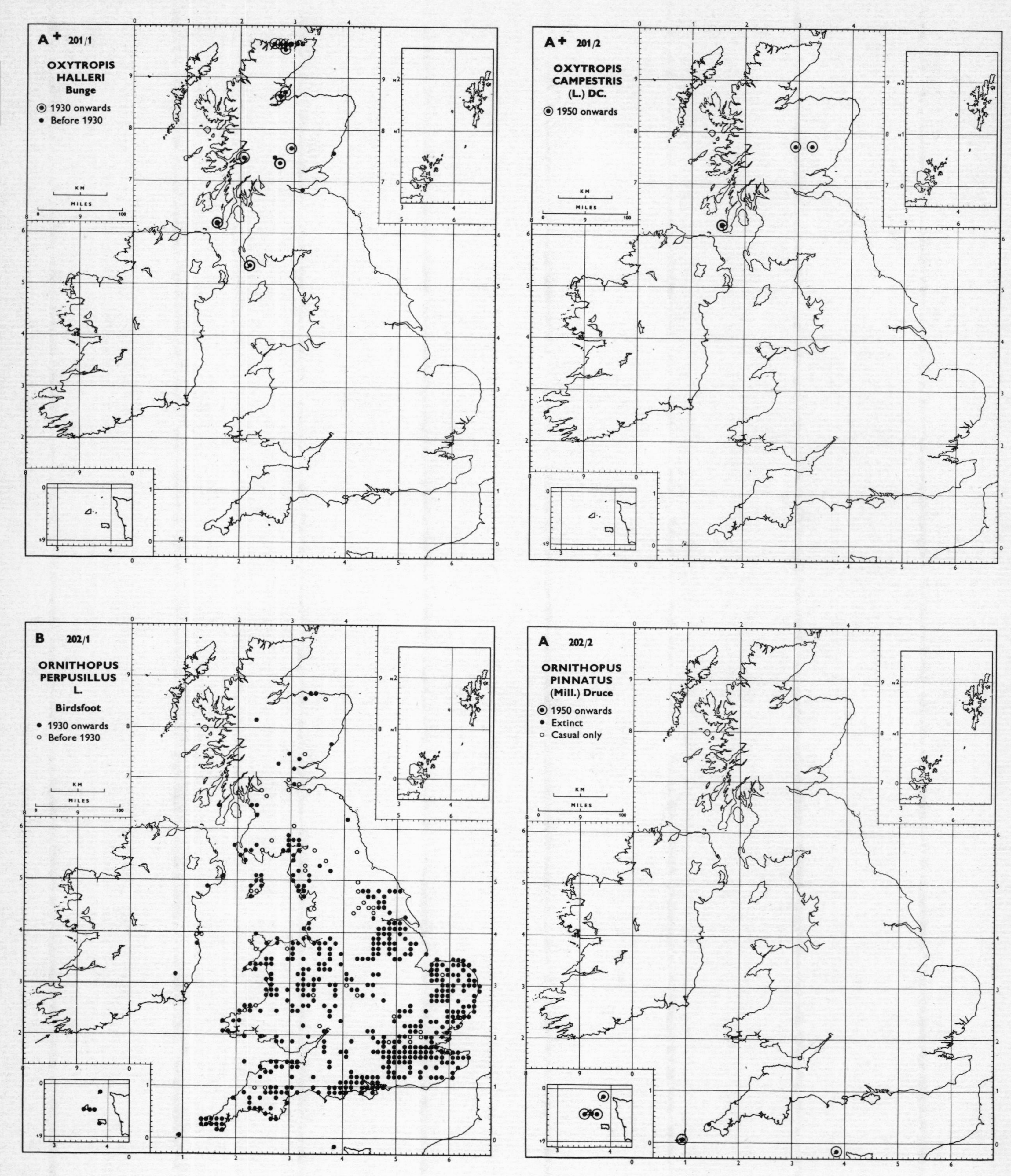
A + 201/1
OXYTROPIS HALLERI Bunge
1930 onwards
Before 1930
A + 201/2
OXYTROPIS CAMPESTRIS (L.) DC.
1950 onwards
B 202/1
ORNITHOPUS PERPUSILLUS L.
Birdsfoot
1930 onwards
Before 1930
A 202/2
ORNITHOPUS PINNATUS (Mill.) Druce
1950 onwards
Extinct
Casual only
KM
MILES

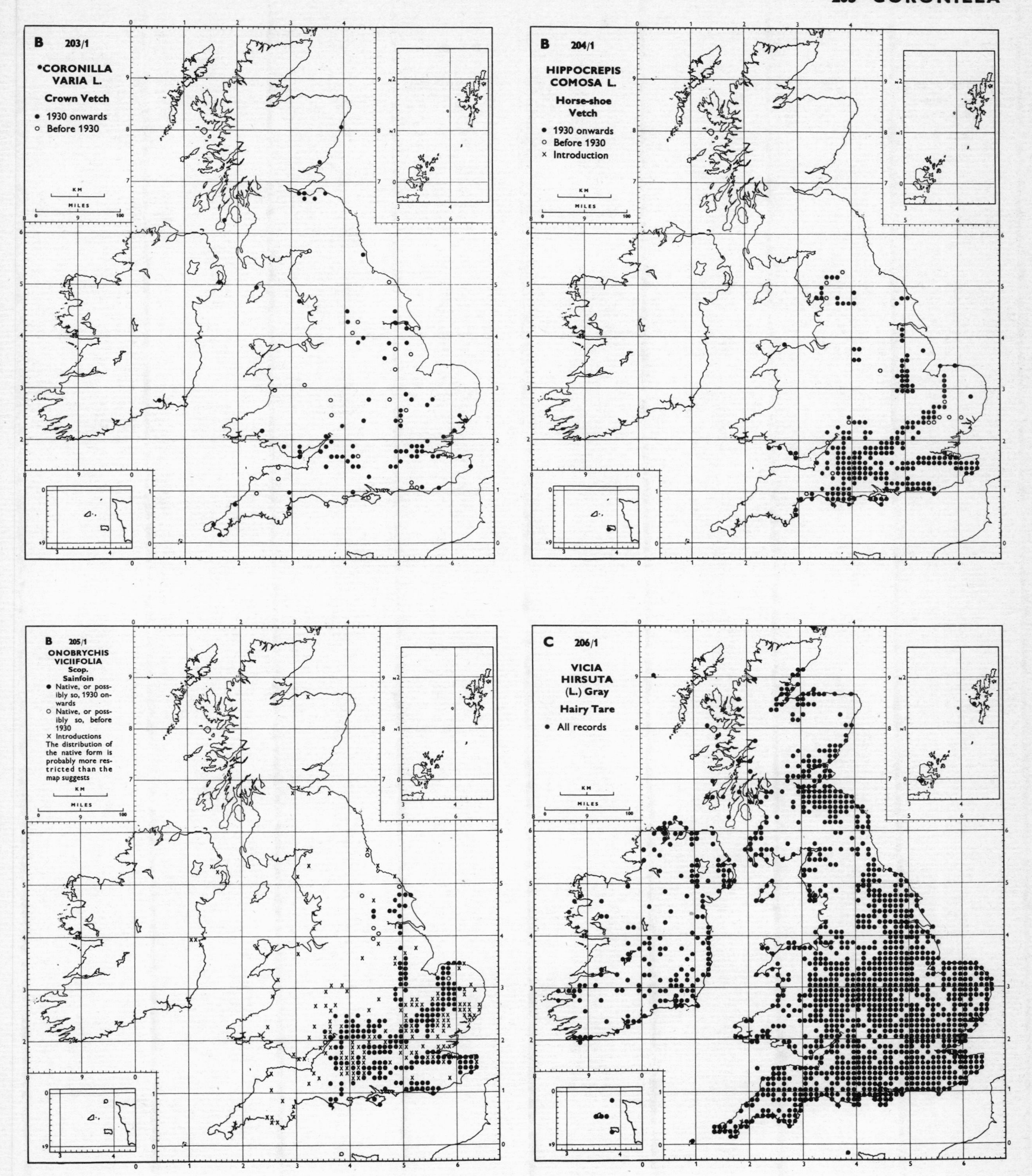
B 203/1
*CORONILLA VARIA L.
Crown Vetch
• 1930 onwards
○ Before 1930
B 204/1
HIPPOCREPIS COMOSA L.
Horse-shoe Vetch
• 1930 onwards
○ Before 1930
× Introduction
B 205/1
ONOBRYCHIS VICIIFOLIA Scop.
Sainfoin
• Native, or possibly so, 1930 onwards
○ Native, or possibly so, before 1930
× Introductions
The distribution of the native form is probably more restricted than the map suggests
C 206/1
VICIA HIRSUTA (L.) Gray
Hairy Tare
• All records

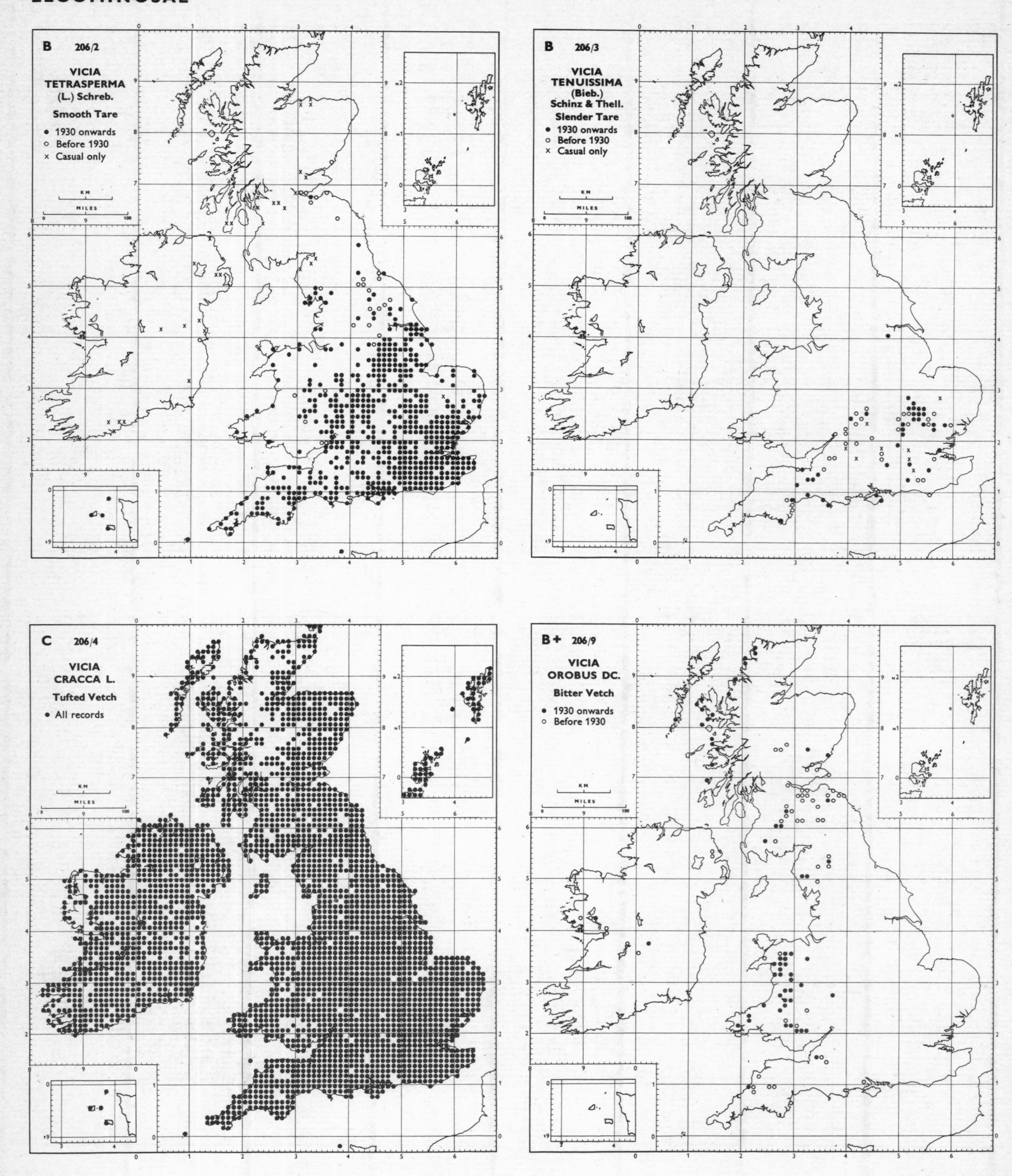
B 206/2
VICIA TETRASPERMA (L.) Schreb.
Smooth Tare
• 1930 onwards
○ Before 1930
× Casual only
B 206/3
VICIA TENUISSIMA (Bieb.) Schinz & Thell.
Slender Tare
• 1930 onwards
○ Before 1930
× Casual only
C 206/4
VICIA CRACCA L.
Tufted Vetch
• All records
B+ 206/9
VICIA OROBUS DC.
Bitter Vetch
• 1930 onwards
○ Before 1930

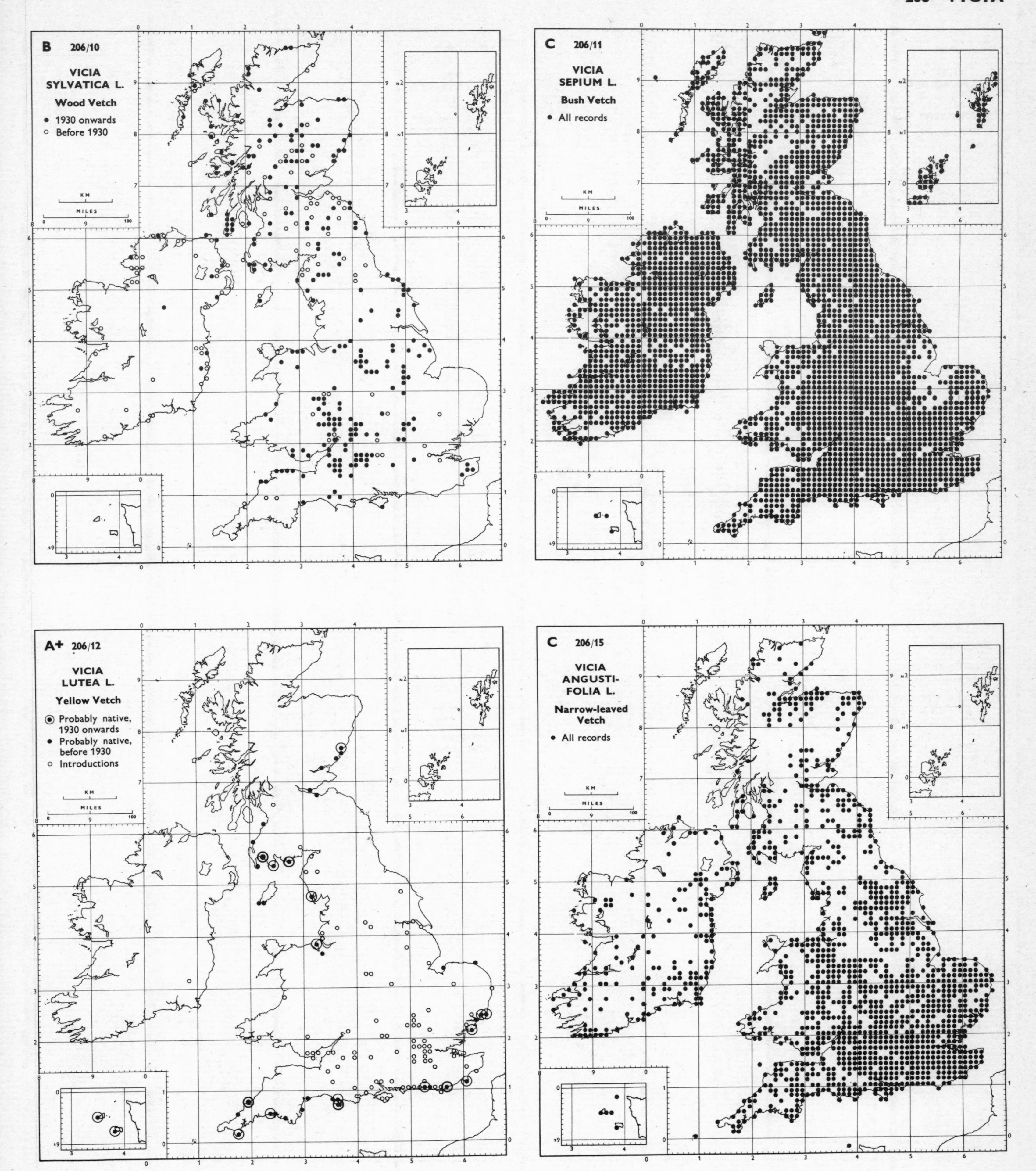
B 206/10
VICIA SYLVATICA L.
Wood Vetch
1930 onwards
Before 1930
C 206/11
VICIA SEPIUM L.
Bush Vetch
All records
A+ 206/12
VICIA LUTEA L.
Yellow Vetch
Probably native, 1930 onwards
Probably native, before 1930
Introductions
C 206/15
VICIA ANGUSTIFOLIA L.
Narrow-leaved Vetch
All records

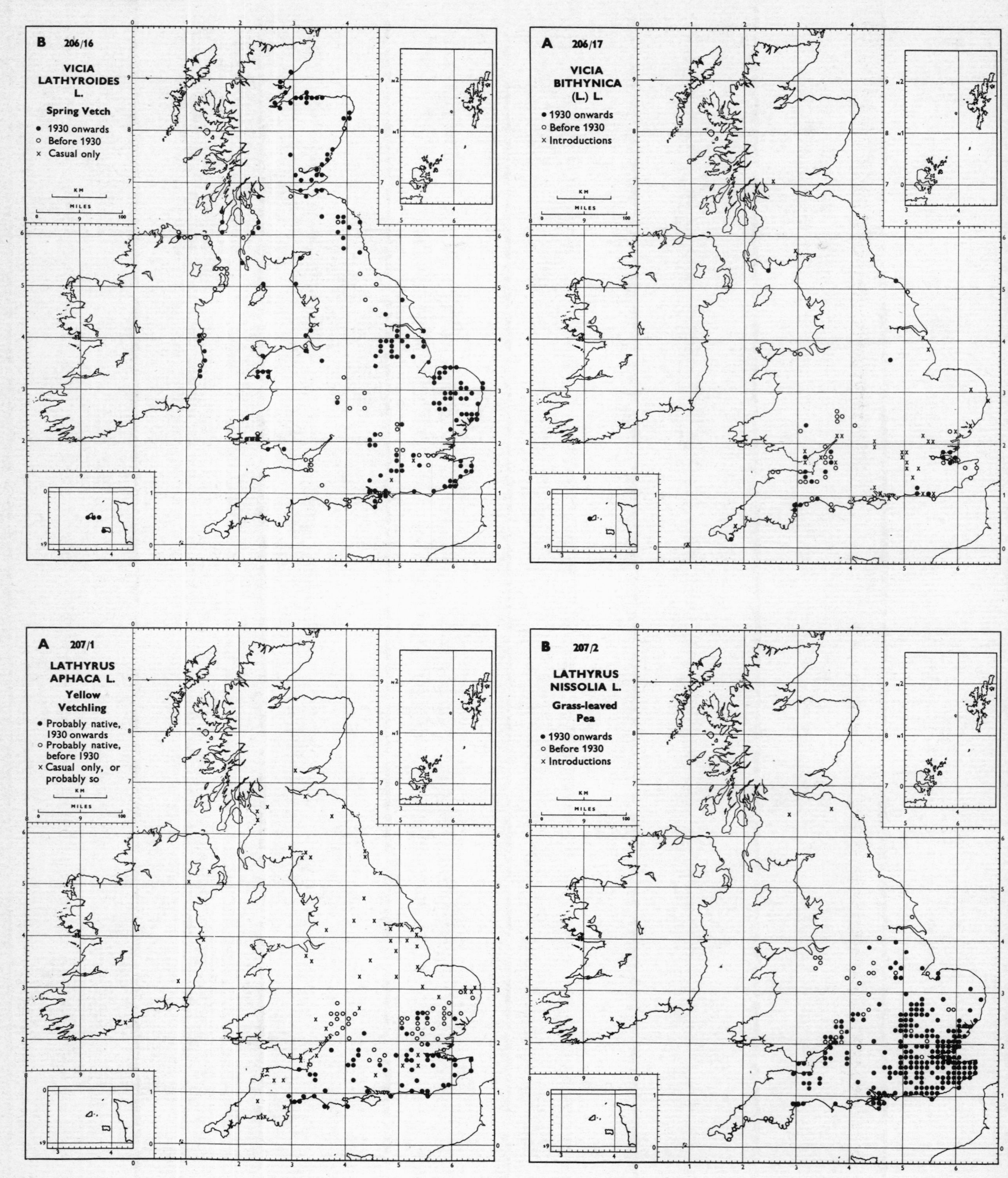
B 206/16
VICIA LATHYROIDES L.
Spring Vetch
• 1930 onwards
○ Before 1930
× Casual only
KM
MILES
A 206/17
VICIA BITHYNICA (L.) L.
• 1930 onwards
○ Before 1930
× Introductions
A 207/1
LATHYRUS APHACA L.
Yellow Vetchling
• Probably native, 1930 onwards
○ Probably native, before 1930
× Casual only, or probably so
B 207/2
LATHYRUS NISSOLIA L.
Grass-leaved Pea
• 1930 onwards
○ Before 1930
× Introductions

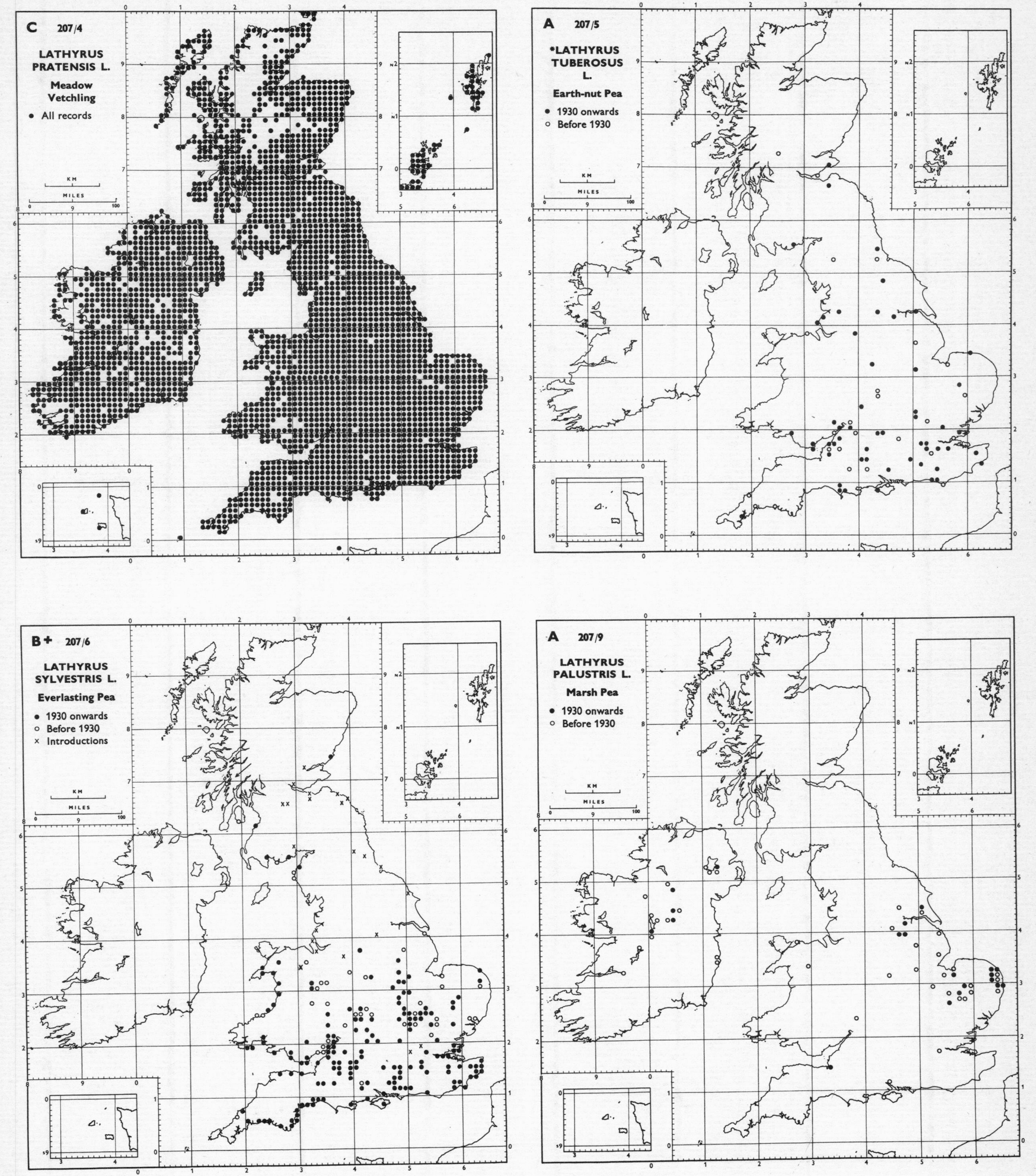

C 207/4
LATHYRUS PRATENSIS L.
Meadow Vetchling
• All records
A 207/5
*LATHYRUS TUBEROSUS L.
Earth-nut Pea
• 1930 onwards
○ Before 1930
B+ 207/6
LATHYRUS SYLVESTRIS L.
Everlasting Pea
• 1930 onwards
○ Before 1930
× Introductions
A 207/9
LATHYRUS PALUSTRIS L.
Marsh Pea
• 1930 onwards
○ Before 1930
KM
MILES

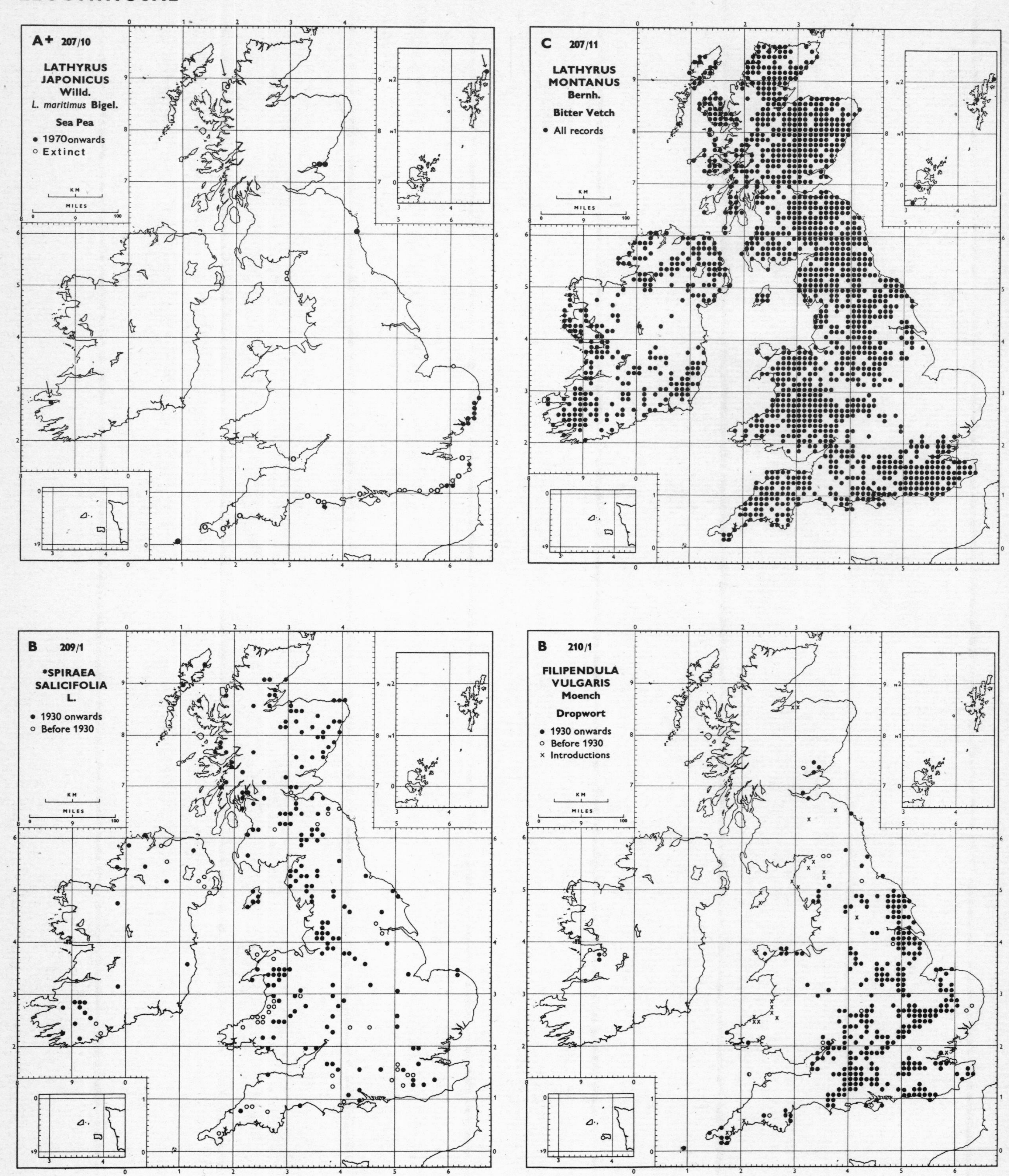
A+ 207/10
LATHYRUS JAPONICUS Willd.
L. maritimus Bigel.
Sea Pea
• 1970 onwards
○ Extinct
C 207/11
LATHYRUS MONTANUS Bernh.
Bitter Vetch
• All records
B 209/1
*SPIRAEA SALICIFOLIA L.
• 1930 onwards
○ Before 1930
B 210/1
FILIPENDULA VULGARIS Moench
Dropwort
• 1930 onwards
○ Before 1930
× Introductions

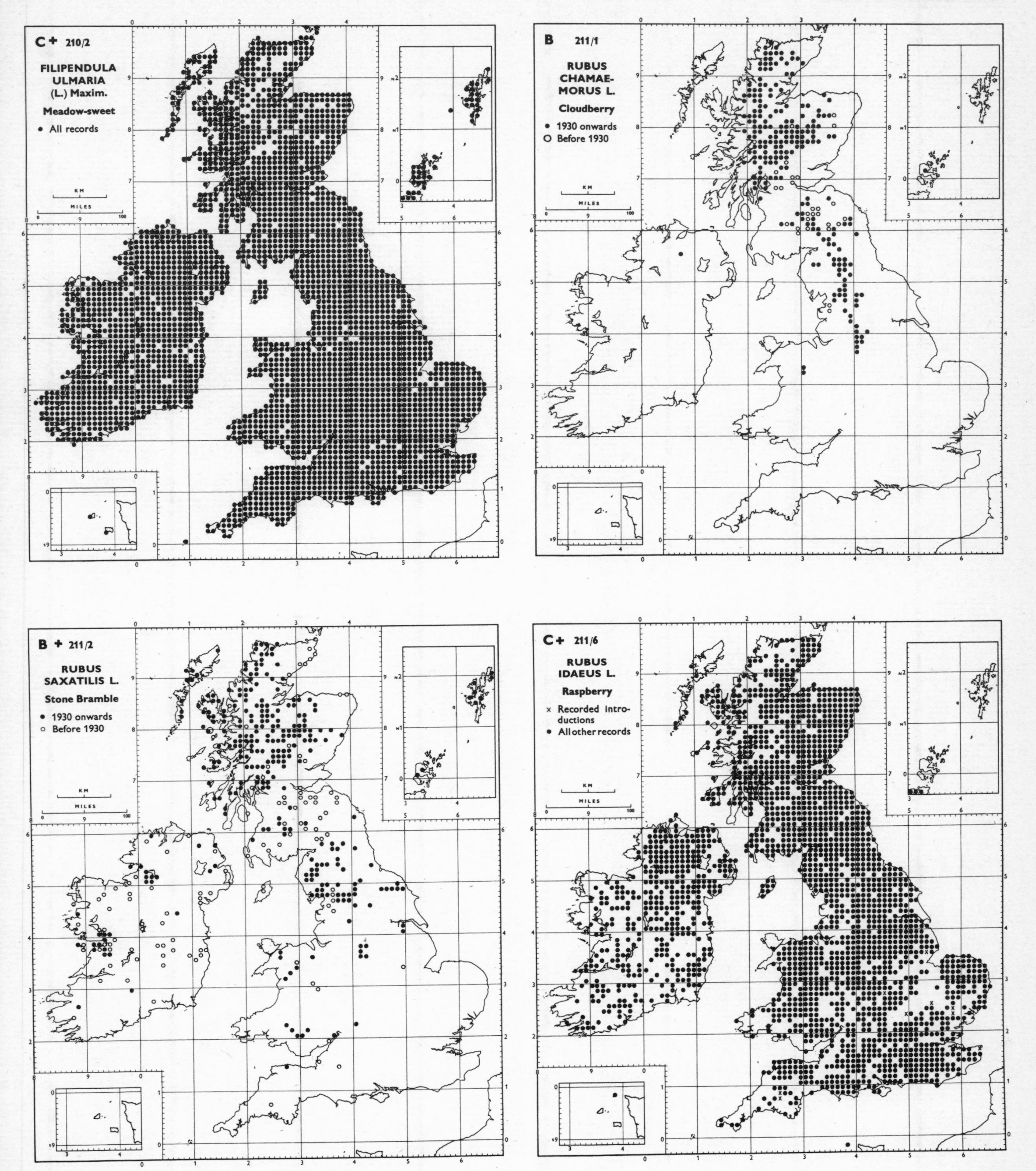
C + 210/2
FILIPENDULA ULMARIA (L.) Maxim.
Meadow-sweet
• All records
B 211/1
RUBUS CHAMAE-MORUS L.
Cloudberry
• 1930 onwards
○ Before 1930
B + 211/2
RUBUS SAXATILIS L.
Stone Bramble
• 1930 onwards
○ Before 1930
C + 211/6
RUBUS IDAEUS L.
Raspberry
x Recorded introductions
• All other records
KM
MILES

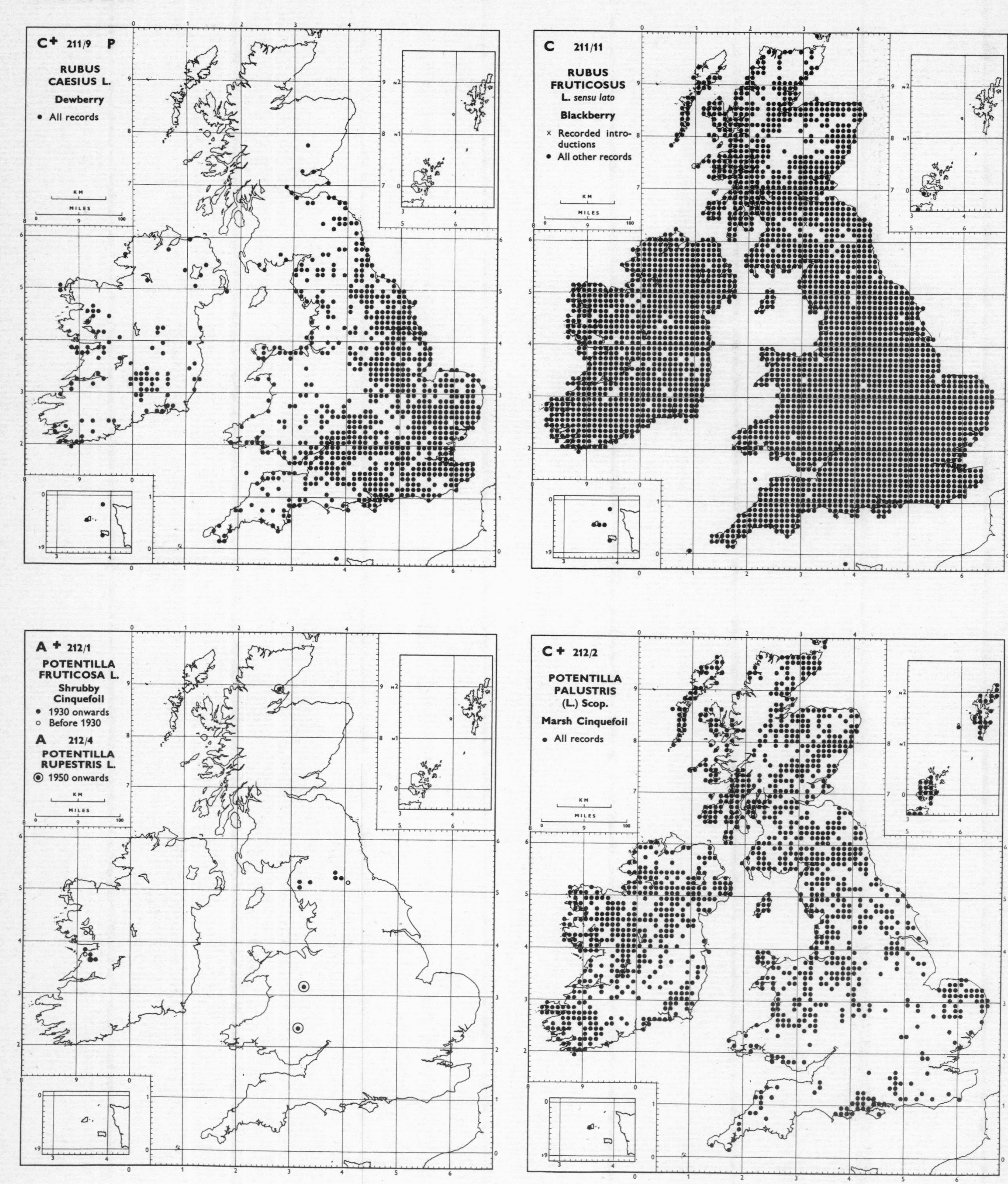

C+ 211/9 P
RUBUS CAESIUS L.
Dewberry
• All records
C 211/11
RUBUS FRUTICOSUS L. sensu lato
Blackberry
× Recorded introductions
• All other records
A + 212/1
POTENTILLA FRUTICOSA L.
Shrubby Cinquefoil
• 1930 onwards
○ Before 1930
A 212/4
POTENTILLA RUPESTRIS L.
◉ 1950 onwards
C + 212/2
POTENTILLA PALUSTRIS (L.) Scop.
Marsh Cinquefoil
• All records
KM
MILES

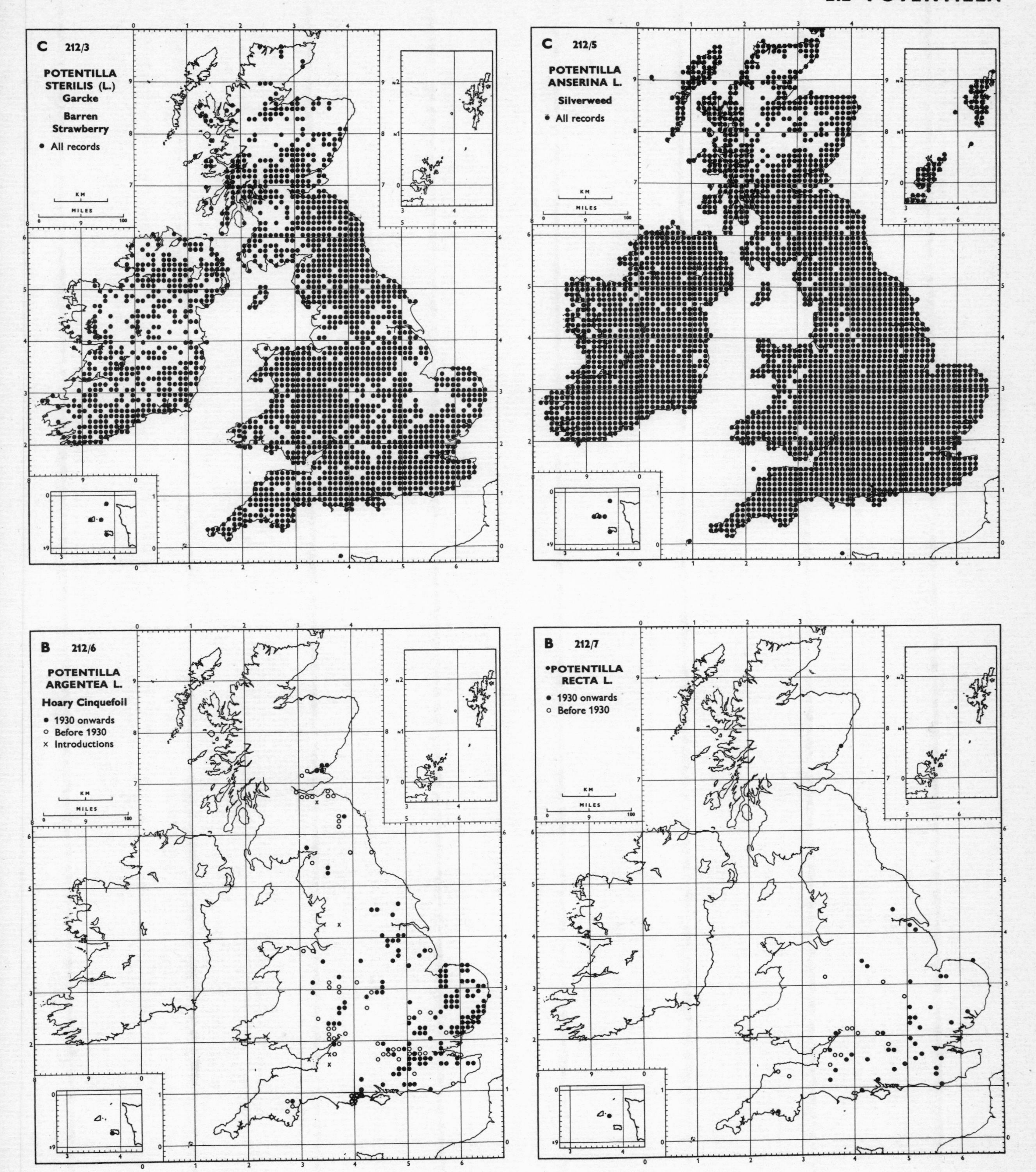
C 212/3
POTENTILLA STERILIS (L.) Garcke
Barren Strawberry
• All records
C 212/5
POTENTILLA ANSERINA L.
Silverweed
• All records
B 212/6
POTENTILLA ARGENTEA L.
Hoary Cinquefoil
• 1930 onwards
○ Before 1930
× Introductions
B 212/7
•POTENTILLA RECTA L.
• 1930 onwards
○ Before 1930
KM
MILES

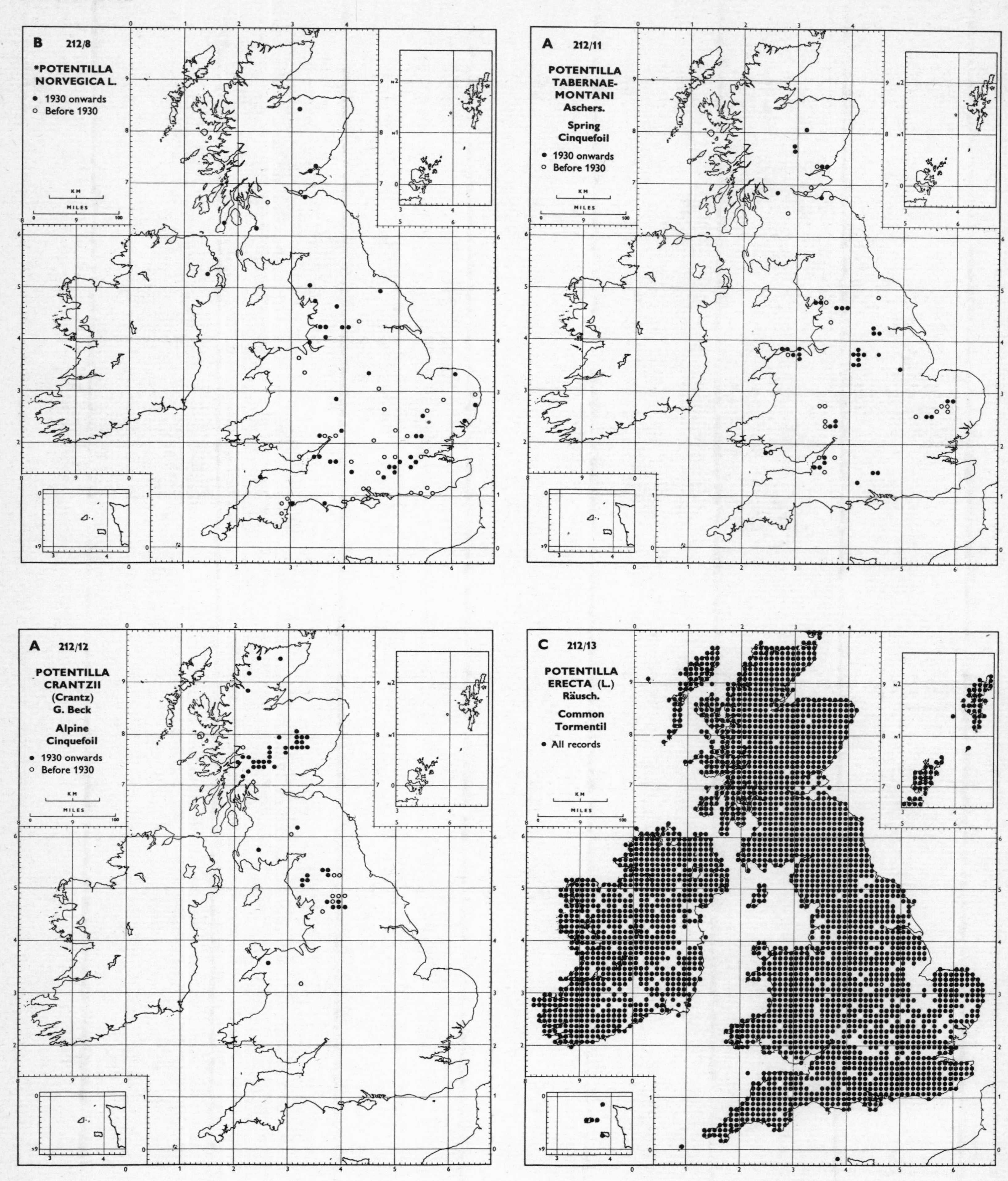
B 212/8
•POTENTILLA NORVEGICA L.
• 1930 onwards
○ Before 1930
A 212/11
POTENTILLA TABERNAE-MONTANI Aschers.
Spring Cinquefoil
• 1930 onwards
○ Before 1930
A 212/12
POTENTILLA CRANTZII (Crantz) G. Beck
Alpine Cinquefoil
• 1930 onwards
○ Before 1930
C 212/13
POTENTILLA ERECTA (L.) Räusch.
Common Tormentil
• All records

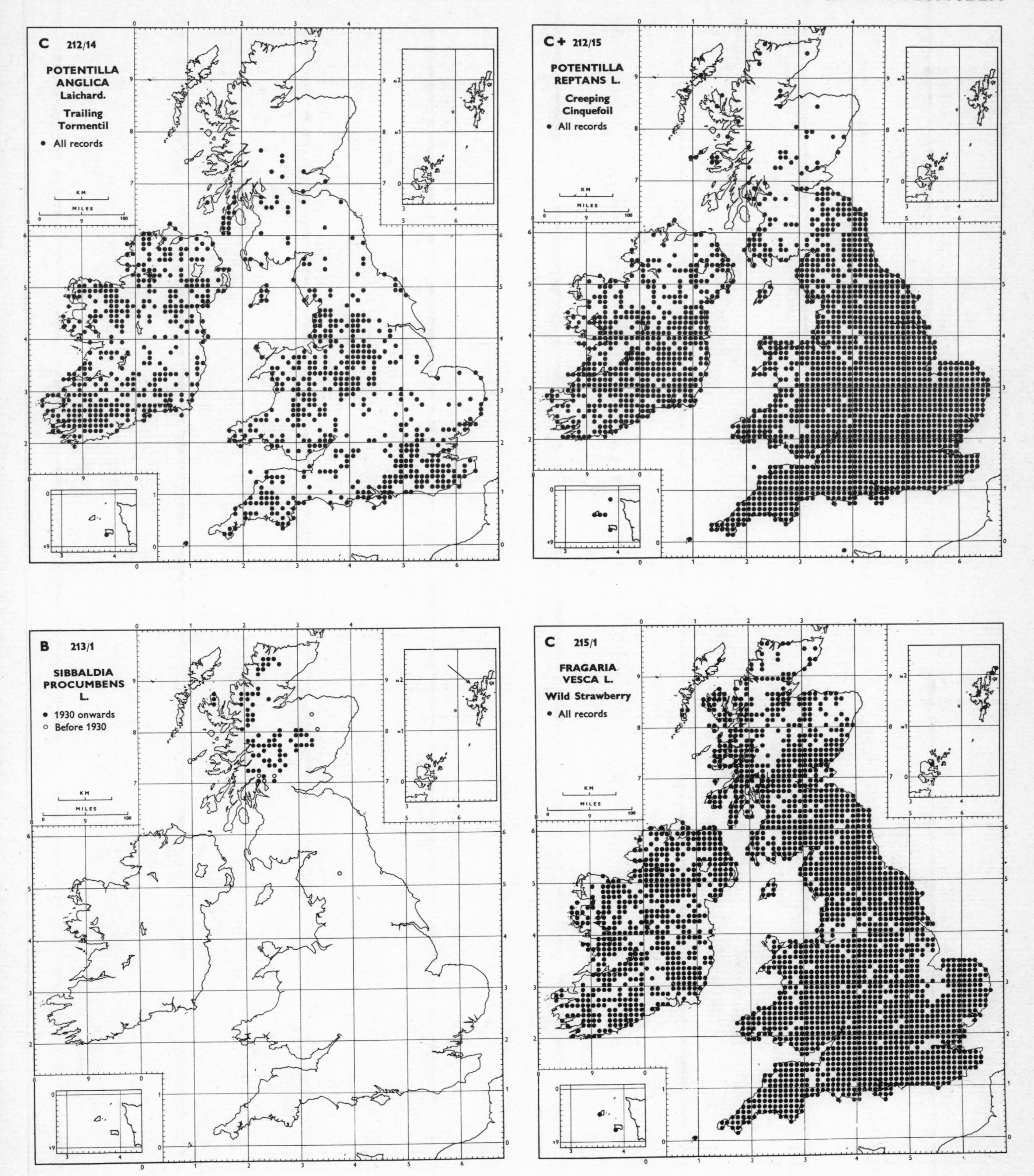
C 212/14
POTENTILLA ANGLICA Laichard.
Trailing Tormentil
• All records
C + 212/15
POTENTILLA REPTANS L.
Creeping Cinquefoil
• All records
B 213/1
SIBBALDIA PROCUMBENS L.
• 1930 onwards
○ Before 1930
C 215/1
FRAGARIA VESCA L.
Wild Strawberry
• All records

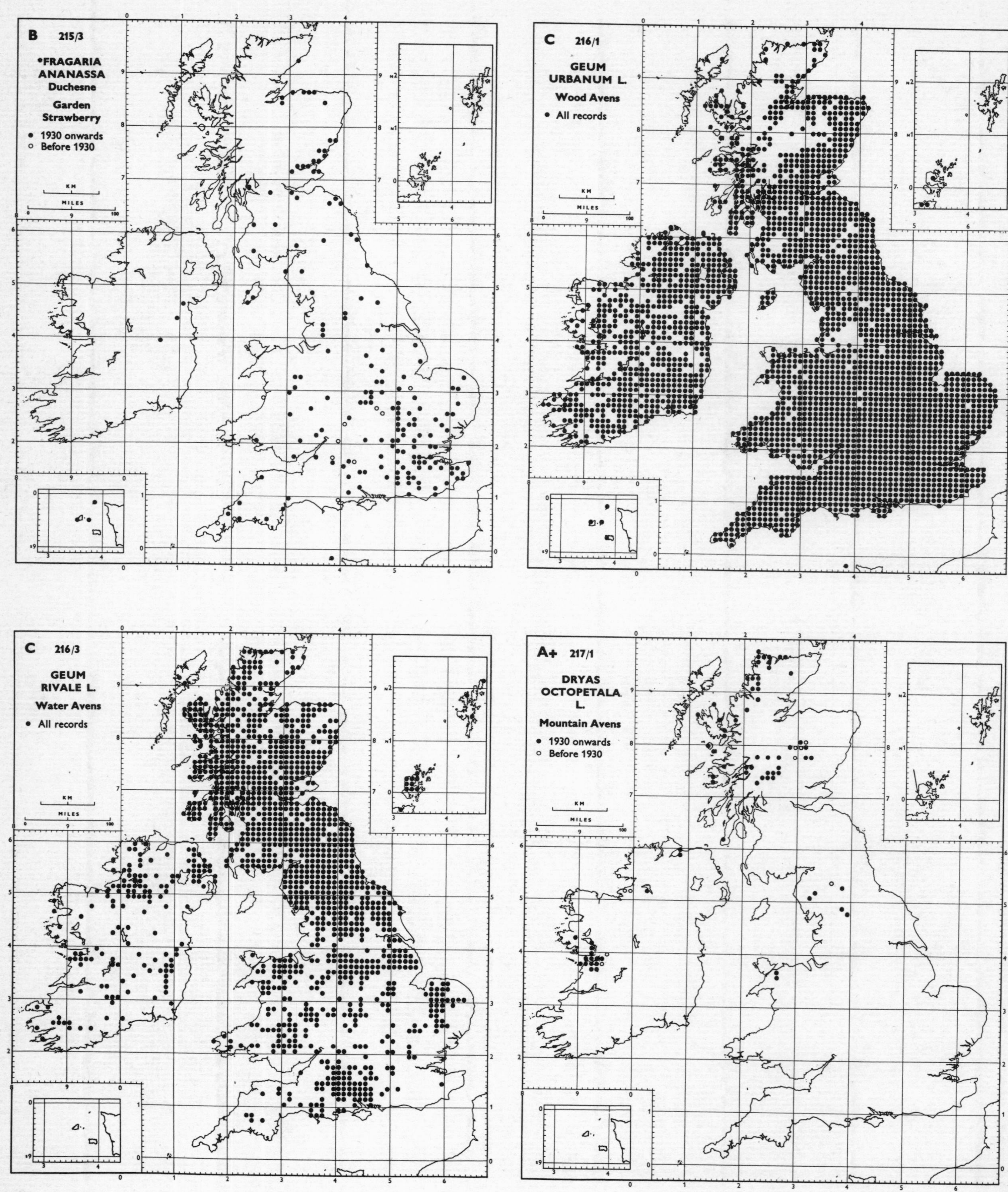
B 215/3
*FRAGARIA ANANASSA Duchesne
Garden Strawberry
• 1930 onwards
○ Before 1930
C 216/1
GEUM URBANUM L.
Wood Avens
• All records
C 216/3
GEUM RIVALE L.
Water Avens
• All records
A+ 217/1
DRYAS OCTOPETALA L.
Mountain Avens
• 1930 onwards
○ Before 1930
KM
MILES

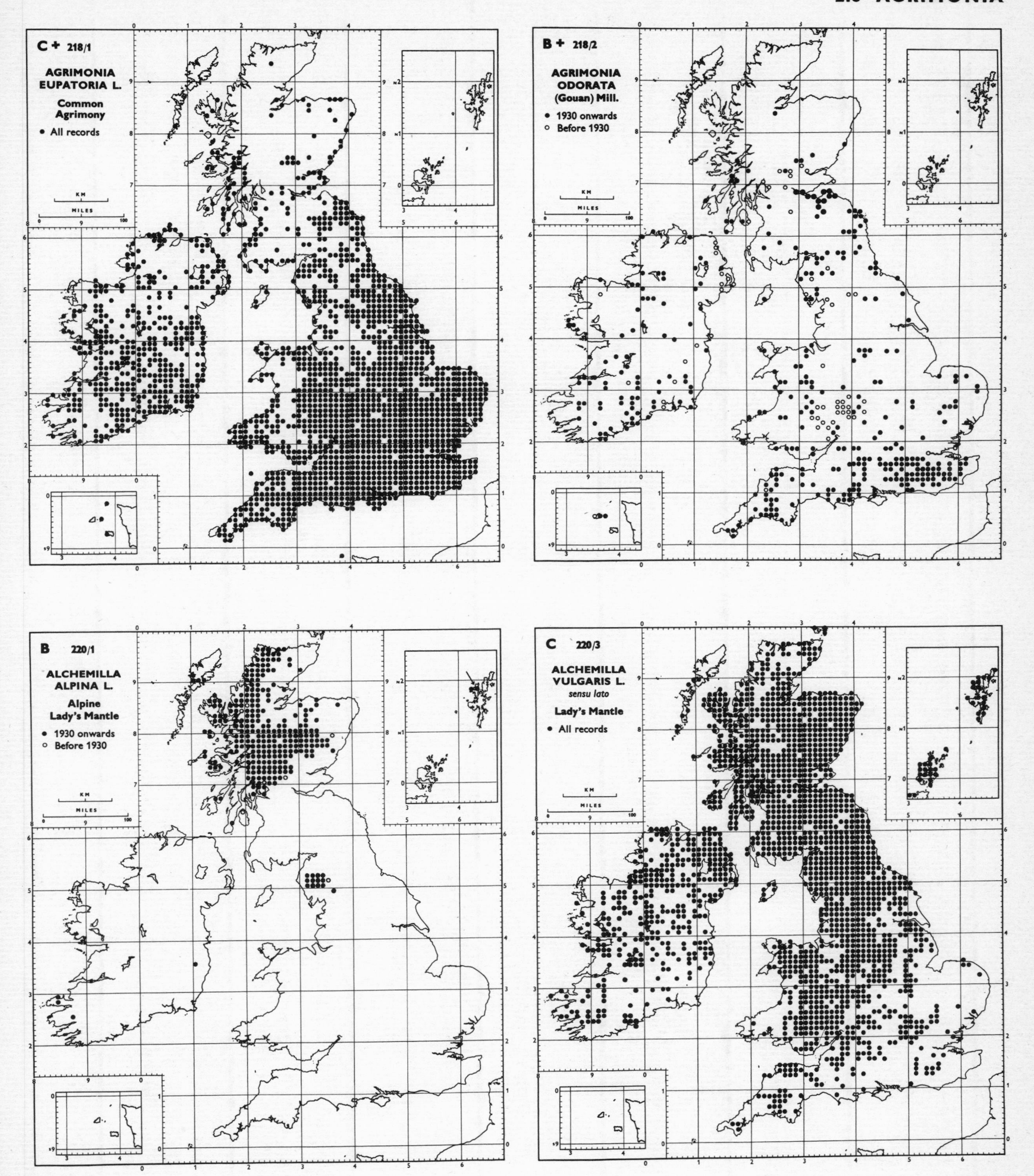
C + 218/1
AGRIMONIA EUPATORIA L.
Common Agrimony
• All records
B + 218/2
AGRIMONIA ODORATA (Gouan) Mill.
• 1930 onwards
○ Before 1930
B 220/1
ALCHEMILLA ALPINA L.
Alpine Lady's Mantle
• 1930 onwards
○ Before 1930
C 220/3
ALCHEMILLA VULGARIS L.
sensu lato
Lady's Mantle
• All records
KM
MILES

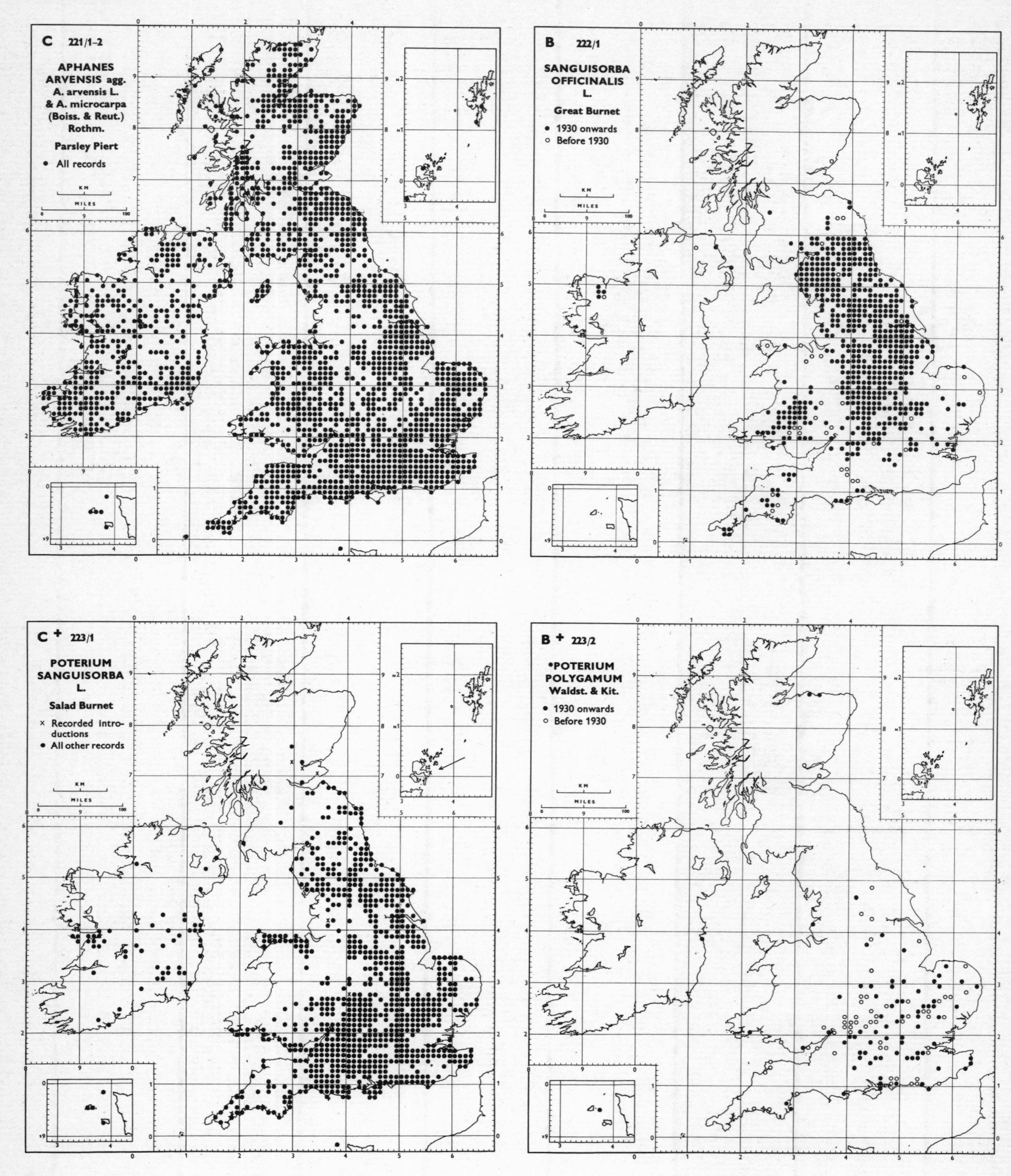

C 221/1–2
APHANES ARVENSIS agg.
A. arvensis L.
& A. microcarpa (Boiss. & Reut.) Rothm.
Parsley Piert
• All records
B 222/1
SANGUISORBA OFFICINALIS L.
Great Burnet
• 1930 onwards
○ Before 1930
C + 223/1
POTERIUM SANGUISORBA L.
Salad Burnet
× Recorded introductions
• All other records
B + 223/2
*POTERIUM POLYGAMUM Waldst. & Kit.
• 1930 onwards
○ Before 1930

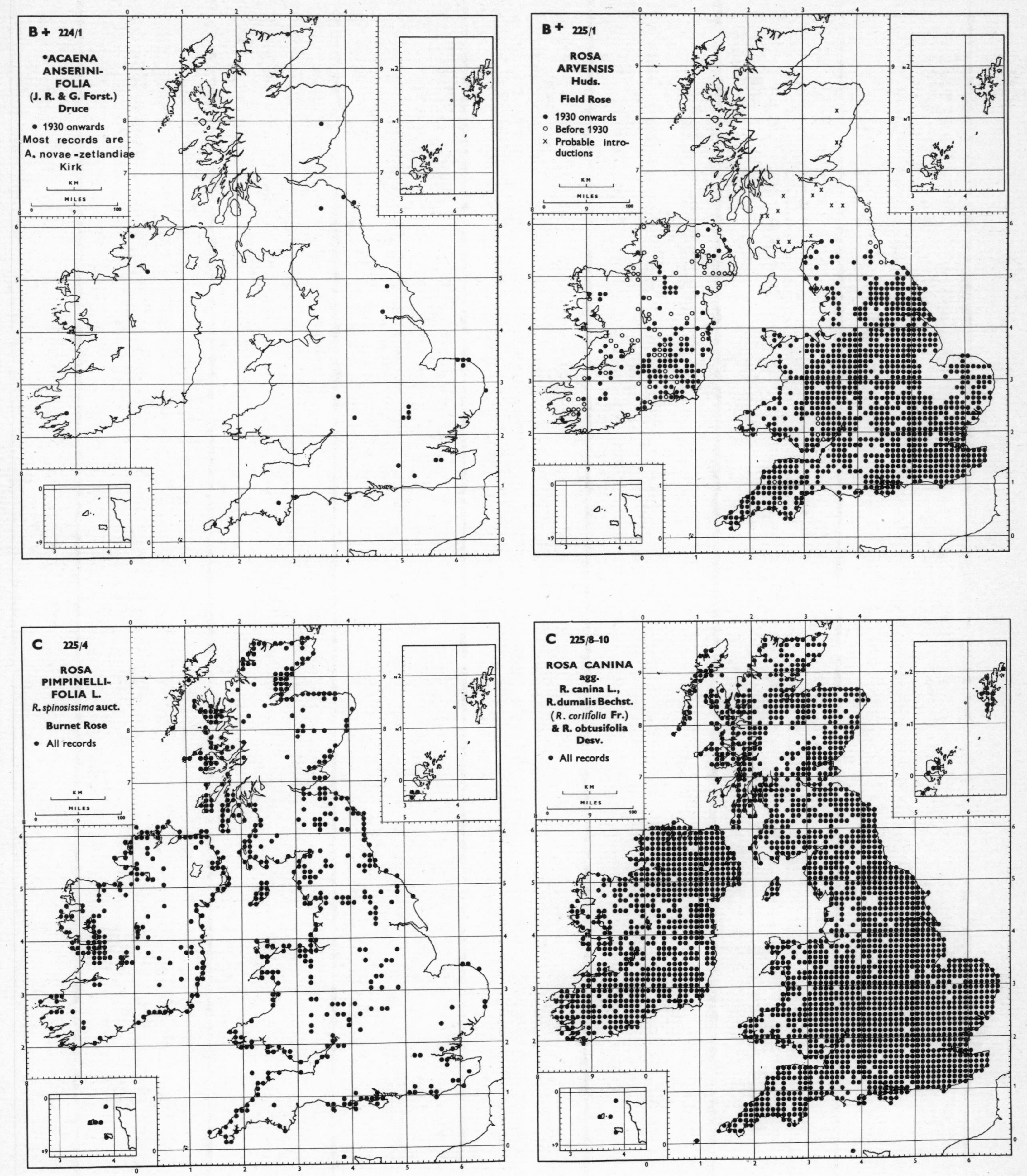
B + 224/1
*ACAENA ANSERINIFOLIA (J. R. & G. Forst.) Druce
• 1930 onwards
Most records are A. novae-zetlandiae Kirk
KM
MILES
B + 225/1
ROSA ARVENSIS Huds.
Field Rose
• 1930 onwards
○ Before 1930
× Probable introductions
C 225/4
ROSA PIMPINELLIFOLIA L.
R. spinosissima auct.
Burnet Rose
• All records
C 225/8–10
ROSA CANINA agg.
R. canina L., R. dumalis Bechst. (R. coriifolia Fr.) & R. obtusifolia Desv.
• All records

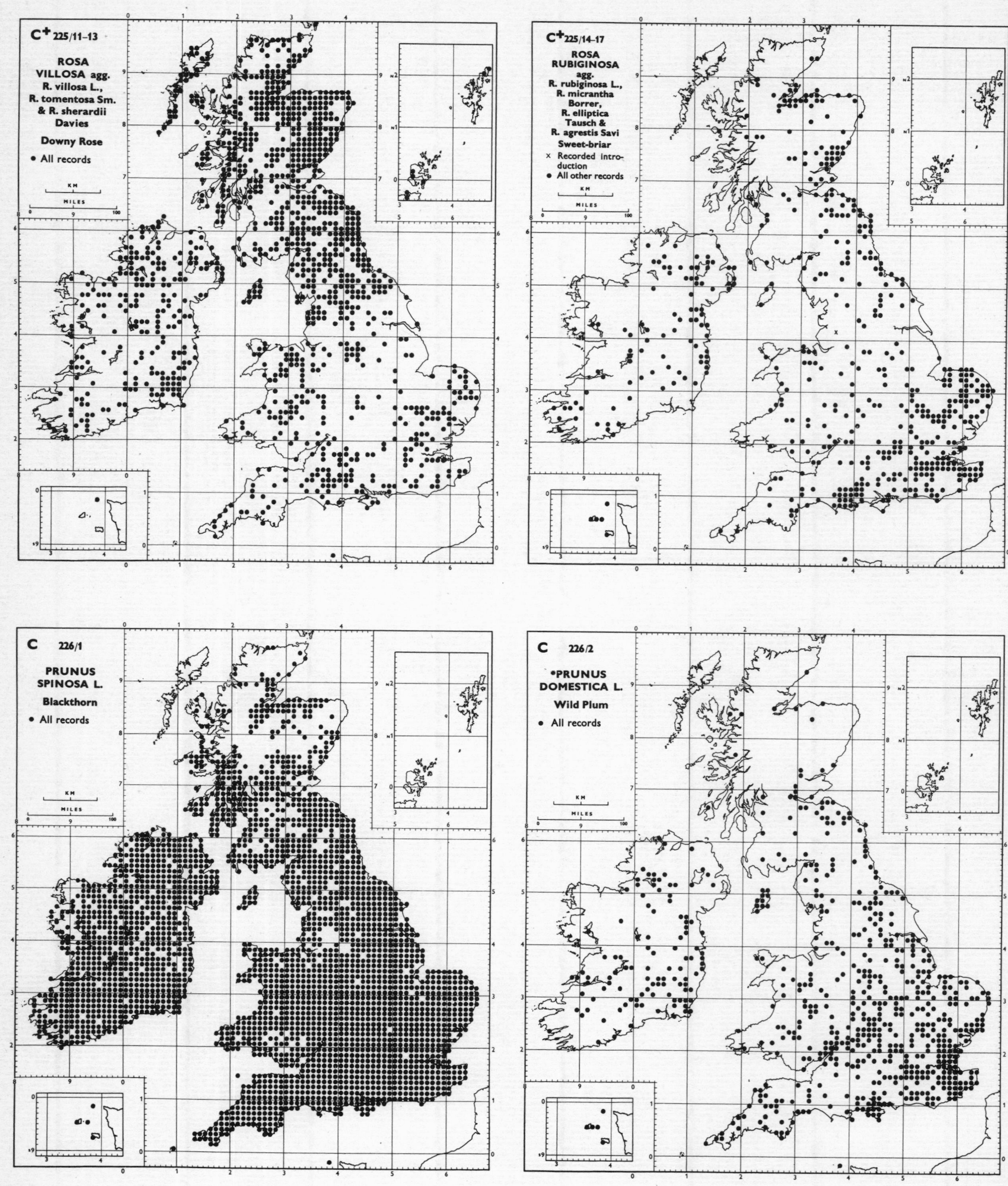
C+ 225/11–13
ROSA VILLOSA agg.
R. villosa L., R. tomentosa Sm. & R. sherardii Davies
Downy Rose
• All records
KM
MILES
C+ 225/14–17
ROSA RUBIGINOSA agg.
R. rubiginosa L., R. micrantha Borrer, R. elliptica Tausch & R. agrestis Savi
Sweet-briar
× Recorded introduction
• All other records
C 226/1
PRUNUS SPINOSA L.
Blackthorn
• All records
C 226/2
•PRUNUS DOMESTICA L.
Wild Plum
• All records

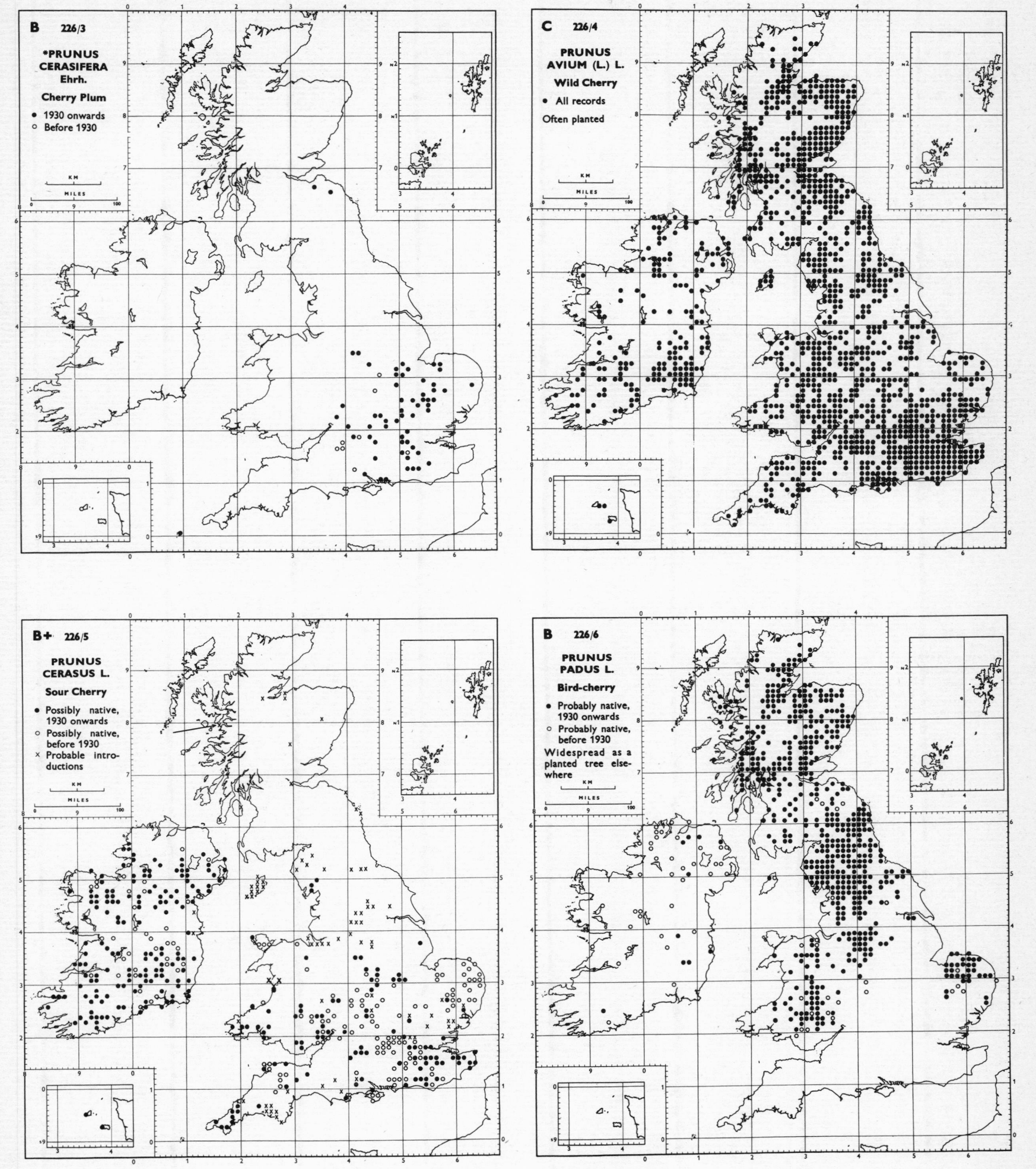
B 226/3
*PRUNUS CERASIFERA Ehrh.
Cherry Plum
● 1930 onwards
○ Before 1930
C 226/4
PRUNUS AVIUM (L.) L.
Wild Cherry
● All records
Often planted
B+ 226/5
PRUNUS CERASUS L.
Sour Cherry
● Possibly native, 1930 onwards
○ Possibly native, before 1930
× Probable introductions
B 226/6
PRUNUS PADUS L.
Bird-cherry
● Probably native, 1930 onwards
○ Probably native, before 1930
Widespread as a planted tree elsewhere

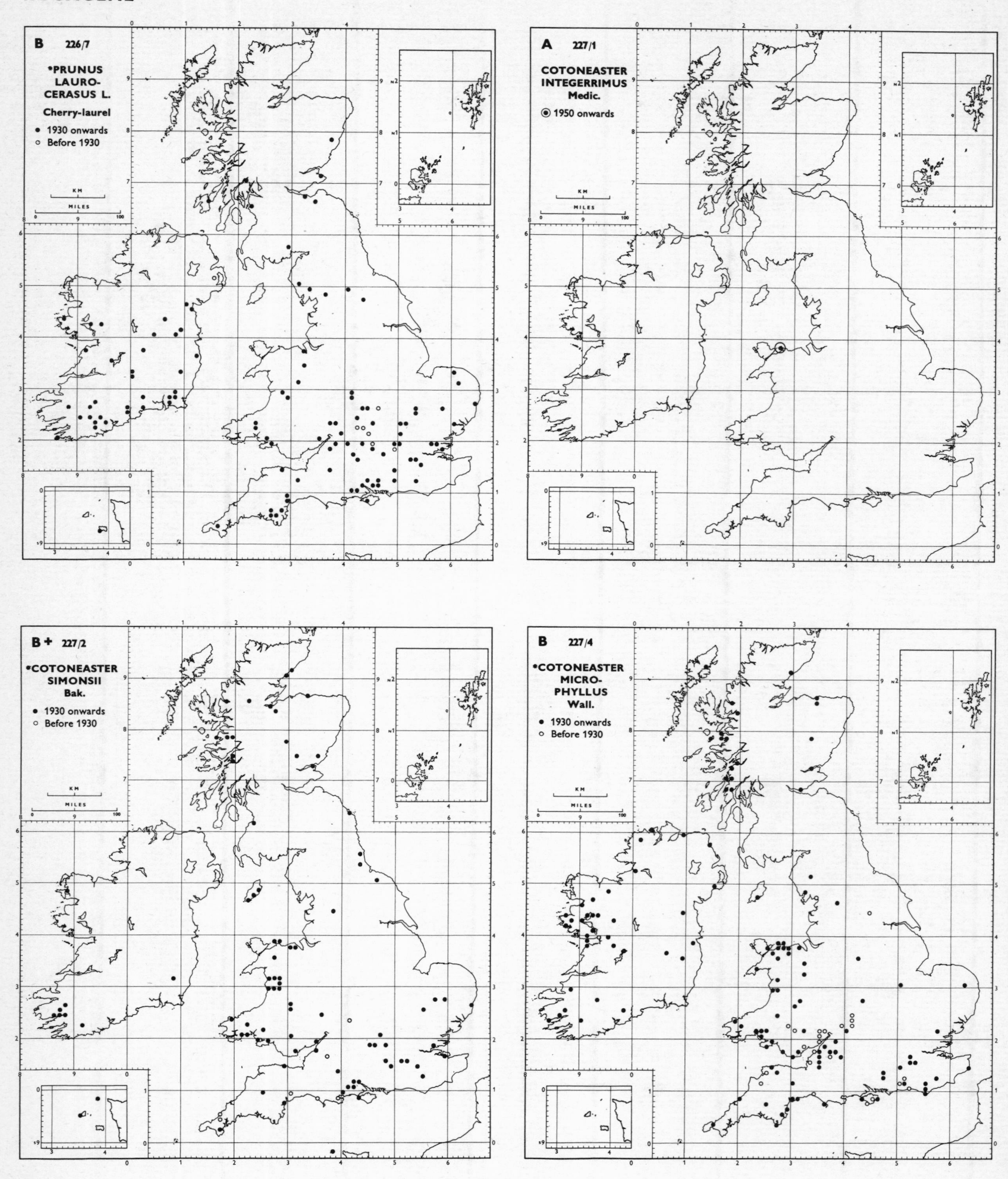
B 226/7
*PRUNUS LAURO-CERASUS L.
Cherry-laurel
● 1930 onwards
○ Before 1930
A 227/1
COTONEASTER INTEGERRIMUS Medic.
◉ 1950 onwards
B+ 227/2
*COTONEASTER SIMONSII Bak.
● 1930 onwards
○ Before 1930
B 227/4
*COTONEASTER MICRO-PHYLLUS Wall.
● 1930 onwards
○ Before 1930
KM
MILES

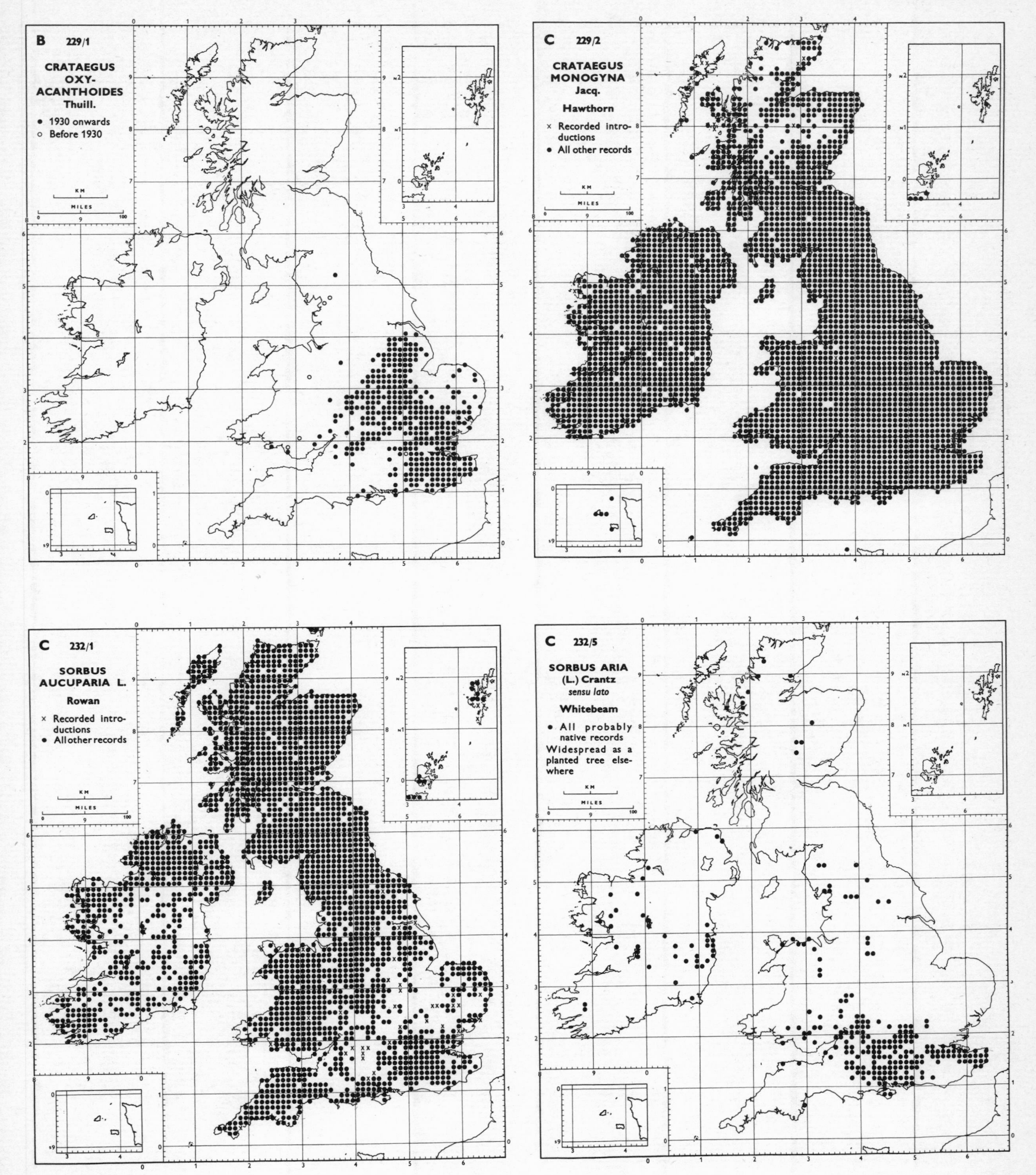
B 229/1
CRATAEGUS OXY-ACANTHOIDES Thuill.
• 1930 onwards
○ Before 1930
C 229/2
CRATAEGUS MONOGYNA Jacq.
Hawthorn
× Recorded introductions
• All other records
C 232/1
SORBUS AUCUPARIA L.
Rowan
× Recorded introductions
• All other records
C 232/5
SORBUS ARIA (L.) Crantz
sensu lato
Whitebeam
• All probably native records
Widespread as a planted tree elsewhere

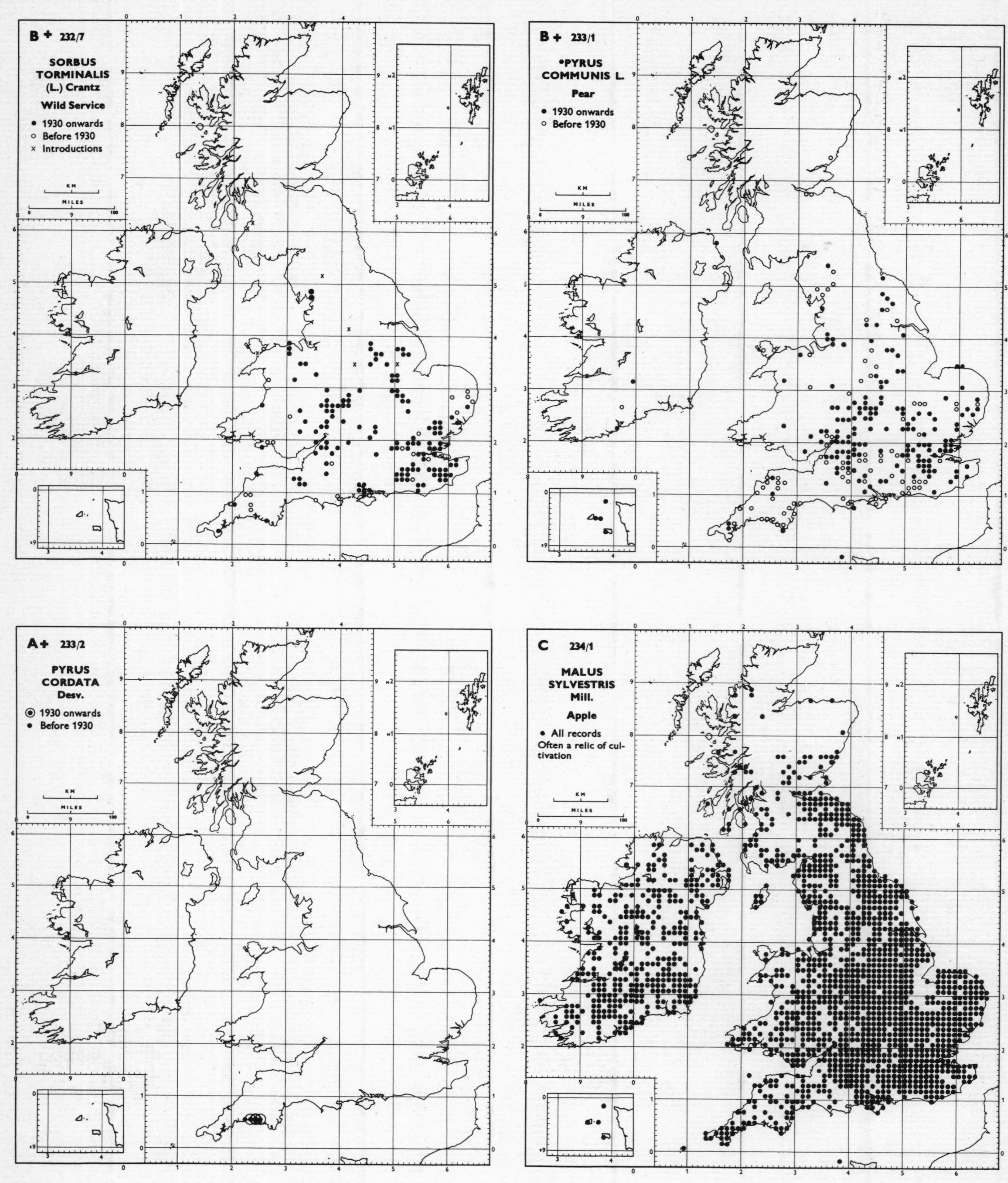
B + 232/7
SORBUS TORMINALIS (L.) Crantz
Wild Service
• 1930 onwards
○ Before 1930
× Introductions
B + 233/1
*PYRUS COMMUNIS L.
Pear
• 1930 onwards
○ Before 1930
A+ 233/2
PYRUS CORDATA Desv.
◉ 1930 onwards
• Before 1930
C 234/1
MALUS SYLVESTRIS Mill.
Apple
• All records
Often a relic of cultivation
KM
MILES

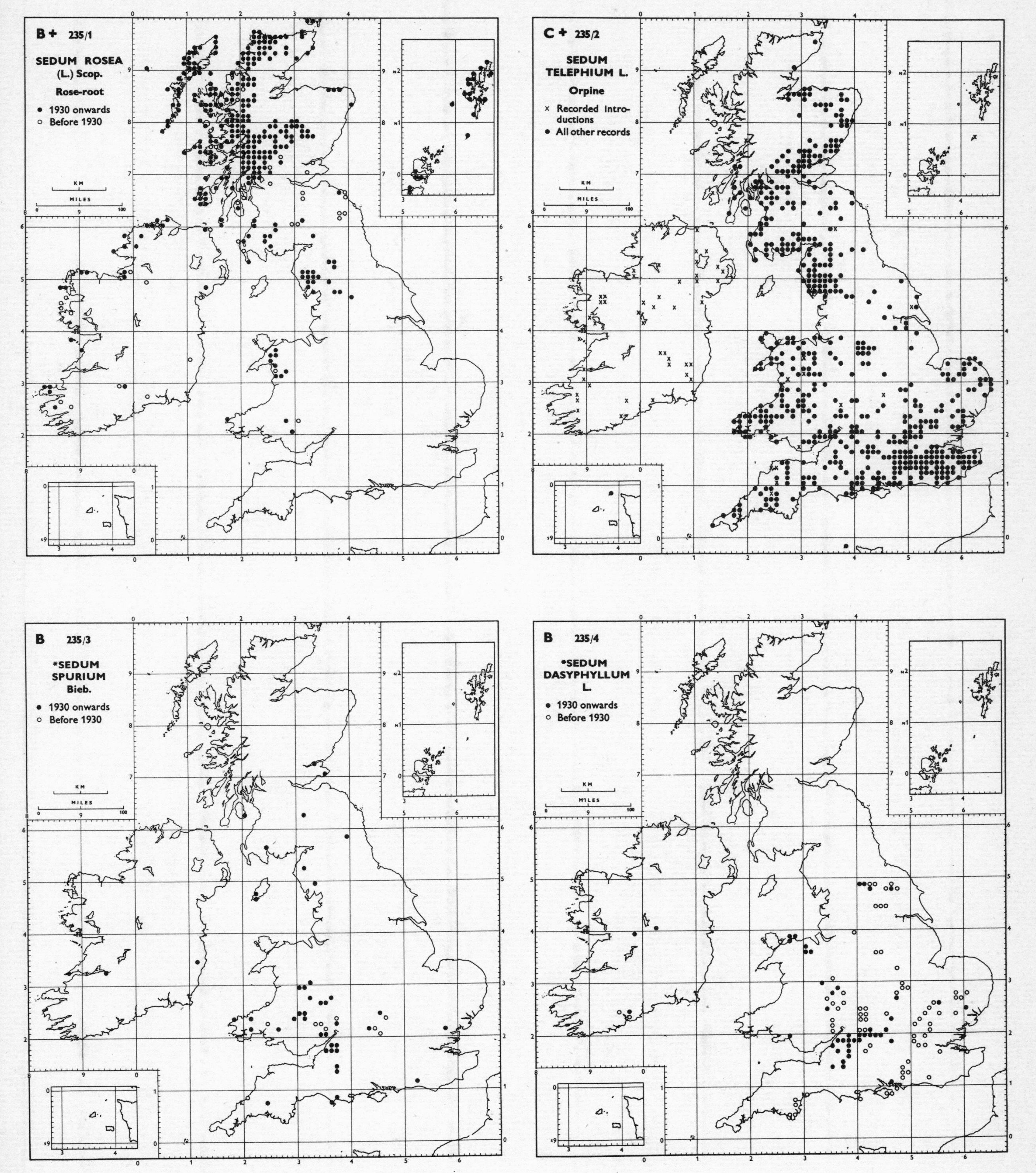
B + 235/1
SEDUM ROSEA (L.) Scop.
Rose-root
1930 onwards
Before 1930
C + 235/2
SEDUM TELEPHIUM L.
Orpine
Recorded introductions
All other records
B 235/3
*SEDUM SPURIUM Bieb.
1930 onwards
Before 1930
B 235/4
*SEDUM DASYPHYLLUM L.
1930 onwards
Before 1930
KM
MILES

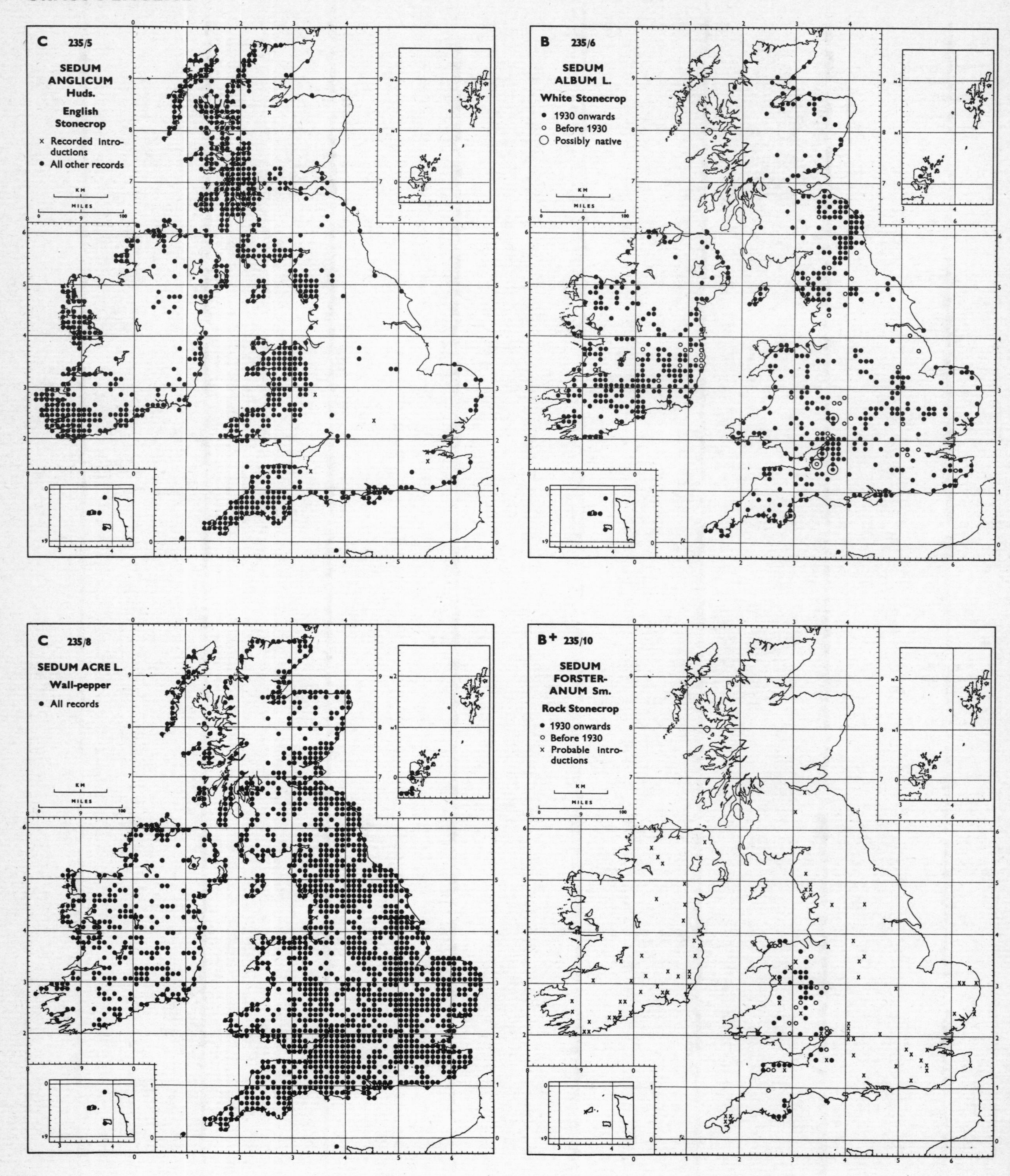
C 235/5
SEDUM ANGLICUM Huds.
English Stonecrop
× Recorded introductions
• All other records
KM
MILES
B 235/6
SEDUM ALBUM L.
White Stonecrop
• 1930 onwards
○ Before 1930
◯ Possibly native
C 235/8
SEDUM ACRE L.
Wall-pepper
• All records
B+ 235/10
SEDUM FORSTERANUM Sm.
Rock Stonecrop
• 1930 onwards
○ Before 1930
× Probable introductions

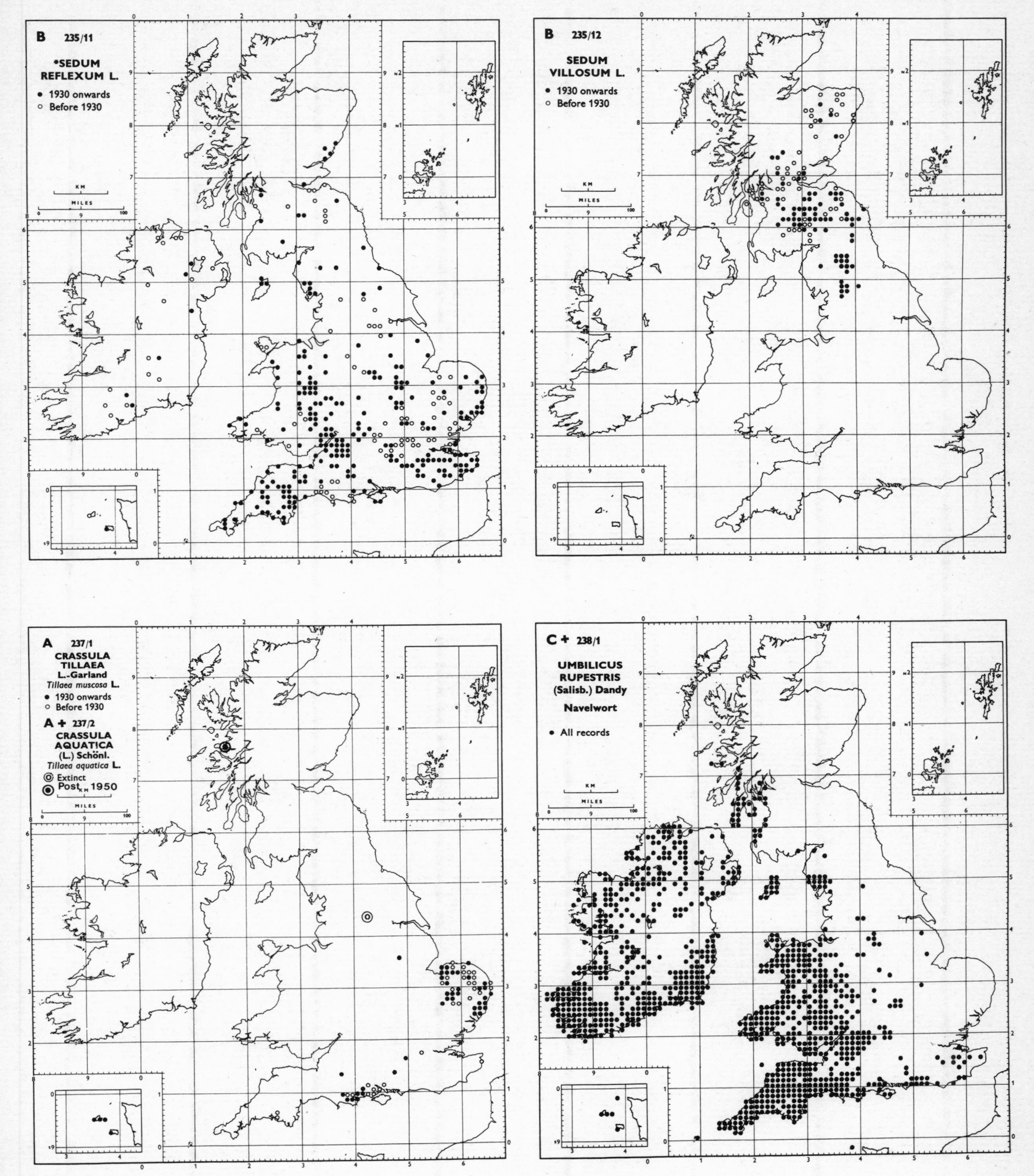
B 235/11
*SEDUM REFLEXUM L.
• 1930 onwards
○ Before 1930
B 235/12
SEDUM VILLOSUM L.
• 1930 onwards
○ Before 1930
A 237/1
CRASSULA TILLAEA L.-Garland
Tillaea muscosa L.
• 1930 onwards
○ Before 1930
A + 237/2
CRASSULA AQUATICA (L.) Schönl.
Tillaea aquatica L.
◎ Extinct
◉ Post 1950
C + 238/1
UMBILICUS RUPESTRIS (Salisb.) Dandy
Navelwort
• All records

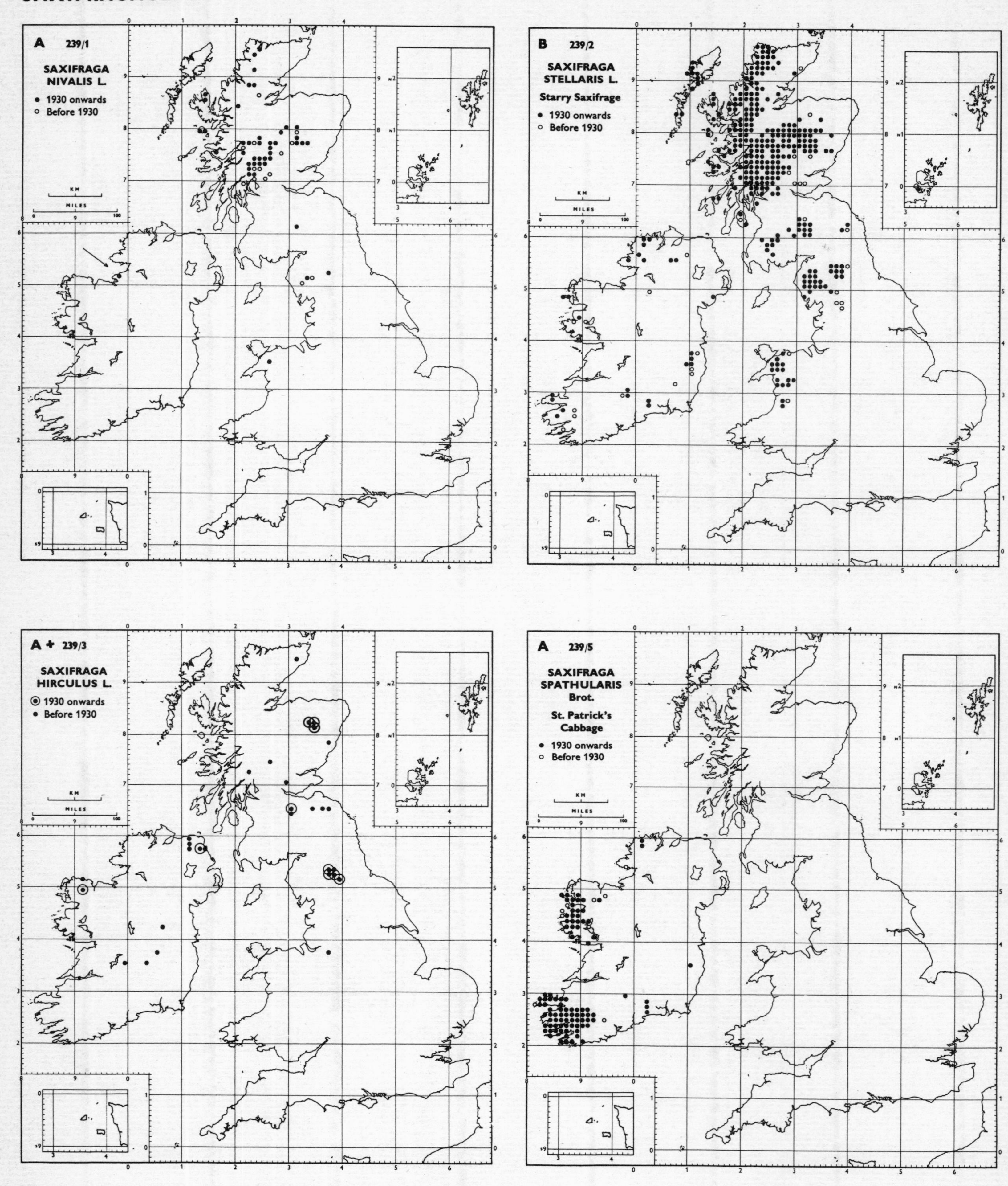
A 239/1
SAXIFRAGA NIVALIS L.
1930 onwards
Before 1930
B 239/2
SAXIFRAGA STELLARIS L.
Starry Saxifrage
1930 onwards
Before 1930
A + 239/3
SAXIFRAGA HIRCULUS L.
1930 onwards
Before 1930
A 239/5
SAXIFRAGA SPATHULARIS Brot.
St. Patrick's Cabbage
1930 onwards
Before 1930
KM
MILES

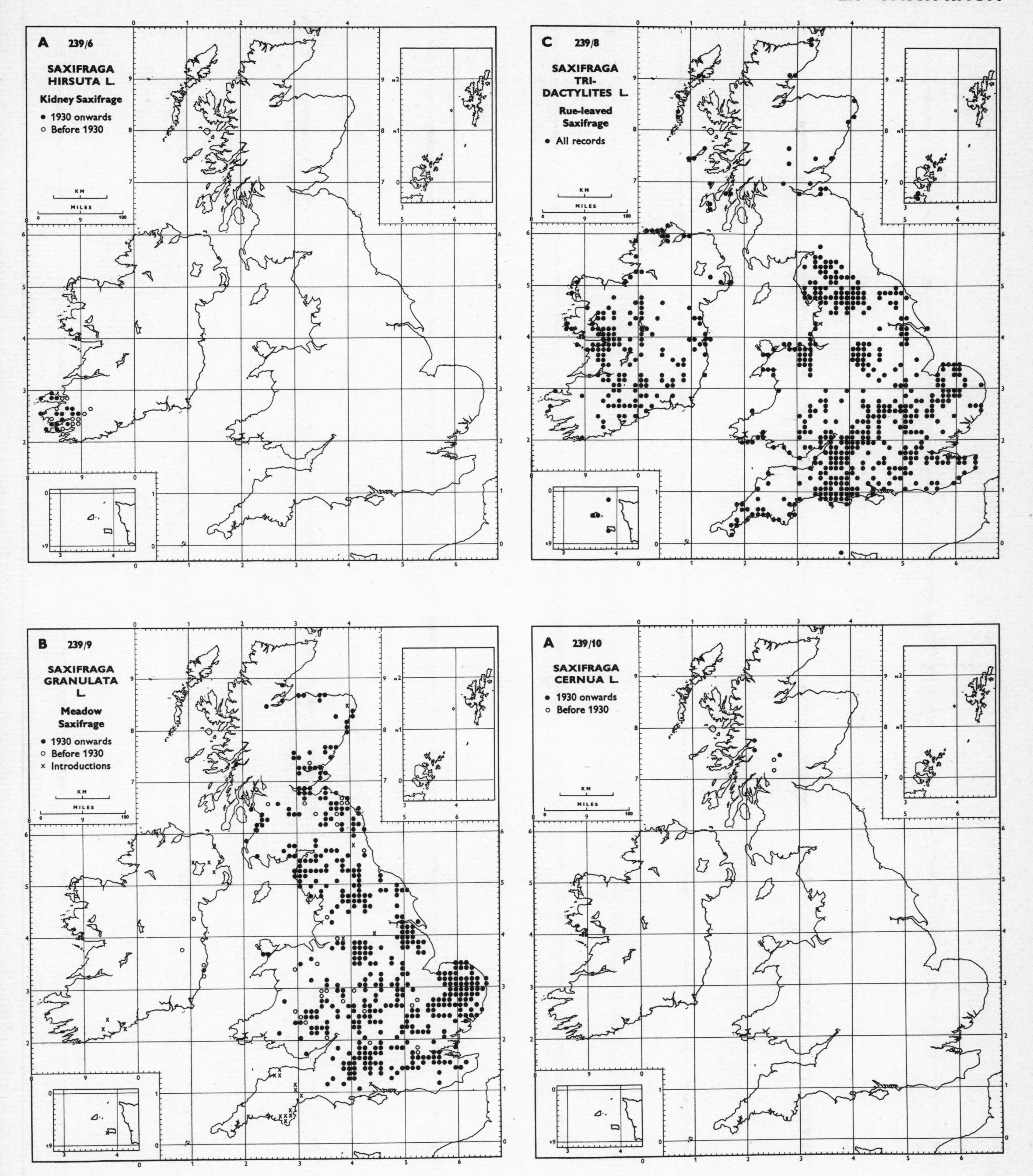
A 239/6
SAXIFRAGA HIRSUTA L.
Kidney Saxifrage
• 1930 onwards
○ Before 1930
C 239/8
SAXIFRAGA TRI-DACTYLITES L.
Rue-leaved Saxifrage
• All records
B 239/9
SAXIFRAGA GRANULATA L.
Meadow Saxifrage
• 1930 onwards
○ Before 1930
x Introductions
A 239/10
SAXIFRAGA CERNUA L.
• 1930 onwards
○ Before 1930
KM
MILES

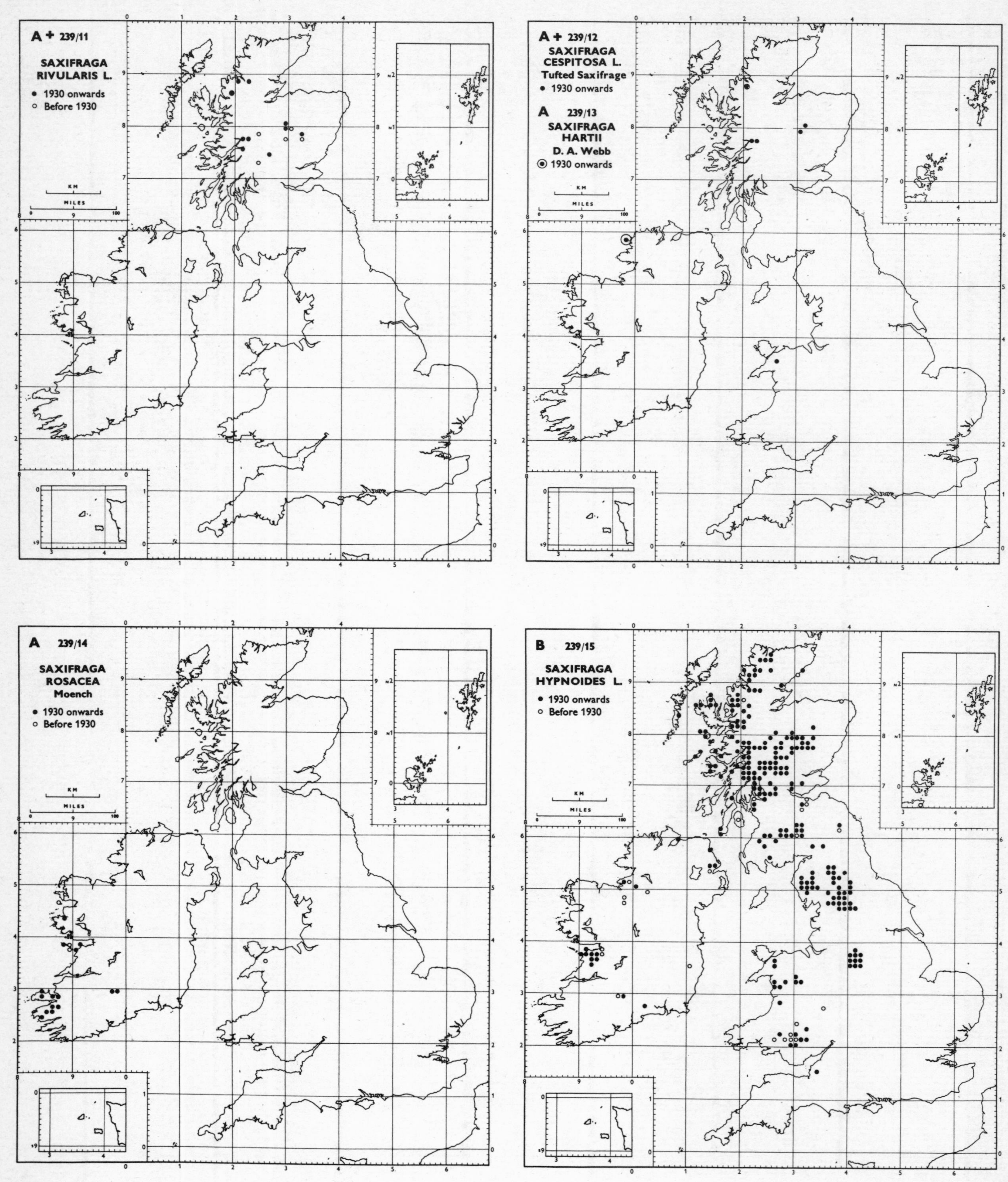
A + 239/11
SAXIFRAGA RIVULARIS L.
• 1930 onwards
○ Before 1930
A + 239/12
SAXIFRAGA CESPITOSA L.
Tufted Saxifrage
• 1930 onwards
A 239/13
SAXIFRAGA HARTII
D. A. Webb
◉ 1930 onwards
A 239/14
SAXIFRAGA ROSACEA
Moench
• 1930 onwards
○ Before 1930
B 239/15
SAXIFRAGA HYPNOIDES L.
• 1930 onwards
○ Before 1930
KM
MILES

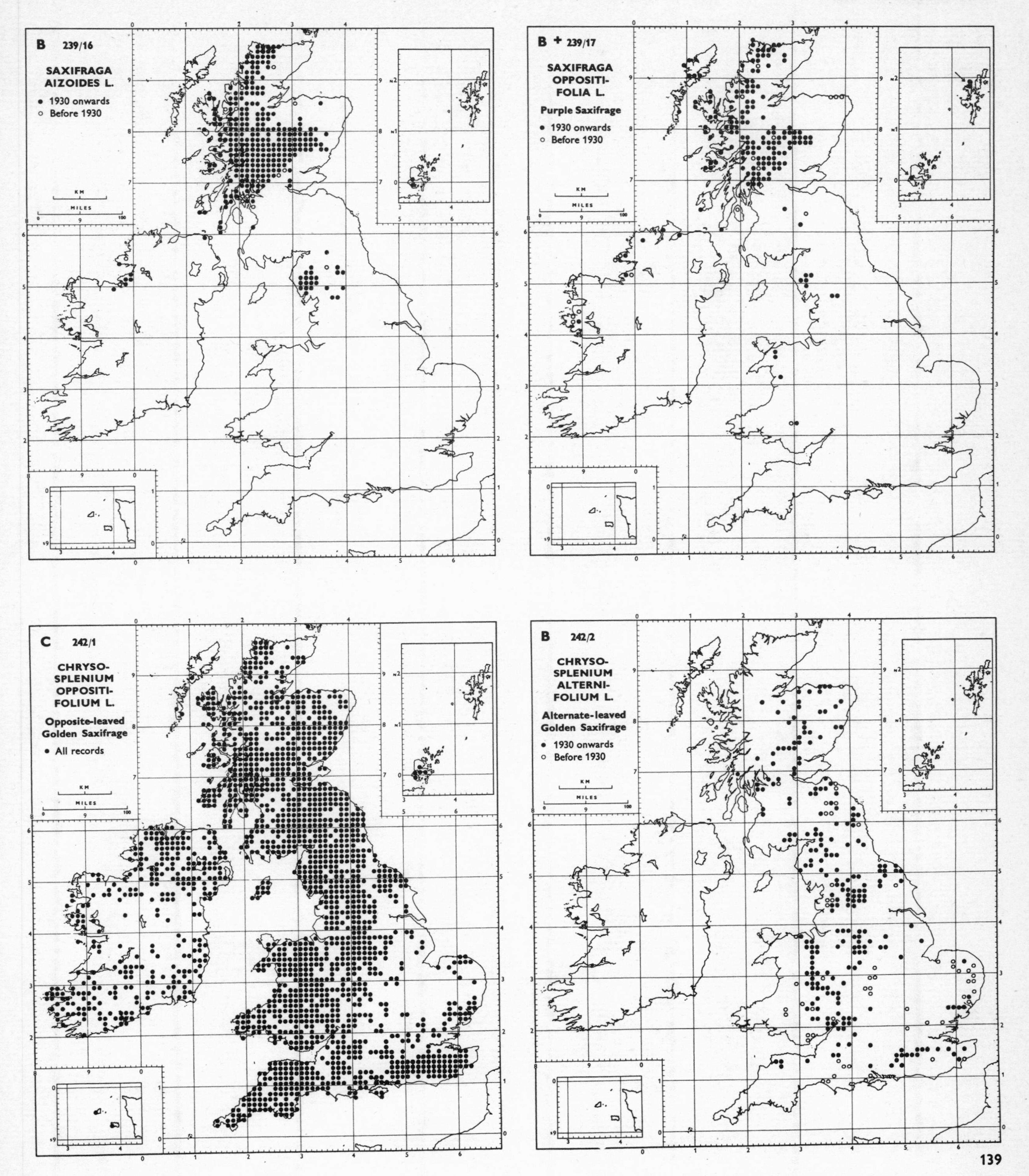
B 239/16
SAXIFRAGA AIZOIDES L.
• 1930 onwards
○ Before 1930
B + 239/17
SAXIFRAGA OPPOSITIFOLIA L.
Purple Saxifrage
• 1930 onwards
○ Before 1930
C 242/1
CHRYSOSPLENIUM OPPOSITIFOLIUM L.
Opposite-leaved Golden Saxifrage
• All records
B 242/2
CHRYSOSPLENIUM ALTERNIFOLIUM L.
Alternate-leaved Golden Saxifrage
• 1930 onwards
○ Before 1930

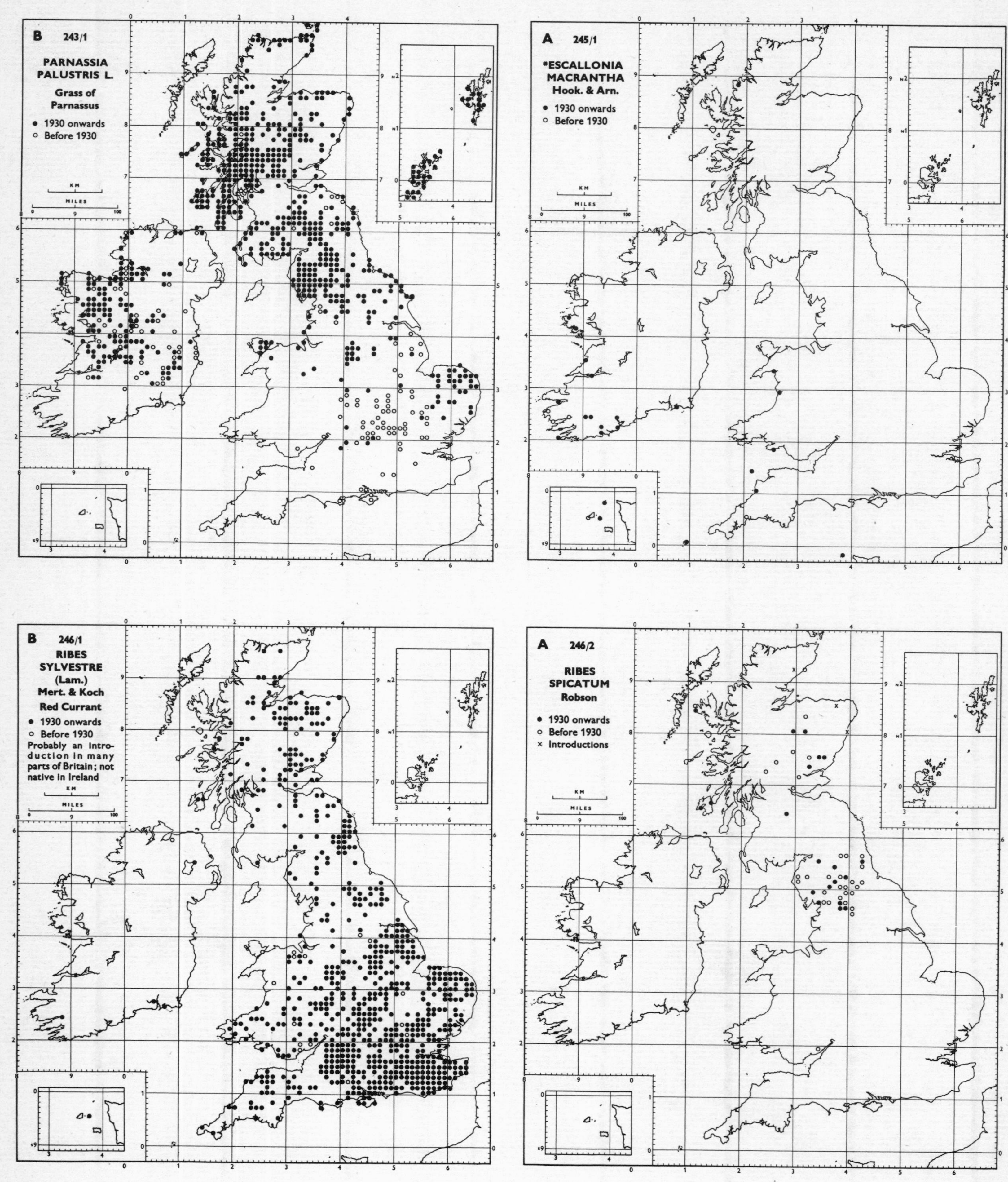
B 243/1
PARNASSIA PALUSTRIS L.
Grass of Parnassus
● 1930 onwards
○ Before 1930
A 245/1
*ESCALLONIA MACRANTHA Hook. & Arn.
● 1930 onwards
○ Before 1930
B 246/1
RIBES SYLVESTRE (Lam.) Mert. & Koch
Red Currant
● 1930 onwards
○ Before 1930
Probably an introduction in many parts of Britain; not native in Ireland
A 246/2
RIBES SPICATUM Robson
● 1930 onwards
○ Before 1930
× Introductions
KM
MILES

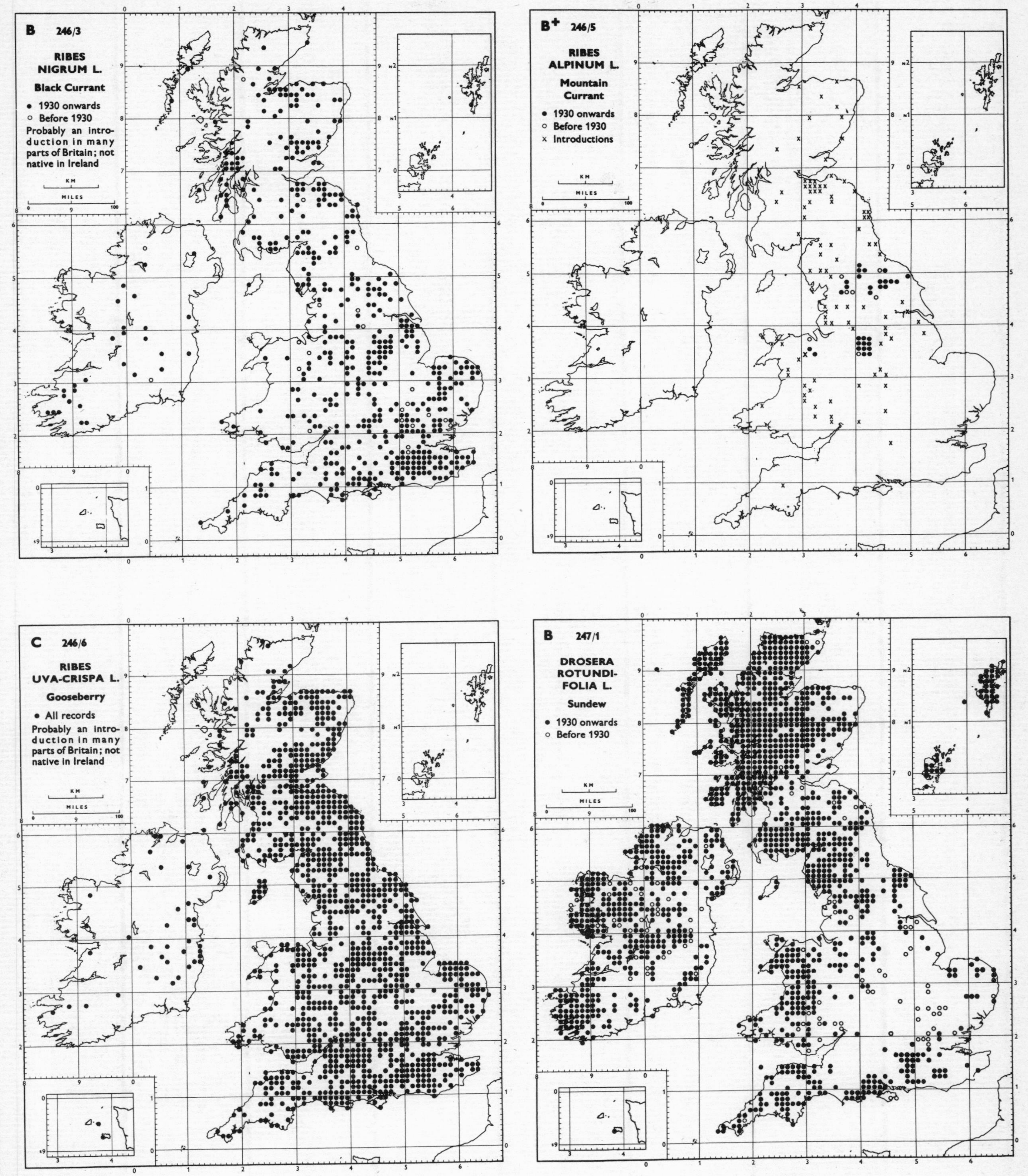
B 246/3
RIBES NIGRUM L.
Black Currant
• 1930 onwards
○ Before 1930
Probably an introduction in many parts of Britain; not native in Ireland
B+ 246/5
RIBES ALPINUM L.
Mountain Currant
• 1930 onwards
○ Before 1930
x Introductions
C 246/6
RIBES UVA-CRISPA L.
Gooseberry
• All records
Probably an introduction in many parts of Britain; not native in Ireland
B 247/1
DROSERA ROTUNDIFOLIA L.
Sundew
• 1930 onwards
○ Before 1930
KM
MILES

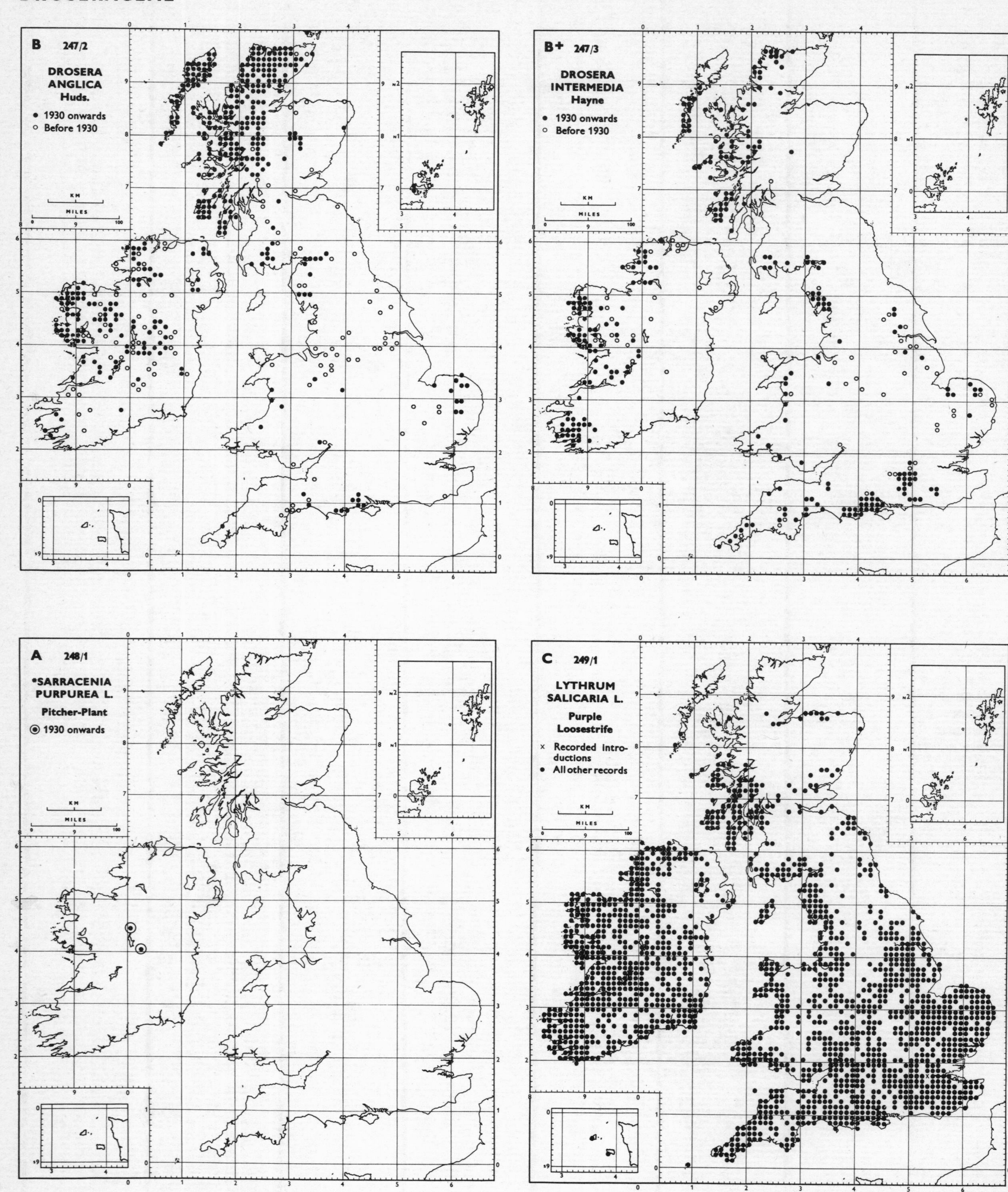
B 247/2
DROSERA ANGLICA Huds.
• 1930 onwards
○ Before 1930
B+ 247/3
DROSERA INTERMEDIA Hayne
• 1930 onwards
○ Before 1930
A 248/1
*SARRACENIA PURPUREA L.
Pitcher-Plant
◉ 1930 onwards
C 249/1
LYTHRUM SALICARIA L.
Purple Loosestrife
× Recorded introductions
• All other records

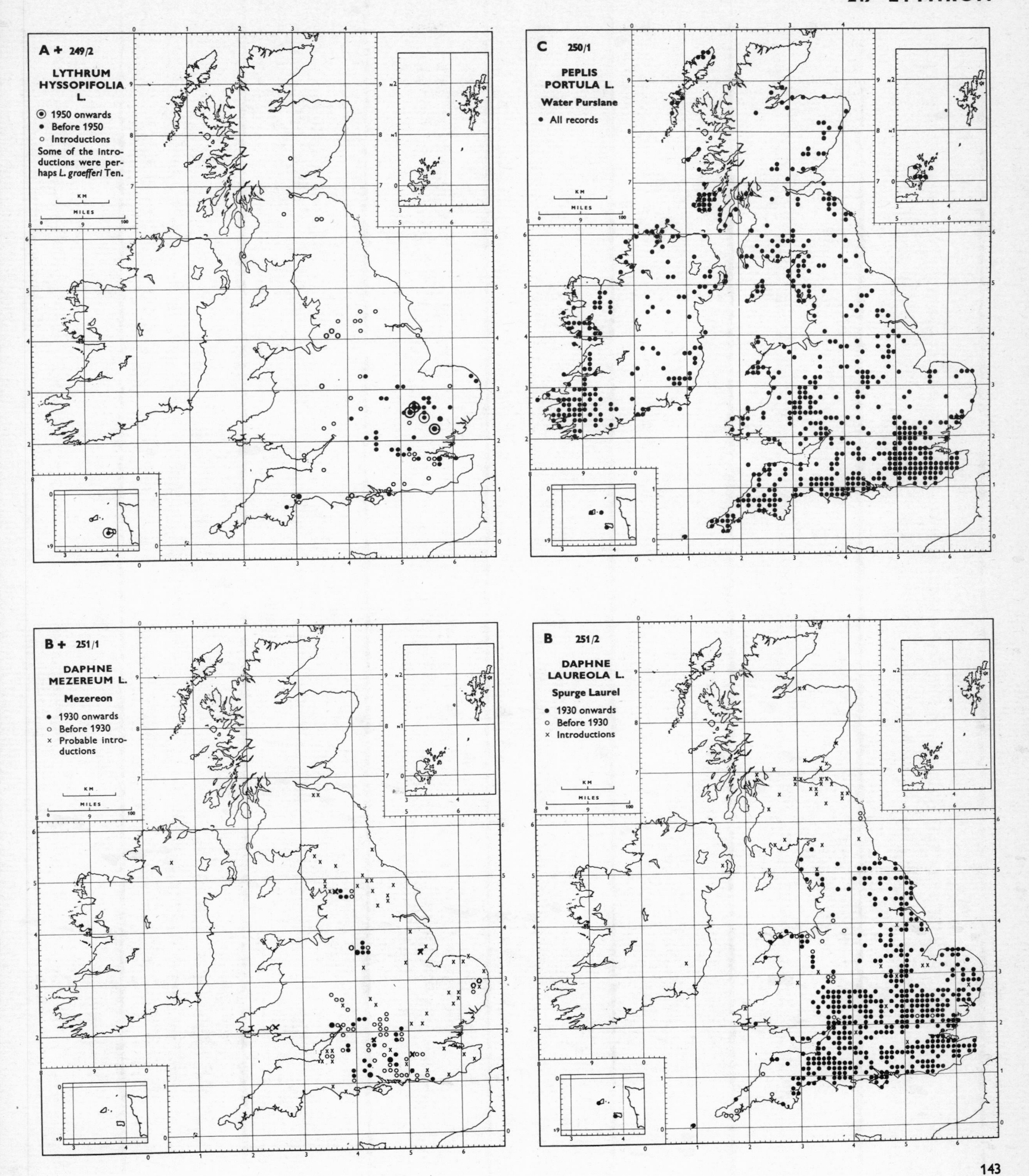
A + 249/2
LYTHRUM HYSSOPIFOLIA L.
1950 onwards
Before 1950
Introductions
Some of the introductions were perhaps L. graefferi Ten.
C 250/1
PEPLIS PORTULA L.
Water Purslane
All records
B + 251/1
DAPHNE MEZEREUM L.
Mezereon
1930 onwards
Before 1930
Probable introductions
B 251/2
DAPHNE LAUREOLA L.
Spurge Laurel
1930 onwards
Before 1930
Introductions

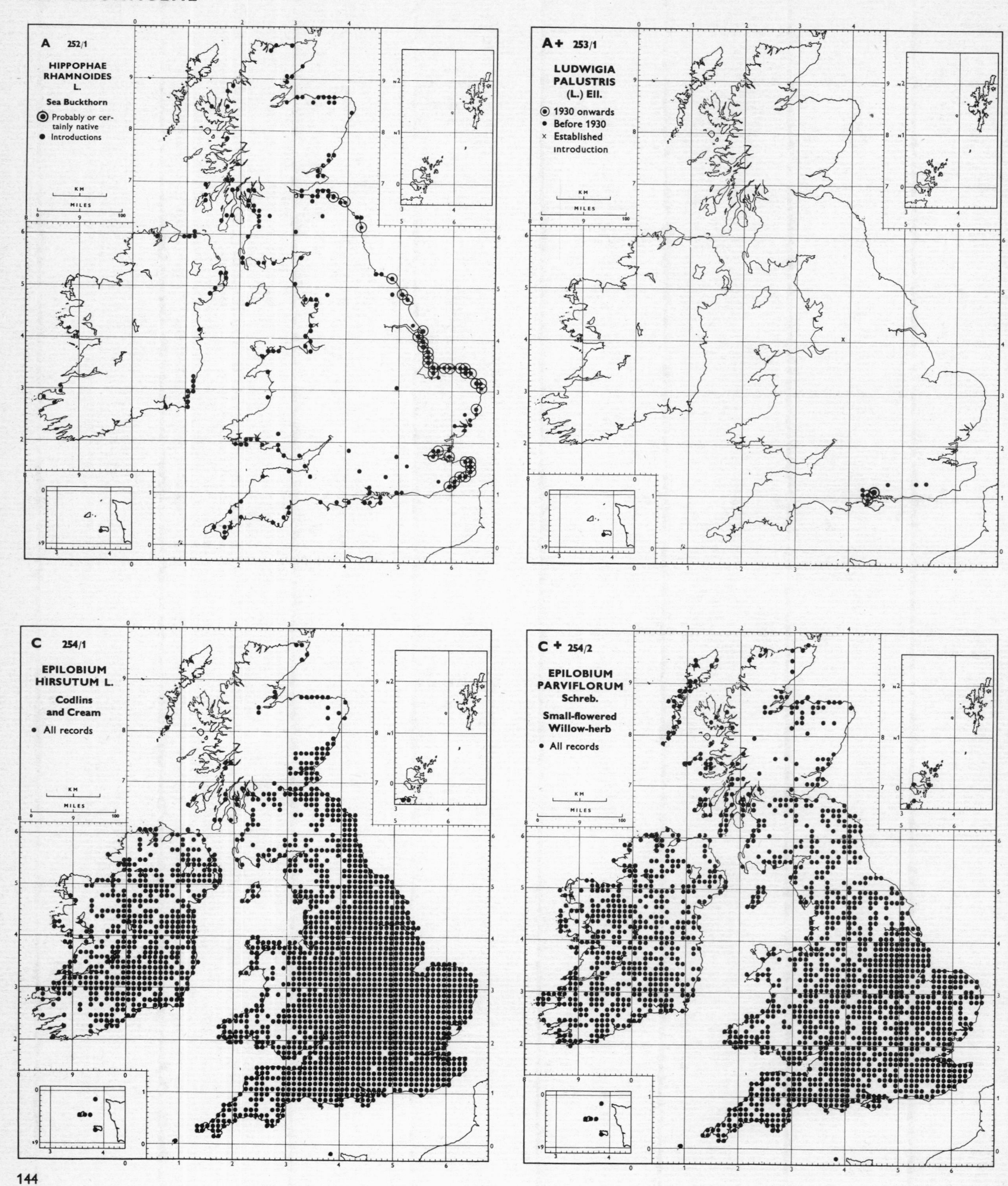
A 252/1
HIPPOPHAE RHAMNOIDES L.
Sea Buckthorn
Probably or certainly native
Introductions
A+ 253/1
LUDWIGIA PALUSTRIS (L.) Ell.
1930 onwards
Before 1930
Established introduction
C 254/1
EPILOBIUM HIRSUTUM L.
Codlins and Cream
All records
C+ 254/2
EPILOBIUM PARVIFLORUM Schreb.
Small-flowered Willow-herb
All records
KM
MILES

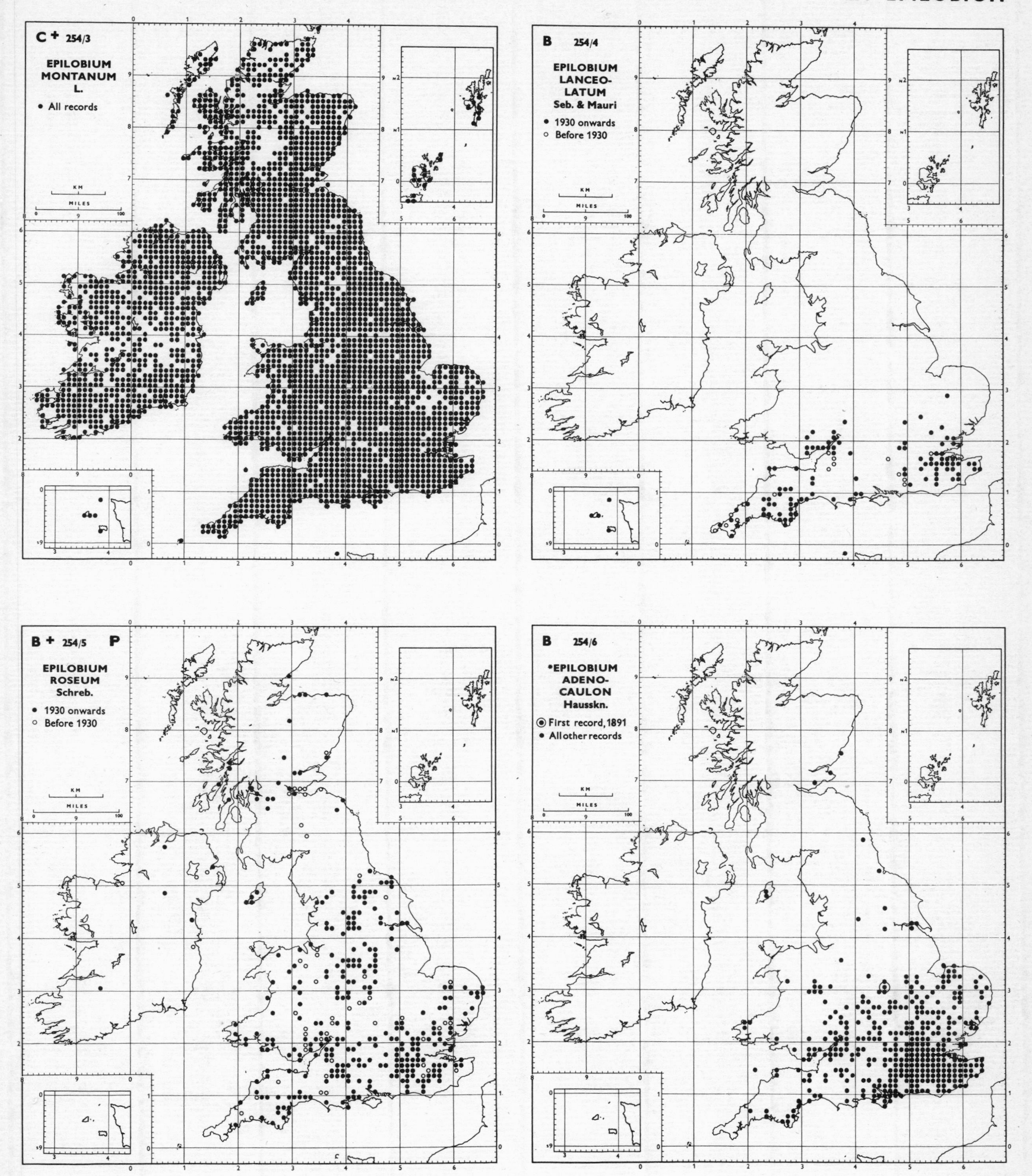
C + 254/3
EPILOBIUM MONTANUM L.
• All records
B 254/4
EPILOBIUM LANCEOLATUM Seb. & Mauri
• 1930 onwards
○ Before 1930
B + 254/5 P
EPILOBIUM ROSEUM Schreb.
• 1930 onwards
○ Before 1930
B 254/6
*EPILOBIUM ADENOCAULON Hausskn.
⦿ First record, 1891
• All other records
KM
MILES

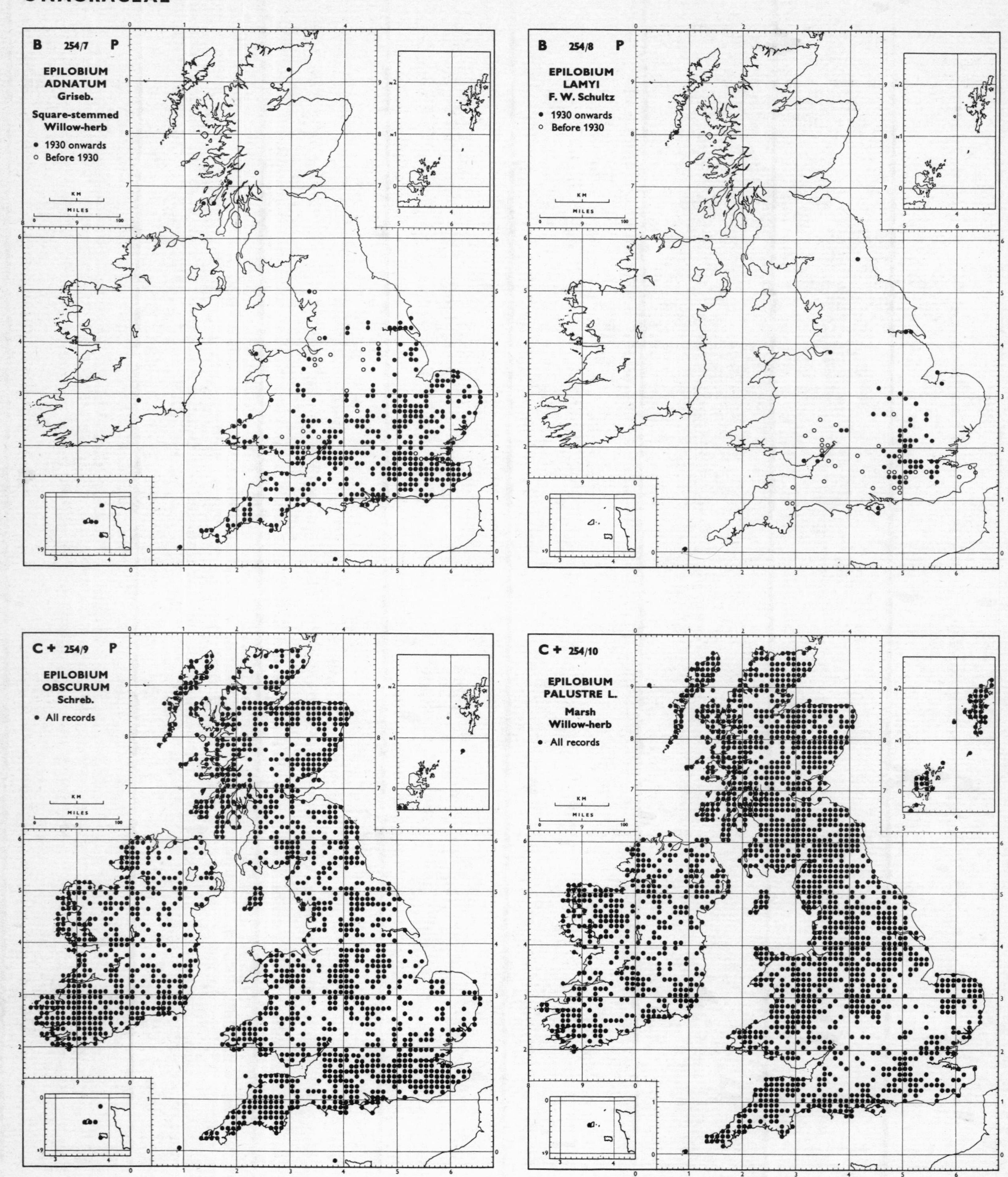
B 254/7 P
EPILOBIUM ADNATUM Griseb.
Square-stemmed Willow-herb
• 1930 onwards
○ Before 1930
KM
MILES
B 254/8 P
EPILOBIUM LAMYI F. W. Schultz
• 1930 onwards
○ Before 1930
C + 254/9 P
EPILOBIUM OBSCURUM Schreb.
• All records
C + 254/10
EPILOBIUM PALUSTRE L.
Marsh Willow-herb
• All records

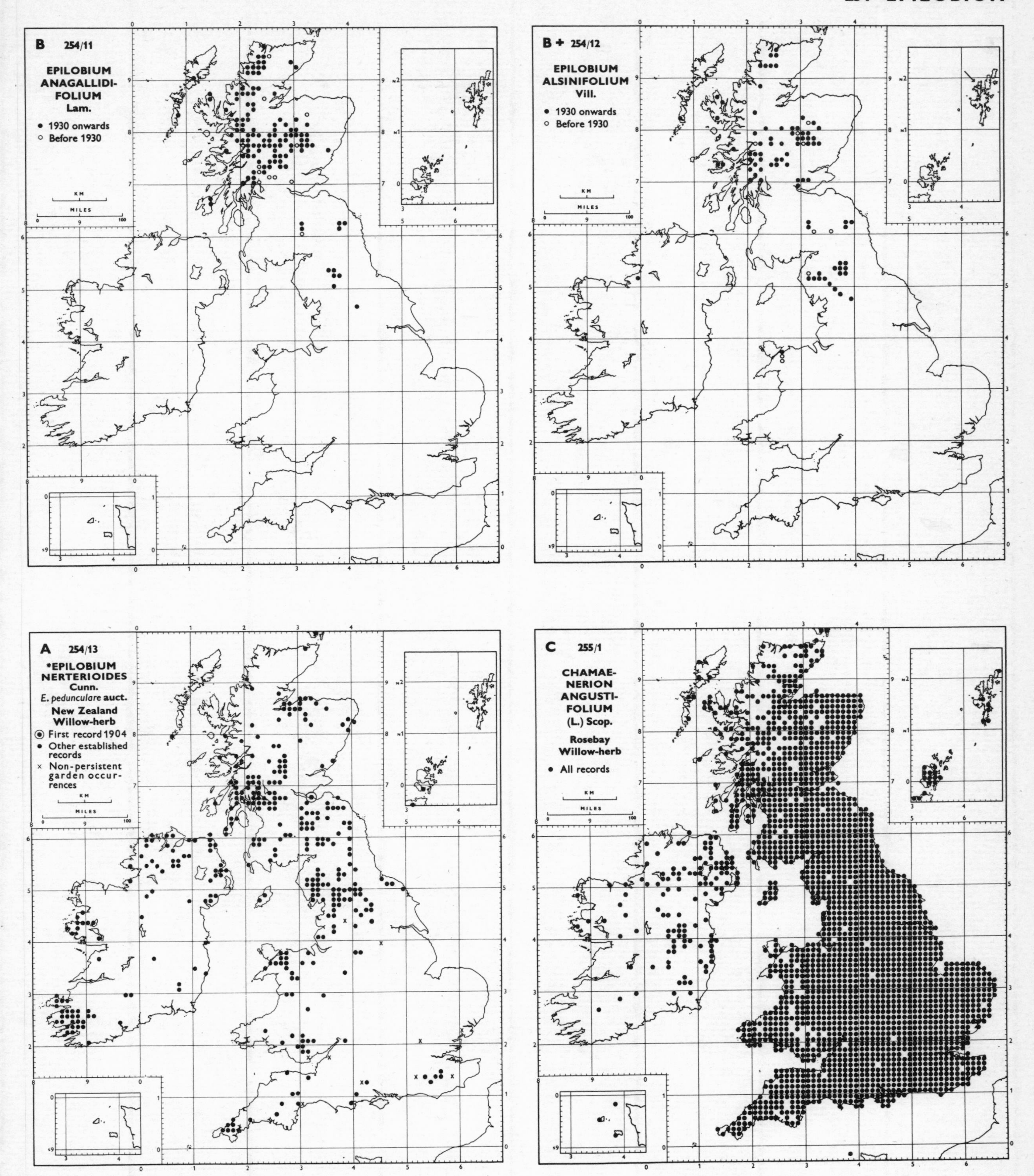
B 254/11
EPILOBIUM ANAGALLIDIFOLIUM Lam.
• 1930 onwards
○ Before 1930
B + 254/12
EPILOBIUM ALSINIFOLIUM Vill.
• 1930 onwards
○ Before 1930
A 254/13
*EPILOBIUM NERTERIOIDES Cunn.
E. pedunculare auct.
New Zealand Willow-herb
First record 1904
Other established records
× Non-persistent garden occurrences
C 255/1
CHAMAENERION ANGUSTIFOLIUM (L.) Scop.
Rosebay Willow-herb
• All records
KM
MILES

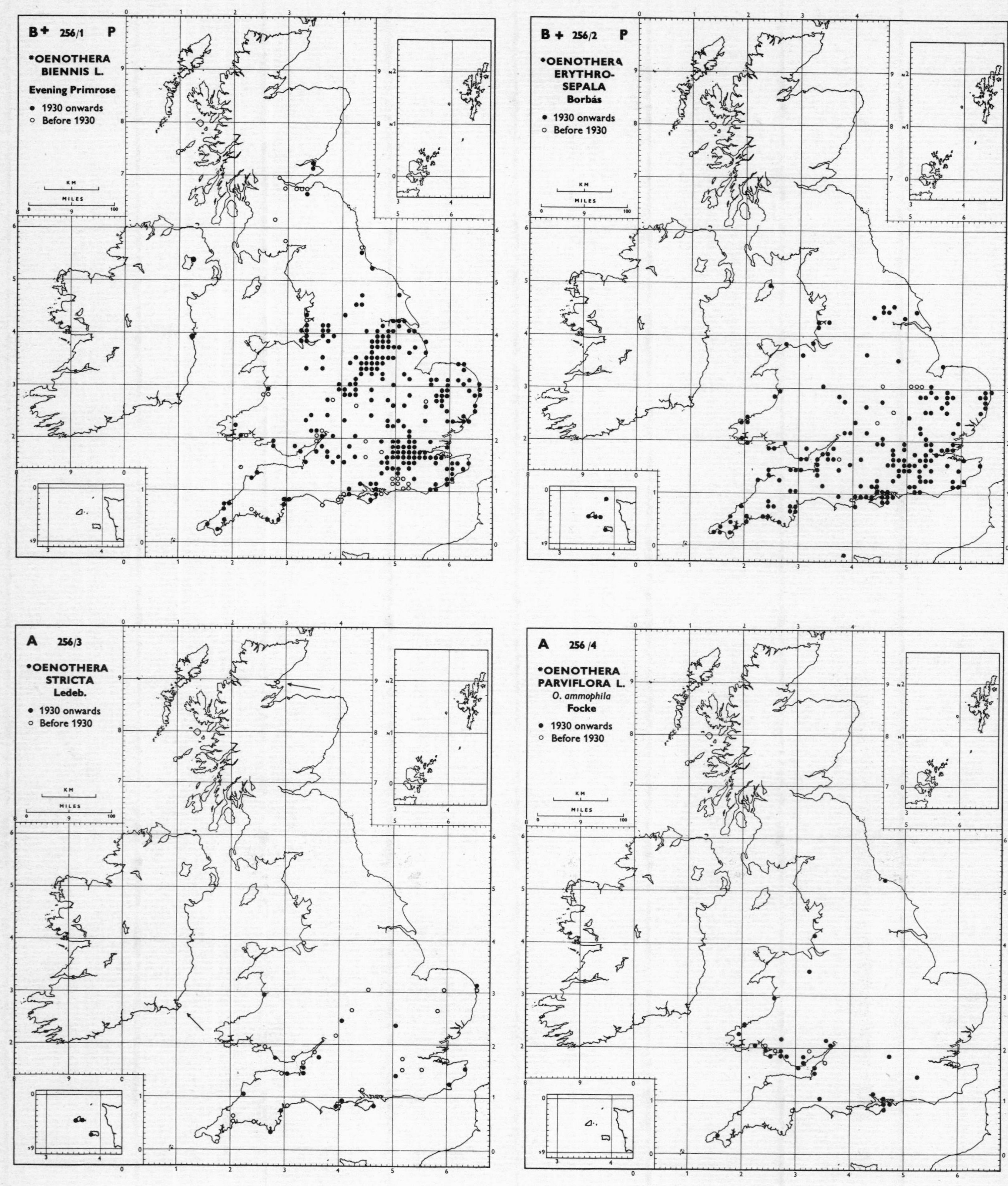
B+ 256/1 P
•OENOTHERA
BIENNIS L.
Evening Primrose
• 1930 onwards
○ Before 1930
KM
MILES
B + 256/2 P
•OENOTHERA
ERYTHRO-
SEPALA
Borbás
• 1930 onwards
○ Before 1930
KM
MILES
A 256/3
•OENOTHERA
STRICTA
Ledeb.
• 1930 onwards
○ Before 1930
KM
MILES
A 256/4
•OENOTHERA
PARVIFLORA L.
O. ammophila
Focke
• 1930 onwards
○ Before 1930
KM
MILES

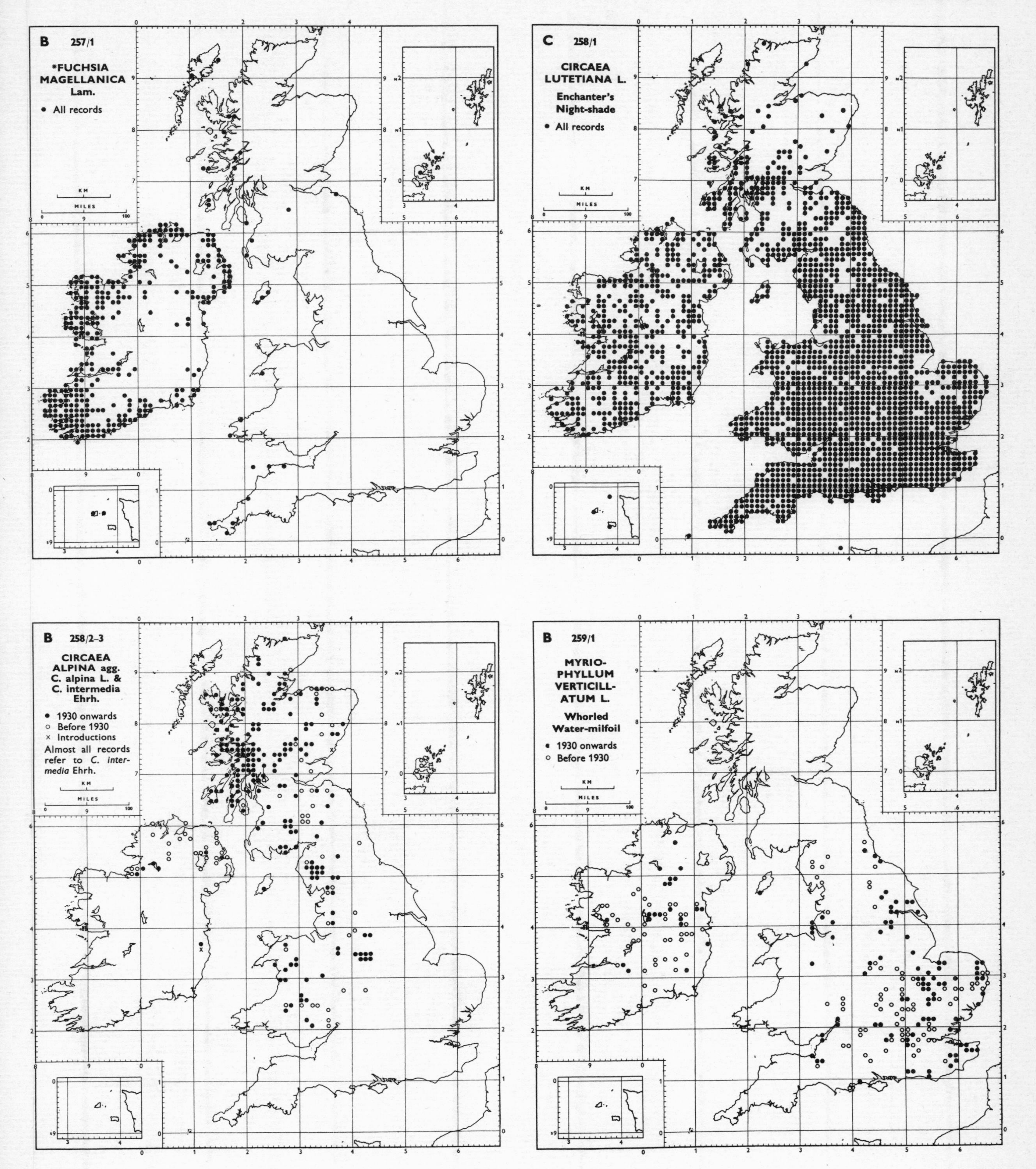
B 257/1
*FUCHSIA MAGELLANICA Lam.
• All records
C 258/1
CIRCAEA LUTETIANA L.
Enchanter's Night-shade
• All records
B 258/2-3
CIRCAEA ALPINA agg.
C. alpina L. & C. intermedia Ehrh.
• 1930 onwards
○ Before 1930
× Introductions
Almost all records refer to C. intermedia Ehrh.
B 259/1
MYRIOPHYLLUM VERTICILLATUM L.
Whorled Water-milfoil
• 1930 onwards
○ Before 1930

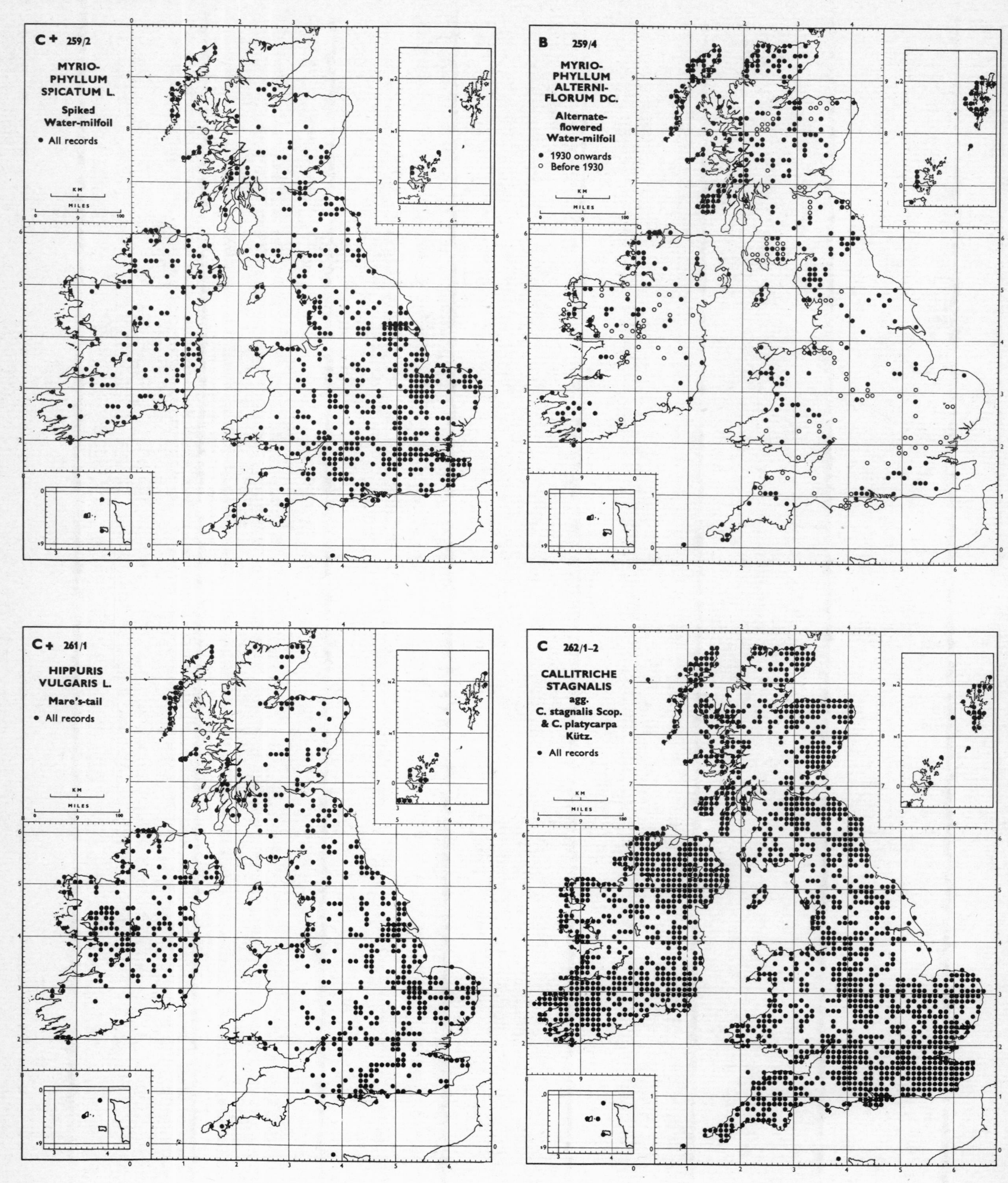

C+ 259/2
MYRIOPHYLLUM SPICATUM L.
Spiked Water-milfoil
• All records
B 259/4
MYRIOPHYLLUM ALTERNIFLORUM DC.
Alternate-flowered Water-milfoil
• 1930 onwards
○ Before 1930
C+ 261/1
HIPPURIS VULGARIS L.
Mare's-tail
• All records
C 262/1–2
CALLITRICHE STAGNALIS agg.
C. stagnalis Scop. & C. platycarpa Kütz.
• All records
KM
MILES

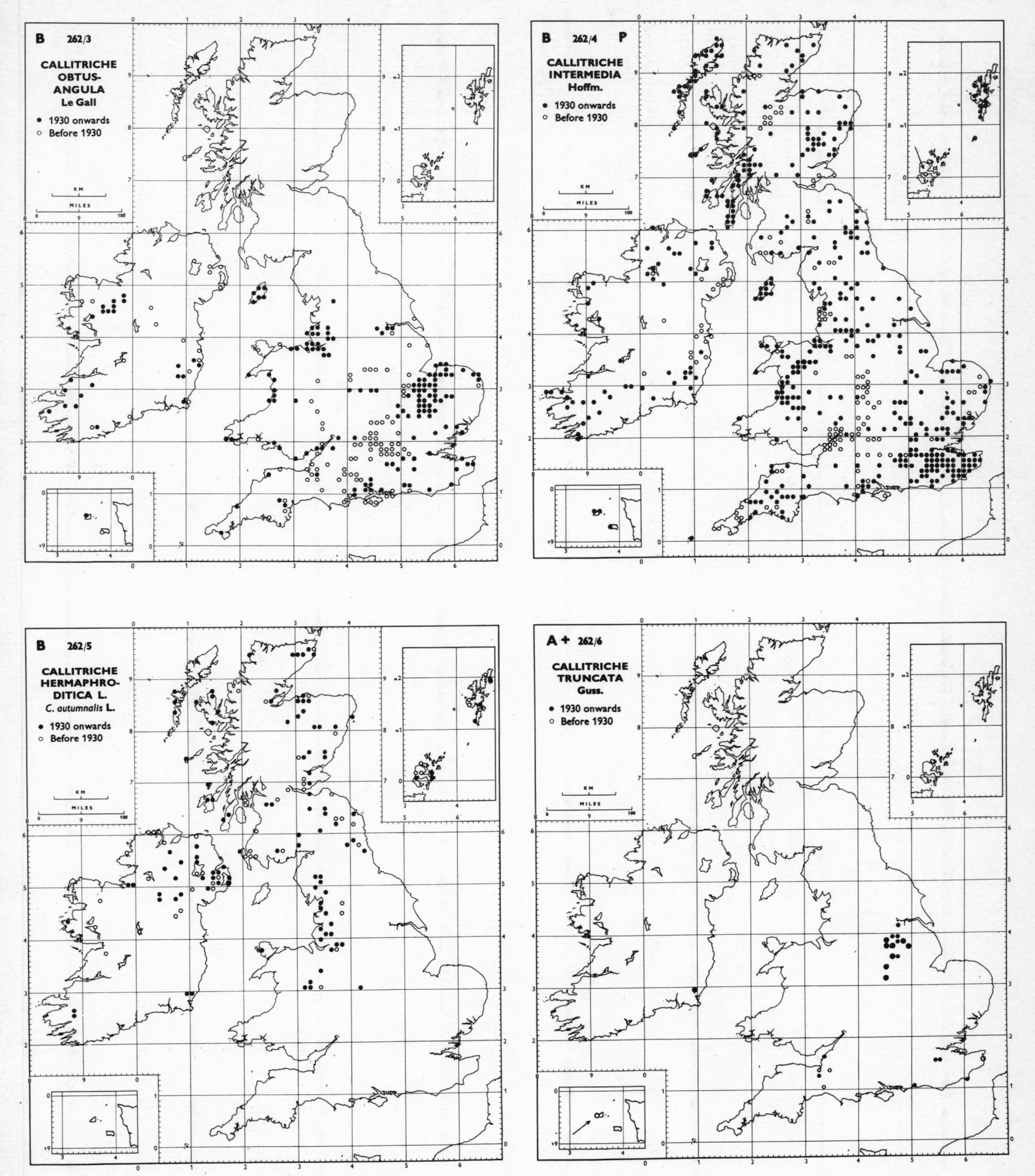
B 262/3
CALLITRICHE OBTUS-ANGULA Le Gall
• 1930 onwards
○ Before 1930
B 262/4 P
CALLITRICHE INTERMEDIA Hoffm.
• 1930 onwards
○ Before 1930
B 262/5
CALLITRICHE HERMAPHRODITICA L.
C. autumnalis L.
• 1930 onwards
○ Before 1930
A + 262/6
CALLITRICHE TRUNCATA Guss.
• 1930 onwards
○ Before 1930

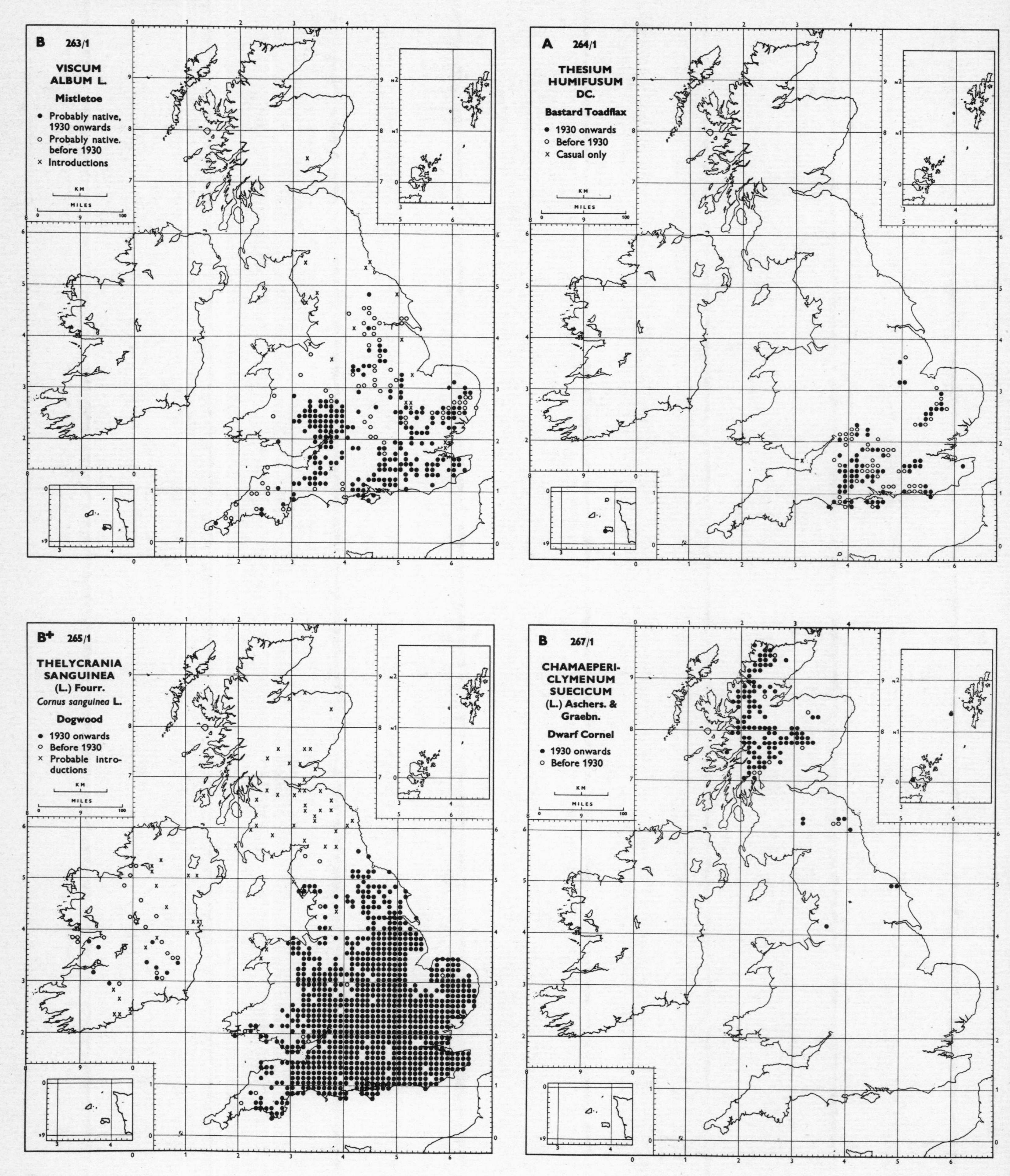
B 263/1
VISCUM ALBUM L.
Mistletoe
• Probably native, 1930 onwards
○ Probably native, before 1930
× Introductions
A 264/1
THESIUM HUMIFUSUM DC.
Bastard Toadflax
• 1930 onwards
○ Before 1930
× Casual only
B+ 265/1
THELYCRANIA SANGUINEA (L.) Fourr.
Cornus sanguinea L.
Dogwood
• 1930 onwards
○ Before 1930
× Probable introductions
B 267/1
CHAMAEPERICLYMENUM SUECICUM (L.) Aschers. & Graebn.
Dwarf Cornel
• 1930 onwards
○ Before 1930

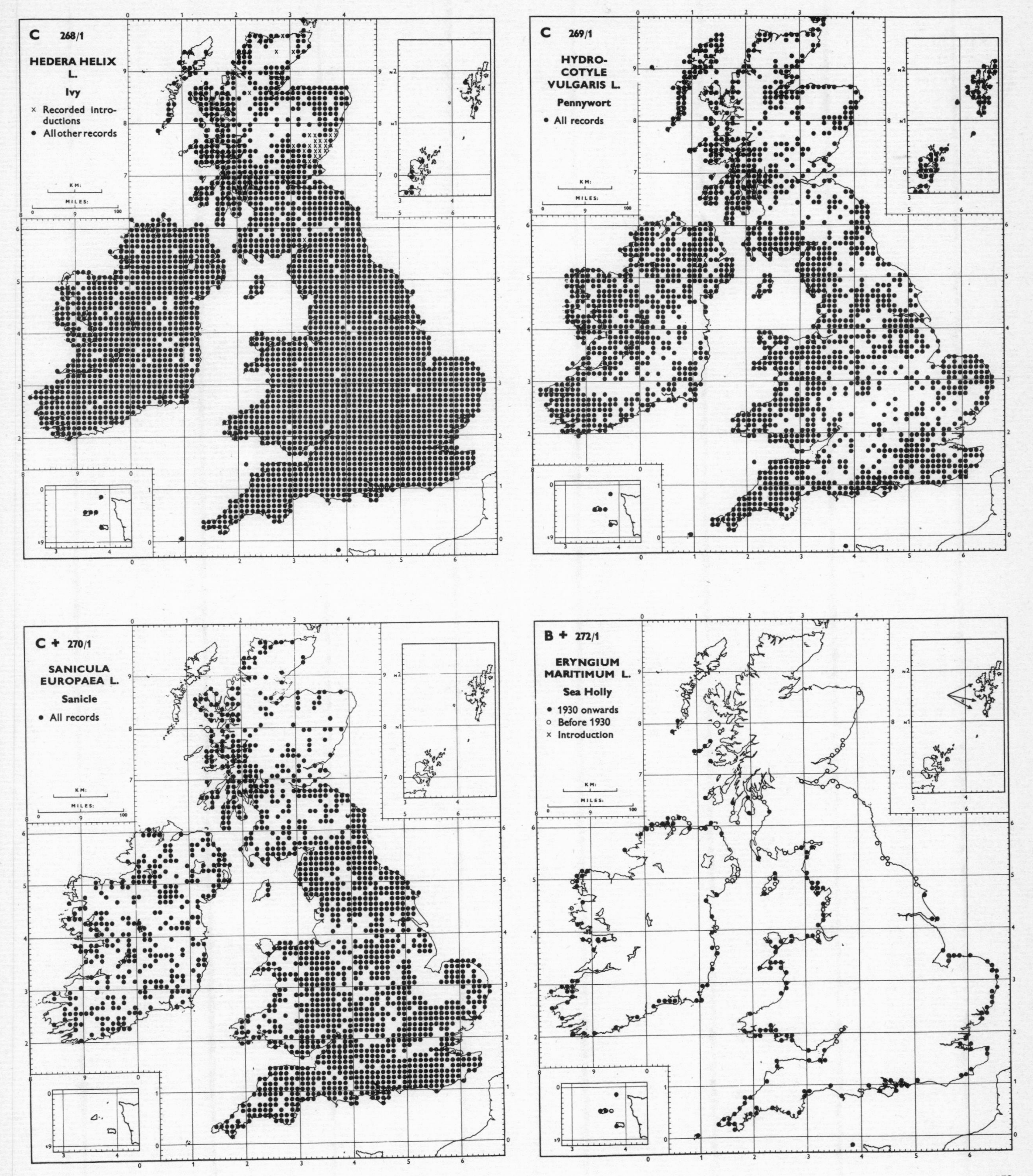
C 268/1
HEDERA HELIX L.
Ivy
× Recorded introductions
• All other records
C 269/1
HYDROCOTYLE VULGARIS L.
Pennywort
• All records
C + 270/1
SANICULA EUROPAEA L.
Sanicle
• All records
B + 272/1
ERYNGIUM MARITIMUM L.
Sea Holly
• 1930 onwards
○ Before 1930
× Introduction

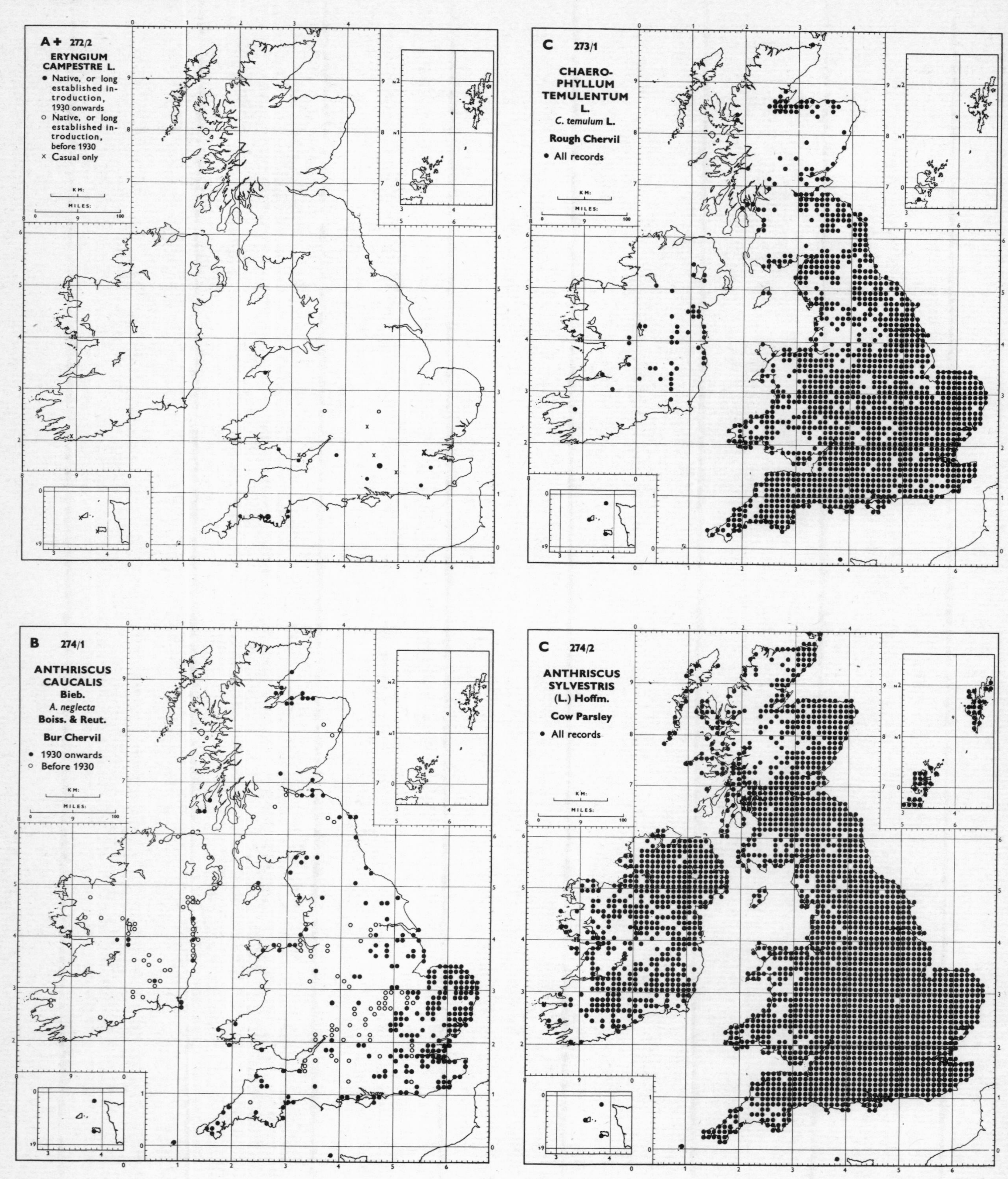
A+ 272/2
ERYNGIUM CAMPESTRE L.
• Native, or long established introduction, 1930 onwards
○ Native, or long established introduction, before 1930
× Casual only
KM:
MILES:
C 273/1
CHAERO-PHYLLUM TEMULENTUM L.
C. temulum L.
Rough Chervil
• All records
B 274/1
ANTHRISCUS CAUCALIS Bieb.
A. neglecta Boiss. & Reut.
Bur Chervil
• 1930 onwards
○ Before 1930
C 274/2
ANTHRISCUS SYLVESTRIS (L.) Hoffm.
Cow Parsley
• All records

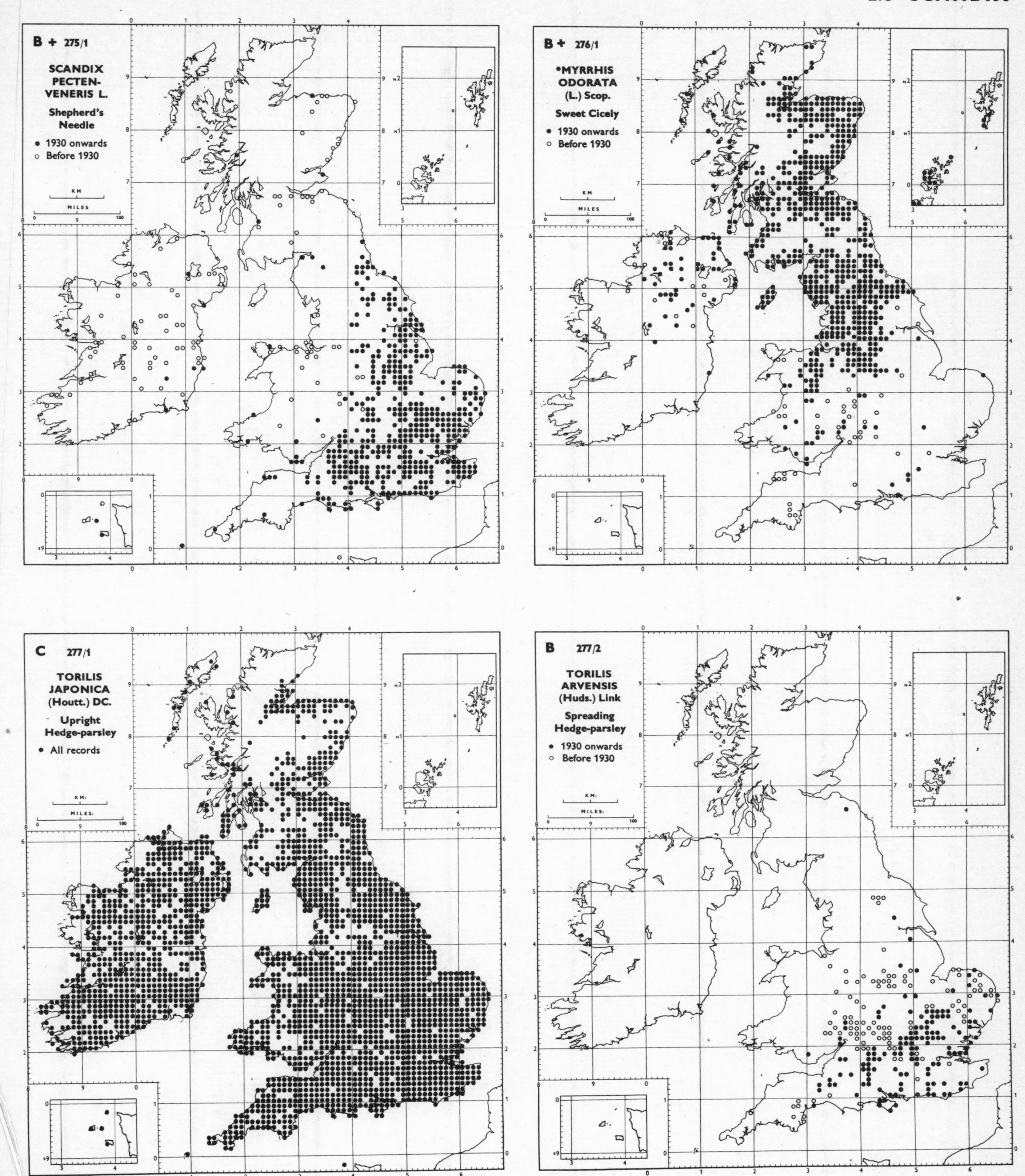
B + 275/1
SCANDIX PECTEN-VENERIS L.
Shepherd's Needle
• 1930 onwards
○ Before 1930
B + 276/1
*MYRRHIS ODORATA (L.) Scop.
Sweet Cicely
• 1930 onwards
○ Before 1930
C 277/1
TORILIS JAPONICA (Houtt.) DC.
Upright Hedge-parsley
• All records
B 277/2
TORILIS ARVENSIS (Huds.) Link
Spreading Hedge-parsley
• 1930 onwards
○ Before 1930

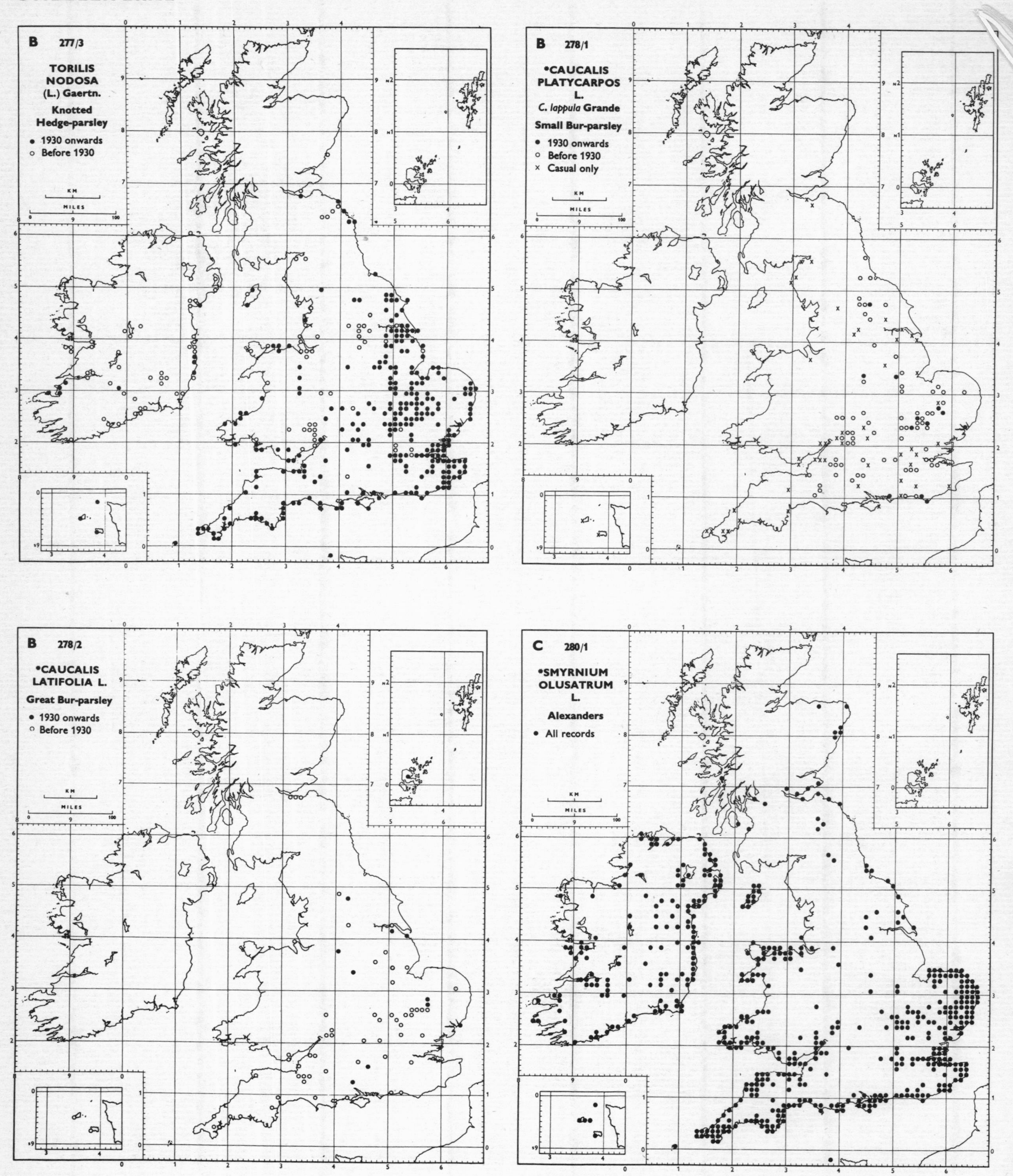
B 277/3
TORILIS NODOSA (L.) Gaertn.
Knotted Hedge-parsley
• 1930 onwards
○ Before 1930
B 278/1
•CAUCALIS PLATYCARPOS L.
C. lappula Grande
Small Bur-parsley
• 1930 onwards
○ Before 1930
× Casual only
B 278/2
•CAUCALIS LATIFOLIA L.
Great Bur-parsley
• 1930 onwards
○ Before 1930
C 280/1
•SMYRNIUM OLUSATRUM L.
Alexanders
• All records
KM
MILES

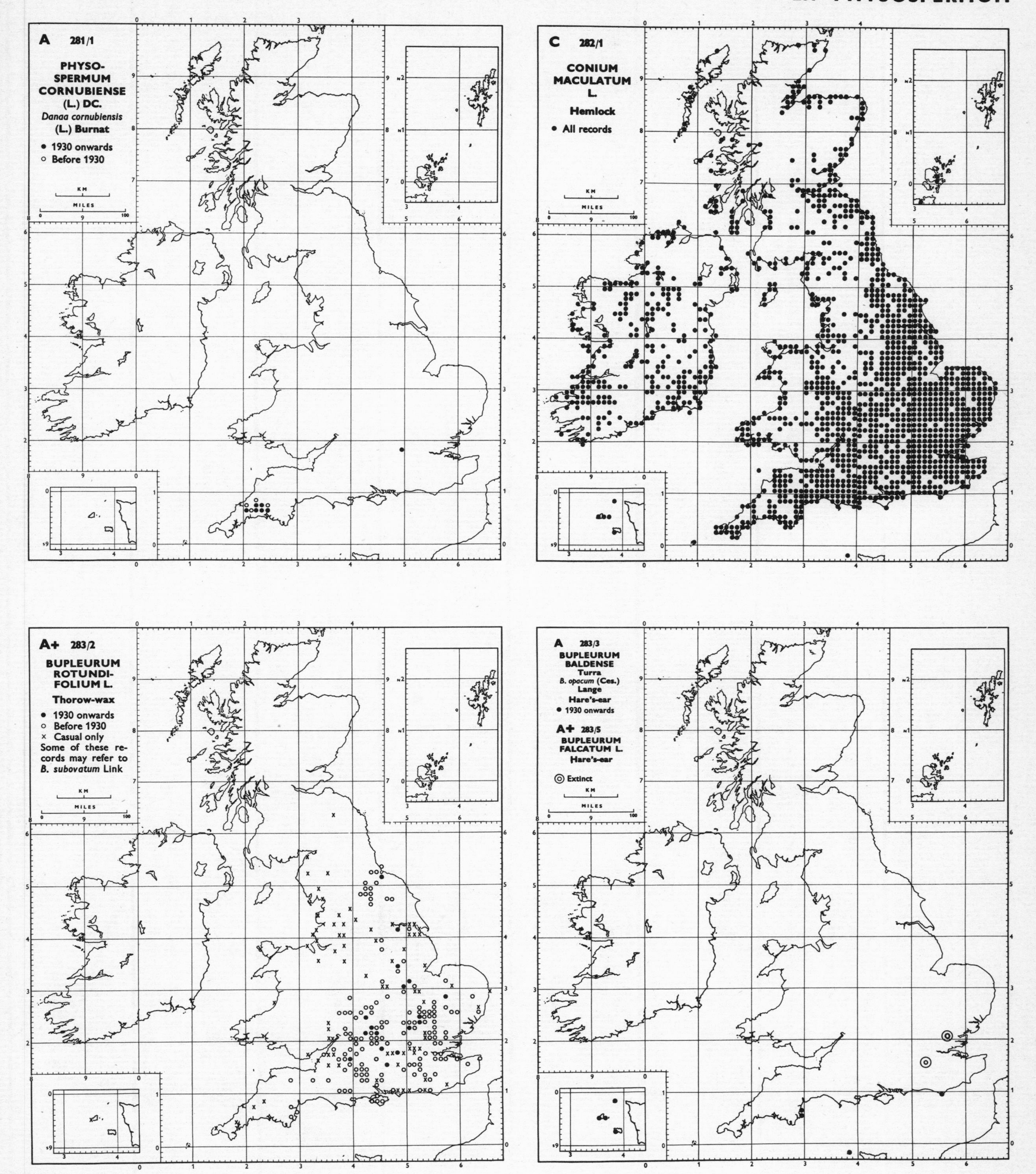
A 281/1
PHYSOSPERMUM CORNUBIENSE (L.) DC.
Danaa cornubiensis (L.) Burnat
• 1930 onwards
○ Before 1930
C 282/1
CONIUM MACULATUM L.
Hemlock
• All records
A+ 283/2
BUPLEURUM ROTUNDIFOLIUM L.
Thorow-wax
• 1930 onwards
○ Before 1930
× Casual only
Some of these records may refer to B. subovatum Link
A 283/3
BUPLEURUM BALDENSE Turra
B. opacum (Ces.) Lange
Hare's-ear
• 1930 onwards
A+ 283/5
BUPLEURUM FALCATUM L.
Hare's-ear
◎ Extinct
KM
MILES

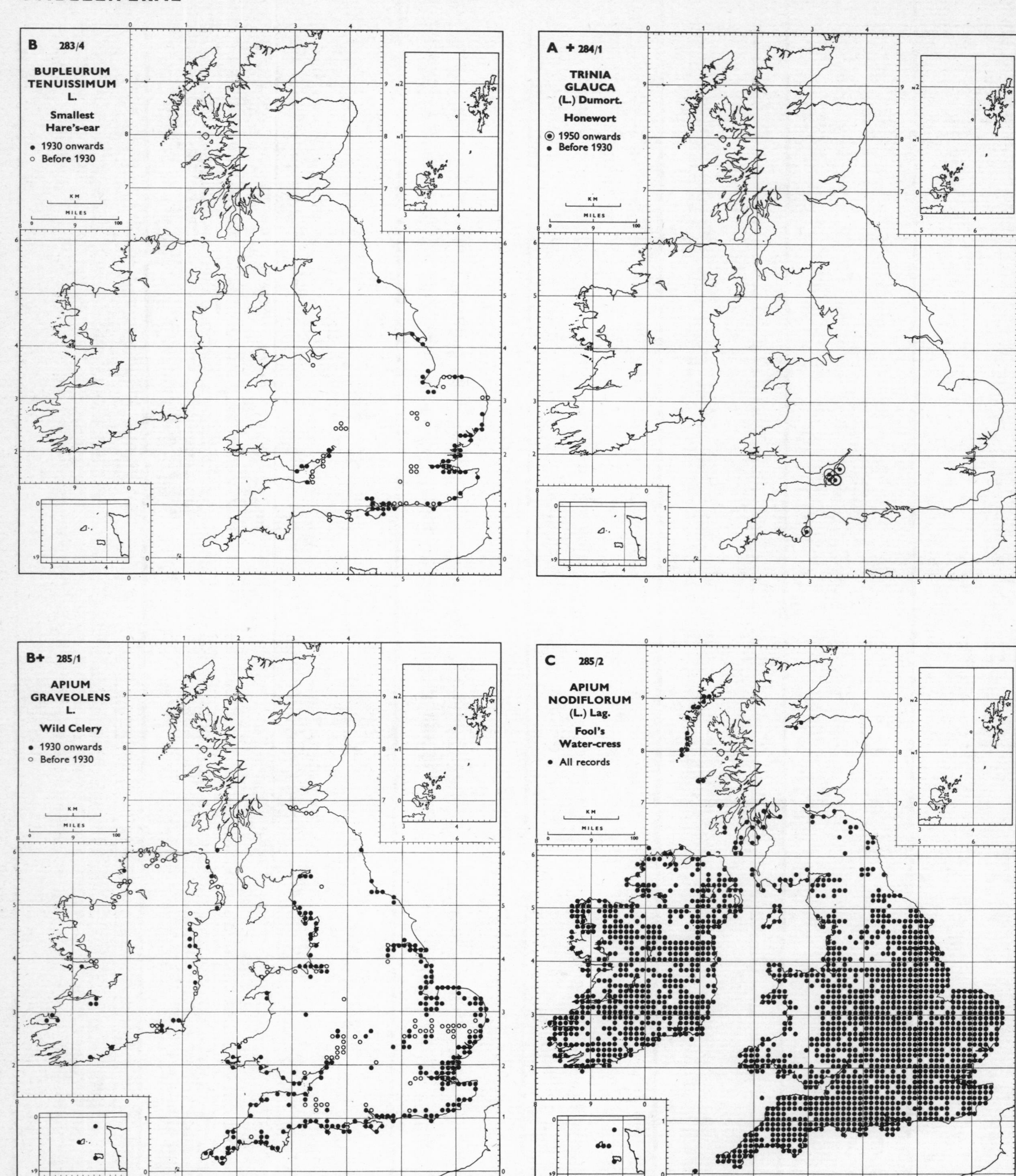
B 283/4
BUPLEURUM TENUISSIMUM L.
Smallest Hare's-ear
• 1930 onwards
○ Before 1930
A + 284/1
TRINIA GLAUCA (L.) Dumort.
Honewort
⊙ 1950 onwards
• Before 1930
B+ 285/1
APIUM GRAVEOLENS L.
Wild Celery
• 1930 onwards
○ Before 1930
C 285/2
APIUM NODIFLORUM (L.) Lag.
Fool's Water-cress
• All records
KM
MILES

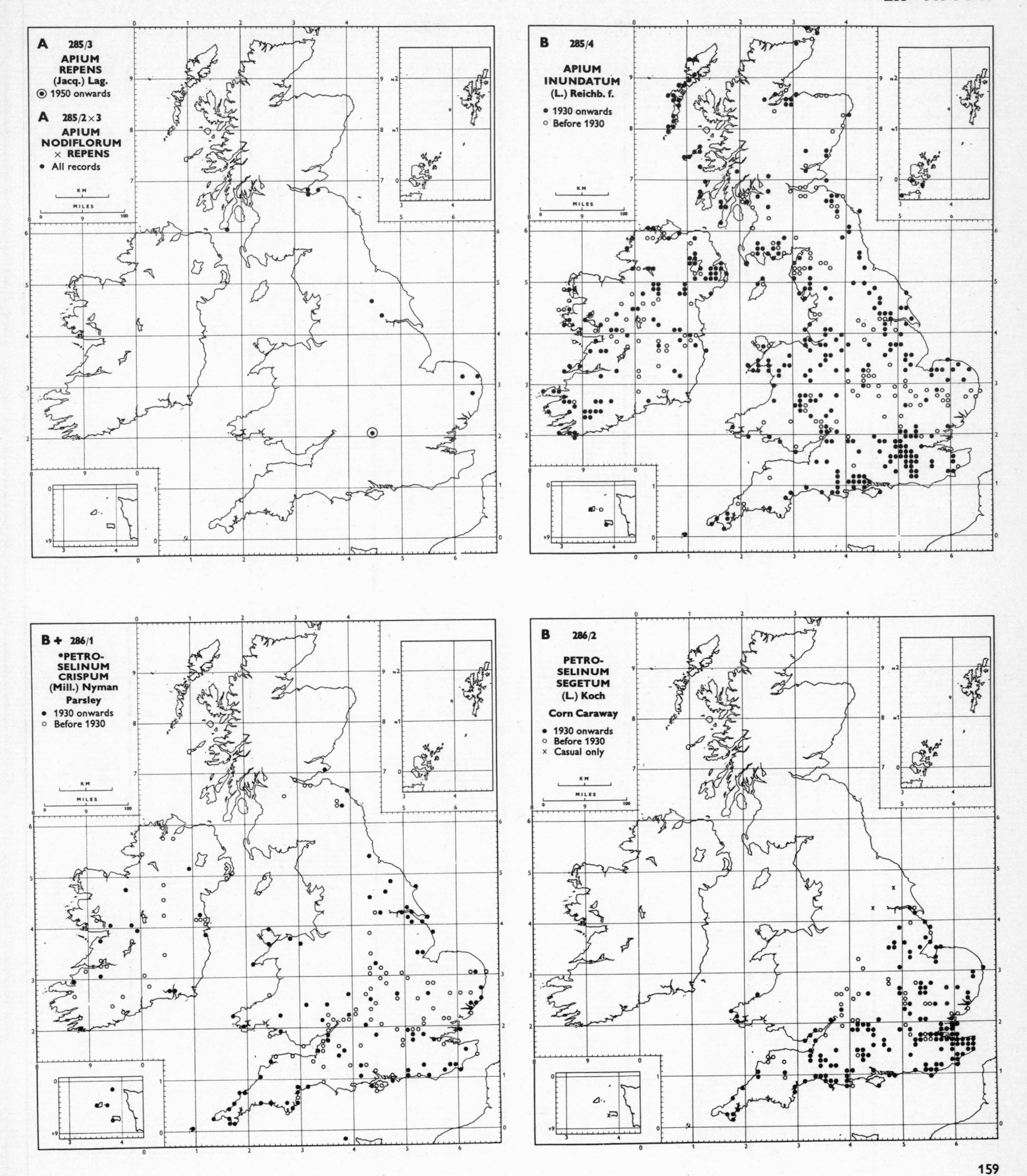
A 285/3
APIUM
REPENS
(Jacq.) Lag.
1950 onwards
A 285/2×3
APIUM
NODIFLORUM
× REPENS
All records
KM
MILES
B 285/4
APIUM
INUNDATUM
(L.) Reichb. f.
1930 onwards
Before 1930
B + 286/1
*PETRO-
SELINUM
CRISPUM
(Mill.) Nyman
Parsley
1930 onwards
Before 1930
B 286/2
PETRO-
SELINUM
SEGETUM
(L.) Koch
Corn Caraway
1930 onwards
Before 1930
Casual only

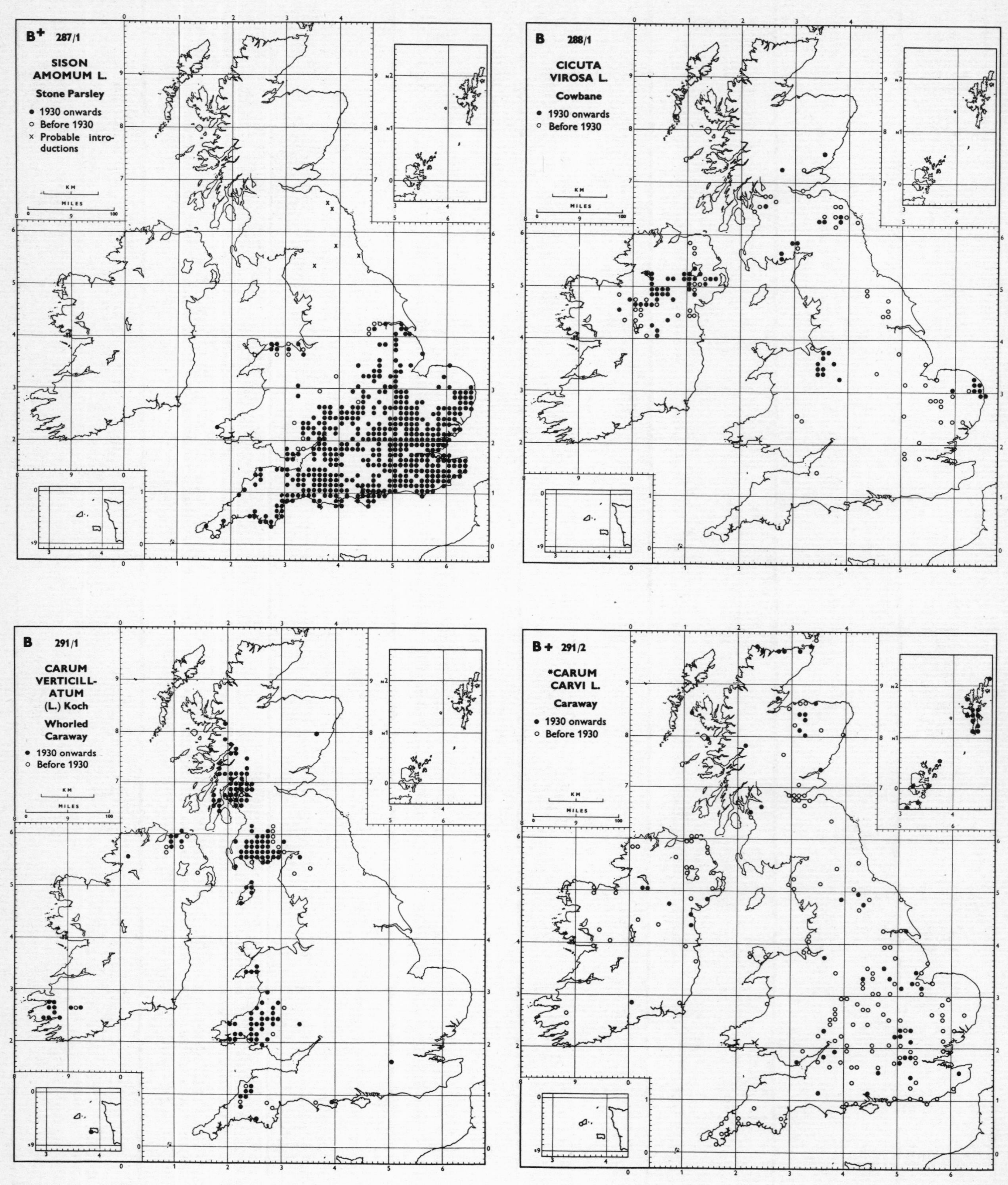
B+ 287/1
SISON AMOMUM L.
Stone Parsley
• 1930 onwards
○ Before 1930
× Probable introductions
B 288/1
CICUTA VIROSA L.
Cowbane
• 1930 onwards
○ Before 1930
B 291/1
CARUM VERTICILLATUM (L.) Koch
Whorled Caraway
• 1930 onwards
○ Before 1930
B+ 291/2
*CARUM CARVI L.
Caraway
• 1930 onwards
○ Before 1930
KM
MILES

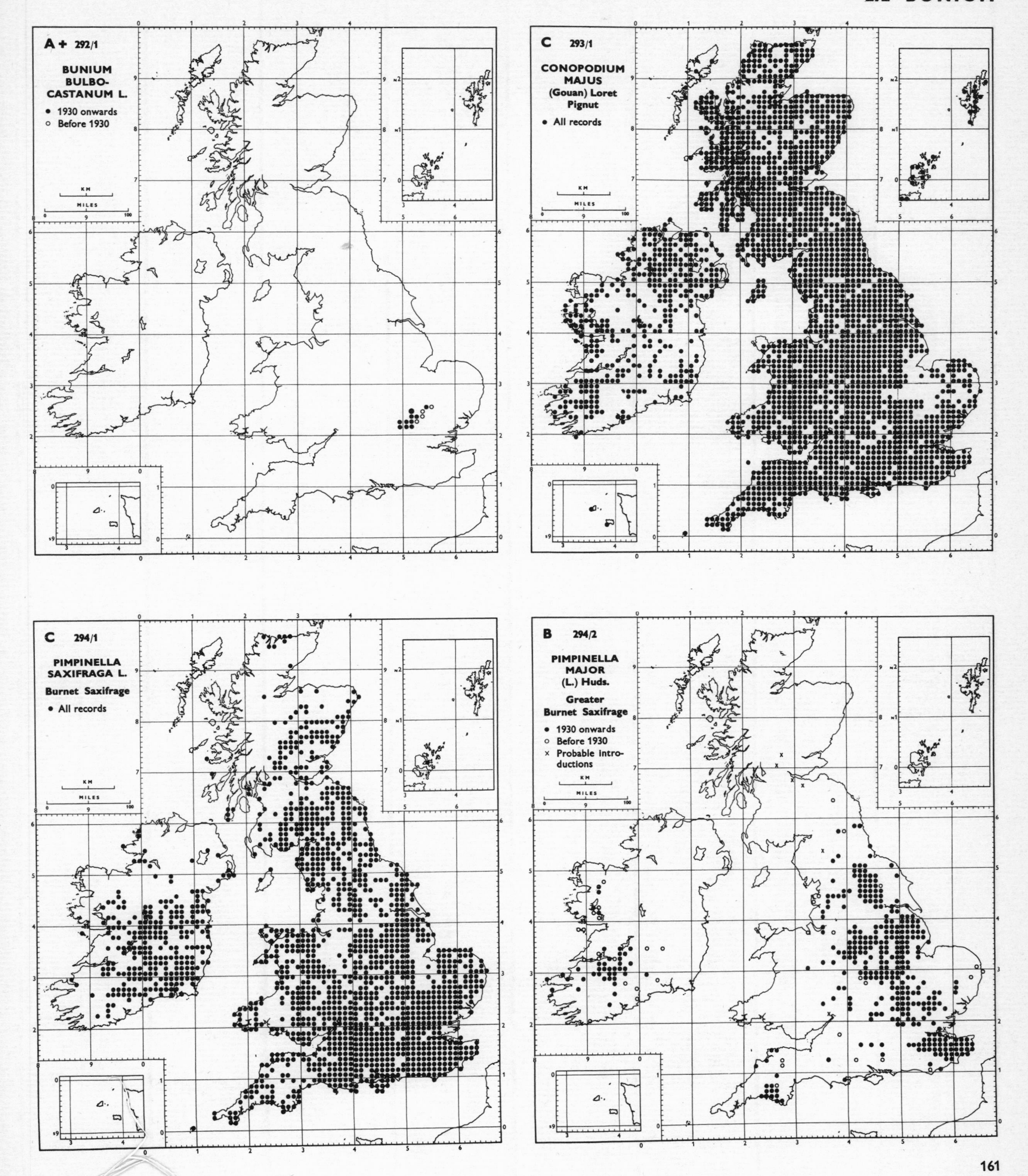
A + 292/1
BUNIUM BULBO-CASTANUM L.
• 1930 onwards
○ Before 1930
C 293/1
CONOPODIUM MAJUS (Gouan) Loret
Pignut
• All records
C 294/1
PIMPINELLA SAXIFRAGA L.
Burnet Saxifrage
• All records
B 294/2
PIMPINELLA MAJOR (L.) Huds.
Greater Burnet Saxifrage
• 1930 onwards
○ Before 1930
× Probable introductions
KM
MILES

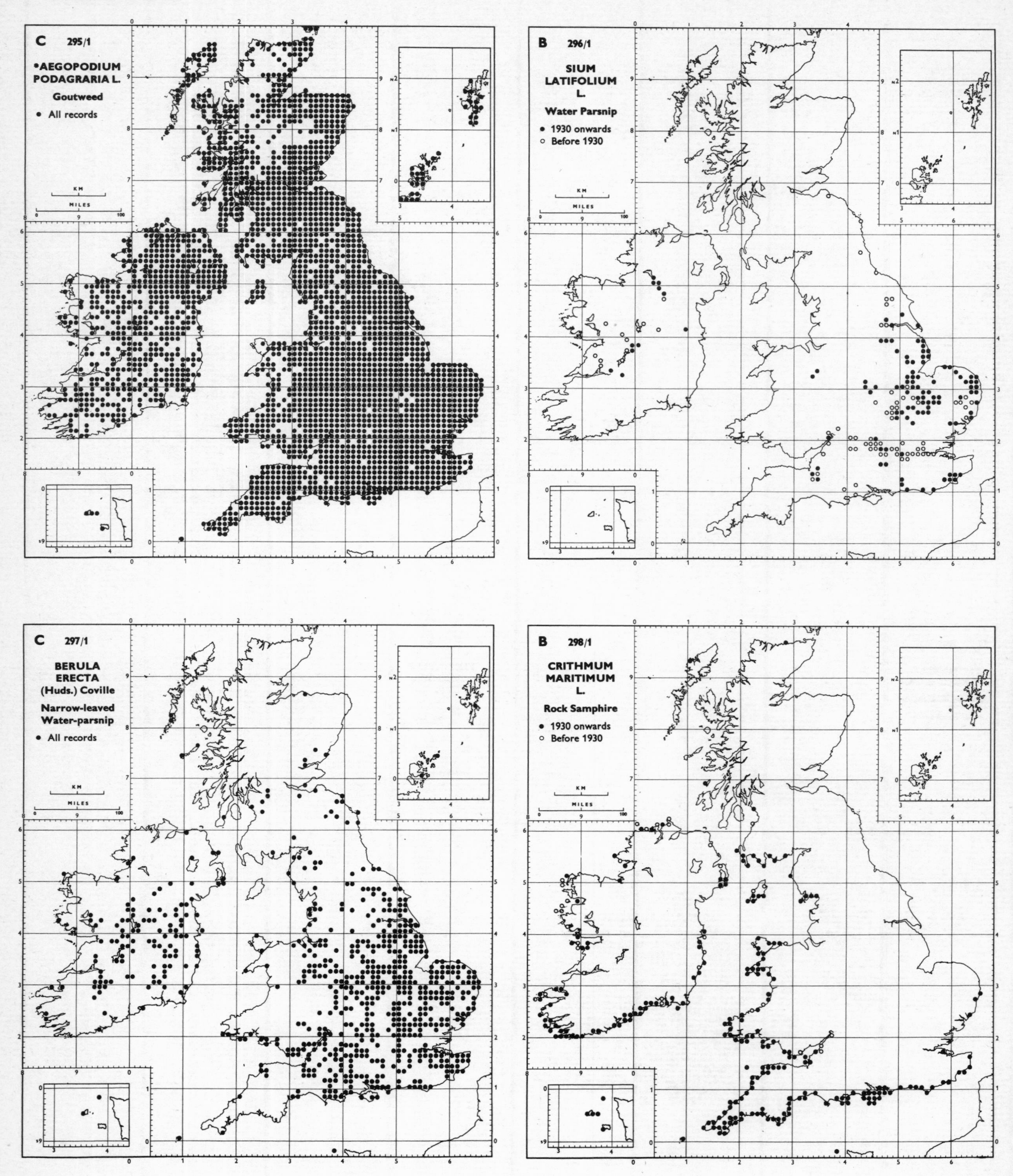
C 295/1
•AEGOPODIUM PODAGRARIA L.
Goutweed
• All records
B 296/1
SIUM LATIFOLIUM L.
Water Parsnip
• 1930 onwards
○ Before 1930
C 297/1
BERULA ERECTA (Huds.) Coville
Narrow-leaved Water-parsnip
• All records
B 298/1
CRITHMUM MARITIMUM L.
Rock Samphire
• 1930 onwards
○ Before 1930
KM
MILES

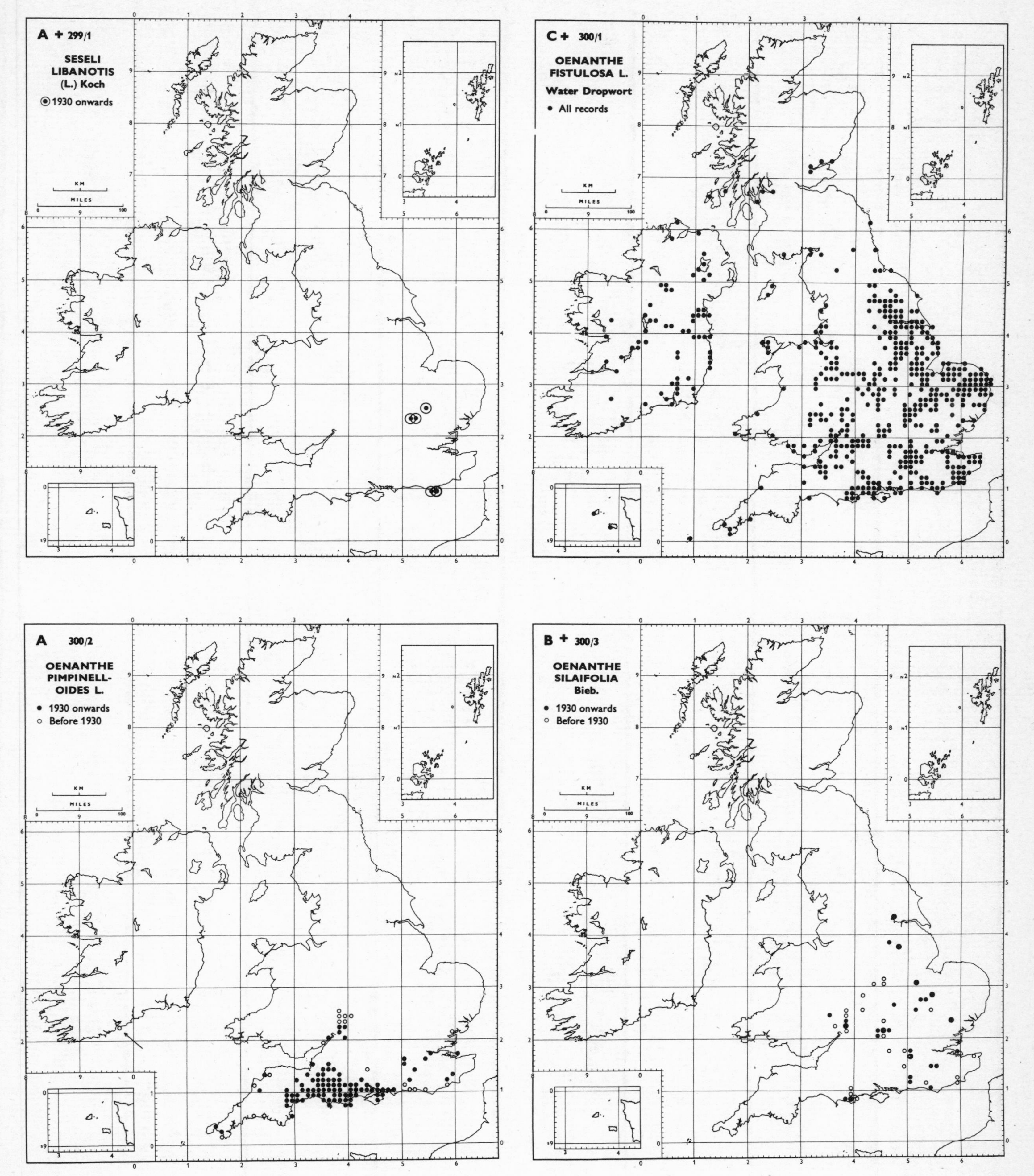
A + 299/1
SESELI LIBANOTIS (L.) Koch
1930 onwards
C + 300/1
OENANTHE FISTULOSA L.
Water Dropwort
All records
A 300/2
OENANTHE PIMPINELL-OIDES L.
1930 onwards
Before 1930
B + 300/3
OENANTHE SILAIFOLIA Bieb.
1930 onwards
Before 1930

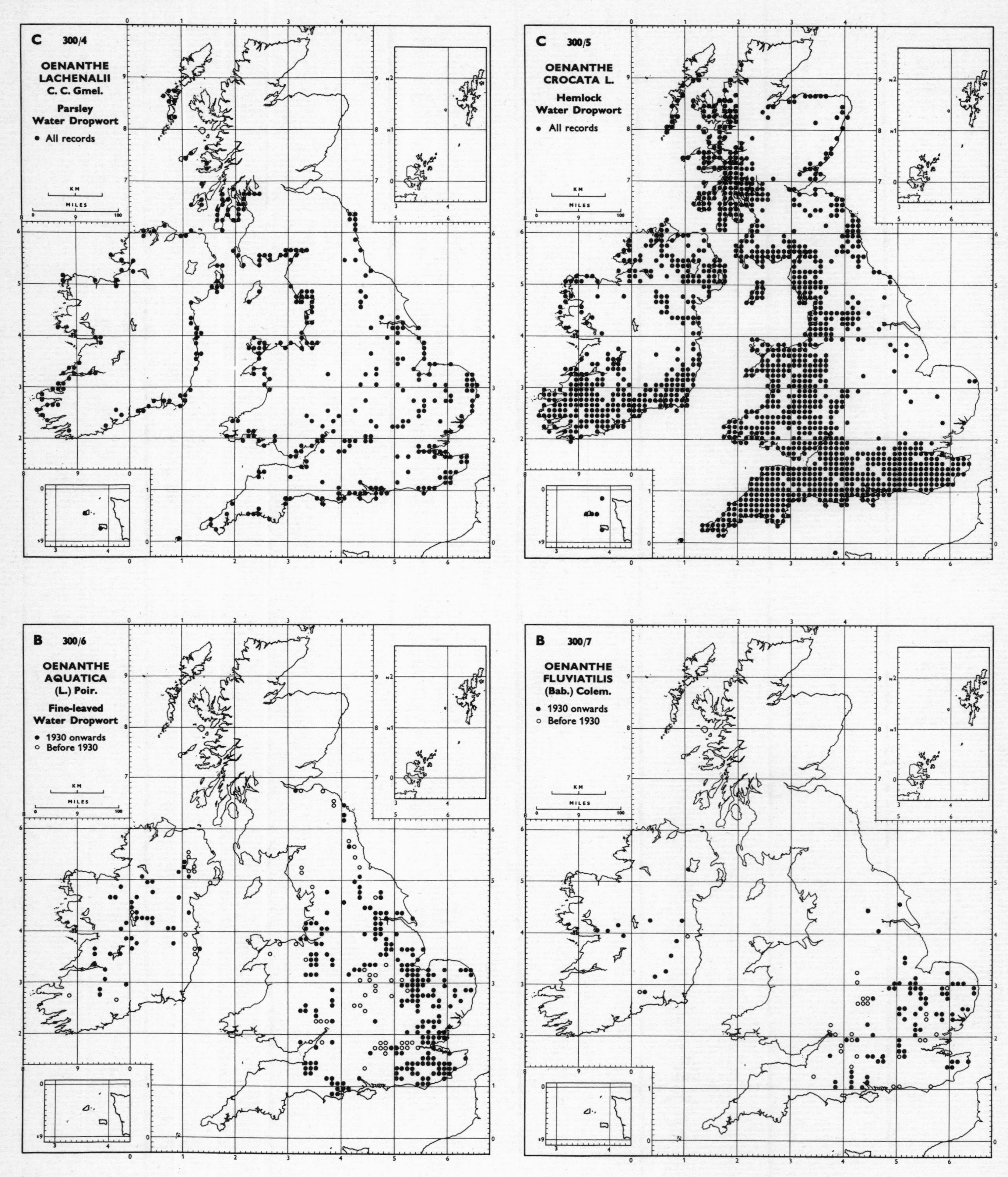
C 300/4
OENANTHE LACHENALII C. C. Gmel.
Parsley Water Dropwort
• All records
C 300/5
OENANTHE CROCATA L.
Hemlock Water Dropwort
• All records
B 300/6
OENANTHE AQUATICA (L.) Poir.
Fine-leaved Water Dropwort
• 1930 onwards
○ Before 1930
B 300/7
OENANTHE FLUVIATILIS (Bab.) Colem.
• 1930 onwards
○ Before 1930

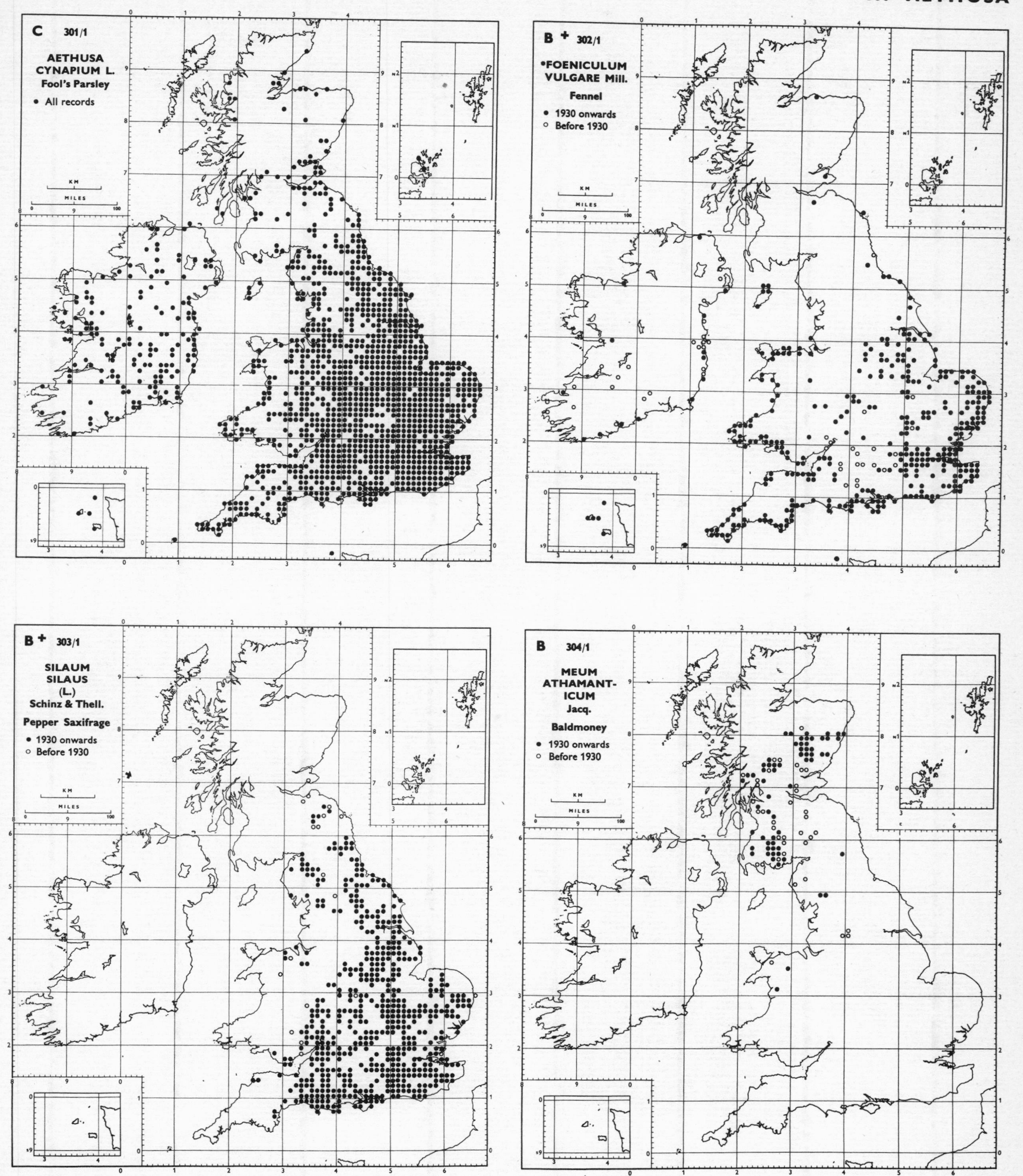
C 301/1
AETHUSA CYNAPIUM L.
Fool's Parsley
• All records
B + 302/1
•FOENICULUM VULGARE Mill.
Fennel
• 1930 onwards
○ Before 1930
B + 303/1
SILAUM SILAUS (L.) Schinz & Thell.
Pepper Saxifrage
• 1930 onwards
○ Before 1930
B 304/1
MEUM ATHAMANTICUM Jacq.
Baldmoney
• 1930 onwards
○ Before 1930
KM
MILES

UMBELLIFERAE

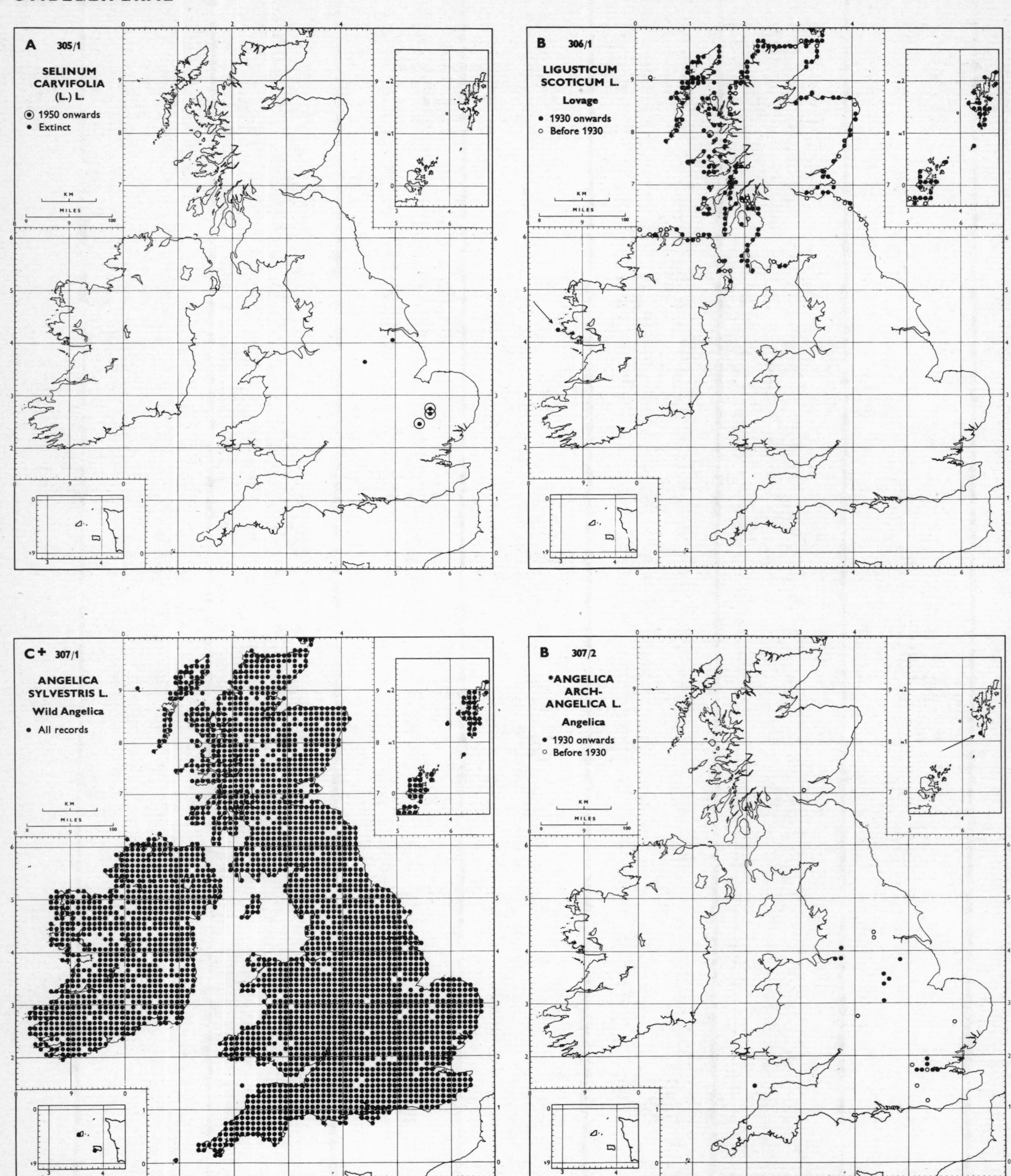

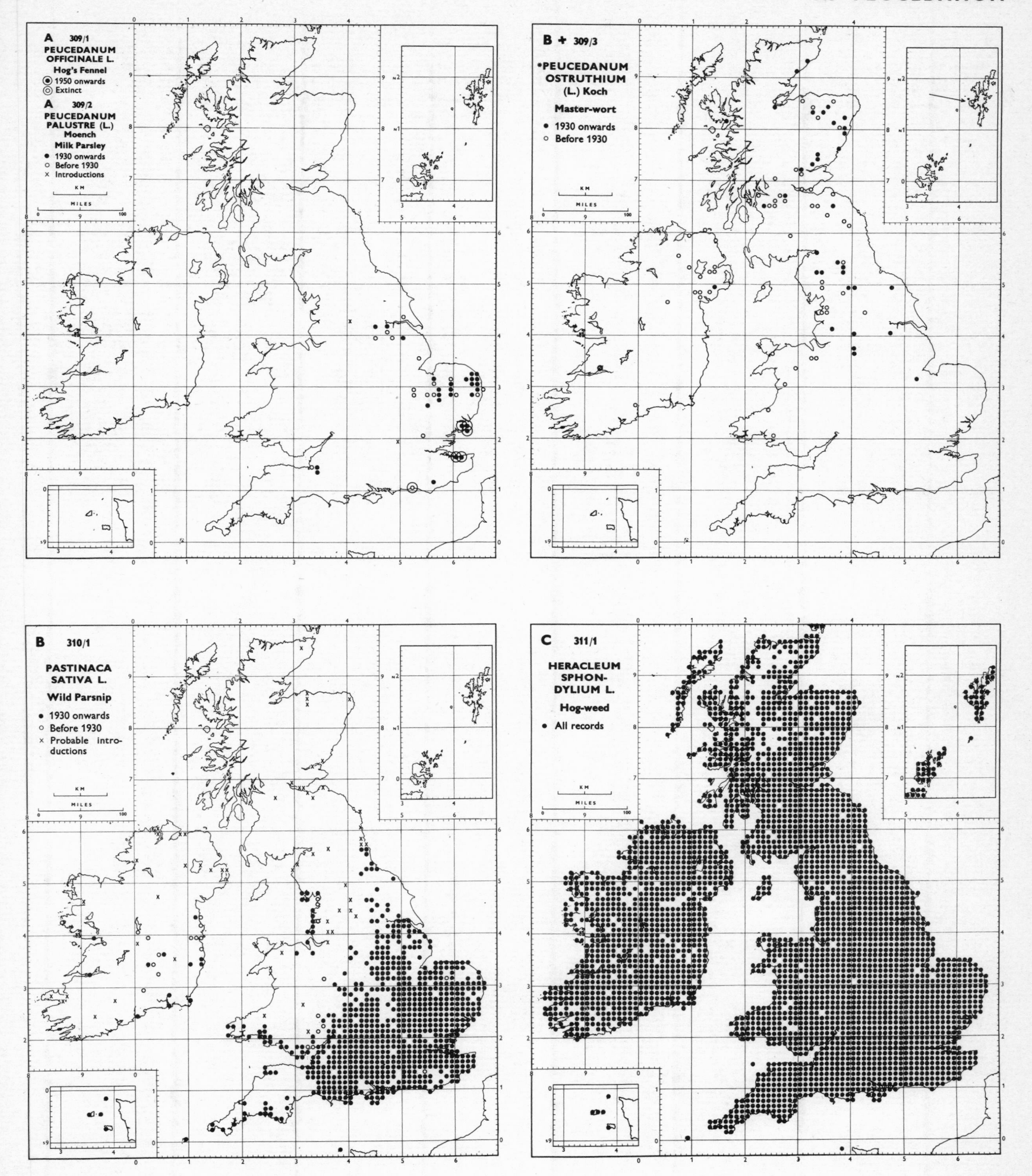
A 309/1
PEUCEDANUM OFFICINALE L.
Hog's Fennel
1950 onwards
Extinct
A 309/2
PEUCEDANUM PALUSTRE (L.) Moench
Milk Parsley
1930 onwards
Before 1930
Introductions
B + 309/3
*PEUCEDANUM OSTRUTHIUM (L.) Koch
Master-wort
1930 onwards
Before 1930
B 310/1
PASTINACA SATIVA L.
Wild Parsnip
1930 onwards
Before 1930
Probable introductions
C 311/1
HERACLEUM SPHONDYLIUM L.
Hog-weed
All records
KM
MILES

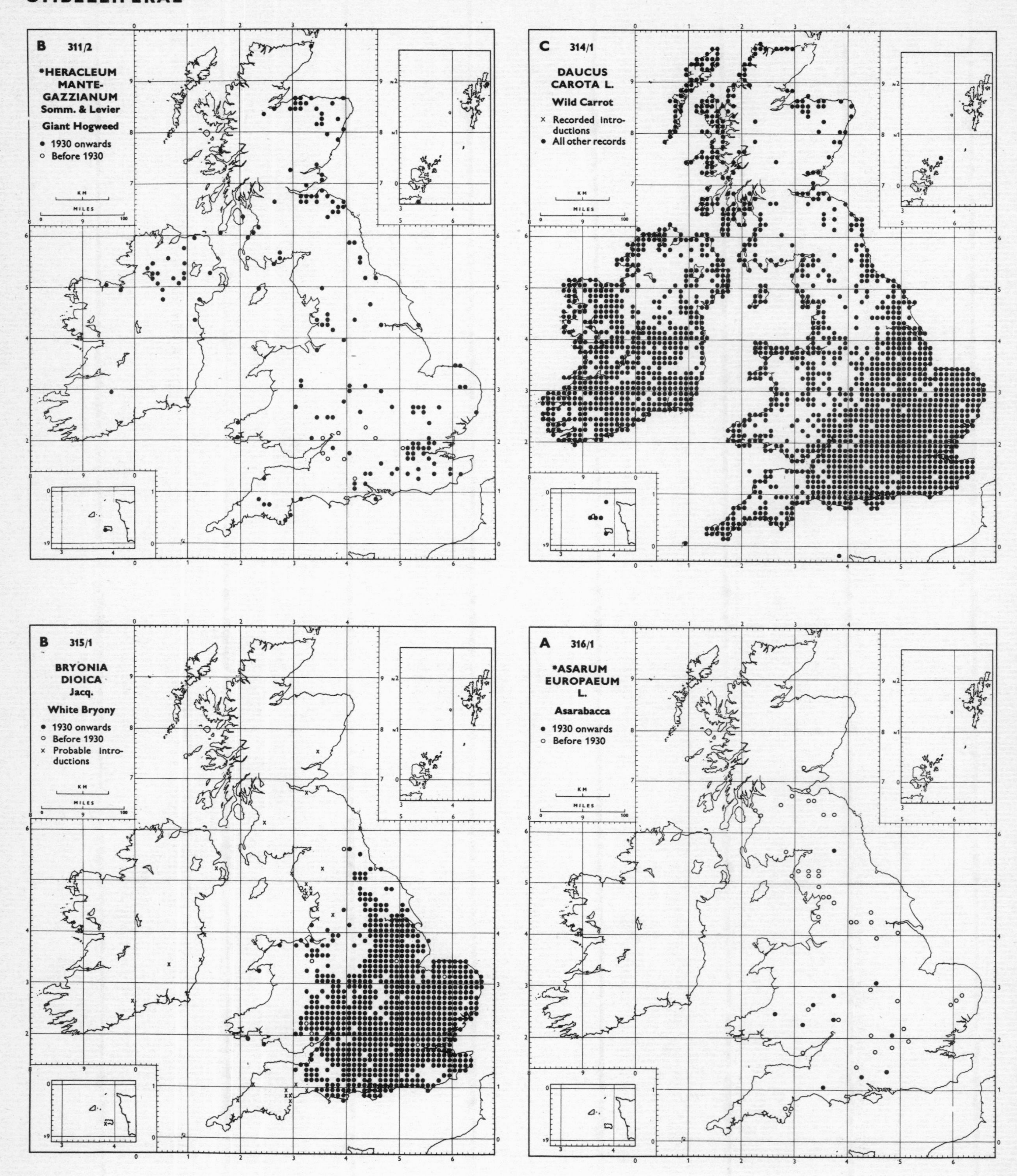
B 311/2
*HERACLEUM MANTE-GAZZIANUM Somm. & Levier
Giant Hogweed
• 1930 onwards
○ Before 1930
C 314/1
DAUCUS CAROTA L.
Wild Carrot
× Recorded introductions
• All other records
B 315/1
BRYONIA DIOICA Jacq.
White Bryony
• 1930 onwards
○ Before 1930
× Probable introductions
A 316/1
*ASARUM EUROPAEUM L.
Asarabacca
• 1930 onwards
○ Before 1930
KM
MILES

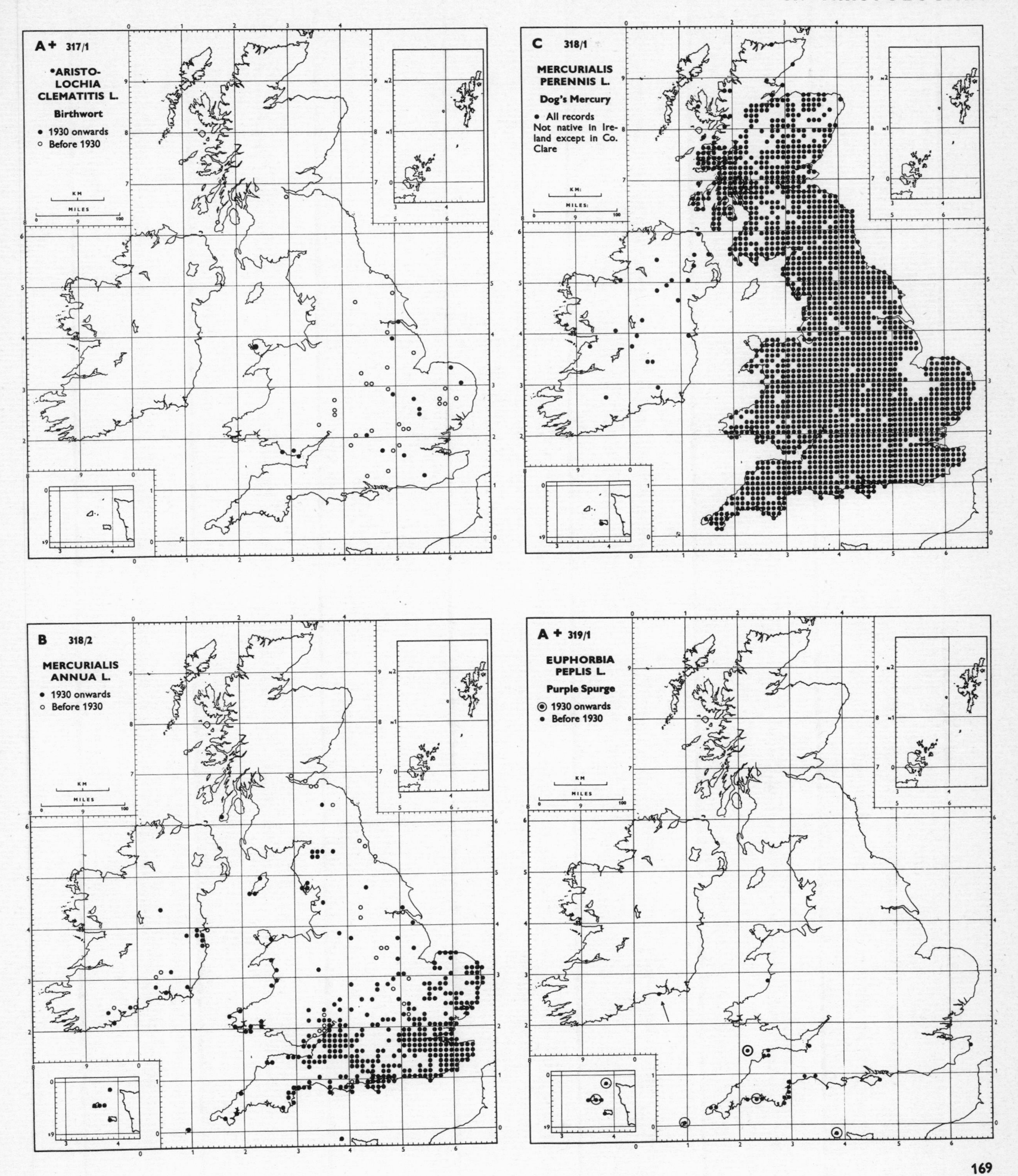
A+ 317/1
*ARISTOLOCHIA CLEMATITIS L.
Birthwort
• 1930 onwards
○ Before 1930
C 318/1
MERCURIALIS PERENNIS L.
Dog's Mercury
• All records
Not native in Ireland except in Co. Clare
B 318/2
MERCURIALIS ANNUA L.
• 1930 onwards
○ Before 1930
A+ 319/1
EUPHORBIA PEPLIS L.
Purple Spurge
⊙ 1930 onwards
• Before 1930
KM
MILES

EUPHORBIACEAE

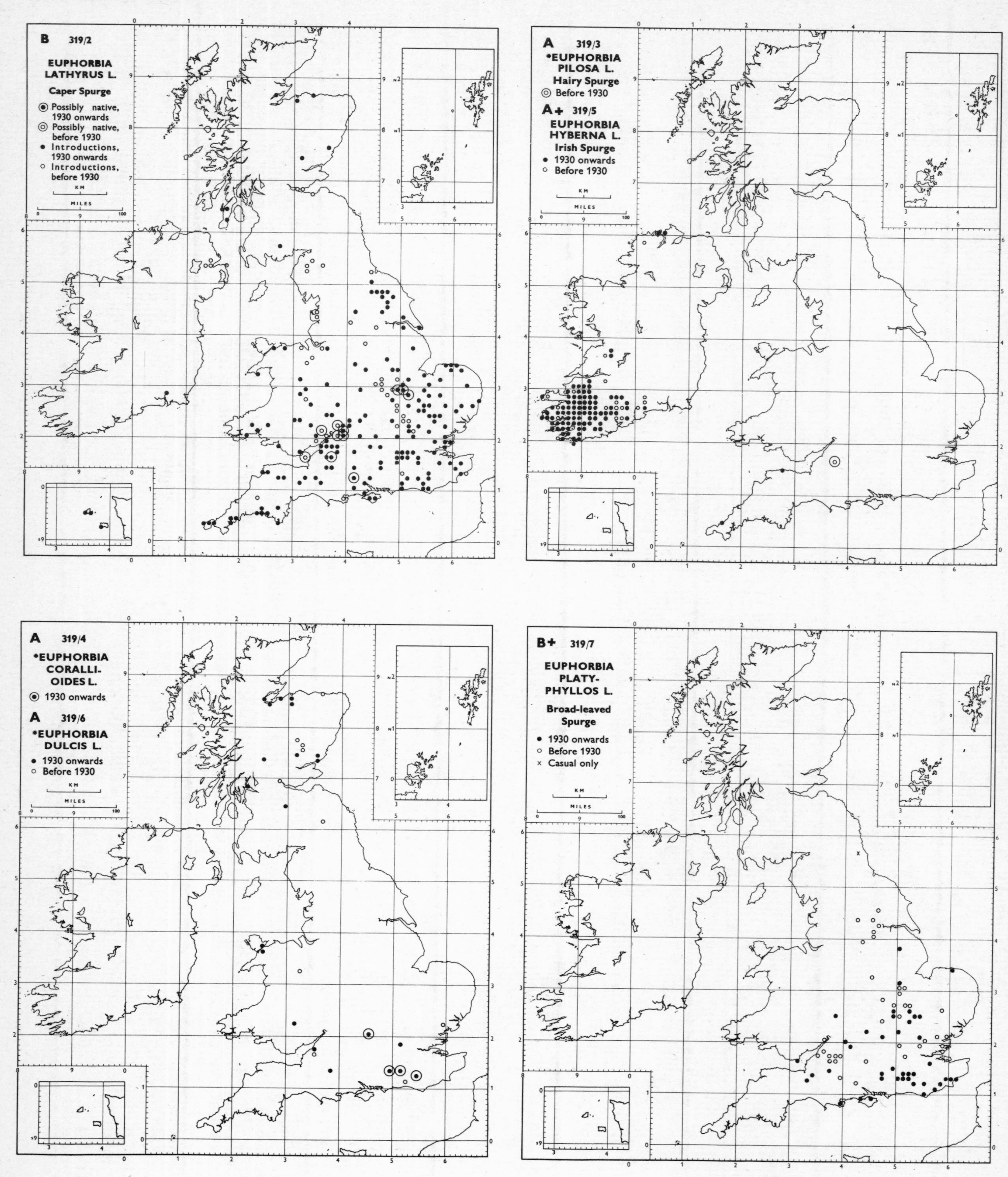

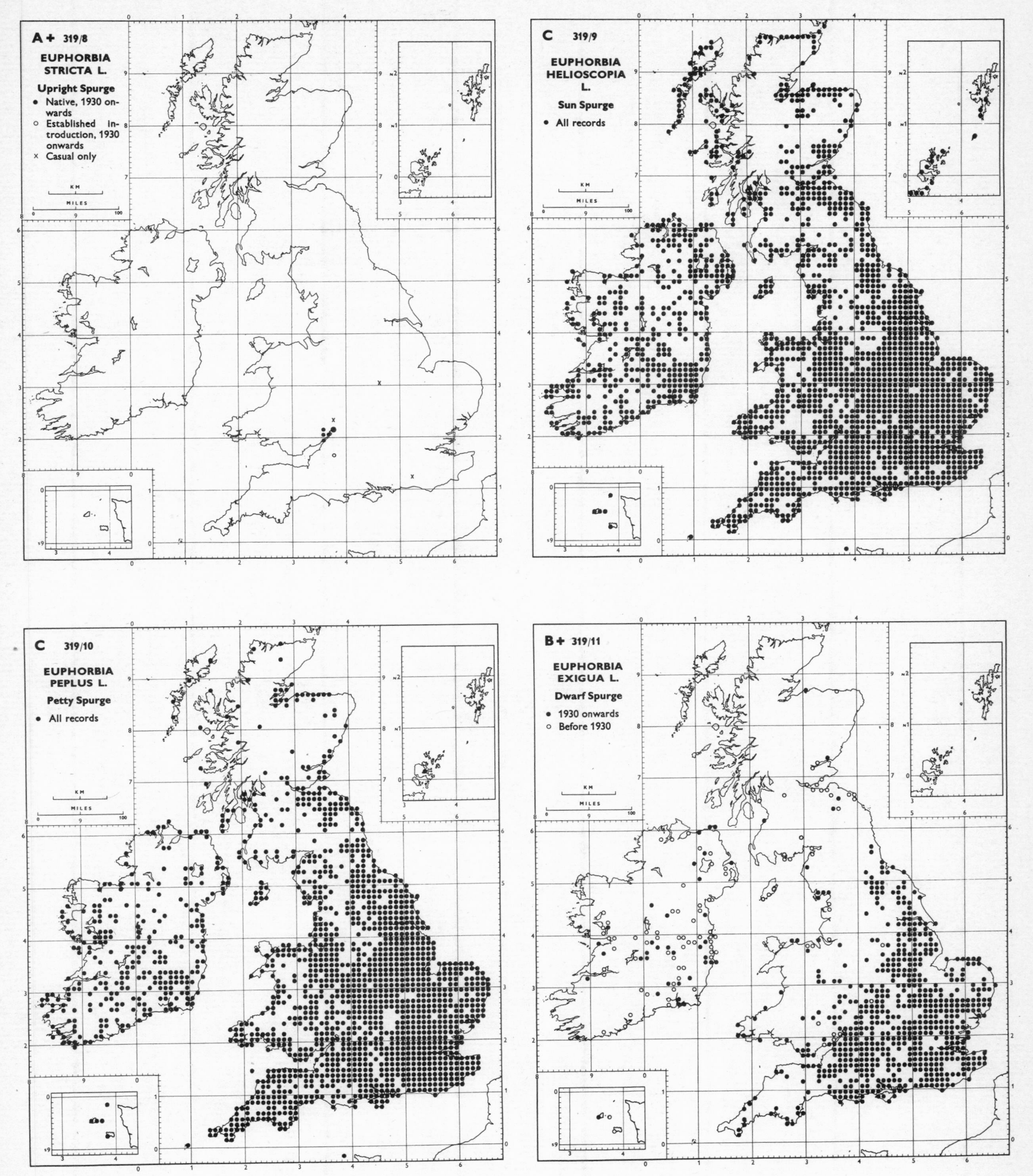
A + 319/8
EUPHORBIA STRICTA L.
Upright Spurge
• Native, 1930 onwards
○ Established introduction, 1930 onwards
× Casual only
C 319/9
EUPHORBIA HELIOSCOPIA L.
Sun Spurge
• All records
C 319/10
EUPHORBIA PEPLUS L.
Petty Spurge
• All records
B + 319/11
EUPHORBIA EXIGUA L.
Dwarf Spurge
• 1930 onwards
○ Before 1930
KM
MILES

EUPHORBIACEAE

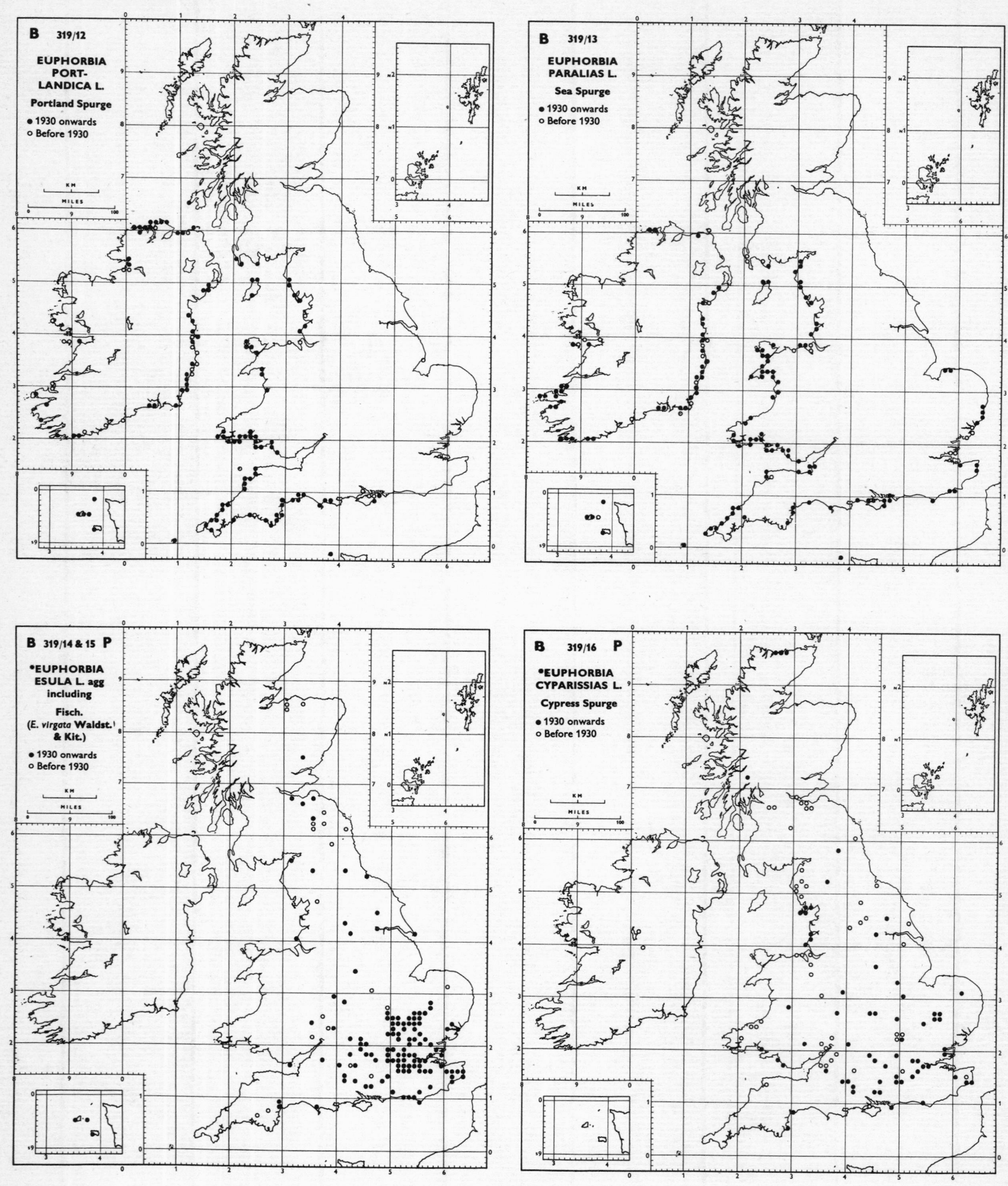

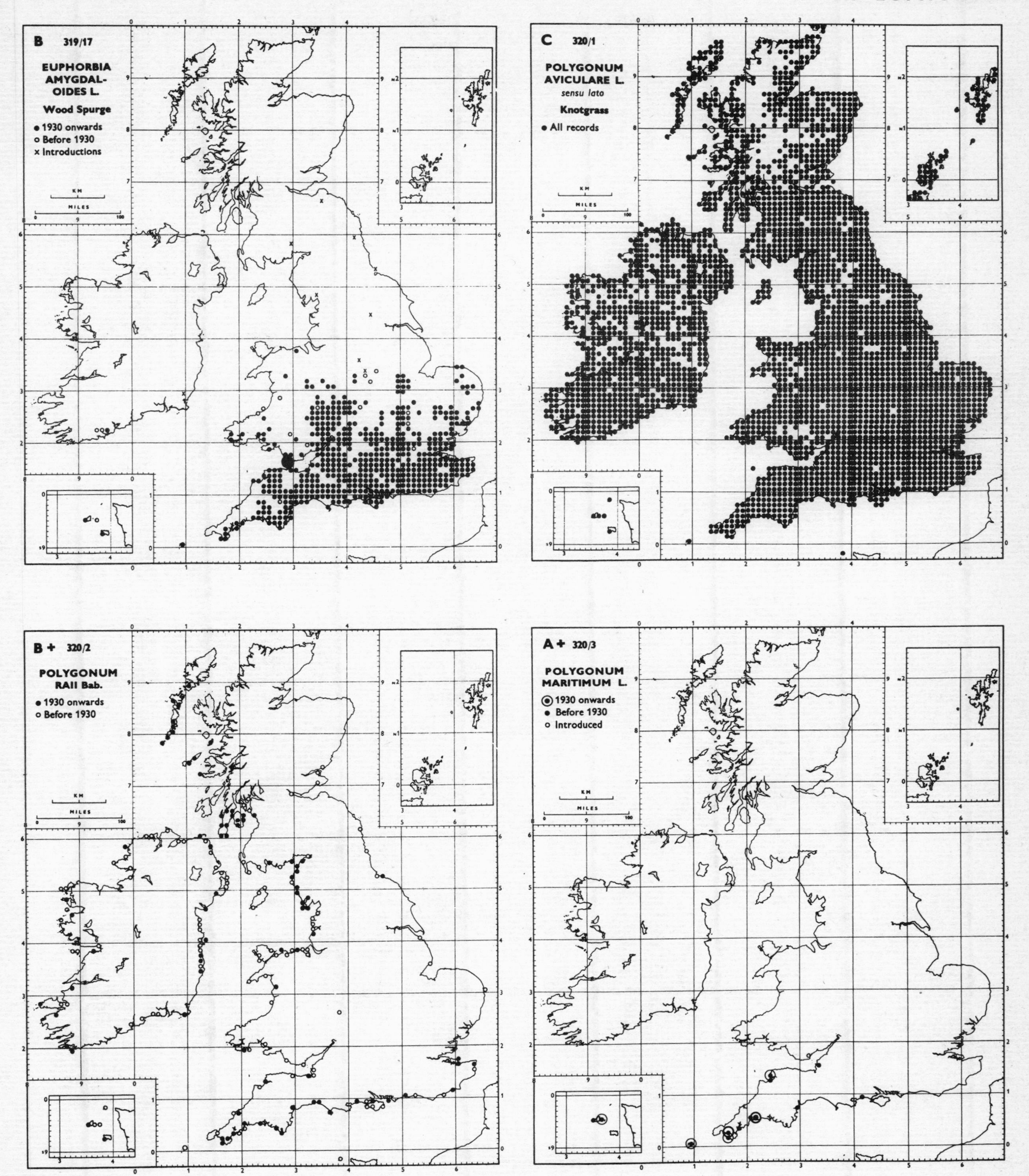
B 319/17
EUPHORBIA AMYGDALOIDES L.
Wood Spurge
● 1930 onwards
○ Before 1930
x Introductions
C 320/1
POLYGONUM AVICULARE L.
sensu lato
Knotgrass
● All records
B + 320/2
POLYGONUM RAII Bab.
● 1930 onwards
○ Before 1930
A + 320/3
POLYGONUM MARITIMUM L.
◉ 1930 onwards
● Before 1930
○ Introduced
KM
MILES

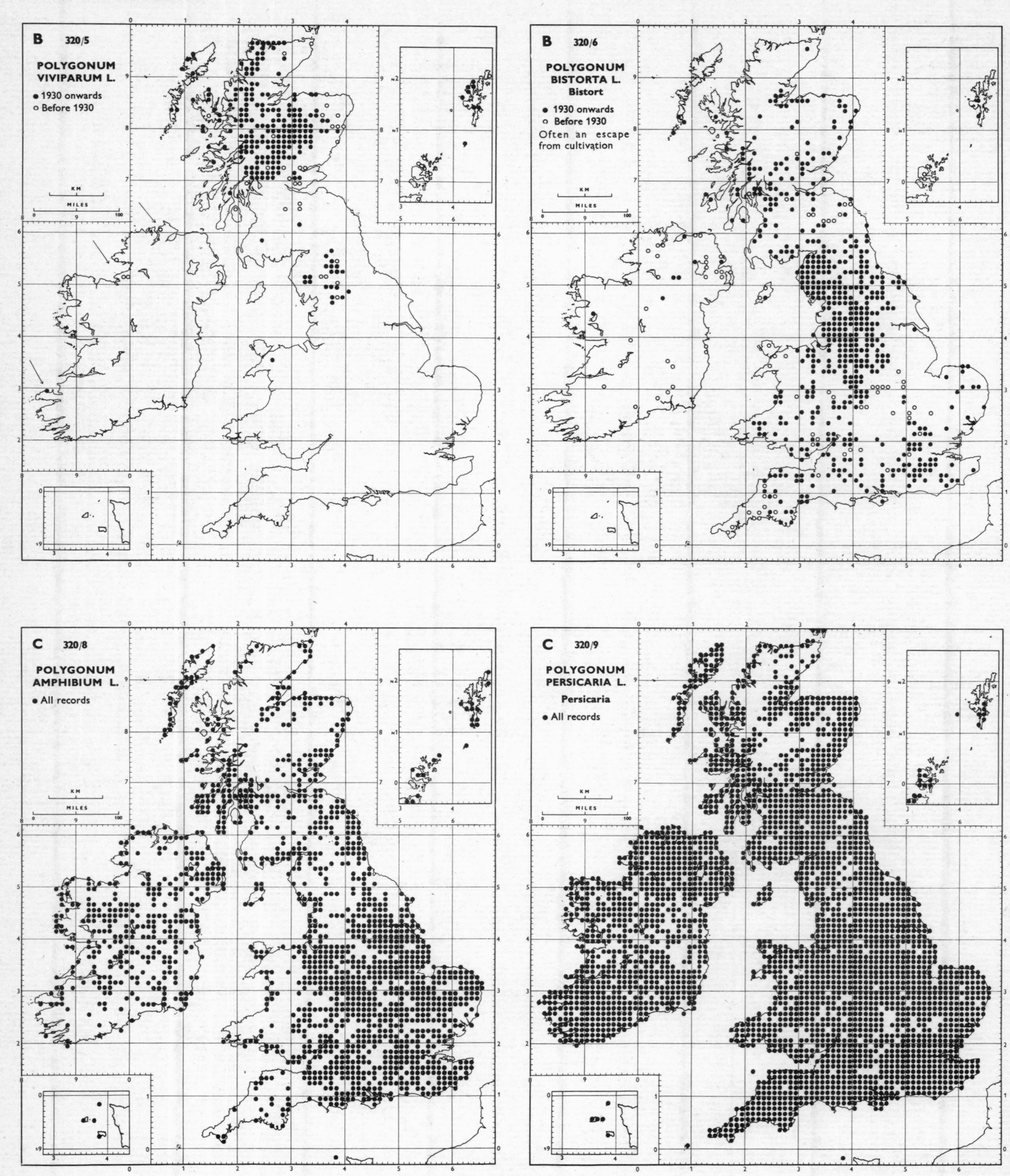
B 320/5
POLYGONUM VIVIPARUM L.
• 1930 onwards
○ Before 1930
B 320/6
POLYGONUM BISTORTA L.
Bistort
• 1930 onwards
○ Before 1930
Often an escape from cultivation
C 320/8
POLYGONUM AMPHIBIUM L.
• All records
C 320/9
POLYGONUM PERSICARIA L.
Persicaria
• All records

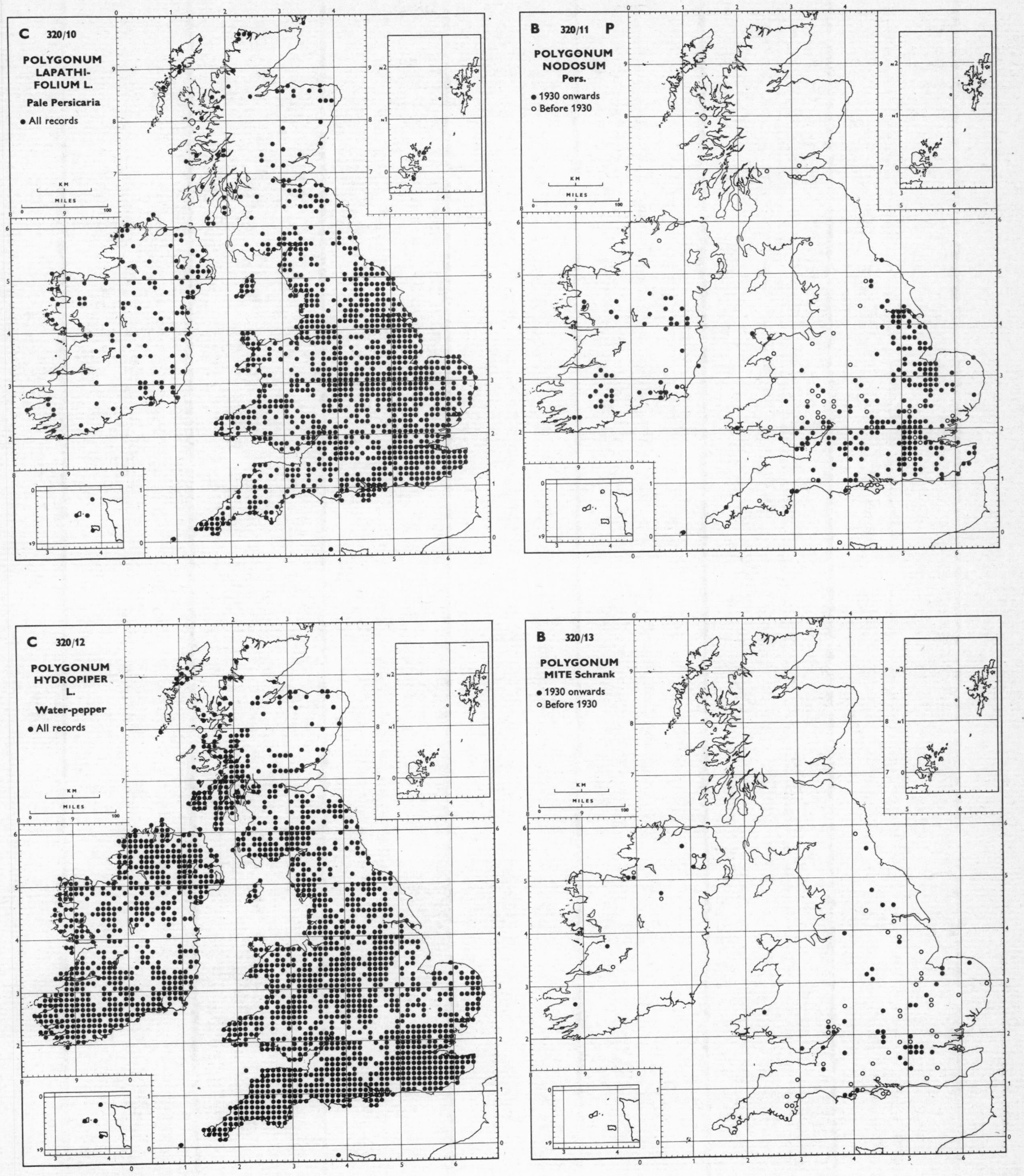
C 320/10
POLYGONUM LAPATHIFOLIUM L.
Pale Persicaria
● All records
B 320/11 P
POLYGONUM NODOSUM Pers.
● 1930 onwards
○ Before 1930
C 320/12
POLYGONUM HYDROPIPER L.
Water-pepper
● All records
B 320/13
POLYGONUM MITE Schrank
● 1930 onwards
○ Before 1930
KM
MILES

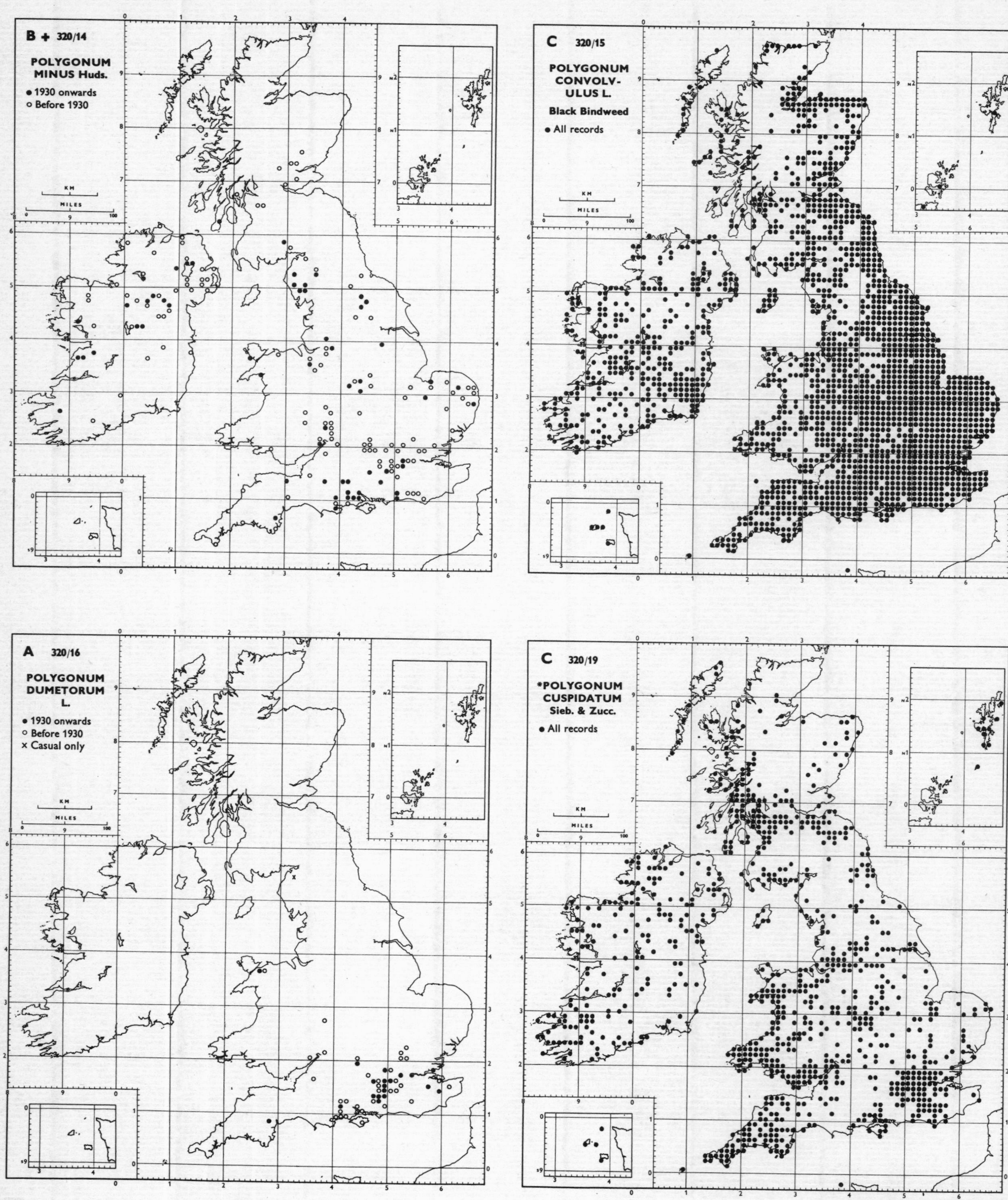
B + 320/14
POLYGONUM
MINUS Huds.
• 1930 onwards
○ Before 1930
C 320/15
POLYGONUM
CONVOLV-
ULUS L.
Black Bindweed
• All records
A 320/16
POLYGONUM
DUMETORUM
L.
• 1930 onwards
○ Before 1930
× Casual only
C 320/19
*POLYGONUM
CUSPIDATUM
Sieb. & Zucc.
• All records
KM
MILES

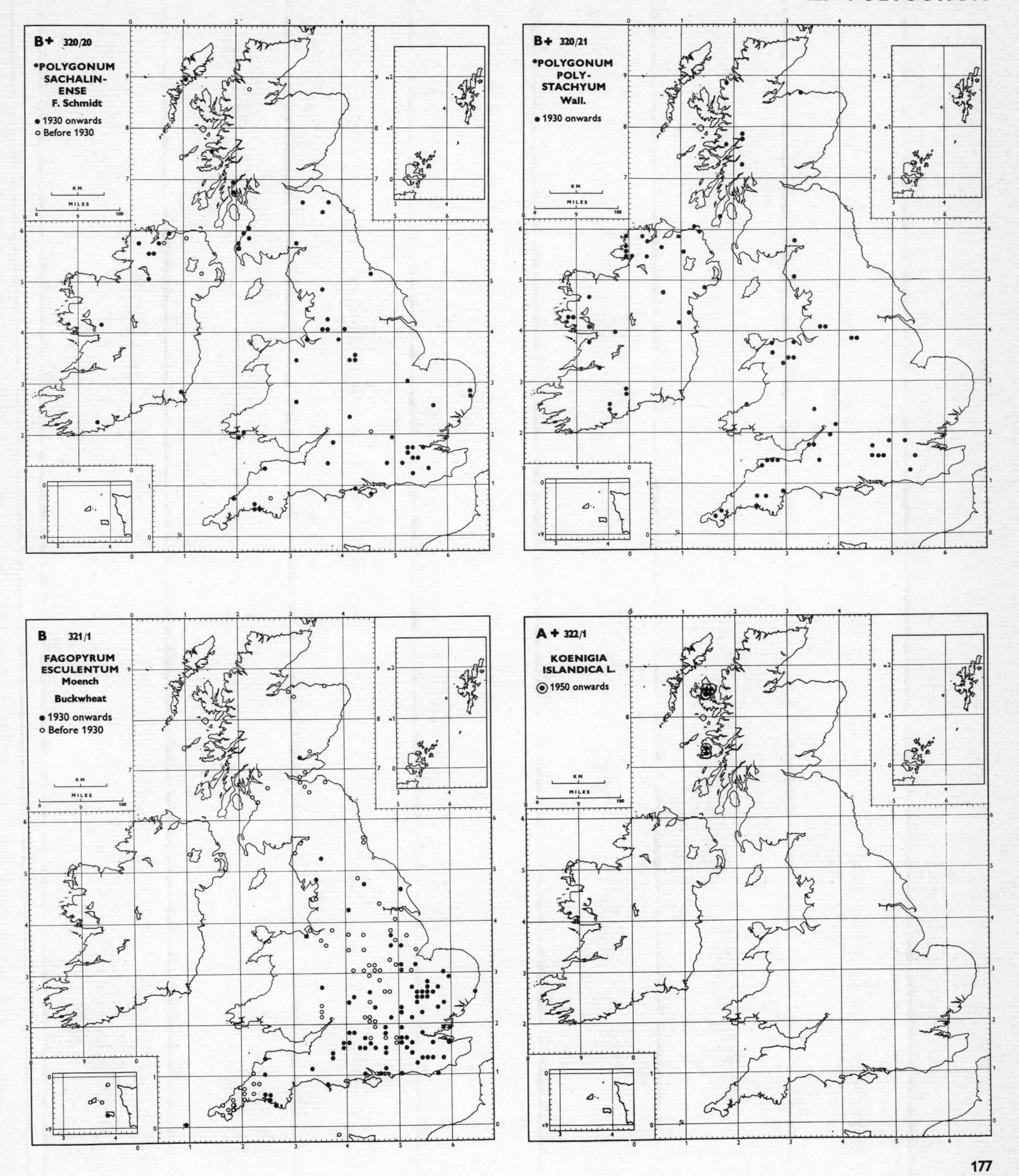
B+ 320/20
*POLYGONUM SACHALINENSE F. Schmidt
● 1930 onwards
○ Before 1930
B+ 320/21
*POLYGONUM POLYSTACHYUM Wall.
● 1930 onwards
B 321/1
FAGOPYRUM ESCULENTUM Moench
Buckwheat
● 1930 onwards
○ Before 1930
A+ 322/1
KOENIGIA ISLANDICA L.
◉ 1950 onwards
KM
MILES

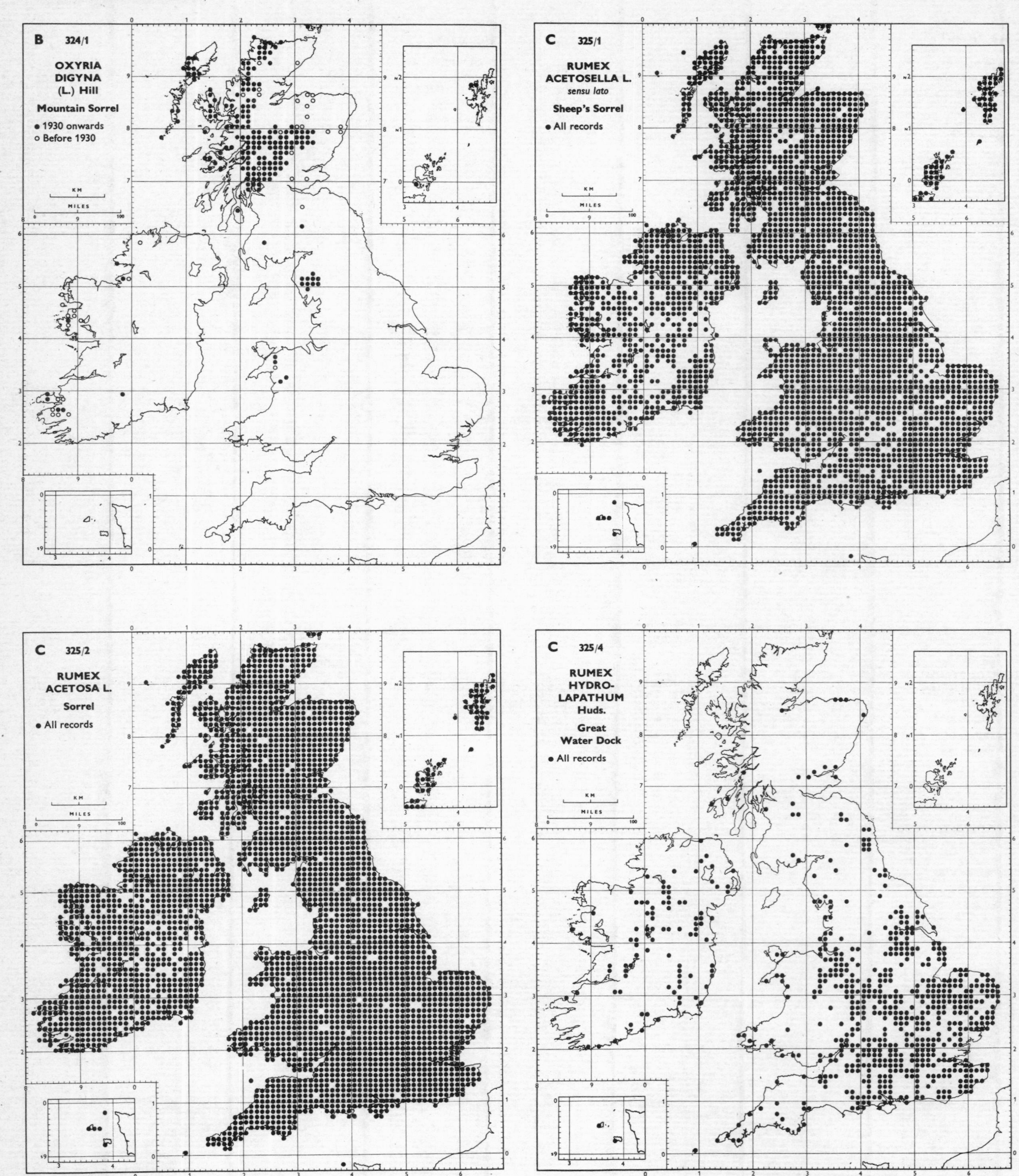
B 324/1
OXYRIA DIGYNA (L.) Hill
Mountain Sorrel
● 1930 onwards
○ Before 1930
C 325/1
RUMEX ACETOSELLA L. sensu lato
Sheep's Sorrel
● All records
C 325/2
RUMEX ACETOSA L.
Sorrel
● All records
C 325/4
RUMEX HYDROLAPATHUM Huds.
Great Water Dock
● All records
KM
MILES

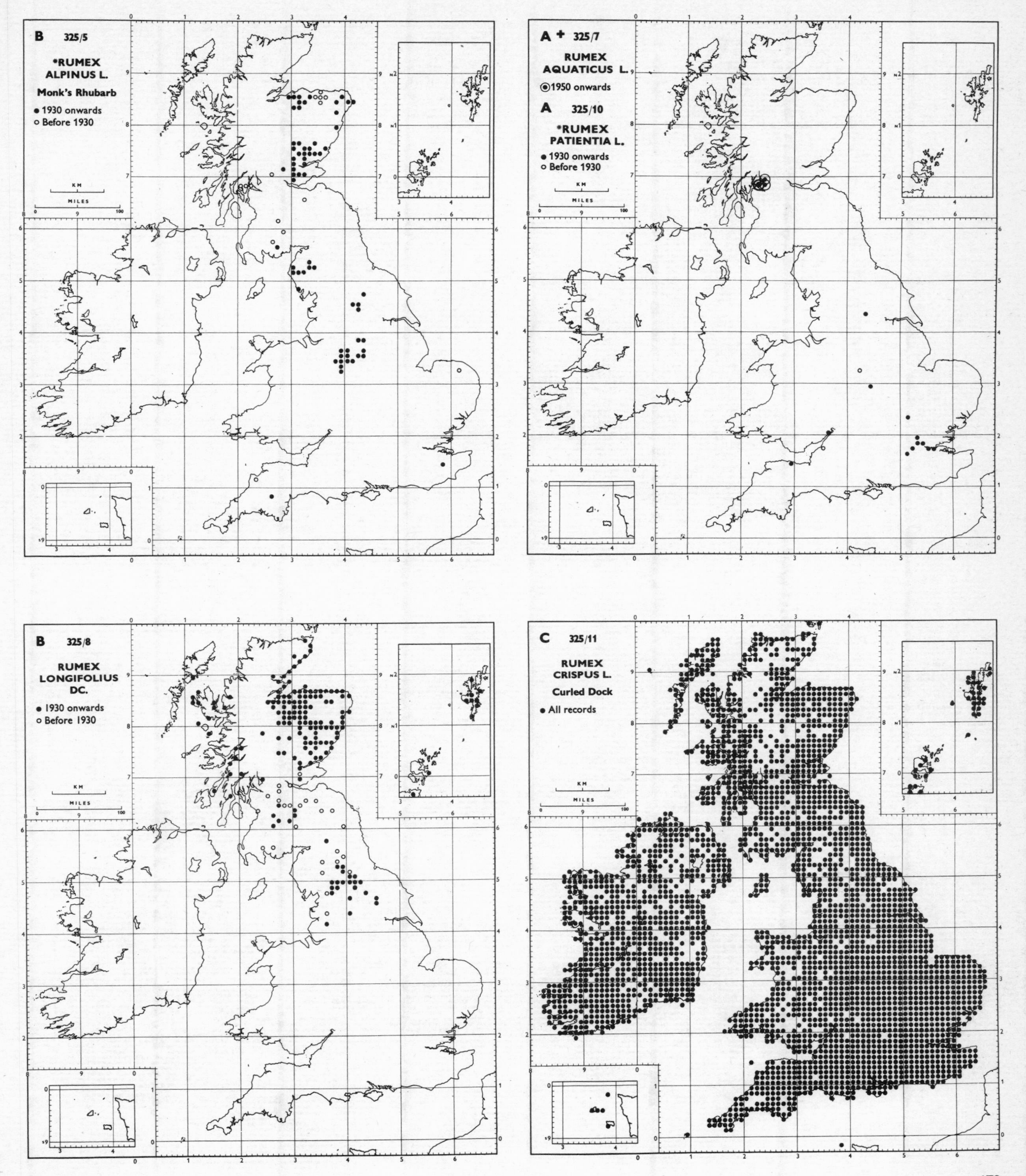
B 325/5
*RUMEX ALPINUS L.
Monk's Rhubarb
● 1930 onwards
○ Before 1930
A + 325/7
RUMEX AQUATICUS L.
⊙ 1950 onwards
A 325/10
*RUMEX PATIENTIA L.
● 1930 onwards
○ Before 1930
B 325/8
RUMEX LONGIFOLIUS DC.
● 1930 onwards
○ Before 1930
C 325/11
RUMEX CRISPUS L.
Curled Dock
● All records
KM
MILES

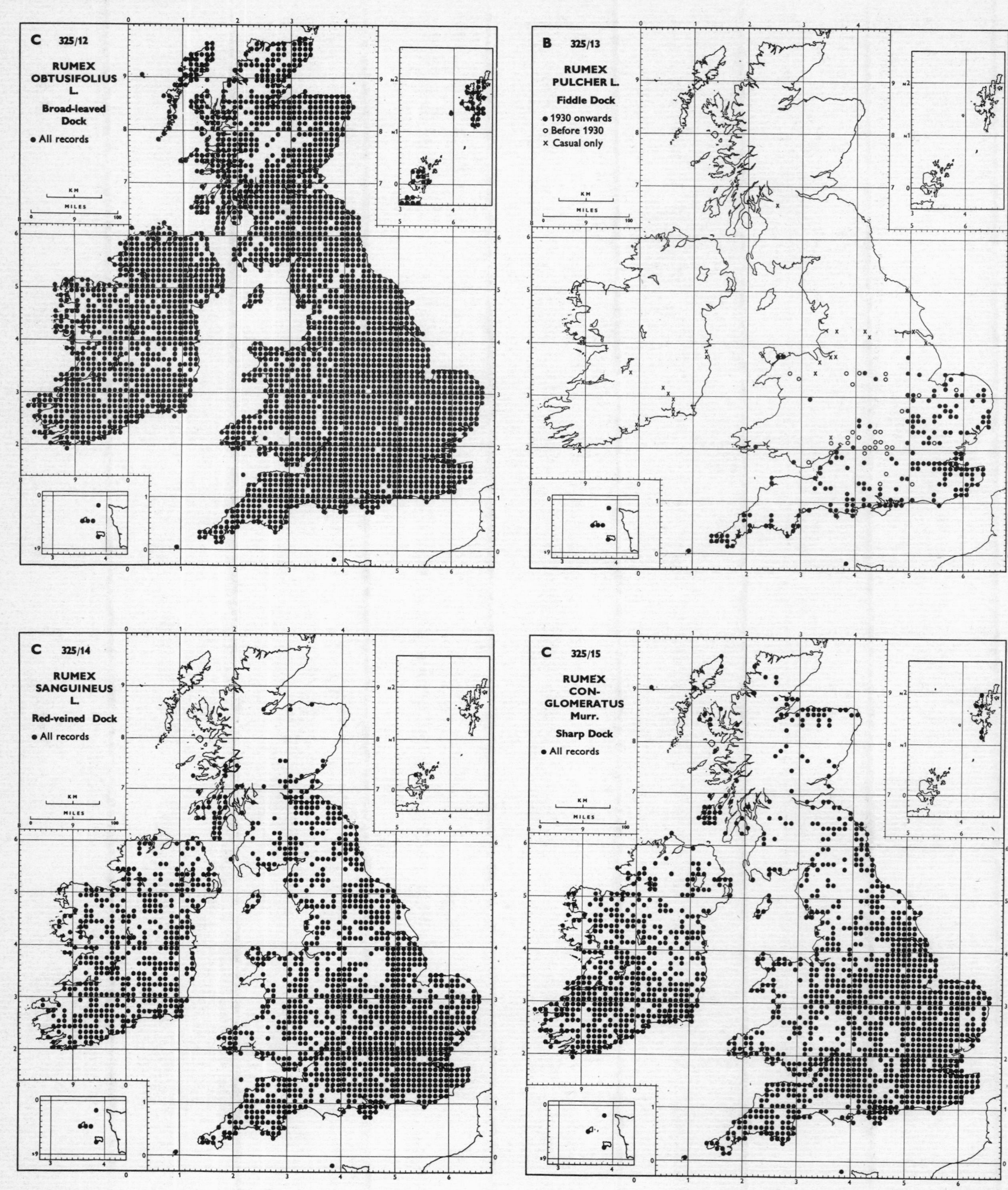
C 325/12
RUMEX OBTUSIFOLIUS L.
Broad-leaved Dock
● All records
B 325/13
RUMEX PULCHER L.
Fiddle Dock
● 1930 onwards
○ Before 1930
× Casual only
C 325/14
RUMEX SANGUINEUS L.
Red-veined Dock
● All records
C 325/15
RUMEX CON-GLOMERATUS Murr.
Sharp Dock
● All records

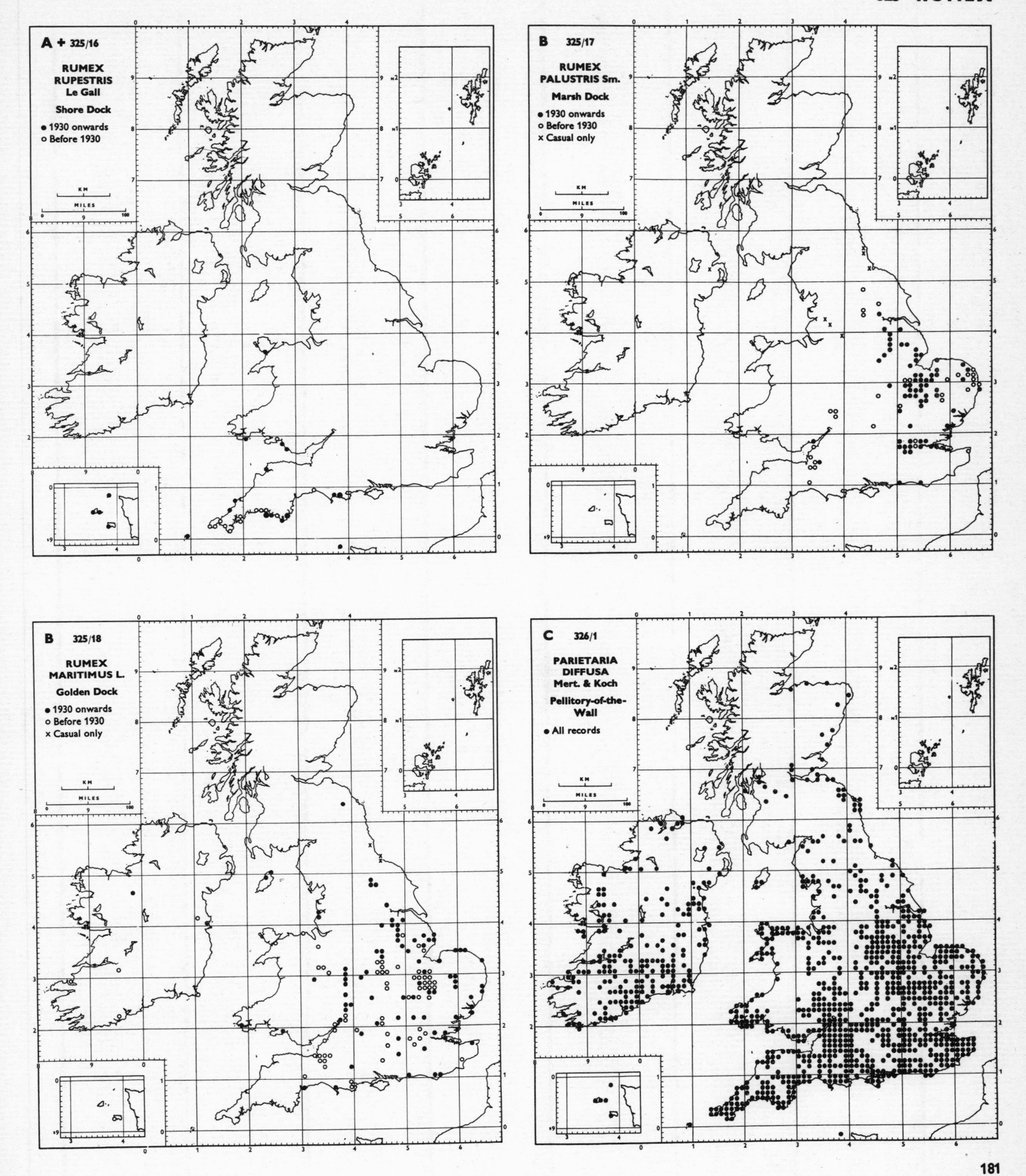
A + 325/16
RUMEX
RUPESTRIS
Le Gall
Shore Dock
• 1930 onwards
○ Before 1930
B 325/17
RUMEX
PALUSTRIS Sm.
Marsh Dock
• 1930 onwards
○ Before 1930
× Casual only
B 325/18
RUMEX
MARITIMUS L.
Golden Dock
• 1930 onwards
○ Before 1930
× Casual only
C 326/1
PARIETARIA
DIFFUSA
Mert. & Koch
Pellitory-of-the-Wall
• All records
KM
MILES

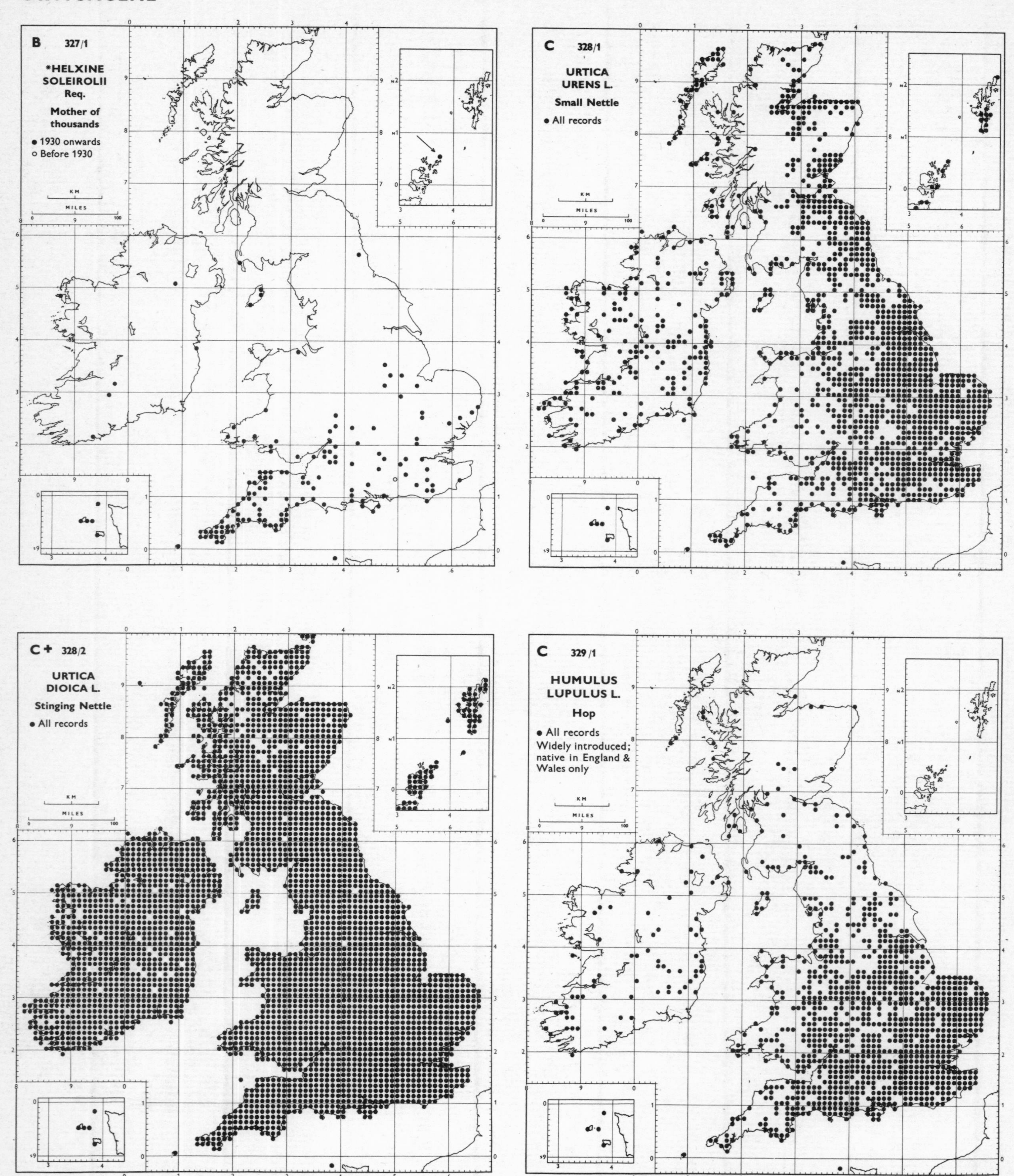
B 327/1
*HELXINE SOLEIROLII Req.
Mother of thousands
● 1930 onwards
○ Before 1930
C 328/1
URTICA URENS L.
Small Nettle
● All records
C+ 328/2
URTICA DIOICA L.
Stinging Nettle
● All records
C 329/1
HUMULUS LUPULUS L.
Hop
● All records
Widely introduced; native in England & Wales only
KM
MILES

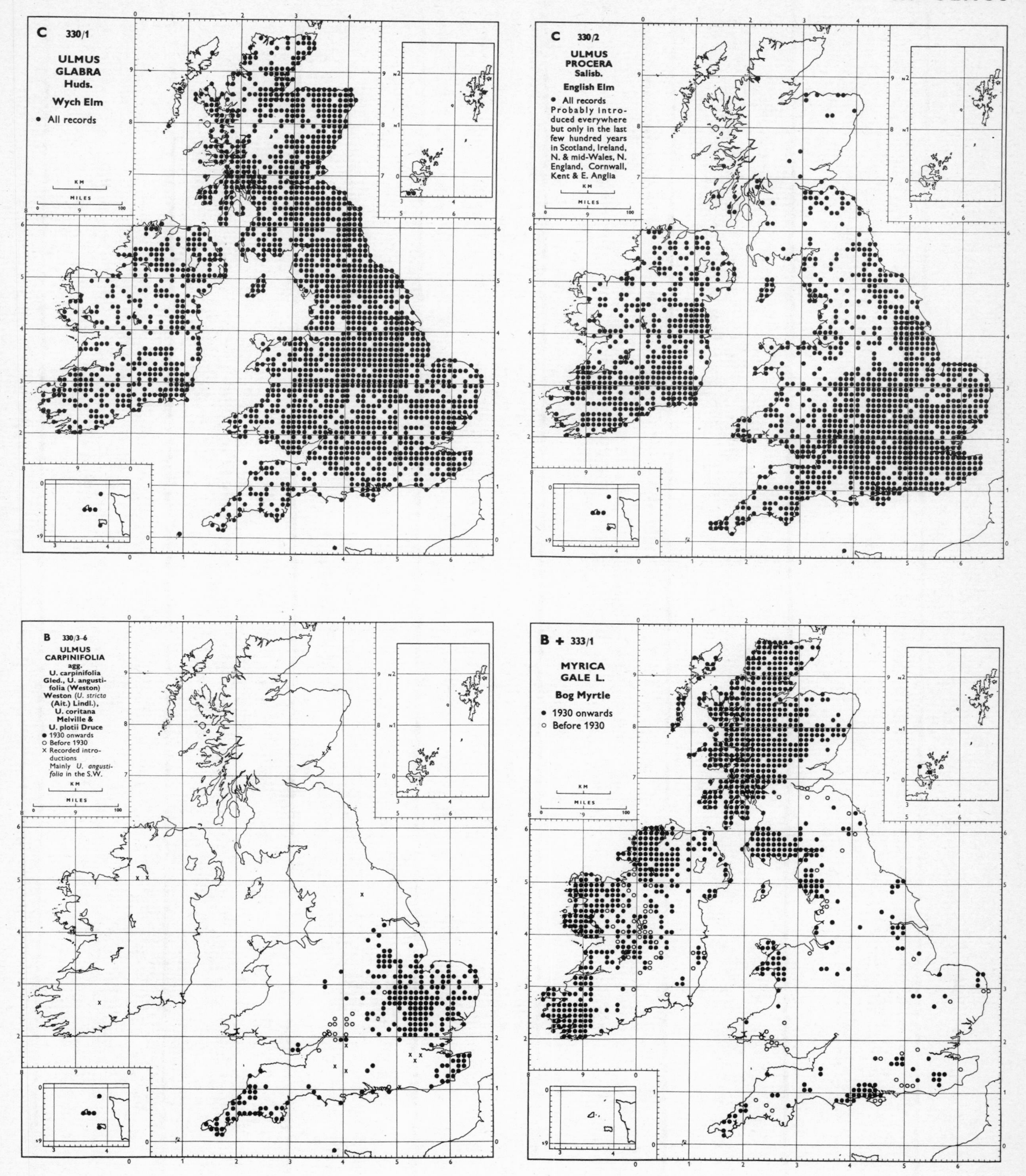
C 330/1
ULMUS GLABRA Huds.
Wych Elm
● All records
KM
MILES
C 330/2
ULMUS PROCERA Salisb.
English Elm
● All records
Probably introduced everywhere but only in the last few hundred years in Scotland, Ireland, N. & mid-Wales, N. England, Cornwall, Kent & E. Anglia
KM
MILES
B 330/3-6
ULMUS CARPINIFOLIA agg.
U. carpinifolia Gled., U. angustifolia (Weston) Weston (U. stricta (Ait.) Lindl.), U. coritana Melville & U. plotii Druce
● 1930 onwards
○ Before 1930
× Recorded introductions
Mainly U. angustifolia in the S.W.
KM
MILES
B + 333/1
MYRICA GALE L.
Bog Myrtle
● 1930 onwards
○ Before 1930
KM
MILES

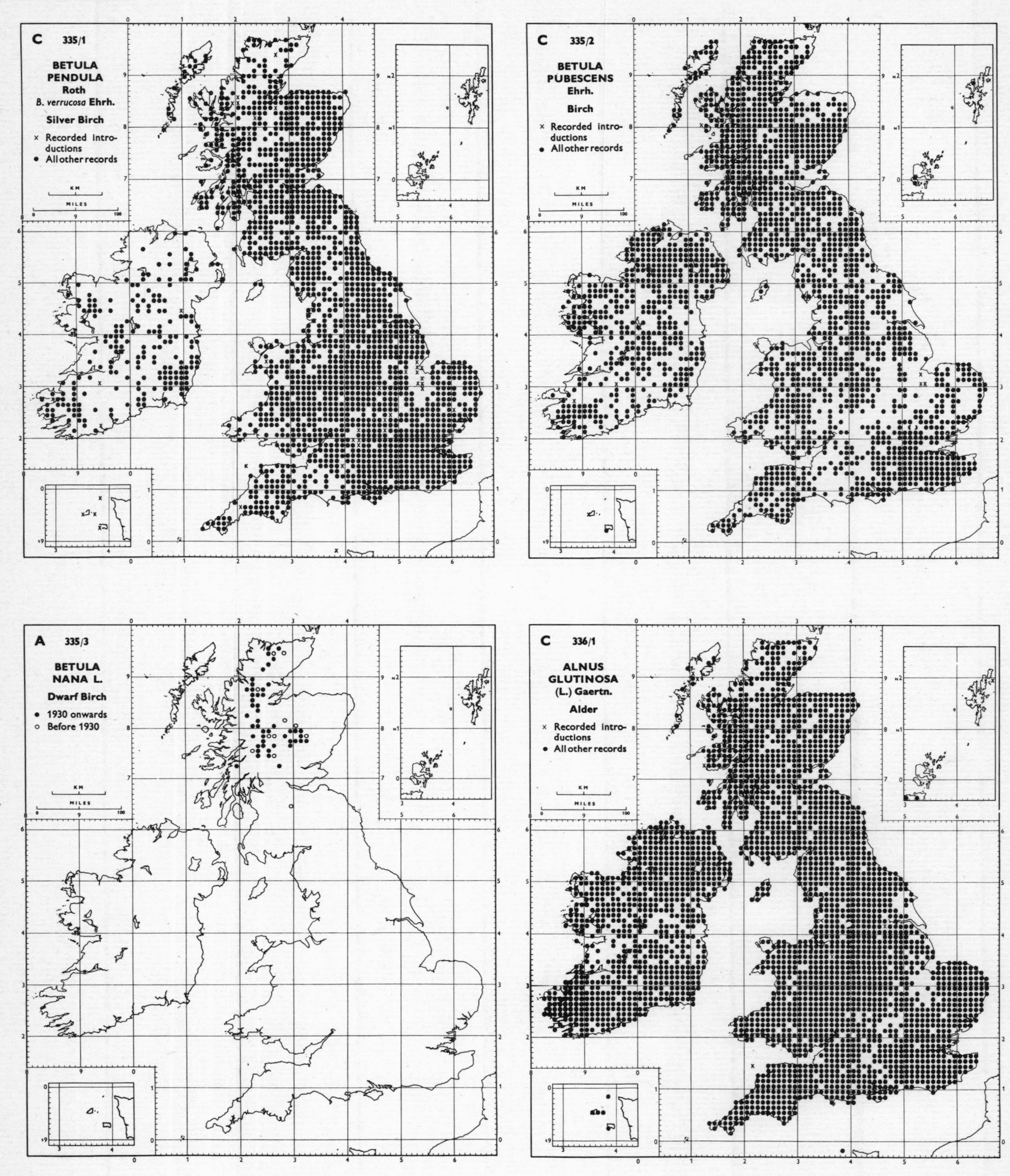
C 335/1
BETULA PENDULA Roth
B. verrucosa Ehrh.
Silver Birch
× Recorded introductions
• All other records
KM
MILES
C 335/2
BETULA PUBESCENS Ehrh.
Birch
× Recorded introductions
• All other records
A 335/3
BETULA NANA L.
Dwarf Birch
• 1930 onwards
○ Before 1930
C 336/1
ALNUS GLUTINOSA (L.) Gaertn.
Alder
× Recorded introductions
• All other records

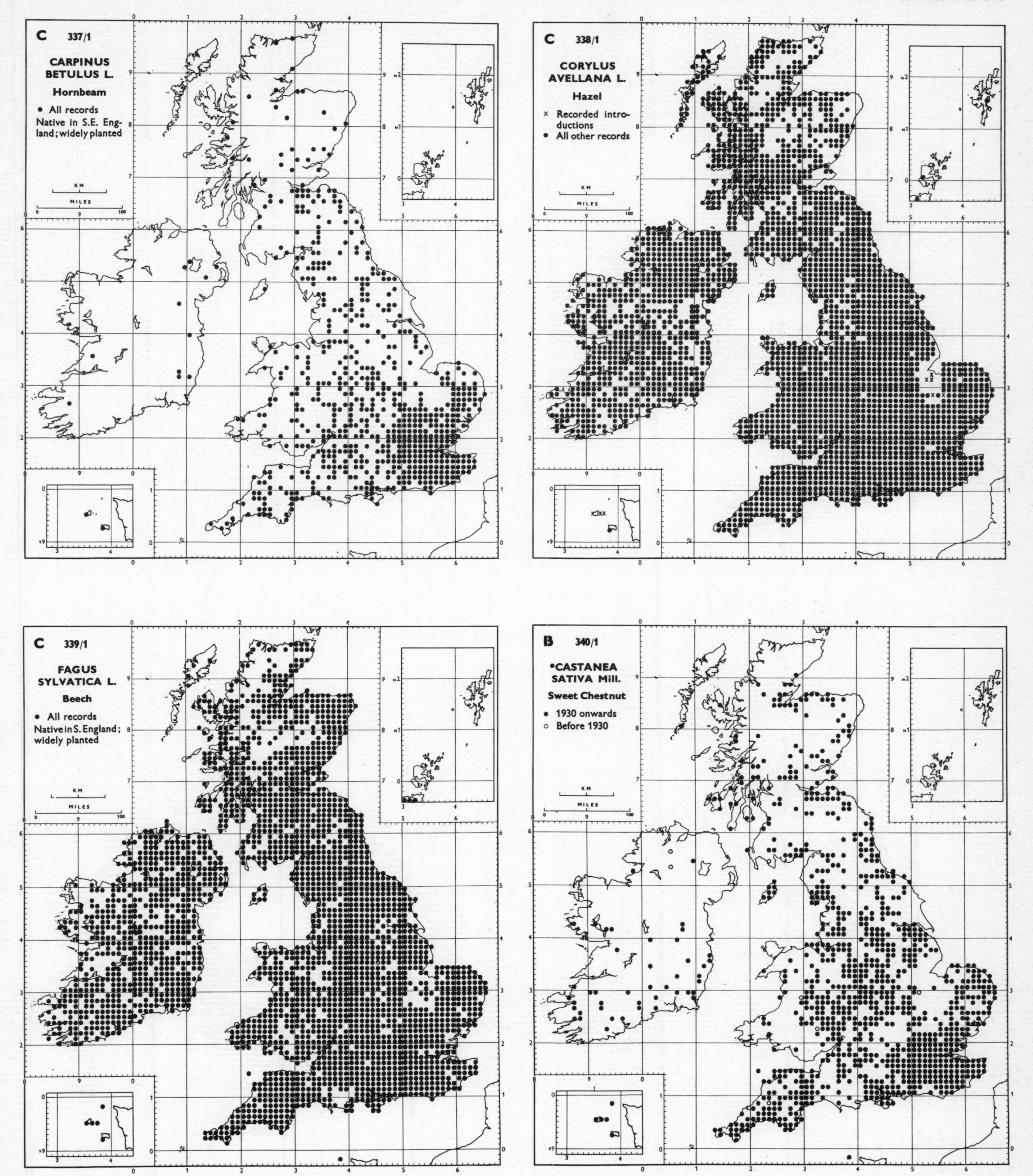
C 337/1
CARPINUS BETULUS L.
Hornbeam
• All records
Native in S.E. England; widely planted
C 338/1
CORYLUS AVELLANA L.
Hazel
× Recorded introductions
• All other records
C 339/1
FAGUS SYLVATICA L.
Beech
• All records
Native in S. England; widely planted
B 340/1
•CASTANEA SATIVA Mill.
Sweet Chestnut
• 1930 onwards
○ Before 1930
KM
MILES

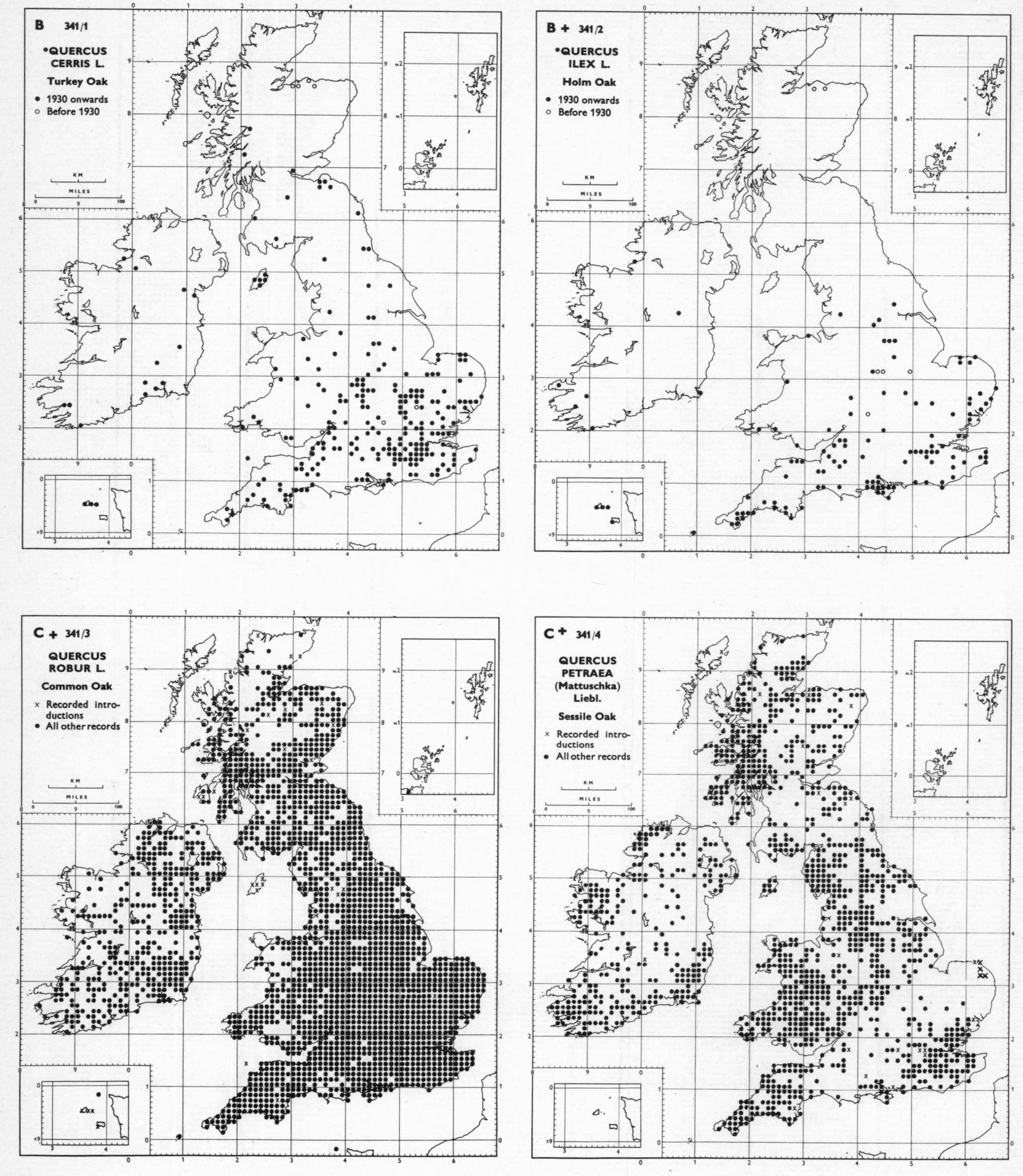
B 341/1
*QUERCUS CERRIS L.
Turkey Oak
● 1930 onwards
○ Before 1930
B + 341/2
*QUERCUS ILEX L.
Holm Oak
● 1930 onwards
○ Before 1930
C + 341/3
QUERCUS ROBUR L.
Common Oak
× Recorded introductions
● All other records
C + 341/4
QUERCUS PETRAEA (Mattuschka) Liebl.
Sessile Oak
× Recorded introductions
● All other records

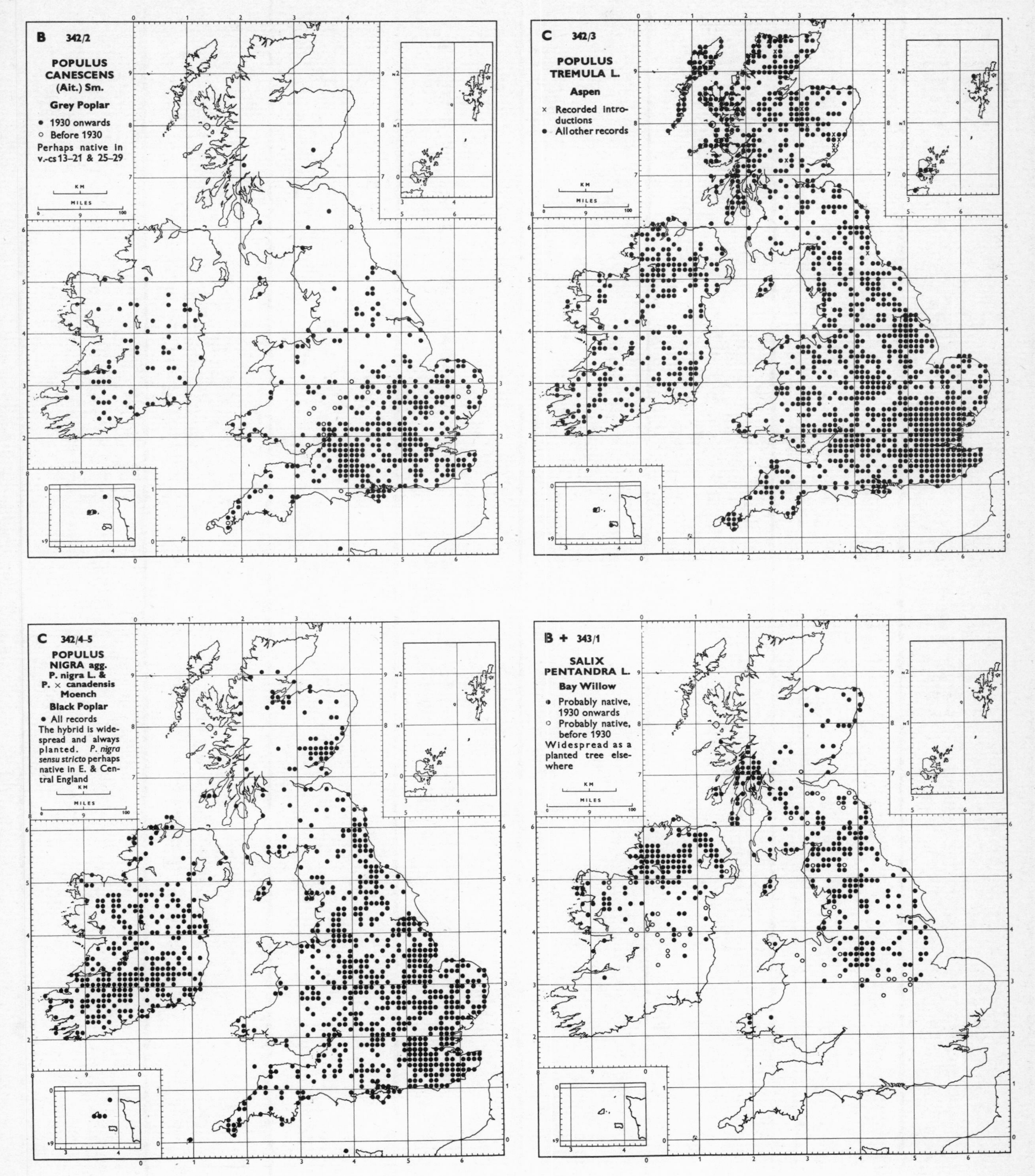
B 342/2
POPULUS CANESCENS (Ait.) Sm.
Grey Poplar
• 1930 onwards
○ Before 1930
Perhaps native in v.-cs 13–21 & 25–29
KM
MILES
C 342/3
POPULUS TREMULA L.
Aspen
× Recorded introductions
• All other records
KM
MILES
C 342/4-5
POPULUS NIGRA agg.
P. nigra L. & P. × canadensis Moench
Black Poplar
• All records
The hybrid is widespread and always planted. P. nigra sensu stricto perhaps native in E. & Central England
KM
MILES
B + 343/1
SALIX PENTANDRA L.
Bay Willow
• Probably native, 1930 onwards
○ Probably native, before 1930
Widespread as a planted tree elsewhere
KM
MILES

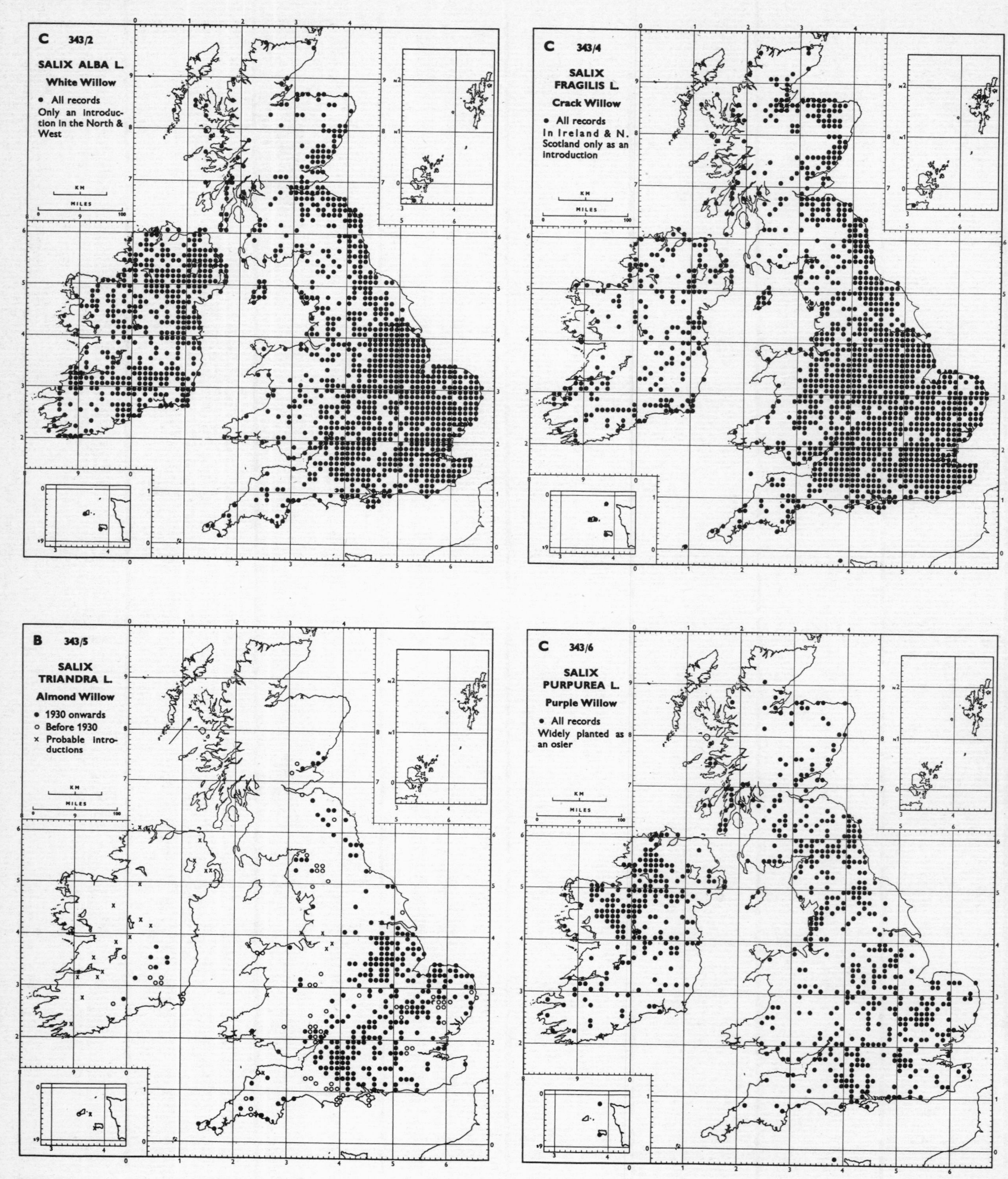
C 343/2
SALIX ALBA L.
White Willow
• All records
Only an introduction in the North & West
C 343/4
SALIX FRAGILIS L.
Crack Willow
• All records
In Ireland & N. Scotland only as an introduction
B 343/5
SALIX TRIANDRA L.
Almond Willow
• 1930 onwards
○ Before 1930
× Probable introductions
C 343/6
SALIX PURPUREA L.
Purple Willow
• All records
Widely planted as an osier

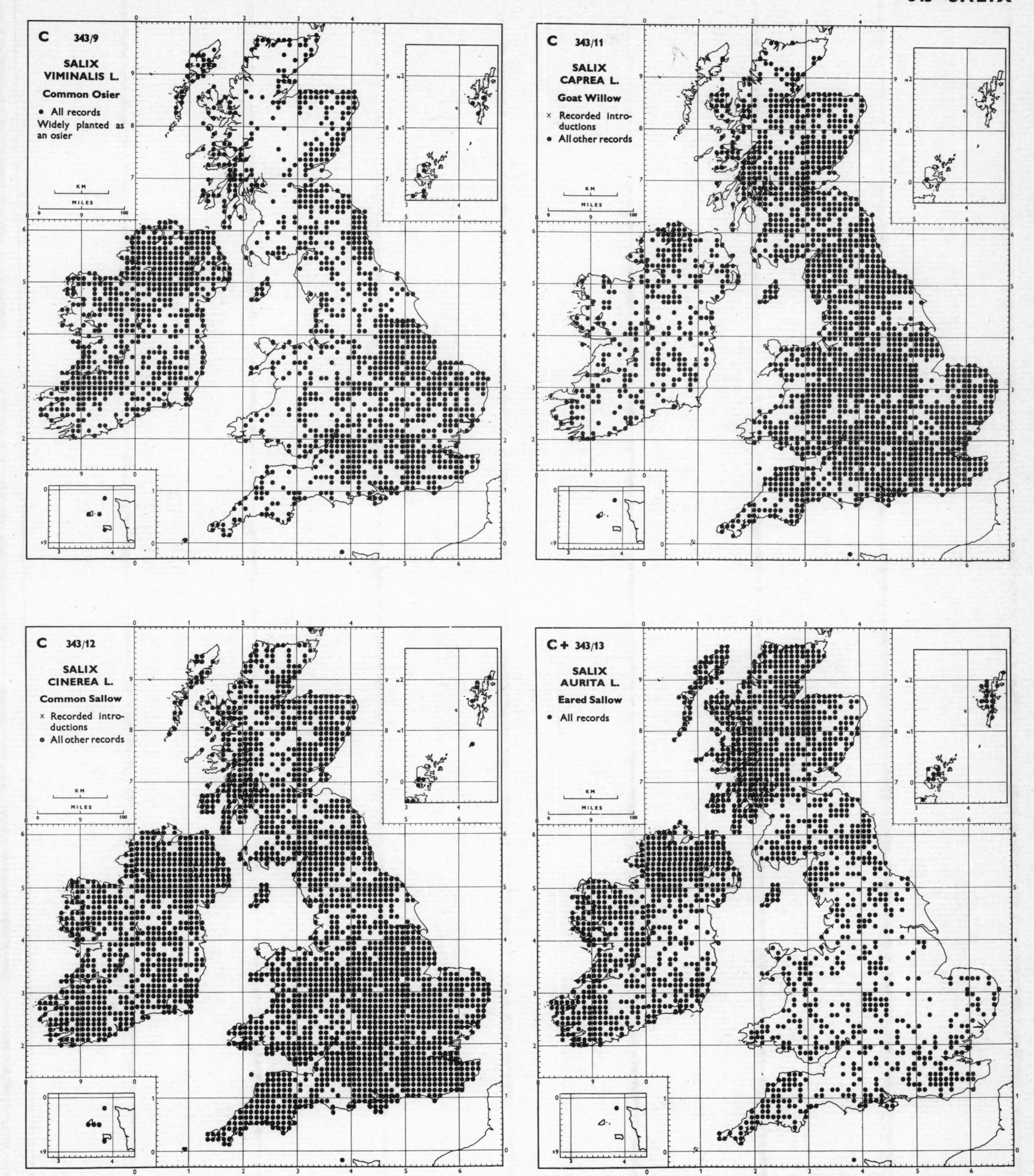
C 343/9
SALIX VIMINALIS L.
Common Osier
• All records
Widely planted as an osier
KM
MILES
C 343/11
SALIX CAPREA L.
Goat Willow
× Recorded introductions
• All other records
C 343/12
SALIX CINEREA L.
Common Sallow
× Recorded introductions
• All other records
C+ 343/13
SALIX AURITA L.
Eared Sallow
• All records

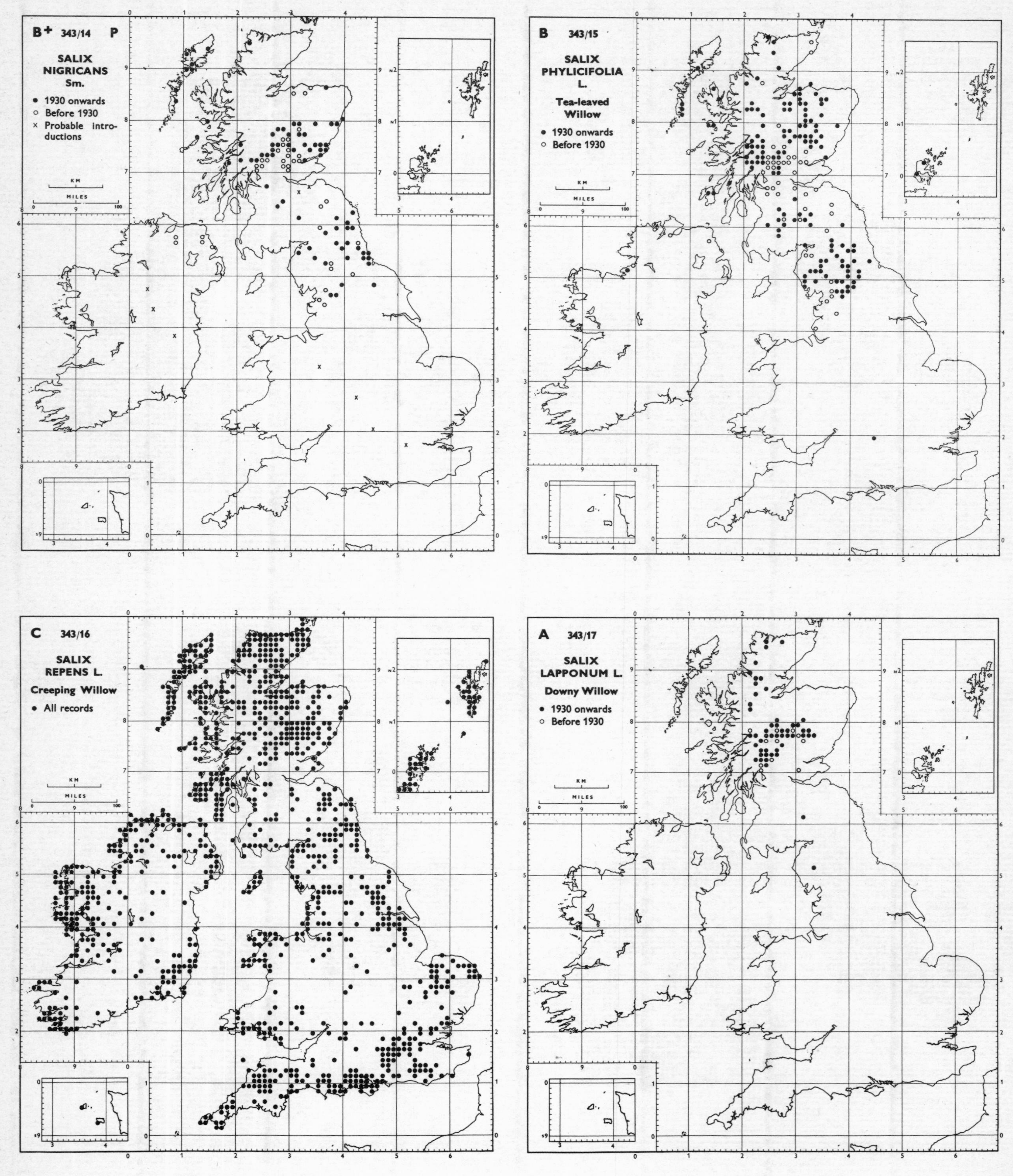
B+ 343/14 P
SALIX
NIGRICANS
Sm.
• 1930 onwards
○ Before 1930
× Probable introductions
B 343/15
SALIX
PHYLICIFOLIA
L.
Tea-leaved
Willow
• 1930 onwards
○ Before 1930
C 343/16
SALIX
REPENS L.
Creeping Willow
• All records
A 343/17
SALIX
LAPPONUM L.
Downy Willow
• 1930 onwards
○ Before 1930

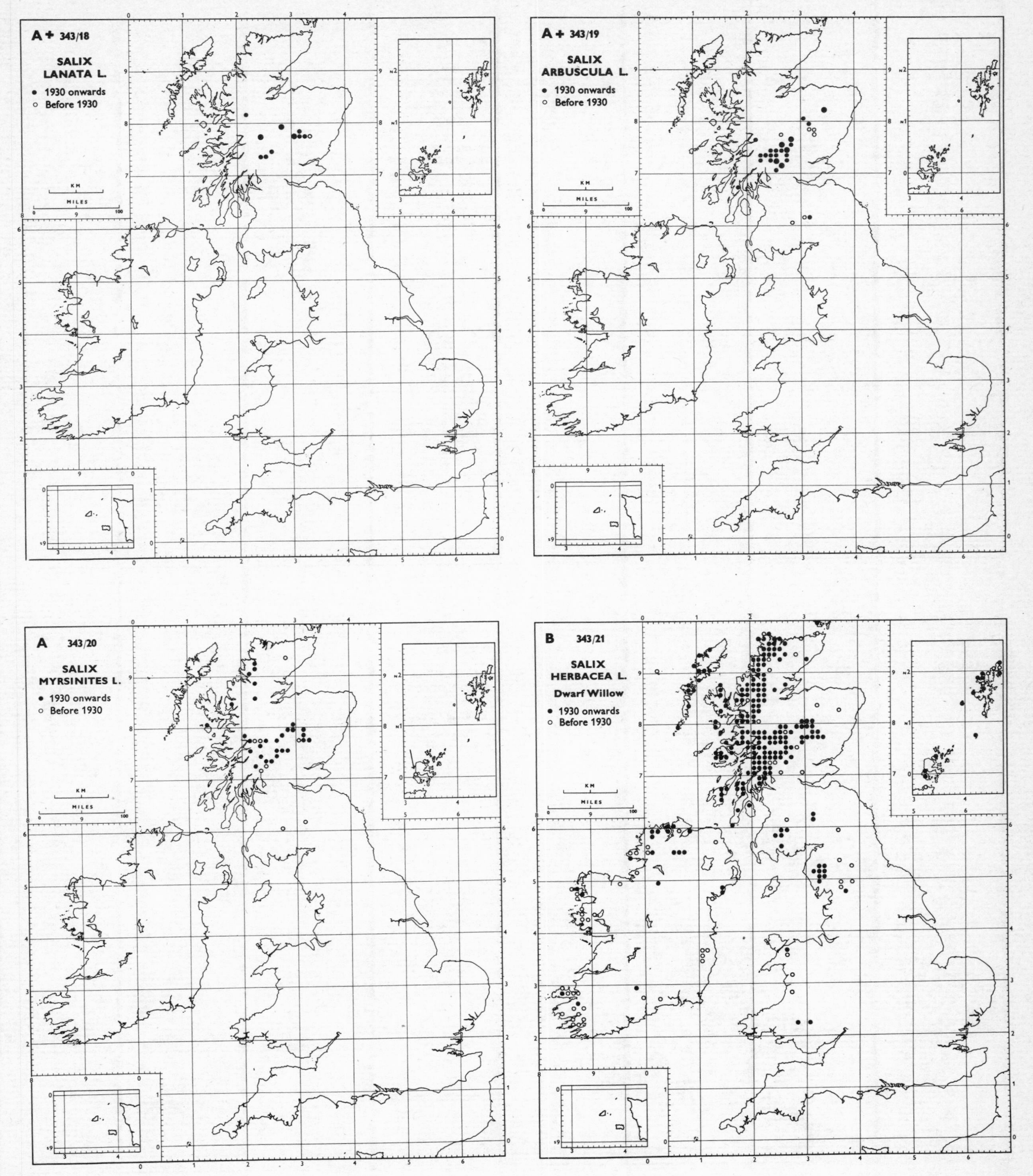
A + 343/18
SALIX
LANATA L.
• 1930 onwards
○ Before 1930
A + 343/19
SALIX
ARBUSCULA L.
• 1930 onwards
○ Before 1930
A 343/20
SALIX
MYRSINITES L.
• 1930 onwards
○ Before 1930
B 343/21
SALIX
HERBACEA L.
Dwarf Willow
• 1930 onwards
○ Before 1930
KM
MILES

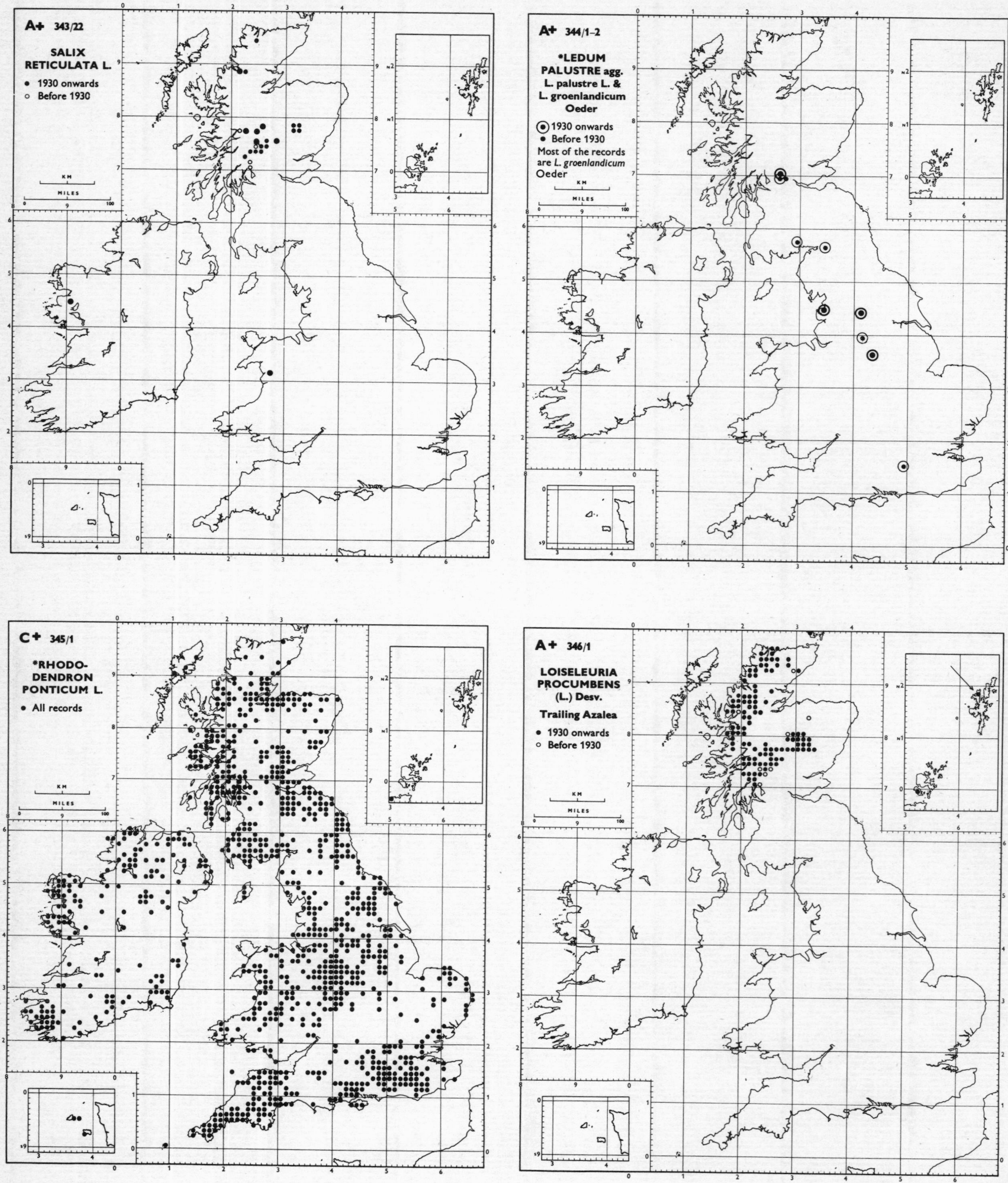
A+ 343/22
SALIX RETICULATA L.
● 1930 onwards
○ Before 1930
KM
MILES
A+ 344/1-2
*LEDUM PALUSTRE agg.
L. palustre L. &
L. groenlandicum Oeder
⊙ 1930 onwards
● Before 1930
Most of the records are L. groenlandicum Oeder
KM
MILES
C+ 345/1
*RHODODENDRON PONTICUM L.
● All records
KM
MILES
A+ 346/1
LOISELEURIA PROCUMBENS (L.) Desv.
Trailing Azalea
● 1930 onwards
○ Before 1930
KM
MILES

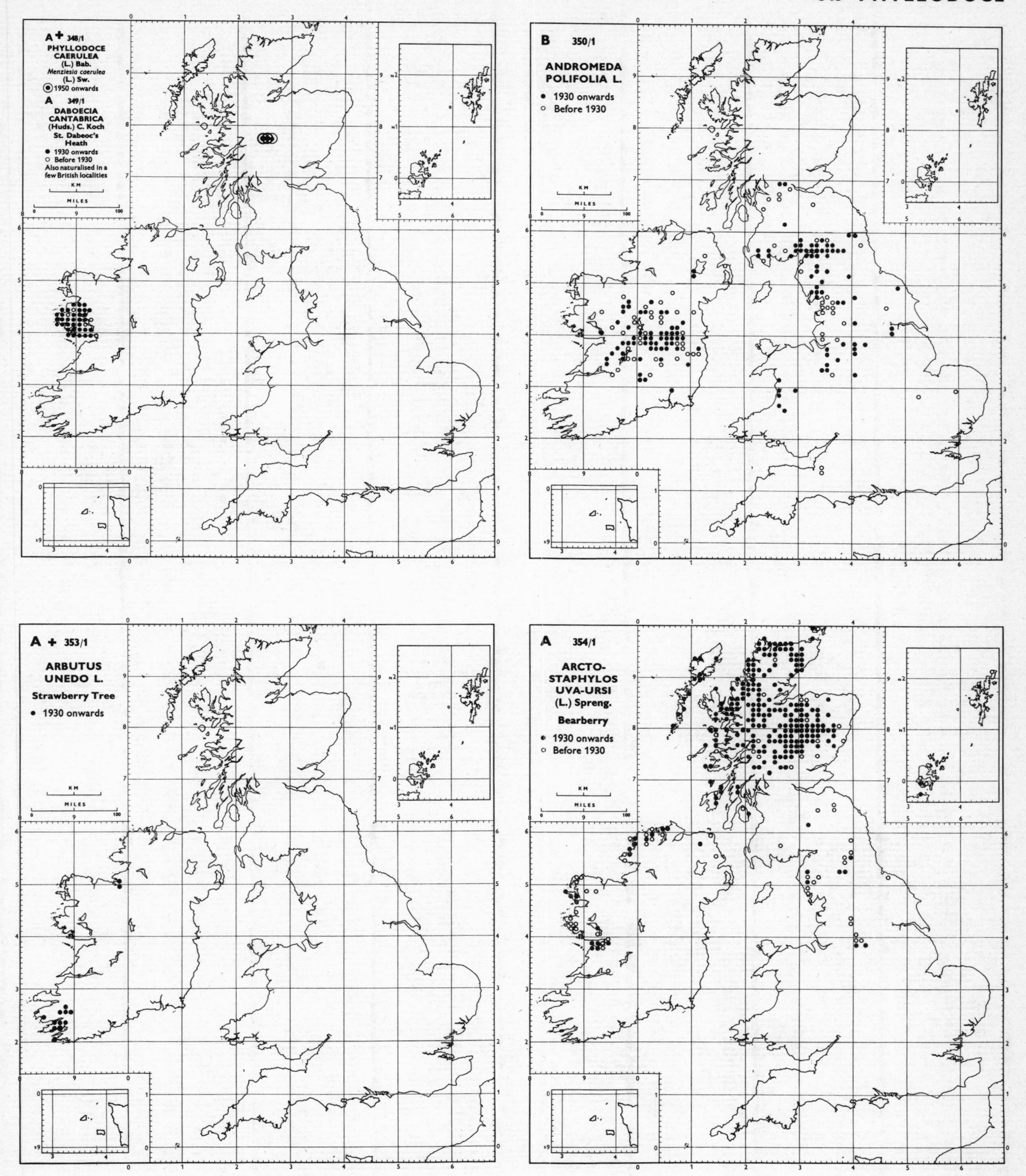
A + 348/1
PHYLLODOCE CAERULEA (L.) Bab.
Menziesia caerulea (L.) Sw.
1950 onwards
A 349/1
DABOECIA CANTABRICA (Huds.) C. Koch
St. Dabeoc's Heath
1930 onwards
Before 1930
Also naturalised in a few British localities
B 350/1
ANDROMEDA POLIFOLIA L.
1930 onwards
Before 1930
A + 353/1
ARBUTUS UNEDO L.
Strawberry Tree
1930 onwards
A 354/1
ARCTOSTAPHYLOS UVA-URSI (L.) Spreng.
Bearberry
1930 onwards
Before 1930
KM
MILES

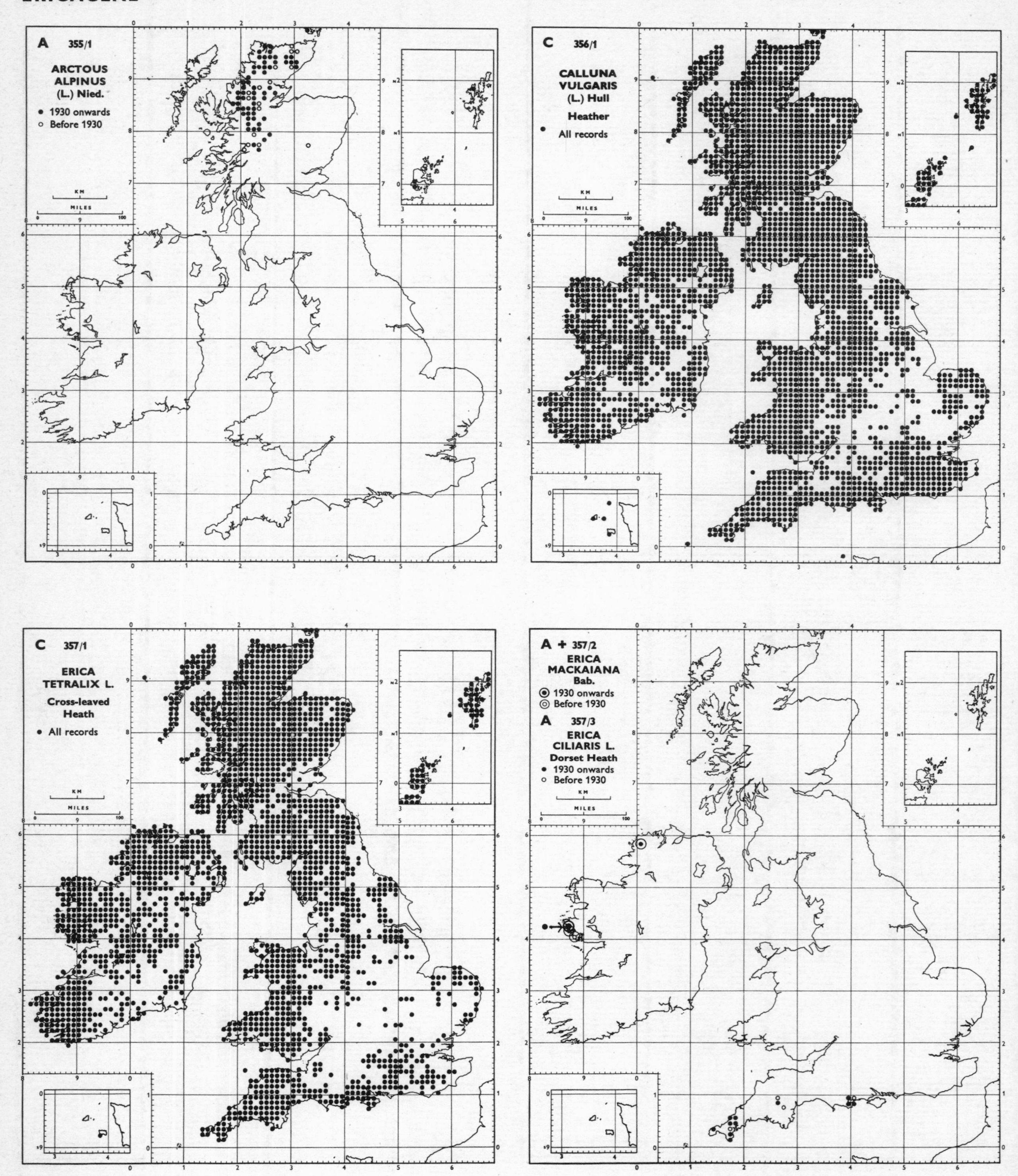
A 355/1
ARCTOUS ALPINUS (L.) Nied.
• 1930 onwards
○ Before 1930
C 356/1
CALLUNA VULGARIS (L.) Hull
Heather
• All records
C 357/1
ERICA TETRALIX L.
Cross-leaved Heath
• All records
A + 357/2
ERICA MACKAIANA Bab.
1930 onwards
Before 1930
A 357/3
ERICA CILIARIS L.
Dorset Heath
• 1930 onwards
○ Before 1930
KM
MILES

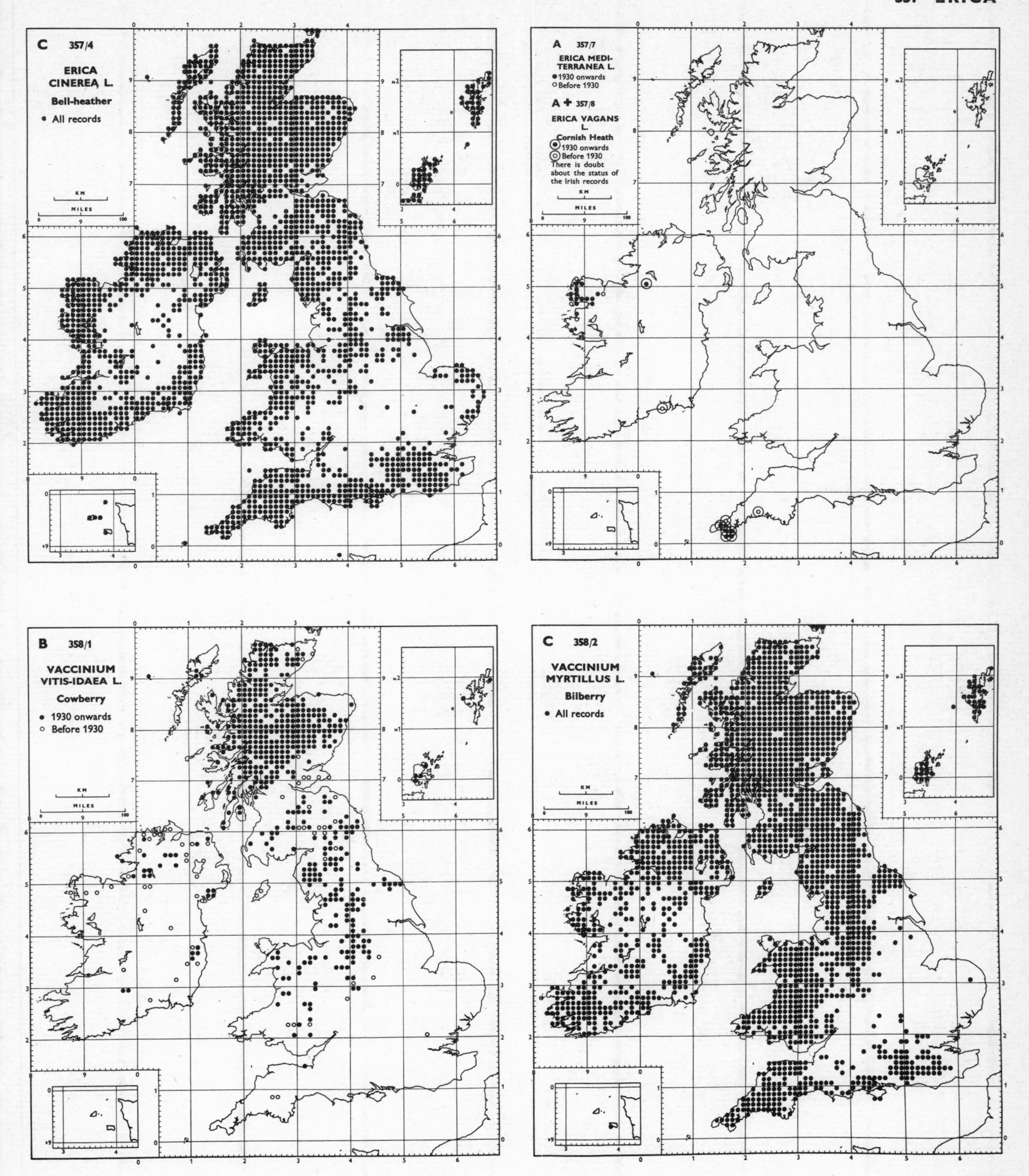
C 357/4
ERICA CINEREA L.
Bell-heather
• All records
A 357/7
ERICA MEDITERRANEA L.
• 1930 onwards
○ Before 1930
A + 357/8
ERICA VAGANS L.
Cornish Heath
1930 onwards
Before 1930
There is doubt about the status of the Irish records
B 358/1
VACCINIUM VITIS-IDAEA L.
Cowberry
• 1930 onwards
○ Before 1930
C 358/2
VACCINIUM MYRTILLUS L.
Bilberry
• All records

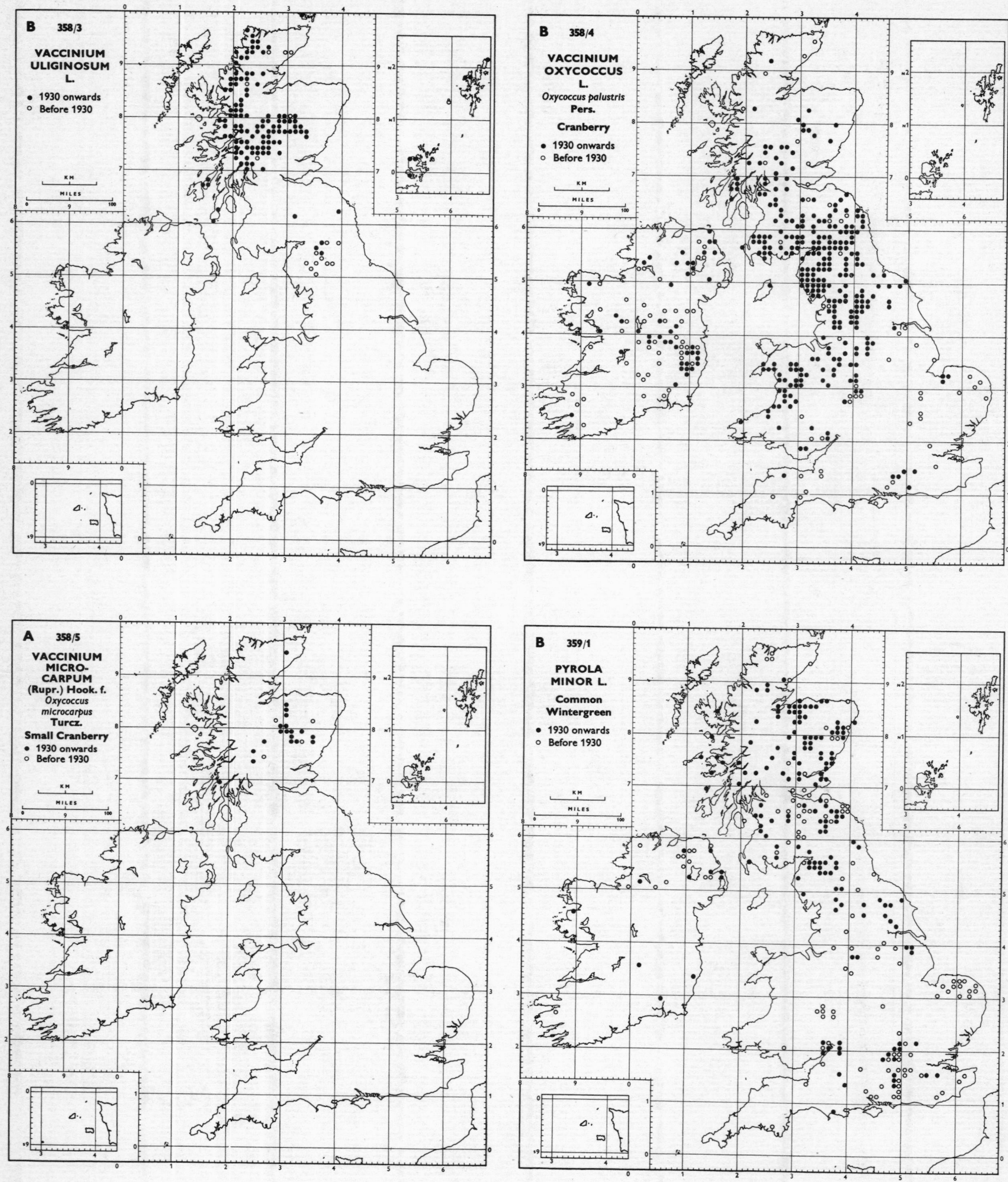
B 358/3
VACCINIUM ULIGINOSUM L.
1930 onwards
Before 1930
B 358/4
VACCINIUM OXYCOCCUS L.
Oxycoccus palustris Pers.
Cranberry
1930 onwards
Before 1930
A 358/5
VACCINIUM MICRO-CARPUM (Rupr.) Hook. f.
Oxycoccus microcarpus Turcz.
Small Cranberry
1930 onwards
Before 1930
B 359/1
PYROLA MINOR L.
Common Wintergreen
1930 onwards
Before 1930

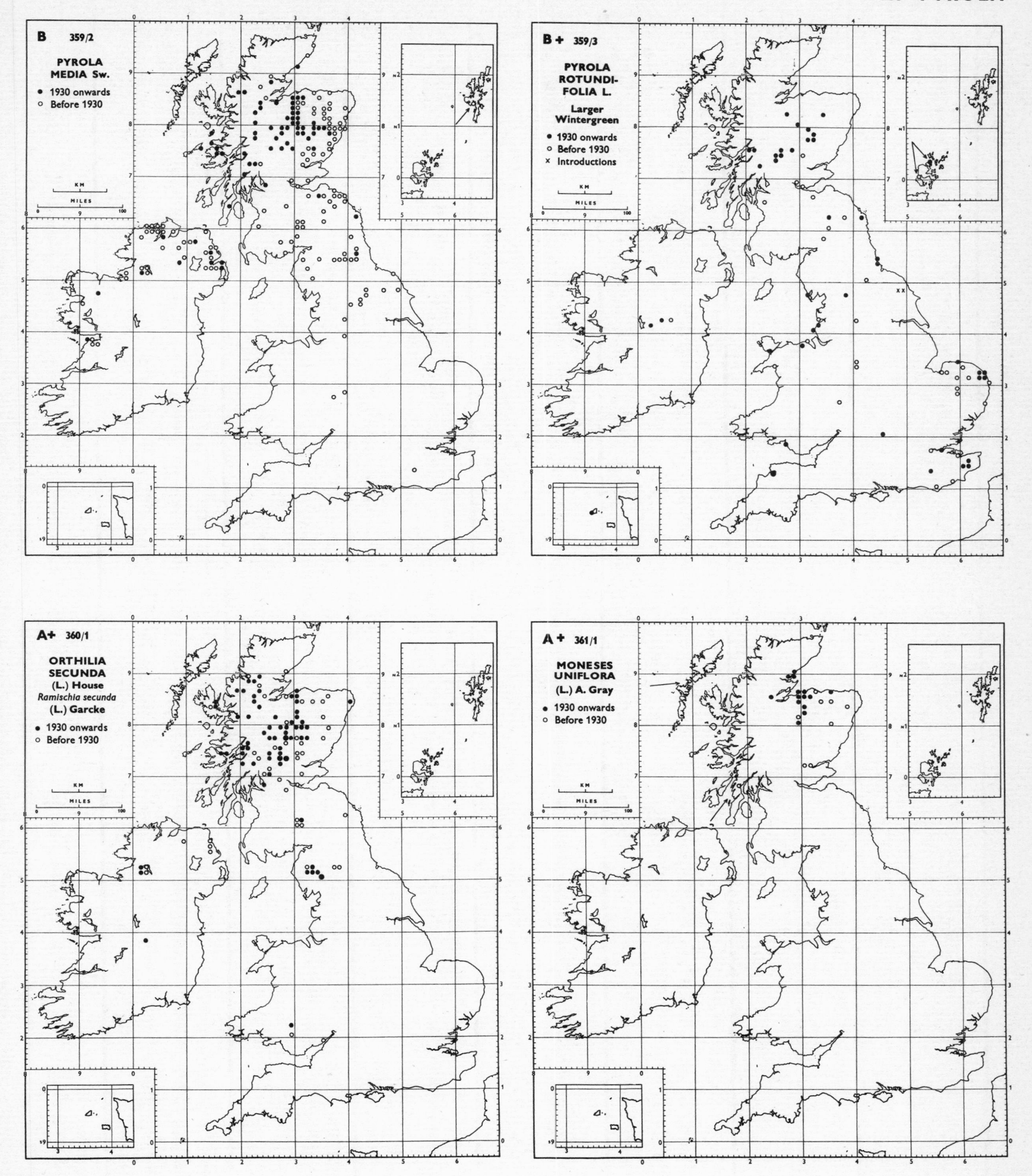
B 359/2
PYROLA
MEDIA Sw.
• 1930 onwards
○ Before 1930
KM
MILES
B + 359/3
PYROLA
ROTUNDI-
FOLIA L.
Larger
Wintergreen
• 1930 onwards
○ Before 1930
× Introductions
A+ 360/1
ORTHILIA
SECUNDA
(L.) House
Ramischia secunda
(L.) Garcke
• 1930 onwards
○ Before 1930
A + 361/1
MONESES
UNIFLORA
(L.) A. Gray
• 1930 onwards
○ Before 1930

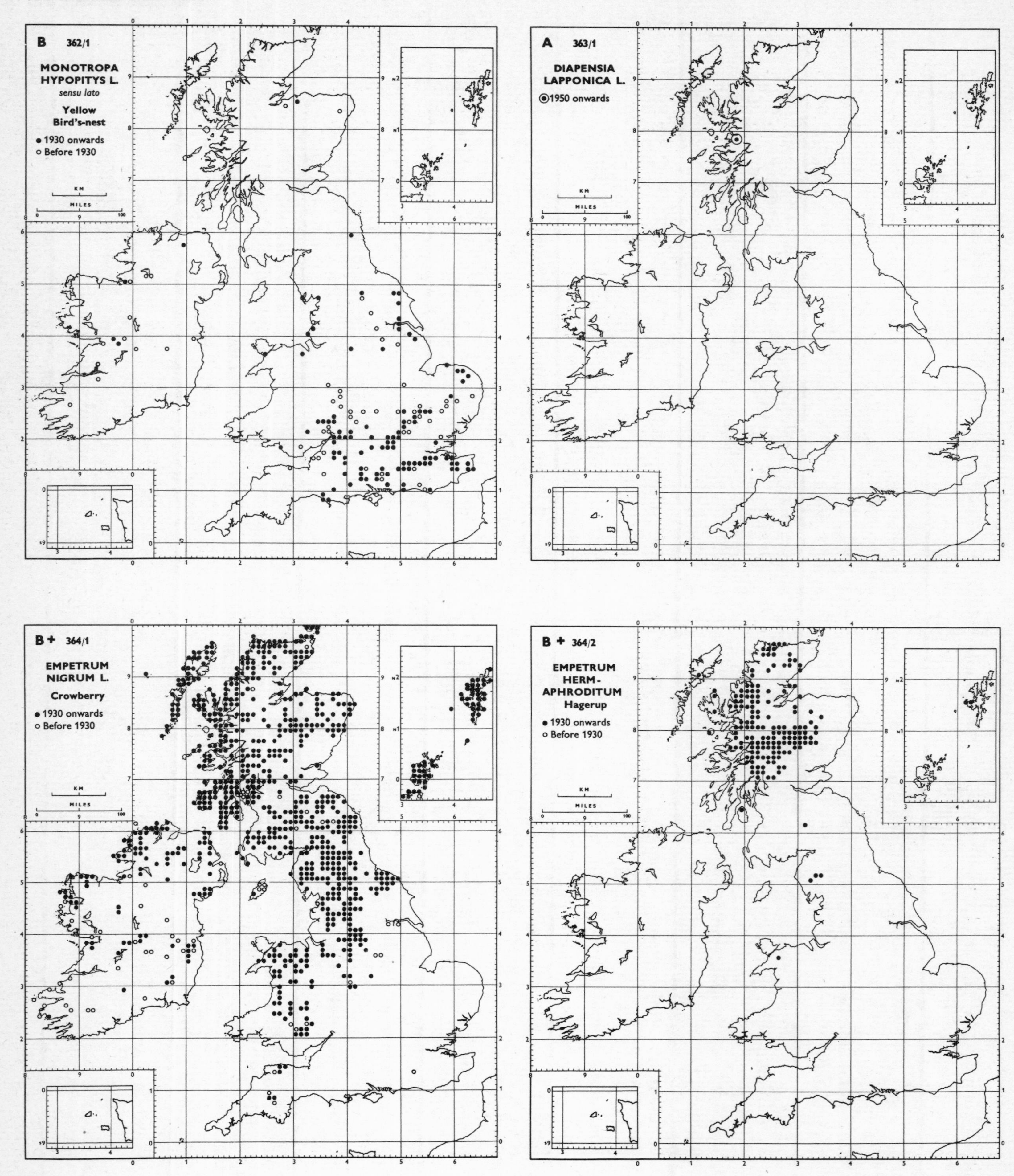
B 362/1
MONOTROPA HYPOPITYS L.
sensu lato
Yellow Bird's-nest
• 1930 onwards
○ Before 1930
A 363/1
DIAPENSIA LAPPONICA L.
⊙ 1950 onwards
B+ 364/1
EMPETRUM NIGRUM L.
Crowberry
• 1930 onwards
○ Before 1930
B+ 364/2
EMPETRUM HERMAPHRODITUM Hagerup
• 1930 onwards
○ Before 1930

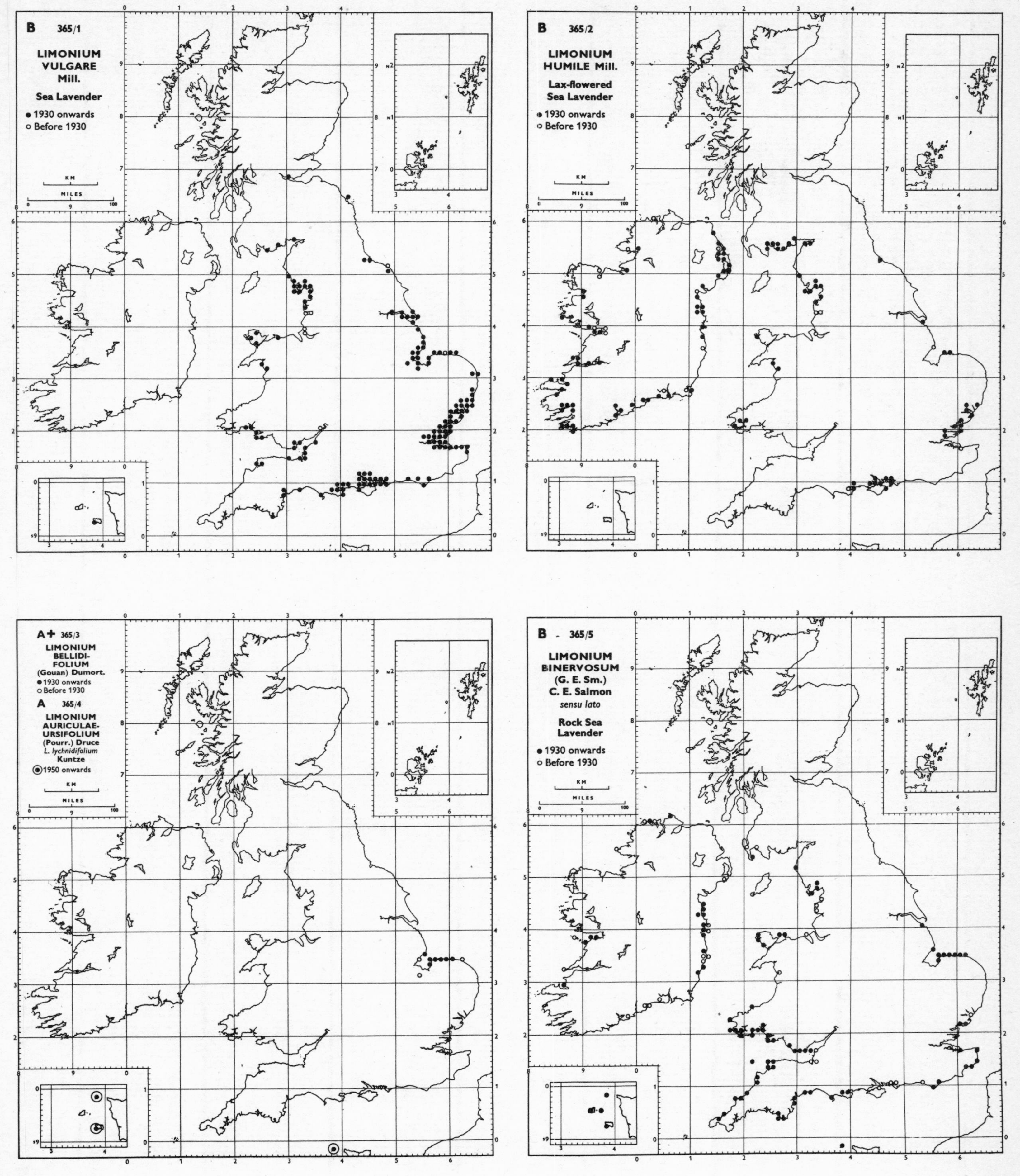
B 365/1
LIMONIUM VULGARE Mill.
Sea Lavender
● 1930 onwards
○ Before 1930
B 365/2
LIMONIUM HUMILE Mill.
Lax-flowered Sea Lavender
● 1930 onwards
○ Before 1930
A+ 365/3
LIMONIUM BELLIDIFOLIUM (Gouan) Dumort.
● 1930 onwards
○ Before 1930
A 365/4
LIMONIUM AURICULAE-URSIFOLIUM (Pourr.) Druce
L. lychnidifolium Kuntze
◉ 1950 onwards
B 365/5
LIMONIUM BINERVOSUM (G. E. Sm.) C. E. Salmon
sensu lato
Rock Sea Lavender
● 1930 onwards
○ Before 1930
KM
MILES

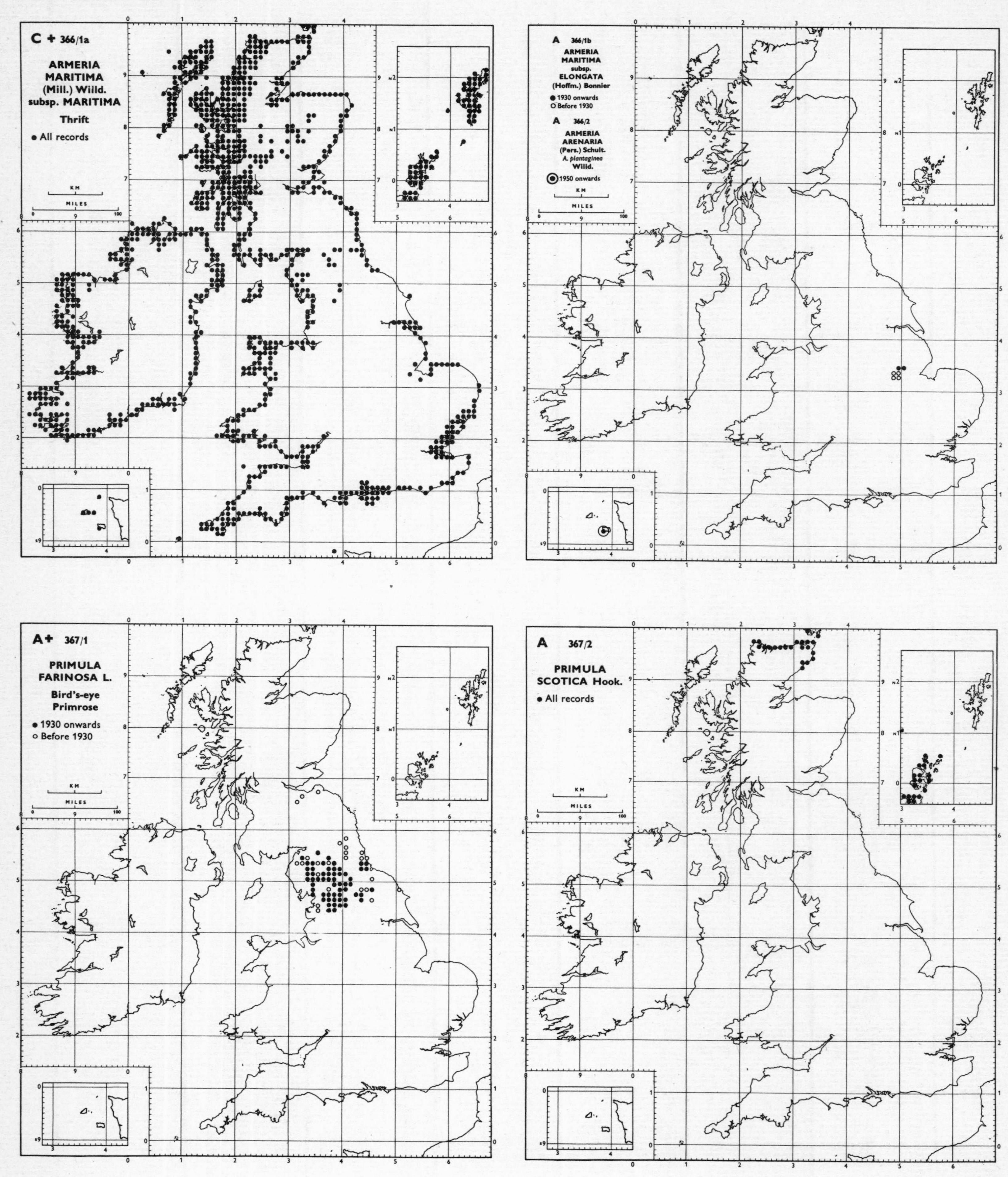
C + 366/1a
ARMERIA MARITIMA (Mill.) Willd. subsp. MARITIMA
Thrift
● All records
KM
MILES
A 366/1b
ARMERIA MARITIMA subsp. ELONGATA (Hoffm.) Bonnier
● 1930 onwards
○ Before 1930
A 366/2
ARMERIA ARENARIA (Pers.) Schult.
A. plantaginea Willd.
◉ 1950 onwards
A+ 367/1
PRIMULA FARINOSA L.
Bird's-eye Primrose
● 1930 onwards
○ Before 1930
A 367/2
PRIMULA SCOTICA Hook.
● All records

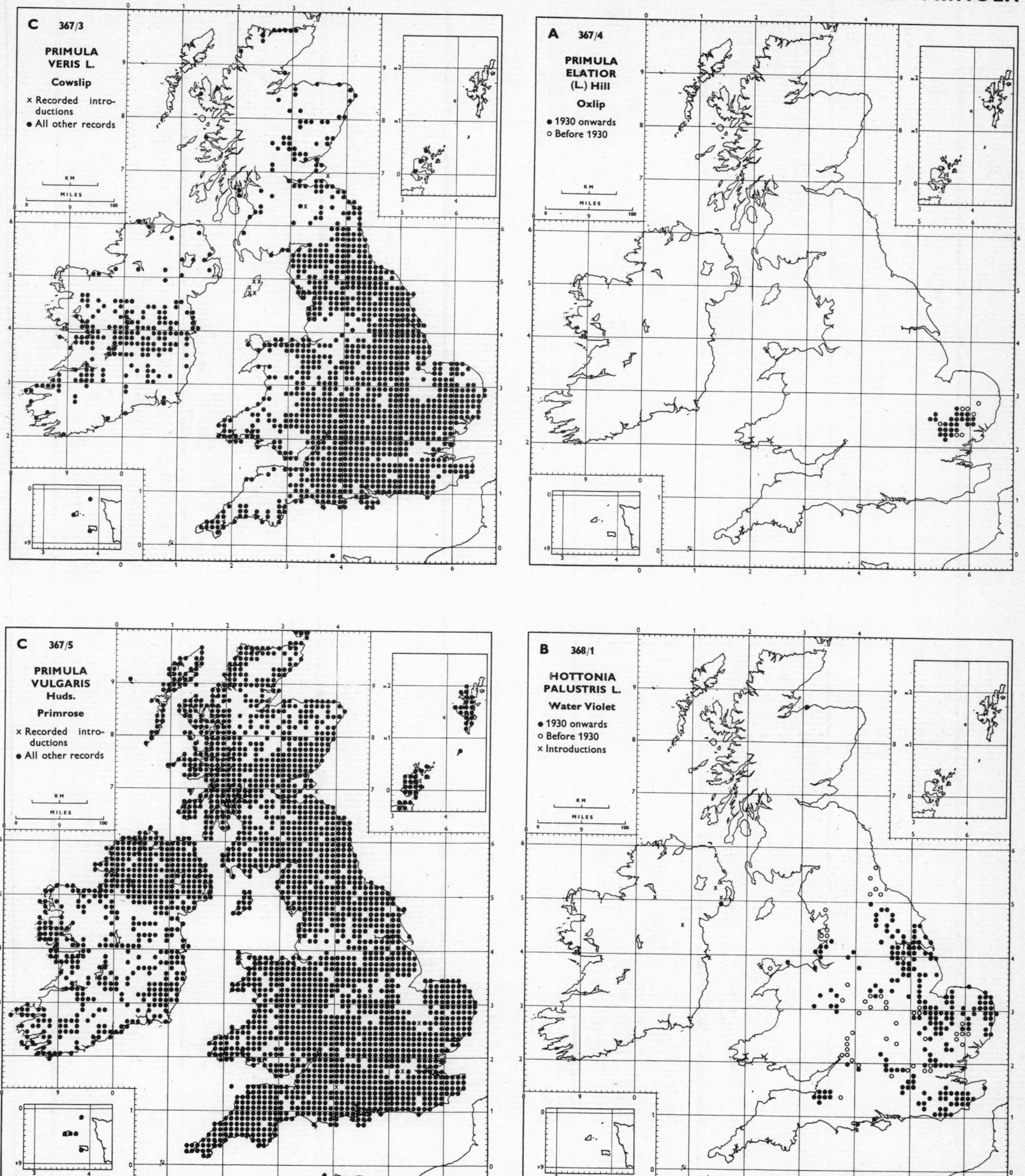
C 367/3
PRIMULA VERIS L.
Cowslip
× Recorded introductions
● All other records
A 367/4
PRIMULA ELATIOR (L.) Hill
Oxlip
● 1930 onwards
○ Before 1930
C 367/5
PRIMULA VULGARIS Huds.
Primrose
× Recorded introductions
● All other records
B 368/1
HOTTONIA PALUSTRIS L.
Water Violet
● 1930 onwards
○ Before 1930
× Introductions
KM
MILES

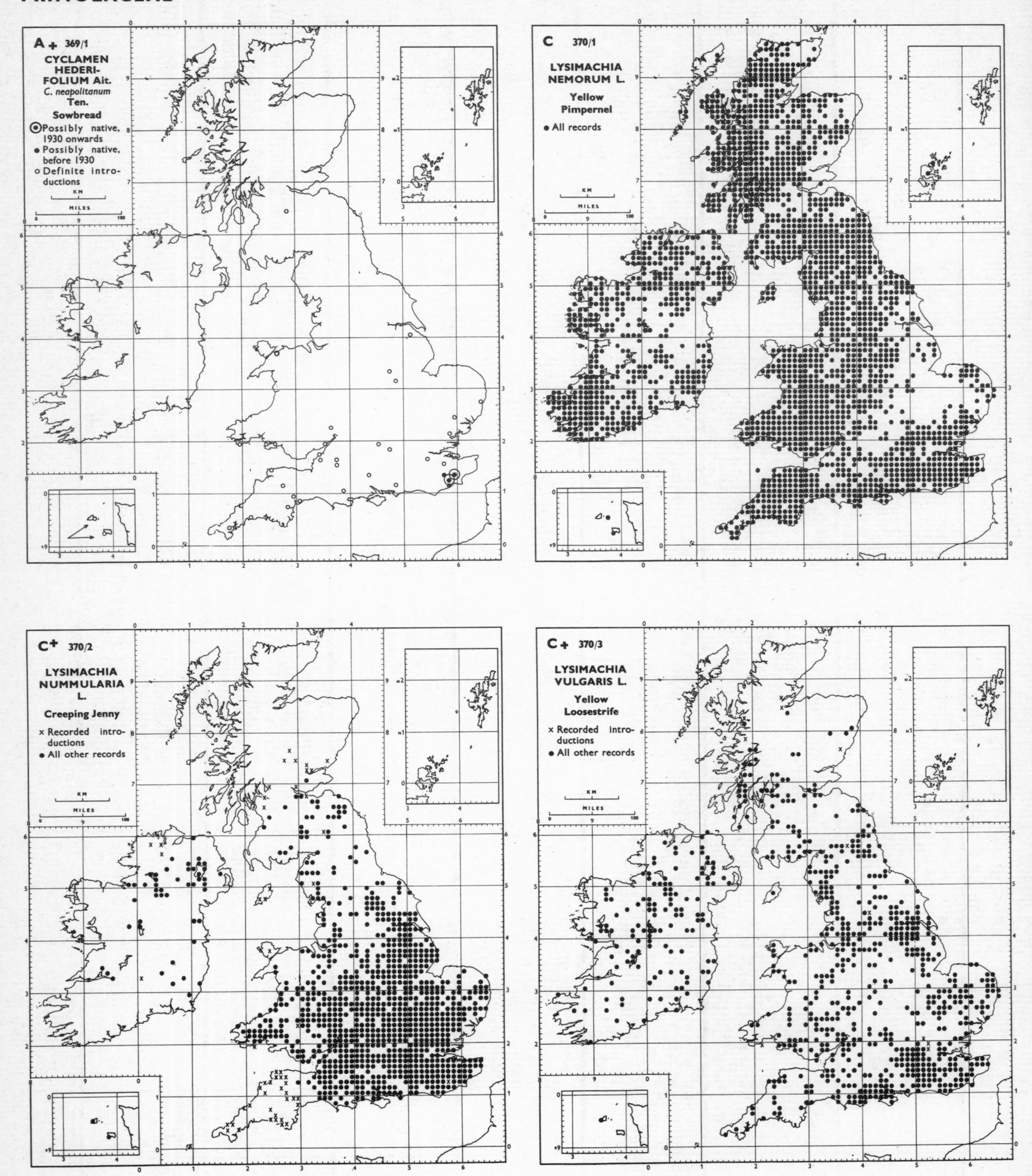
A + 369/1
CYCLAMEN HEDERIFOLIUM Ait.
C. neapolitanum Ten.
Sowbread
Possibly native, 1930 onwards
Possibly native, before 1930
Definite introductions
C 370/1
LYSIMACHIA NEMORUM L.
Yellow Pimpernel
All records
C+ 370/2
LYSIMACHIA NUMMULARIA L.
Creeping Jenny
× Recorded introductions
All other records
C + 370/3
LYSIMACHIA VULGARIS L.
Yellow Loosestrife
× Recorded introductions
All other records
KM
MILES

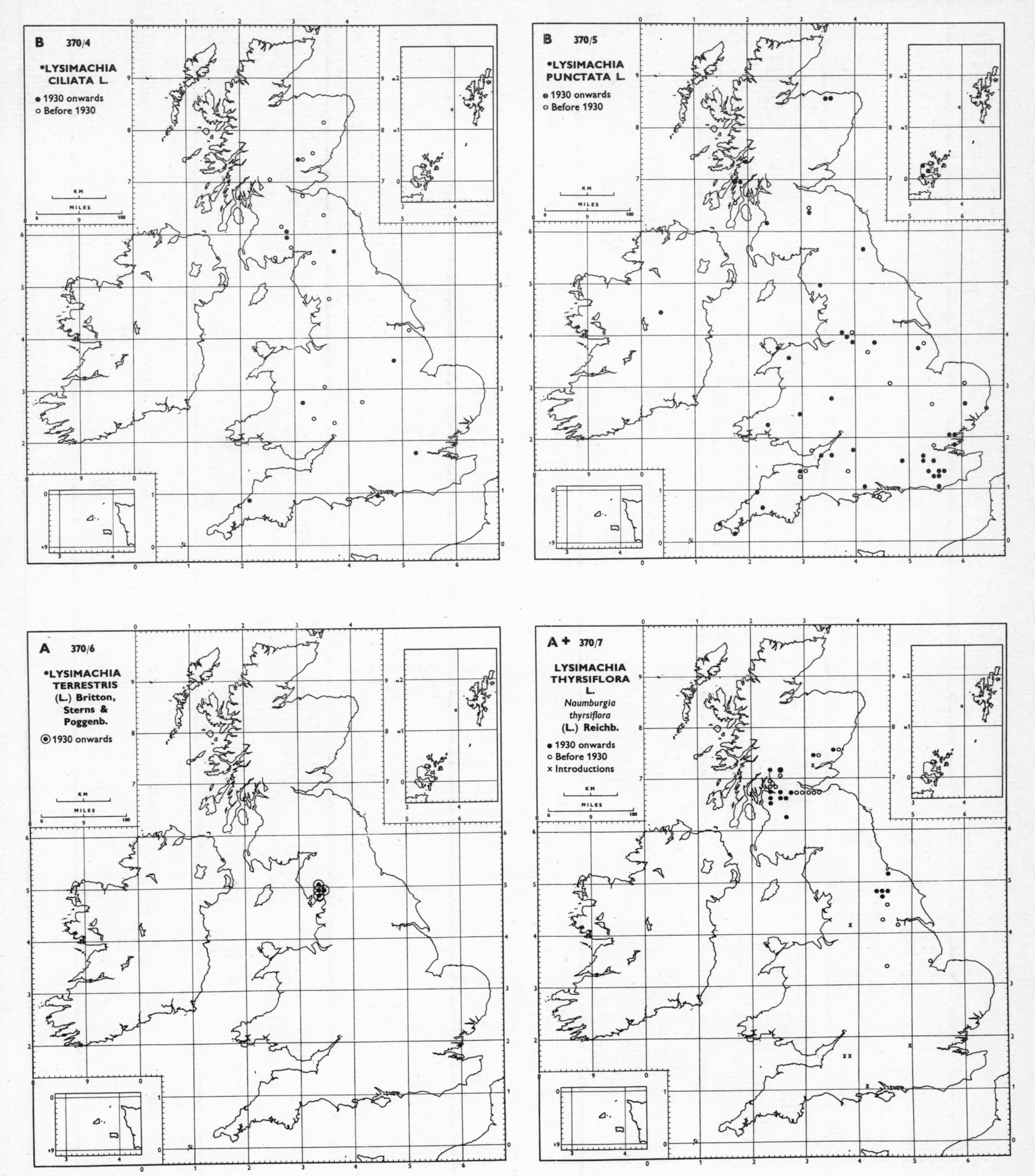
B 370/4
•LYSIMACHIA CILIATA L.
• 1930 onwards
○ Before 1930
B 370/5
•LYSIMACHIA PUNCTATA L.
• 1930 onwards
○ Before 1930
A 370/6
•LYSIMACHIA TERRESTRIS (L.) Britton, Sterns & Poggenb.
⊙ 1930 onwards
A+ 370/7
LYSIMACHIA THYRSIFLORA L.
Naumburgia thyrsiflora (L.) Reichb.
• 1930 onwards
○ Before 1930
x Introductions
KM
MILES

PRIMULACEAE

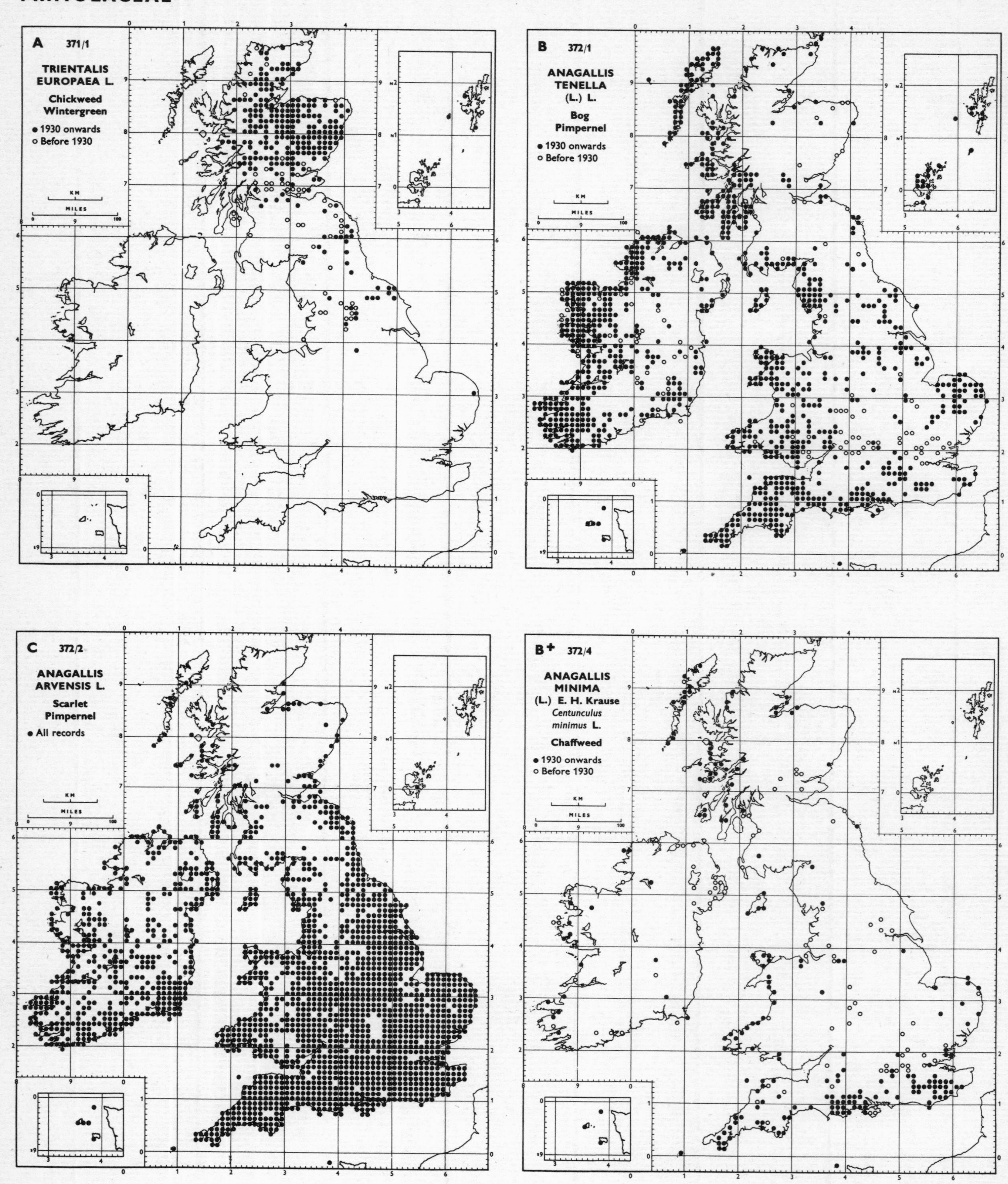

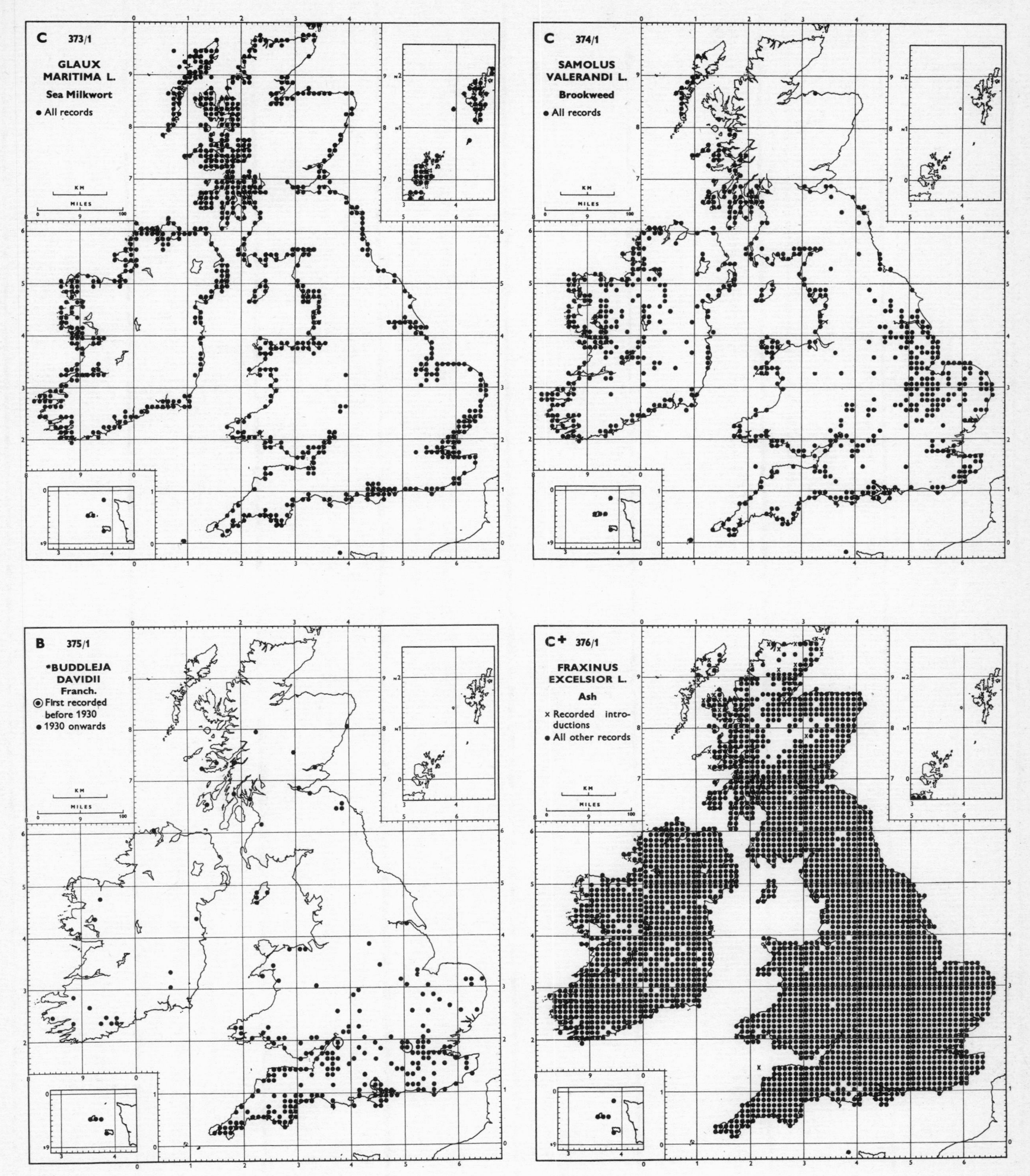
C 373/1
GLAUX MARITIMA L.
Sea Milkwort
• All records
C 374/1
SAMOLUS VALERANDI L.
Brookweed
• All records
B 375/1
*BUDDLEJA DAVIDII Franch.
◉ First recorded before 1930
• 1930 onwards
C+ 376/1
FRAXINUS EXCELSIOR L.
Ash
× Recorded introductions
• All other records
KM
MILES

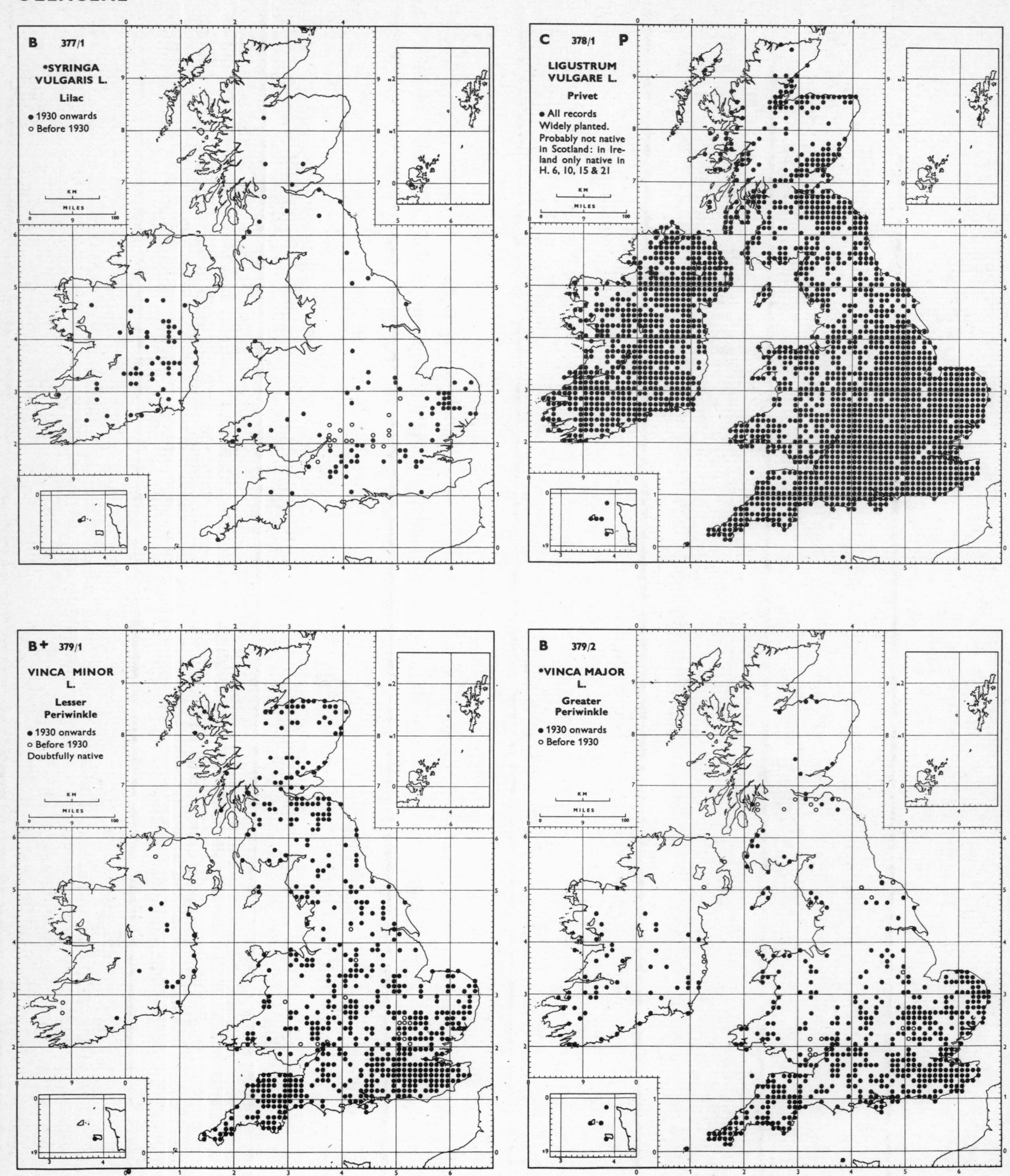
B 377/1
*SYRINGA VULGARIS L.
Lilac
● 1930 onwards
○ Before 1930
C 378/1 P
LIGUSTRUM VULGARE L.
Privet
● All records
Widely planted.
Probably not native in Scotland: in Ireland only native in H. 6, 10, 15 & 21
B+ 379/1
VINCA MINOR L.
Lesser Periwinkle
● 1930 onwards
○ Before 1930
Doubtfully native
B 379/2
*VINCA MAJOR L.
Greater Periwinkle
● 1930 onwards
○ Before 1930

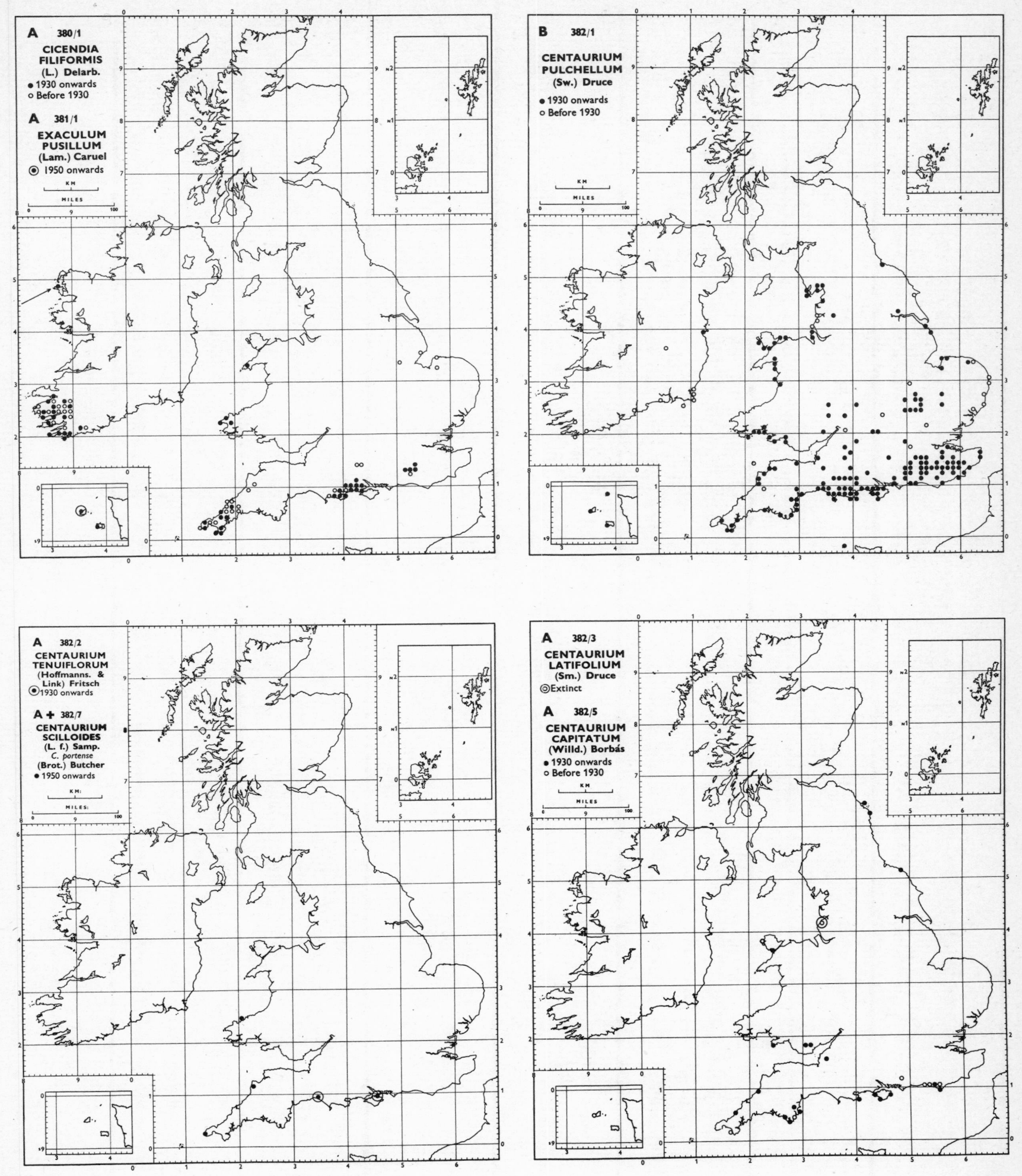
A 380/1
CICENDIA FILIFORMIS (L.) Delarb.
• 1930 onwards
○ Before 1930
A 381/1
EXACULUM PUSILLUM (Lam.) Caruel
◉ 1950 onwards
B 382/1
CENTAURIUM PULCHELLUM (Sw.) Druce
• 1930 onwards
○ Before 1930
A 382/2
CENTAURIUM TENUIFLORUM (Hoffmanns. & Link) Fritsch
◉ 1930 onwards
A + 382/7
CENTAURIUM SCILLOIDES (L. f.) Samp.
C. portense (Brot.) Butcher
• 1950 onwards
A 382/3
CENTAURIUM LATIFOLIUM (Sm.) Druce
◎ Extinct
A 382/5
CENTAURIUM CAPITATUM (Willd.) Borbás
• 1930 onwards
○ Before 1930

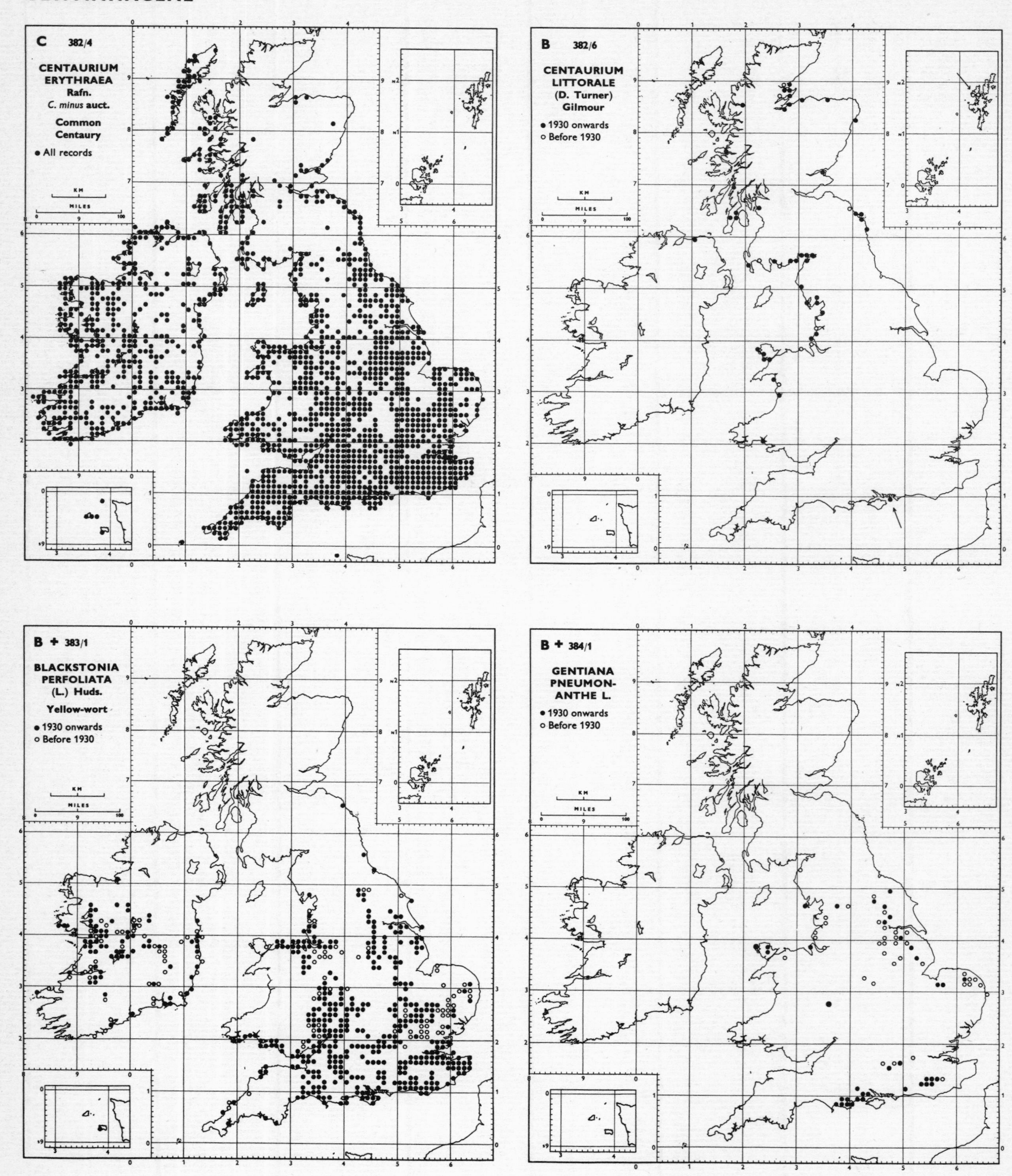
C 382/4
CENTAURIUM ERYTHRAEA Rafn.
C. minus auct.
Common Centaury
● All records
B 382/6
CENTAURIUM LITTORALE (D. Turner) Gilmour
● 1930 onwards
○ Before 1930
B + 383/1
BLACKSTONIA PERFOLIATA (L.) Huds.
Yellow-wort
● 1930 onwards
○ Before 1930
B + 384/1
GENTIANA PNEUMONANTHE L.
● 1930 onwards
○ Before 1930

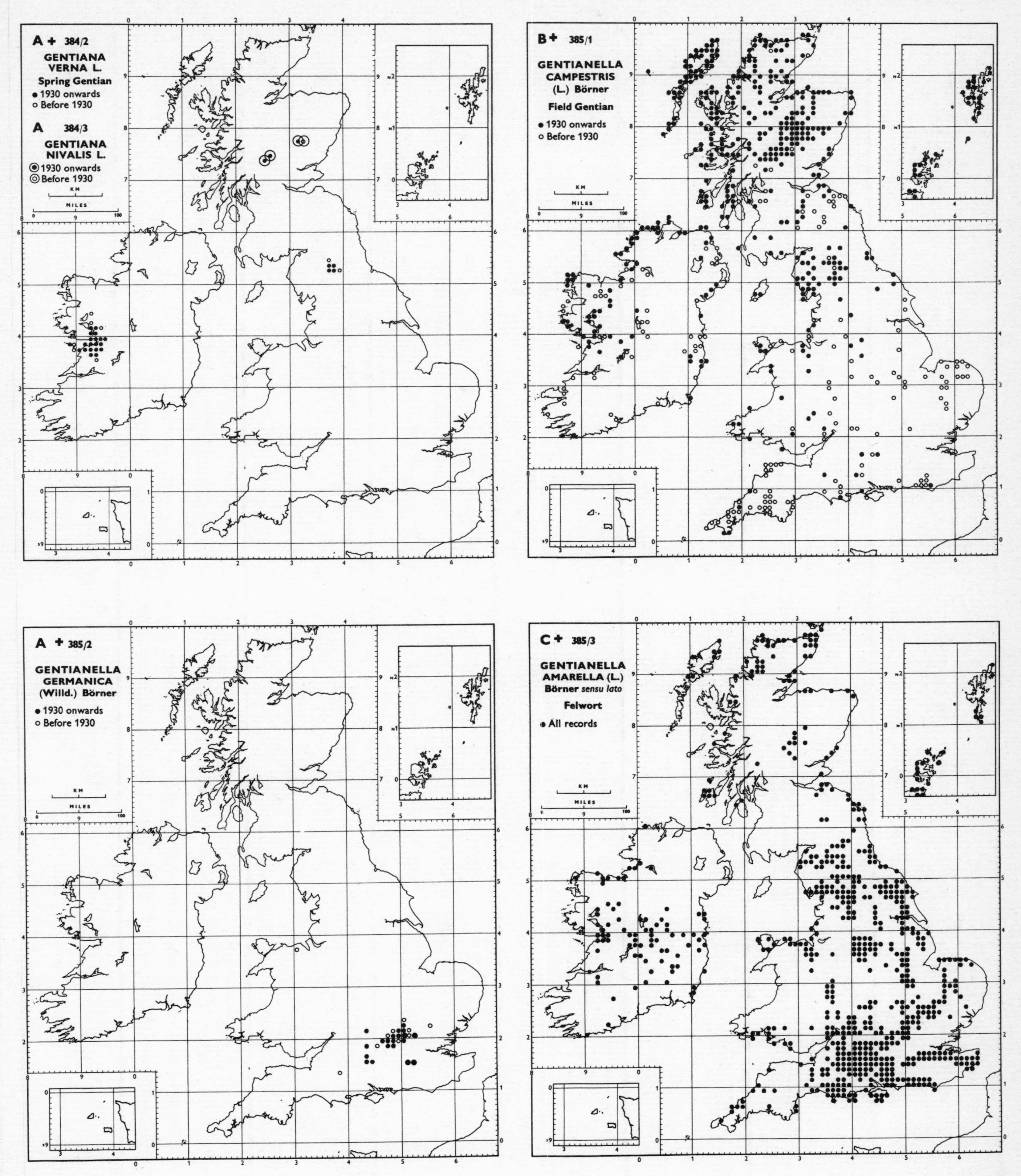
A + 384/2
GENTIANA VERNA L.
Spring Gentian
● 1930 onwards
○ Before 1930
A 384/3
GENTIANA NIVALIS L.
◉ 1930 onwards
◎ Before 1930
B + 385/1
GENTIANELLA CAMPESTRIS (L.) Börner
Field Gentian
● 1930 onwards
○ Before 1930
A + 385/2
GENTIANELLA GERMANICA (Willd.) Börner
● 1930 onwards
○ Before 1930
C + 385/3
GENTIANELLA AMARELLA (L.) Börner sensu lato
Felwort
● All records
KM
MILES

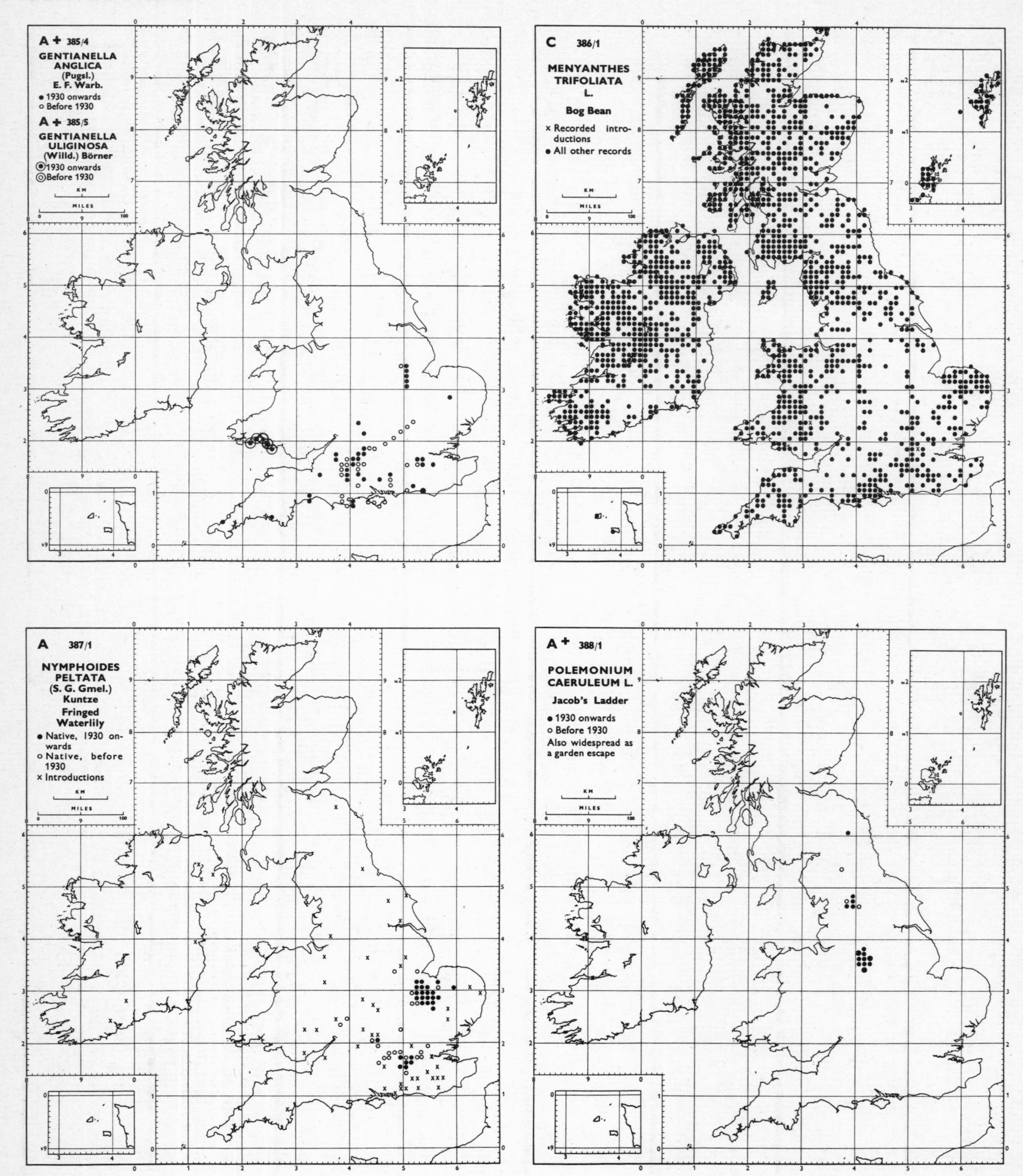
A + 385/4
GENTIANELLA ANGLICA (Pugsl.) E. F. Warb.
• 1930 onwards
○ Before 1930
A + 385/5
GENTIANELLA ULIGINOSA (Willd.) Börner
◉ 1930 onwards
◎ Before 1930
KM
MILES
C 386/1
MENYANTHES TRIFOLIATA L.
Bog Bean
× Recorded introductions
• All other records
A 387/1
NYMPHOIDES PELTATA (S. G. Gmel.) Kuntze
Fringed Waterlily
• Native, 1930 onwards
○ Native, before 1930
× Introductions
A + 388/1
POLEMONIUM CAERULEUM L.
Jacob's Ladder
• 1930 onwards
○ Before 1930
Also widespread as a garden escape

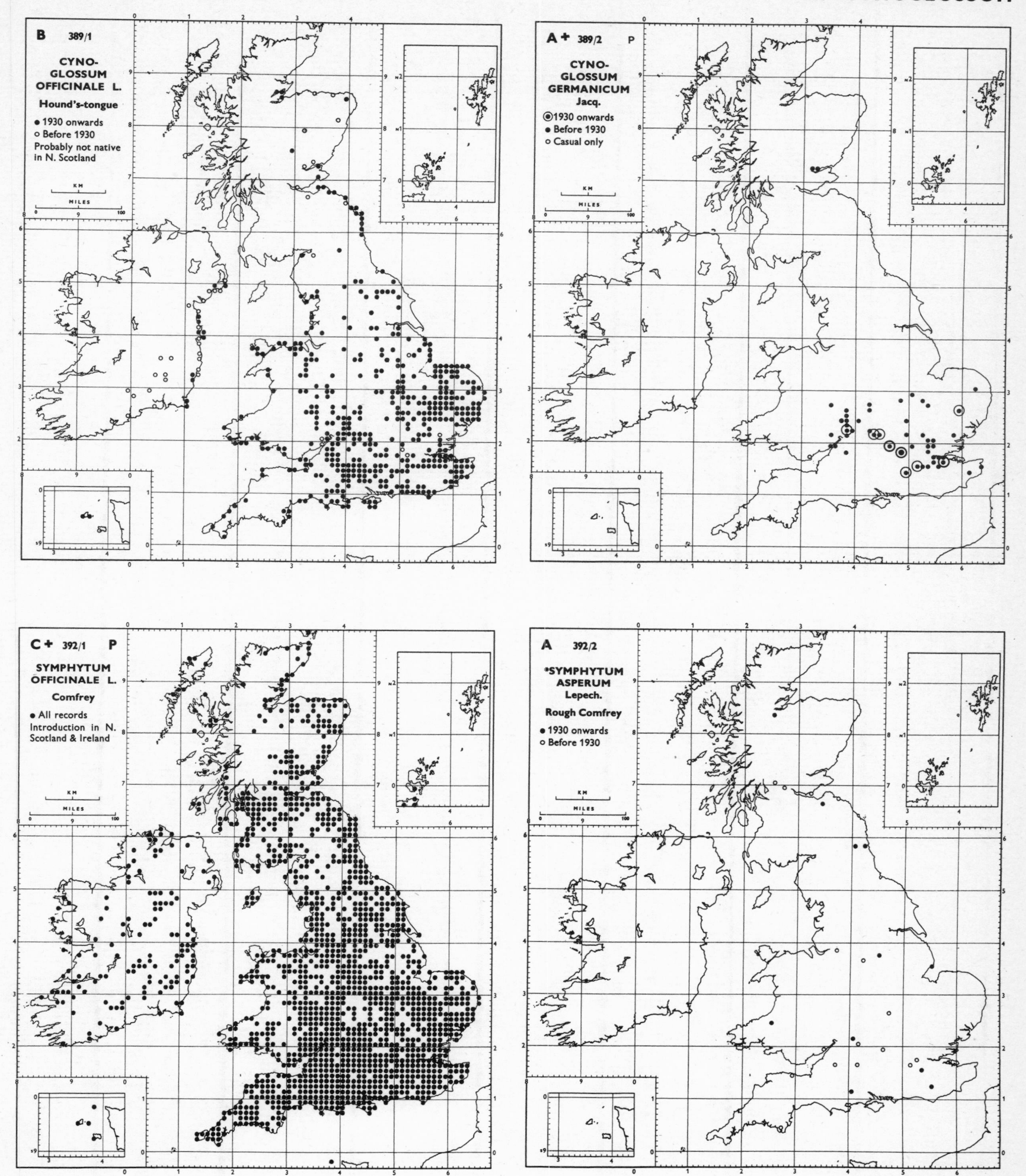

B 389/1
CYNO-GLOSSUM OFFICINALE L.
Hound's-tongue
● 1930 onwards
○ Before 1930
Probably not native in N. Scotland
A + 389/2 P
CYNO-GLOSSUM GERMANICUM Jacq.
◉ 1930 onwards
● Before 1930
○ Casual only
C + 392/1 P
SYMPHYTUM OFFICINALE L.
Comfrey
● All records
Introduction in N. Scotland & Ireland
A 392/2
*SYMPHYTUM ASPERUM Lepech.
Rough Comfrey
● 1930 onwards
○ Before 1930
KM
MILES

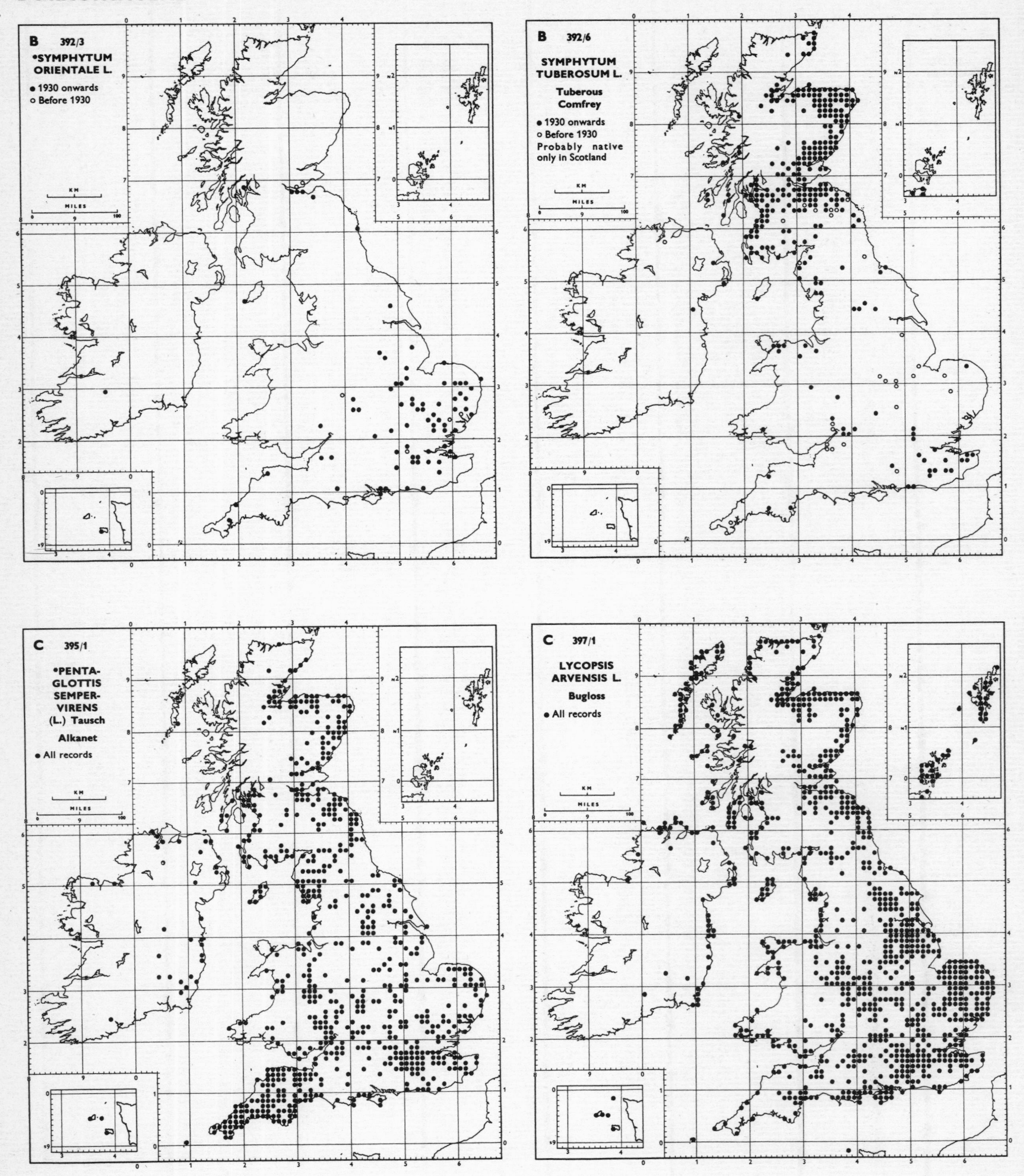
B 392/3
•SYMPHYTUM ORIENTALE L.
● 1930 onwards
○ Before 1930
B 392/6
SYMPHYTUM TUBEROSUM L.
Tuberous Comfrey
● 1930 onwards
○ Before 1930
Probably native only in Scotland
C 395/1
•PENTAGLOTTIS SEMPERVIRENS (L.) Tausch
Alkanet
● All records
C 397/1
LYCOPSIS ARVENSIS L.
Bugloss
● All records
KM
MILES

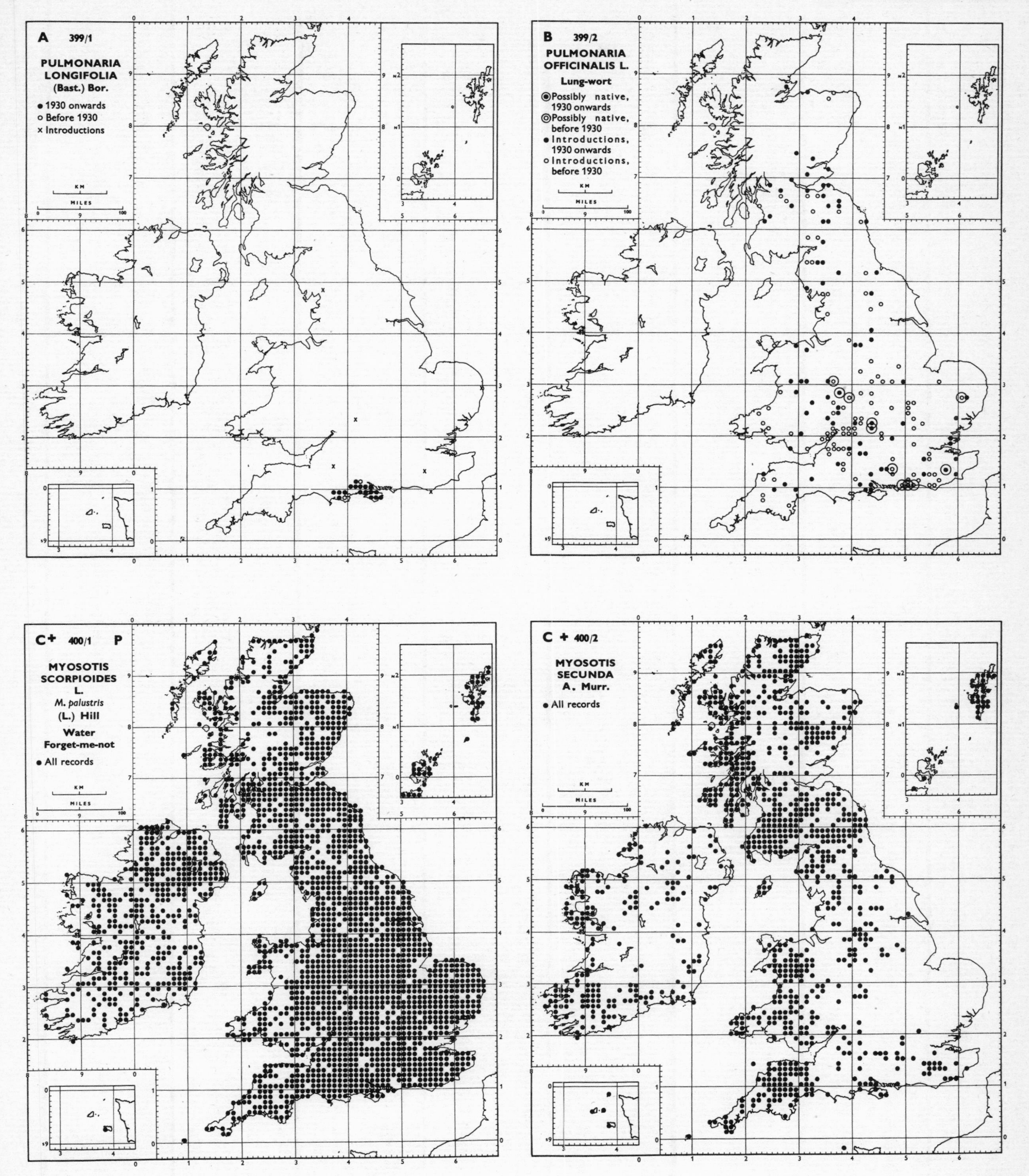
A 399/1
PULMONARIA LONGIFOLIA (Bast.) Bor.
• 1930 onwards
○ Before 1930
× Introductions
B 399/2
PULMONARIA OFFICINALIS L.
Lung-wort
Possibly native, 1930 onwards
Possibly native, before 1930
• Introductions, 1930 onwards
○ Introductions, before 1930
C+ 400/1 P
MYOSOTIS SCORPIOIDES L.
M. palustris (L.) Hill
Water Forget-me-not
• All records
C + 400/2
MYOSOTIS SECUNDA A. Murr.
• All records
KM
MILES

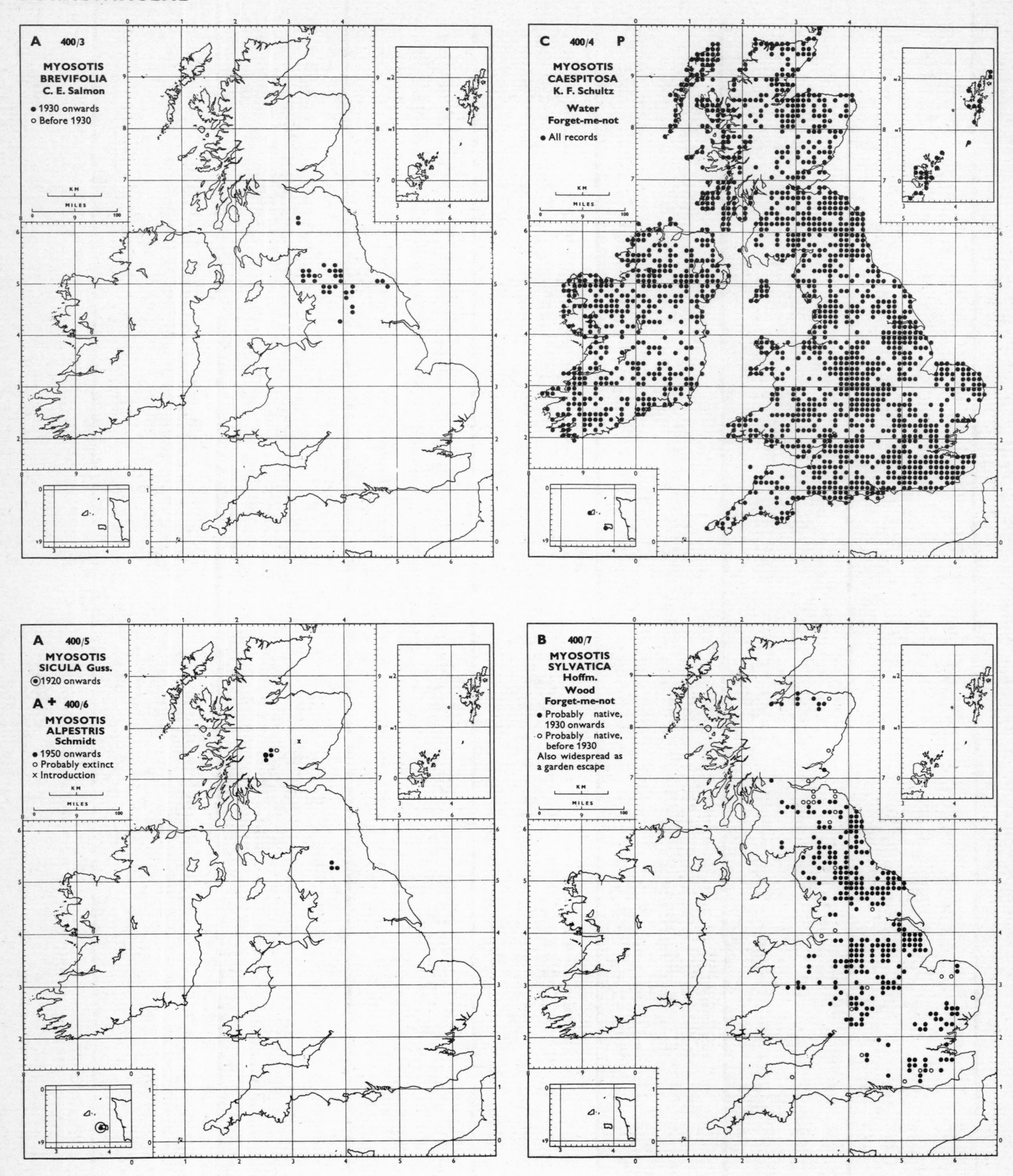
A 400/3
MYOSOTIS
BREVIFOLIA
C. E. Salmon
• 1930 onwards
○ Before 1930
C 400/4 P
MYOSOTIS
CAESPITOSA
K. F. Schultz
Water
Forget-me-not
• All records
A 400/5
MYOSOTIS
SICULA Guss.
⊙1920 onwards
A + 400/6
MYOSOTIS
ALPESTRIS
Schmidt
• 1950 onwards
○ Probably extinct
× Introduction
B 400/7
MYOSOTIS
SYLVATICA
Hoffm.
Wood
Forget-me-not
• Probably native, 1930 onwards
○ Probably native, before 1930
Also widespread as a garden escape
KM
MILES

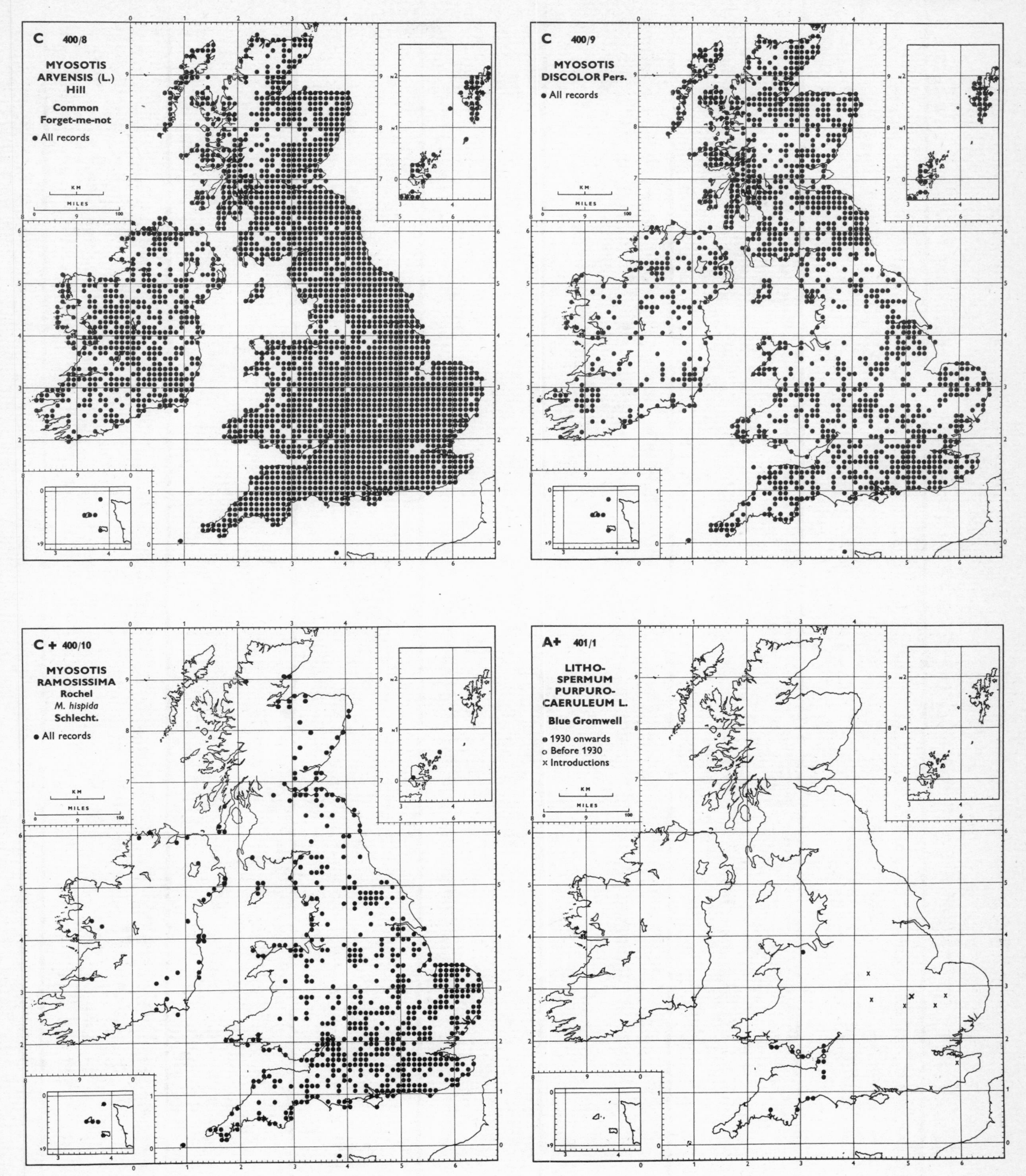
C 400/8
MYOSOTIS ARVENSIS (L.) Hill
Common Forget-me-not
● All records
C 400/9
MYOSOTIS DISCOLOR Pers.
● All records
C + 400/10
MYOSOTIS RAMOSISSIMA Rochel
M. hispida Schlecht.
● All records
A+ 401/1
LITHO-SPERMUM PURPURO-CAERULEUM L.
Blue Gromwell
● 1930 onwards
○ Before 1930
x Introductions
KM
MILES

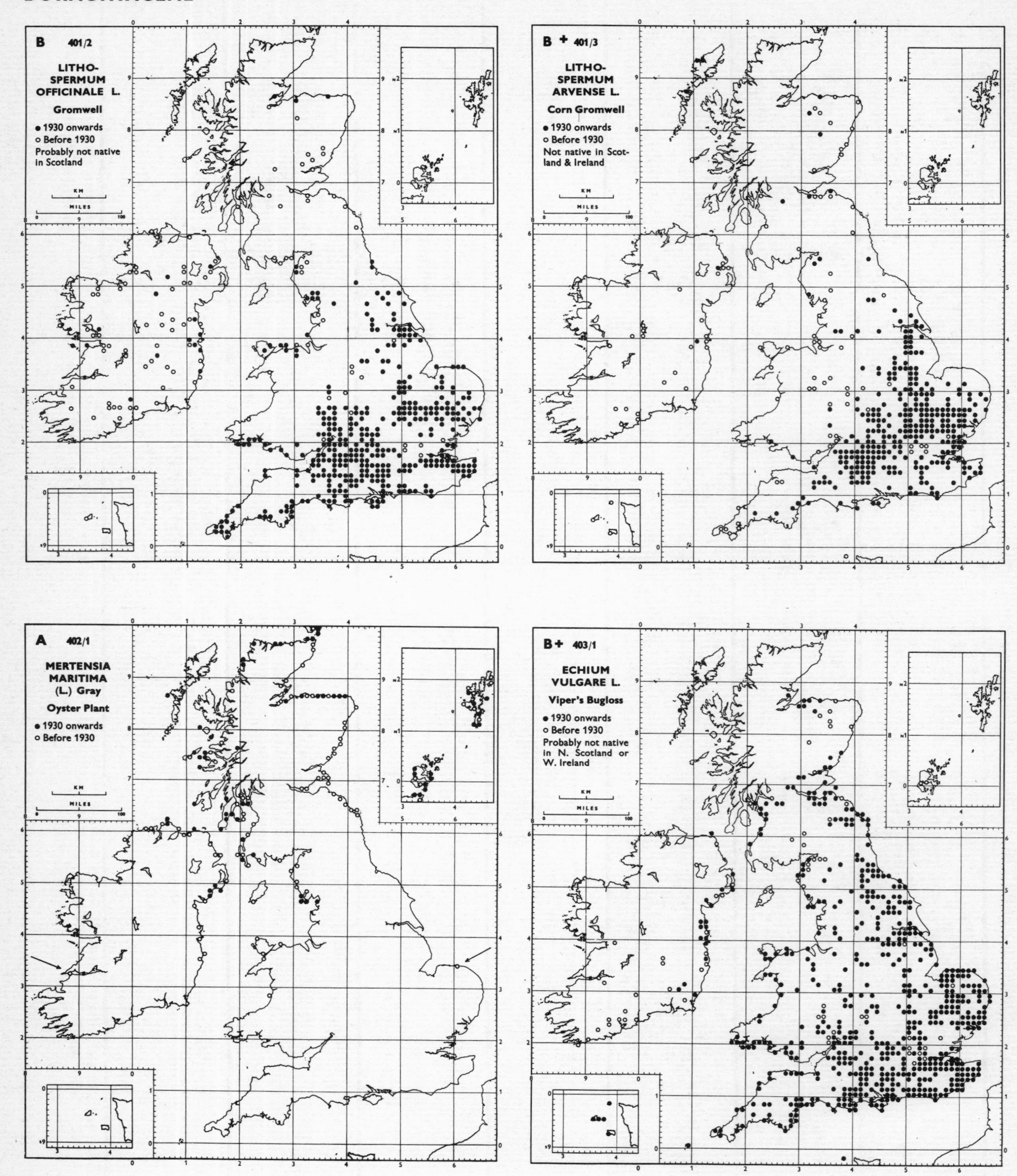
B 401/2
LITHO-SPERMUM OFFICINALE L.
Gromwell
• 1930 onwards
○ Before 1930
Probably not native in Scotland
B + 401/3
LITHO-SPERMUM ARVENSE L.
Corn Gromwell
• 1930 onwards
○ Before 1930
Not native in Scotland & Ireland
A 402/1
MERTENSIA MARITIMA (L.) Gray
Oyster Plant
• 1930 onwards
○ Before 1930
B+ 403/1
ECHIUM VULGARE L.
Viper's Bugloss
• 1930 onwards
○ Before 1930
Probably not native in N. Scotland or W. Ireland
KM
MILES

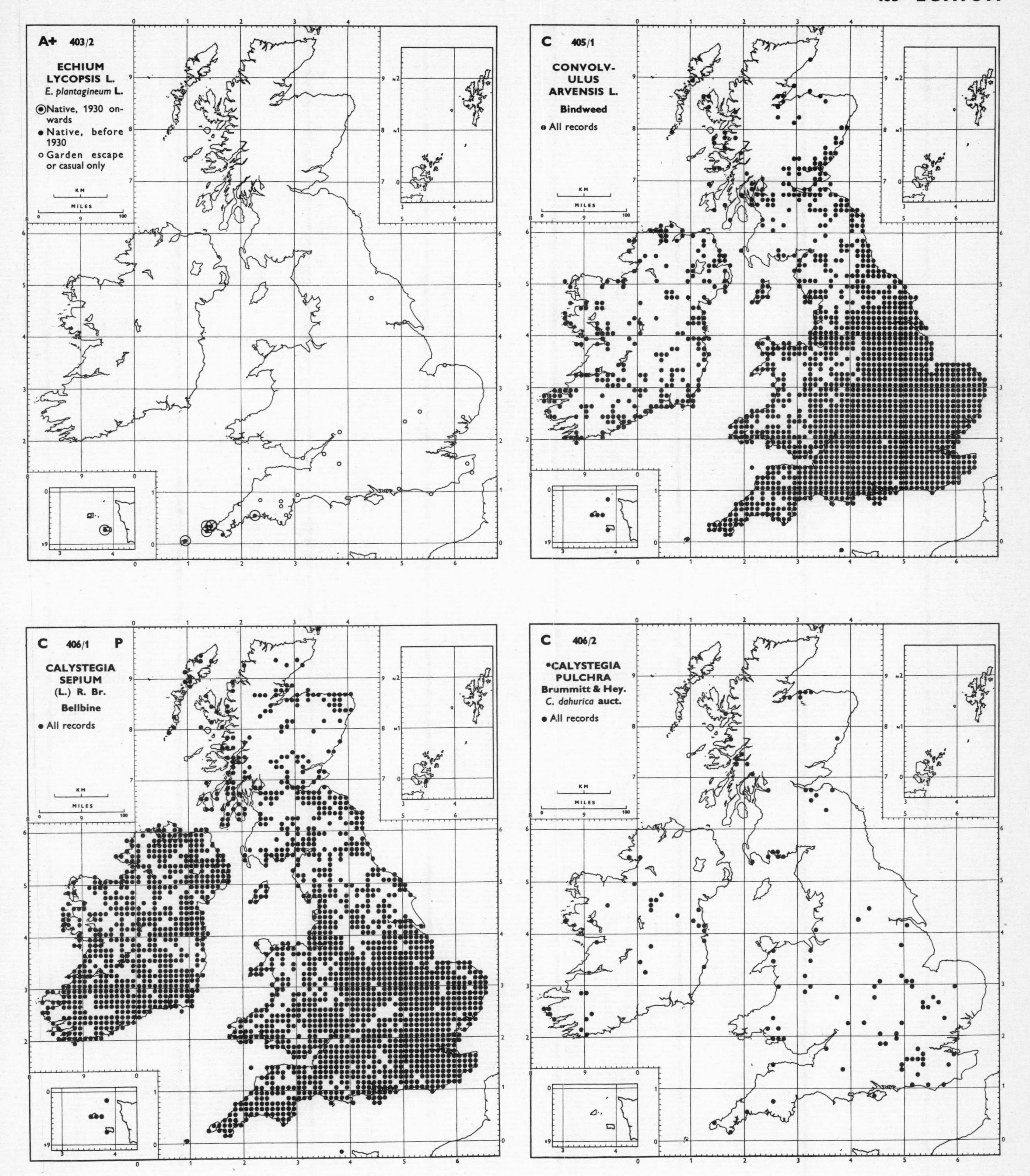
A+ 403/2
ECHIUM LYCOPSIS L.
E. plantagineum L.
Native, 1930 onwards
Native, before 1930
Garden escape or casual only
C 405/1
CONVOLVULUS ARVENSIS L.
Bindweed
All records
C 406/1 P
CALYSTEGIA SEPIUM (L.) R. Br.
Bellbine
All records
C 406/2
•CALYSTEGIA PULCHRA Brummitt & Hey.
C. dahurica auct.
All records
KM
MILES

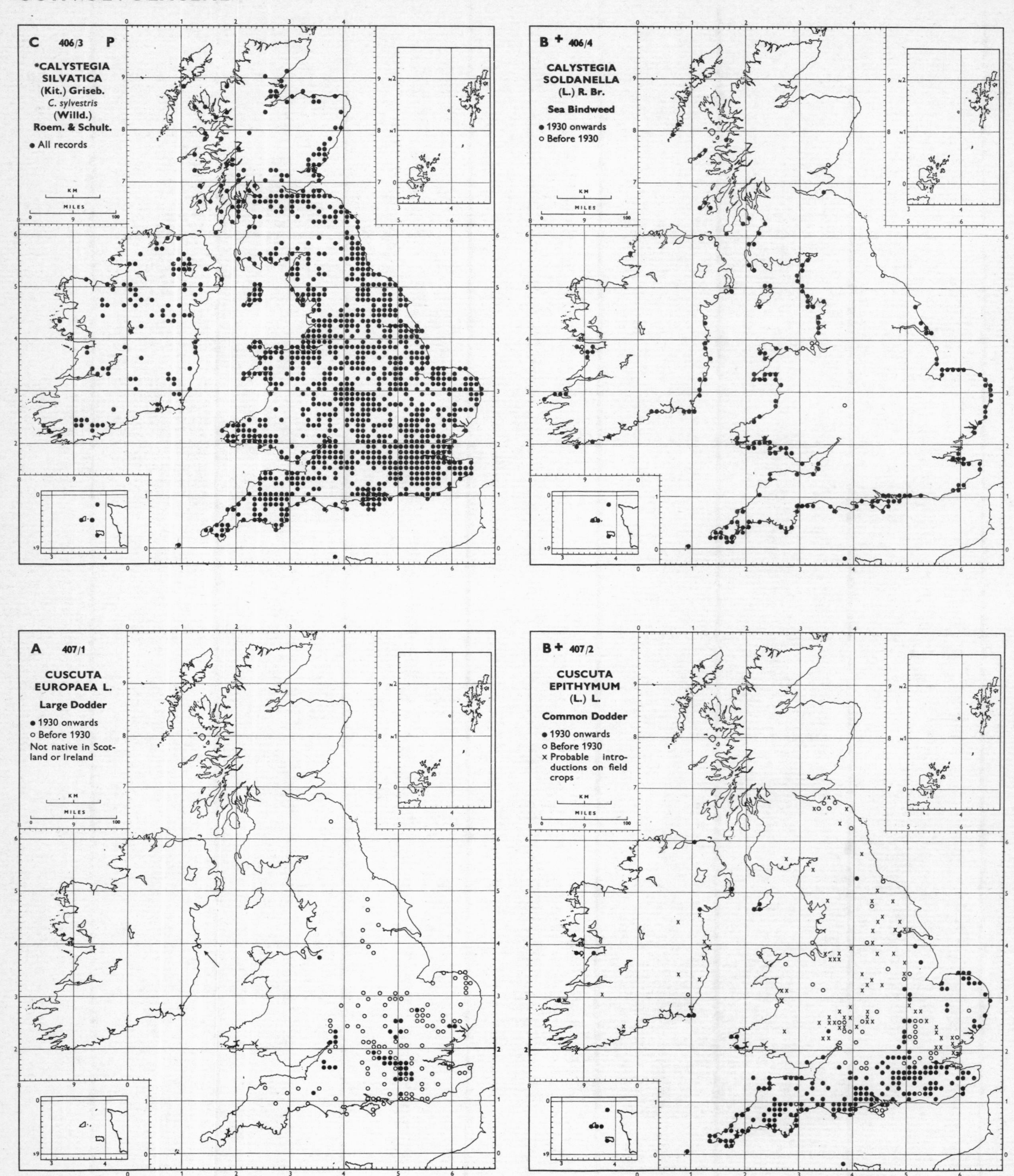
C 406/3 P
*CALYSTEGIA SILVATICA (Kit.) Griseb.
C. sylvestris (Willd.) Roem. & Schult.
● All records
KM
MILES
B + 406/4
CALYSTEGIA SOLDANELLA (L.) R. Br.
Sea Bindweed
● 1930 onwards
○ Before 1930
A 407/1
CUSCUTA EUROPAEA L.
Large Dodder
● 1930 onwards
○ Before 1930
Not native in Scotland or Ireland
B + 407/2
CUSCUTA EPITHYMUM (L.) L.
Common Dodder
● 1930 onwards
○ Before 1930
× Probable introductions on field crops

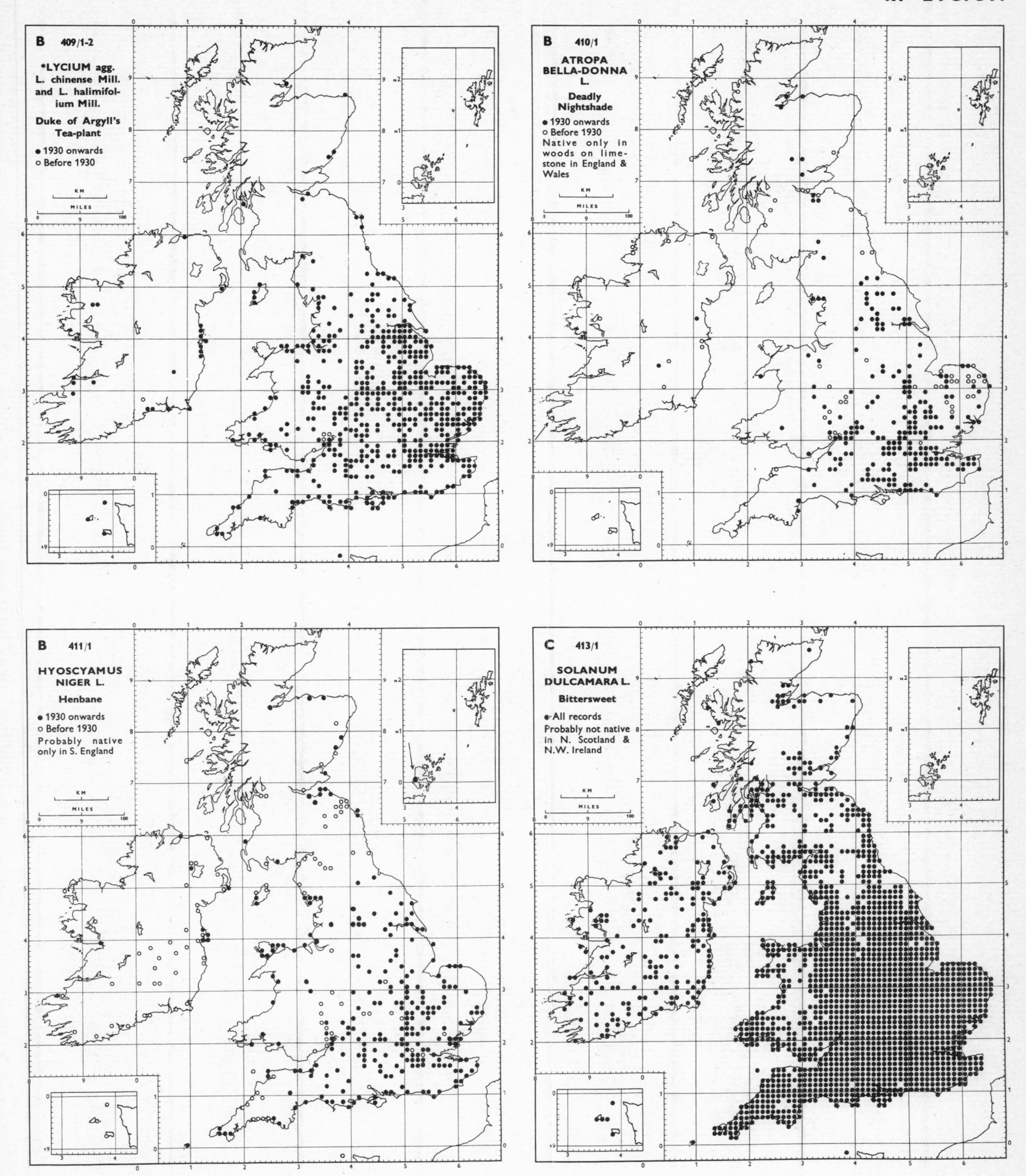
B 409/1-2
*LYCIUM agg.
L. chinense Mill.
and L. halimifolium Mill.
Duke of Argyll's Tea-plant
● 1930 onwards
○ Before 1930
B 410/1
ATROPA BELLA-DONNA L.
Deadly Nightshade
● 1930 onwards
○ Before 1930
Native only in woods on limestone in England & Wales
B 411/1
HYOSCYAMUS NIGER L.
Henbane
● 1930 onwards
○ Before 1930
Probably native only in S. England
C 413/1
SOLANUM DULCAMARA L.
Bittersweet
● All records
Probably not native in N. Scotland & N.W. Ireland
KM
MILES

SOLANACEAE

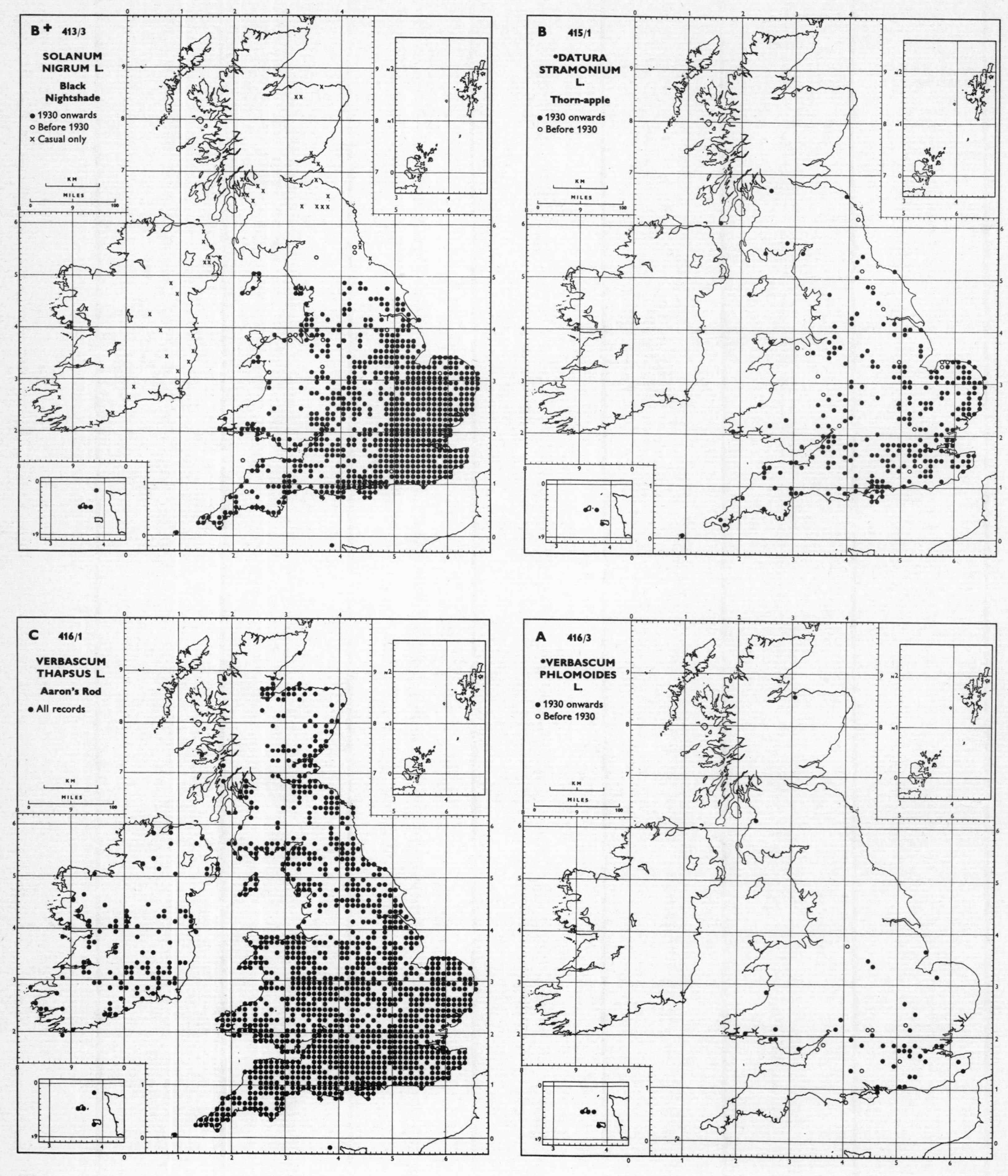
B+ 413/3
SOLANUM NIGRUM L.
Black Nightshade
● 1930 onwards
○ Before 1930
× Casual only
B 415/1
•DATURA STRAMONIUM L.
Thorn-apple
● 1930 onwards
○ Before 1930
C 416/1
VERBASCUM THAPSUS L.
Aaron's Rod
● All records
A 416/3
•VERBASCUM PHLOMOIDES L.
● 1930 onwards
○ Before 1930

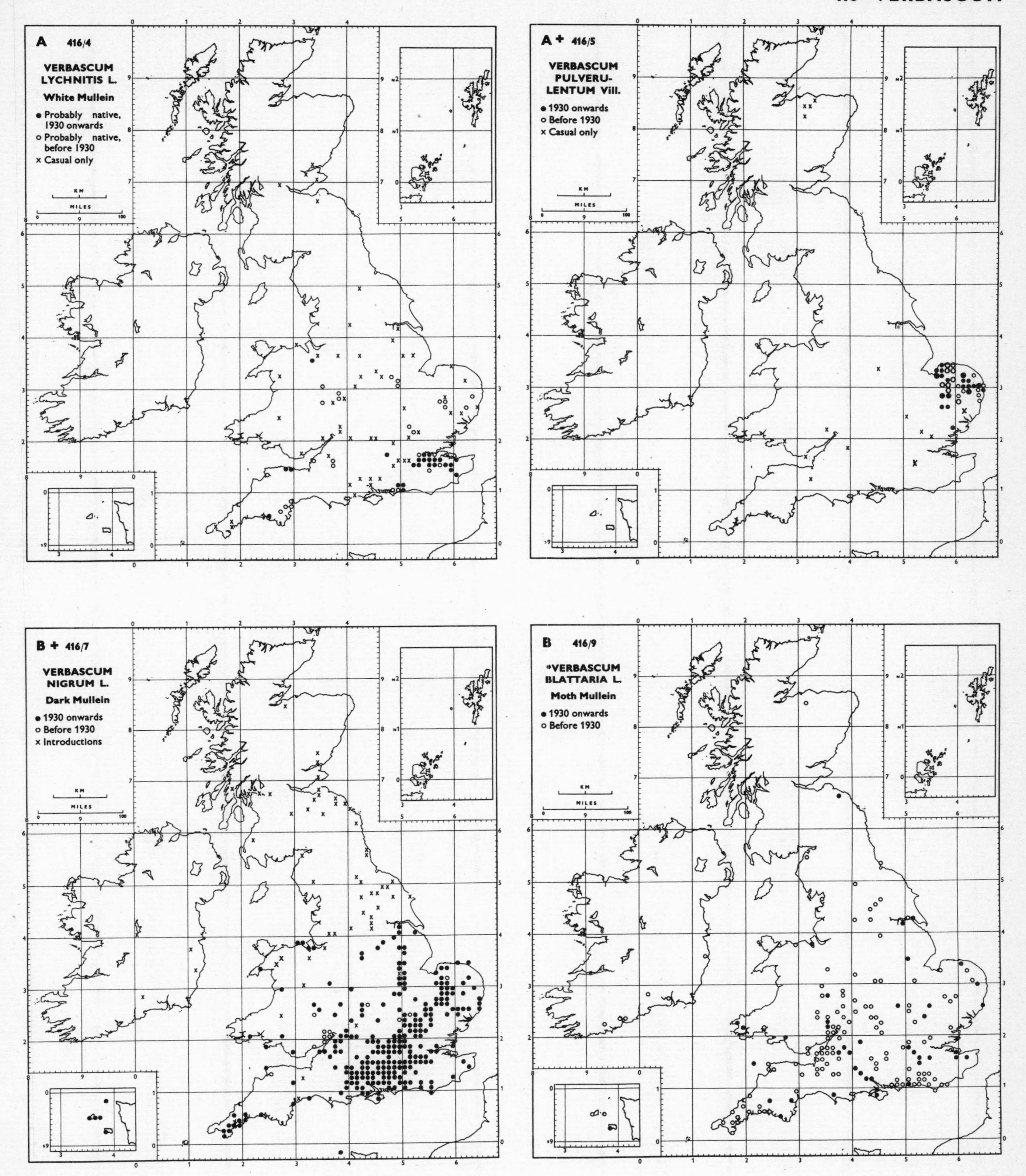
A 416/4
VERBASCUM LYCHNITIS L.
White Mullein
● Probably native, 1930 onwards
○ Probably native, before 1930
x Casual only
A + 416/5
VERBASCUM PULVERU-LENTUM Vill.
● 1930 onwards
○ Before 1930
x Casual only
B + 416/7
VERBASCUM NIGRUM L.
Dark Mullein
● 1930 onwards
○ Before 1930
x Introductions
B 416/9
°VERBASCUM BLATTARIA L.
Moth Mullein
● 1930 onwards
○ Before 1930
KM
MILES

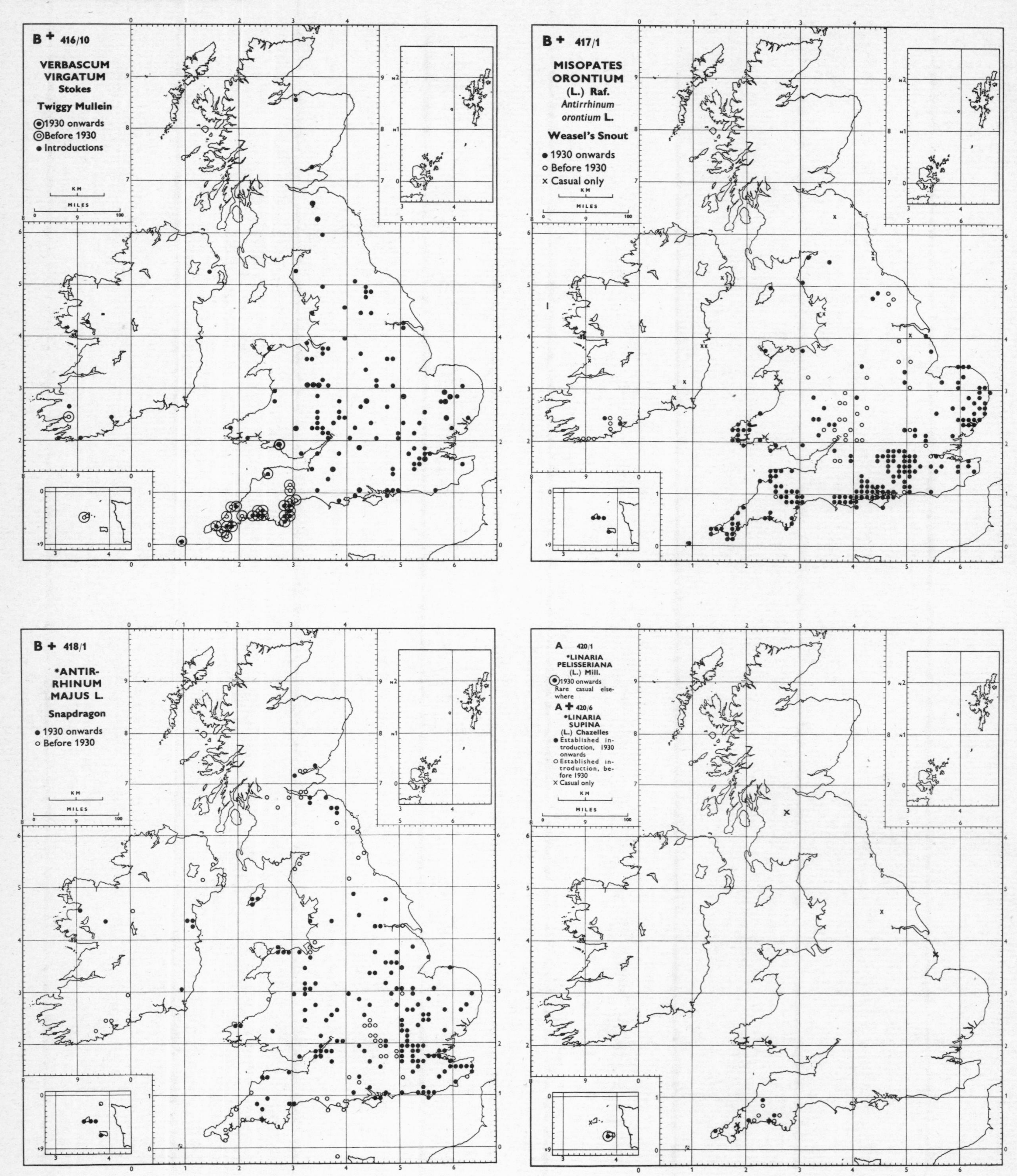
B + 416/10
VERBASCUM VIRGATUM Stokes
Twiggy Mullein
1930 onwards
Before 1930
Introductions
B + 417/1
MISOPATES ORONTIUM (L.) Raf.
Antirrhinum orontium L.
Weasel's Snout
1930 onwards
Before 1930
Casual only
B + 418/1
*ANTIRRHINUM MAJUS L.
Snapdragon
1930 onwards
Before 1930
A 420/1
*LINARIA PELISSERIANA (L.) Mill.
1930 onwards
Rare casual elsewhere
A + 420/6
*LINARIA SUPINA (L.) Chazelles
Established introduction, 1930 onwards
Established introduction, before 1930
Casual only
KM
MILES

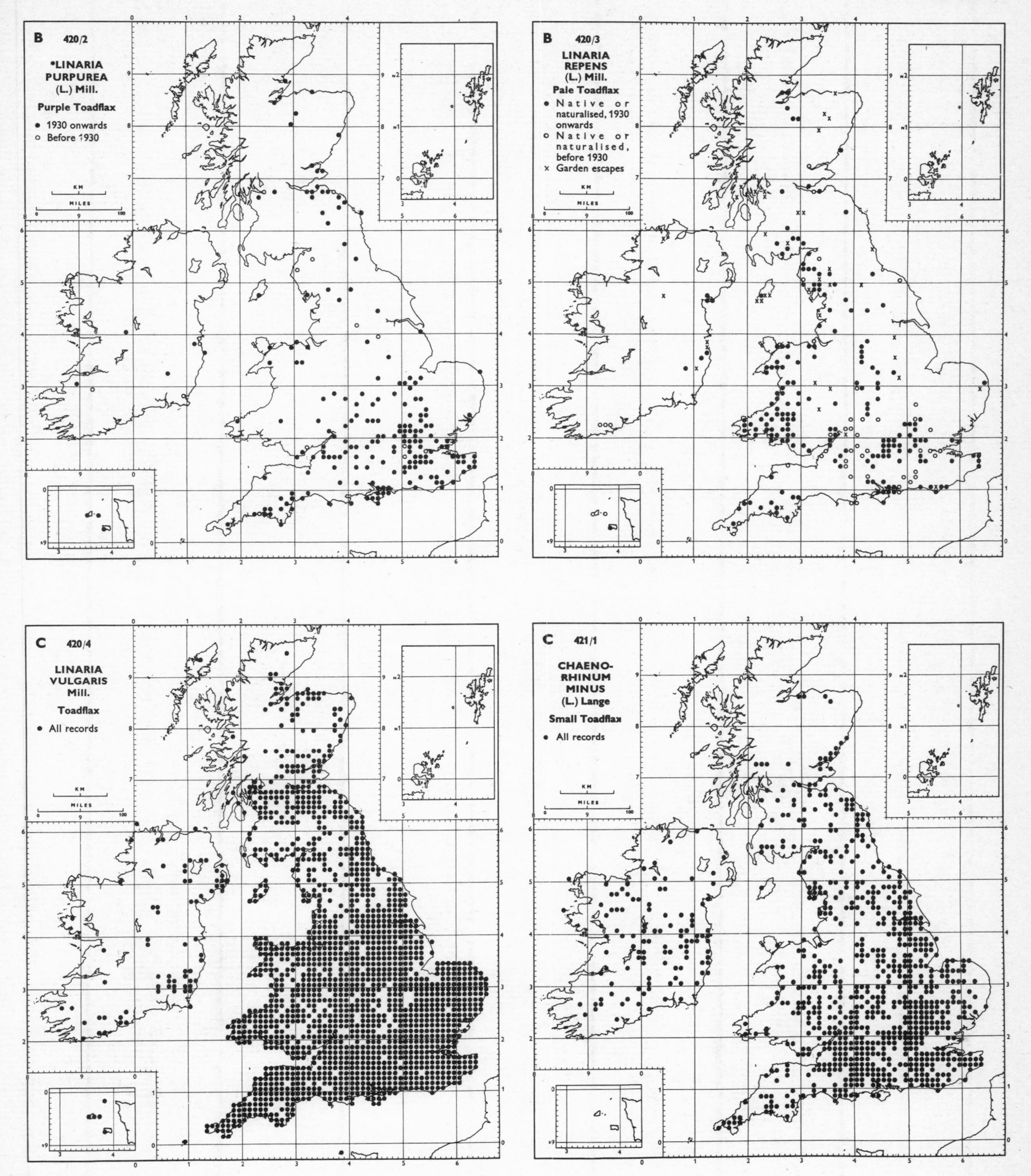
B 420/2
*LINARIA PURPUREA (L.) Mill.
Purple Toadflax
• 1930 onwards
○ Before 1930
B 420/3
LINARIA REPENS (L.) Mill.
Pale Toadflax
• Native or naturalised, 1930 onwards
○ Native or naturalised, before 1930
× Garden escapes
C 420/4
LINARIA VULGARIS Mill.
Toadflax
• All records
C 421/1
CHAENORHINUM MINUS (L.) Lange
Small Toadflax
• All records

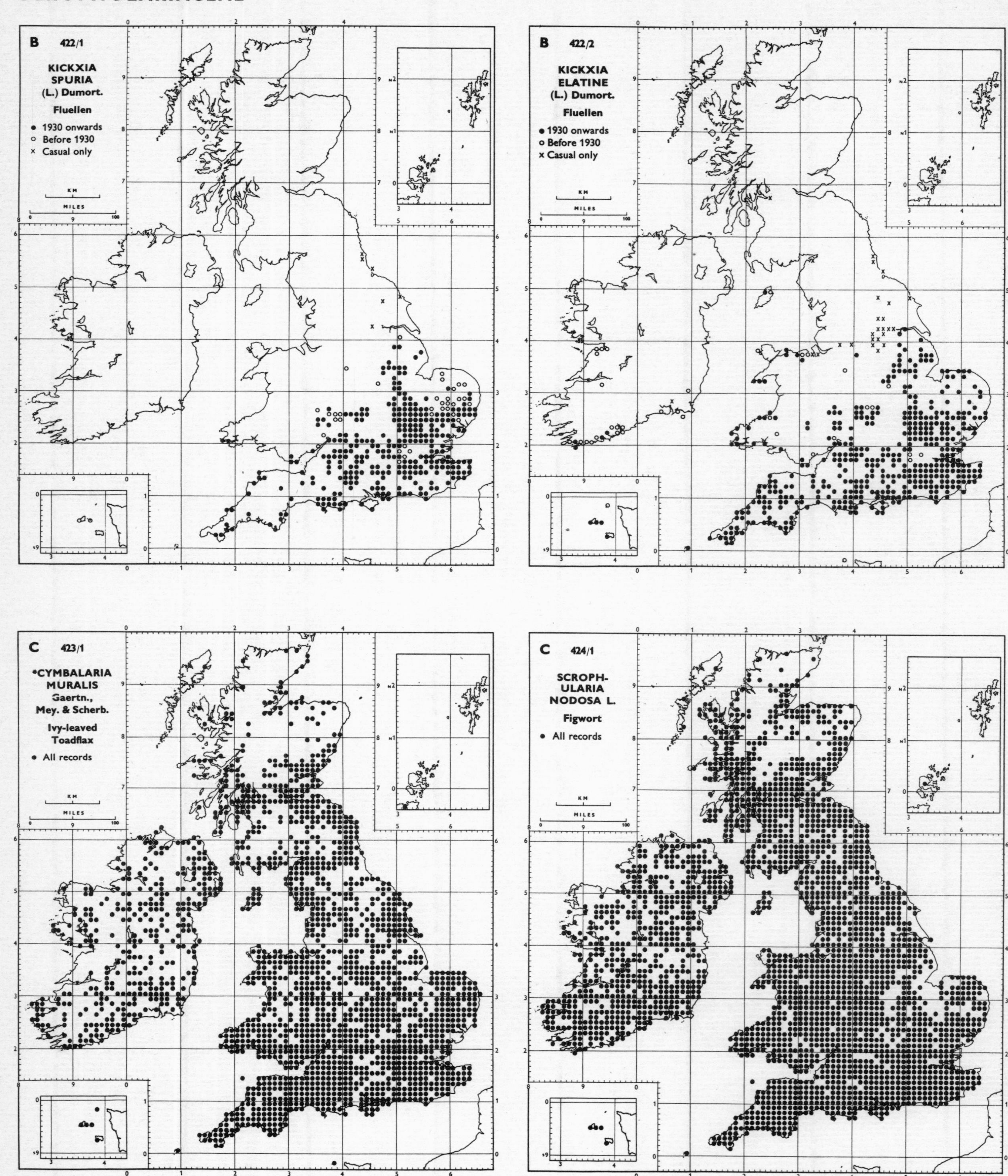
B 422/1
KICKXIA SPURIA (L.) Dumort.
Fluellen
1930 onwards
Before 1930
Casual only
B 422/2
KICKXIA ELATINE (L.) Dumort.
Fluellen
1930 onwards
Before 1930
Casual only
C 423/1
*CYMBALARIA MURALIS Gaertn., Mey. & Scherb.
Ivy-leaved Toadflax
All records
C 424/1
SCROPHULARIA NODOSA L.
Figwort
All records

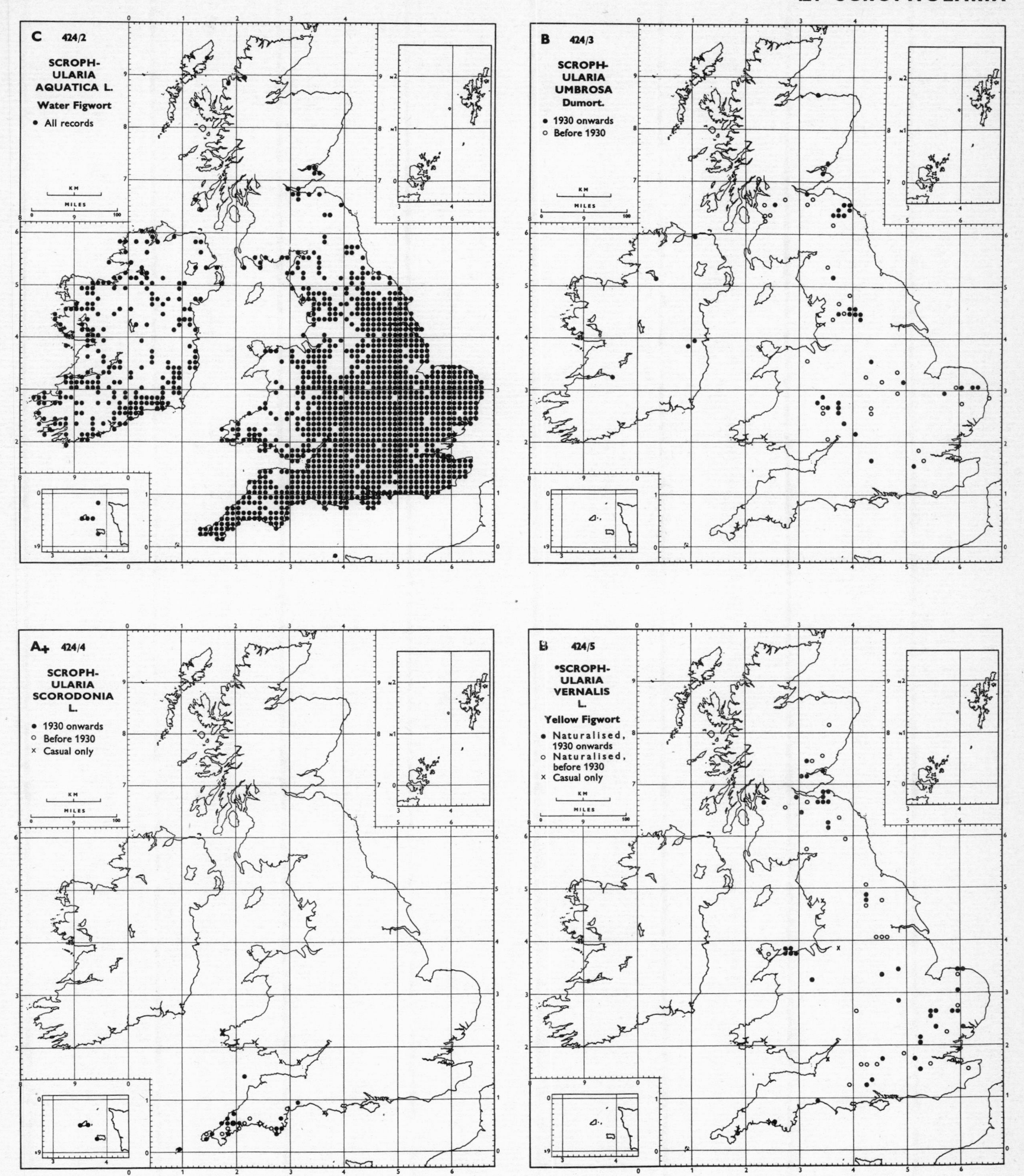
C 424/2
SCROPH-ULARIA AQUATICA L.
Water Figwort
• All records
B 424/3
SCROPH-ULARIA UMBROSA Dumort.
• 1930 onwards
○ Before 1930
A+ 424/4
SCROPH-ULARIA SCORODONIA L.
• 1930 onwards
○ Before 1930
× Casual only
B 424/5
*SCROPH-ULARIA VERNALIS L.
Yellow Figwort
• Naturalised, 1930 onwards
○ Naturalised, before 1930
× Casual only
KM
MILES

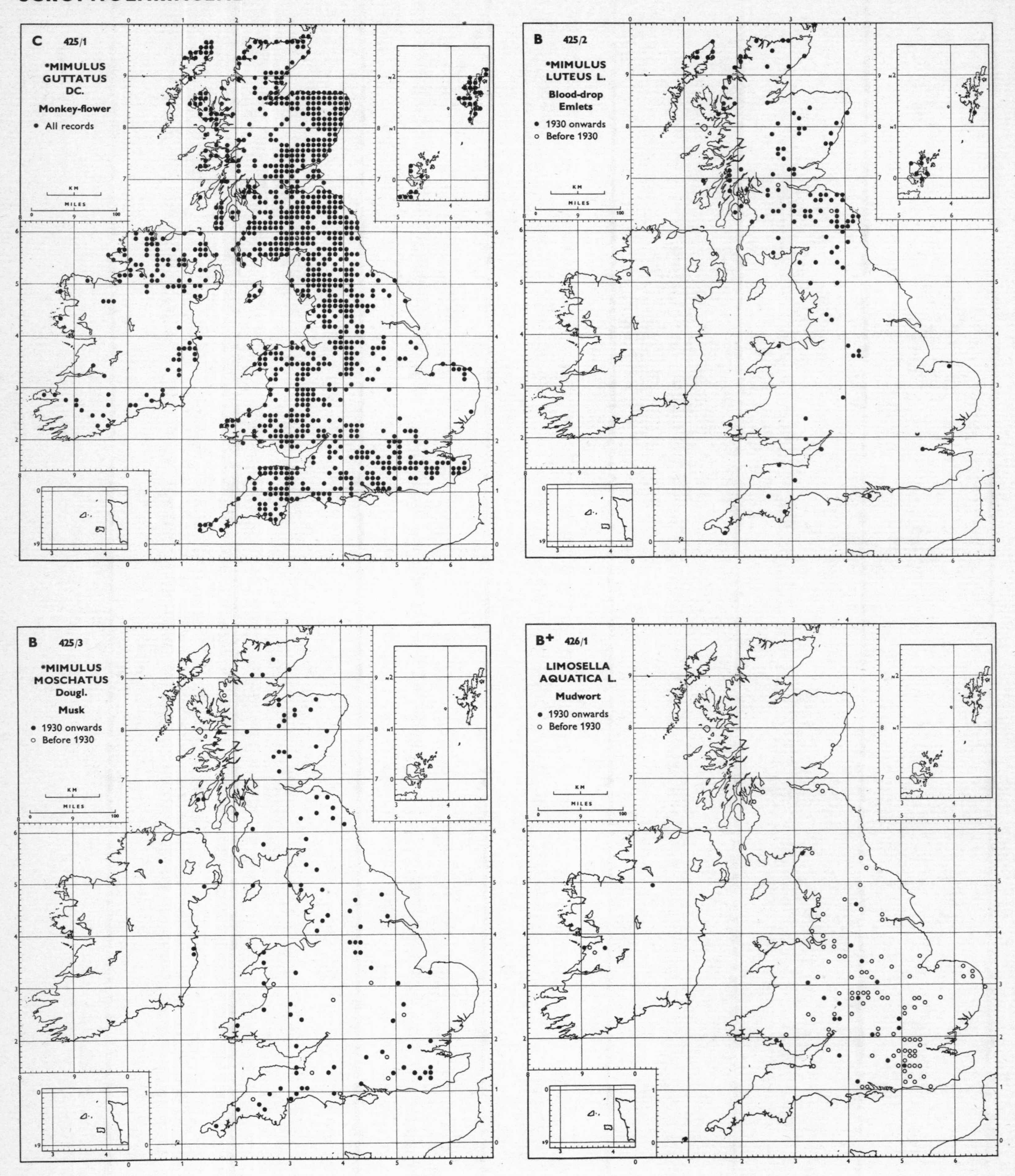
C 425/1
*MIMULUS GUTTATUS DC.
Monkey-flower
• All records
B 425/2
*MIMULUS LUTEUS L.
Blood-drop Emlets
• 1930 onwards
○ Before 1930
B 425/3
*MIMULUS MOSCHATUS Dougl.
Musk
• 1930 onwards
○ Before 1930
B+ 426/1
LIMOSELLA AQUATICA L.
Mudwort
• 1930 onwards
○ Before 1930
KM
MILES

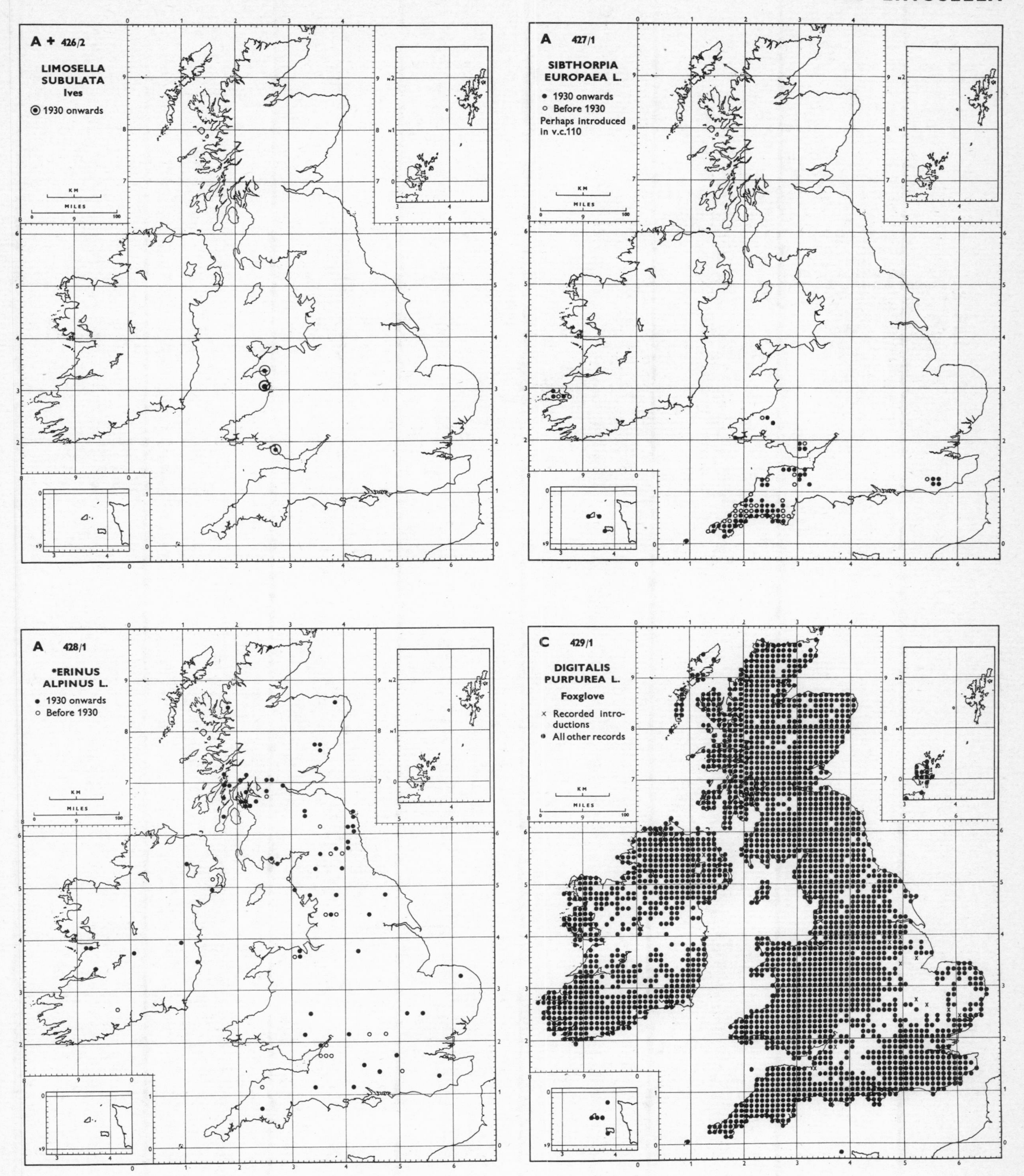
A + 426/2
LIMOSELLA SUBULATA Ives
1930 onwards
A 427/1
SIBTHORPIA EUROPAEA L.
1930 onwards
Before 1930
Perhaps introduced in v.c.110
A 428/1
*ERINUS ALPINUS L.
1930 onwards
Before 1930
C 429/1
DIGITALIS PURPUREA L.
Foxglove
Recorded introductions
All other records
KM
MILES

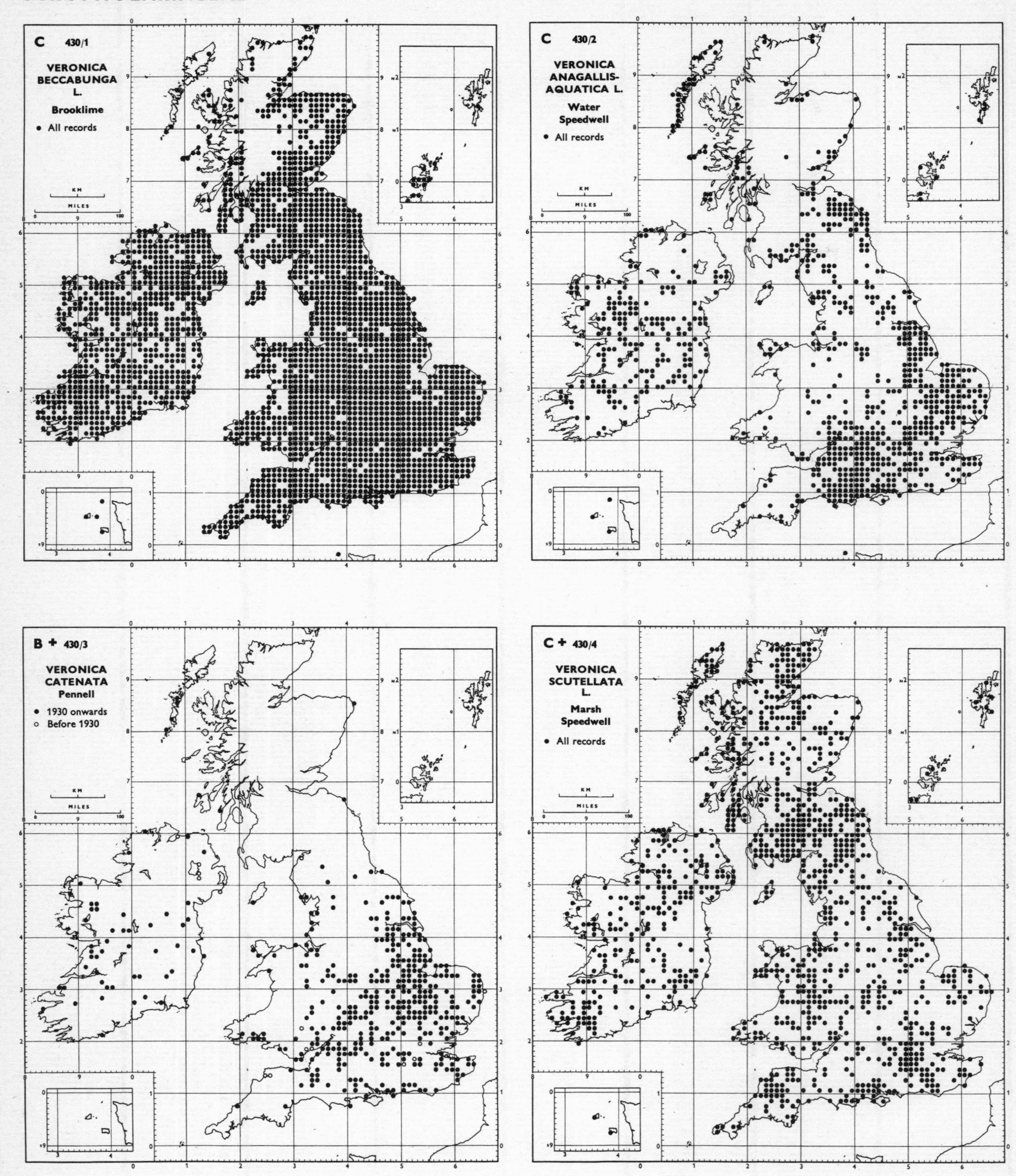
C 430/1
VERONICA BECCABUNGA L.
Brooklime
• All records
C 430/2
VERONICA ANAGALLIS-AQUATICA L.
Water Speedwell
• All records
B + 430/3
VERONICA CATENATA Pennell
• 1930 onwards
○ Before 1930
C + 430/4
VERONICA SCUTELLATA L.
Marsh Speedwell
• All records

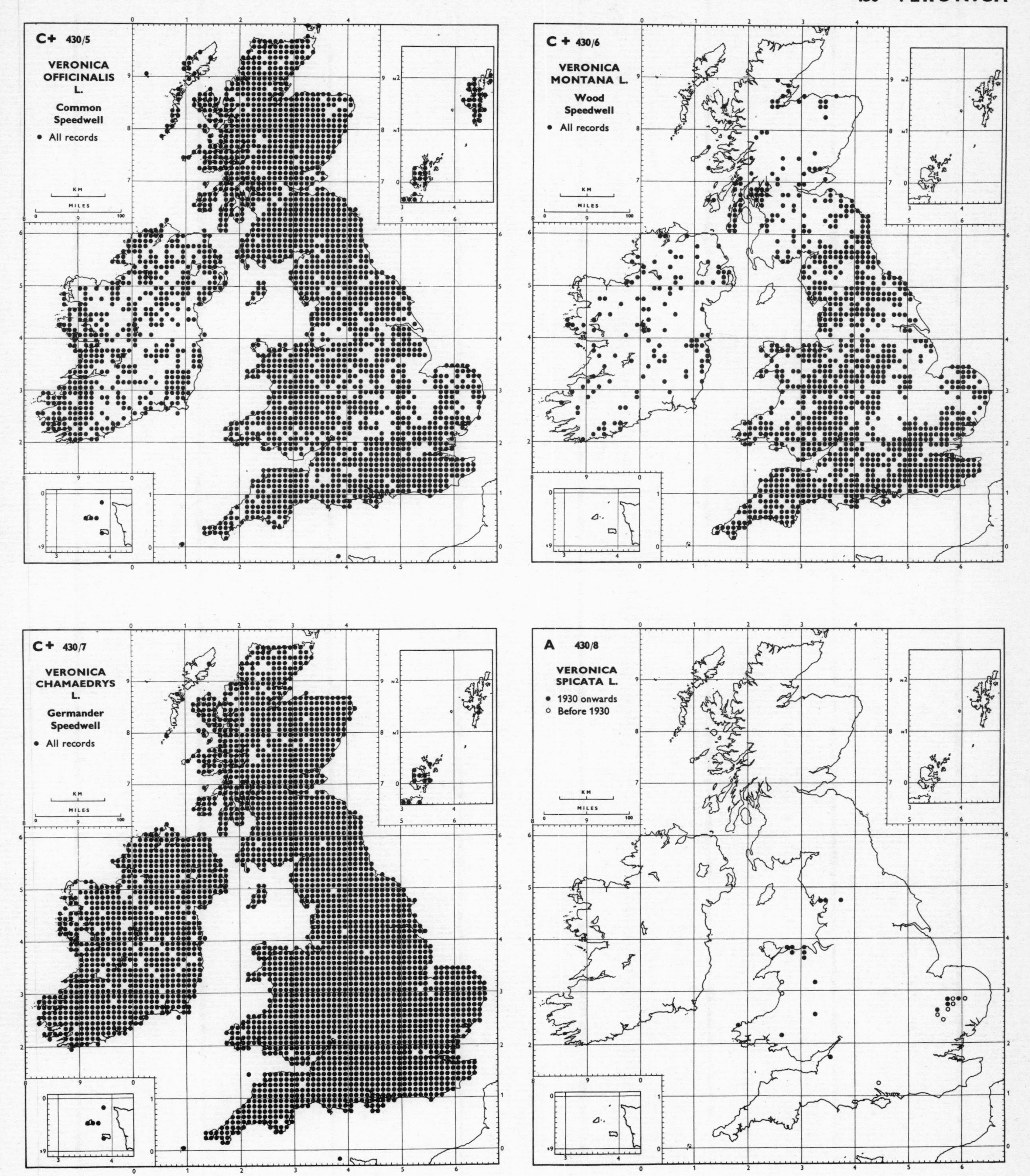
C+ 430/5
VERONICA OFFICINALIS L.
Common Speedwell
• All records
C + 430/6
VERONICA MONTANA L.
Wood Speedwell
• All records
C+ 430/7
VERONICA CHAMAEDRYS L.
Germander Speedwell
• All records
A 430/8
VERONICA SPICATA L.
• 1930 onwards
○ Before 1930

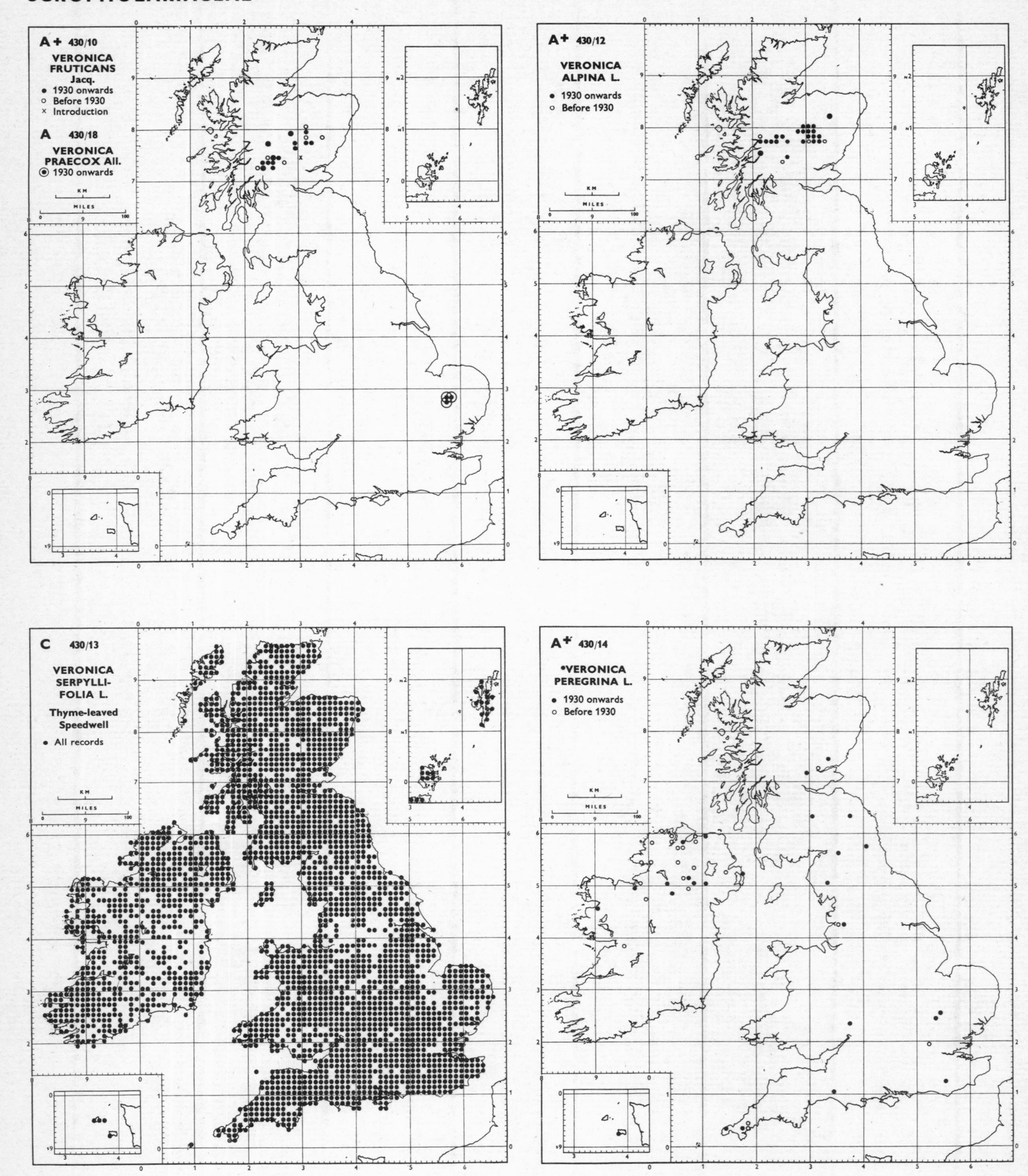
A + 430/10
VERONICA FRUTICANS Jacq.
● 1930 onwards
○ Before 1930
× Introduction
A 430/18
VERONICA PRAECOX All.
⊙ 1930 onwards
A + 430/12
VERONICA ALPINA L.
● 1930 onwards
○ Before 1930
C 430/13
VERONICA SERPYLLIFOLIA L.
Thyme-leaved Speedwell
● All records
A + 430/14
*VERONICA PEREGRINA L.
● 1930 onwards
○ Before 1930
KM
MILES

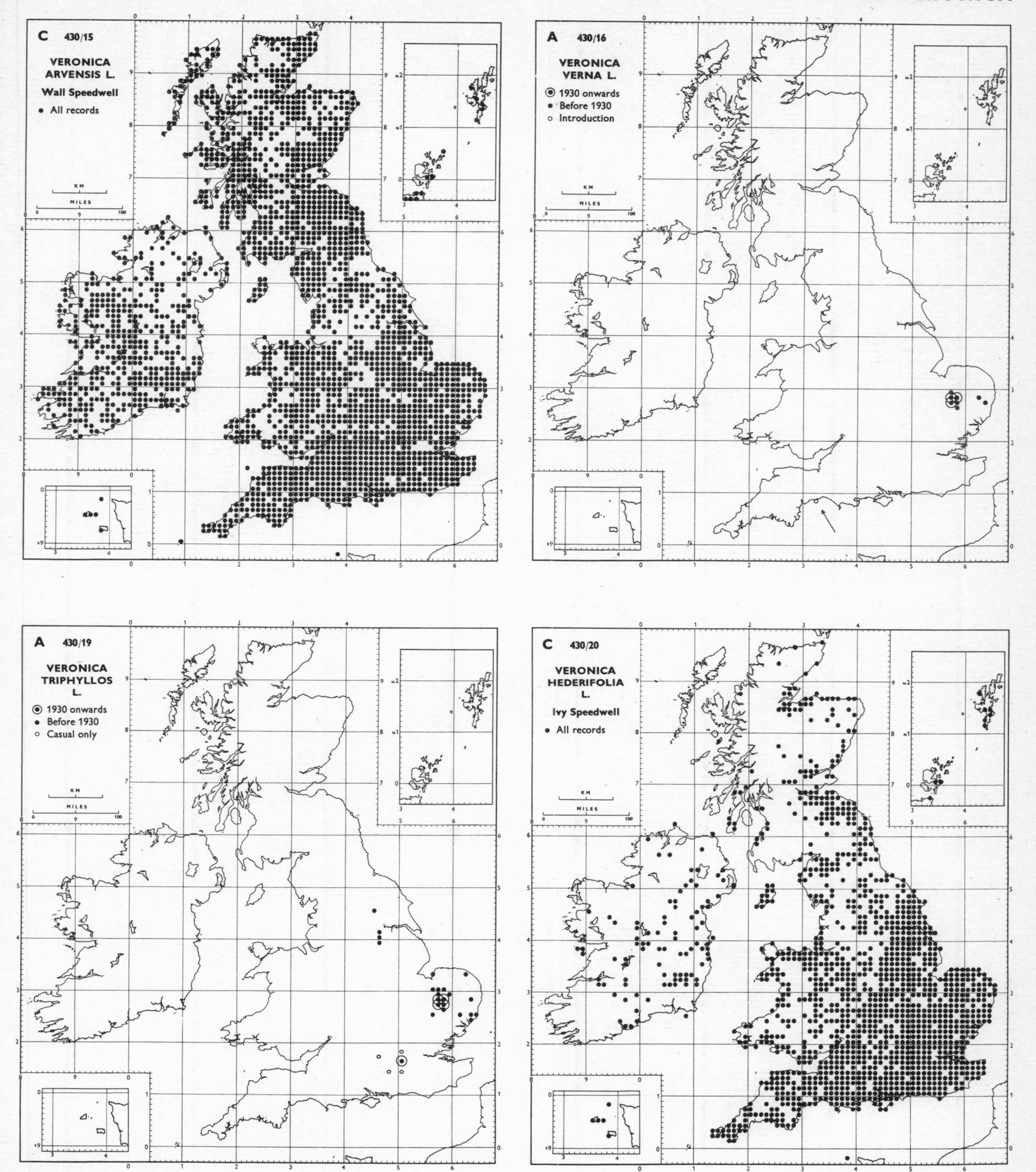
C 430/15
VERONICA ARVENSIS L.
Wall Speedwell
● All records
A 430/16
VERONICA VERNA L.
◉ 1930 onwards
● Before 1930
○ Introduction
A 430/19
VERONICA TRIPHYLLOS L.
◉ 1930 onwards
● Before 1930
○ Casual only
C 430/20
VERONICA HEDERIFOLIA L.
Ivy Speedwell
● All records
KM
MILES

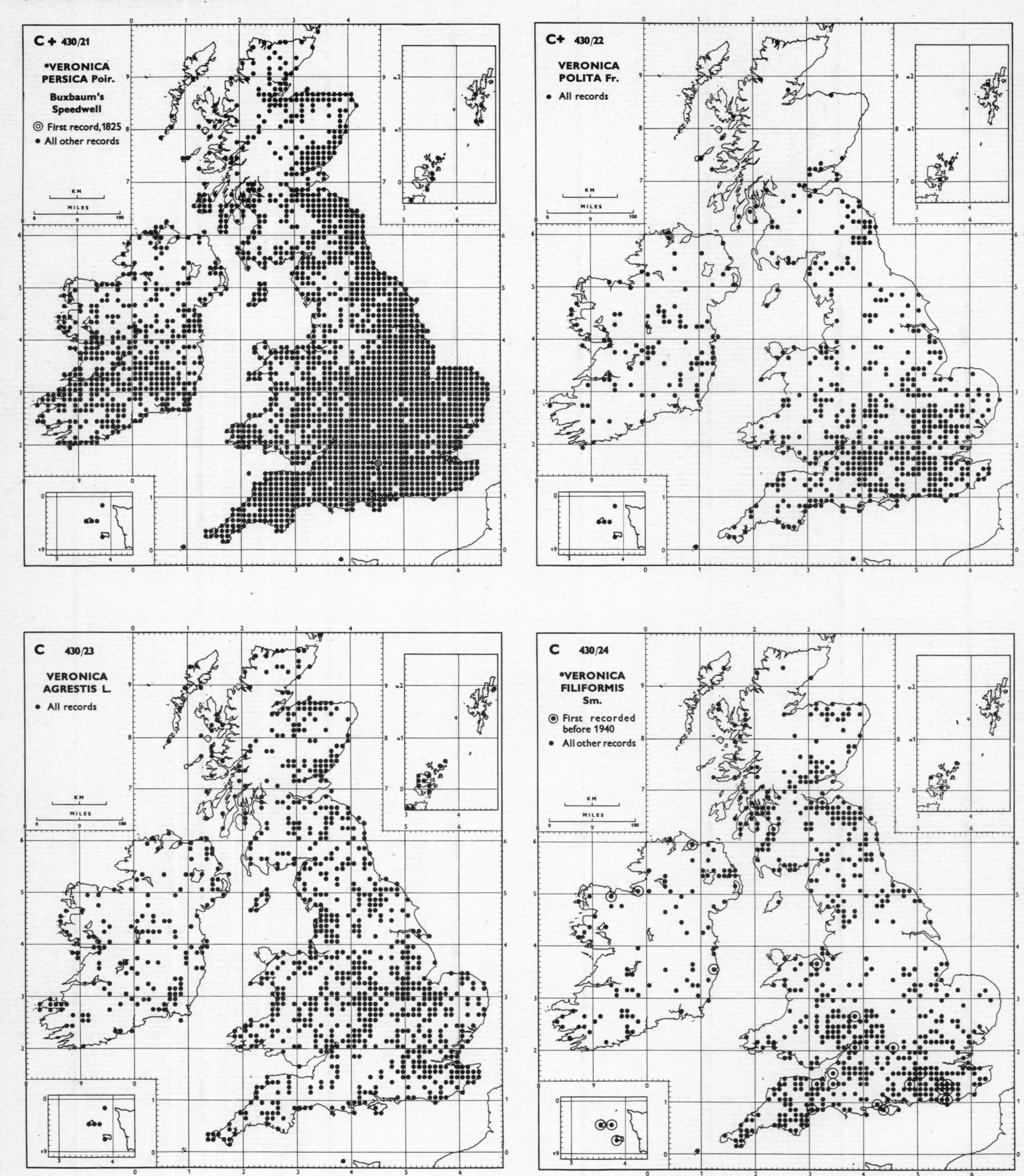

C + 430/21
•VERONICA PERSICA Poir.
Buxbaum's Speedwell
First record, 1825
All other records
C+ 430/22
VERONICA POLITA Fr.
All records
C 430/23
VERONICA AGRESTIS L.
All records
C 430/24
•VERONICA FILIFORMIS Sm.
First recorded before 1940
All other records

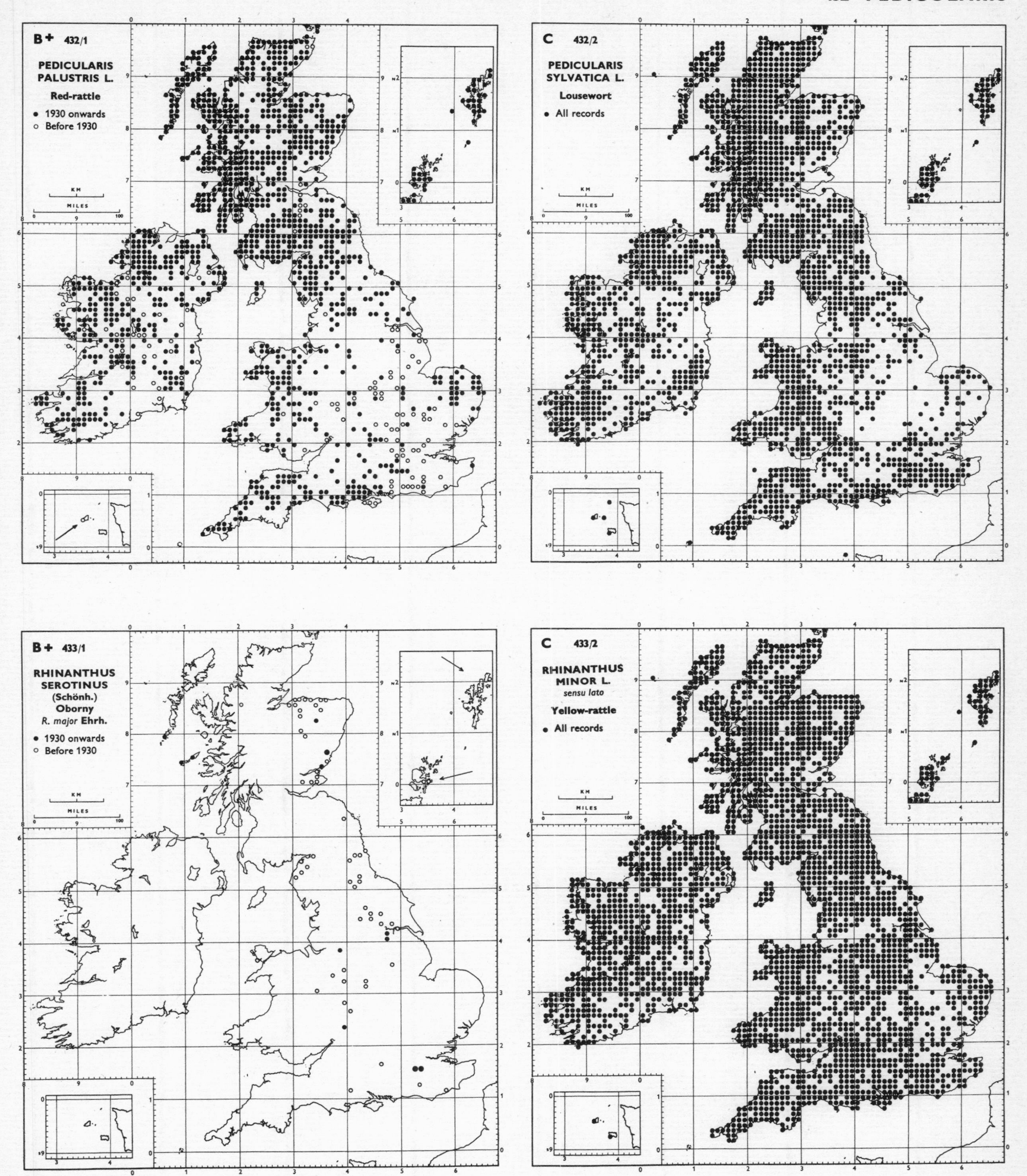
B+ 432/1
PEDICULARIS PALUSTRIS L.
Red-rattle
• 1930 onwards
○ Before 1930
C 432/2
PEDICULARIS SYLVATICA L.
Lousewort
• All records
B+ 433/1
RHINANTHUS SEROTINUS (Schönh.) Oborny
R. major Ehrh.
• 1930 onwards
○ Before 1930
C 433/2
RHINANTHUS MINOR L.
sensu lato
Yellow-rattle
• All records

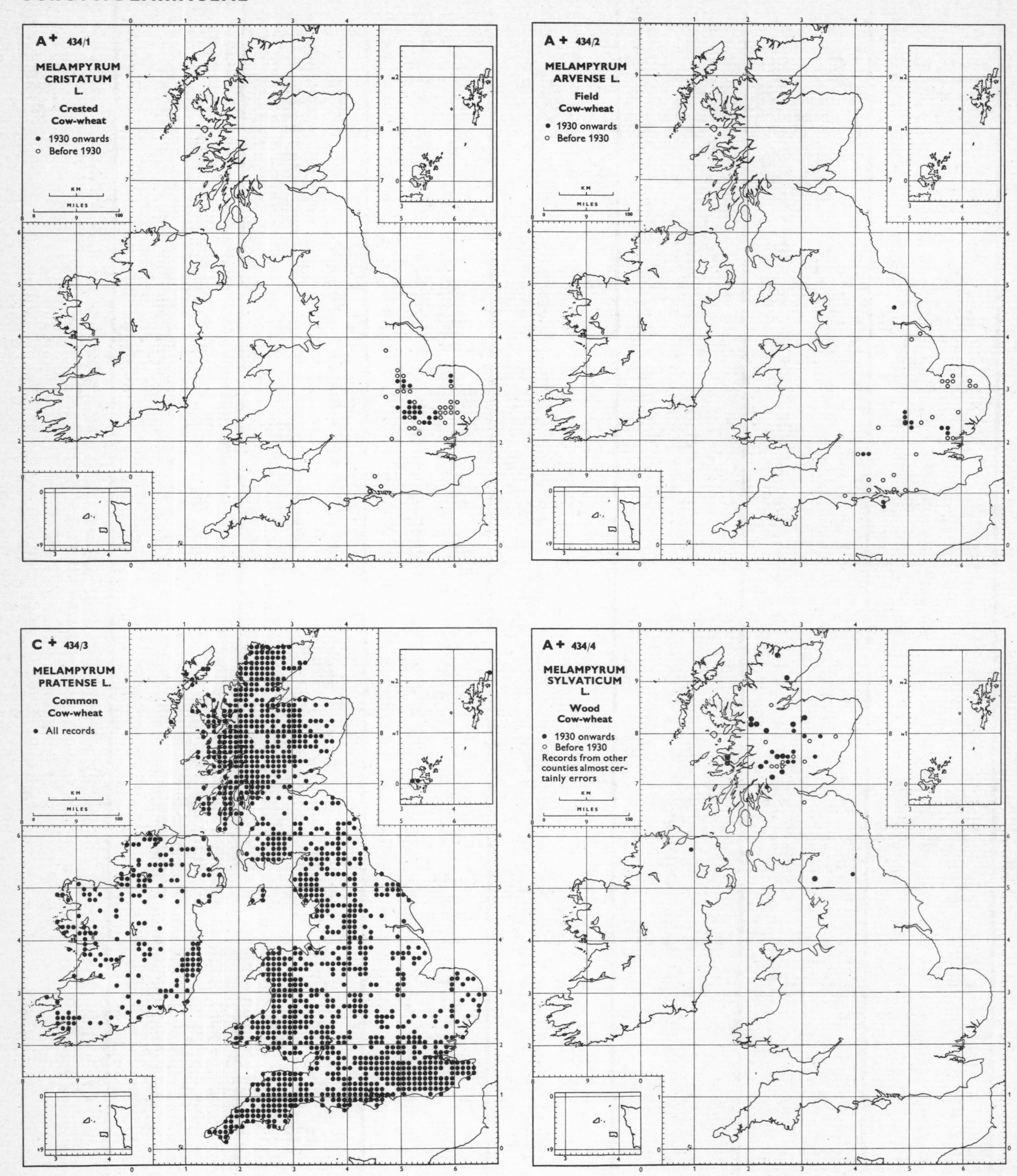
A+ 434/1
MELAMPYRUM CRISTATUM L.
Crested Cow-wheat
• 1930 onwards
○ Before 1930
A+ 434/2
MELAMPYRUM ARVENSE L.
Field Cow-wheat
• 1930 onwards
○ Before 1930
C+ 434/3
MELAMPYRUM PRATENSE L.
Common Cow-wheat
• All records
A+ 434/4
MELAMPYRUM SYLVATICUM L.
Wood Cow-wheat
• 1930 onwards
○ Before 1930
Records from other counties almost certainly errors

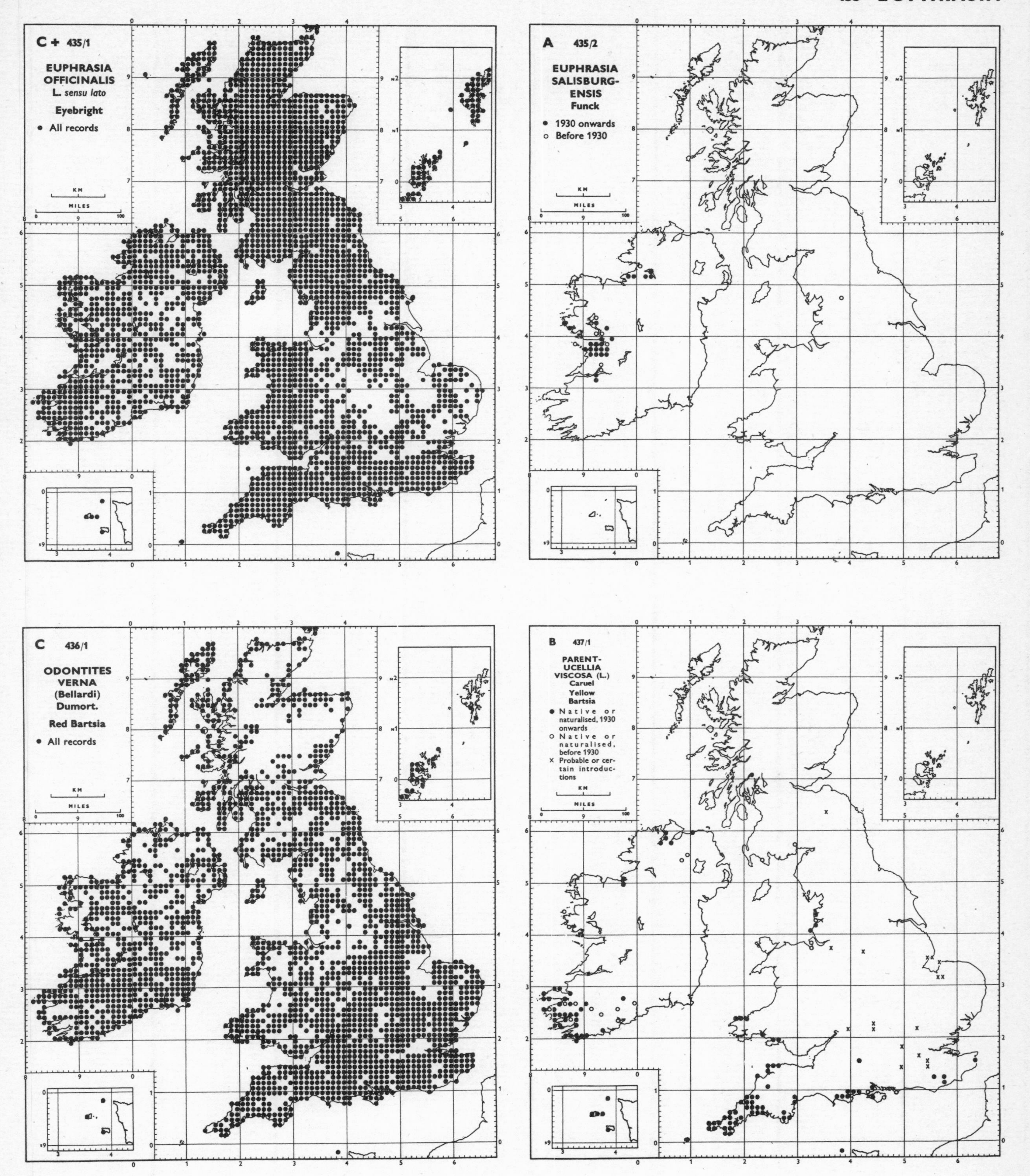

C + 435/1
EUPHRASIA OFFICINALIS L. sensu lato
Eyebright
● All records
A 435/2
EUPHRASIA SALISBURGENSIS Funck
● 1930 onwards
○ Before 1930
C 436/1
ODONTITES VERNA (Bellardi) Dumort.
Red Bartsia
● All records
B 437/1
PARENTUCELLIA VISCOSA (L.) Caruel
Yellow Bartsia
● Native or naturalised, 1930 onwards
○ Native or naturalised, before 1930
× Probable or certain introductions
KM
MILES

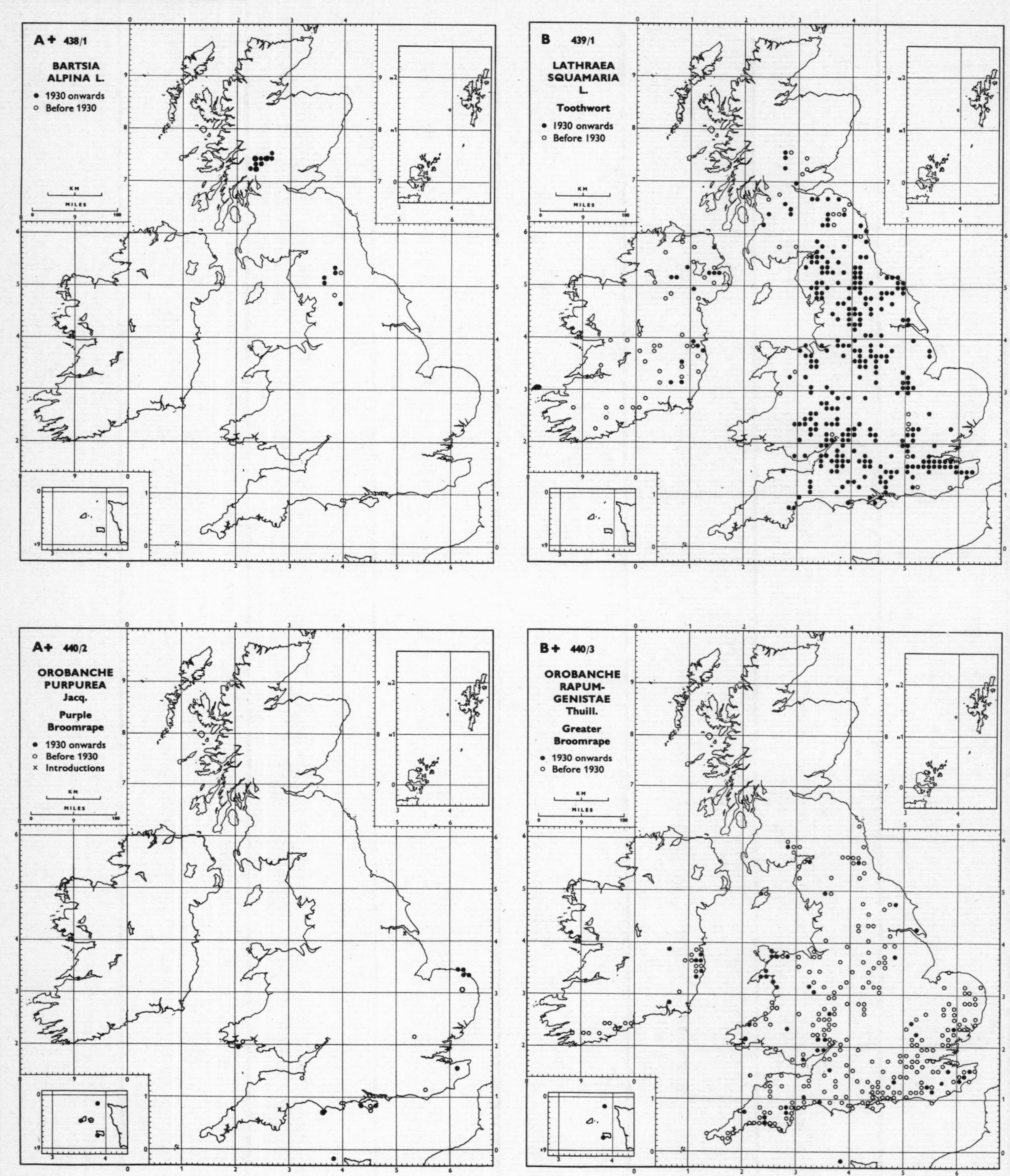
A+ 438/1
BARTSIA ALPINA L.
• 1930 onwards
○ Before 1930
B 439/1
LATHRAEA SQUAMARIA L.
Toothwort
• 1930 onwards
○ Before 1930
A+ 440/2
OROBANCHE PURPUREA Jacq.
Purple Broomrape
• 1930 onwards
○ Before 1930
× Introductions
B+ 440/3
OROBANCHE RAPUM-GENISTAE Thuill.
Greater Broomrape
• 1930 onwards
○ Before 1930
KM
MILES

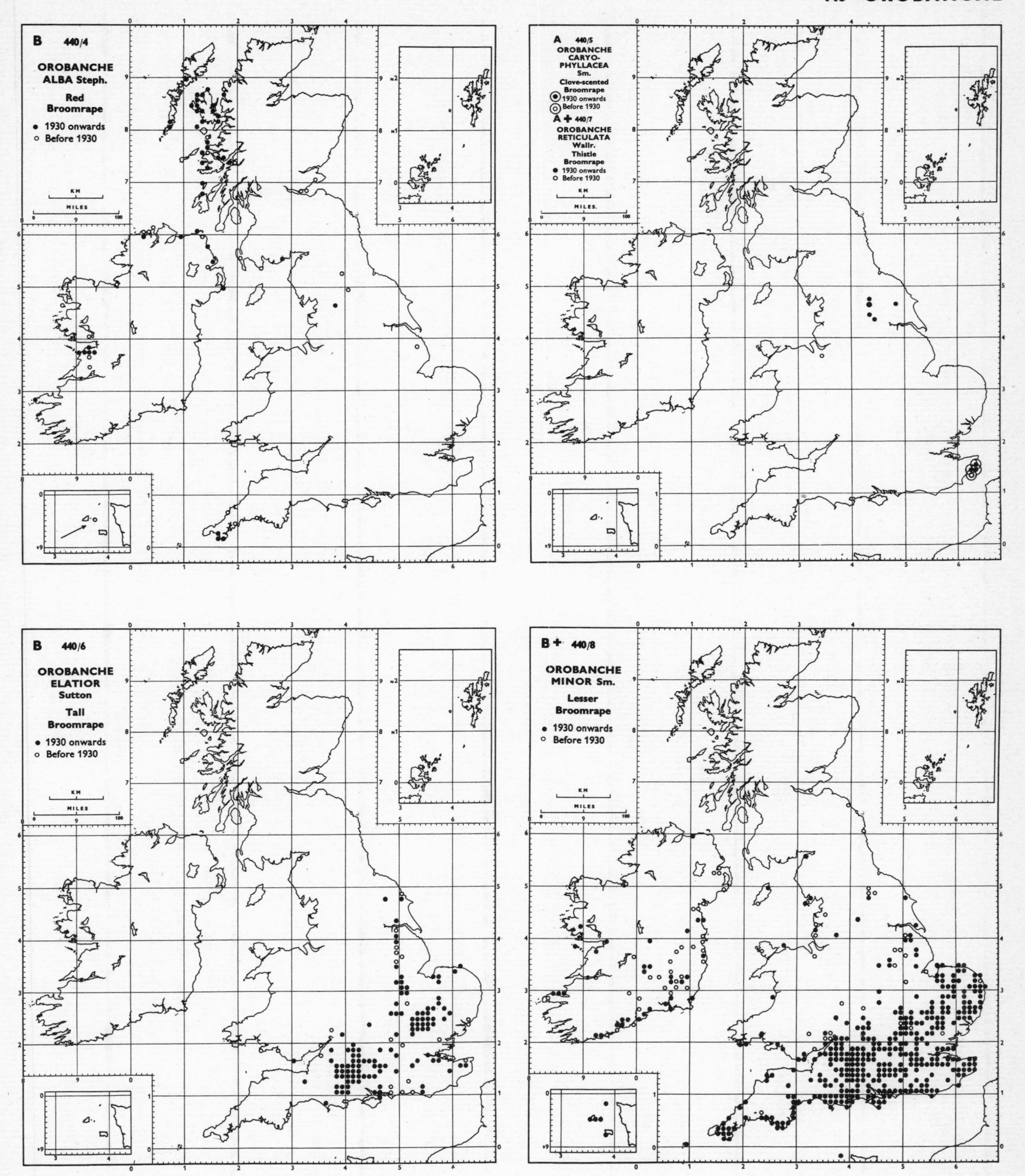
B 440/4
OROBANCHE ALBA Steph.
Red Broomrape
1930 onwards
Before 1930
A 440/5
OROBANCHE CARYOPHYLLACEA Sm.
Clove-scented Broomrape
1930 onwards
Before 1930
A + 440/7
OROBANCHE RETICULATA Wallr.
Thistle Broomrape
1930 onwards
Before 1930
B 440/6
OROBANCHE ELATIOR Sutton
Tall Broomrape
1930 onwards
Before 1930
B + 440/8
OROBANCHE MINOR Sm.
Lesser Broomrape
1930 onwards
Before 1930
KM
MILES

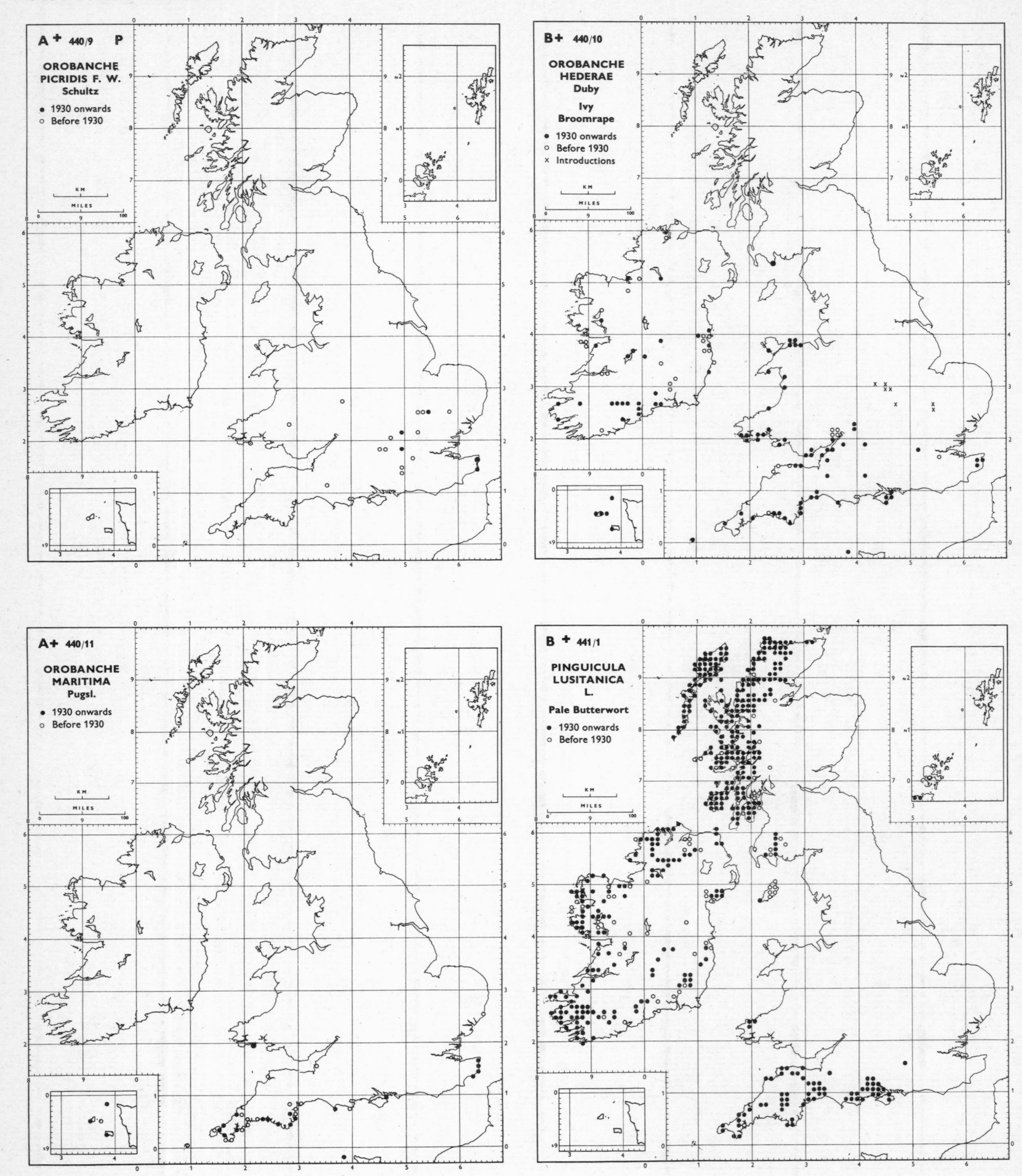
A + 440/9 P
OROBANCHE
PICRIDIS F. W.
Schultz
● 1930 onwards
○ Before 1930
KM
MILES
B + 440/10
OROBANCHE
HEDERAE
Duby
Ivy
Broomrape
● 1930 onwards
○ Before 1930
× Introductions
A + 440/11
OROBANCHE
MARITIMA
Pugsl.
● 1930 onwards
○ Before 1930
B + 441/1
PINGUICULA
LUSITANICA
L.
Pale Butterwort
● 1930 onwards
○ Before 1930

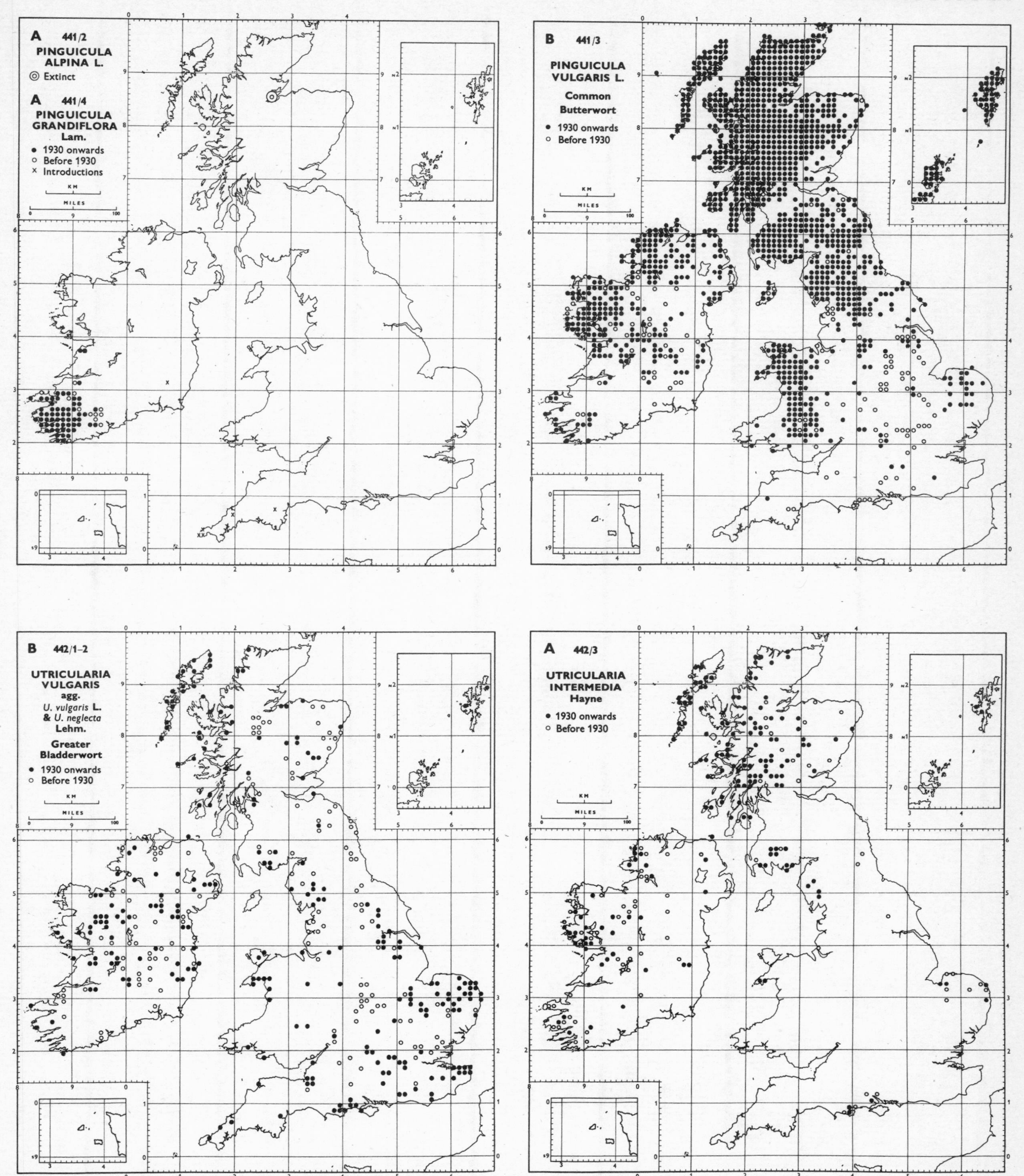
A 441/2
PINGUICULA ALPINA L.
Extinct
A 441/4
PINGUICULA GRANDIFLORA Lam.
1930 onwards
Before 1930
Introductions
B 441/3
PINGUICULA VULGARIS L.
Common Butterwort
1930 onwards
Before 1930
B 442/1-2
UTRICULARIA VULGARIS agg.
U. vulgaris L. & U. neglecta Lehm.
Greater Bladderwort
1930 onwards
Before 1930
A 442/3
UTRICULARIA INTERMEDIA Hayne
1930 onwards
Before 1930

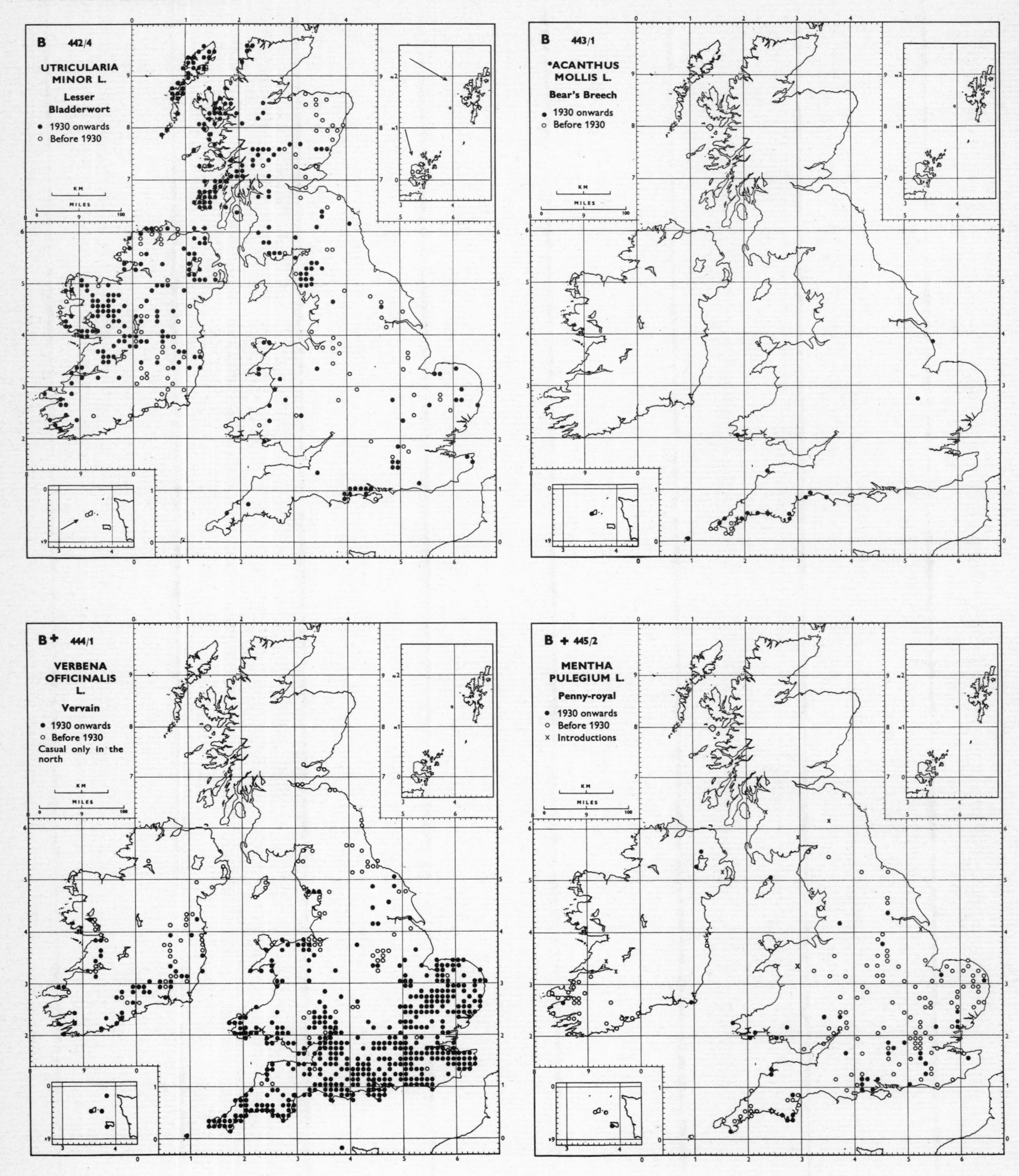
B 442/4
UTRICULARIA MINOR L.
Lesser Bladderwort
● 1930 onwards
○ Before 1930
B 443/1
*ACANTHUS MOLLIS L.
Bear's Breech
● 1930 onwards
○ Before 1930
B+ 444/1
VERBENA OFFICINALIS L.
Vervain
● 1930 onwards
○ Before 1930
Casual only in the north
B + 445/2
MENTHA PULEGIUM L.
Penny-royal
● 1930 onwards
○ Before 1930
× Introductions

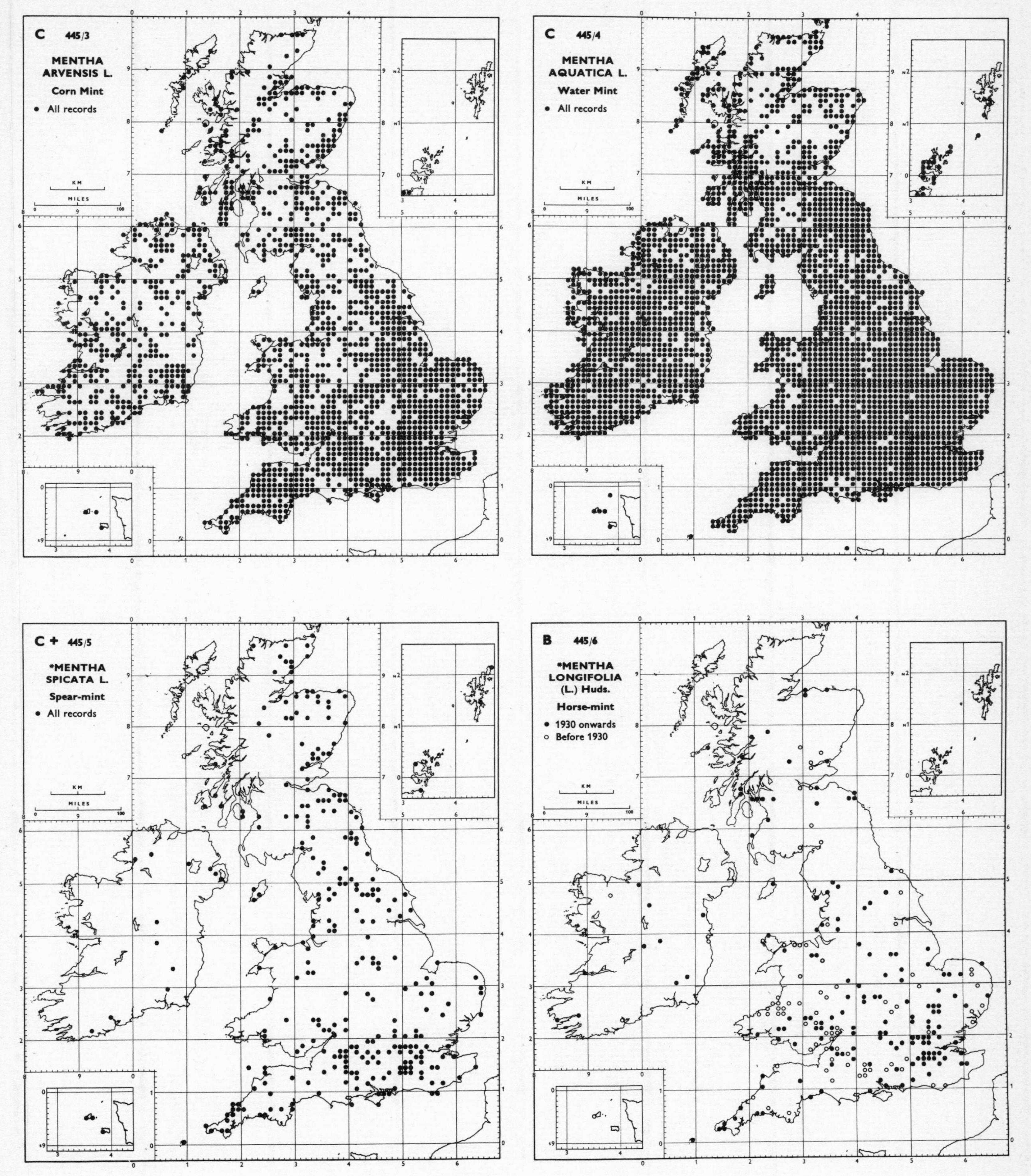
C 445/3
MENTHA
ARVENSIS L.
Corn Mint
• All records
C 445/4
MENTHA
AQUATICA L.
Water Mint
• All records
C + 445/5
*MENTHA
SPICATA L.
Spear-mint
• All records
B 445/6
*MENTHA
LONGIFOLIA
(L.) Huds.
Horse-mint
• 1930 onwards
○ Before 1930
KM
MILES

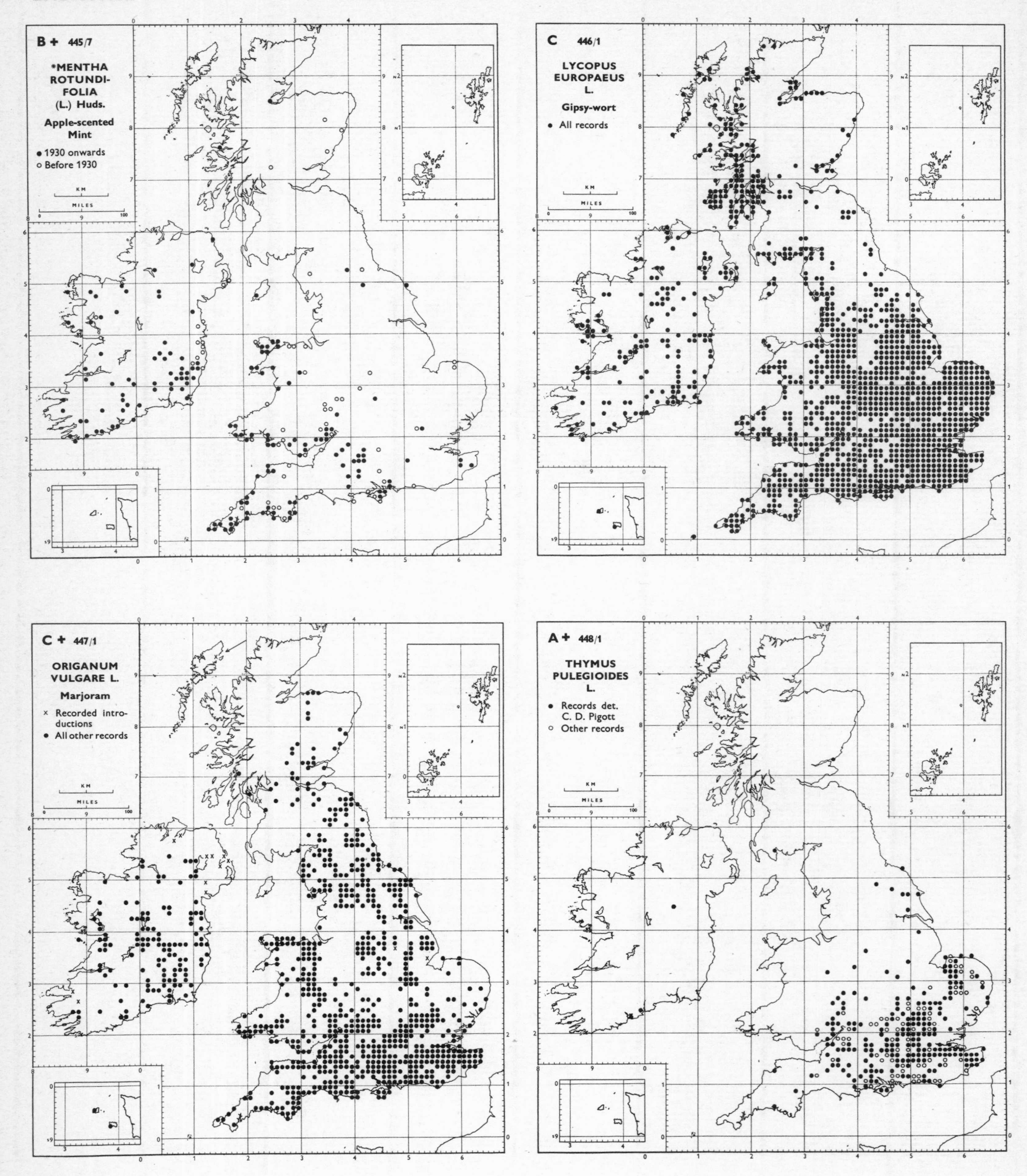
B + 445/7
*MENTHA ROTUNDIFOLIA (L.) Huds.
Apple-scented Mint
● 1930 onwards
○ Before 1930
KM
MILES
C 446/1
LYCOPUS EUROPAEUS L.
Gipsy-wort
● All records
C + 447/1
ORIGANUM VULGARE L.
Marjoram
× Recorded introductions
● All other records
A + 448/1
THYMUS PULEGIOIDES L.
● Records det. C. D. Pigott
○ Other records

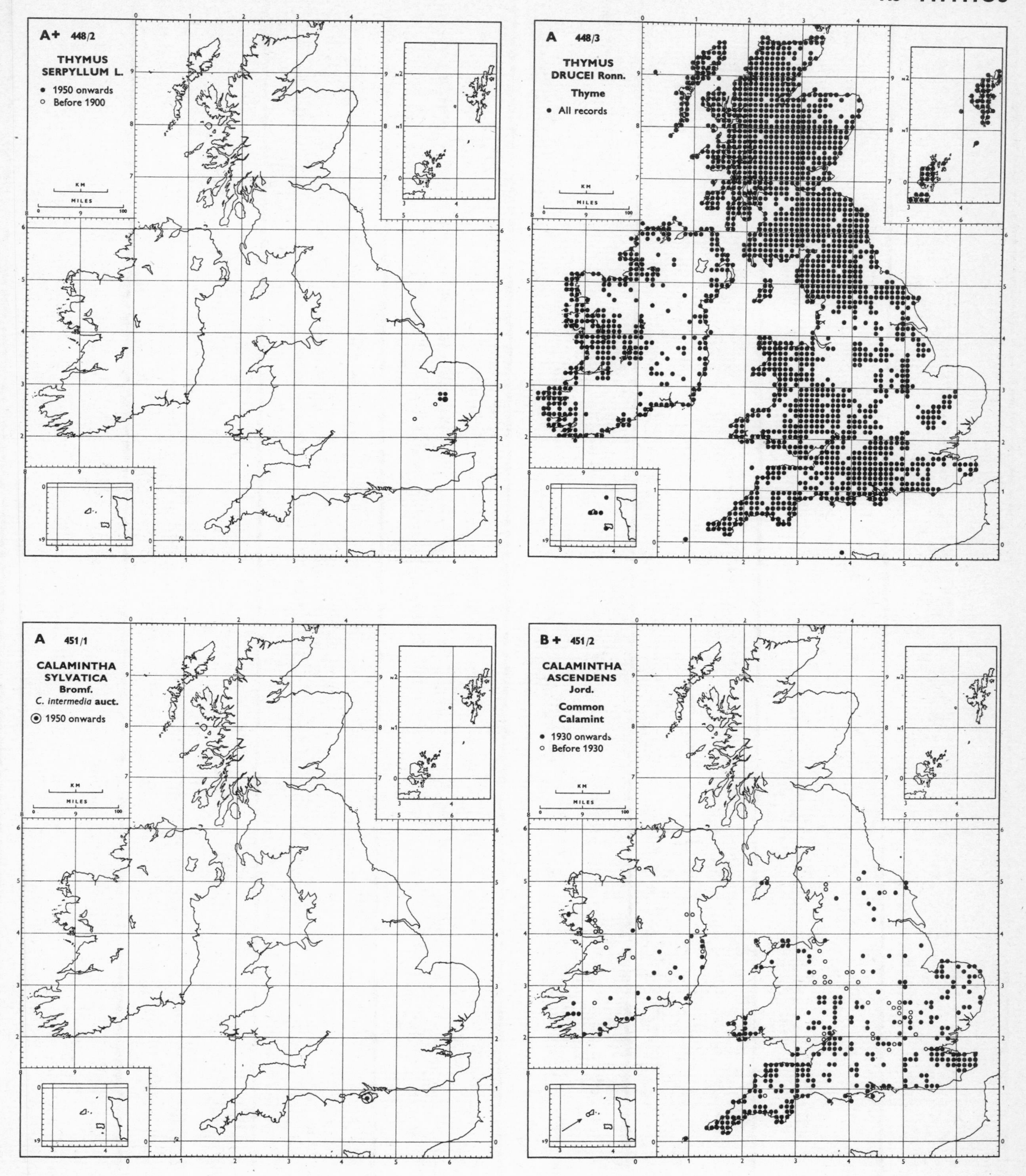
A+ 448/2
THYMUS SERPYLLUM L.
1950 onwards
Before 1900
A 448/3
THYMUS DRUCEI Ronn.
Thyme
All records
A 451/1
CALAMINTHA SYLVATICA Bromf.
C. intermedia auct.
1950 onwards
B+ 451/2
CALAMINTHA ASCENDENS Jord.
Common Calamint
1930 onwards
Before 1930
KM
MILES

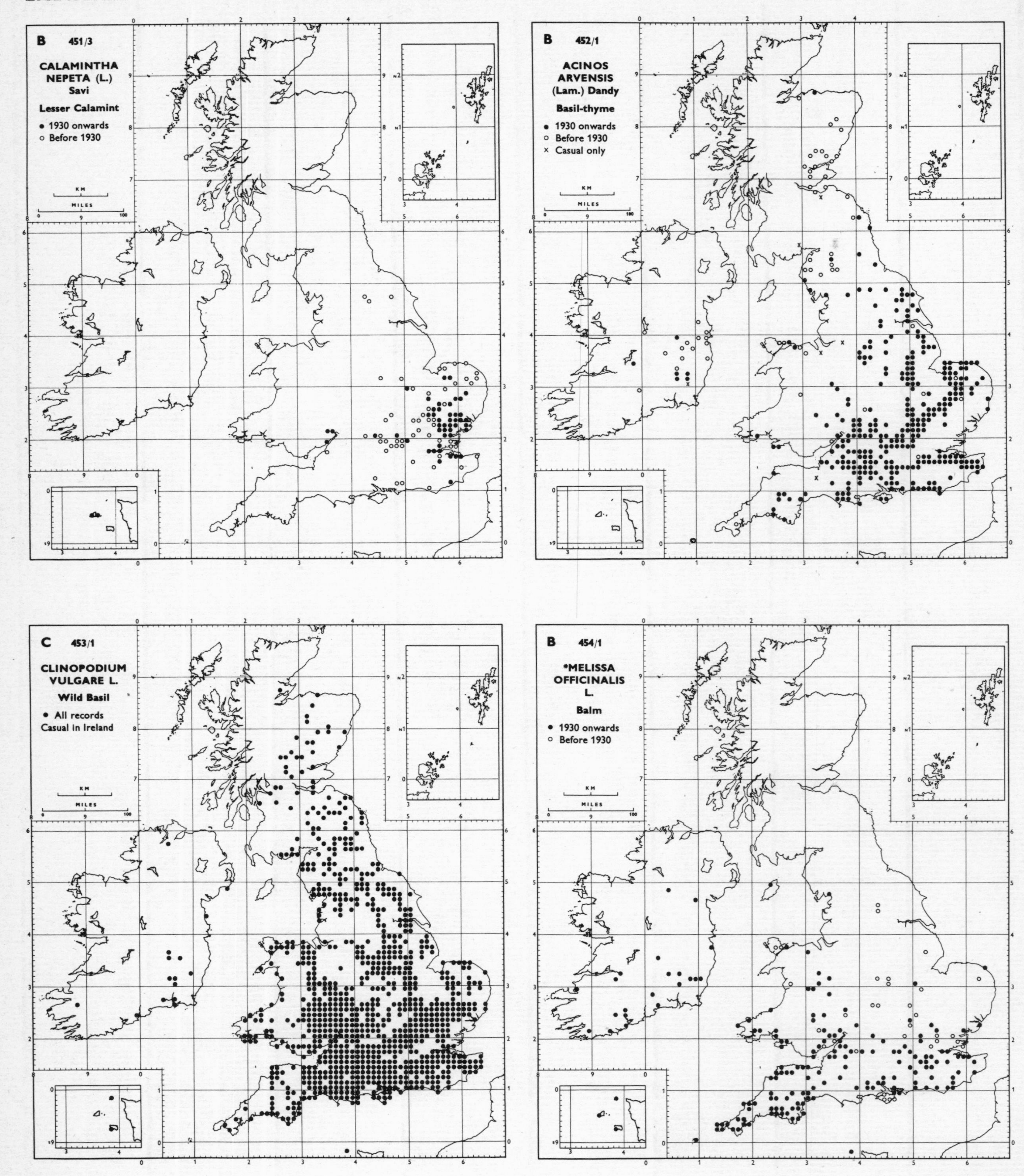
B 451/3
CALAMINTHA NEPETA (L.) Savi
Lesser Calamint
• 1930 onwards
○ Before 1930
B 452/1
ACINOS ARVENSIS (Lam.) Dandy
Basil-thyme
• 1930 onwards
○ Before 1930
× Casual only
C 453/1
CLINOPODIUM VULGARE L.
Wild Basil
• All records
Casual in Ireland
B 454/1
*MELISSA OFFICINALIS L.
Balm
• 1930 onwards
○ Before 1930

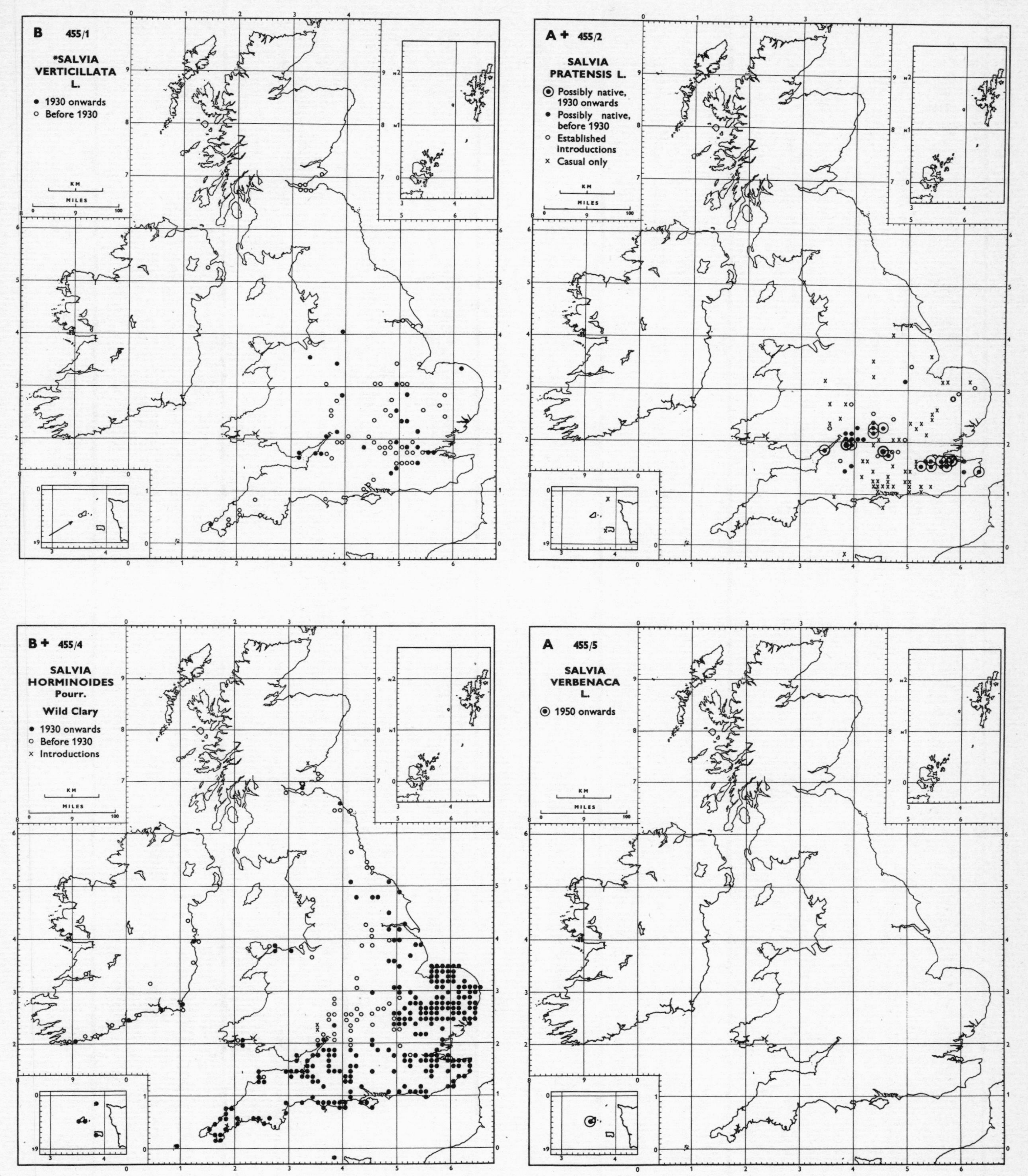
B 455/1
*SALVIA VERTICILLATA L.
1930 onwards
Before 1930
A + 455/2
SALVIA PRATENSIS L.
Possibly native, 1930 onwards
Possibly native, before 1930
Established introductions
Casual only
B + 455/4
SALVIA HORMINOIDES Pourr.
Wild Clary
1930 onwards
Before 1930
Introductions
A 455/5
SALVIA VERBENACA L.
1950 onwards
KM
MILES

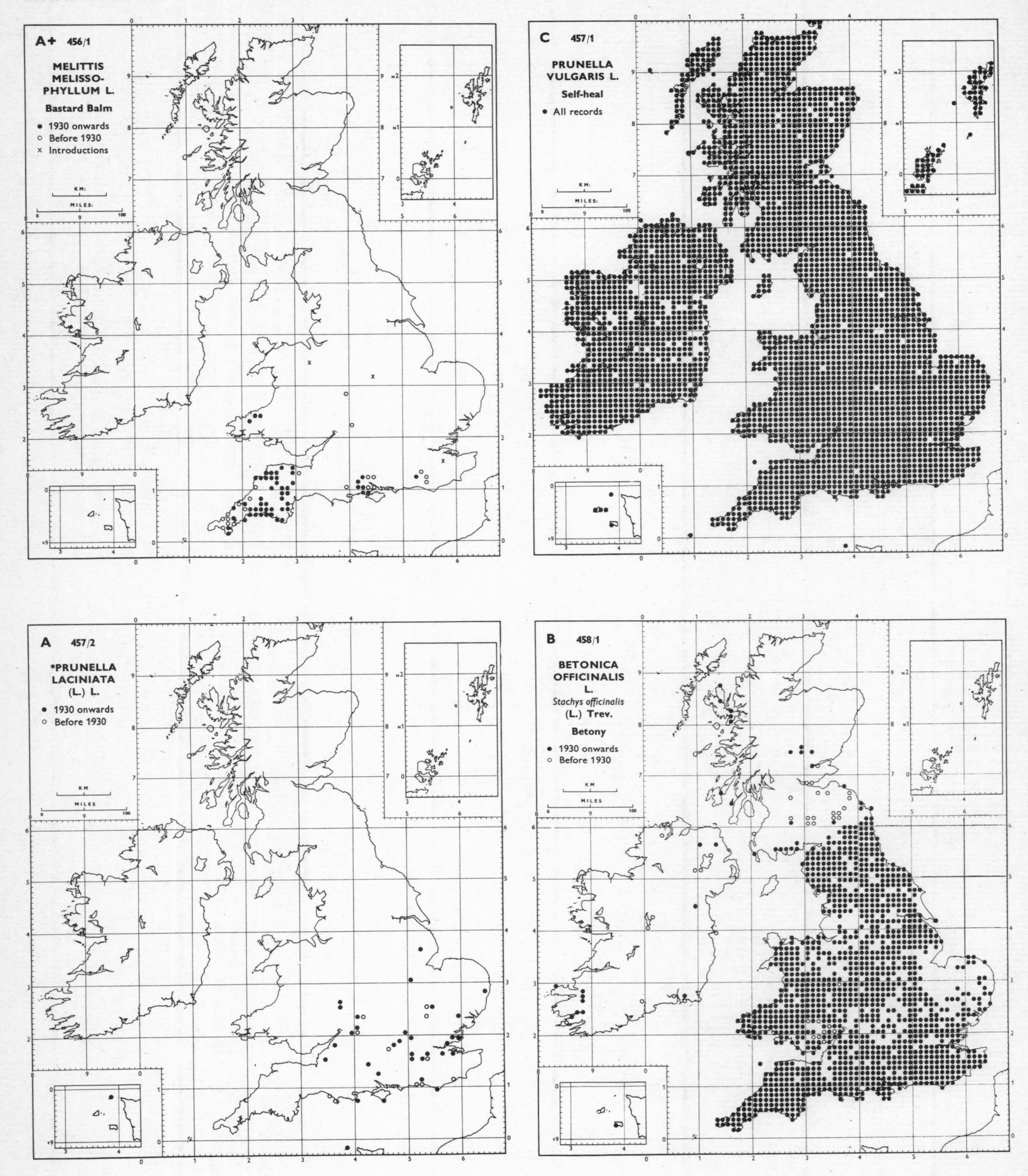
A+ 456/1
MELITTIS MELISSOPHYLLUM L.
Bastard Balm
• 1930 onwards
○ Before 1930
× Introductions
C 457/1
PRUNELLA VULGARIS L.
Self-heal
• All records
A 457/2
*PRUNELLA LACINIATA (L.) L.
• 1930 onwards
○ Before 1930
B 458/1
BETONICA OFFICINALIS L.
Stachys officinalis (L.) Trev.
Betony
• 1930 onwards
○ Before 1930

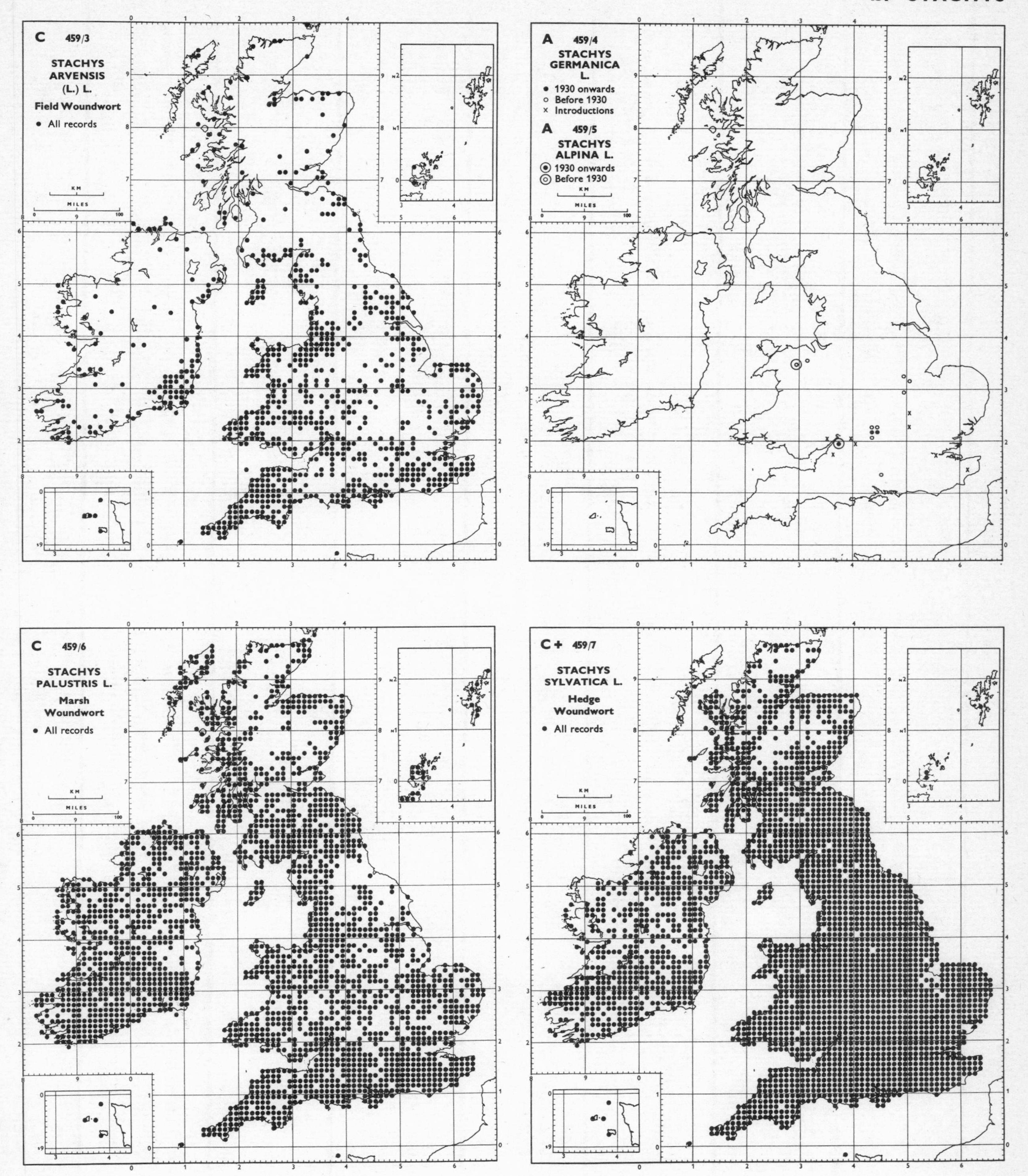
C 459/3
STACHYS ARVENSIS (L.) L.
Field Woundwort
• All records
A 459/4
STACHYS GERMANICA L.
• 1930 onwards
○ Before 1930
x Introductions
A 459/5
STACHYS ALPINA L.
◉ 1930 onwards
◎ Before 1930
C 459/6
STACHYS PALUSTRIS L.
Marsh Woundwort
• All records
C+ 459/7
STACHYS SYLVATICA L.
Hedge Woundwort
• All records
KM
MILES

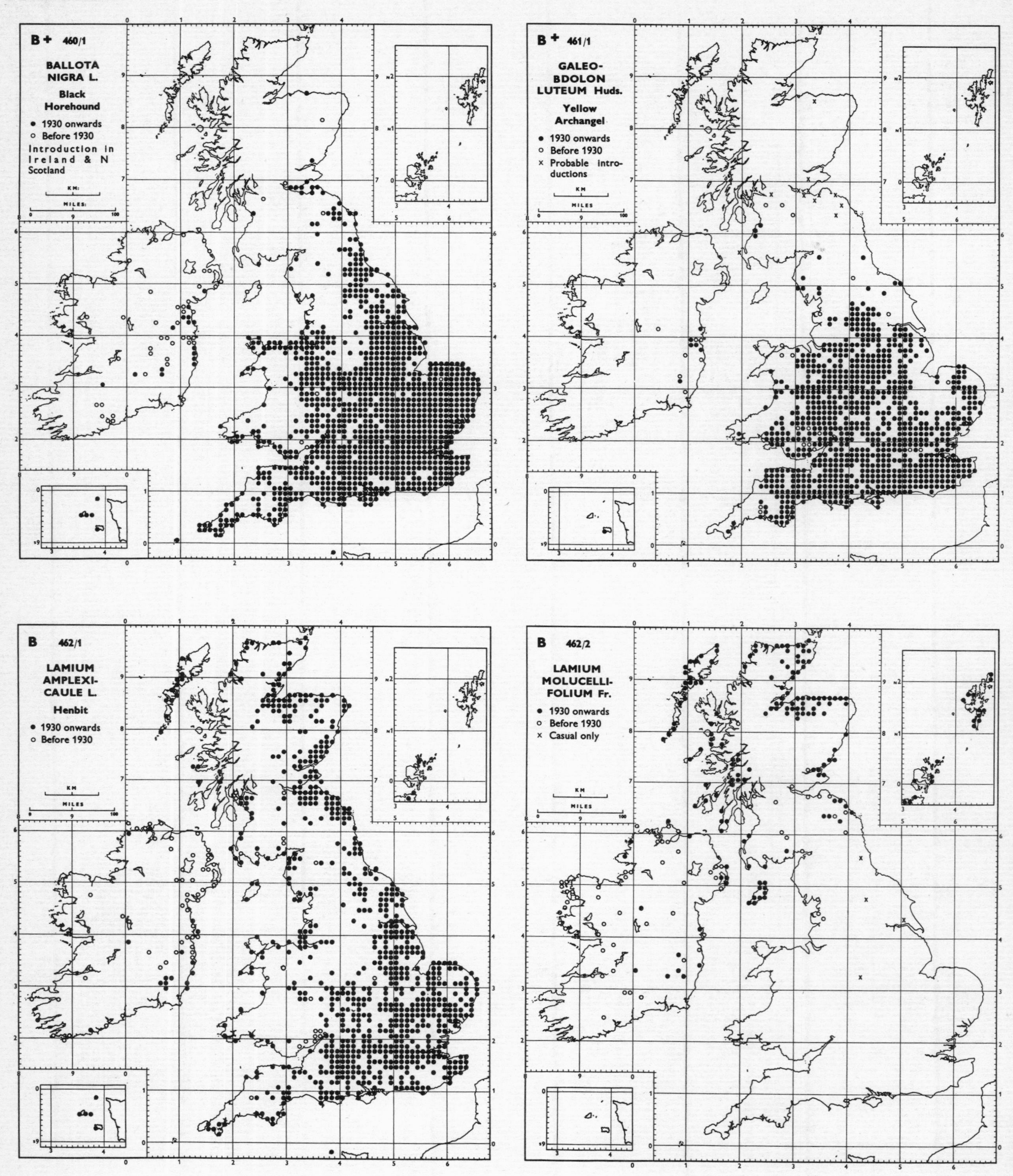
B+ 460/1
BALLOTA NIGRA L.
Black Horehound
• 1930 onwards
○ Before 1930
Introduction in Ireland & N Scotland
B+ 461/1
GALEO-BDOLON LUTEUM Huds.
Yellow Archangel
• 1930 onwards
○ Before 1930
× Probable introductions
B 462/1
LAMIUM AMPLEXI-CAULE L.
Henbit
• 1930 onwards
○ Before 1930
B 462/2
LAMIUM MOLUCELLI-FOLIUM Fr.
• 1930 onwards
○ Before 1930
× Casual only

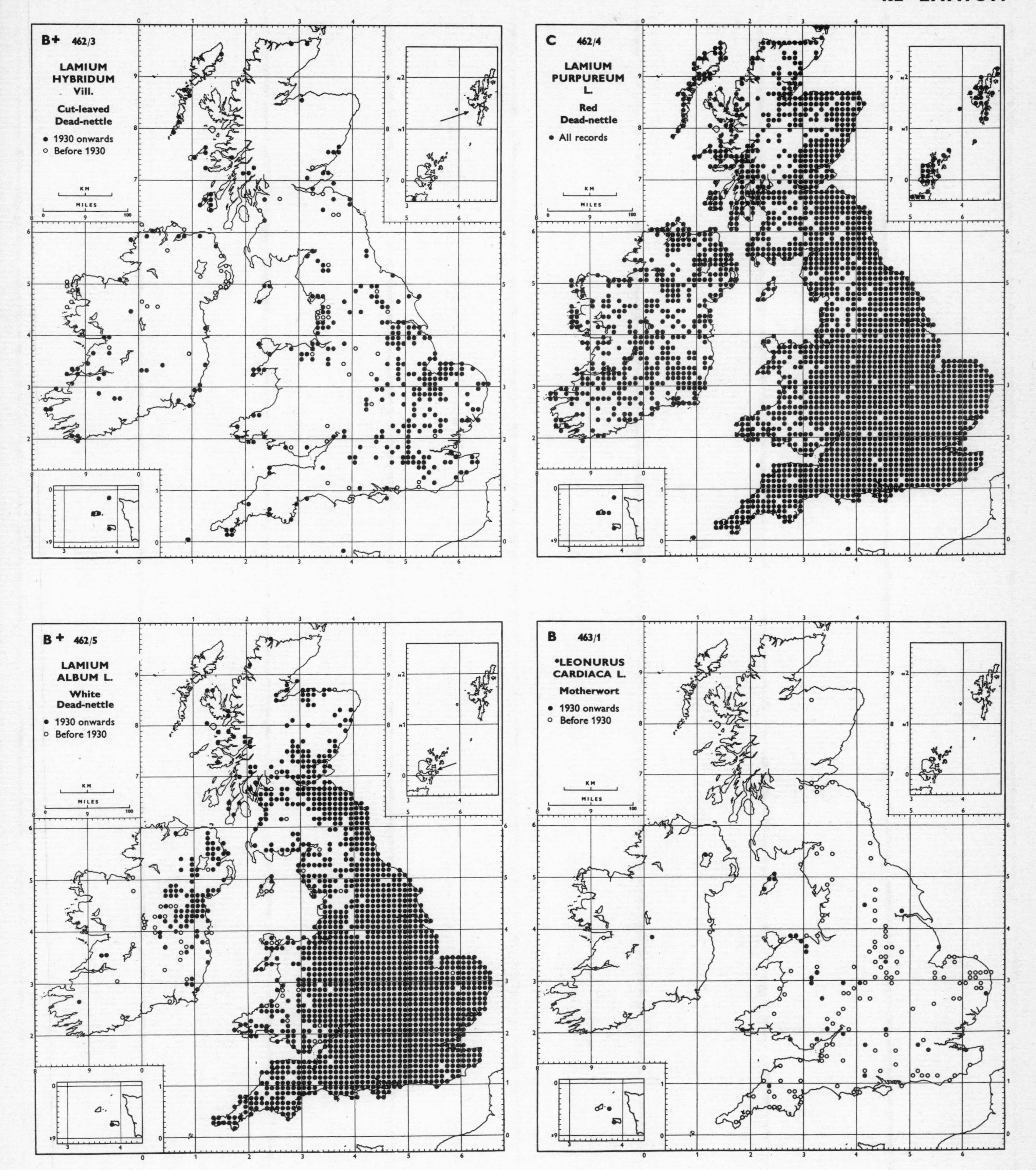

B+ 462/3
LAMIUM HYBRIDUM Vill.
Cut-leaved Dead-nettle
• 1930 onwards
○ Before 1930
C 462/4
LAMIUM PURPUREUM L.
Red Dead-nettle
• All records
B+ 462/5
LAMIUM ALBUM L.
White Dead-nettle
• 1930 onwards
○ Before 1930
B 463/1
*LEONURUS CARDIACA L.
Motherwort
• 1930 onwards
○ Before 1930

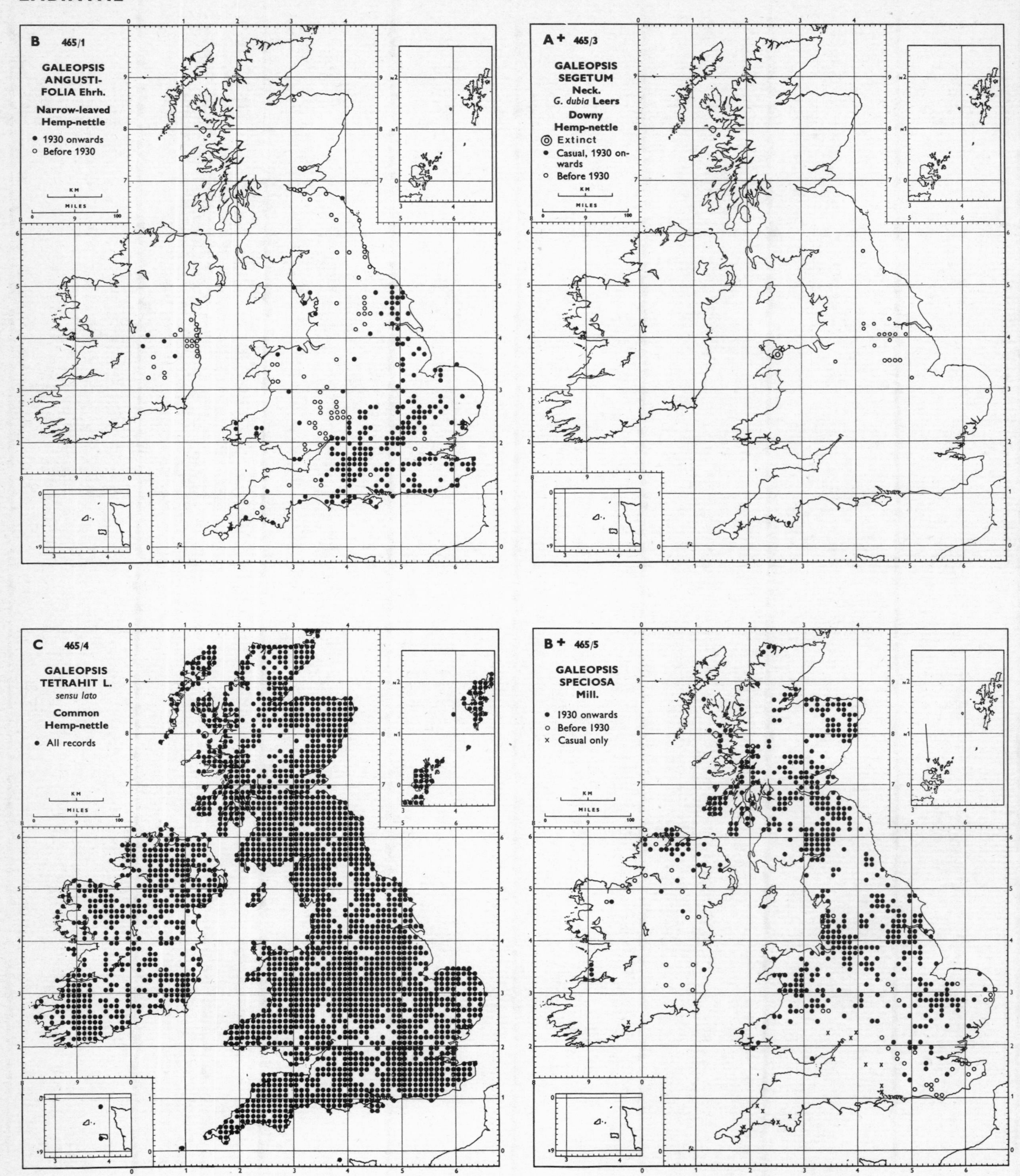
B 465/1
GALEOPSIS ANGUSTIFOLIA Ehrh.
Narrow-leaved Hemp-nettle
• 1930 onwards
○ Before 1930
A+ 465/3
GALEOPSIS SEGETUM Neck.
G. dubia Leers
Downy Hemp-nettle
◎ Extinct
• Casual, 1930 onwards
○ Before 1930
C 465/4
GALEOPSIS TETRAHIT L. sensu lato
Common Hemp-nettle
• All records
B+ 465/5
GALEOPSIS SPECIOSA Mill.
• 1930 onwards
○ Before 1930
× Casual only
KM
MILES

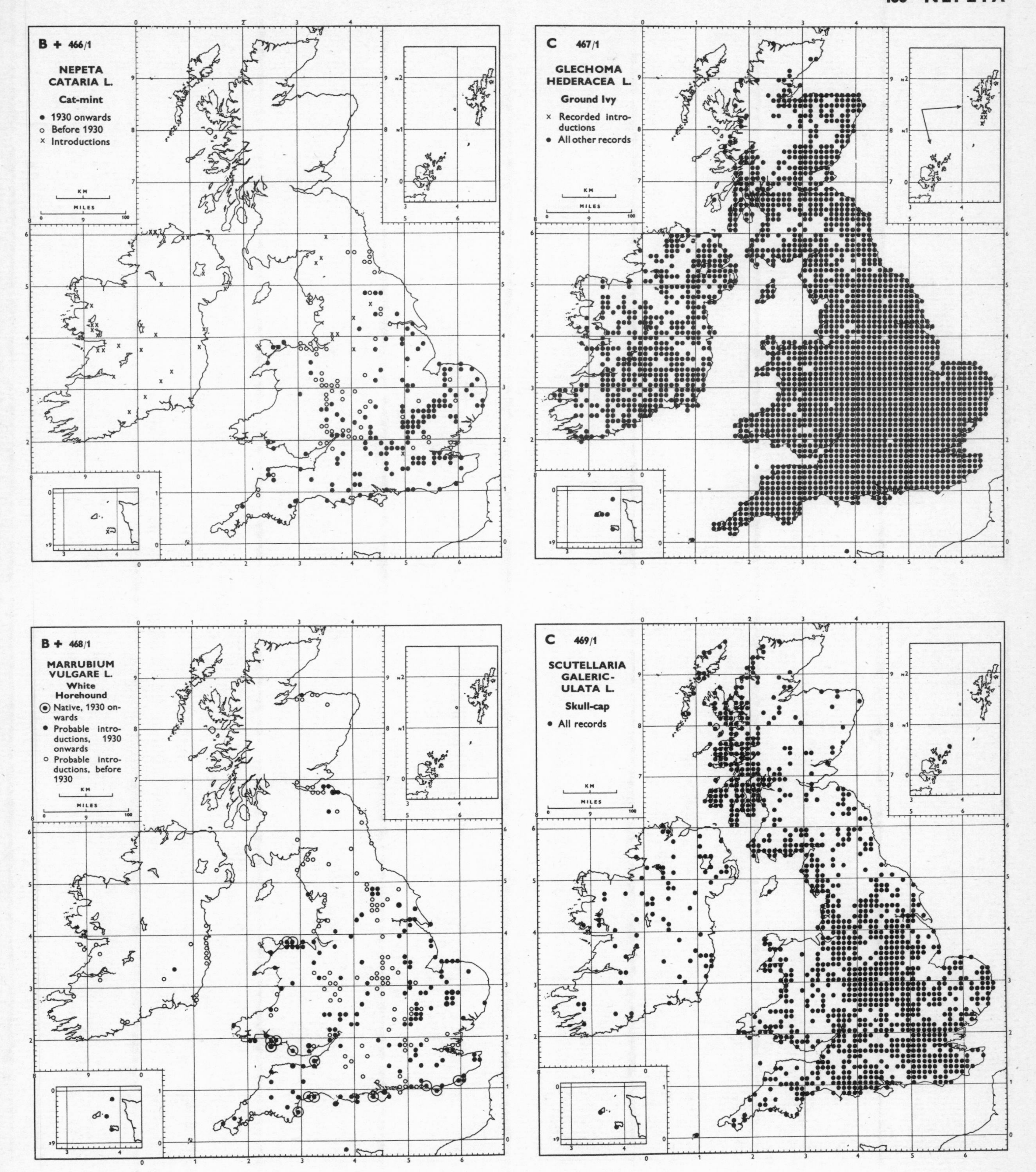
B + 466/1
NEPETA CATARIA L.
Cat-mint
• 1930 onwards
○ Before 1930
× Introductions
C 467/1
GLECHOMA HEDERACEA L.
Ground Ivy
× Recorded introductions
• All other records
B + 468/1
MARRUBIUM VULGARE L.
White Horehound
Native, 1930 onwards
• Probable introductions, 1930 onwards
○ Probable introductions, before 1930
C 469/1
SCUTELLARIA GALERICULATA L.
Skull-cap
• All records

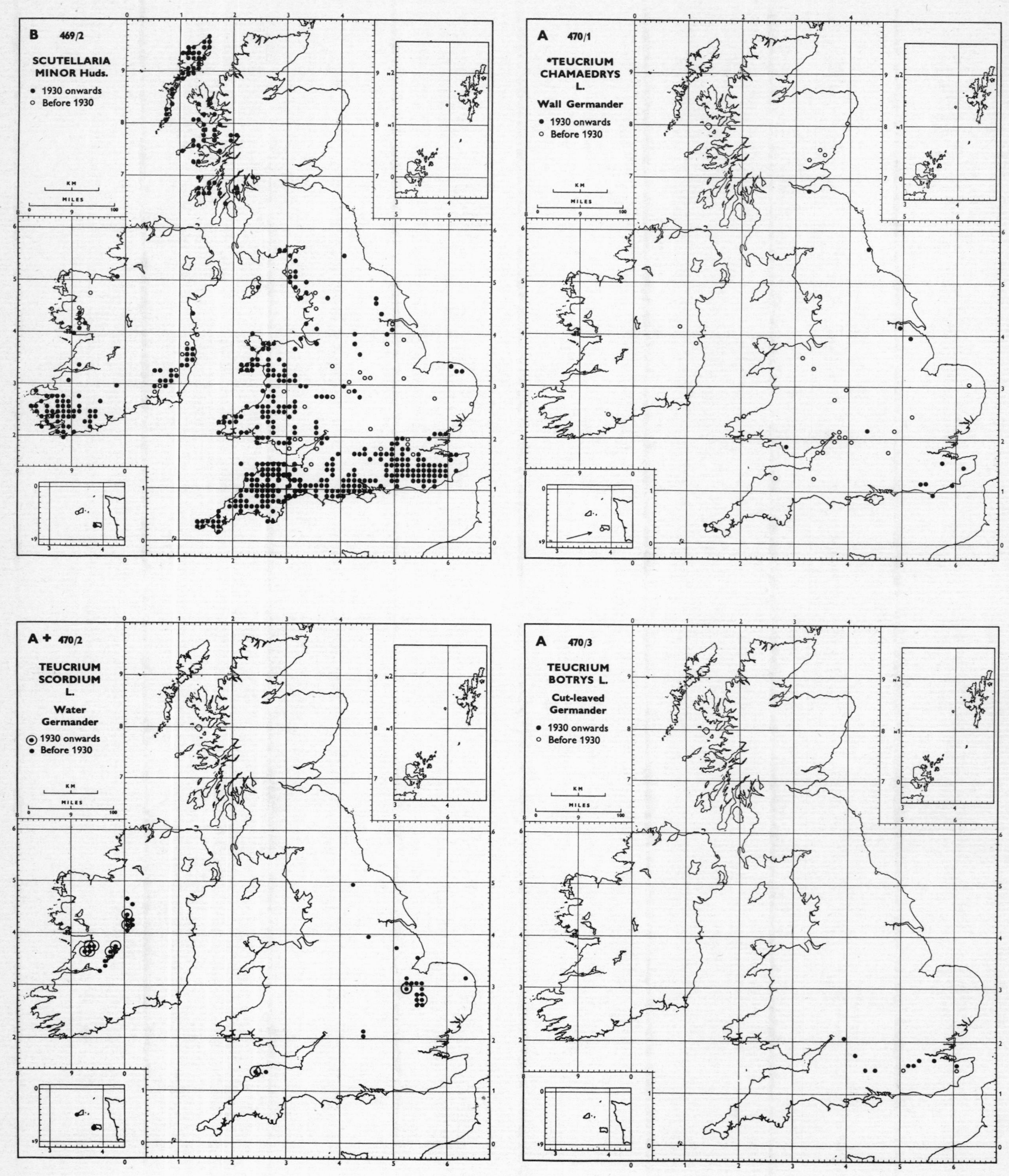

B 469/2
SCUTELLARIA MINOR Huds.
• 1930 onwards
○ Before 1930
A 470/1
*TEUCRIUM CHAMAEDRYS L.
Wall Germander
• 1930 onwards
○ Before 1930
A + 470/2
TEUCRIUM SCORDIUM L.
Water Germander
⊙ 1930 onwards
• Before 1930
A 470/3
TEUCRIUM BOTRYS L.
Cut-leaved Germander
• 1930 onwards
○ Before 1930
KM
MILES

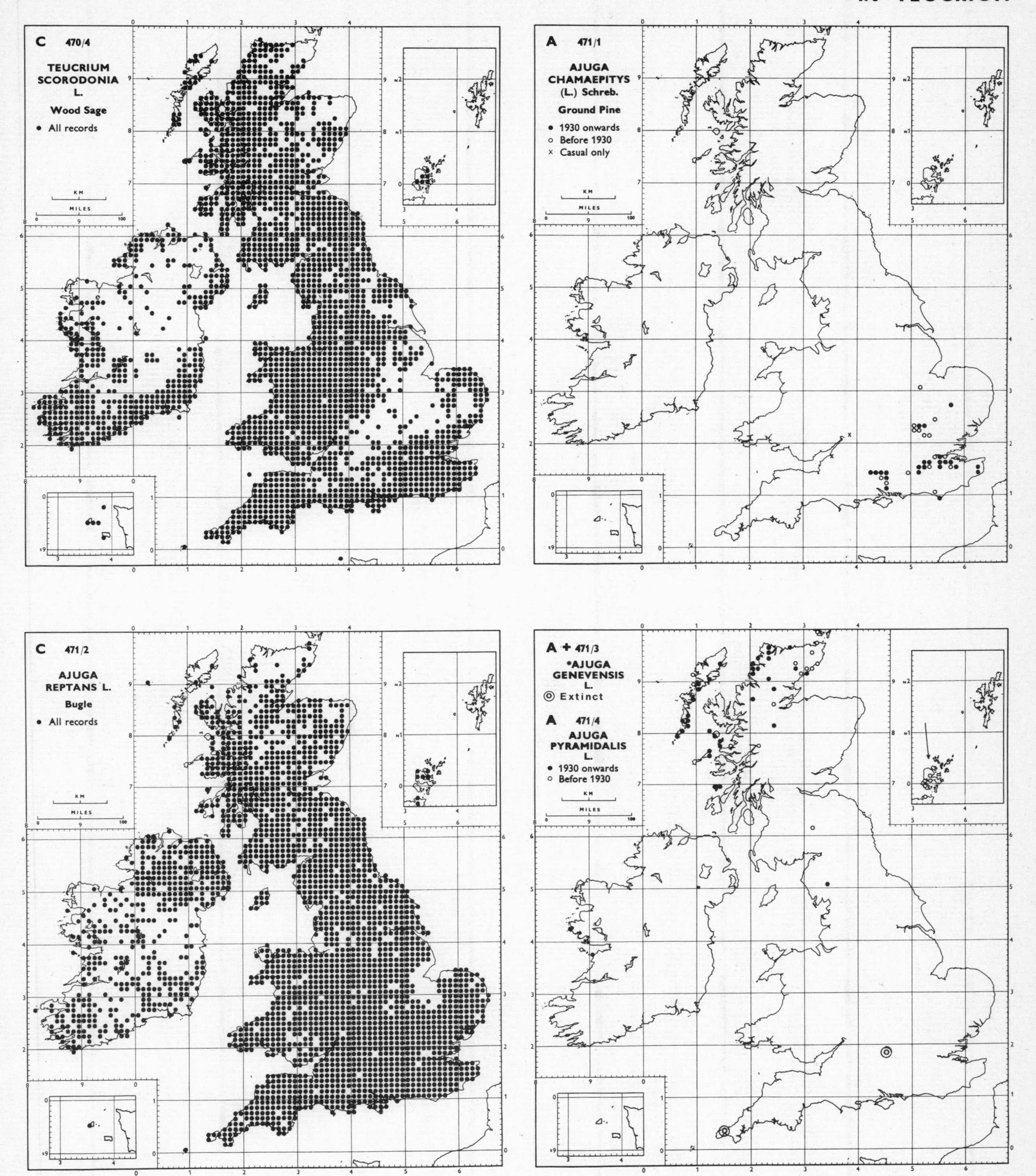

C 470/4
TEUCRIUM SCORODONIA L.
Wood Sage
• All records
A 471/1
AJUGA CHAMAEPITYS (L.) Schreb.
Ground Pine
• 1930 onwards
○ Before 1930
× Casual only
C 471/2
AJUGA REPTANS L.
Bugle
• All records
A + 471/3
*AJUGA GENEVENSIS L.
◎ Extinct
A 471/4
AJUGA PYRAMIDALIS L.
• 1930 onwards
○ Before 1930
KM
MILES

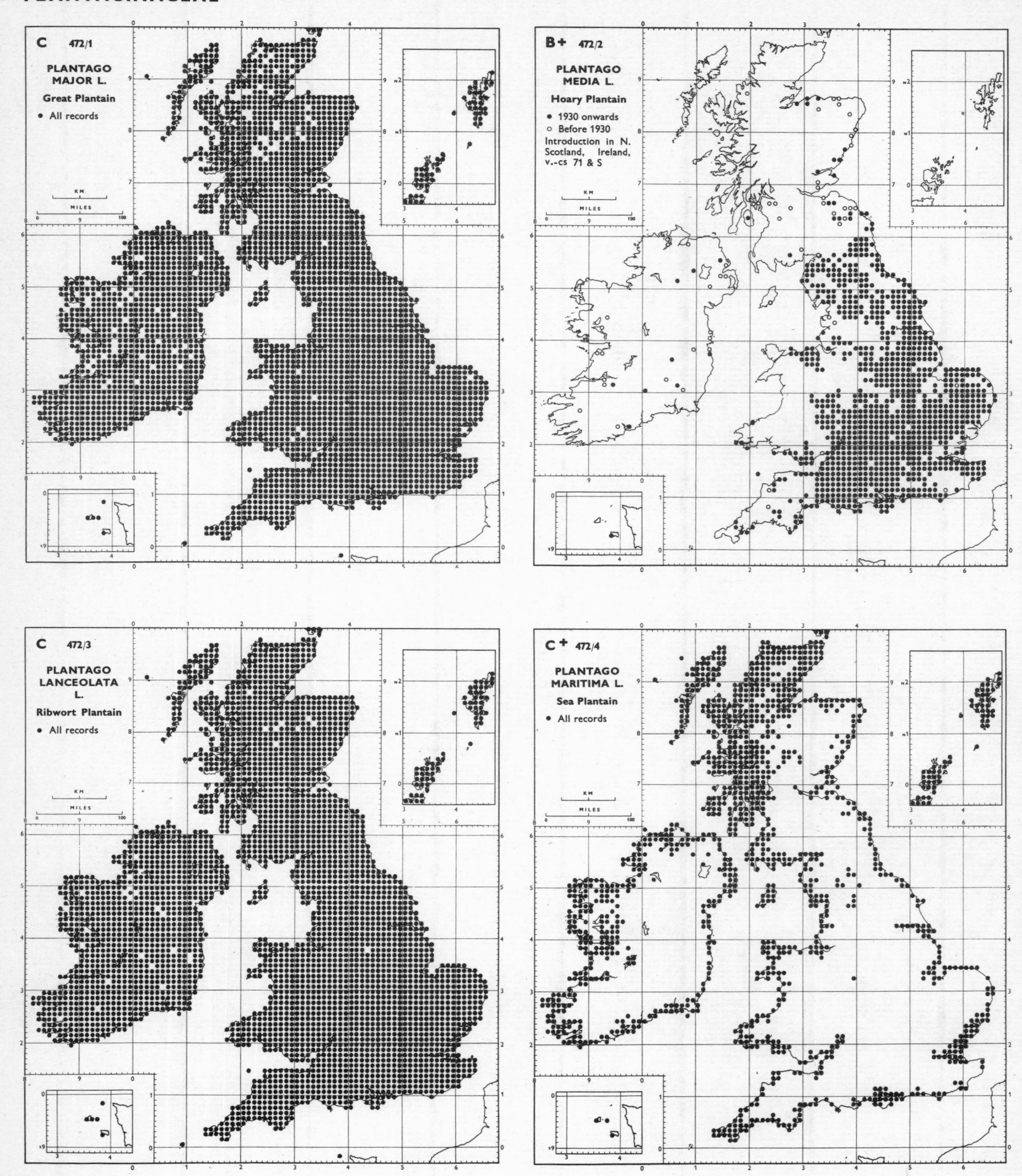
C 472/1
PLANTAGO MAJOR L.
Great Plantain
• All records
B+ 472/2
PLANTAGO MEDIA L.
Hoary Plantain
• 1930 onwards
○ Before 1930
Introduction in N. Scotland, Ireland, v.-cs 71 & S
C 472/3
PLANTAGO LANCEOLATA L.
Ribwort Plantain
• All records
C+ 472/4
PLANTAGO MARITIMA L.
Sea Plantain
• All records
KM
MILES

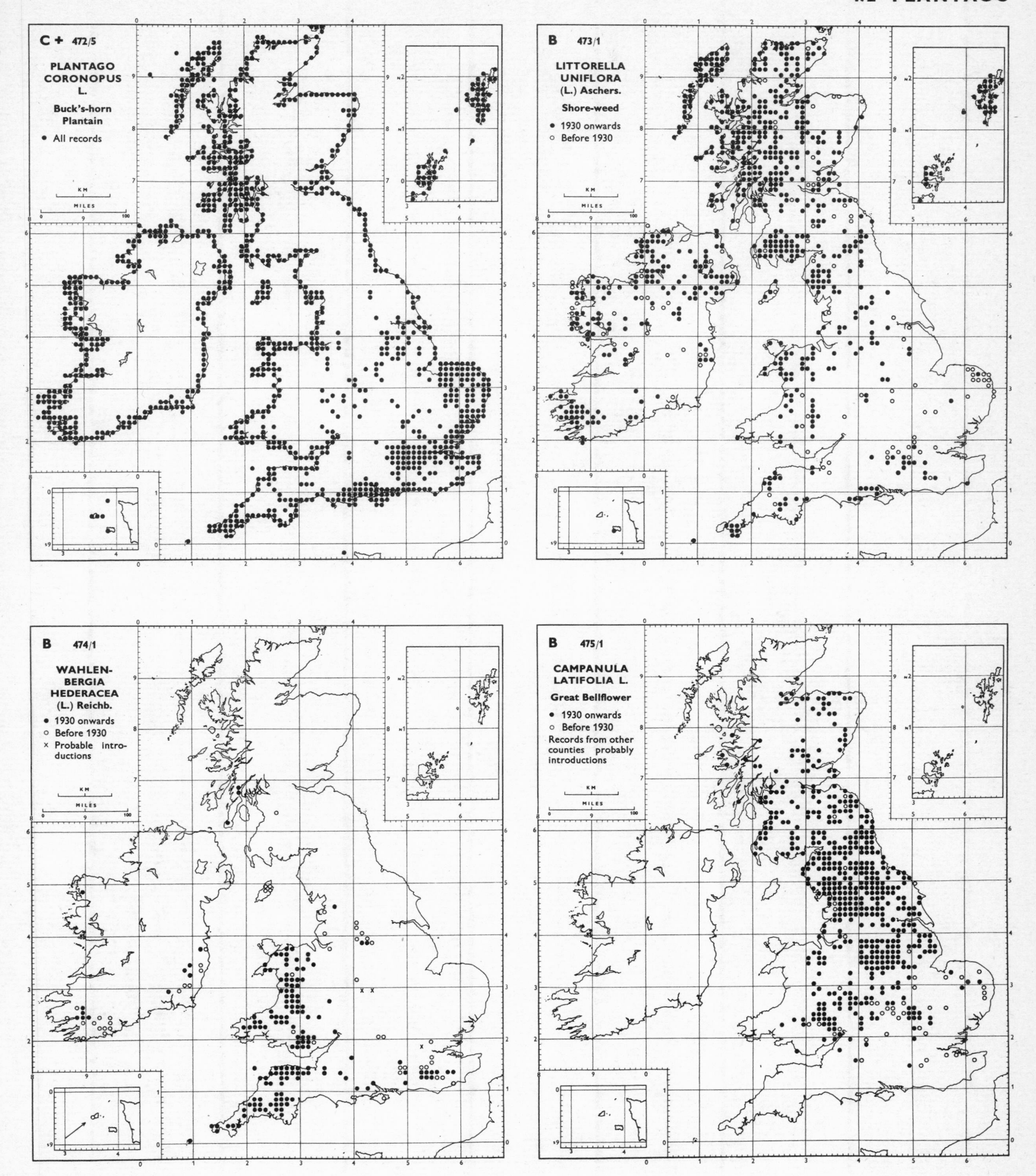
C + 472/5
PLANTAGO CORONOPUS L.
Buck's-horn Plantain
• All records
B 473/1
LITTORELLA UNIFLORA (L.) Aschers.
Shore-weed
• 1930 onwards
○ Before 1930
B 474/1
WAHLEN-BERGIA HEDERACEA (L.) Reichb.
• 1930 onwards
○ Before 1930
× Probable introductions
B 475/1
CAMPANULA LATIFOLIA L.
Great Bellflower
• 1930 onwards
○ Before 1930
Records from other counties probably introductions
KM
MILES

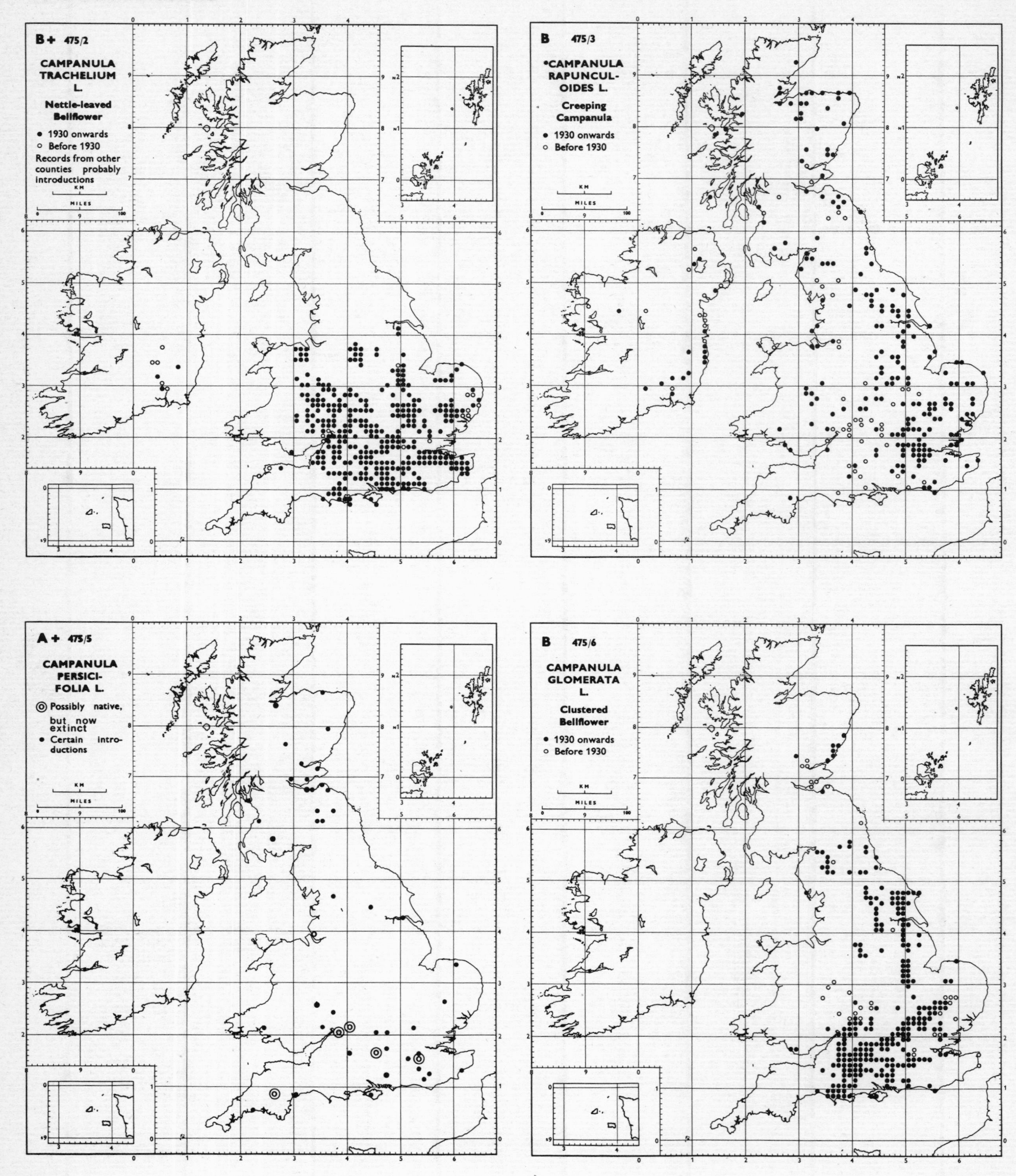
B+ 475/2
CAMPANULA TRACHELIUM L.
Nettle-leaved Bellflower
• 1930 onwards
○ Before 1930
Records from other counties probably introductions
KM
MILES
B 475/3
*CAMPANULA RAPUNCULOIDES L.
Creeping Campanula
• 1930 onwards
○ Before 1930
A+ 475/5
CAMPANULA PERSICIFOLIA L.
◎ Possibly native, but now extinct
• Certain introductions
B 475/6
CAMPANULA GLOMERATA L.
Clustered Bellflower
• 1930 onwards
○ Before 1930

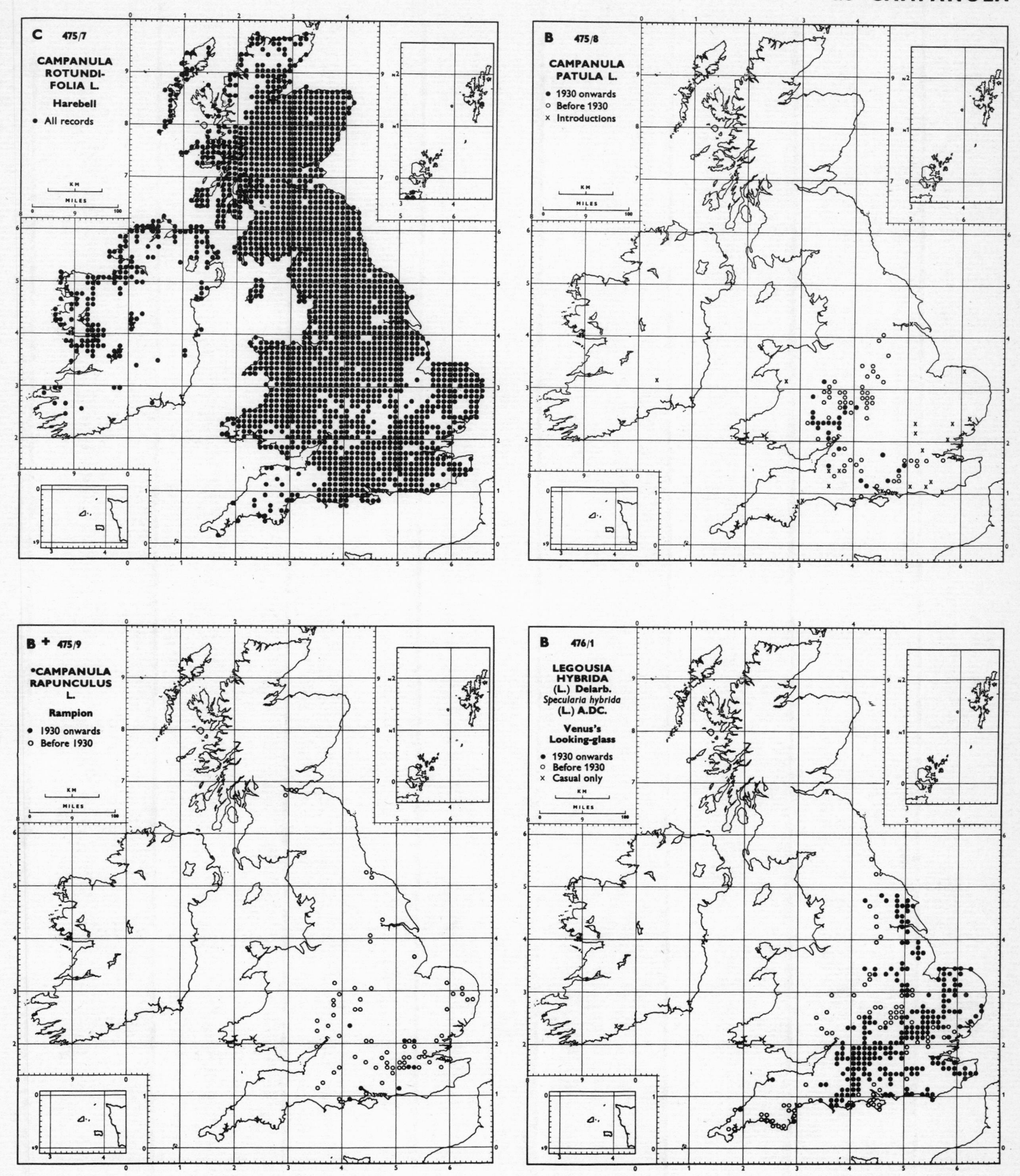
C 475/7
CAMPANULA ROTUNDIFOLIA L.
Harebell
• All records
KM
MILES
B 475/8
CAMPANULA PATULA L.
• 1930 onwards
○ Before 1930
× Introductions
B + 475/9
*CAMPANULA RAPUNCULUS L.
Rampion
• 1930 onwards
○ Before 1930
B 476/1
LEGOUSIA HYBRIDA (L.) Delarb.
Specularia hybrida (L.) A.DC.
Venus's Looking-glass
• 1930 onwards
○ Before 1930
× Casual only

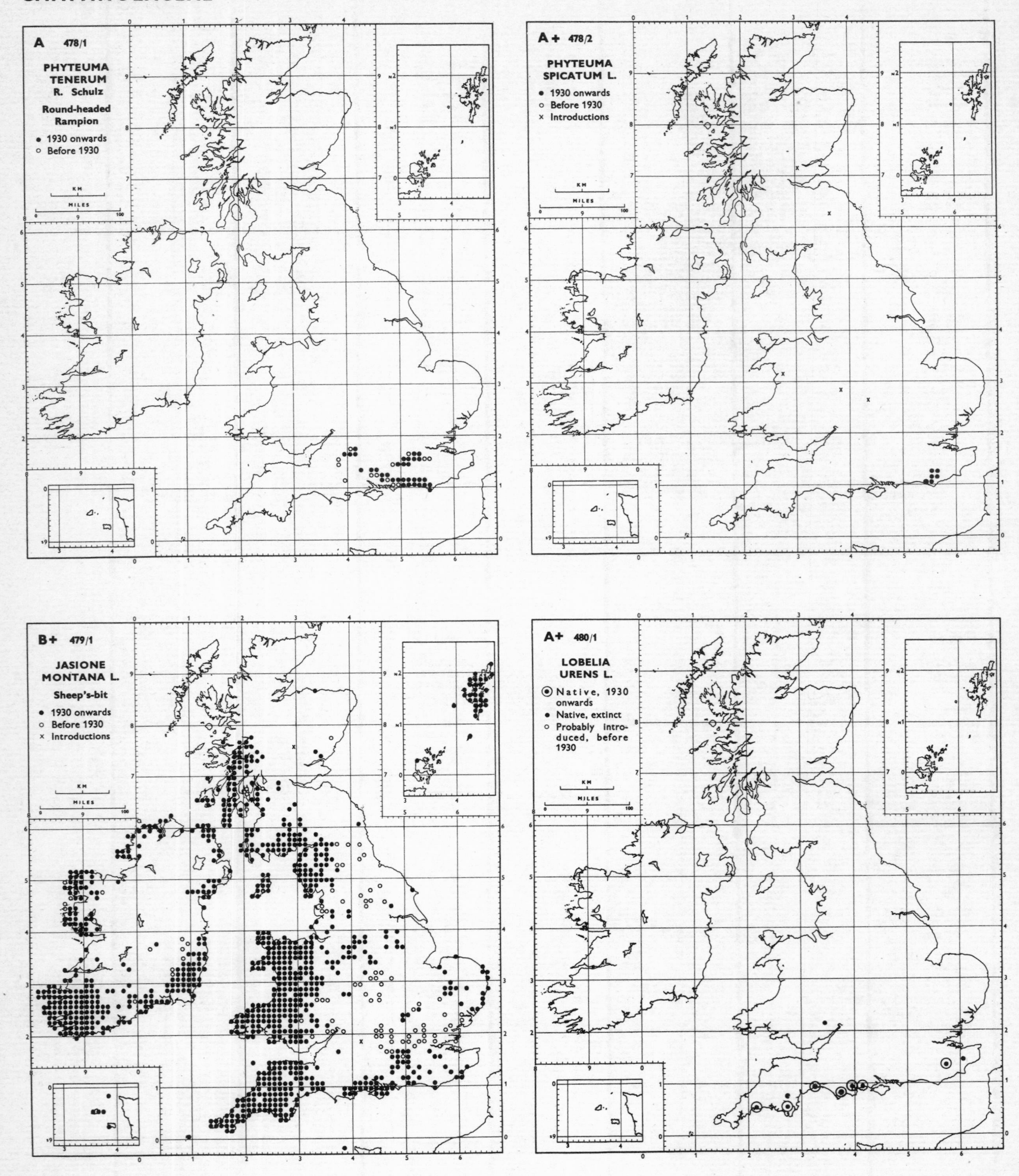
A 478/1
PHYTEUMA TENERUM R. Schulz
Round-headed Rampion
• 1930 onwards
○ Before 1930
A+ 478/2
PHYTEUMA SPICATUM L.
• 1930 onwards
○ Before 1930
× Introductions
B+ 479/1
JASIONE MONTANA L.
Sheep's-bit
• 1930 onwards
○ Before 1930
× Introductions
A+ 480/1
LOBELIA URENS L.
Native, 1930 onwards
• Native, extinct
○ Probably introduced, before 1930
KM
MILES

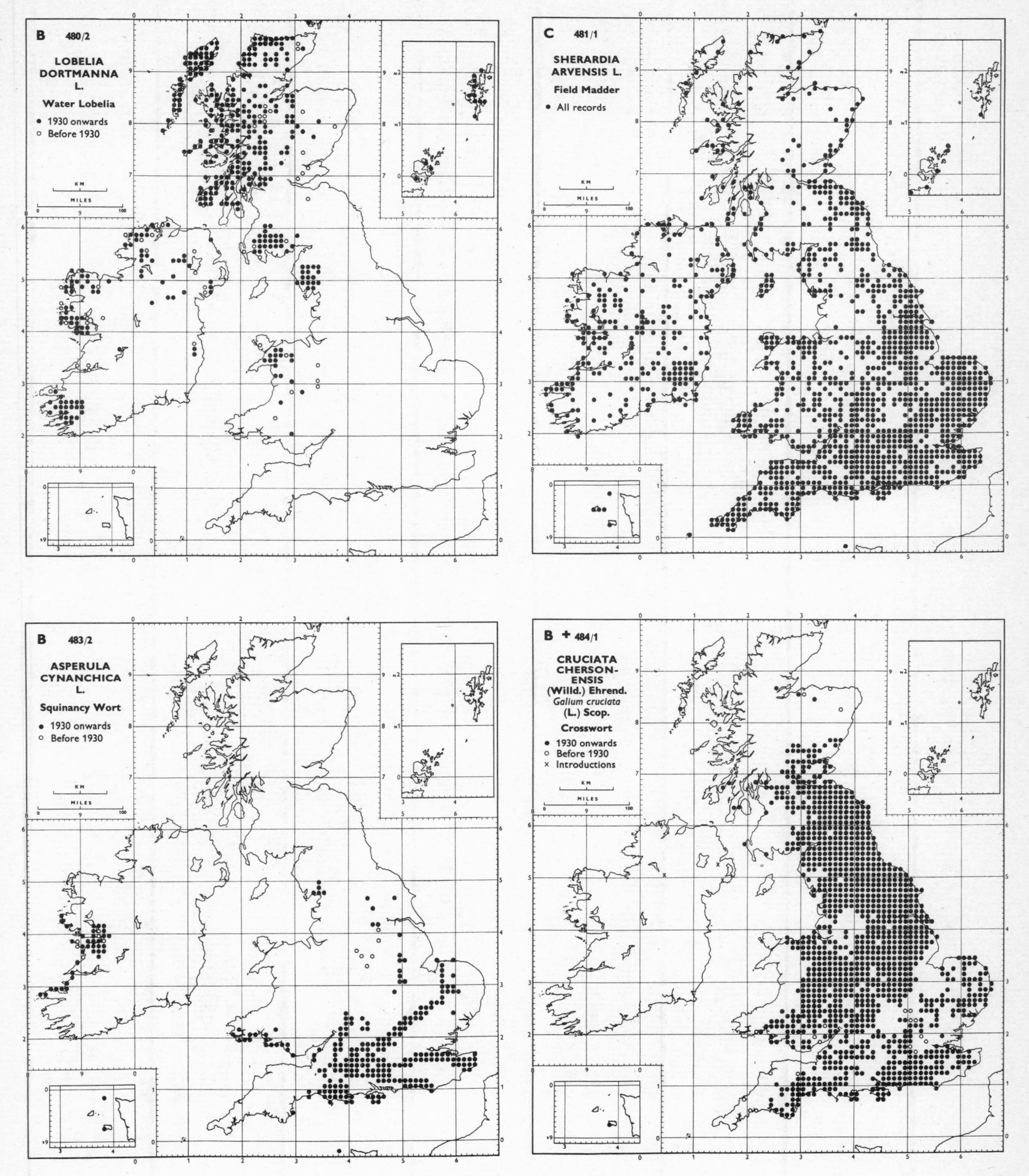
B 480/2
LOBELIA DORTMANNA L.
Water Lobelia
1930 onwards
Before 1930
C 481/1
SHERARDIA ARVENSIS L.
Field Madder
All records
B 483/2
ASPERULA CYNANCHICA L.
Squinancy Wort
1930 onwards
Before 1930
B + 484/1
CRUCIATA CHERSON-ENSIS (Willd.) Ehrend.
Galium cruciata (L.) Scop.
Crosswort
1930 onwards
Before 1930
Introductions
KM
MILES

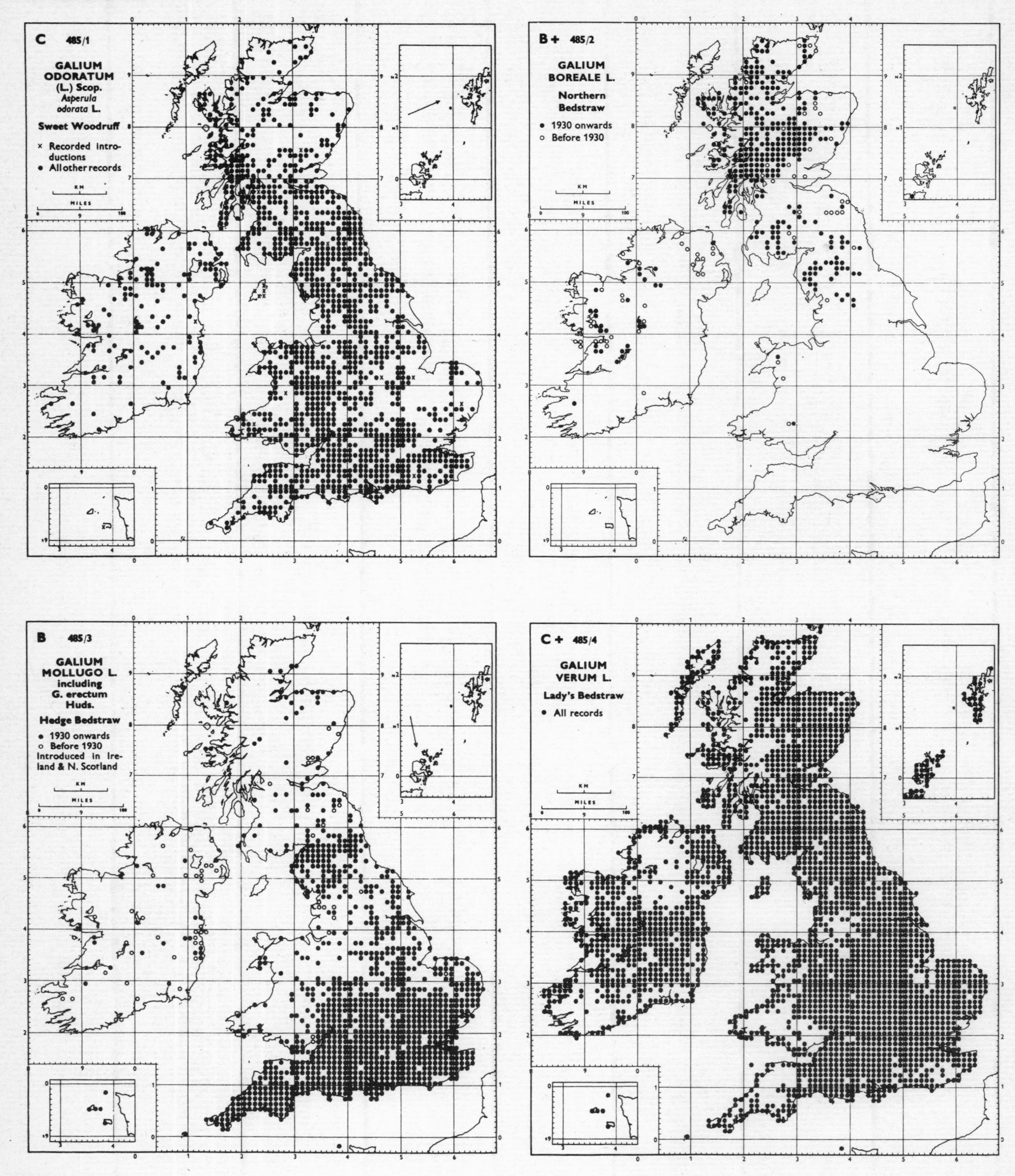
C 485/1
GALIUM ODORATUM (L.) Scop.
Asperula odorata L.
Sweet Woodruff
× Recorded introductions
• All other records
B + 485/2
GALIUM BOREALE L.
Northern Bedstraw
• 1930 onwards
○ Before 1930
B 485/3
GALIUM MOLLUGO L. including G. erectum Huds.
Hedge Bedstraw
• 1930 onwards
○ Before 1930
Introduced in Ireland & N. Scotland
C + 485/4
GALIUM VERUM L.
Lady's Bedstraw
• All records

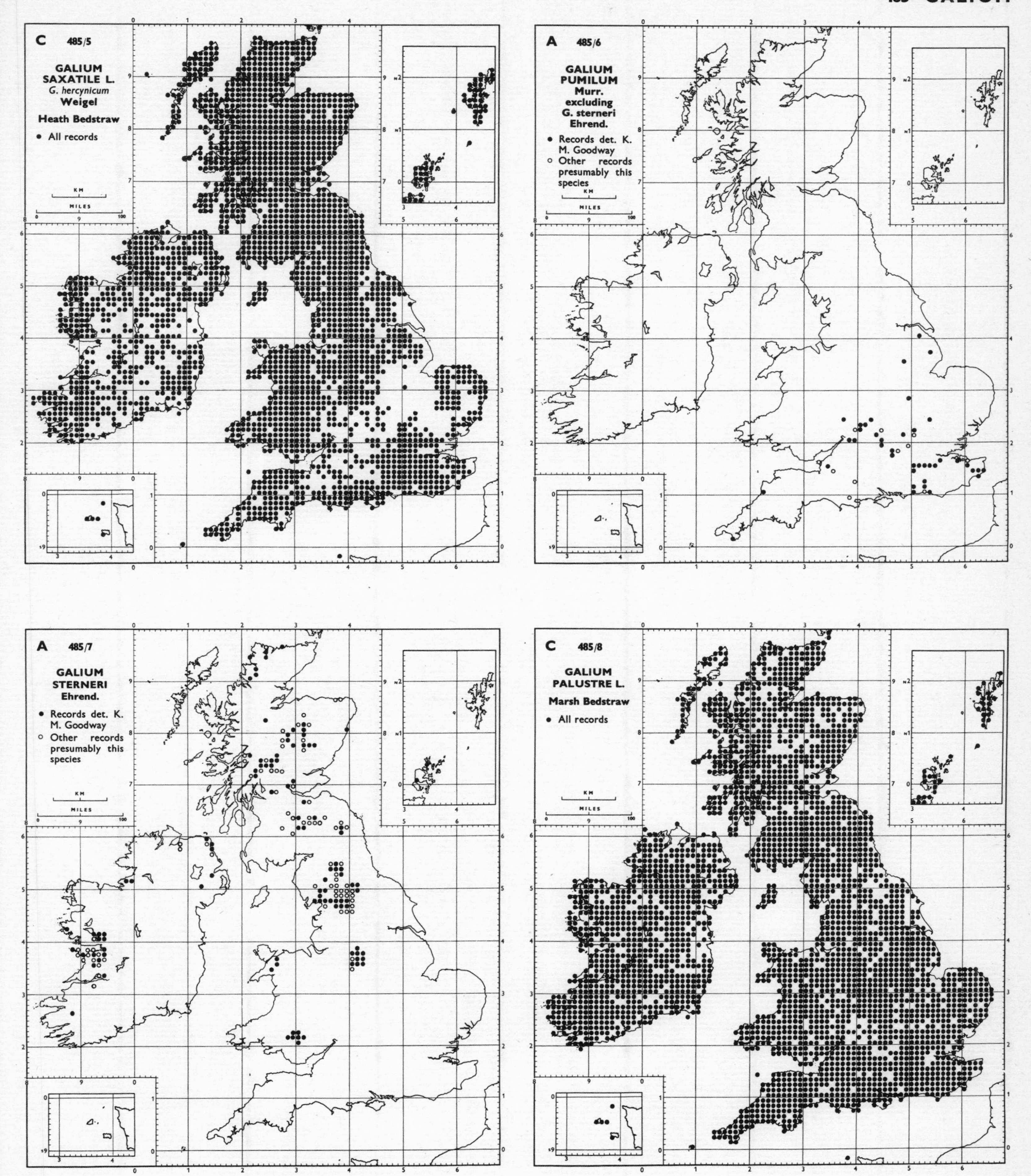
C 485/5
GALIUM SAXATILE L.
G. hercynicum
Weigel
Heath Bedstraw
• All records
A 485/6
GALIUM PUMILUM Murr.
excluding G. sterneri Ehrend.
• Records det. K. M. Goodway
○ Other records presumably this species
A 485/7
GALIUM STERNERI Ehrend.
• Records det. K. M. Goodway
○ Other records presumably this species
C 485/8
GALIUM PALUSTRE L.
Marsh Bedstraw
• All records

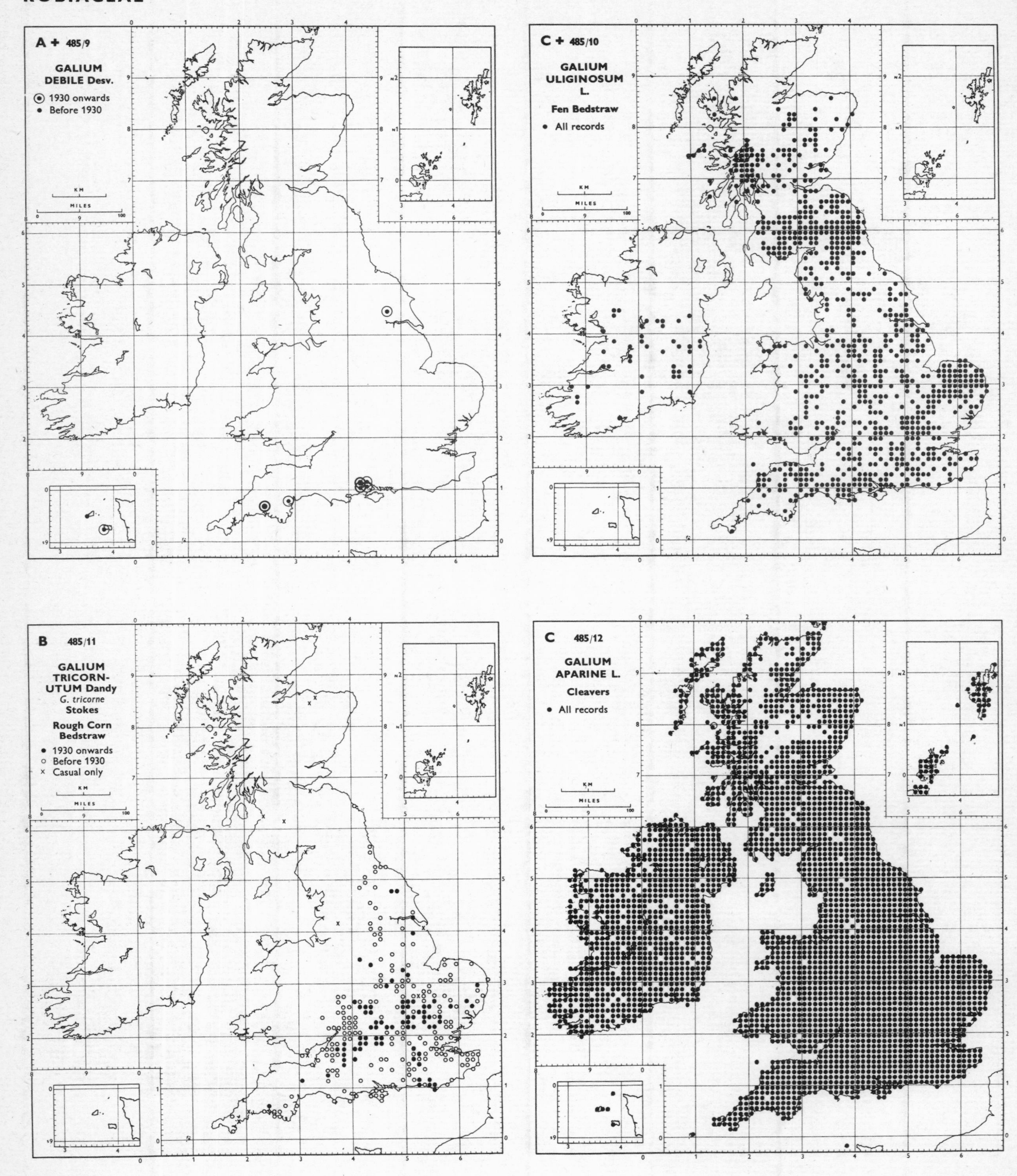
A + 485/9
GALIUM DEBILE Desv.
1930 onwards
Before 1930
C + 485/10
GALIUM ULIGINOSUM L.
Fen Bedstraw
All records
B 485/11
GALIUM TRICORN-UTUM Dandy
G. tricorne Stokes
Rough Corn Bedstraw
1930 onwards
Before 1930
Casual only
C 485/12
GALIUM APARINE L.
Cleavers
All records
KM
MILES

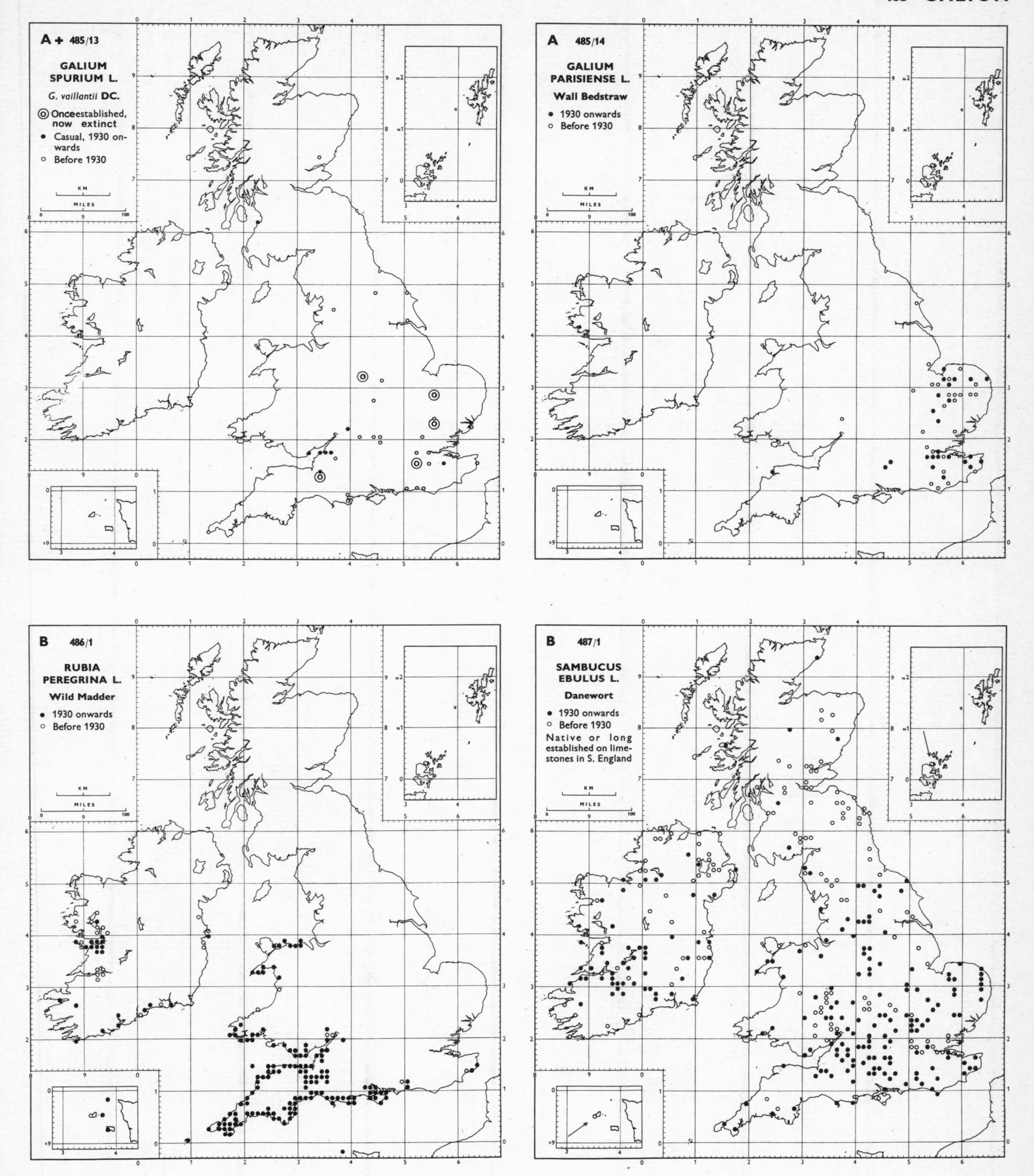
A + 485/13
GALIUM SPURIUM L.
G. vaillantii DC.
Once established, now extinct
Casual, 1930 onwards
Before 1930
A 485/14
GALIUM PARISIENSE L.
Wall Bedstraw
1930 onwards
Before 1930
B 486/1
RUBIA PEREGRINA L.
Wild Madder
1930 onwards
Before 1930
B 487/1
SAMBUCUS EBULUS L.
Danewort
1930 onwards
Before 1930
Native or long established on limestones in S. England

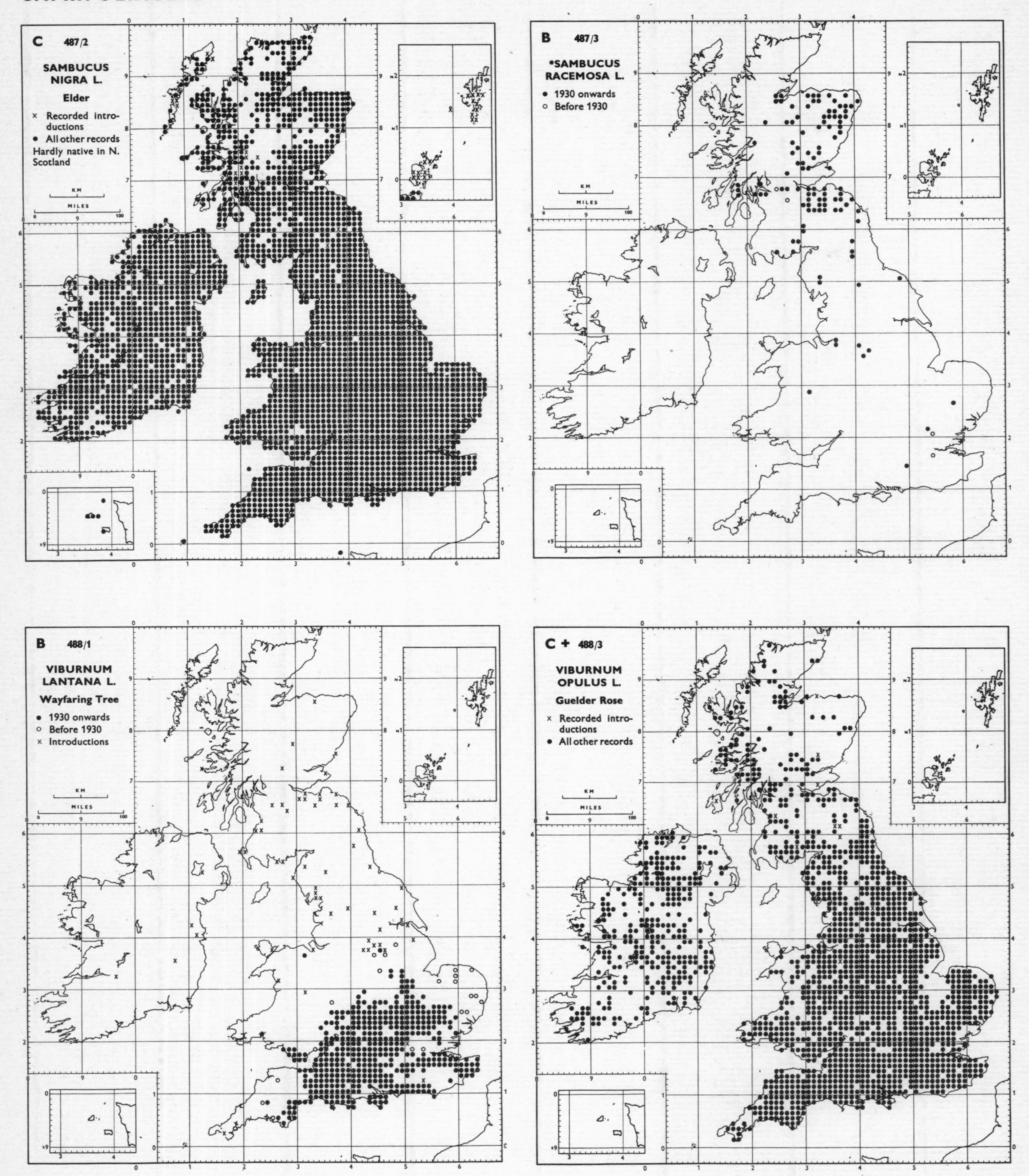
C 487/2
SAMBUCUS NIGRA L.
Elder
× Recorded introductions
• All other records
Hardly native in N. Scotland
B 487/3
*SAMBUCUS RACEMOSA L.
• 1930 onwards
○ Before 1930
B 488/1
VIBURNUM LANTANA L.
Wayfaring Tree
• 1930 onwards
○ Before 1930
× Introductions
C + 488/3
VIBURNUM OPULUS L.
Guelder Rose
× Recorded introductions
• All other records
KM
MILES

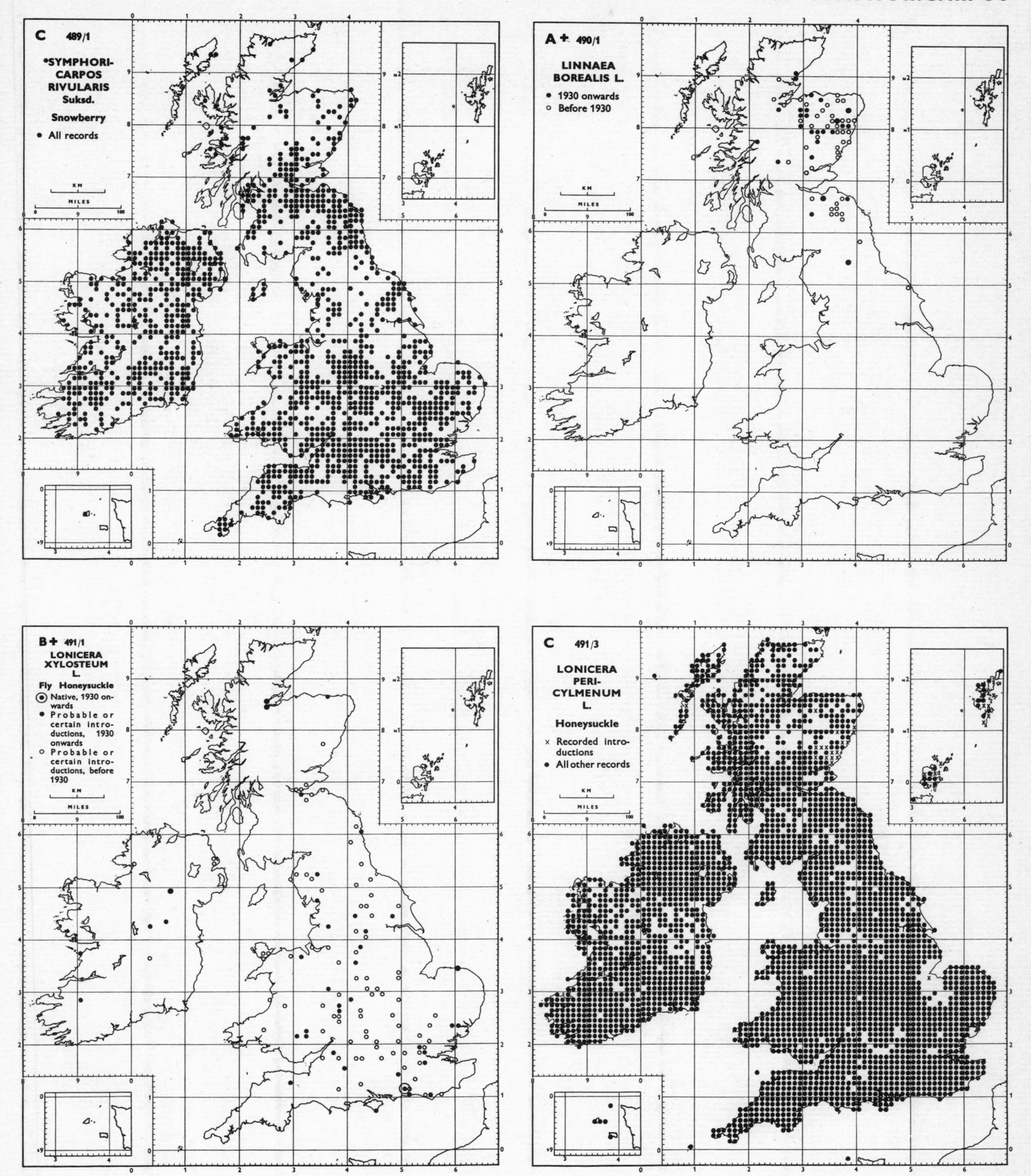
C 489/1
*SYMPHORICARPOS RIVULARIS Suksd.
Snowberry
• All records
A+ 490/1
LINNAEA BOREALIS L.
• 1930 onwards
○ Before 1930
B+ 491/1
LONICERA XYLOSTEUM L.
Fly Honeysuckle
Native, 1930 onwards
• Probable or certain introductions, 1930 onwards
○ Probable or certain introductions, before 1930
C 491/3
LONICERA PERICYLMENUM L.
Honeysuckle
× Recorded introductions
• All other records
KM
MILES

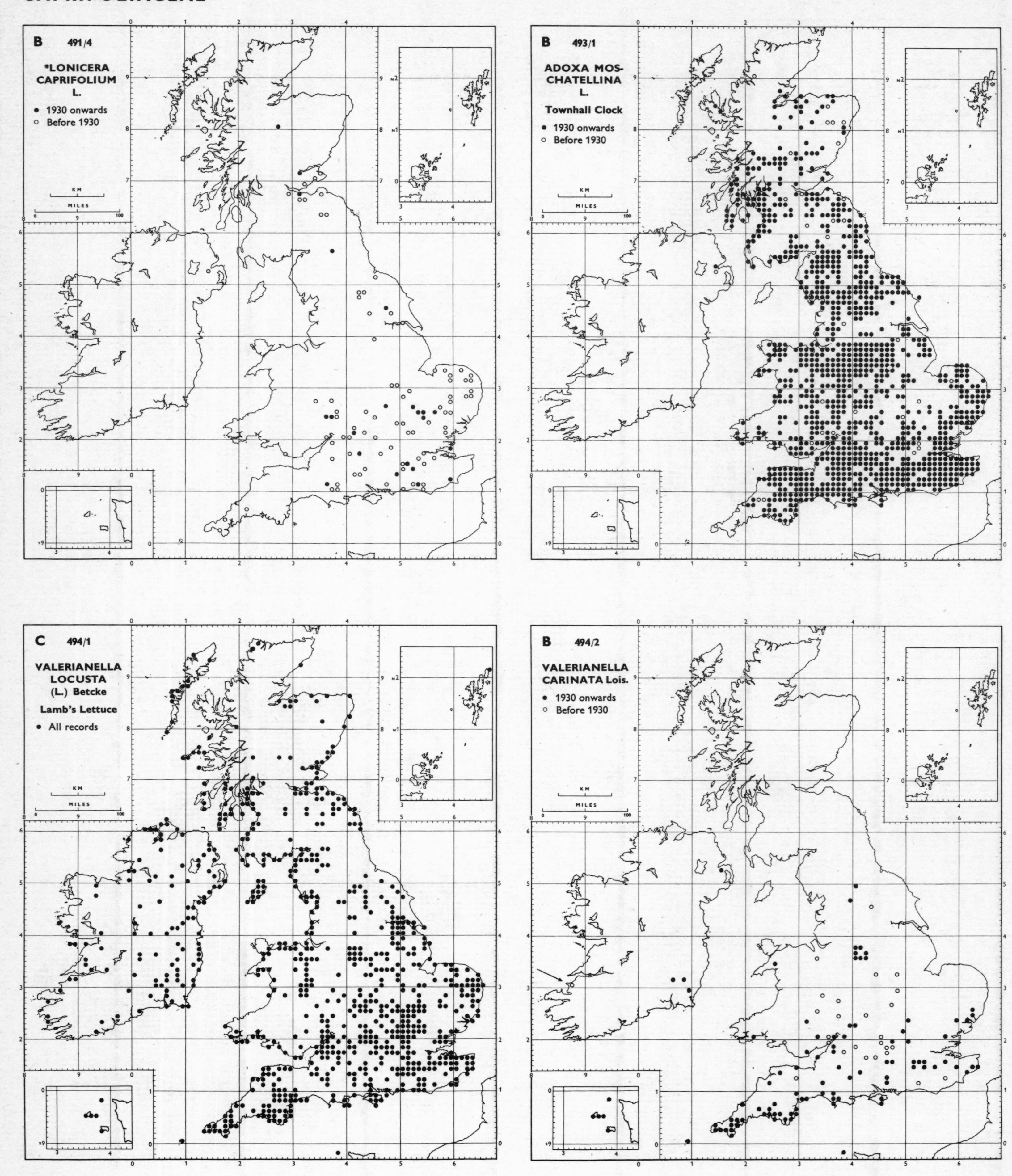

B 491/4
*LONICERA CAPRIFOLIUM L.
● 1930 onwards
○ Before 1930
B 493/1
ADOXA MOSCHATELLINA L.
Townhall Clock
● 1930 onwards
○ Before 1930
C 494/1
VALERIANELLA LOCUSTA (L.) Betcke
Lamb's Lettuce
● All records
B 494/2
VALERIANELLA CARINATA Lois.
● 1930 onwards
○ Before 1930

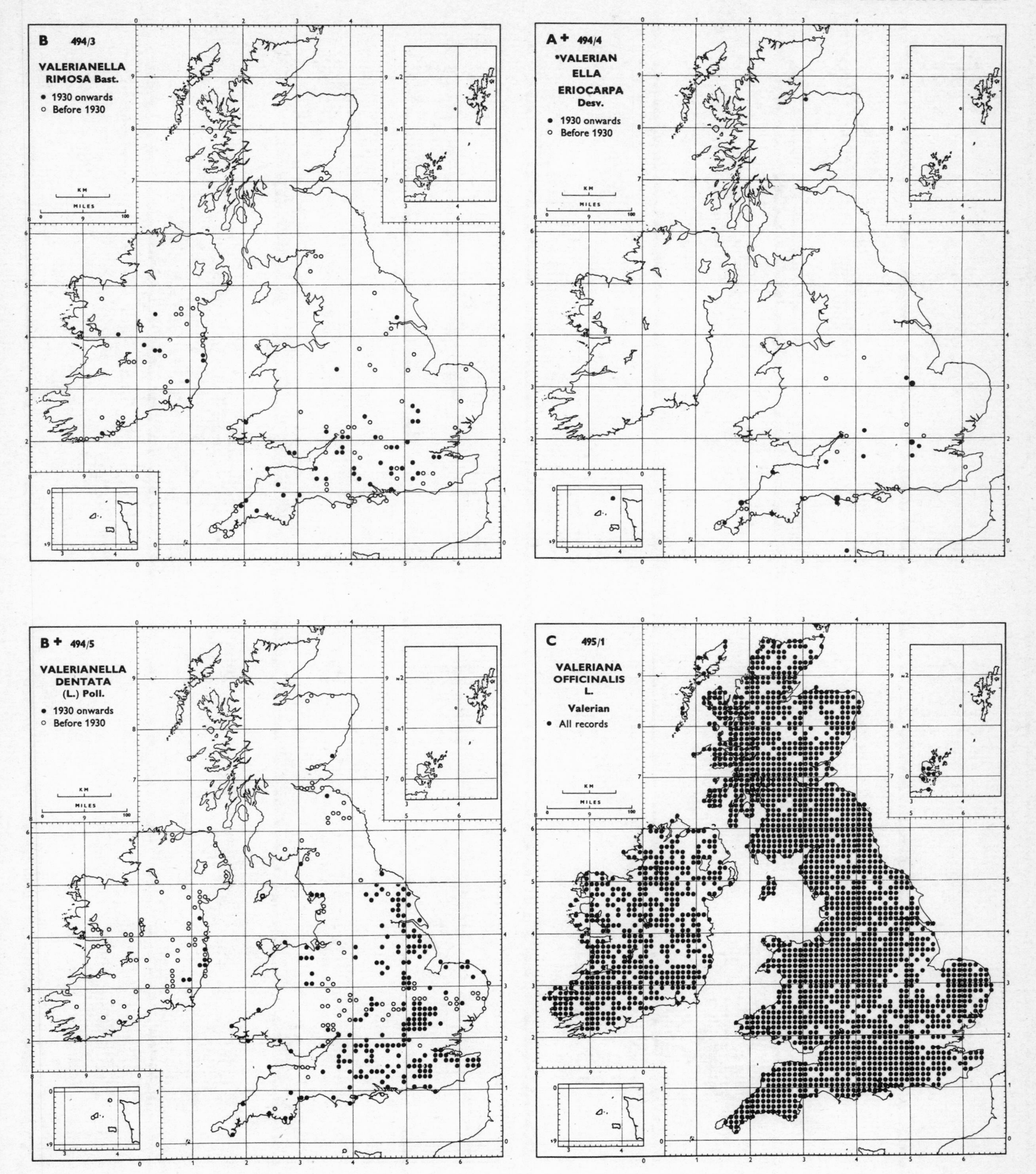
B 494/3
VALERIANELLA RIMOSA Bast.
• 1930 onwards
○ Before 1930
A + 494/4
*VALERIANELLA ERIOCARPA Desv.
• 1930 onwards
○ Before 1930
B + 494/5
VALERIANELLA DENTATA (L.) Poll.
• 1930 onwards
○ Before 1930
C 495/1
VALERIANA OFFICINALIS L.
Valerian
• All records
KM
MILES

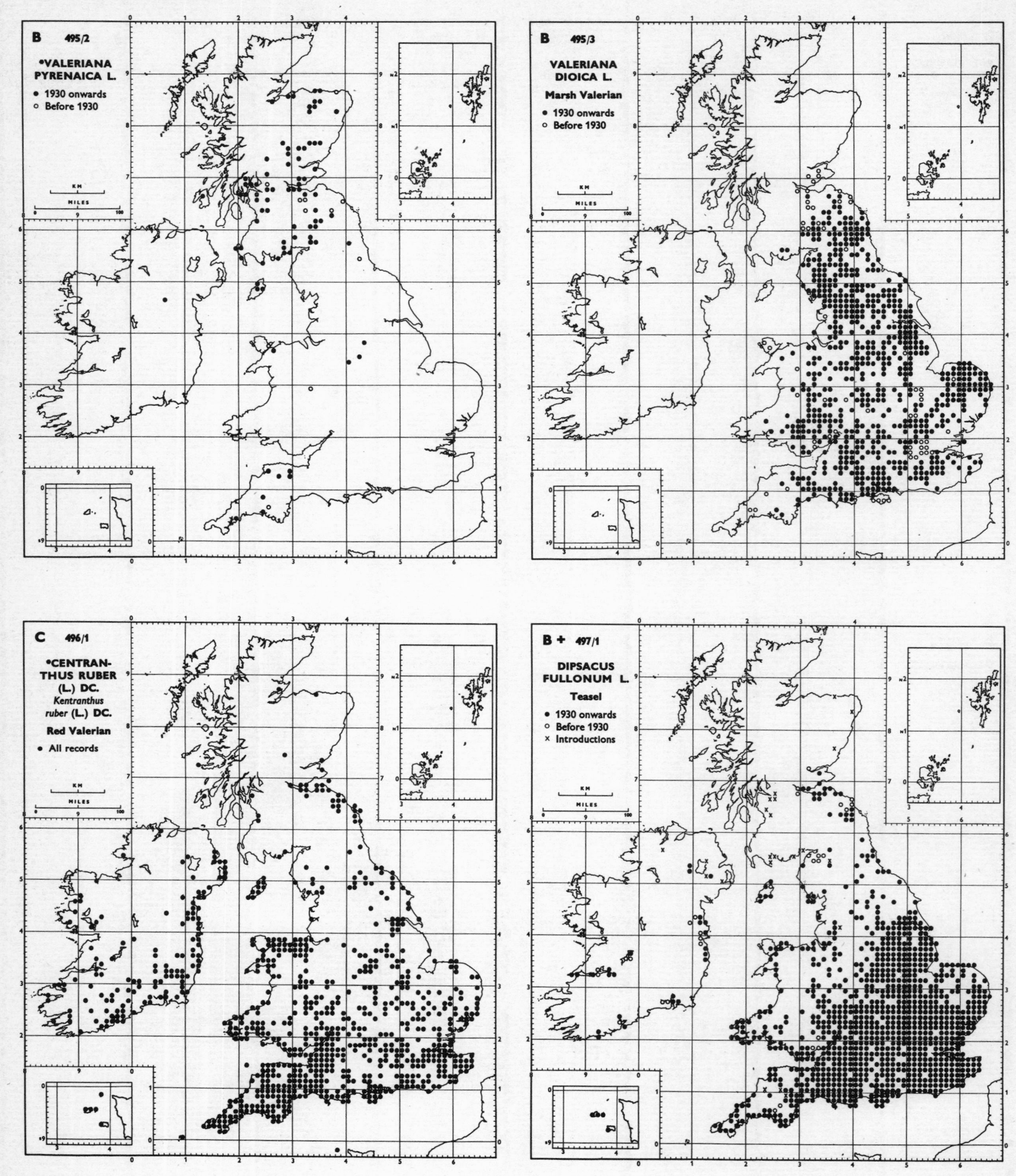
B 495/2
•VALERIANA PYRENAICA L.
• 1930 onwards
○ Before 1930
B 495/3
VALERIANA DIOICA L.
Marsh Valerian
• 1930 onwards
○ Before 1930
C 496/1
•CENTRANTHUS RUBER (L.) DC.
Kentranthus ruber (L.) DC.
Red Valerian
• All records
B + 497/1
DIPSACUS FULLONUM L.
Teasel
• 1930 onwards
○ Before 1930
× Introductions

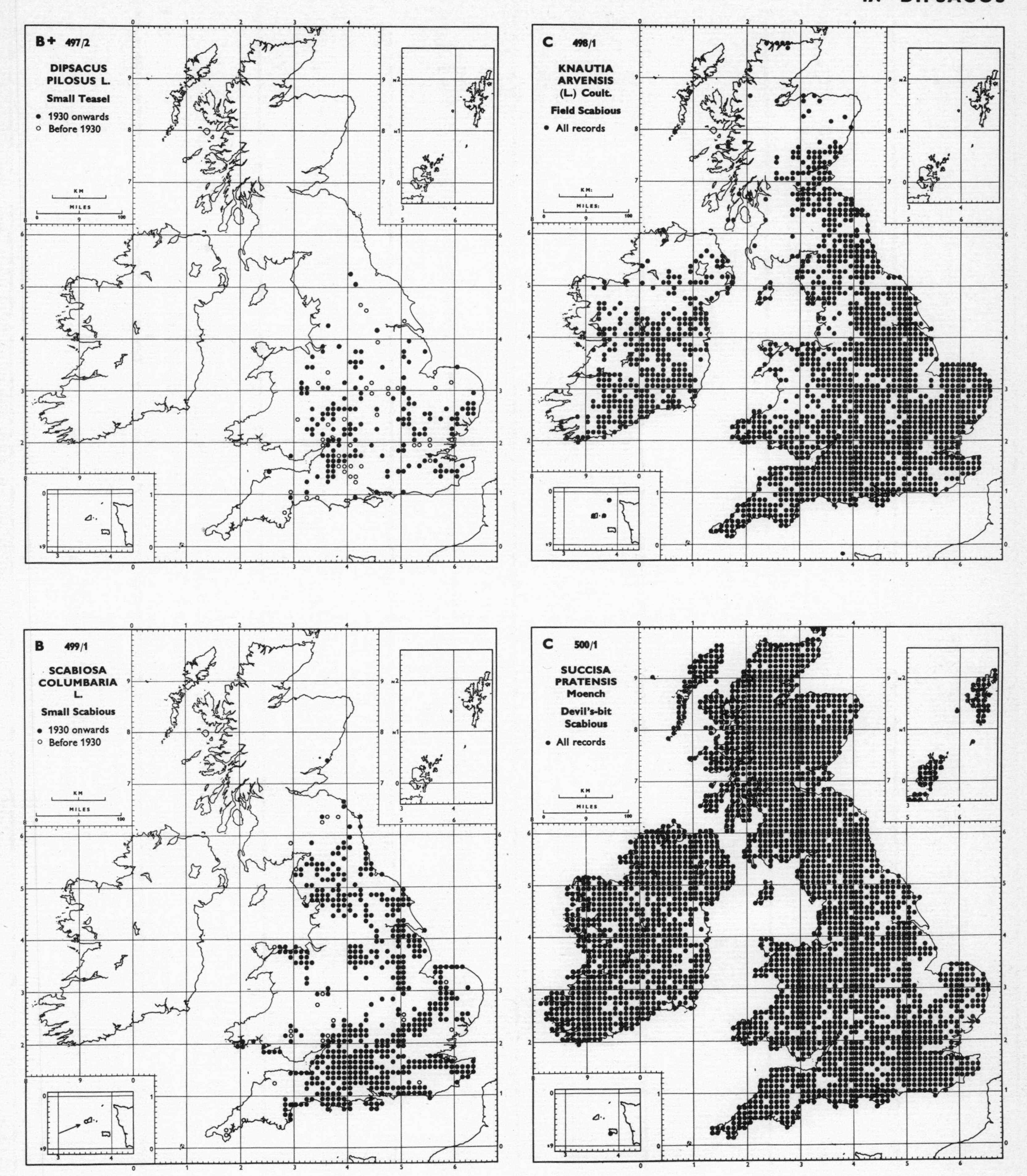
B+ 497/2
DIPSACUS PILOSUS L.
Small Teasel
1930 onwards
Before 1930
C 498/1
KNAUTIA ARVENSIS (L.) Coult.
Field Scabious
All records
B 499/1
SCABIOSA COLUMBARIA L.
Small Scabious
1930 onwards
Before 1930
C 500/1
SUCCISA PRATENSIS Moench
Devil's-bit Scabious
All records

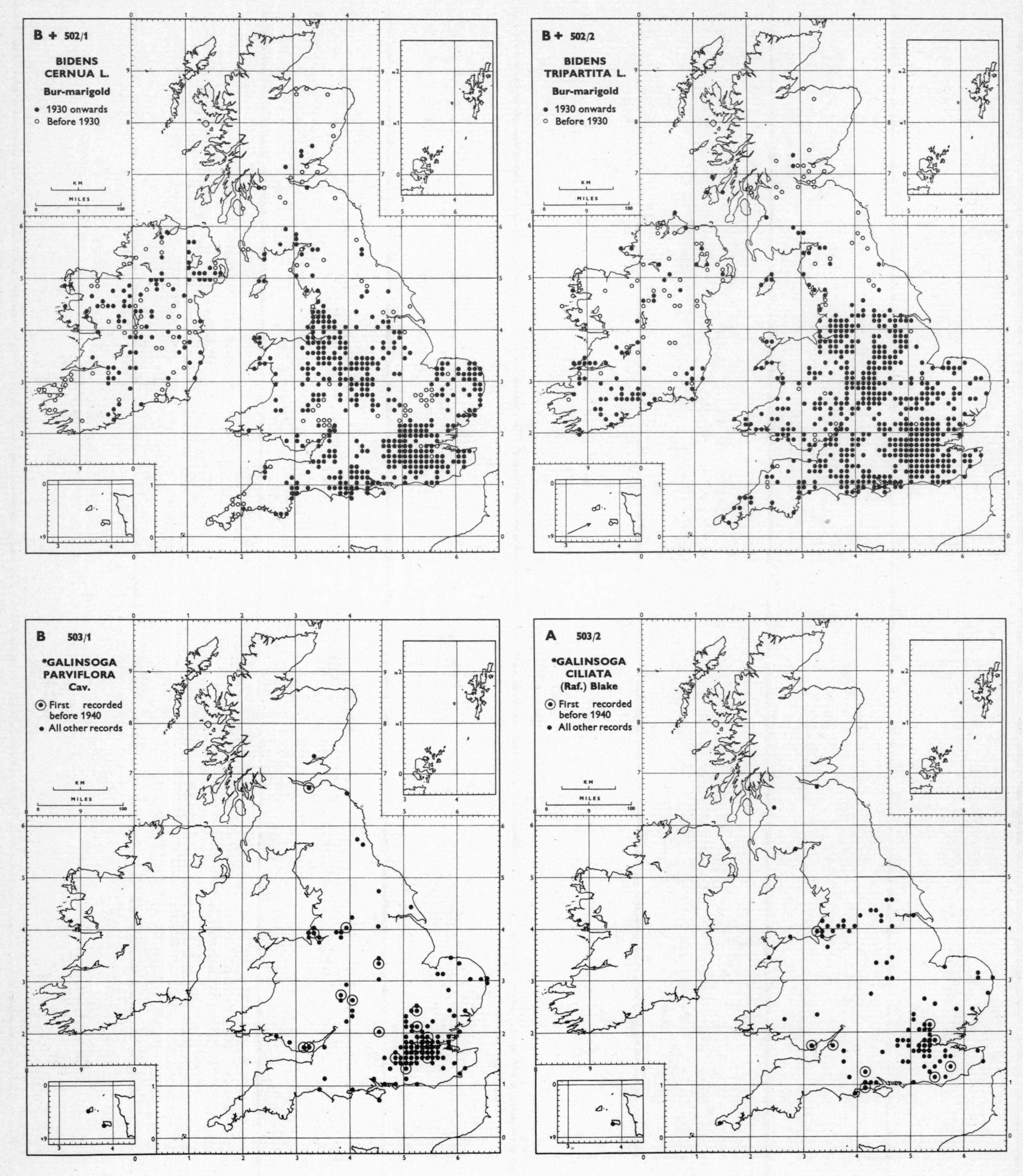
B + 502/1
BIDENS
CERNUA L.
Bur-marigold
1930 onwards
Before 1930
B + 502/2
BIDENS
TRIPARTITA L.
Bur-marigold
1930 onwards
Before 1930
B 503/1
*GALINSOGA
PARVIFLORA
Cav.
First recorded before 1940
All other records
A 503/2
*GALINSOGA
CILIATA
(Raf.) Blake
First recorded before 1940
All other records

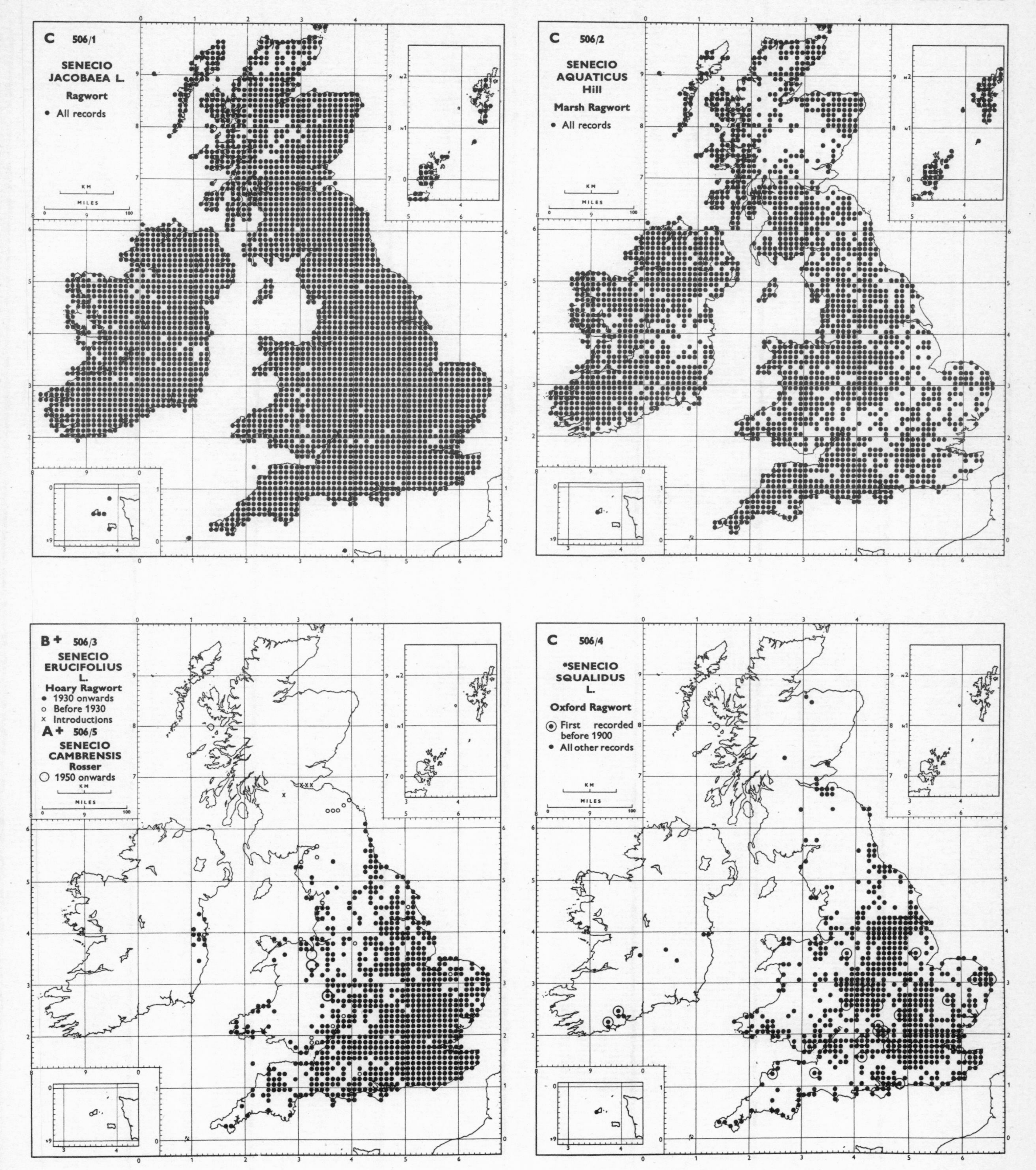
C 506/1
SENECIO JACOBAEA L.
Ragwort
• All records
C 506/2
SENECIO AQUATICUS Hill
Marsh Ragwort
• All records
B+ 506/3
SENECIO ERUCIFOLIUS L.
Hoary Ragwort
• 1930 onwards
○ Before 1930
× Introductions
A+ 506/5
SENECIO CAMBRENSIS Rosser
○ 1950 onwards
C 506/4
*SENECIO SQUALIDUS L.
Oxford Ragwort
⊙ First recorded before 1900
• All other records
KM
MILES

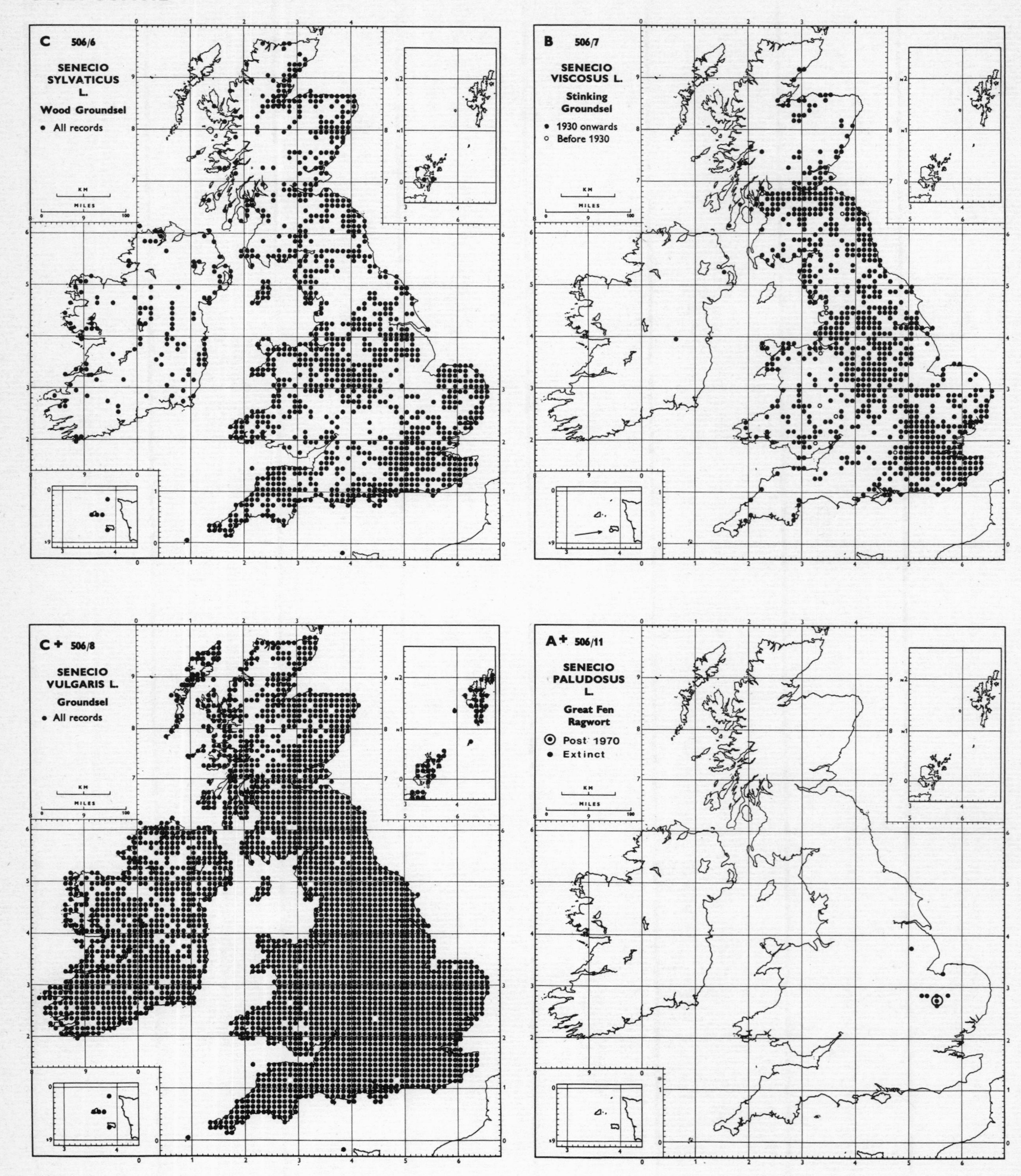
C 506/6
SENECIO SYLVATICUS L.
Wood Groundsel
• All records
B 506/7
SENECIO VISCOSUS L.
Stinking Groundsel
• 1930 onwards
○ Before 1930
C+ 506/8
SENECIO VULGARIS L.
Groundsel
• All records
A+ 506/11
SENECIO PALUDOSUS L.
Great Fen Ragwort
⊙ Post 1970
• Extinct

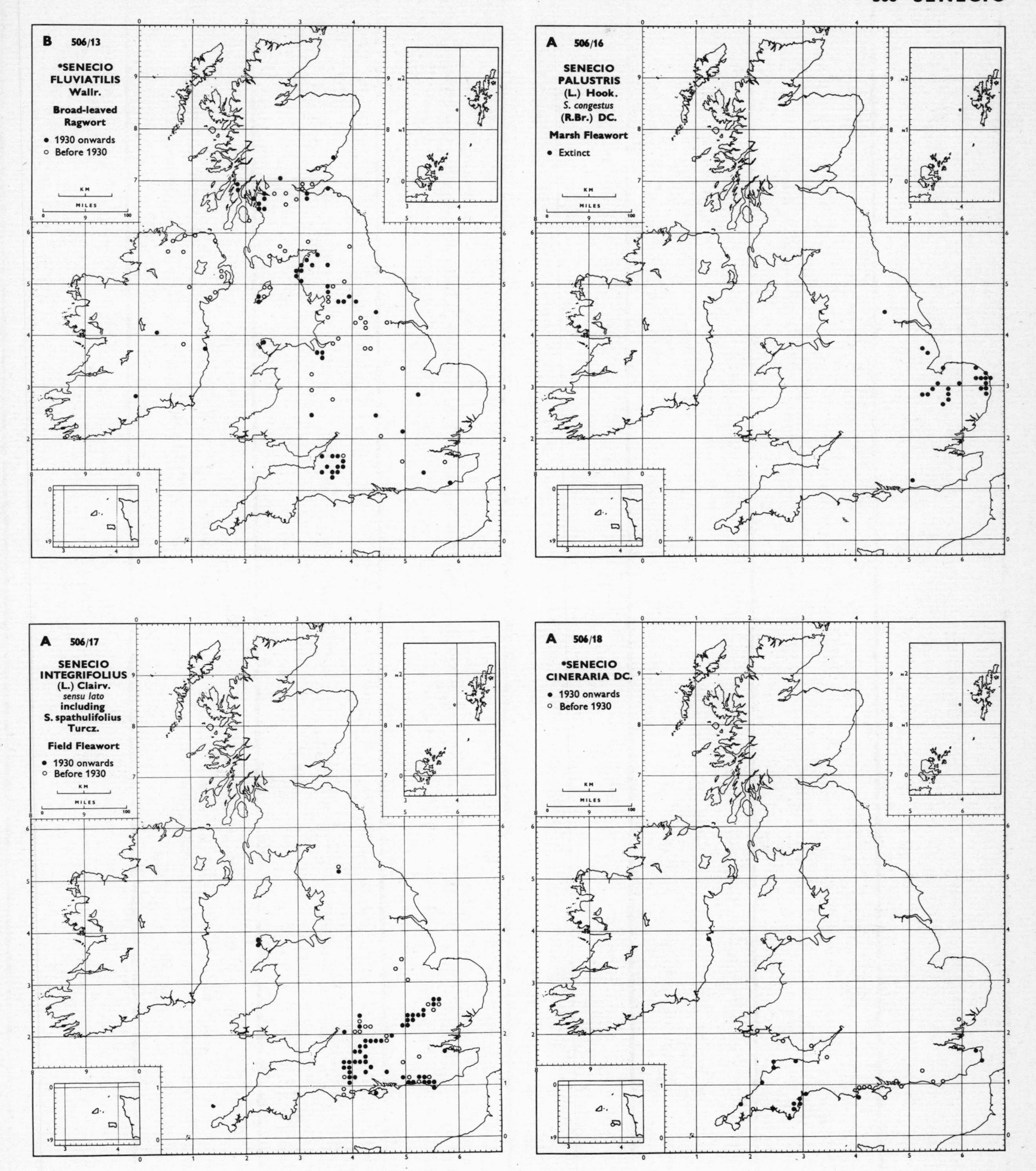

B 506/13
*SENECIO FLUVIATILIS Wallr.
Broad-leaved Ragwort
• 1930 onwards
○ Before 1930
A 506/16
SENECIO PALUSTRIS (L.) Hook.
S. congestus (R.Br.) DC.
Marsh Fleawort
• Extinct
A 506/17
SENECIO INTEGRIFOLIUS (L.) Clairv.
sensu lato
including
S. spathulifolius Turcz.
Field Fleawort
• 1930 onwards
○ Before 1930
A 506/18
*SENECIO CINERARIA DC.
• 1930 onwards
○ Before 1930
KM
MILES

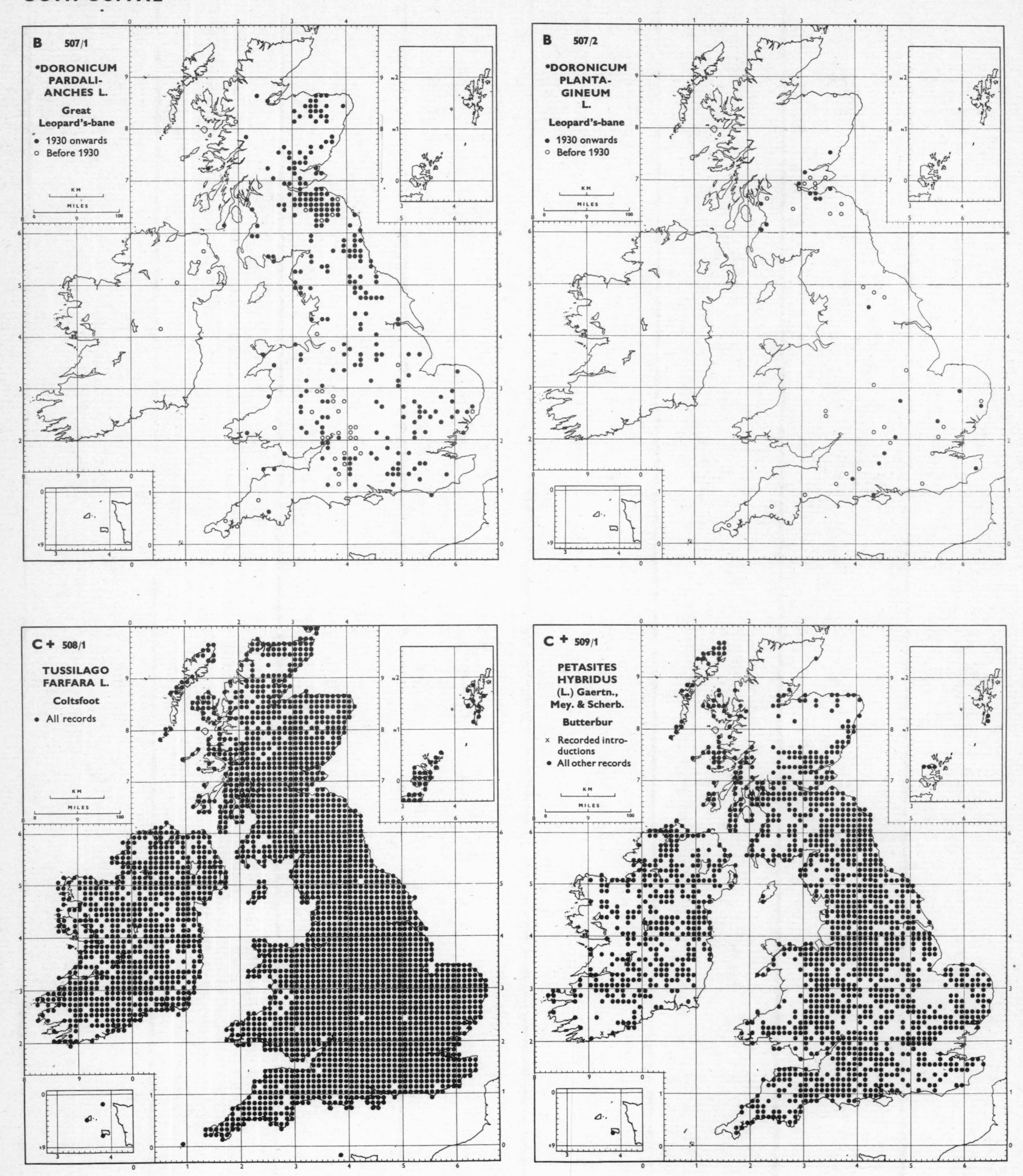
B 507/1
*DORONICUM PARDALIANCHES L.
Great Leopard's-bane
• 1930 onwards
○ Before 1930
B 507/2
*DORONICUM PLANTAGINEUM L.
Leopard's-bane
• 1930 onwards
○ Before 1930
C + 508/1
TUSSILAGO FARFARA L.
Coltsfoot
• All records
C + 509/1
PETASITES HYBRIDUS (L.) Gaertn., Mey. & Scherb.
Butterbur
× Recorded introductions
• All other records

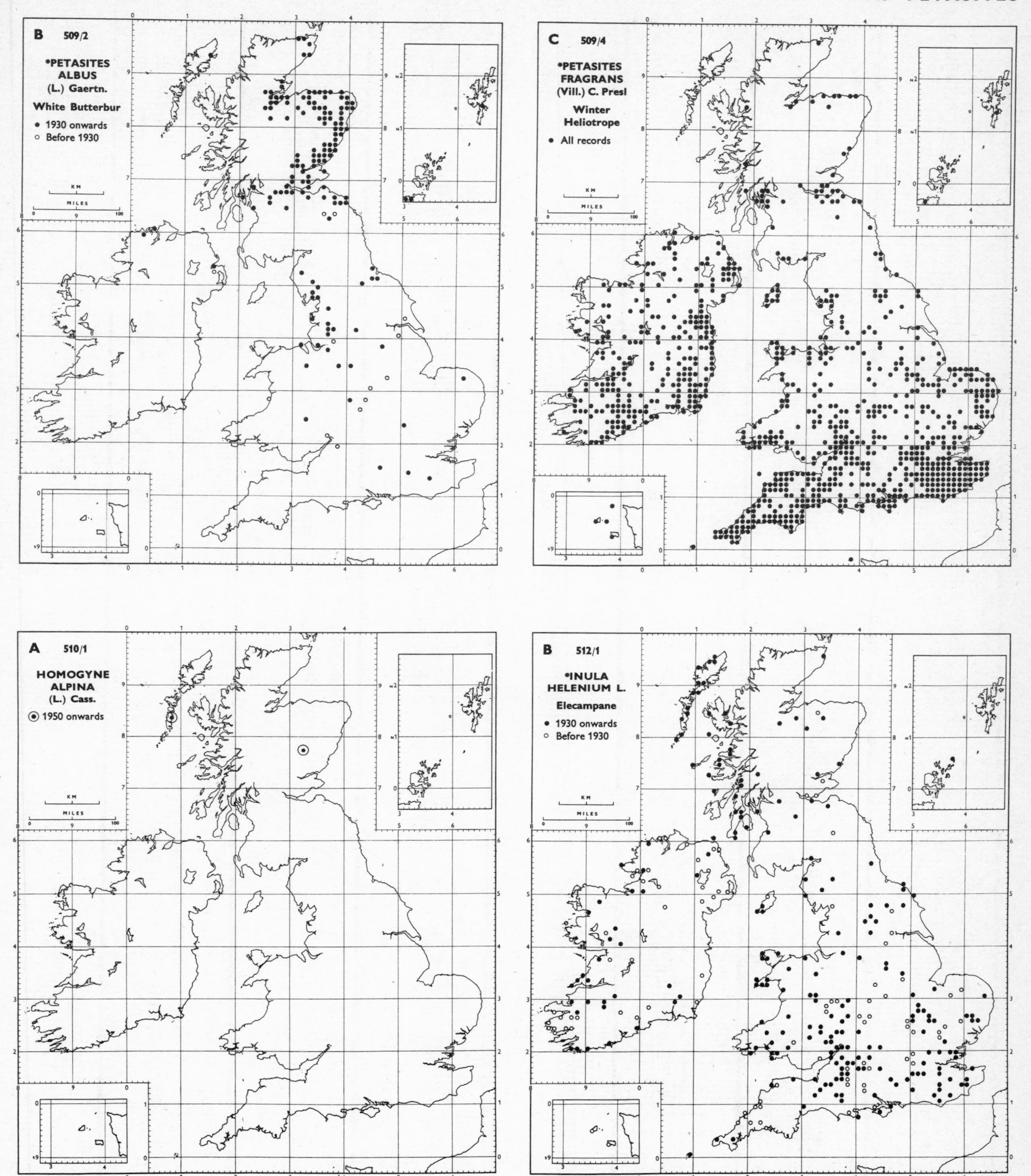
B 509/2
*PETASITES ALBUS (L.) Gaertn.
White Butterbur
1930 onwards
Before 1930
C 509/4
*PETASITES FRAGRANS (Vill.) C. Presl
Winter Heliotrope
All records
A 510/1
HOMOGYNE ALPINA (L.) Cass.
1950 onwards
B 512/1
*INULA HELENIUM L.
Elecampane
1930 onwards
Before 1930

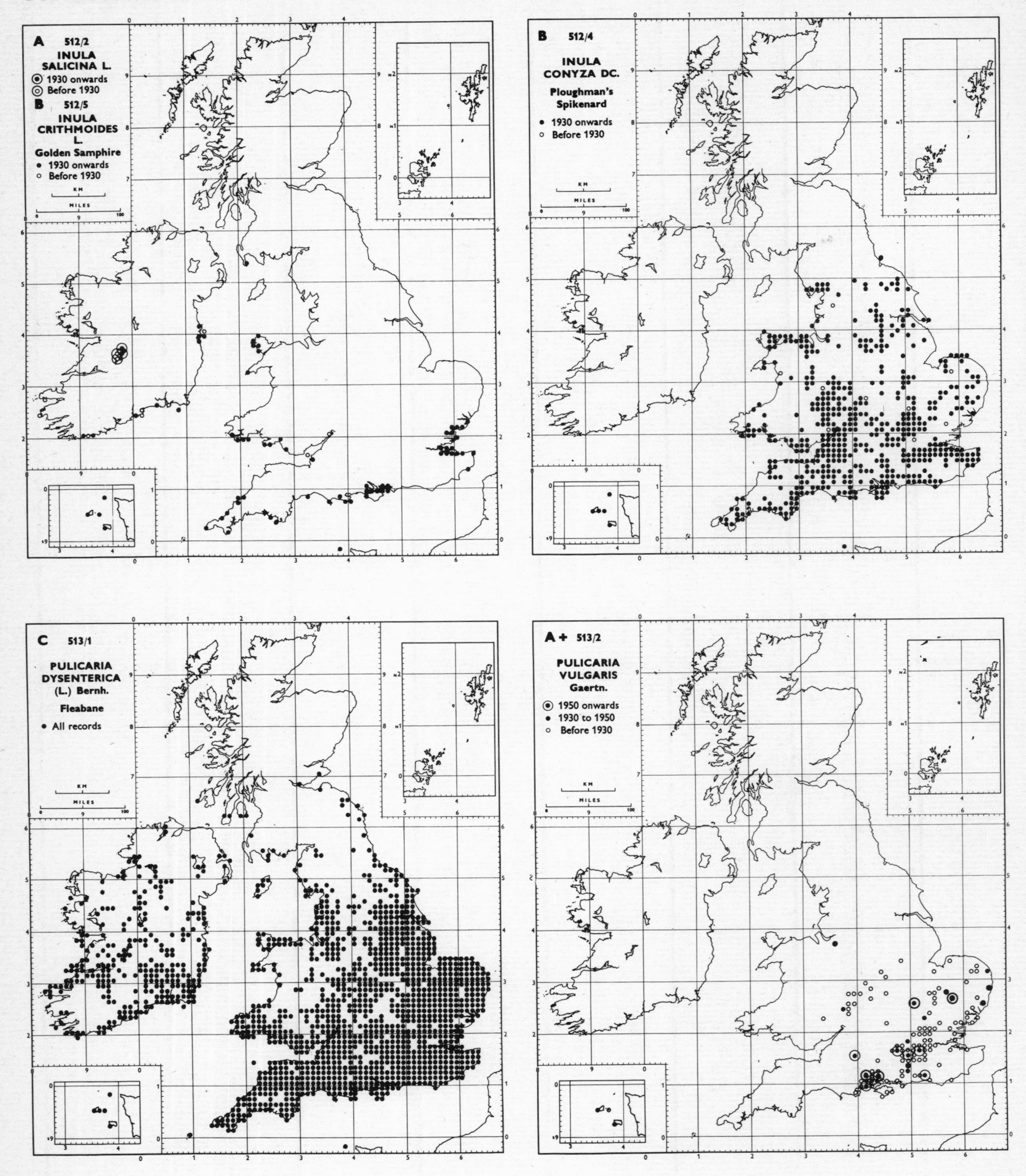
A 512/2
INULA SALICINA L.
1930 onwards
Before 1930
B 512/5
INULA CRITHMOIDES L.
Golden Samphire
1930 onwards
Before 1930
B 512/4
INULA CONYZA DC.
Ploughman's Spikenard
1930 onwards
Before 1930
C 513/1
PULICARIA DYSENTERICA (L.) Bernh.
Fleabane
All records
A+ 513/2
PULICARIA VULGARIS Gaertn.
1950 onwards
1930 to 1950
Before 1930

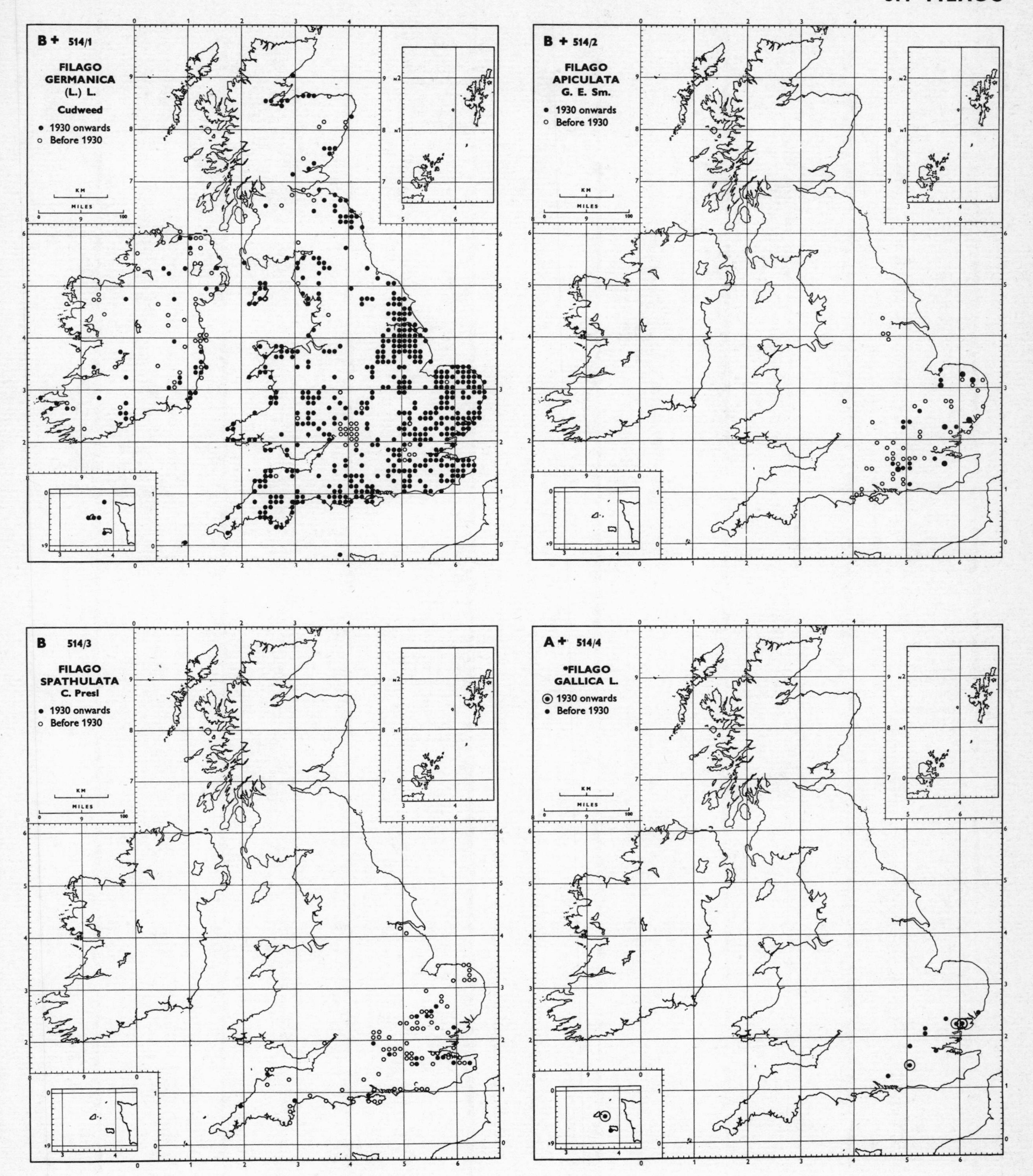
B + 514/1
FILAGO GERMANICA (L.) L.
Cudweed
1930 onwards
Before 1930
KM
MILES
B + 514/2
FILAGO APICULATA G. E. Sm.
1930 onwards
Before 1930
B 514/3
FILAGO SPATHULATA C. Presl
1930 onwards
Before 1930
A + 514/4
°FILAGO GALLICA L.
1930 onwards
Before 1930

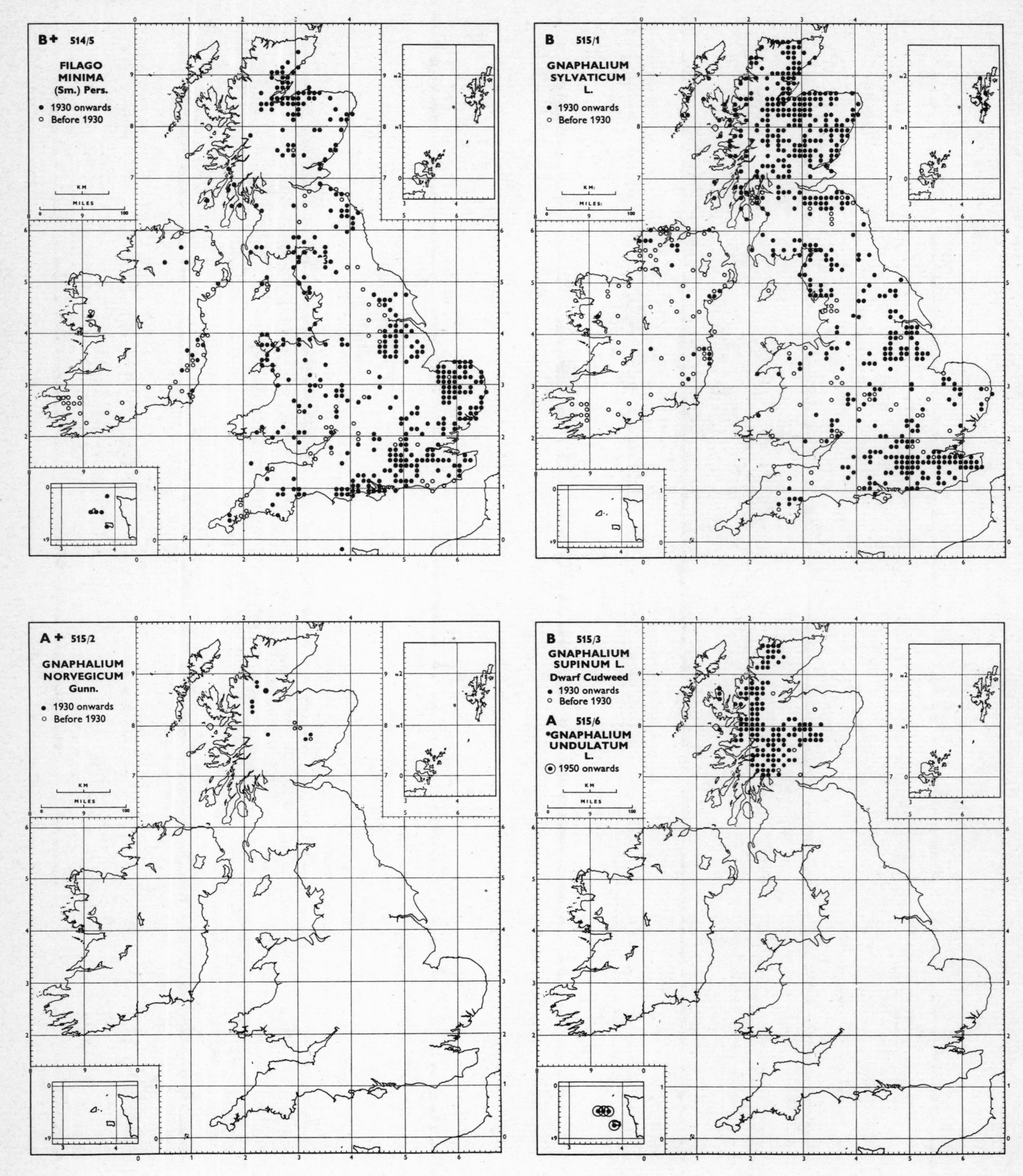
B+ 514/5
FILAGO
MINIMA
(Sm.) Pers.
• 1930 onwards
○ Before 1930
B 515/1
GNAPHALIUM
SYLVATICUM
L.
• 1930 onwards
○ Before 1930
A+ 515/2
GNAPHALIUM
NORVEGICUM
Gunn.
• 1930 onwards
○ Before 1930
B 515/3
GNAPHALIUM
SUPINUM L.
Dwarf Cudweed
• 1930 onwards
○ Before 1930
A 515/6
*GNAPHALIUM
UNDULATUM
L.
⊙ 1950 onwards
KM
MILES

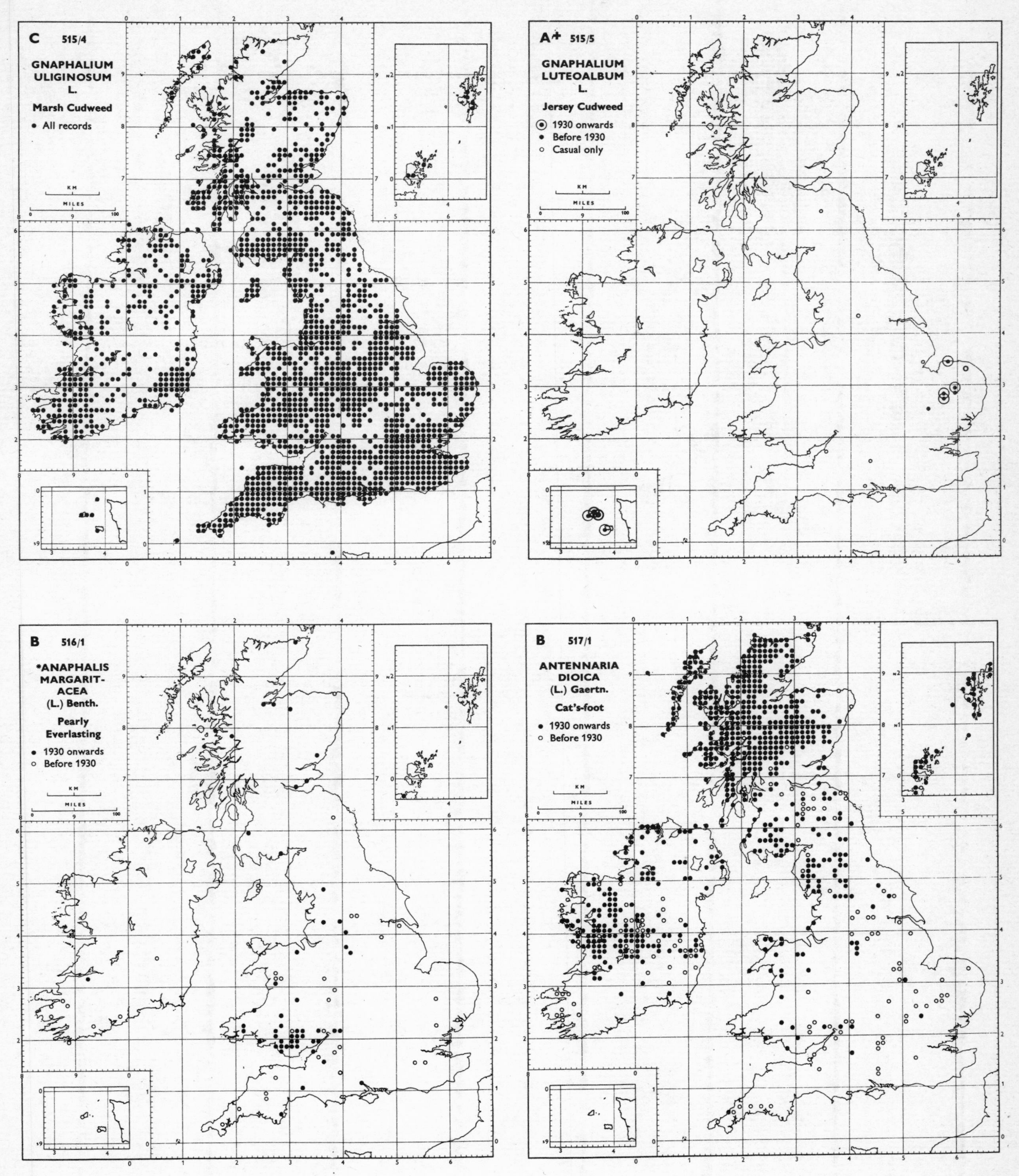
C 515/4
GNAPHALIUM ULIGINOSUM L.
Marsh Cudweed
• All records
KM
MILES
A+ 515/5
GNAPHALIUM LUTEOALBUM L.
Jersey Cudweed
⊙ 1930 onwards
• Before 1930
○ Casual only
KM
MILES
B 516/1
*ANAPHALIS MARGARITACEA (L.) Benth.
Pearly Everlasting
• 1930 onwards
○ Before 1930
KM
MILES
B 517/1
ANTENNARIA DIOICA (L.) Gaertn.
Cat's-foot
• 1930 onwards
○ Before 1930
KM
MILES

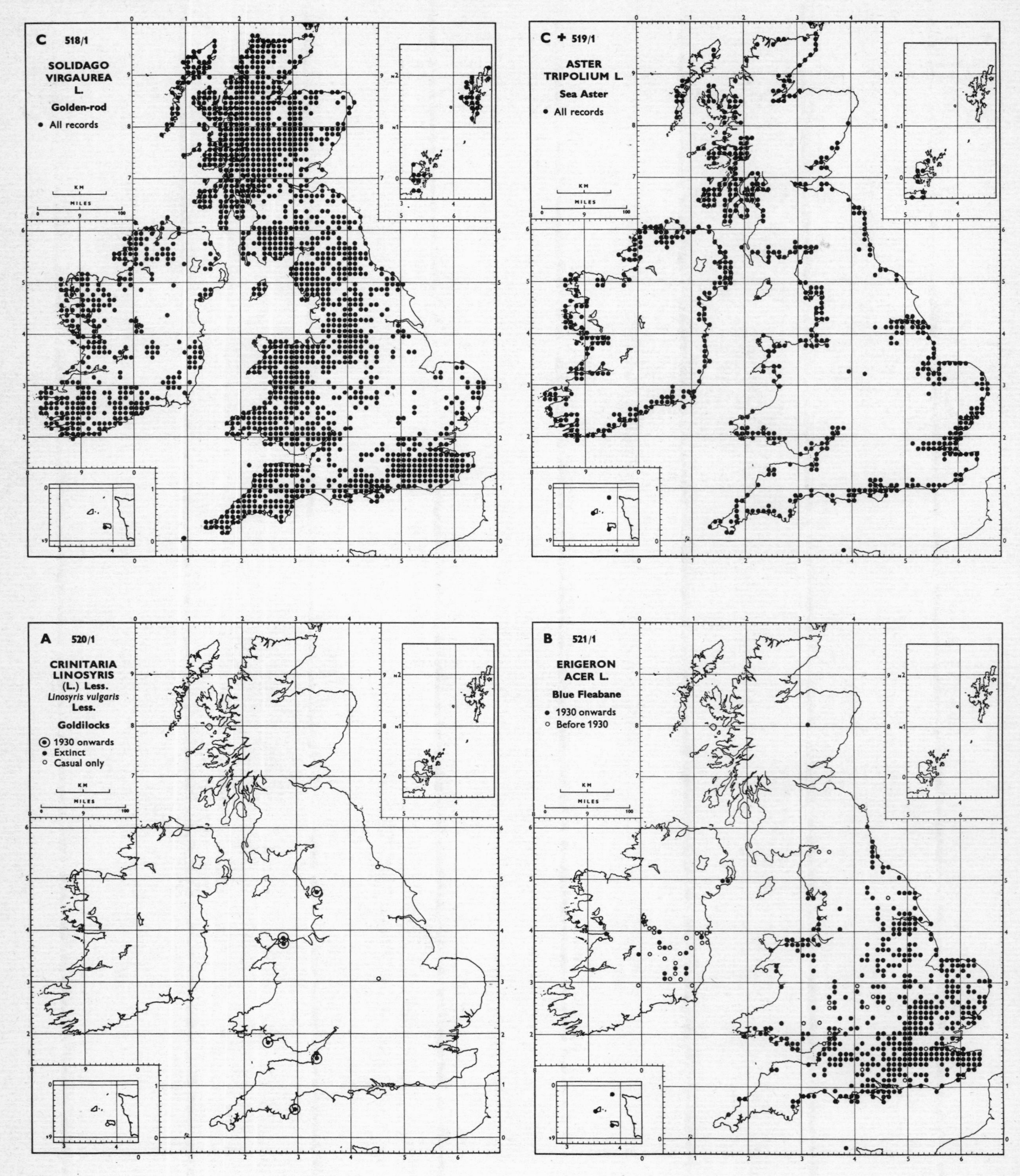
C 518/1
SOLIDAGO VIRGAUREA L.
Golden-rod
• All records
C + 519/1
ASTER TRIPOLIUM L.
Sea Aster
• All records
A 520/1
CRINITARIA LINOSYRIS (L.) Less.
Linosyris vulgaris Less.
Goldilocks
⊙ 1930 onwards
• Extinct
○ Casual only
B 521/1
ERIGERON ACER L.
Blue Fleabane
• 1930 onwards
○ Before 1930
KM
MILES

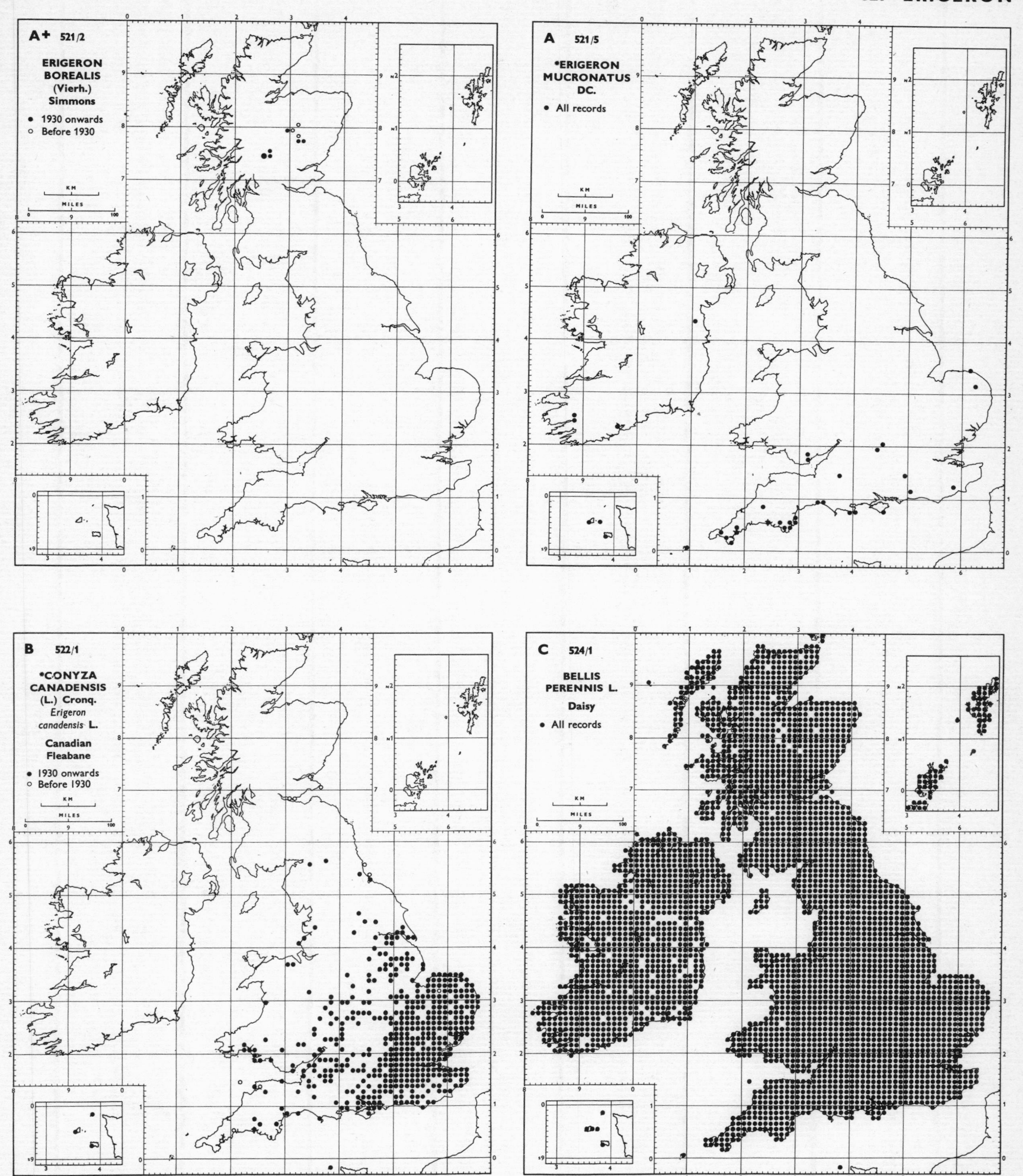
A+ 521/2
ERIGERON BOREALIS (Vierh.) Simmons
● 1930 onwards
○ Before 1930
A 521/5
*ERIGERON MUCRONATUS DC.
● All records
B 522/1
*CONYZA CANADENSIS (L.) Cronq.
Erigeron canadensis L.
Canadian Fleabane
● 1930 onwards
○ Before 1930
C 524/1
BELLIS PERENNIS L.
Daisy
● All records
KM
MILES

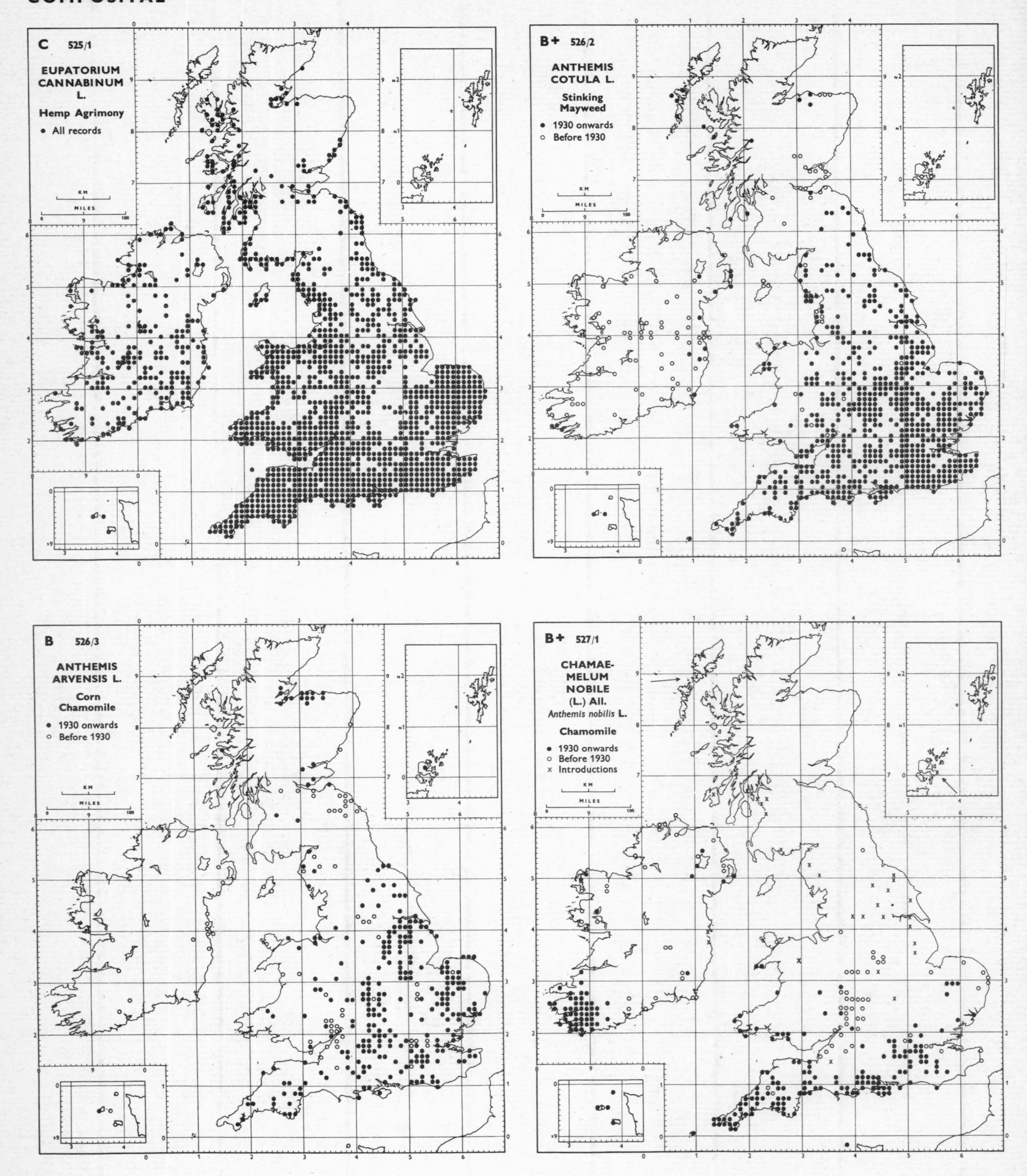

C 525/1
EUPATORIUM CANNABINUM L.
Hemp Agrimony
• All records
B+ 526/2
ANTHEMIS COTULA L.
Stinking Mayweed
• 1930 onwards
○ Before 1930
B 526/3
ANTHEMIS ARVENSIS L.
Corn Chamomile
• 1930 onwards
○ Before 1930
B+ 527/1
CHAMAEMELUM NOBILE (L.) All.
Anthemis nobilis L.
Chamomile
• 1930 onwards
○ Before 1930
x Introductions
KM
MILES

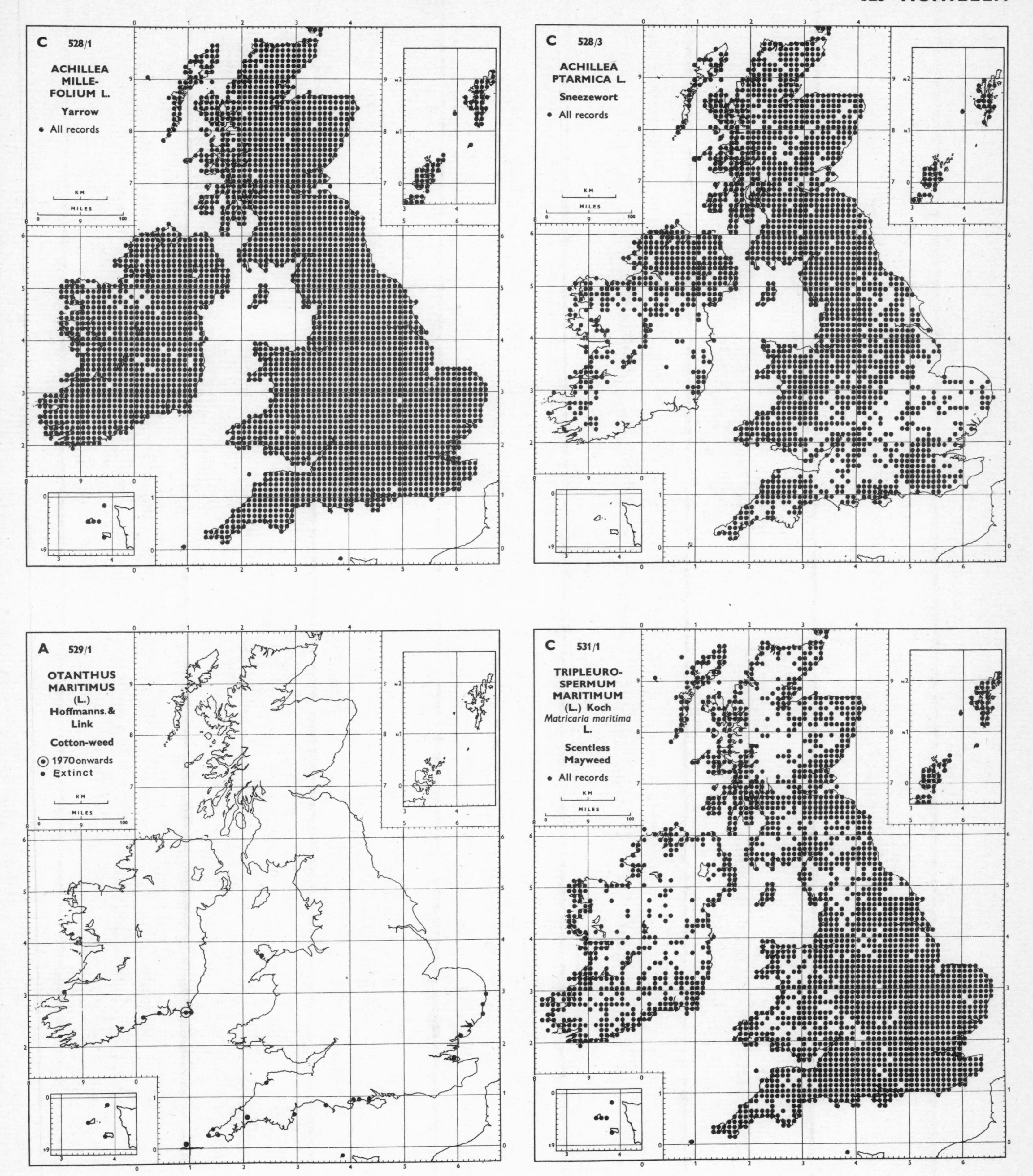
C 528/1
ACHILLEA MILLE-FOLIUM L.
Yarrow
• All records
C 528/3
ACHILLEA PTARMICA L.
Sneezewort
• All records
A 529/1
OTANTHUS MARITIMUS (L.) Hoffmanns. & Link
Cotton-weed
⊙ 1970 onwards
• Extinct
C 531/1
TRIPLEURO-SPERMUM MARITIMUM (L.) Koch
Matricaria maritima L.
Scentless Mayweed
• All records

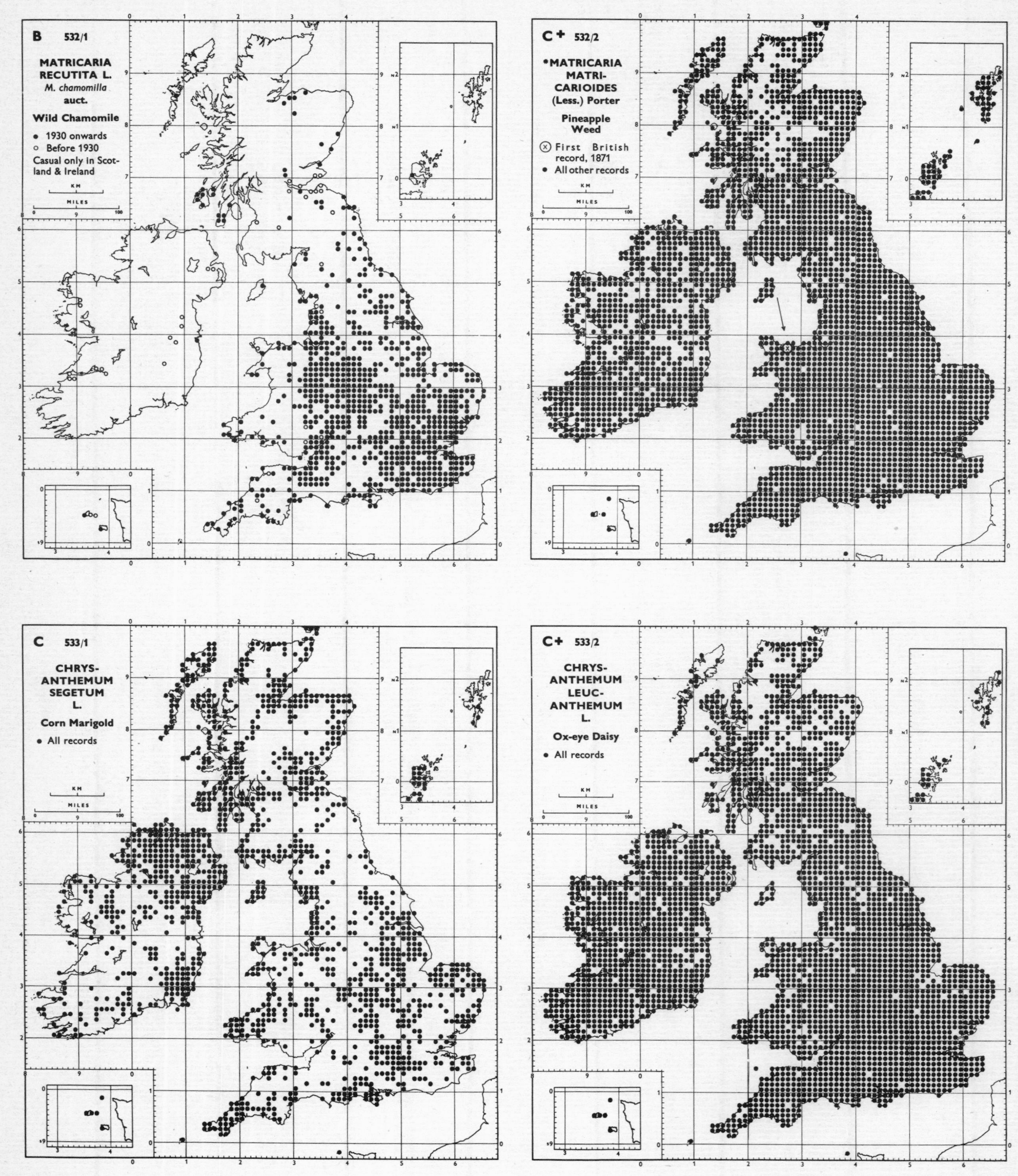
B 532/1
MATRICARIA RECUTITA L.
M. chamomilla auct.
Wild Chamomile
• 1930 onwards
○ Before 1930
Casual only in Scotland & Ireland
C+ 532/2
•MATRICARIA MATRICARIOIDES (Less.) Porter
Pineapple Weed
First British record, 1871
• All other records
C 533/1
CHRYSANTHEMUM SEGETUM L.
Corn Marigold
• All records
C+ 533/2
CHRYSANTHEMUM LEUCANTHEMUM L.
Ox-eye Daisy
• All records

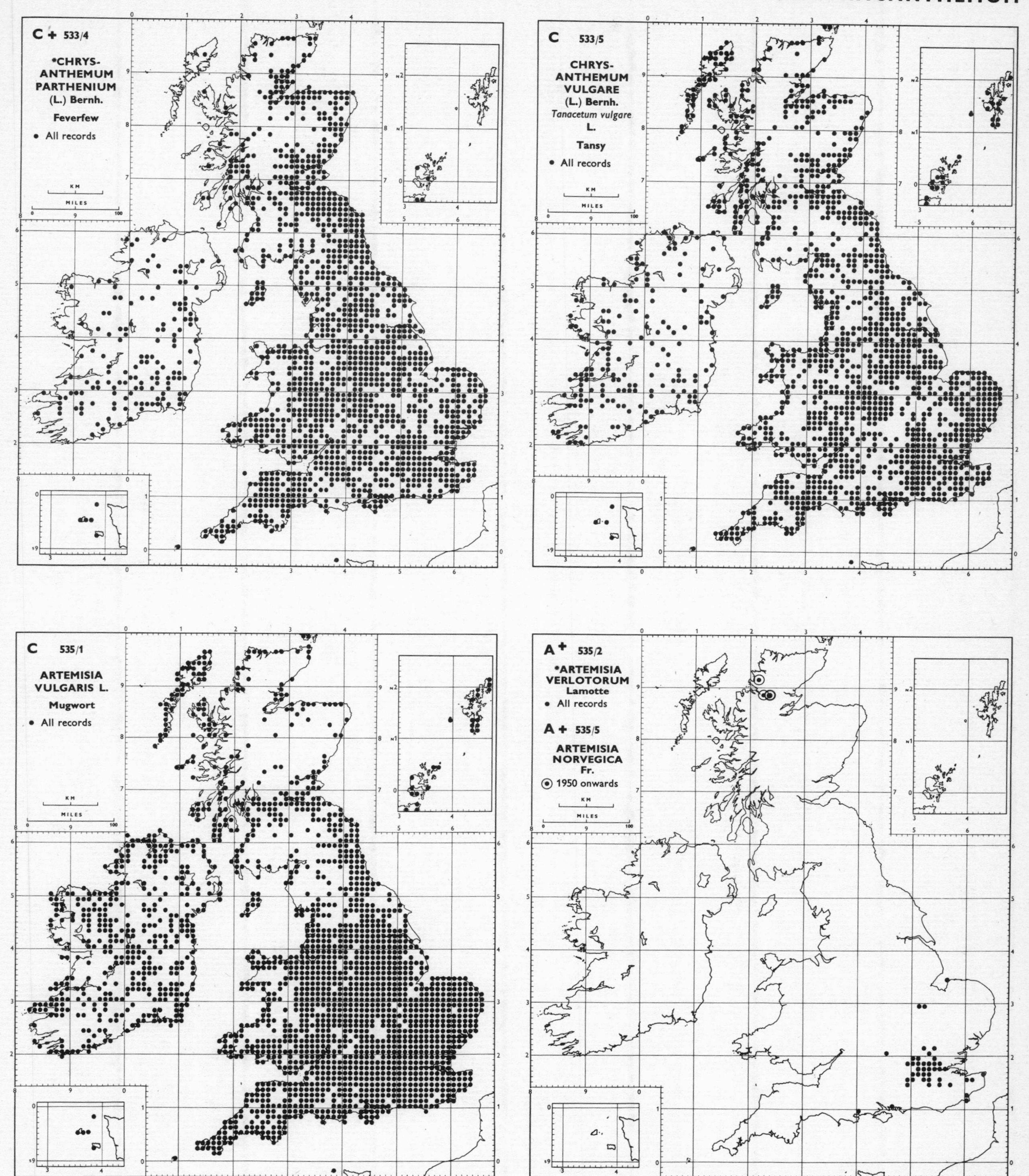
C + 533/4
*CHRYSANTHEMUM PARTHENIUM (L.) Bernh.
Feverfew
• All records
C 533/5
CHRYSANTHEMUM VULGARE (L.) Bernh.
Tanacetum vulgare L.
Tansy
• All records
C 535/1
ARTEMISIA VULGARIS L.
Mugwort
• All records
A+ 535/2
*ARTEMISIA VERLOTORUM Lamotte
• All records
A + 535/5
ARTEMISIA NORVEGICA Fr.
1950 onwards
KM
MILES

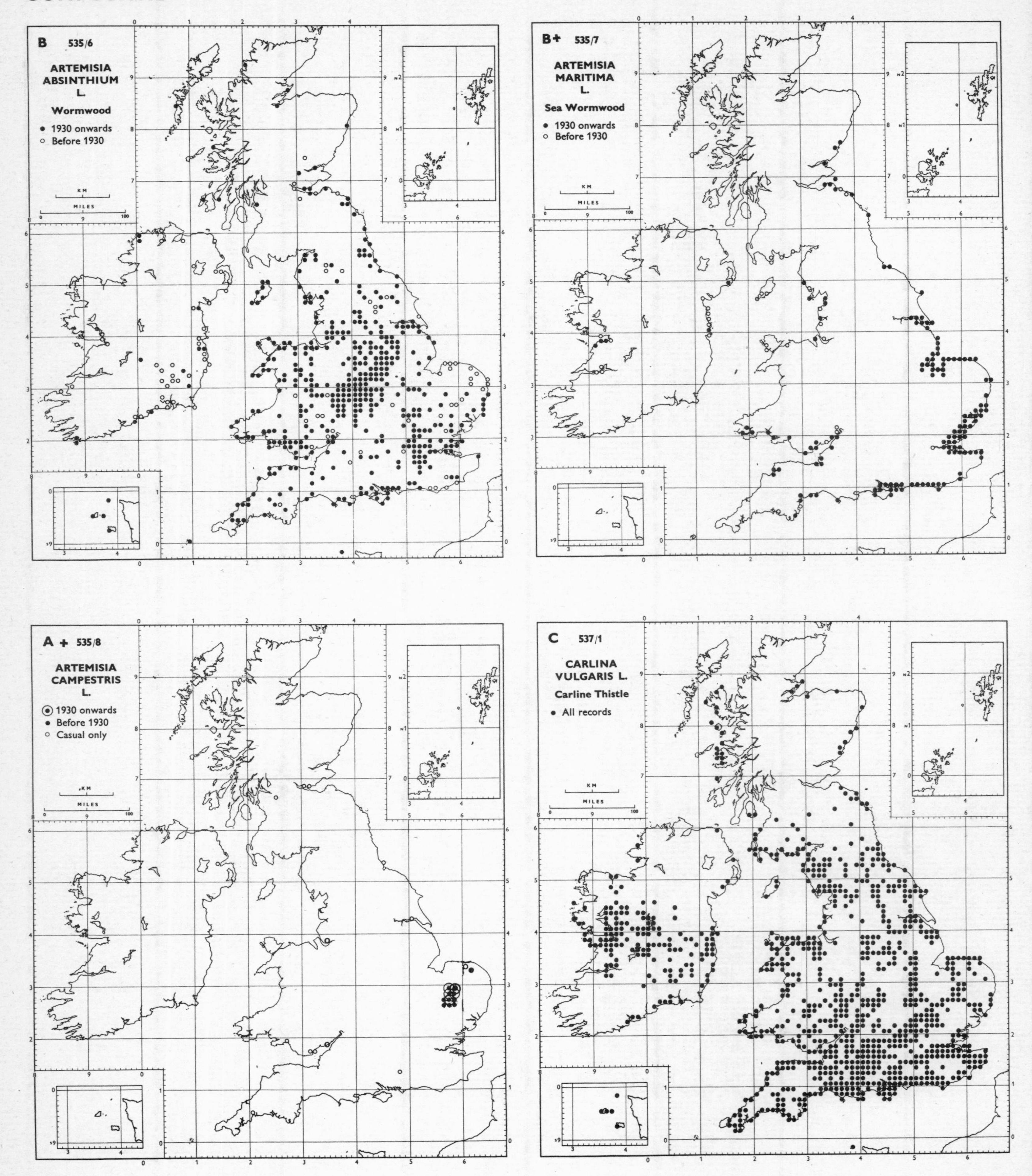

B 535/6
ARTEMISIA ABSINTHIUM L.
Wormwood
• 1930 onwards
○ Before 1930
B+ 535/7
ARTEMISIA MARITIMA L.
Sea Wormwood
• 1930 onwards
○ Before 1930
A + 535/8
ARTEMISIA CAMPESTRIS L.
⊙ 1930 onwards
• Before 1930
○ Casual only
C 537/1
CARLINA VULGARIS L.
Carline Thistle
• All records

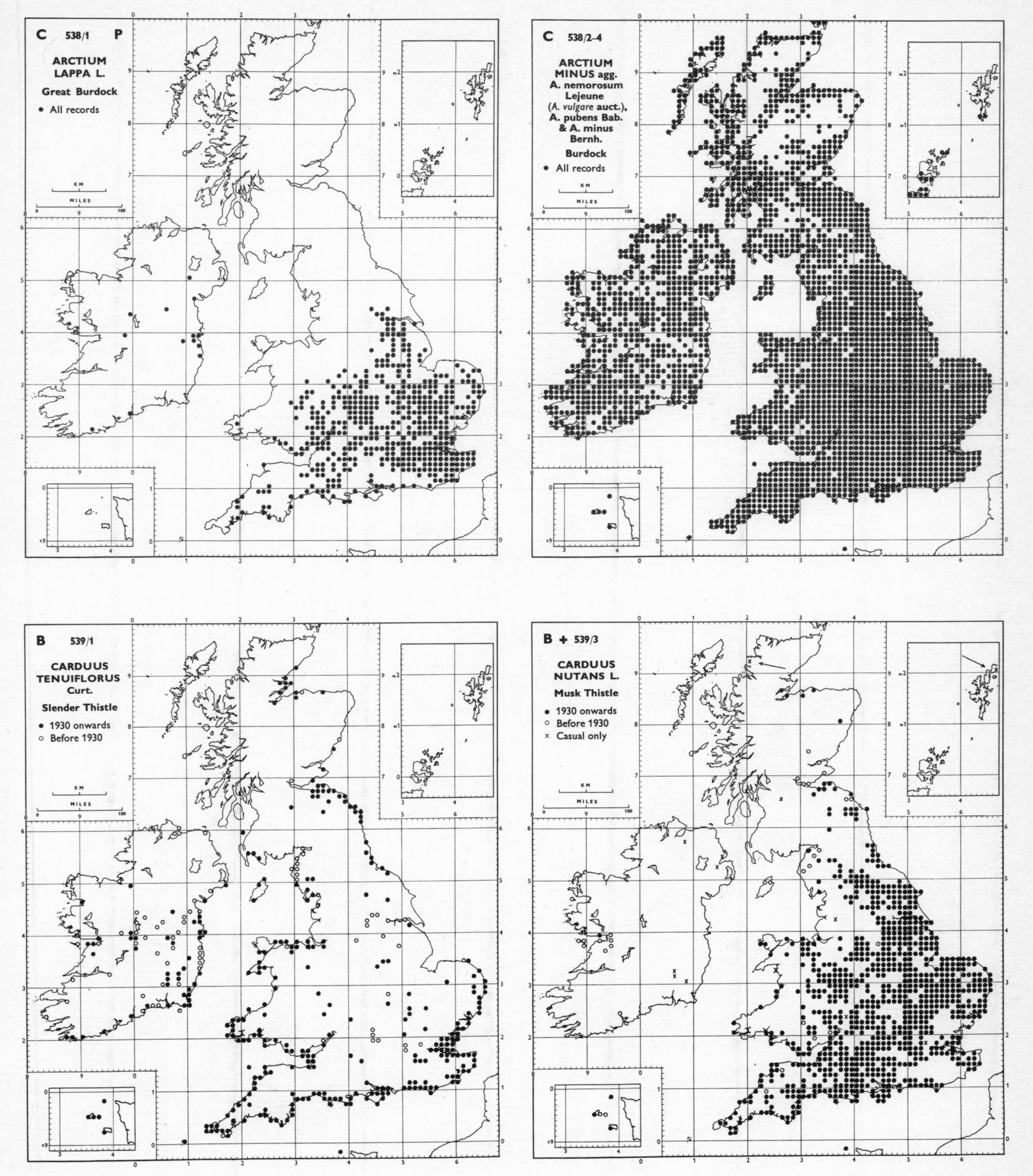
C 538/1 P
ARCTIUM LAPPA L.
Great Burdock
• All records
C 538/2–4
ARCTIUM MINUS agg.
A. nemorosum Lejeune
(A. vulgare auct.),
A. pubens Bab.
& A. minus Bernh.
Burdock
• All records
B 539/1
CARDUUS TENUIFLORUS Curt.
Slender Thistle
• 1930 onwards
○ Before 1930
B + 539/3
CARDUUS NUTANS L.
Musk Thistle
• 1930 onwards
○ Before 1930
× Casual only
KM
MILES

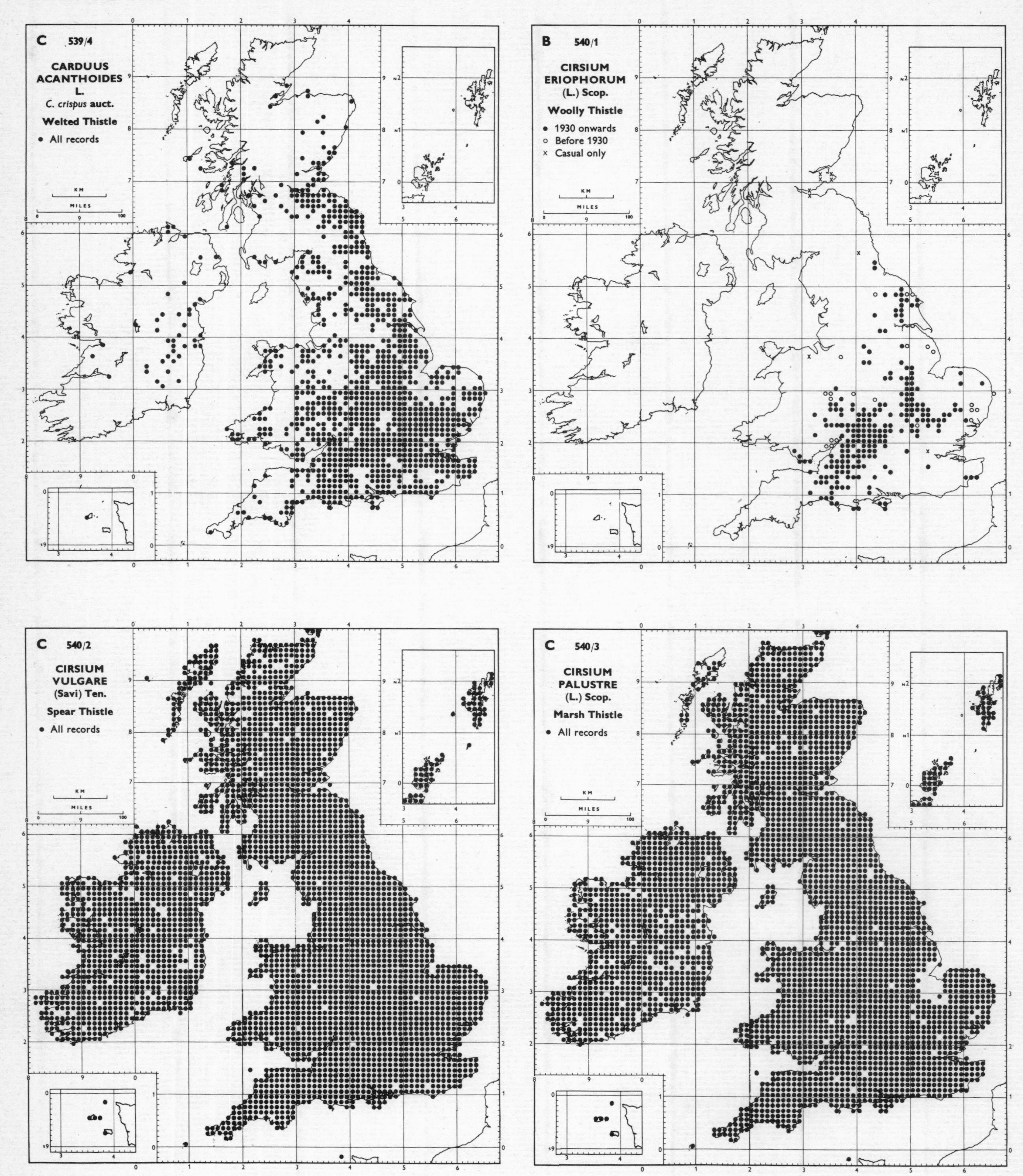
C 539/4
CARDUUS ACANTHOIDES L.
C. crispus auct.
Welted Thistle
• All records
B 540/1
CIRSIUM ERIOPHORUM (L.) Scop.
Woolly Thistle
• 1930 onwards
○ Before 1930
× Casual only
C 540/2
CIRSIUM VULGARE (Savi) Ten.
Spear Thistle
• All records
C 540/3
CIRSIUM PALUSTRE (L.) Scop.
Marsh Thistle
• All records
KM
MILES

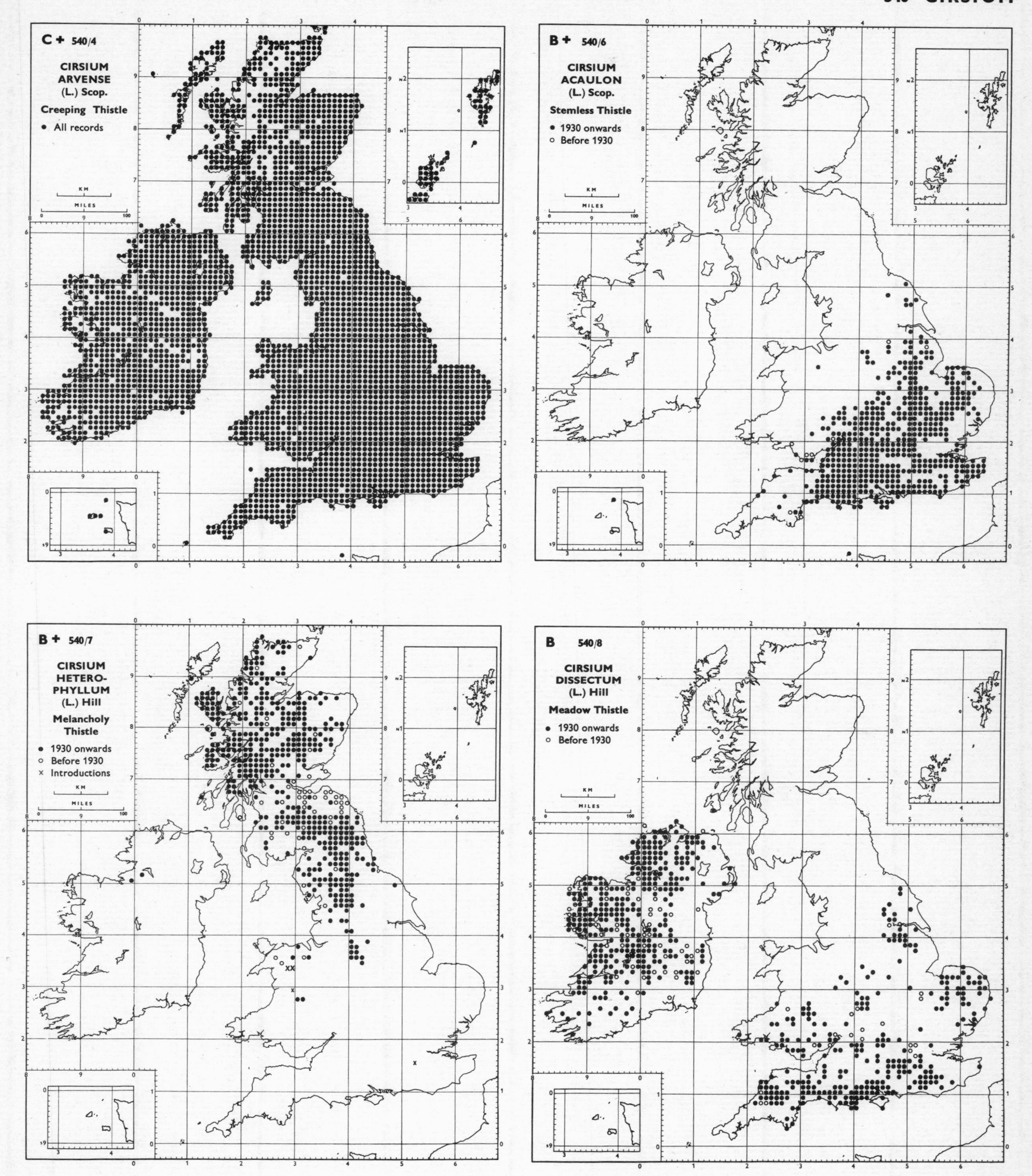
C + 540/4
CIRSIUM ARVENSE (L.) Scop.
Creeping Thistle
• All records
B + 540/6
CIRSIUM ACAULON (L.) Scop.
Stemless Thistle
• 1930 onwards
○ Before 1930
B + 540/7
CIRSIUM HETERO-PHYLLUM (L.) Hill
Melancholy Thistle
• 1930 onwards
○ Before 1930
× Introductions
B 540/8
CIRSIUM DISSECTUM (L.) Hill
Meadow Thistle
• 1930 onwards
○ Before 1930
KM
MILES

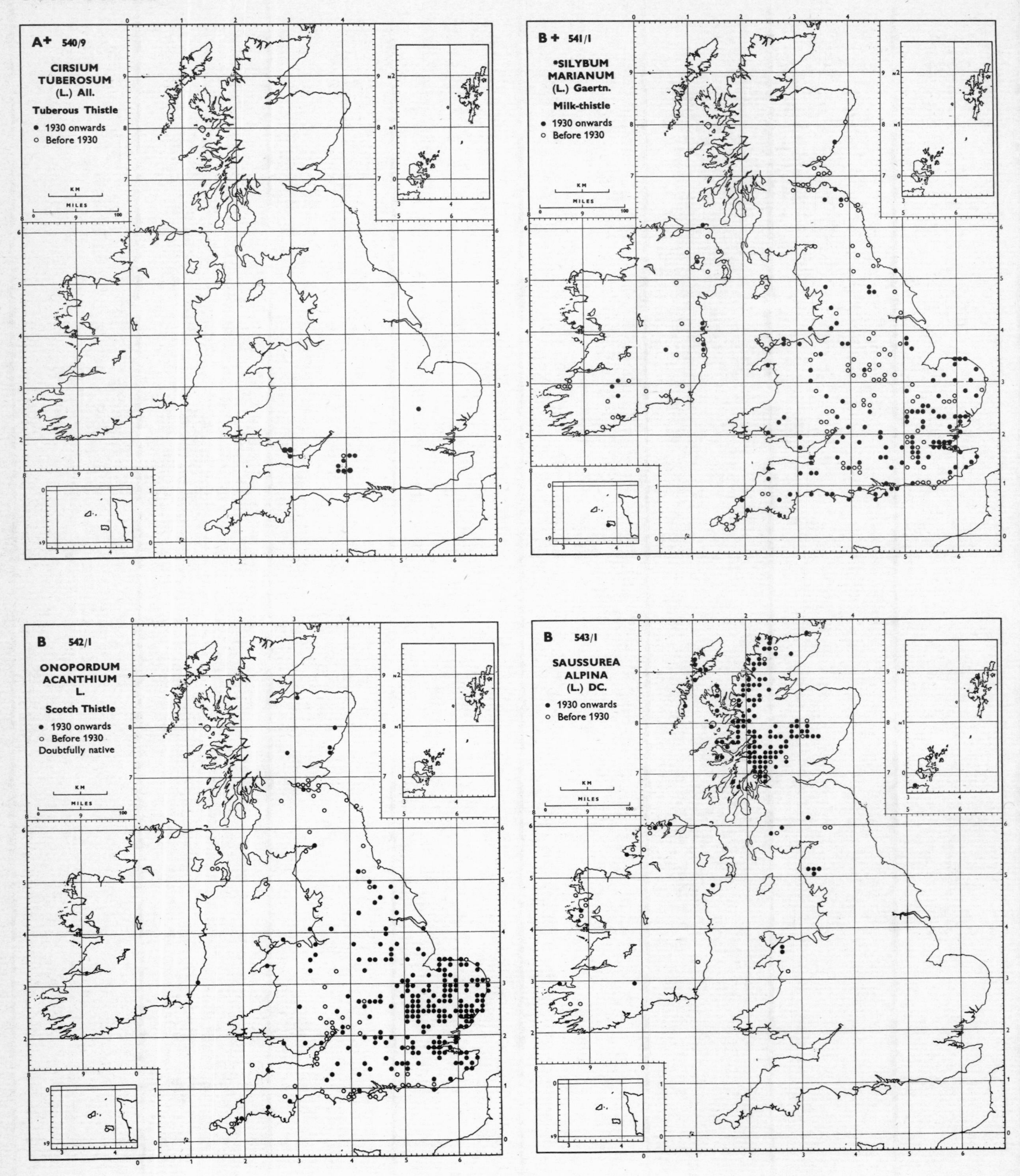
A+ 540/9
CIRSIUM TUBEROSUM (L.) All.
Tuberous Thistle
• 1930 onwards
○ Before 1930
B+ 541/1
*SILYBUM MARIANUM (L.) Gaertn.
Milk-thistle
• 1930 onwards
○ Before 1930
B 542/1
ONOPORDUM ACANTHIUM L.
Scotch Thistle
• 1930 onwards
○ Before 1930
Doubtfully native
B 543/1
SAUSSUREA ALPINA (L.) DC.
• 1930 onwards
○ Before 1930
KM
MILES

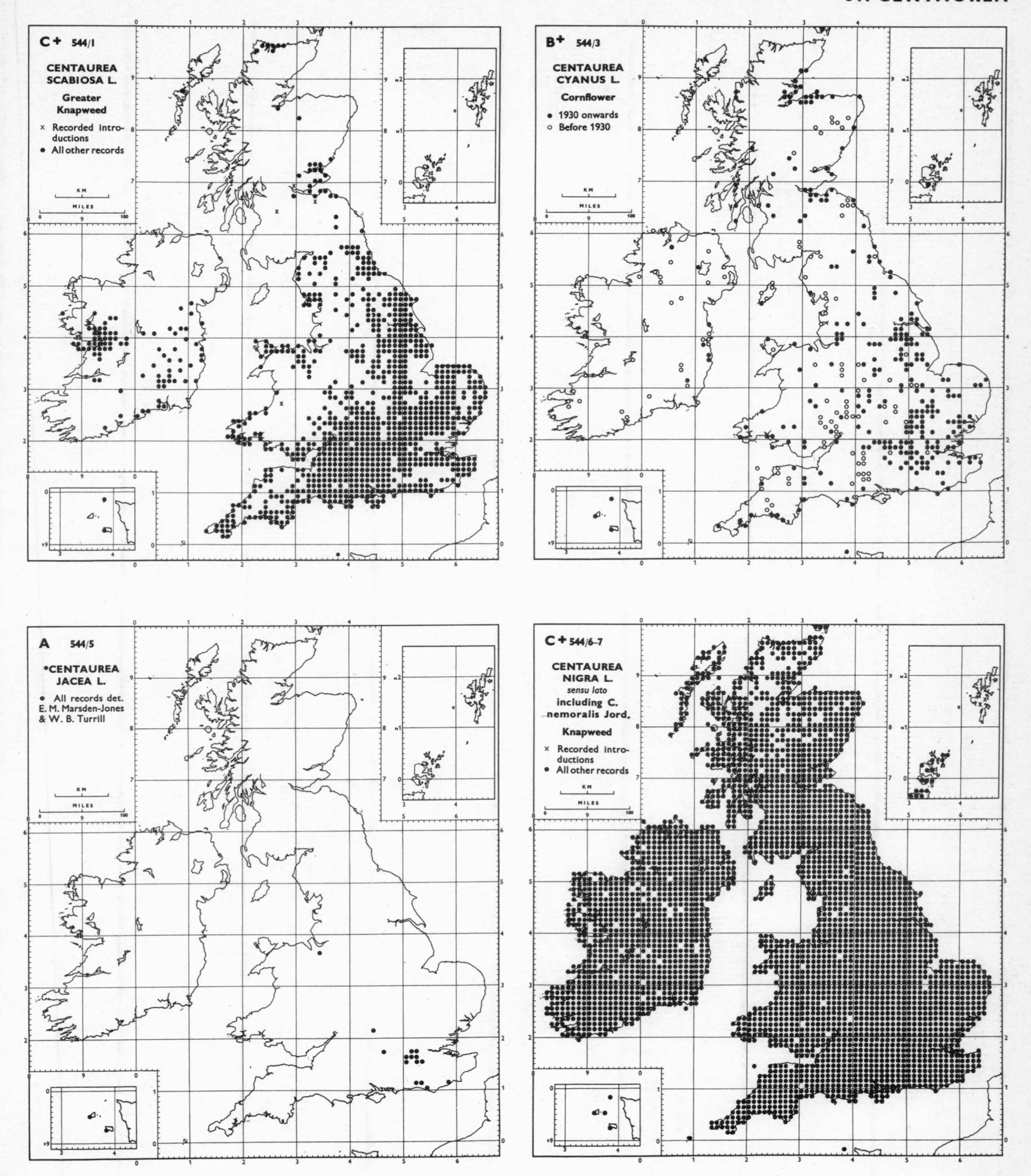
C+ 544/1
CENTAUREA SCABIOSA L.
Greater Knapweed
× Recorded introductions
• All other records
KM
MILES
B+ 544/3
CENTAUREA CYANUS L.
Cornflower
• 1930 onwards
○ Before 1930
A 544/5
•CENTAUREA JACEA L.
• All records det. E. M. Marsden-Jones & W. B. Turrill
C+ 544/6–7
CENTAUREA NIGRA L.
sensu lato
including C. nemoralis Jord.
Knapweed
× Recorded introductions
• All other records

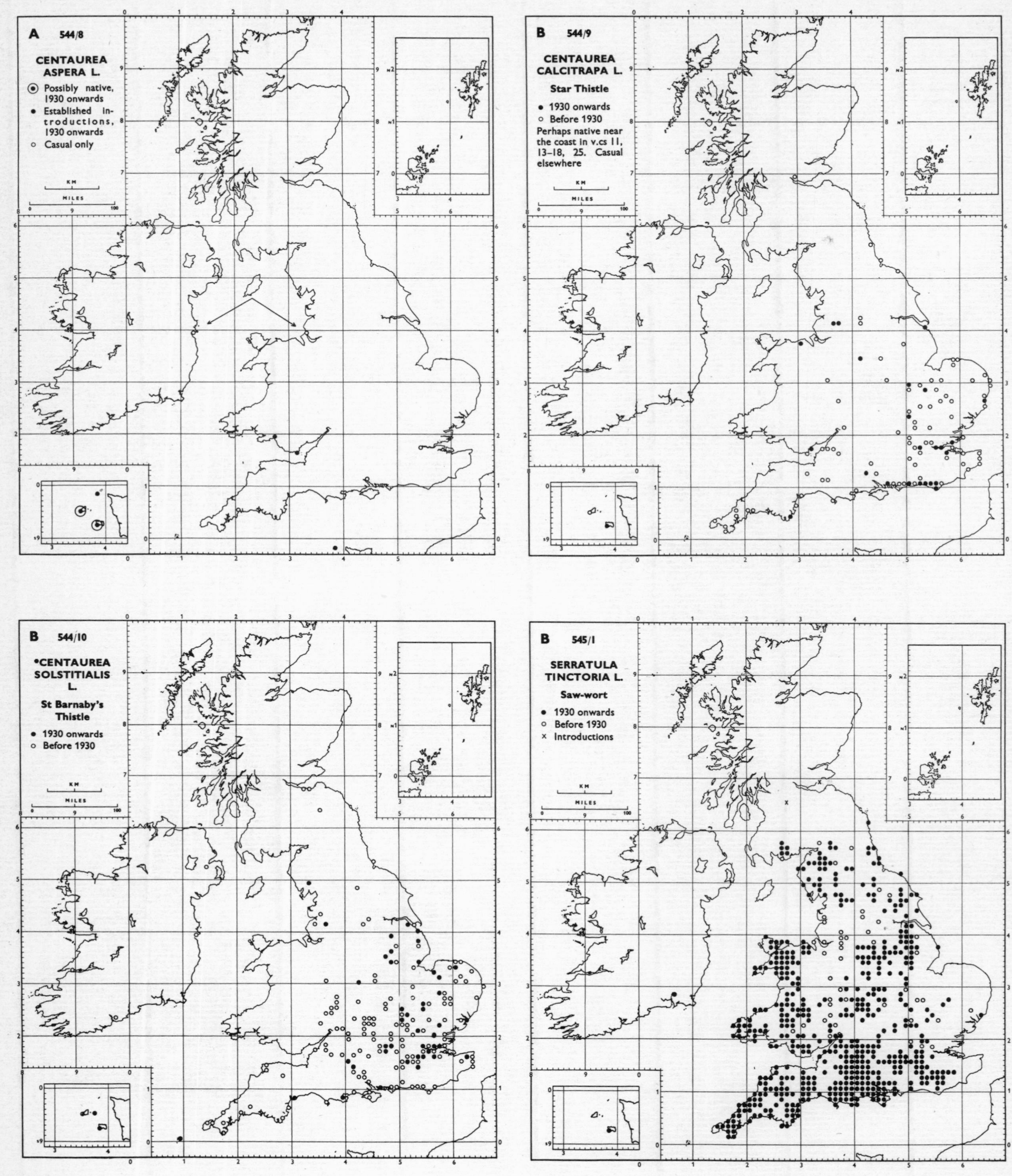
A 544/8
CENTAUREA ASPERA L.
Possibly native, 1930 onwards
Established introductions, 1930 onwards
Casual only
KM
MILES
B 544/9
CENTAUREA CALCITRAPA L.
Star Thistle
1930 onwards
Before 1930
Perhaps native near the coast in v.cs 11, 13–18, 25. Casual elsewhere
KM
MILES
B 544/10
•CENTAUREA SOLSTITIALIS L.
St Barnaby's Thistle
1930 onwards
Before 1930
KM
MILES
B 545/1
SERRATULA TINCTORIA L.
Saw-wort
1930 onwards
Before 1930
× Introductions
KM
MILES

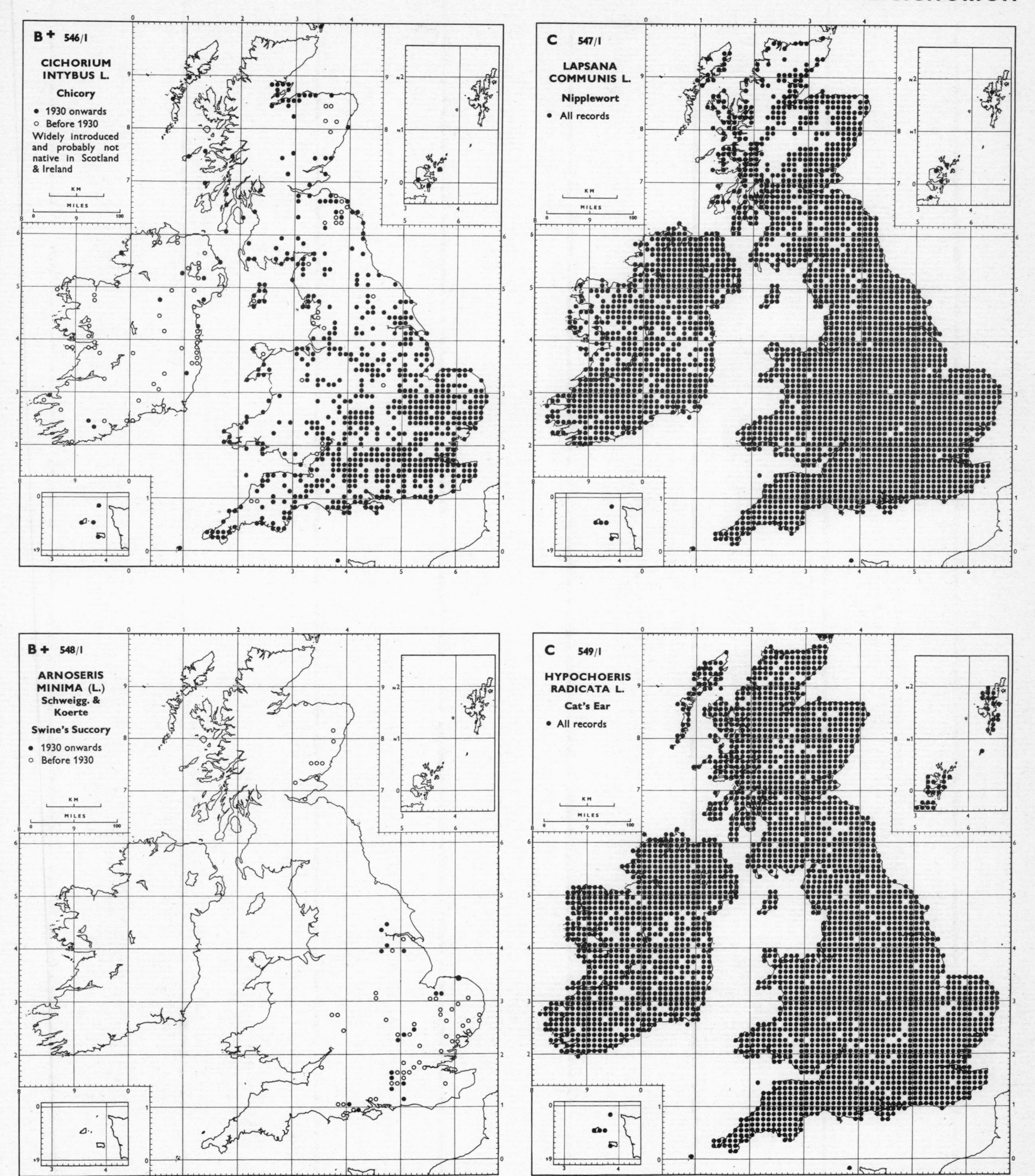
B + 546/1
CICHORIUM INTYBUS L.
Chicory
• 1930 onwards
○ Before 1930
Widely introduced and probably not native in Scotland & Ireland
C 547/1
LAPSANA COMMUNIS L.
Nipplewort
• All records
B + 548/1
ARNOSERIS MINIMA (L.) Schweigg. & Koerte
Swine's Succory
• 1930 onwards
○ Before 1930
C 549/1
HYPOCHOERIS RADICATA L.
Cat's Ear
• All records

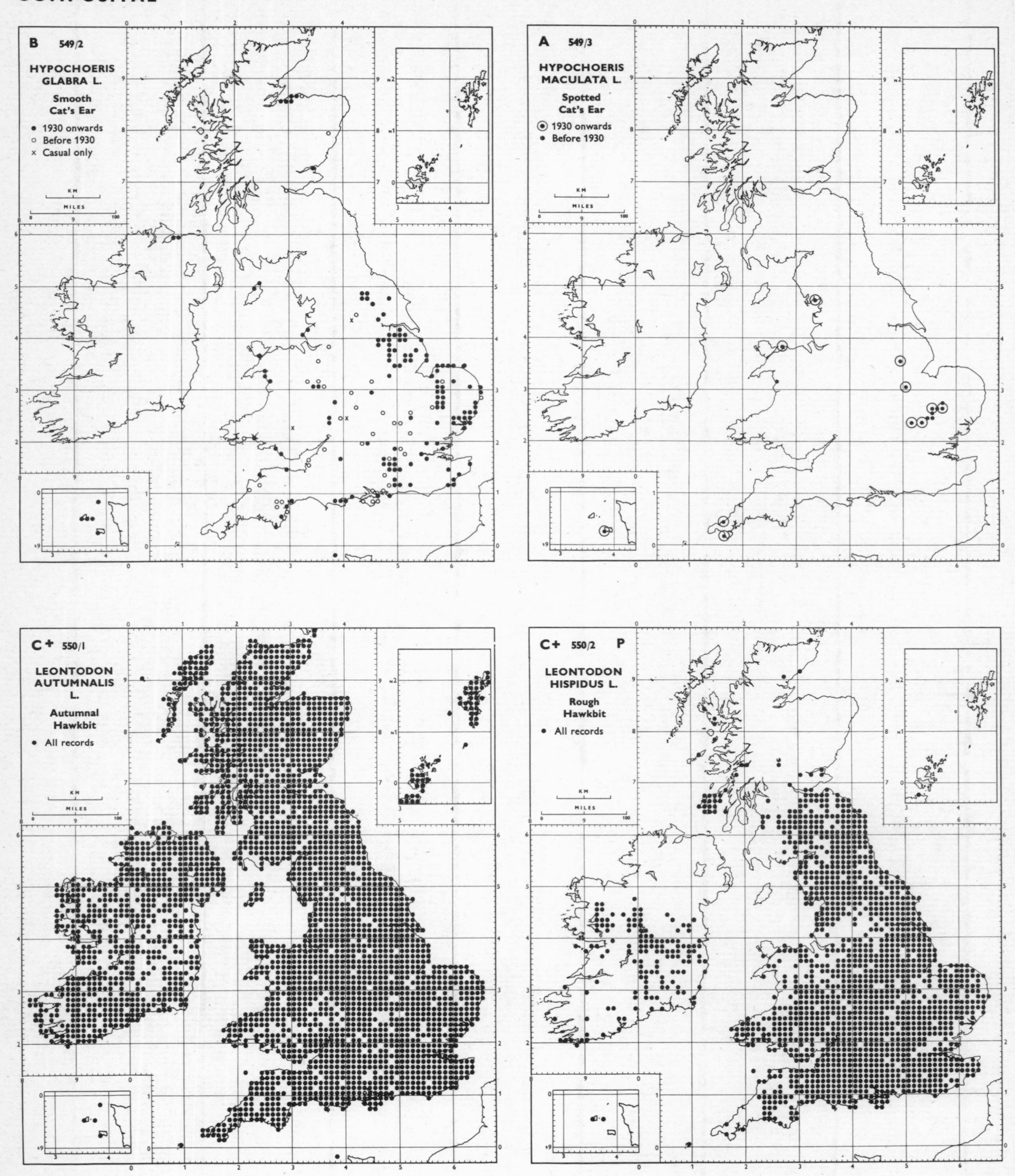
B 549/2
HYPOCHOERIS GLABRA L.
Smooth Cat's Ear
• 1930 onwards
○ Before 1930
× Casual only
A 549/3
HYPOCHOERIS MACULATA L.
Spotted Cat's Ear
⊙ 1930 onwards
• Before 1930
C+ 550/1
LEONTODON AUTUMNALIS L.
Autumnal Hawkbit
• All records
C+ 550/2 P
LEONTODON HISPIDUS L.
Rough Hawkbit
• All records
KM
MILES

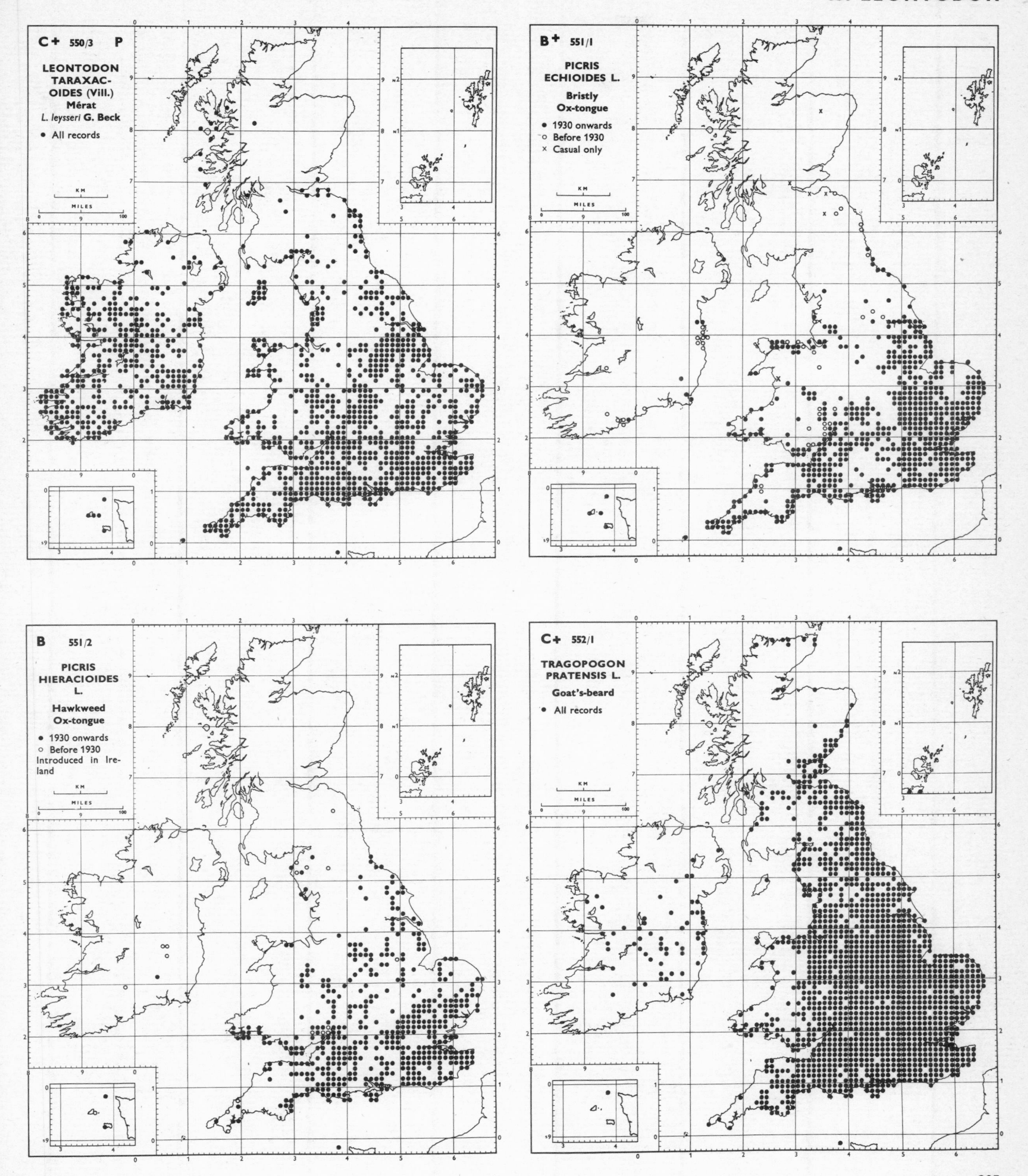

C+ 550/3 P
LEONTODON TARAXACOIDES (Vill.) Mérat
L. leysseri G. Beck
• All records
B+ 551/1
PICRIS ECHIOIDES L.
Bristly Ox-tongue
• 1930 onwards
○ Before 1930
× Casual only
B 551/2
PICRIS HIERACIOIDES L.
Hawkweed Ox-tongue
• 1930 onwards
○ Before 1930
Introduced in Ireland
C+ 552/1
TRAGOPOGON PRATENSIS L.
Goat's-beard
• All records
KM
MILES

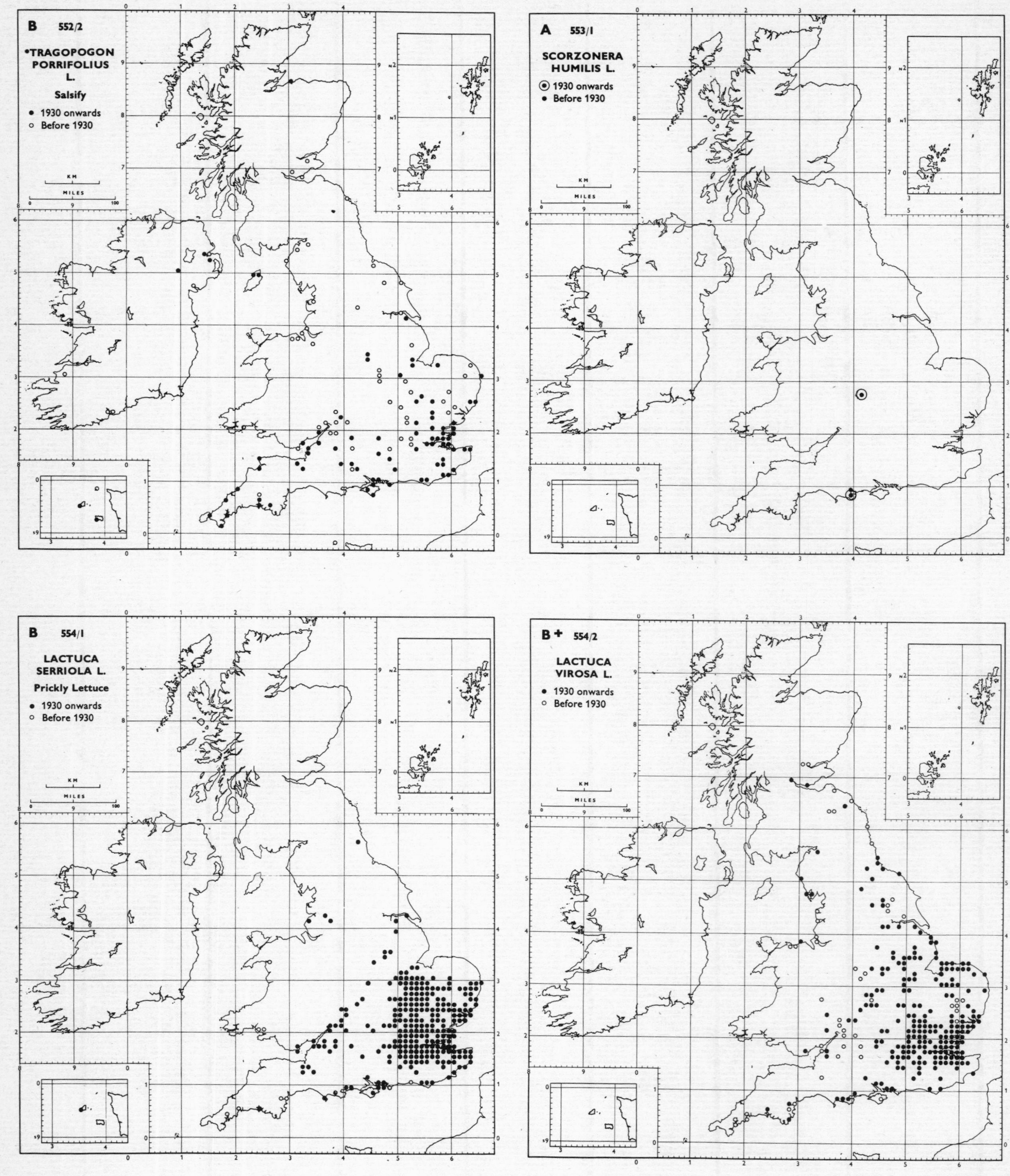
B 552/2
*TRAGOPOGON PORRIFOLIUS L.
Salsify
• 1930 onwards
○ Before 1930
A 553/1
SCORZONERA HUMILIS L.
⊙ 1930 onwards
• Before 1930
B 554/1
LACTUCA SERRIOLA L.
Prickly Lettuce
• 1930 onwards
○ Before 1930
B+ 554/2
LACTUCA VIROSA L.
• 1930 onwards
○ Before 1930

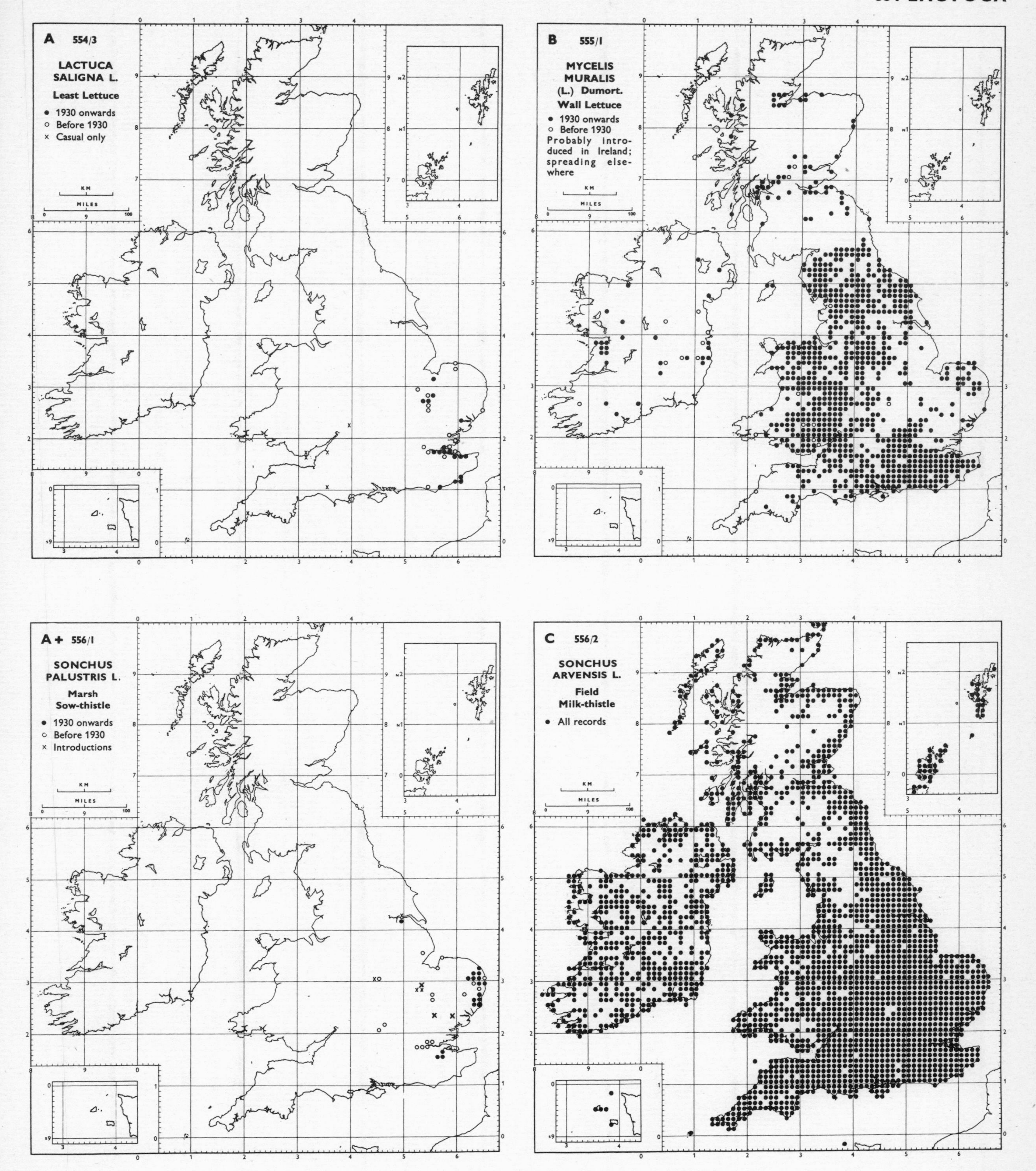
A 554/3
LACTUCA SALIGNA L.
Least Lettuce
• 1930 onwards
○ Before 1930
× Casual only
B 555/1
MYCELIS MURALIS (L.) Dumort.
Wall Lettuce
• 1930 onwards
○ Before 1930
Probably introduced in Ireland; spreading elsewhere
A + 556/1
SONCHUS PALUSTRIS L.
Marsh Sow-thistle
• 1930 onwards
○ Before 1930
× Introductions
C 556/2
SONCHUS ARVENSIS L.
Field Milk-thistle
• All records

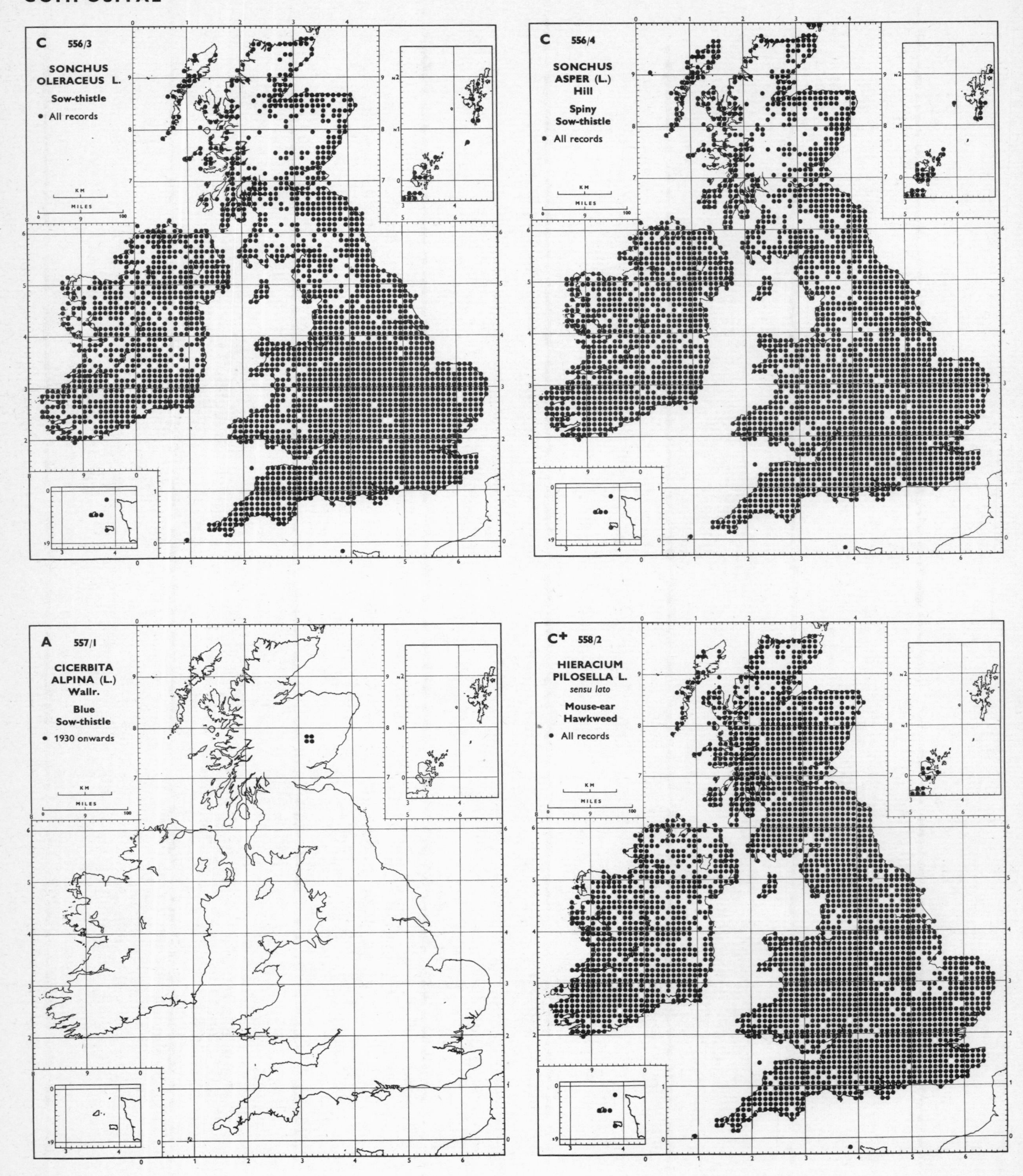
C 556/3
SONCHUS OLERACEUS L.
Sow-thistle
• All records
C 556/4
SONCHUS ASPER (L.) Hill
Spiny Sow-thistle
• All records
A 557/1
CICERBITA ALPINA (L.) Wallr.
Blue Sow-thistle
• 1930 onwards
C+ 558/2
HIERACIUM PILOSELLA L.
sensu lato
Mouse-ear Hawkweed
• All records

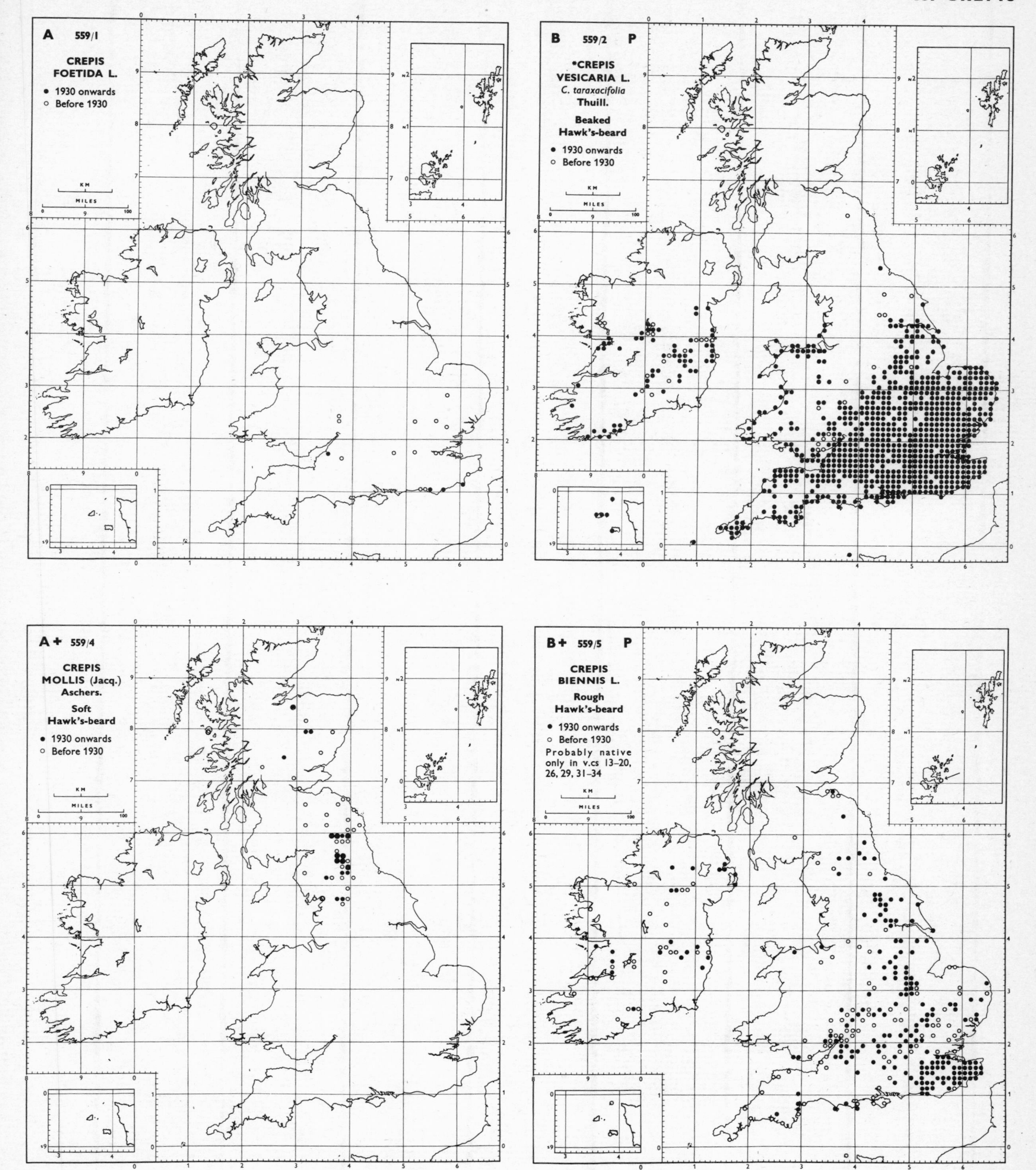
A 559/1
CREPIS FOETIDA L.
• 1930 onwards
○ Before 1930
B 559/2 P
*CREPIS VESICARIA L.
C. taraxacifolia Thuill.
Beaked Hawk's-beard
• 1930 onwards
○ Before 1930
A+ 559/4
CREPIS MOLLIS (Jacq.) Aschers.
Soft Hawk's-beard
• 1930 onwards
○ Before 1930
B+ 559/5 P
CREPIS BIENNIS L.
Rough Hawk's-beard
• 1930 onwards
○ Before 1930
Probably native only in v.cs 13–20, 26, 29, 31–34
KM
MILES

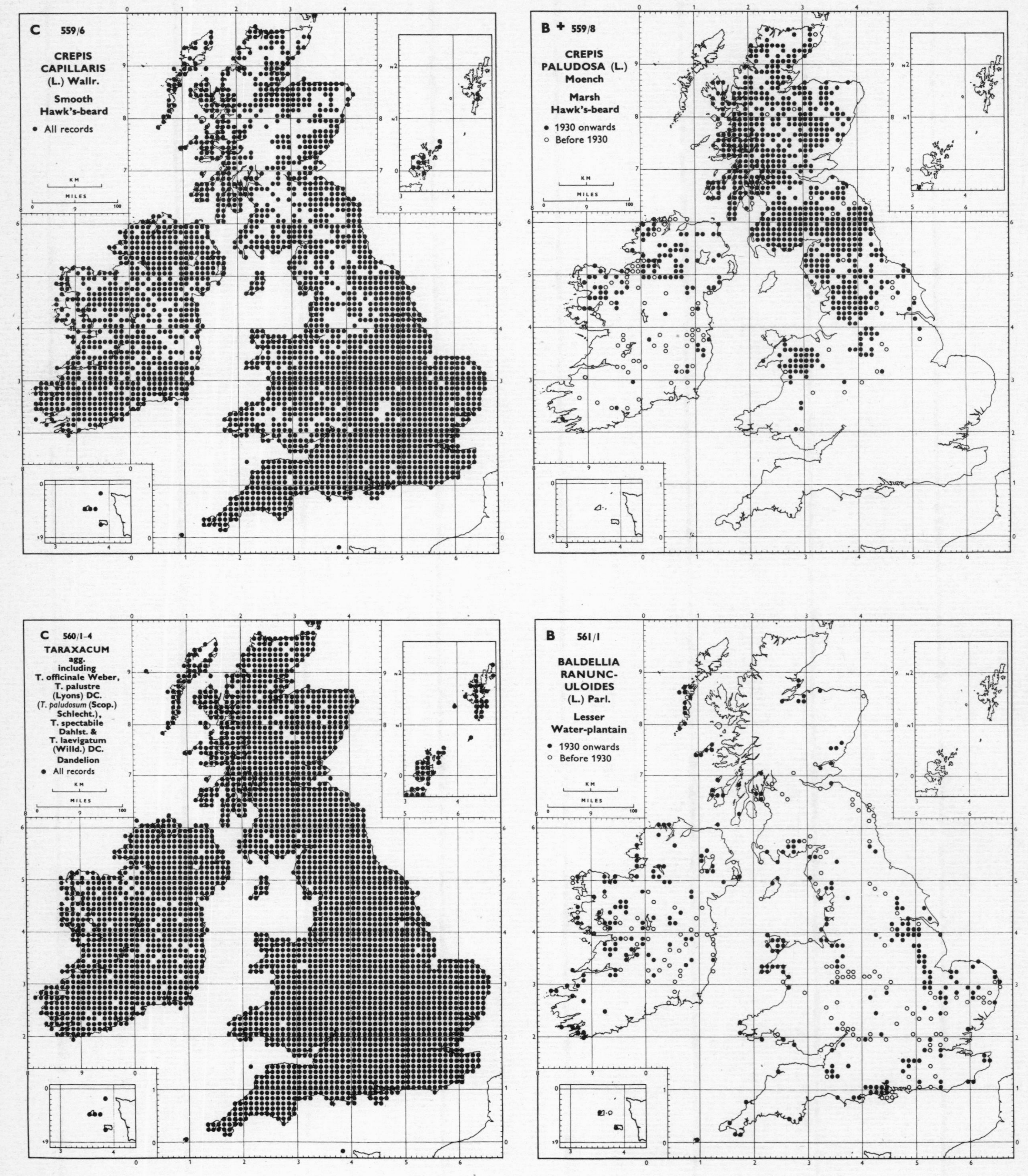
C 559/6
CREPIS CAPILLARIS (L.) Wallr.
Smooth Hawk's-beard
All records
KM
MILES
B + 559/8
CREPIS PALUDOSA (L.) Moench
Marsh Hawk's-beard
1930 onwards
Before 1930
C 560/1-4
TARAXACUM agg. including T. officinale Weber, T. palustre (Lyons) DC. (T. paludosum (Scop.) Schlecht.), T. spectabile Dahlst. & T. laevigatum (Willd.) DC.
Dandelion
All records
B 561/1
BALDELLIA RANUNCULOIDES (L.) Parl.
Lesser Water-plantain
1930 onwards
Before 1930

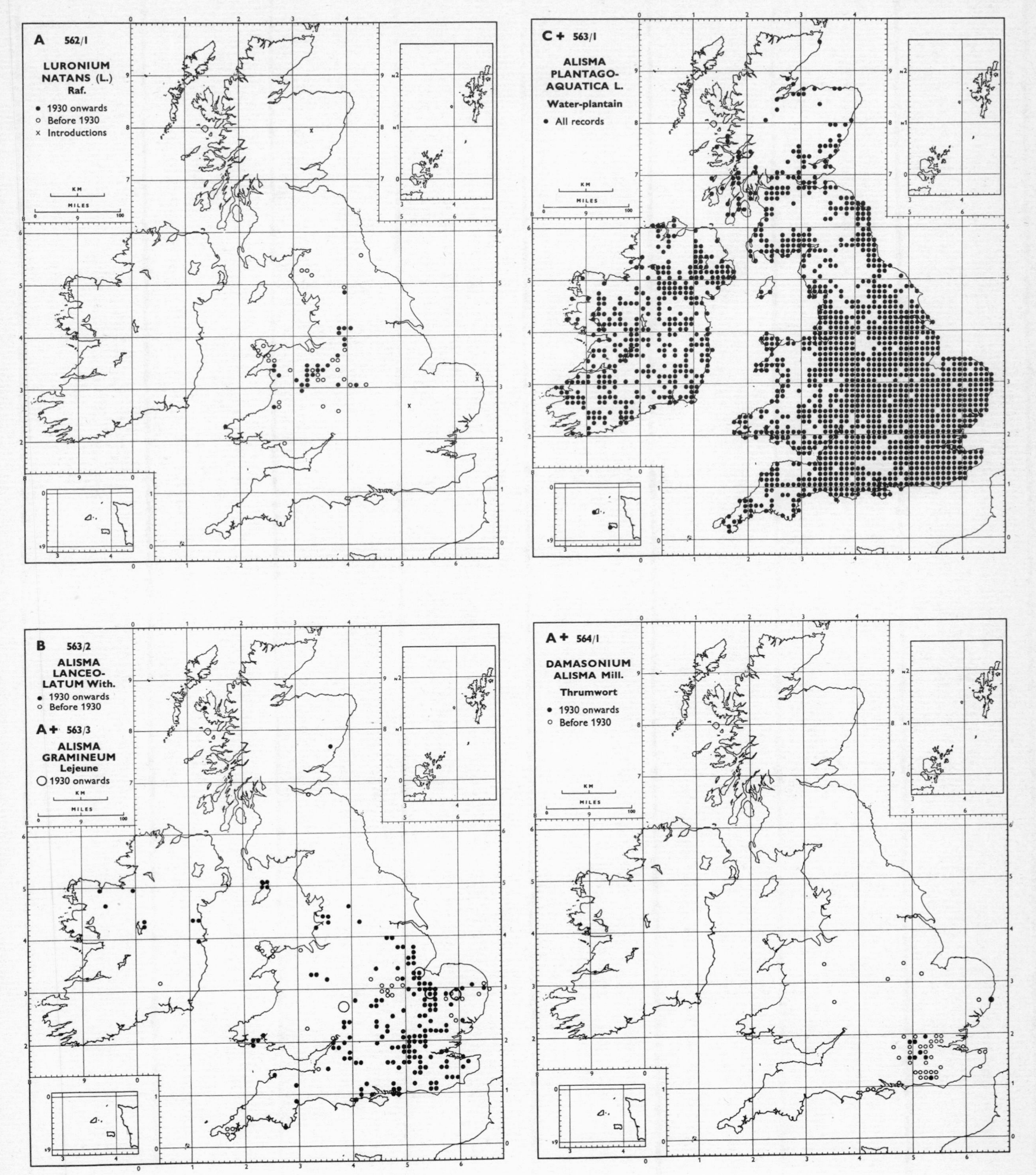
A 562/1
LURONIUM NATANS (L.) Raf.
1930 onwards
Before 1930
Introductions
KM
MILES
C + 563/1
ALISMA PLANTAGO-AQUATICA L.
Water-plantain
All records
B 563/2
ALISMA LANCEOLATUM With.
1930 onwards
Before 1930
A + 563/3
ALISMA GRAMINEUM Lejeune
1930 onwards
A + 564/1
DAMASONIUM ALISMA Mill.
Thrumwort
1930 onwards
Before 1930

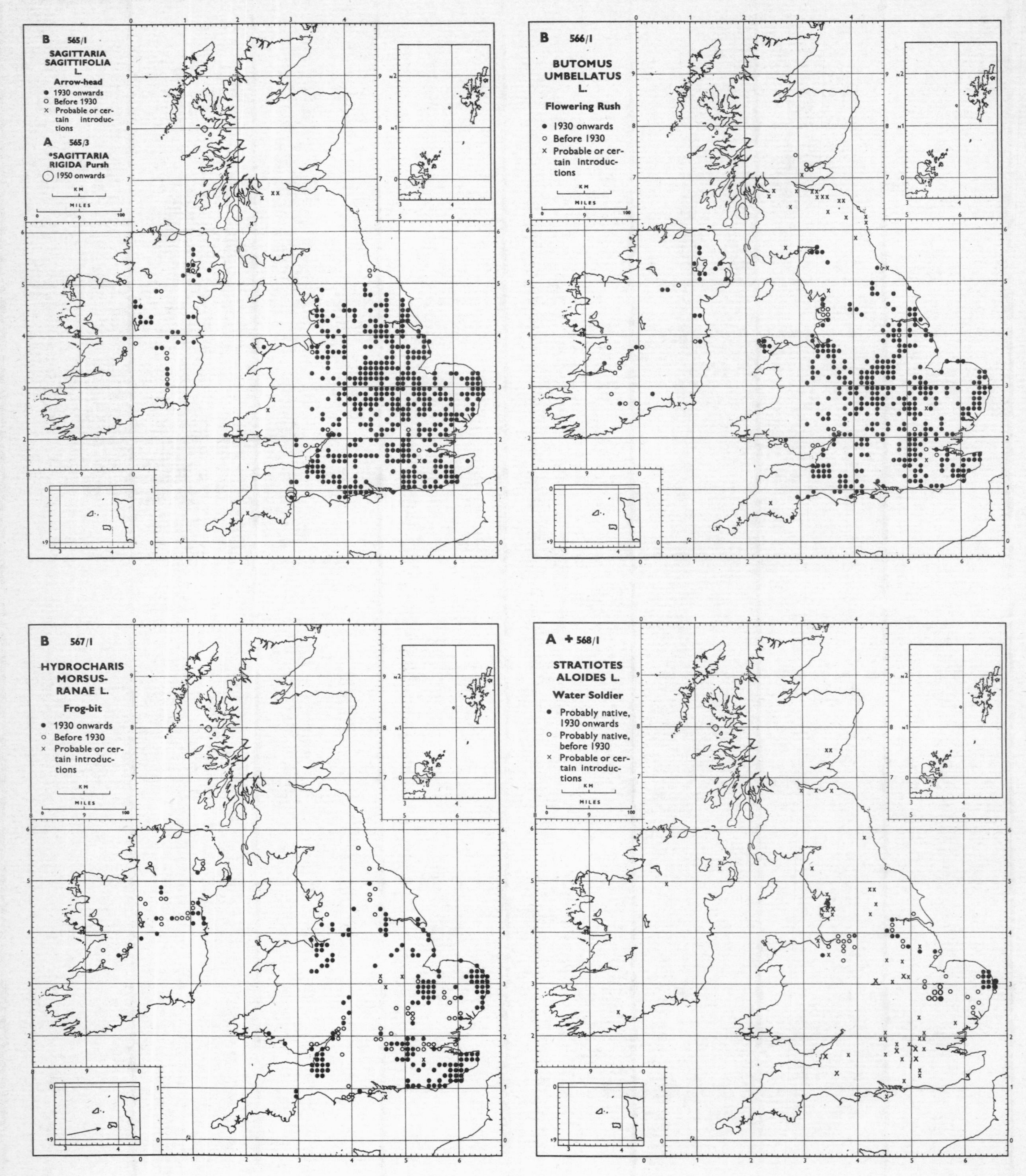
B 565/1
SAGITTARIA SAGITTIFOLIA L.
Arrow-head
1930 onwards
Before 1930
Probable or certain introductions
A 565/3
*SAGITTARIA RIGIDA Pursh
1950 onwards
KM
MILES
B 566/1
BUTOMUS UMBELLATUS L.
Flowering Rush
1930 onwards
Before 1930
Probable or certain introductions
KM
MILES
B 567/1
HYDROCHARIS MORSUS-RANAE L.
Frog-bit
1930 onwards
Before 1930
Probable or certain introductions
KM
MILES
A + 568/1
STRATIOTES ALOIDES L.
Water Soldier
Probably native, 1930 onwards
Probably native, before 1930
Probable or certain introductions
KM
MILES

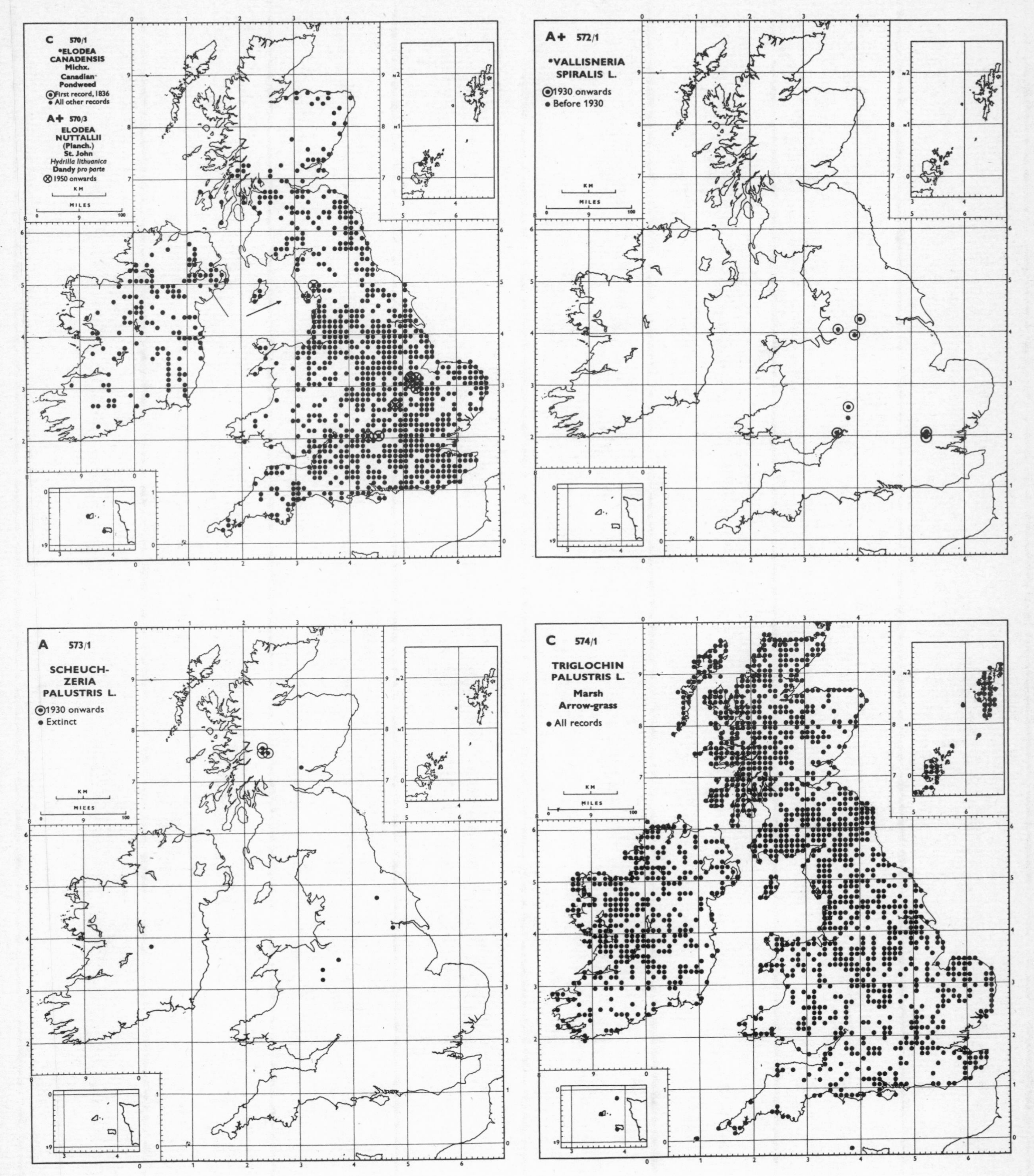
C 570/1
•ELODEA CANADENSIS Michx.
Canadian Pondweed
First record, 1836
All other records
A+ 570/3
ELODEA NUTTALLII (Planch.) St. John
Hydrilla lithuanica Dandy *pro parte*
1950 onwards
KM
MILES
A+ 572/1
•VALLISNERIA SPIRALIS L.
1930 onwards
Before 1930
A 573/1
SCHEUCH-ZERIA PALUSTRIS L.
1930 onwards
Extinct
C 574/1
TRIGLOCHIN PALUSTRIS L.
Marsh Arrow-grass
All records

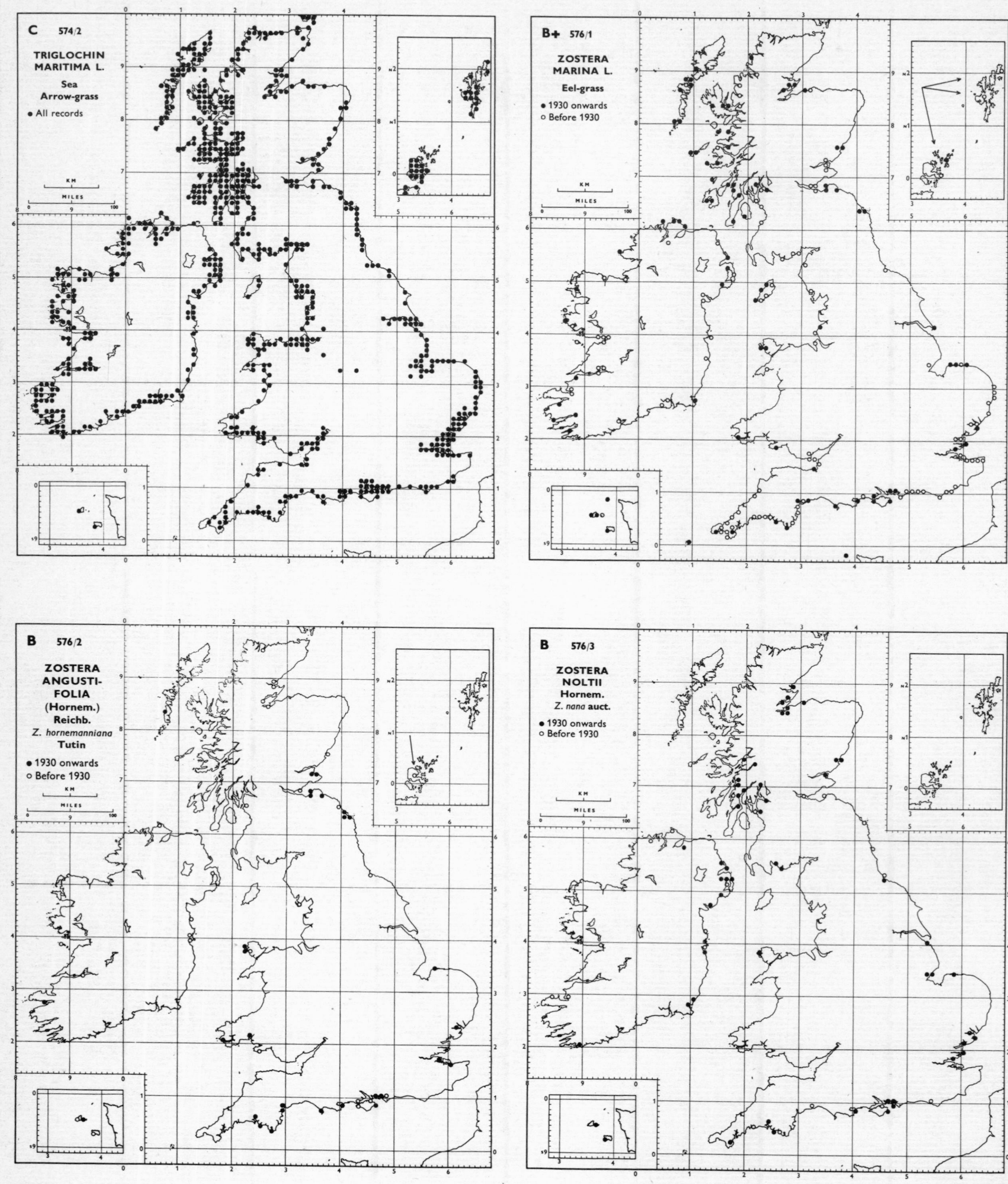
C 574/2
TRIGLOCHIN MARITIMA L.
Sea Arrow-grass
• All records
KM
MILES
B+ 576/1
ZOSTERA MARINA L.
Eel-grass
• 1930 onwards
○ Before 1930
B 576/2
ZOSTERA ANGUSTIFOLIA (Hornem.) Reichb.
Z. hornemanniana Tutin
• 1930 onwards
○ Before 1930
B 576/3
ZOSTERA NOLTII Hornem.
Z. nana auct.
• 1930 onwards
○ Before 1930

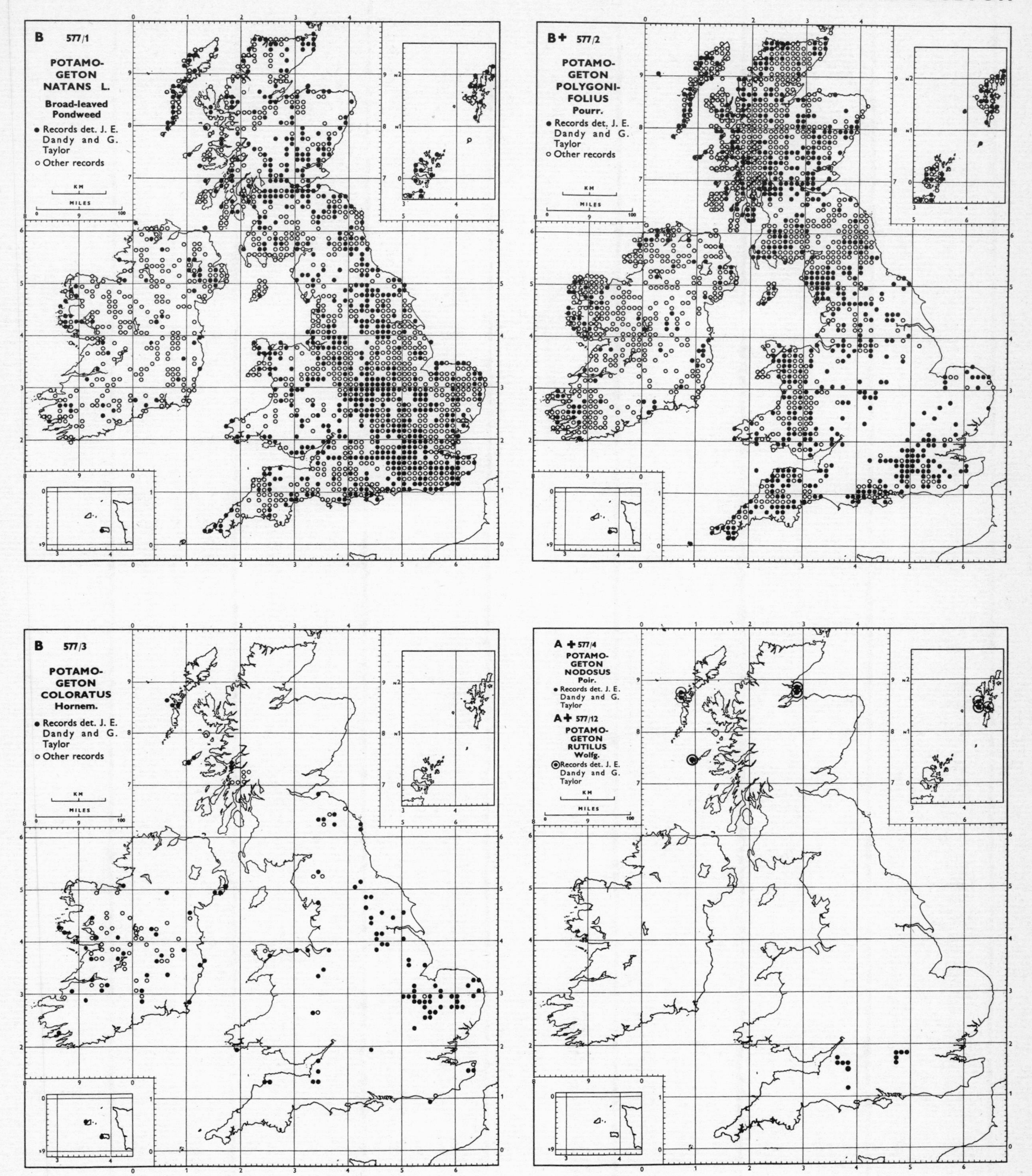
B 577/1
POTAMOGETON NATANS L.
Broad-leaved Pondweed
● Records det. J. E. Dandy and G. Taylor
○ Other records
B+ 577/2
POTAMOGETON POLYGONIFOLIUS Pourr.
● Records det. J. E. Dandy and G. Taylor
○ Other records
B 577/3
POTAMOGETON COLORATUS Hornem.
● Records det. J. E. Dandy and G. Taylor
○ Other records
A+ 577/4
POTAMOGETON NODOSUS Poir.
● Records det. J. E. Dandy and G. Taylor
A+ 577/12
POTAMOGETON RUTILUS Wolfg.
◉ Records det. J. E. Dandy and G. Taylor
KM
MILES

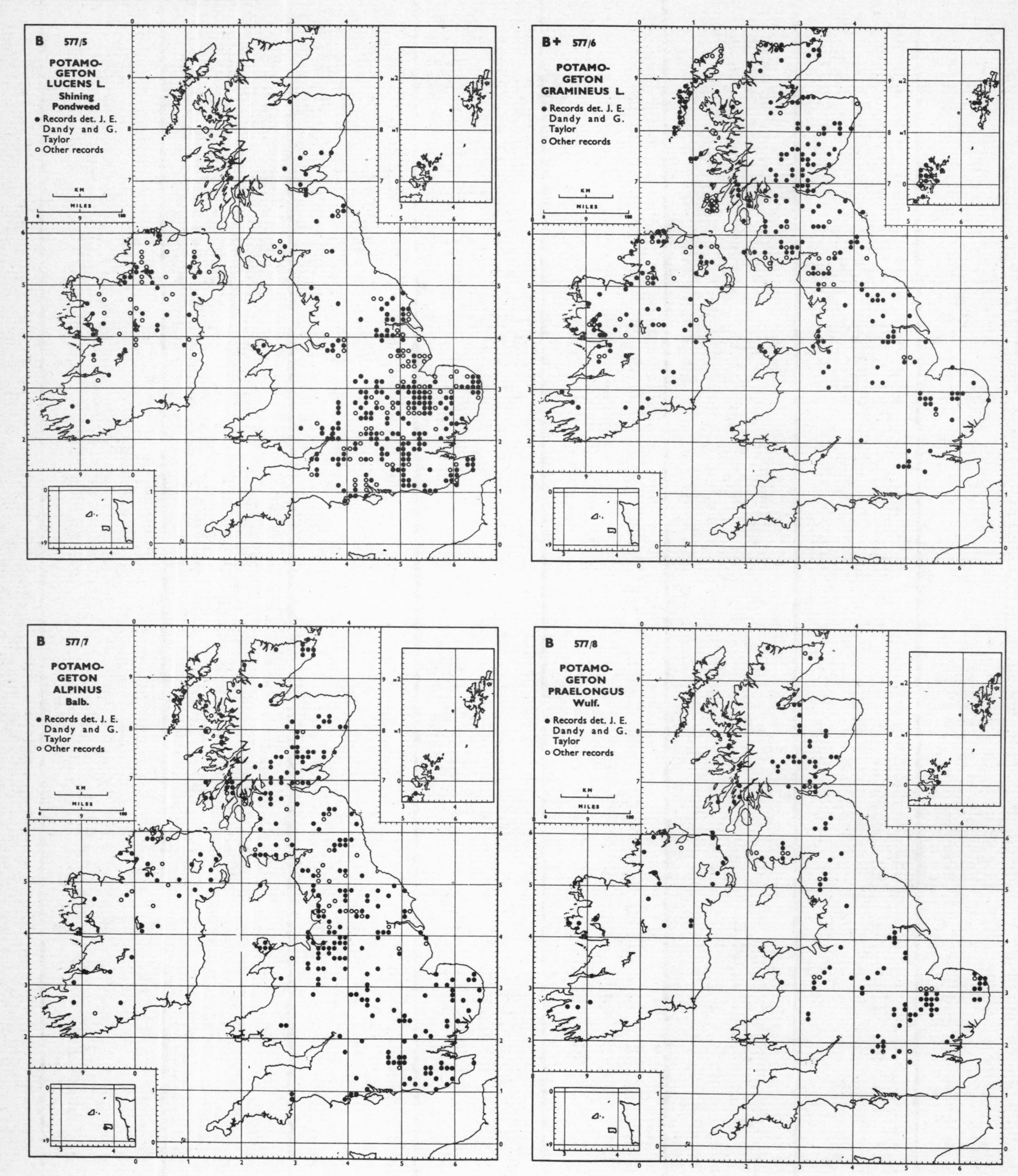
B 577/5
POTAMOGETON LUCENS L.
Shining Pondweed
● Records det. J. E. Dandy and G. Taylor
○ Other records
B+ 577/6
POTAMOGETON GRAMINEUS L.
● Records det. J. E. Dandy and G. Taylor
○ Other records
B 577/7
POTAMOGETON ALPINUS Balb.
● Records det. J. E. Dandy and G. Taylor
○ Other records
B 577/8
POTAMOGETON PRAELONGUS Wulf.
● Records det. J. E. Dandy and G. Taylor
○ Other records

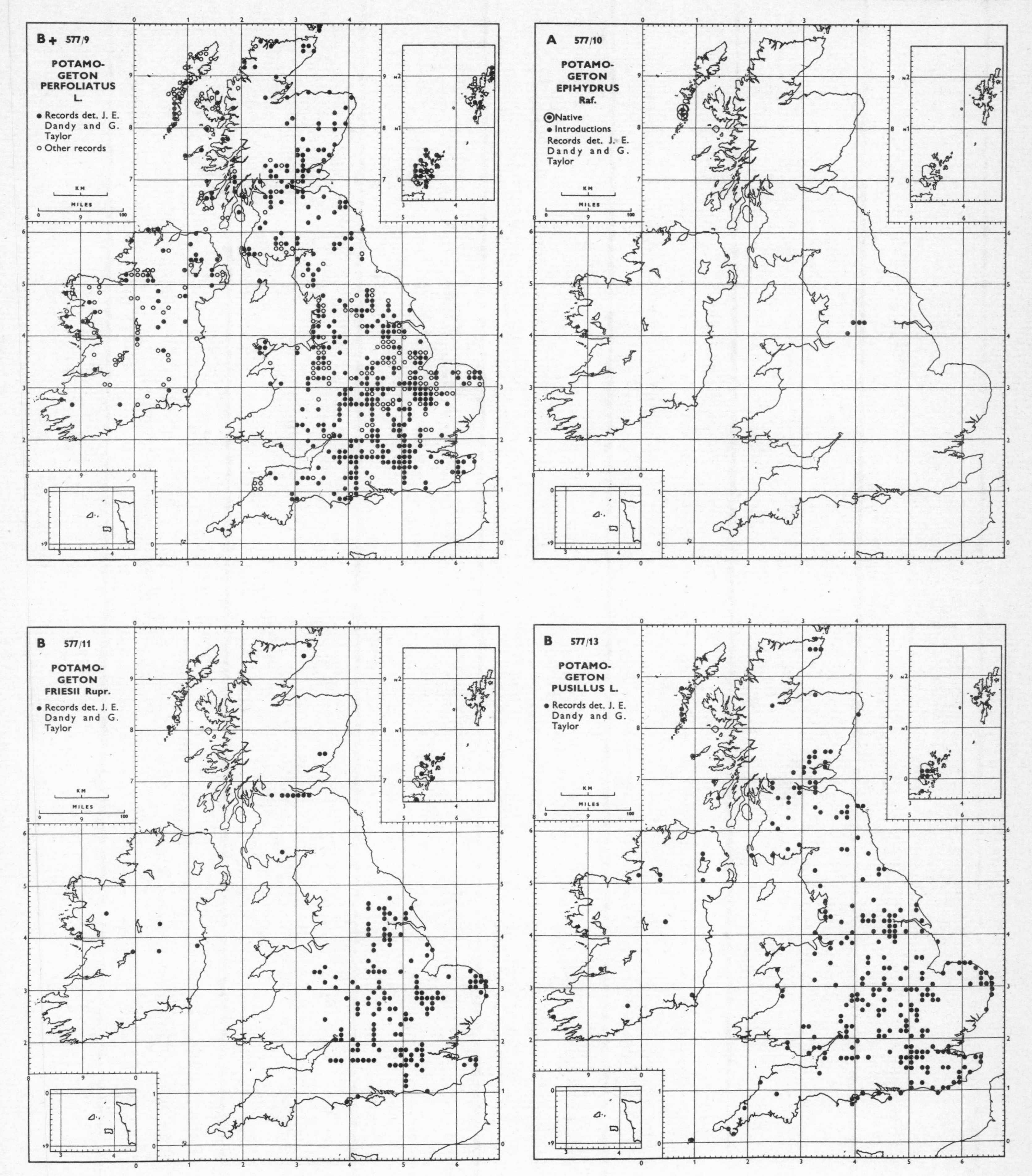
B + 577/9
POTAMOGETON PERFOLIATUS L.
• Records det. J. E. Dandy and G. Taylor
○ Other records
KM
MILES
A 577/10
POTAMOGETON EPIHYDRUS Raf.
Native
• Introductions
Records det. J. E. Dandy and G. Taylor
B 577/11
POTAMOGETON FRIESII Rupr.
• Records det. J. E. Dandy and G. Taylor
B 577/13
POTAMOGETON PUSILLUS L.
• Records det. J. E. Dandy and G. Taylor

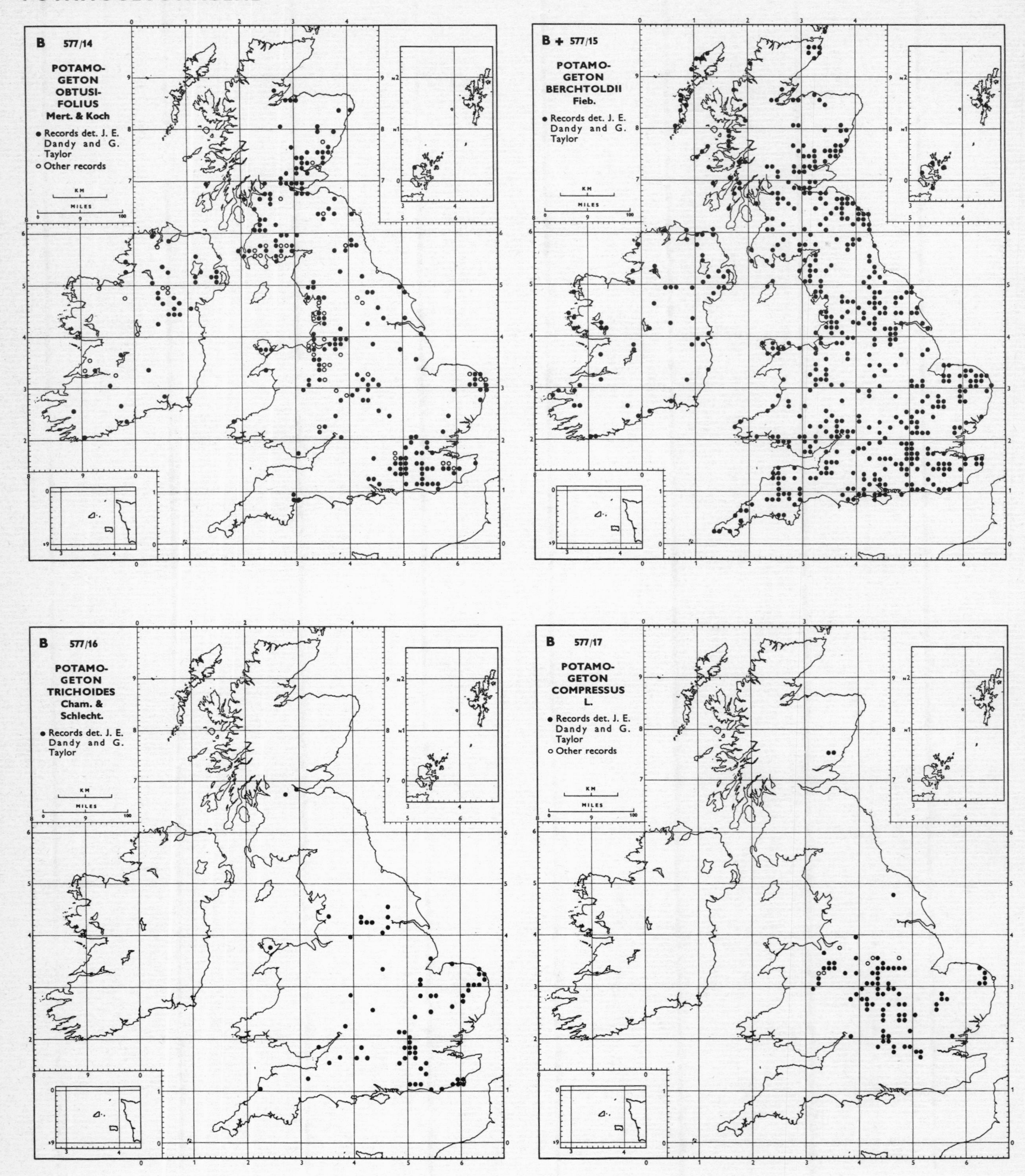
B 577/14
POTAMOGETON OBTUSIFOLIUS Mert. & Koch
● Records det. J. E. Dandy and G. Taylor
○ Other records
KM
MILES
B + 577/15
POTAMOGETON BERCHTOLDII Fieb.
● Records det. J. E. Dandy and G. Taylor
B 577/16
POTAMOGETON TRICHOIDES Cham. & Schlecht.
● Records det. J. E. Dandy and G. Taylor
B 577/17
POTAMOGETON COMPRESSUS L.
● Records det. J. E. Dandy and G. Taylor
○ Other records

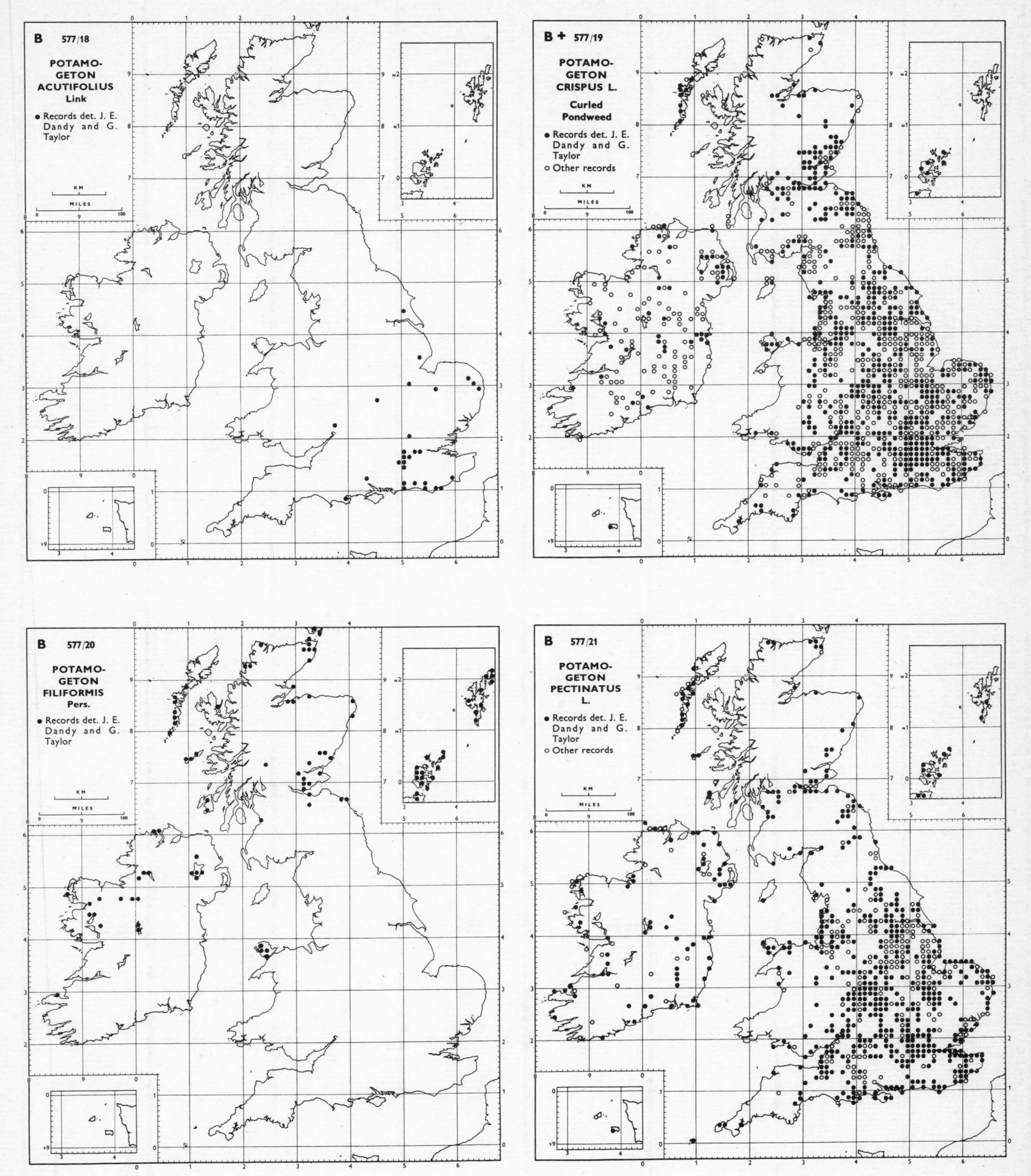
B 577/18
POTAMOGETON ACUTIFOLIUS Link
● Records det. J. E. Dandy and G. Taylor
B + 577/19
POTAMOGETON CRISPUS L.
Curled Pondweed
● Records det. J. E. Dandy and G. Taylor
○ Other records
B 577/20
POTAMOGETON FILIFORMIS Pers.
● Records det. J. E. Dandy and G. Taylor
B 577/21
POTAMOGETON PECTINATUS L.
● Records det. J. E. Dandy and G. Taylor
○ Other records

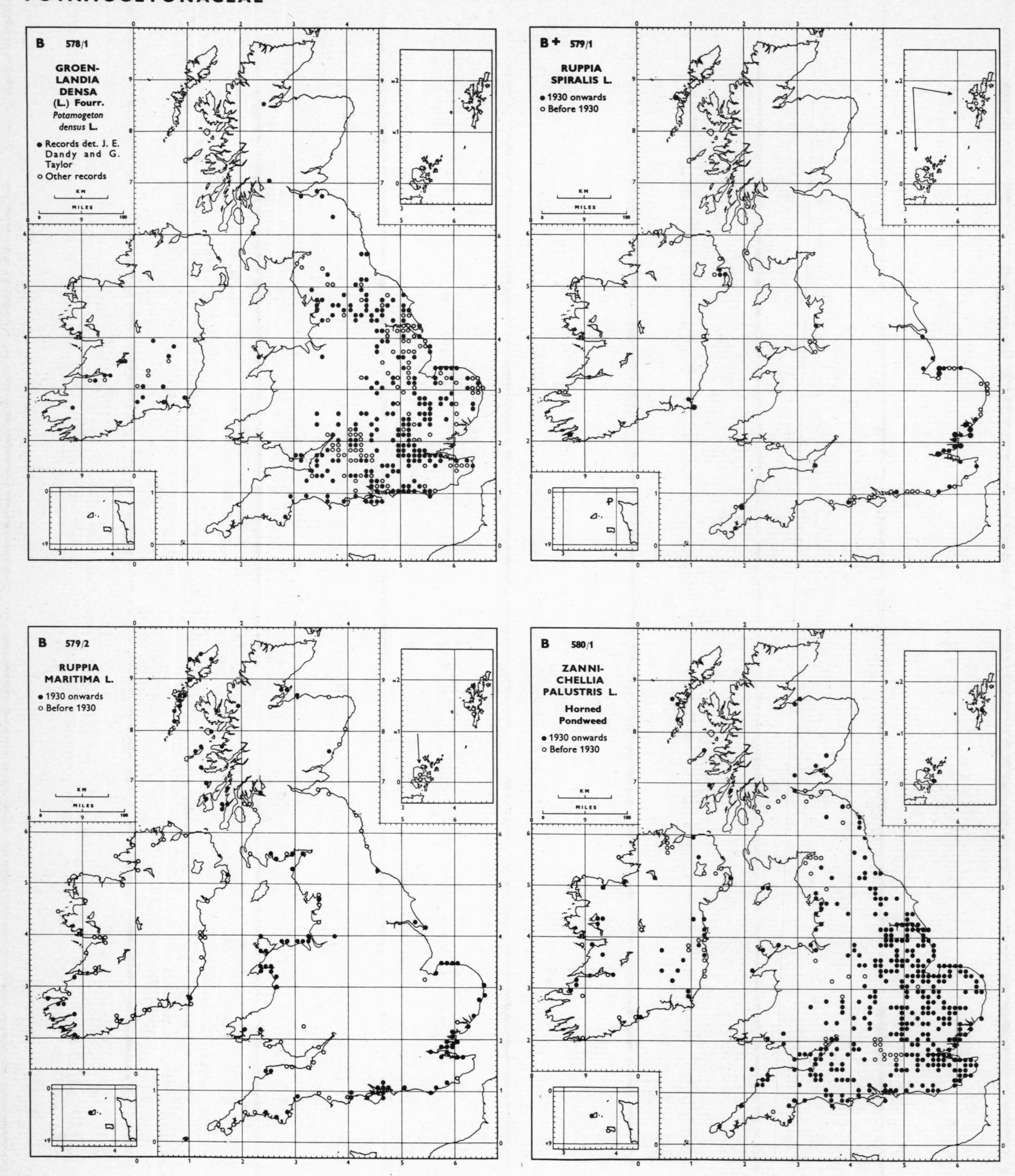
B 578/1
GROENLANDIA DENSA (L.) Fourr.
Potamogeton densus L.
● Records det. J. E. Dandy and G. Taylor
○ Other records
KM
MILES
B+ 579/1
RUPPIA SPIRALIS L.
● 1930 onwards
○ Before 1930
B 579/2
RUPPIA MARITIMA L.
● 1930 onwards
○ Before 1930
B 580/1
ZANNICHELLIA PALUSTRIS L.
Horned Pondweed
● 1930 onwards
○ Before 1930

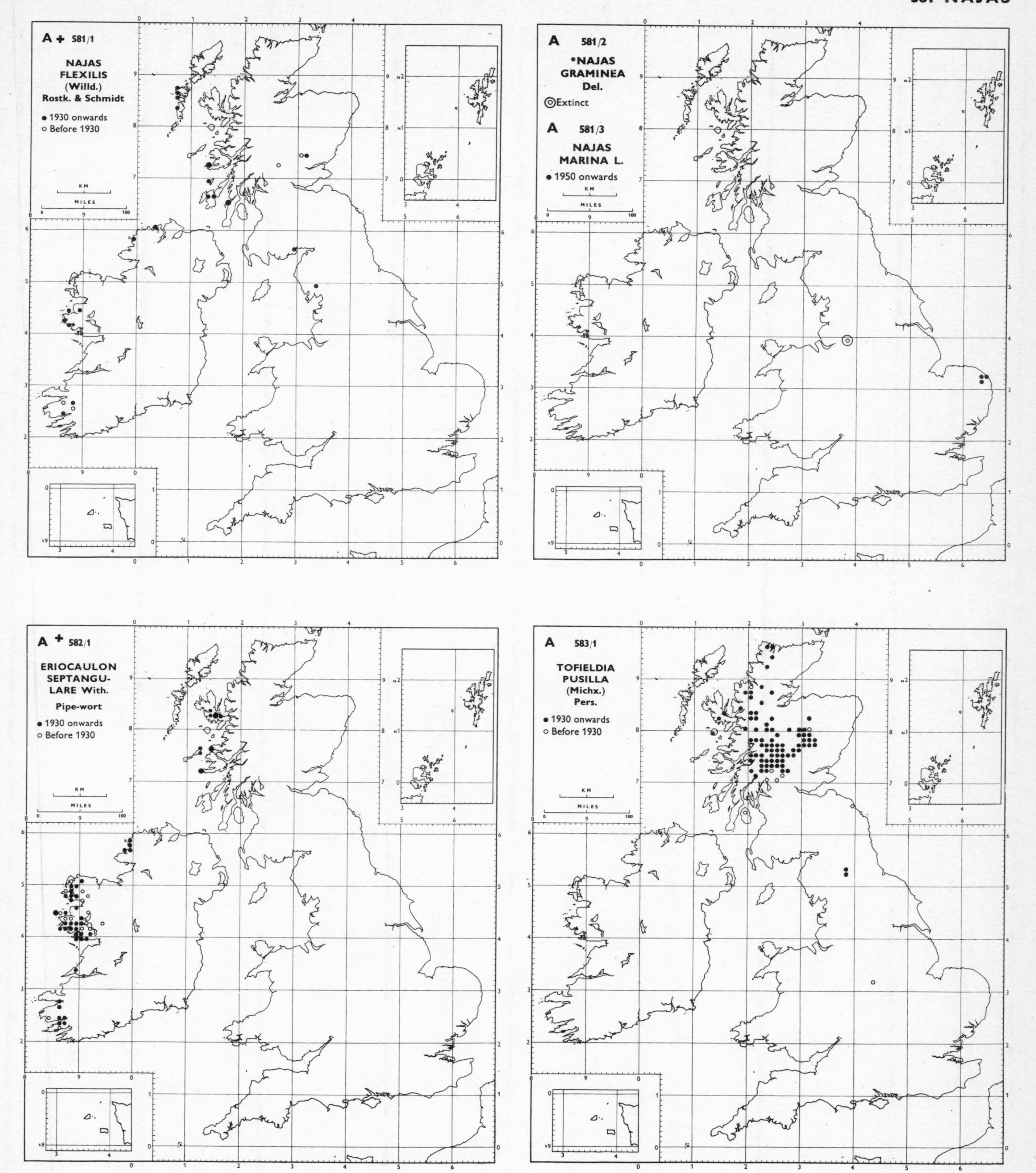
A + 581/1
NAJAS FLEXILIS (Willd.) Rostk. & Schmidt
● 1930 onwards
○ Before 1930
A 581/2
*NAJAS GRAMINEA Del.
◎Extinct
A 581/3
NAJAS MARINA L.
● 1950 onwards
A + 582/1
ERIOCAULON SEPTANGULARE With.
Pipe-wort
● 1930 onwards
○ Before 1930
A 583/1
TOFIELDIA PUSILLA (Michx.) Pers.
● 1930 onwards
○ Before 1930
KM
MILES

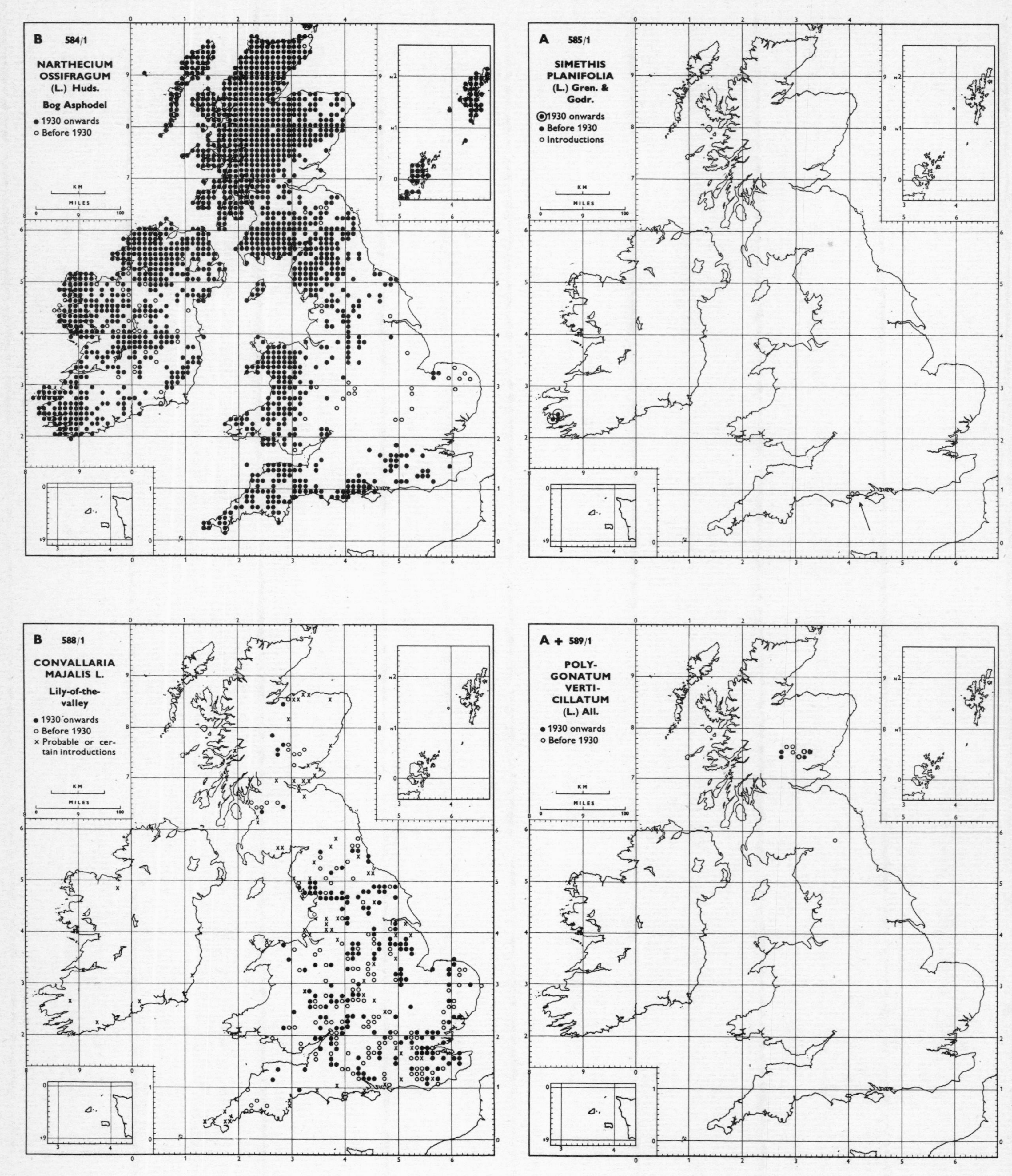
B 584/1
NARTHECIUM OSSIFRAGUM (L.) Huds.
Bog Asphodel
● 1930 onwards
○ Before 1930
A 585/1
SIMETHIS PLANIFOLIA (L.) Gren. & Godr.
◉ 1930 onwards
● Before 1930
○ Introductions
B 588/1
CONVALLARIA MAJALIS L.
Lily-of-the-valley
● 1930 onwards
○ Before 1930
× Probable or certain introductions
A + 589/1
POLYGONATUM VERTICILLATUM (L.) All.
● 1930 onwards
○ Before 1930
KM
MILES

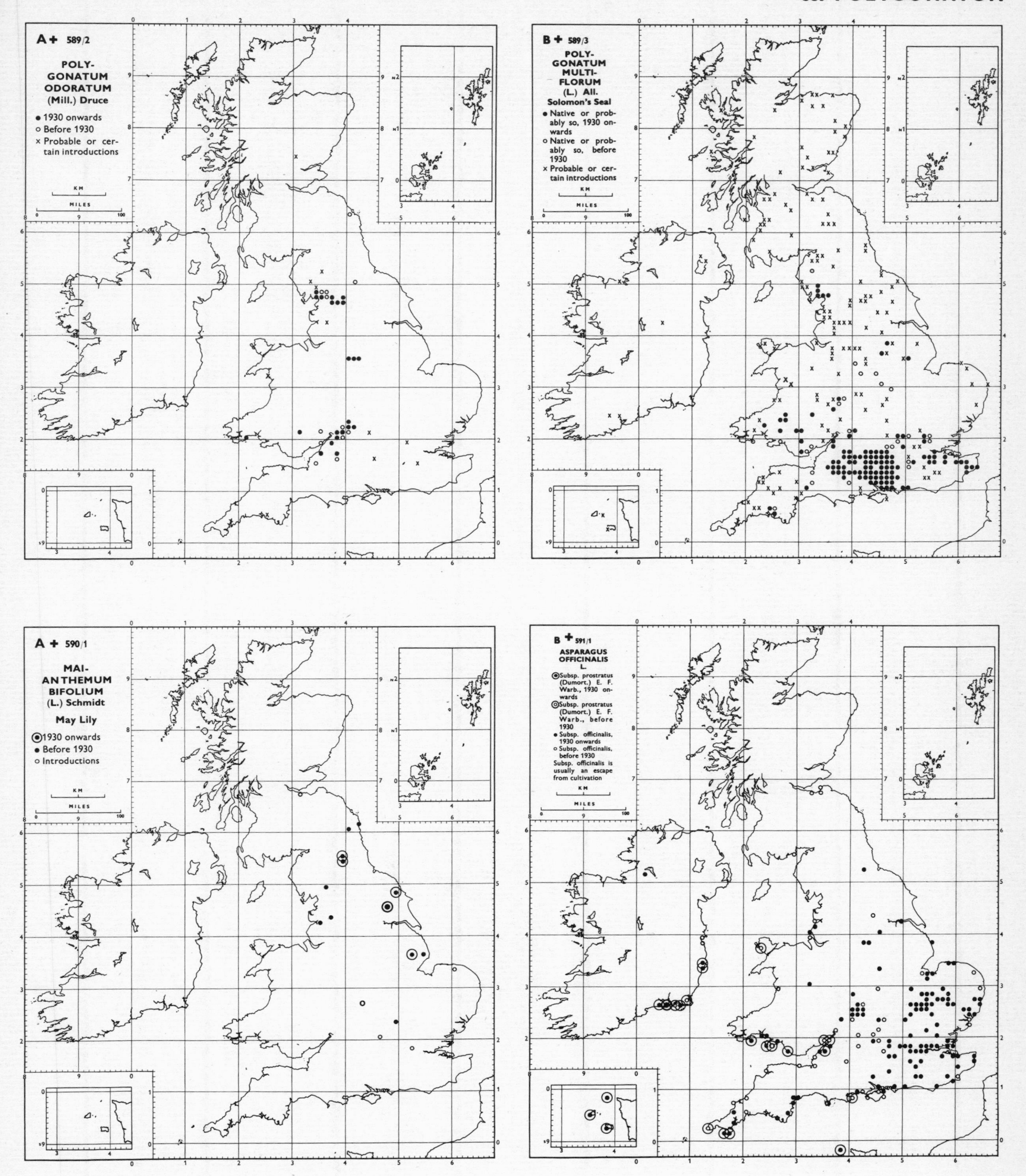
A + 589/2
POLYGONATUM ODORATUM (Mill.) Druce
• 1930 onwards
○ Before 1930
× Probable or certain introductions
KM
MILES
0 9 100
B + 589/3
POLYGONATUM MULTIFLORUM (L.) All.
Solomon's Seal
• Native or probably so, 1930 onwards
○ Native or probably so, before 1930
× Probable or certain introductions
KM
MILES
0 9 100
A + 590/1
MAIANTHEMUM BIFOLIUM (L.) Schmidt
May Lily
⦿ 1930 onwards
• Before 1930
○ Introductions
KM
MILES
0 9 100
B + 591/1
ASPARAGUS OFFICINALIS L.
⦿ Subsp. prostratus (Dumort.) E. F. Warb., 1930 onwards
◎ Subsp. prostratus (Dumort.) E. F. Warb., before 1930
• Subsp. officinalis, 1930 onwards
○ Subsp. officinalis, before 1930
Subsp. officinalis is usually an escape from cultivation
KM
MILES
0 9 100

LILIACEAE

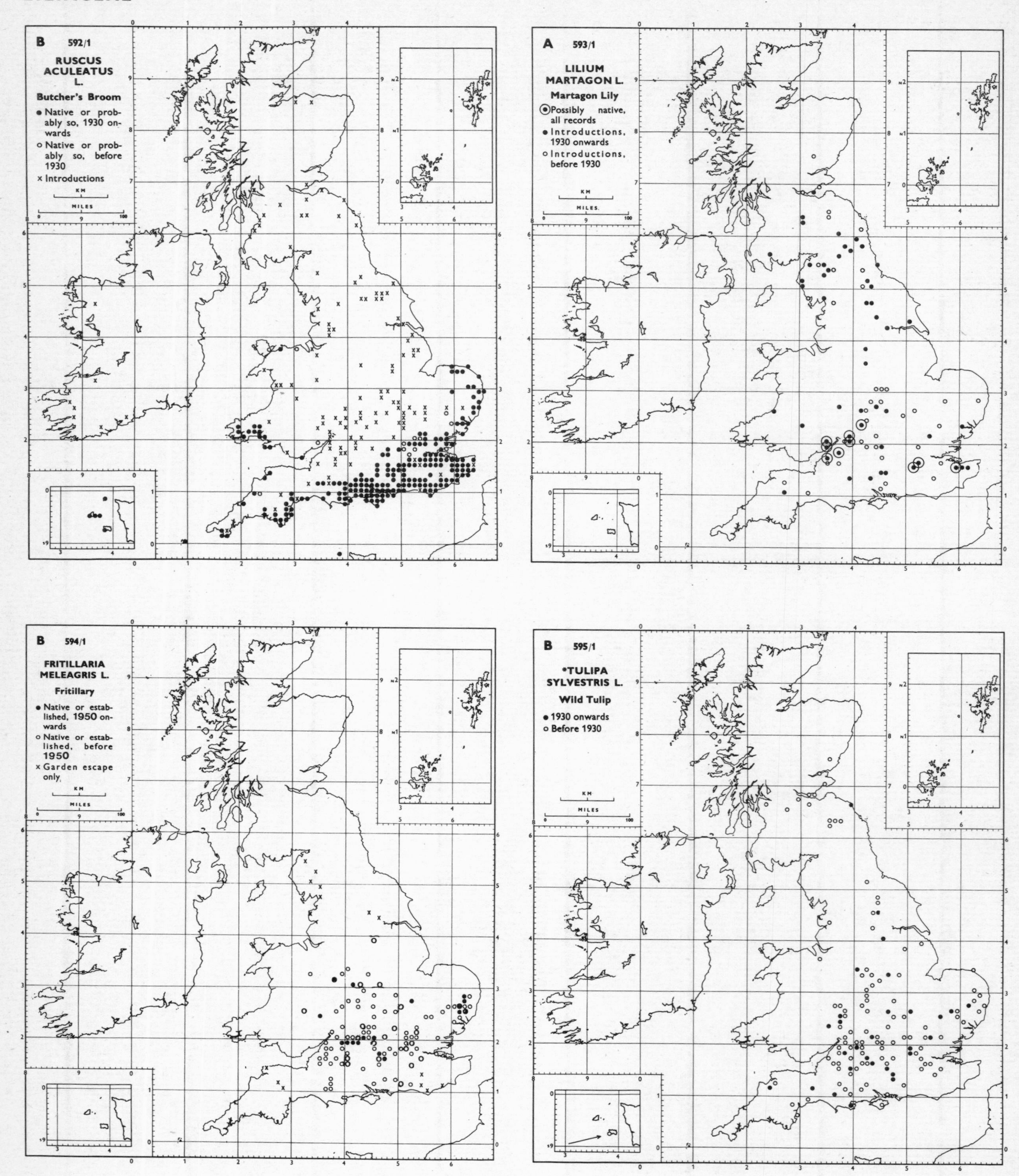

B 592/1
RUSCUS ACULEATUS L.
Butcher's Broom
● Native or probably so, 1930 onwards
○ Native or probably so, before 1930
× Introductions
KM
MILES
A 593/1
LILIUM MARTAGON L.
Martagon Lily
⊙ Possibly native, all records
● Introductions, 1930 onwards
○ Introductions, before 1930
KM
MILES
B 594/1
FRITILLARIA MELEAGRIS L.
Fritillary
● Native or established, 1950 onwards
○ Native or established, before 1950
× Garden escape only
KM
MILES
B 595/1
*TULIPA SYLVESTRIS L.
Wild Tulip
● 1930 onwards
○ Before 1930
KM
MILES

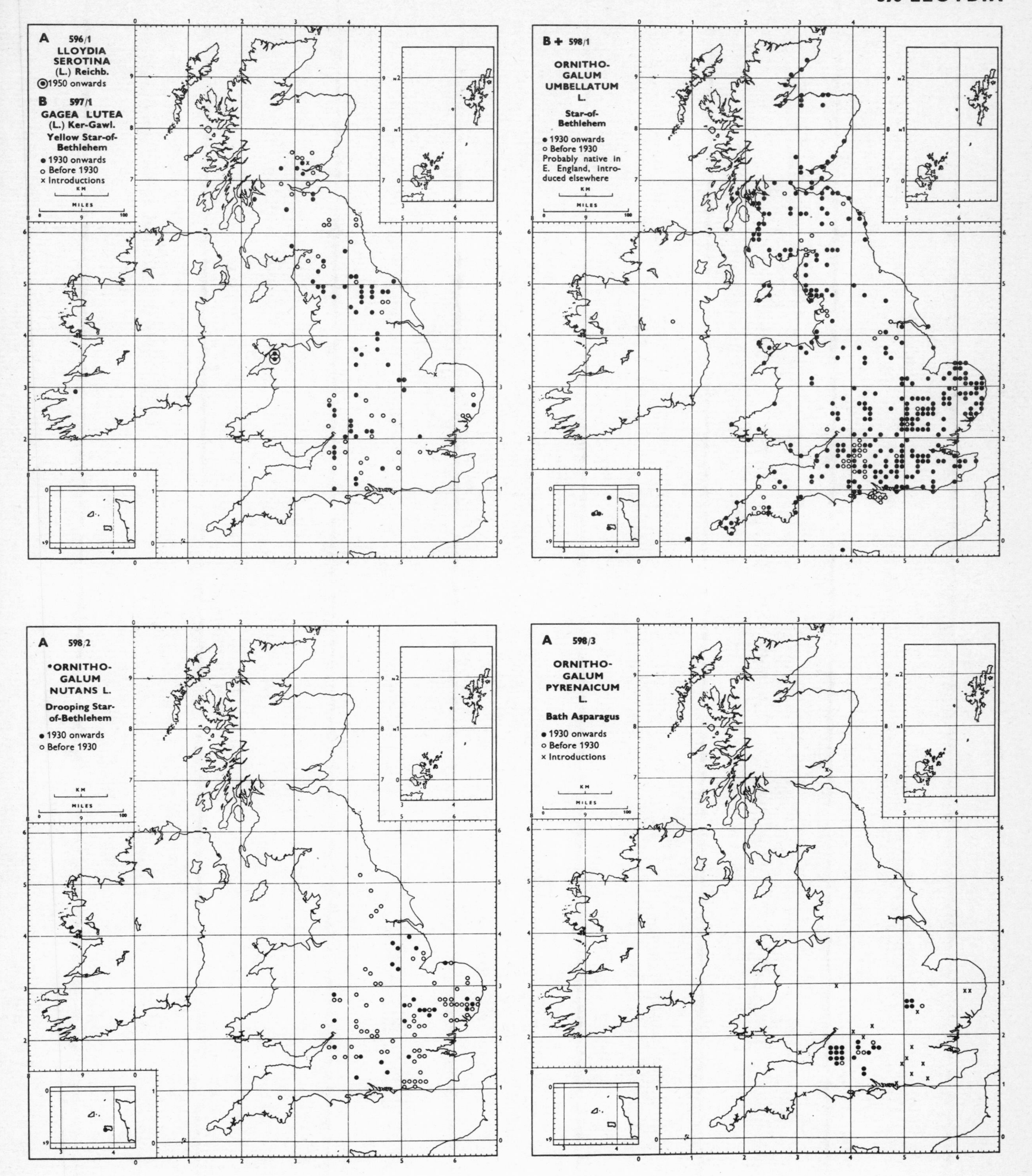
A 596/1
LLOYDIA SEROTINA (L.) Reichb.
1950 onwards
B 597/1
GAGEA LUTEA (L.) Ker-Gawl.
Yellow Star-of-Bethlehem
• 1930 onwards
○ Before 1930
× Introductions
B + 598/1
ORNITHOGALUM UMBELLATUM L.
Star-of-Bethlehem
• 1930 onwards
○ Before 1930
Probably native in E. England, introduced elsewhere
A 598/2
*ORNITHOGALUM NUTANS L.
Drooping Star-of-Bethlehem
• 1930 onwards
○ Before 1930
A 598/3
ORNITHOGALUM PYRENAICUM L.
Bath Asparagus
• 1930 onwards
○ Before 1930
× Introductions

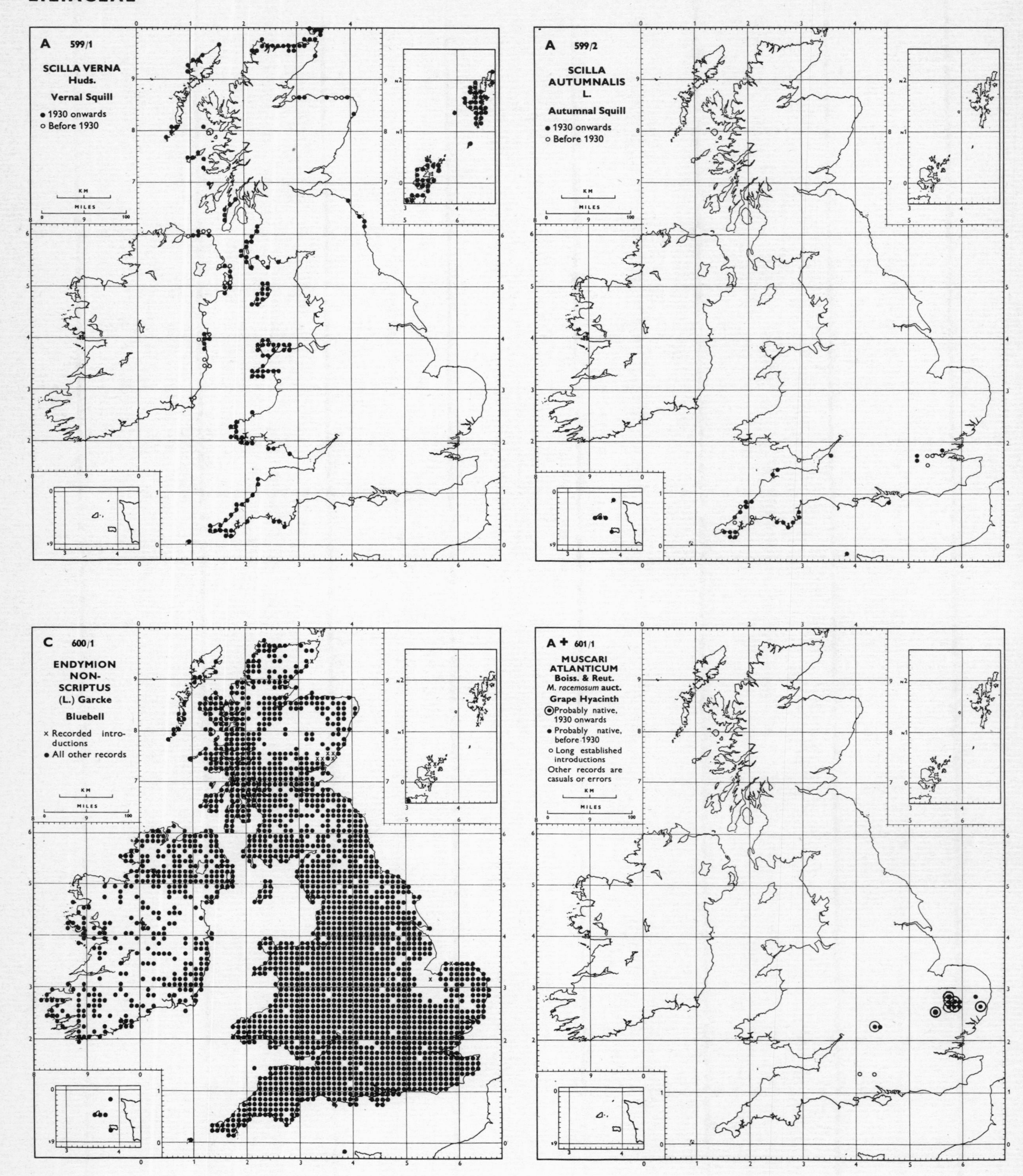
A 599/1
SCILLA VERNA
Huds.
Vernal Squill
● 1930 onwards
○ Before 1930
A 599/2
SCILLA
AUTUMNALIS
L.
Autumnal Squill
● 1930 onwards
○ Before 1930
C 600/1
ENDYMION
NON-
SCRIPTUS
(L.) Garcke
Bluebell
× Recorded introductions
● All other records
A + 601/1
MUSCARI
ATLANTICUM
Boiss. & Reut.
M. racemosum auct.
Grape Hyacinth
◉ Probably native, 1930 onwards
● Probably native, before 1930
○ Long established introductions
Other records are casuals or errors
KM
MILES

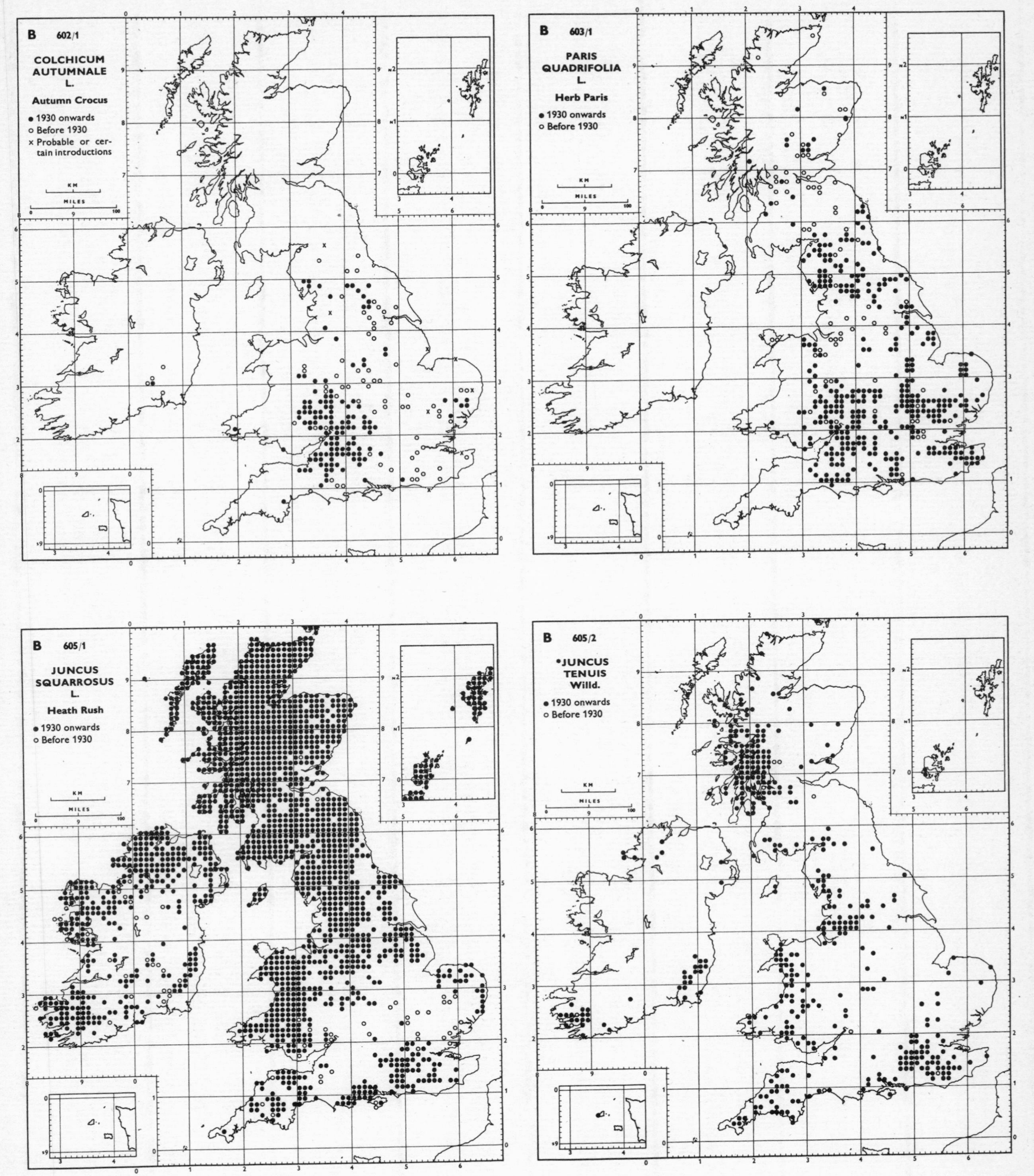
B 602/1
COLCHICUM AUTUMNALE L.
Autumn Crocus
● 1930 onwards
○ Before 1930
x Probable or certain introductions
KM
MILES
B 603/1
PARIS QUADRIFOLIA L.
Herb Paris
● 1930 onwards
○ Before 1930
B 605/1
JUNCUS SQUARROSUS L.
Heath Rush
● 1930 onwards
○ Before 1930
B 605/2
•JUNCUS TENUIS Willd.
● 1930 onwards
○ Before 1930

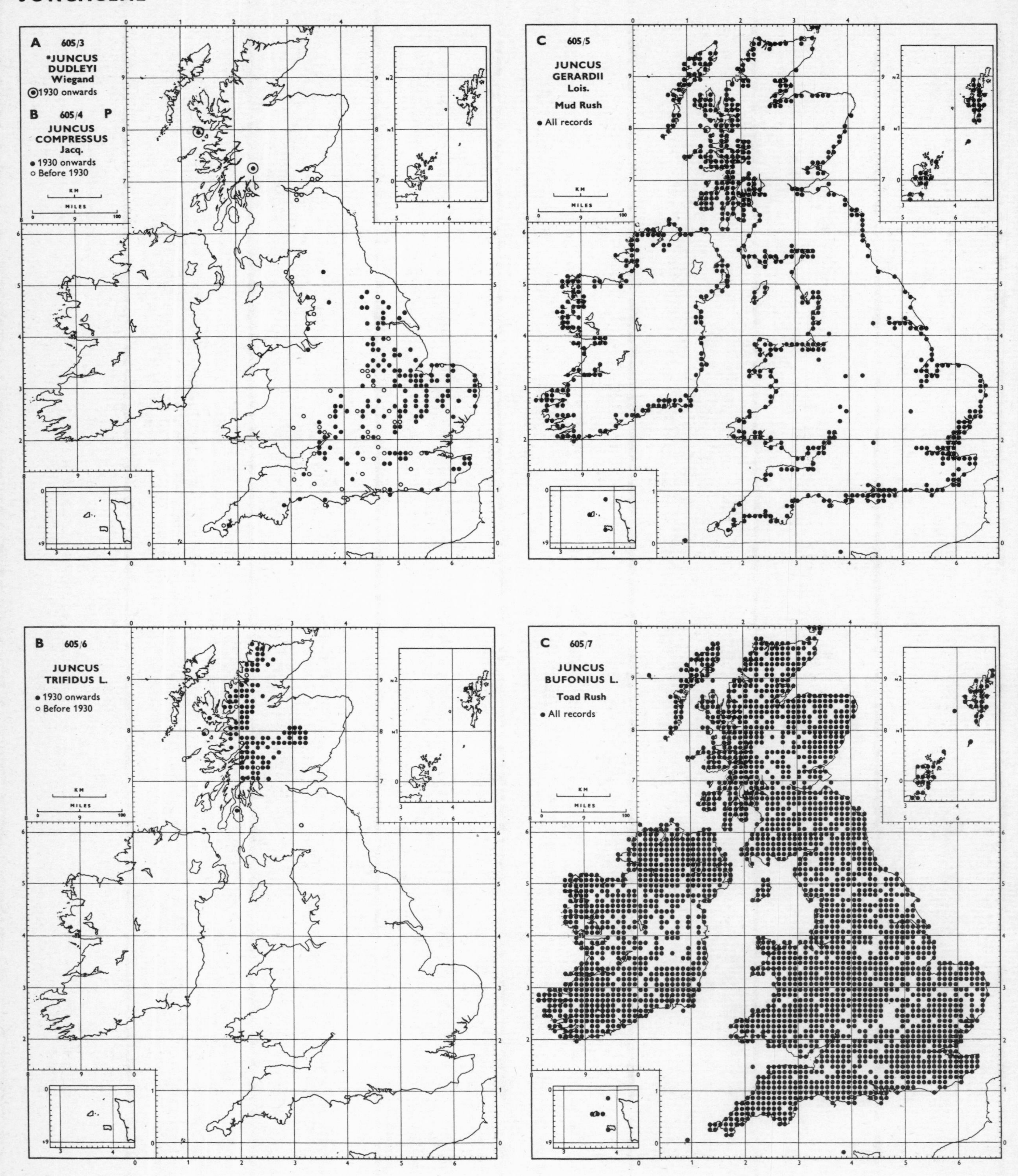
A 605/3
*JUNCUS DUDLEYI Wiegand
1930 onwards
B 605/4 P
JUNCUS COMPRESSUS Jacq.
• 1930 onwards
○ Before 1930
C 605/5
JUNCUS GERARDII Lois.
Mud Rush
• All records
B 605/6
JUNCUS TRIFIDUS L.
• 1930 onwards
○ Before 1930
C 605/7
JUNCUS BUFONIUS L.
Toad Rush
• All records
KM
MILES

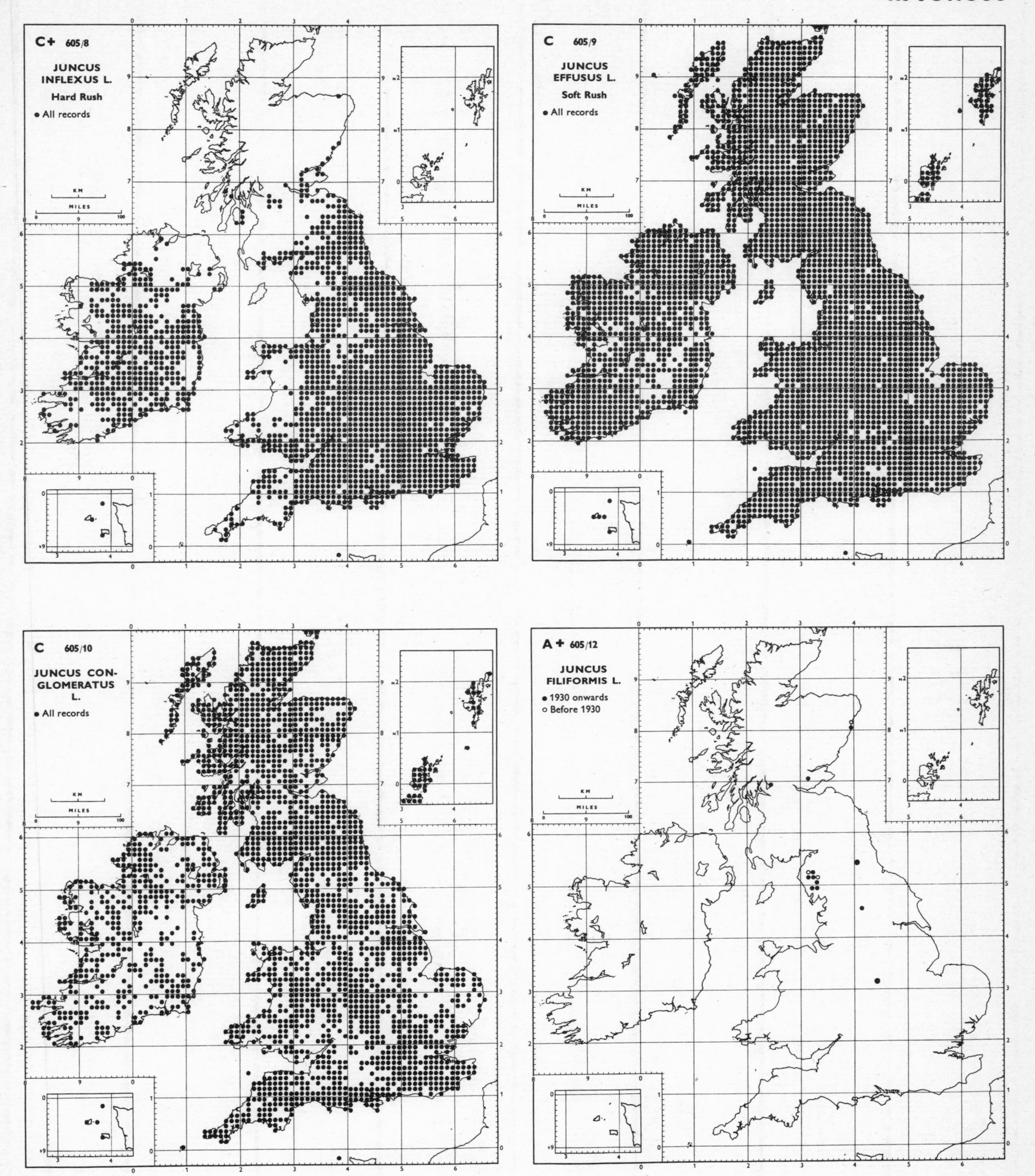
C+ 605/8
JUNCUS
INFLEXUS L.
Hard Rush
● All records
C 605/9
JUNCUS
EFFUSUS L.
Soft Rush
● All records
C 605/10
JUNCUS CON-
GLOMERATUS
L.
● All records
A+ 605/12
JUNCUS
FILIFORMIS L.
● 1930 onwards
○ Before 1930
KM
MILES

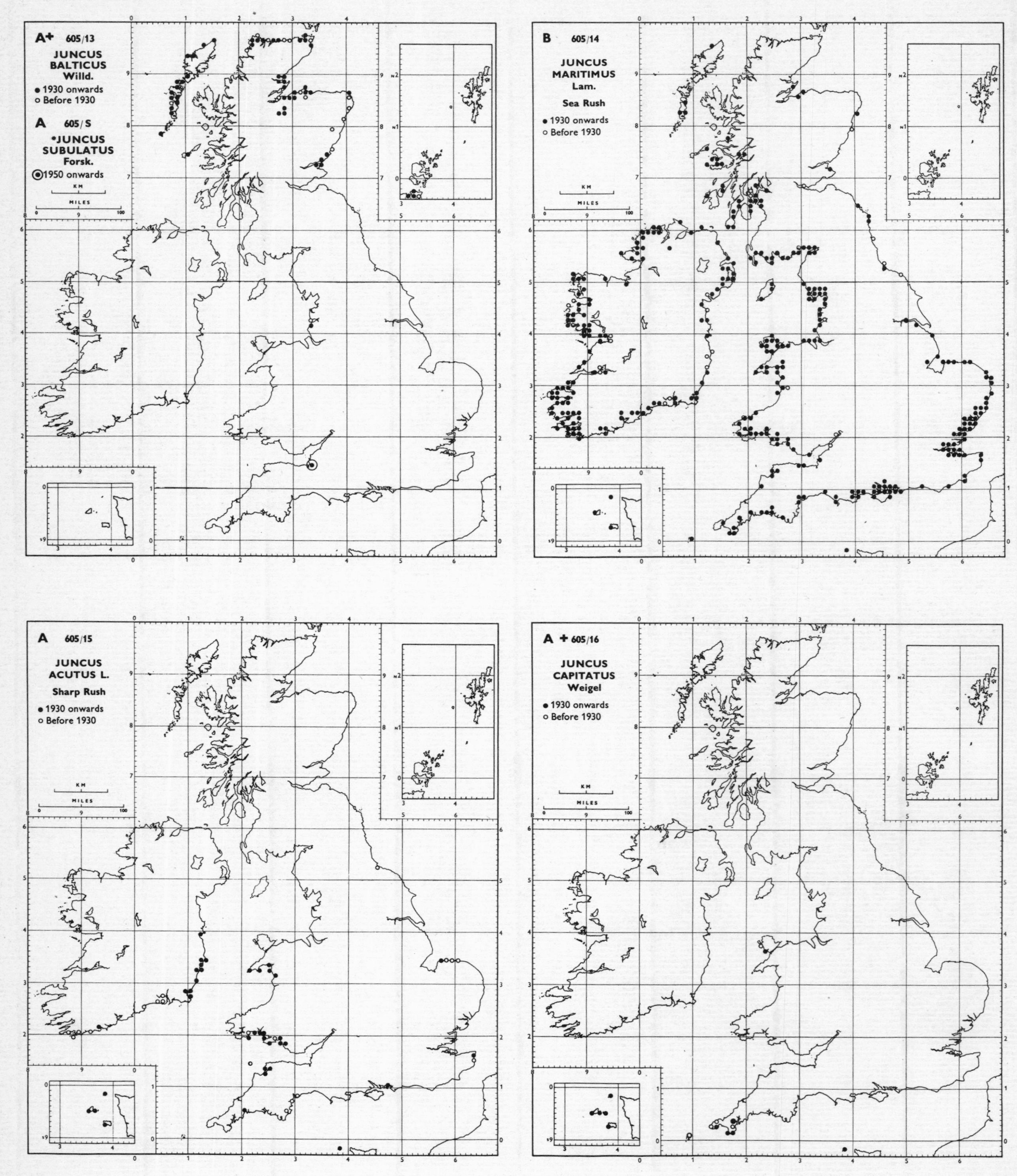
A+ 605/13
JUNCUS BALTICUS Willd.
● 1930 onwards
○ Before 1930
A 605/S
*JUNCUS SUBULATUS Forsk.
⊙ 1950 onwards
B 605/14
JUNCUS MARITIMUS Lam.
Sea Rush
● 1930 onwards
○ Before 1930
A 605/15
JUNCUS ACUTUS L.
Sharp Rush
● 1930 onwards
○ Before 1930
A + 605/16
JUNCUS CAPITATUS Weigel
● 1930 onwards
○ Before 1930
KM
MILES

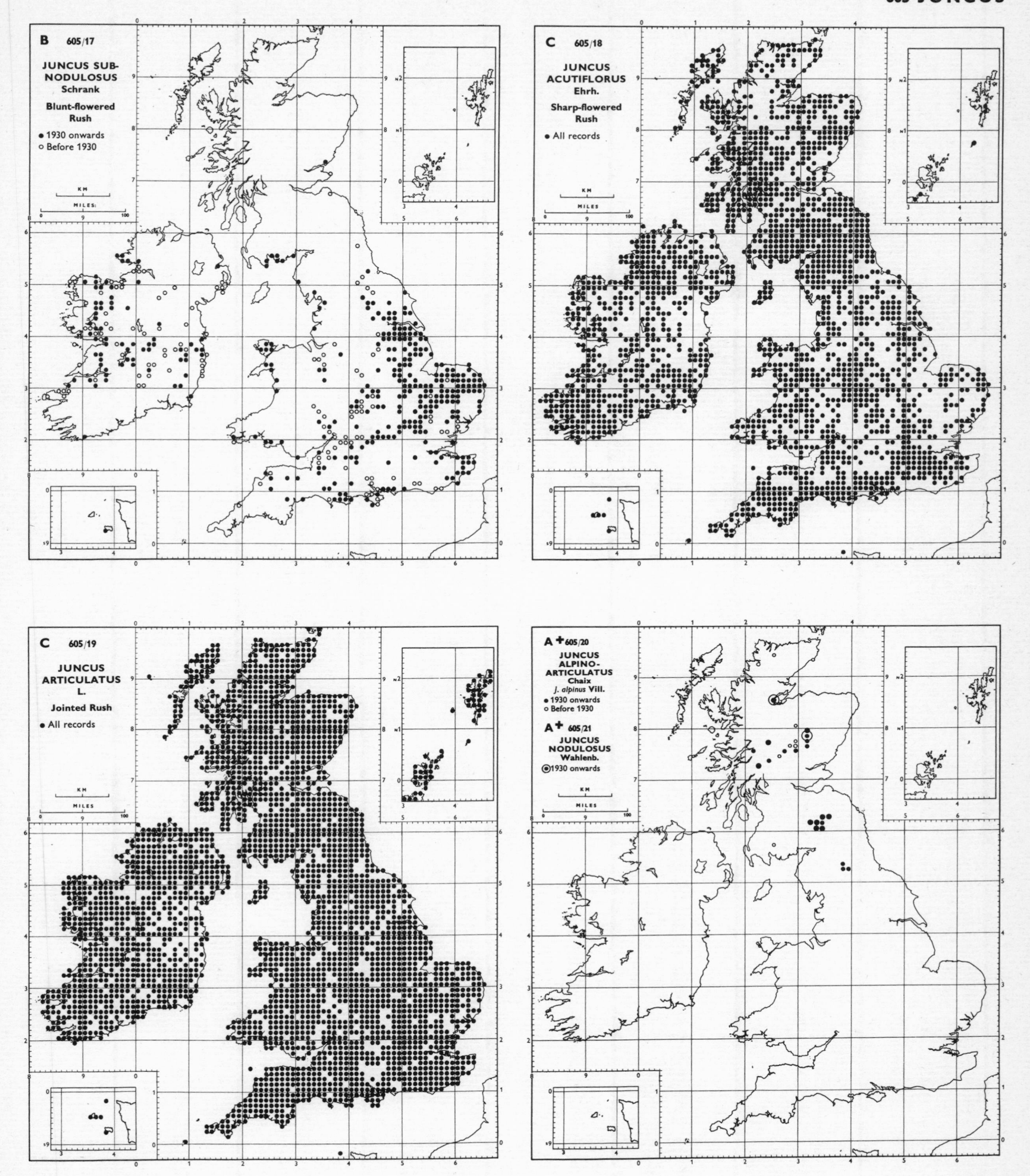
B 605/17
JUNCUS SUB-NODULOSUS
Schrank
Blunt-flowered Rush
1930 onwards
Before 1930
C 605/18
JUNCUS ACUTIFLORUS
Ehrh.
Sharp-flowered Rush
All records
C 605/19
JUNCUS ARTICULATUS
L.
Jointed Rush
All records
A + 605/20
JUNCUS ALPINO-ARTICULATUS
Chaix
J. alpinus Vill.
1930 onwards
Before 1930
A + 605/21
JUNCUS NODULOSUS
Wahlenb.
1930 onwards
KM
MILES

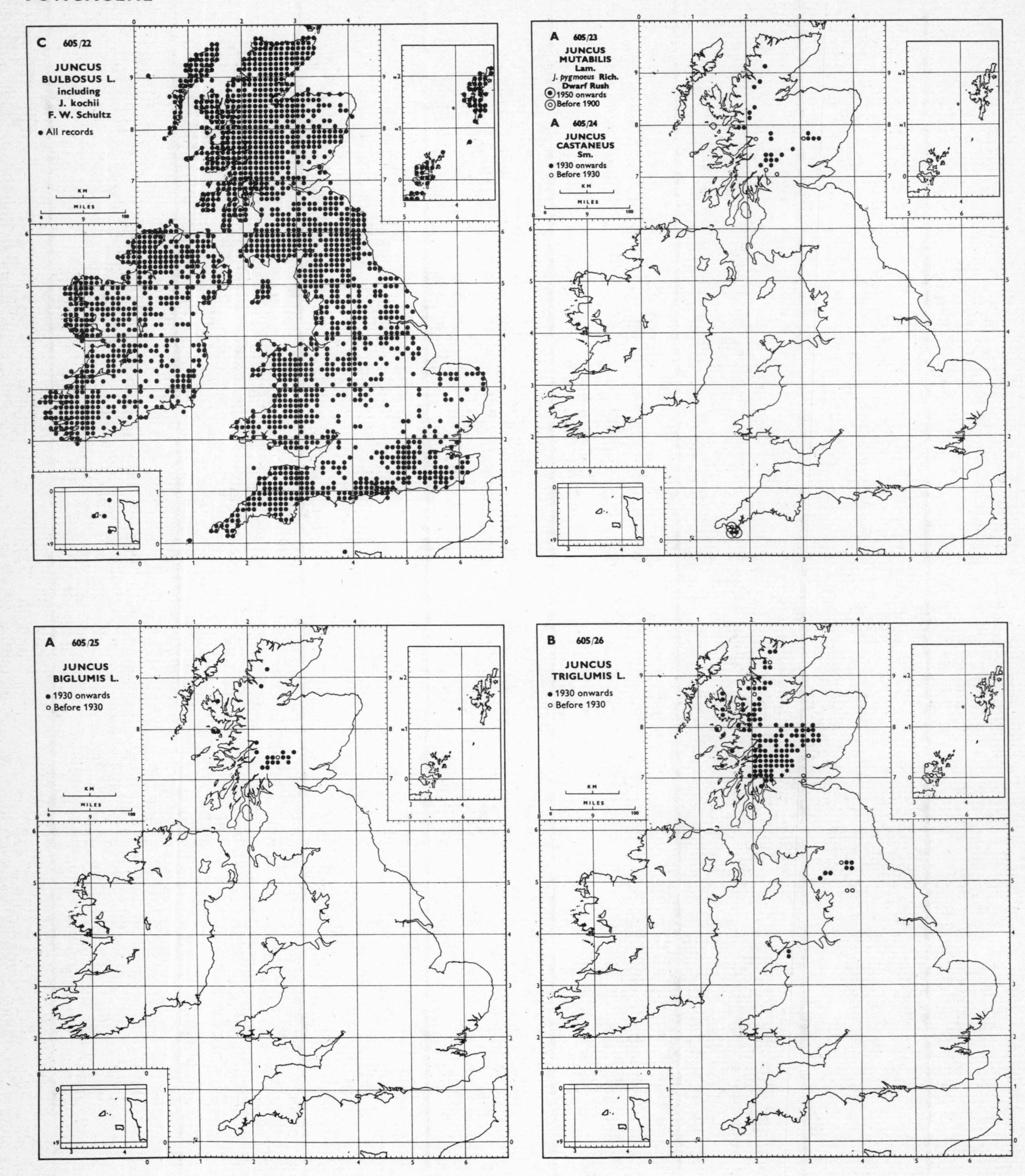
C 605/22
JUNCUS BULBOSUS L. including J. kochii F. W. Schultz
• All records
KM
MILES
A 605/23
JUNCUS MUTABILIS Lam.
J. pygmaeus Rich.
Dwarf Rush
1950 onwards
Before 1900
A 605/24
JUNCUS CASTANEUS Sm.
• 1930 onwards
○ Before 1930
A 605/25
JUNCUS BIGLUMIS L.
• 1930 onwards
○ Before 1930
B 605/26
JUNCUS TRIGLUMIS L.
• 1930 onwards
○ Before 1930

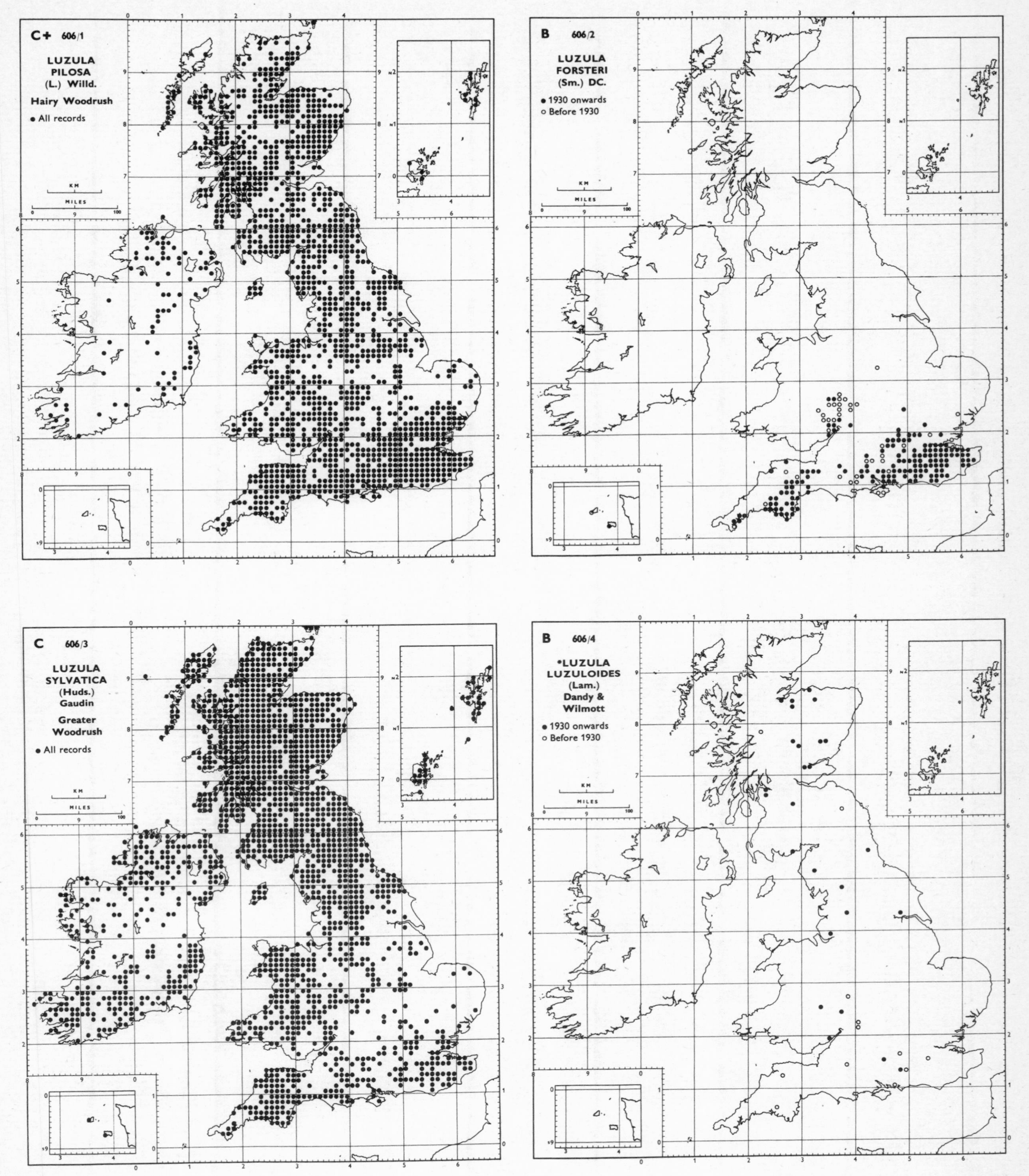
C+ 606/1
LUZULA PILOSA (L.) Willd.
Hairy Woodrush
All records
B 606/2
LUZULA FORSTERI (Sm.) DC.
1930 onwards
Before 1930
C 606/3
LUZULA SYLVATICA (Huds.) Gaudin
Greater Woodrush
All records
B 606/4
*LUZULA LUZULOIDES (Lam.) Dandy & Wilmott
1930 onwards
Before 1930
KM
MILES

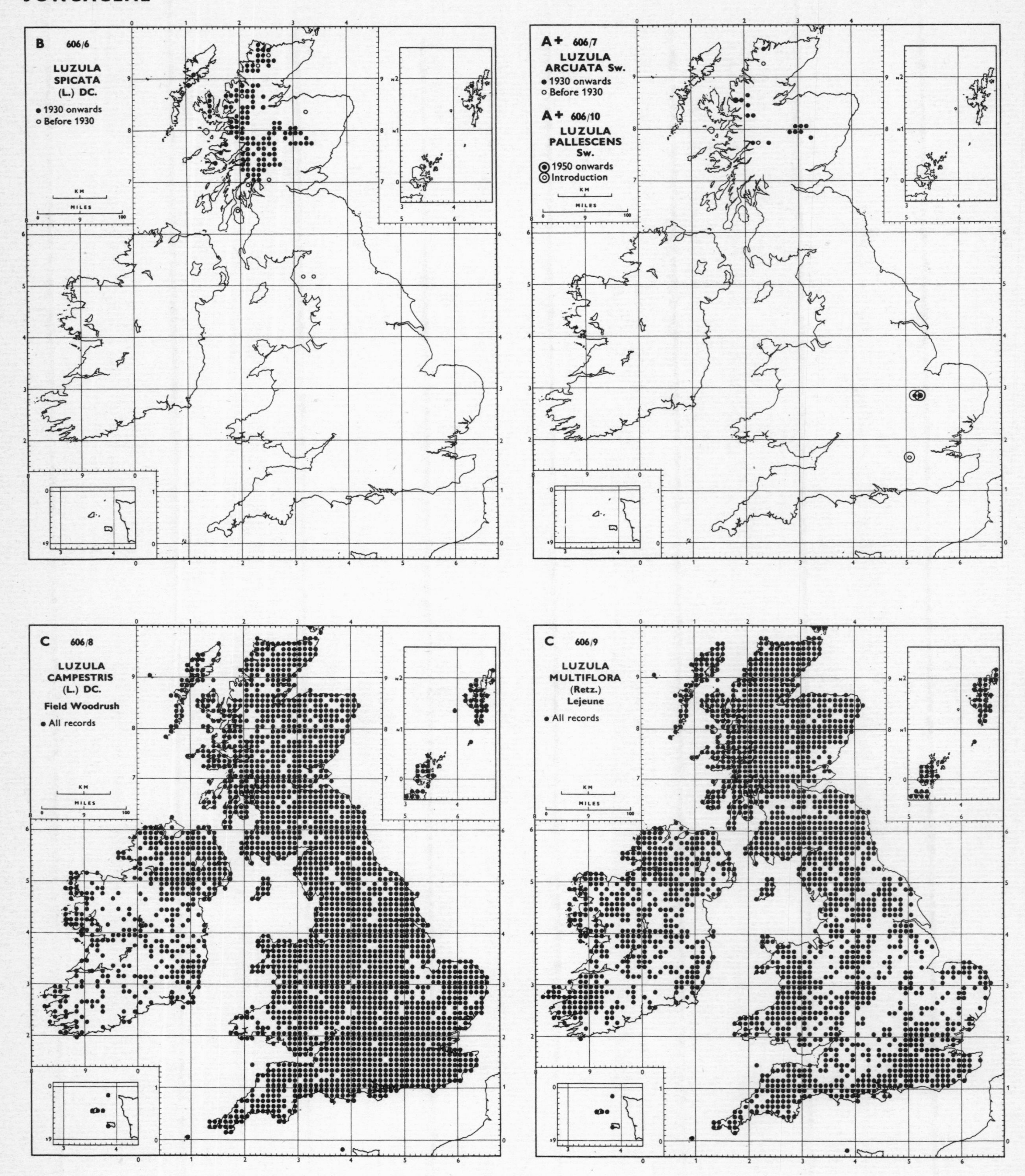
B 606/6
LUZULA SPICATA (L.) DC.
● 1930 onwards
○ Before 1930
A+ 606/7
LUZULA ARCUATA Sw.
● 1930 onwards
○ Before 1930
A+ 606/10
LUZULA PALLESCENS Sw.
1950 onwards
Introduction
C 606/8
LUZULA CAMPESTRIS (L.) DC.
Field Woodrush
● All records
C 606/9
LUZULA MULTIFLORA (Retz.) Lejeune
● All records
KM
MILES

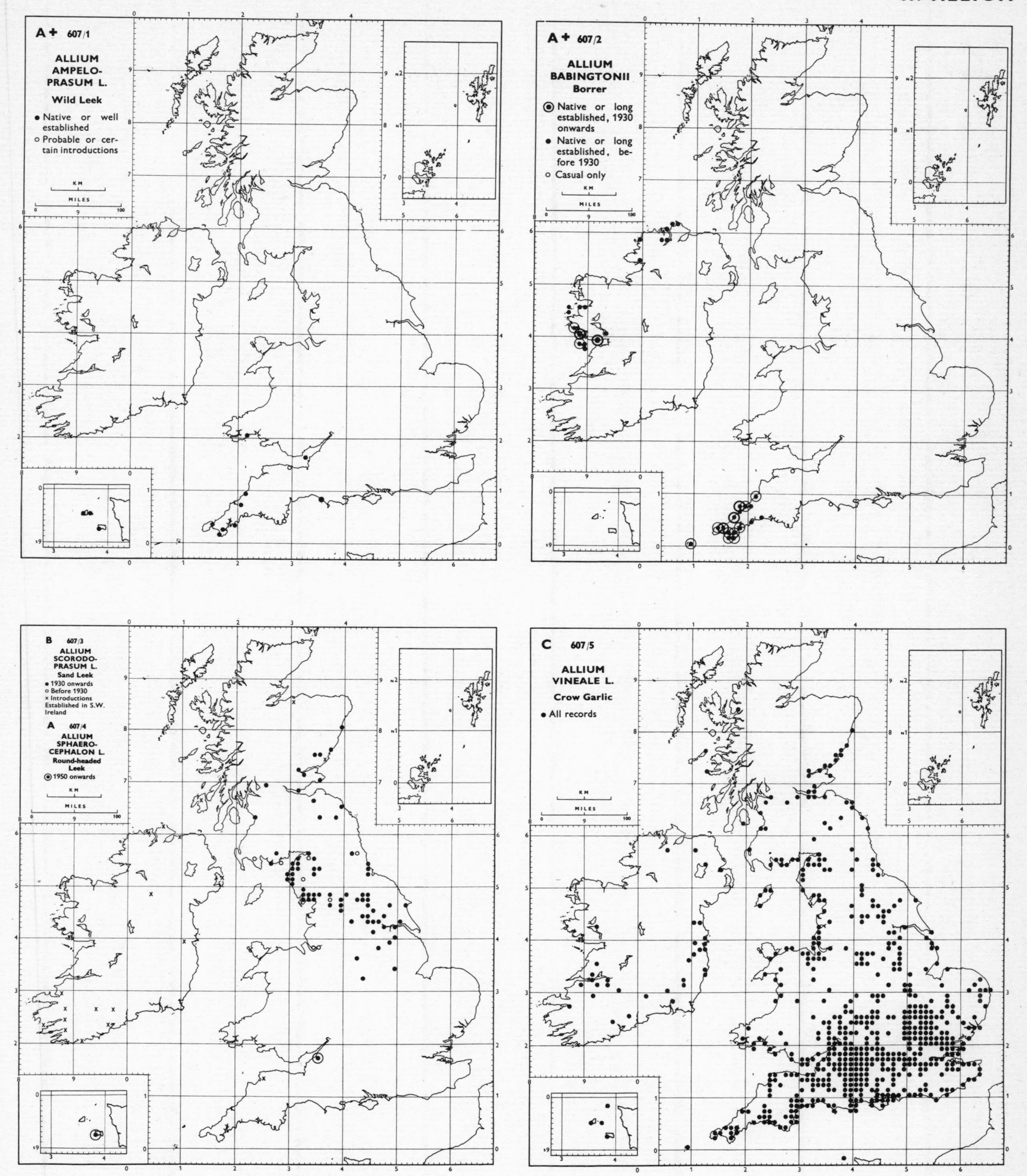
A+ 607/1
ALLIUM AMPELO-PRASUM L.
Wild Leek
• Native or well established
○ Probable or certain introductions
A+ 607/2
ALLIUM BABINGTONII Borrer
◉ Native or long established, 1930 onwards
• Native or long established, before 1930
○ Casual only
B 607/3
ALLIUM SCORODO-PRASUM L.
Sand Leek
• 1930 onwards
○ Before 1930
× Introductions
Established in S.W. Ireland
A 607/4
ALLIUM SPHAERO-CEPHALON L.
Round-headed Leek
◉ 1950 onwards
C 607/5
ALLIUM VINEALE L.
Crow Garlic
• All records

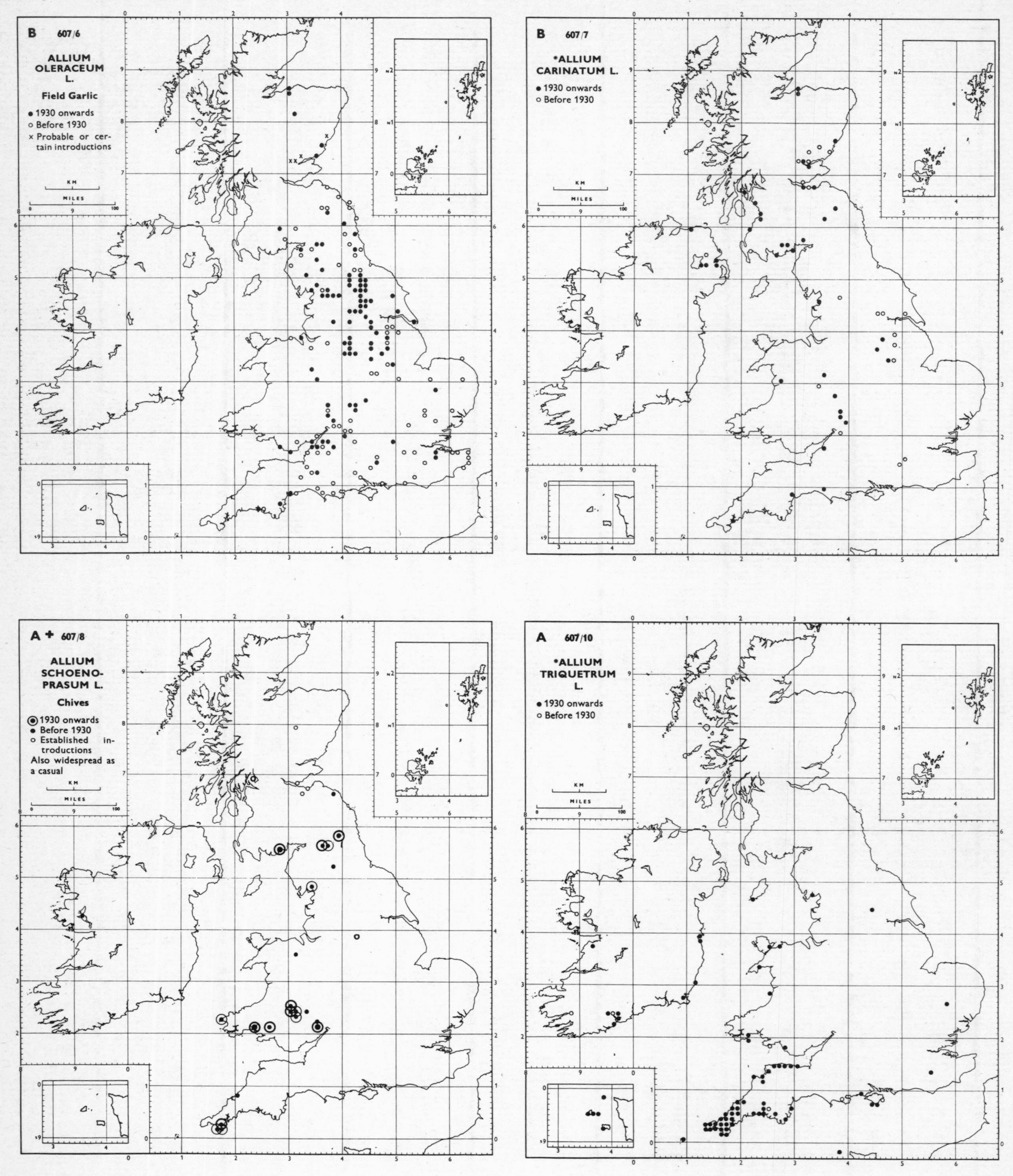
B 607/6
ALLIUM OLERACEUM L.
Field Garlic
• 1930 onwards
○ Before 1930
× Probable or certain introductions
KM
MILES
B 607/7
*ALLIUM CARINATUM L.
• 1930 onwards
○ Before 1930
A + 607/8
ALLIUM SCHOENOPRASUM L.
Chives
1930 onwards
• Before 1930
○ Established introductions
Also widespread as a casual
A 607/10
*ALLIUM TRIQUETRUM L.
• 1930 onwards
○ Before 1930

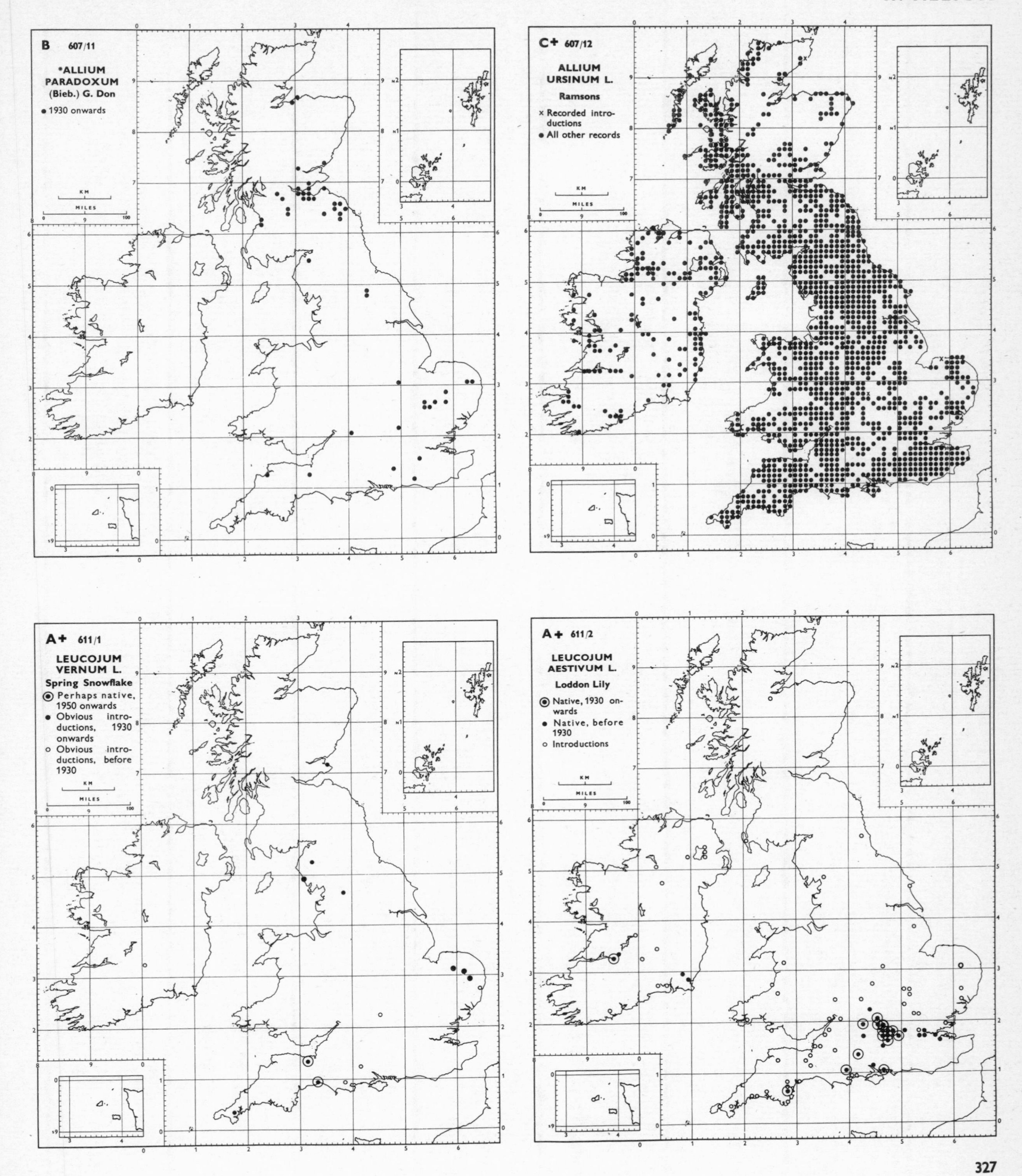
B 607/11
*ALLIUM
PARADOXUM
(Bieb.) G. Don
● 1930 onwards
C+ 607/12
ALLIUM
URSINUM L.
Ramsons
× Recorded introductions
● All other records
A+ 611/1
LEUCOJUM
VERNUM L.
Spring Snowflake
⊙ Perhaps native, 1950 onwards
● Obvious introductions, 1930 onwards
○ Obvious introductions, before 1930
A+ 611/2
LEUCOJUM
AESTIVUM L.
Loddon Lily
⊙ Native, 1930 onwards
● Native, before 1930
○ Introductions
KM
MILES

AMARYLLIDACEAE

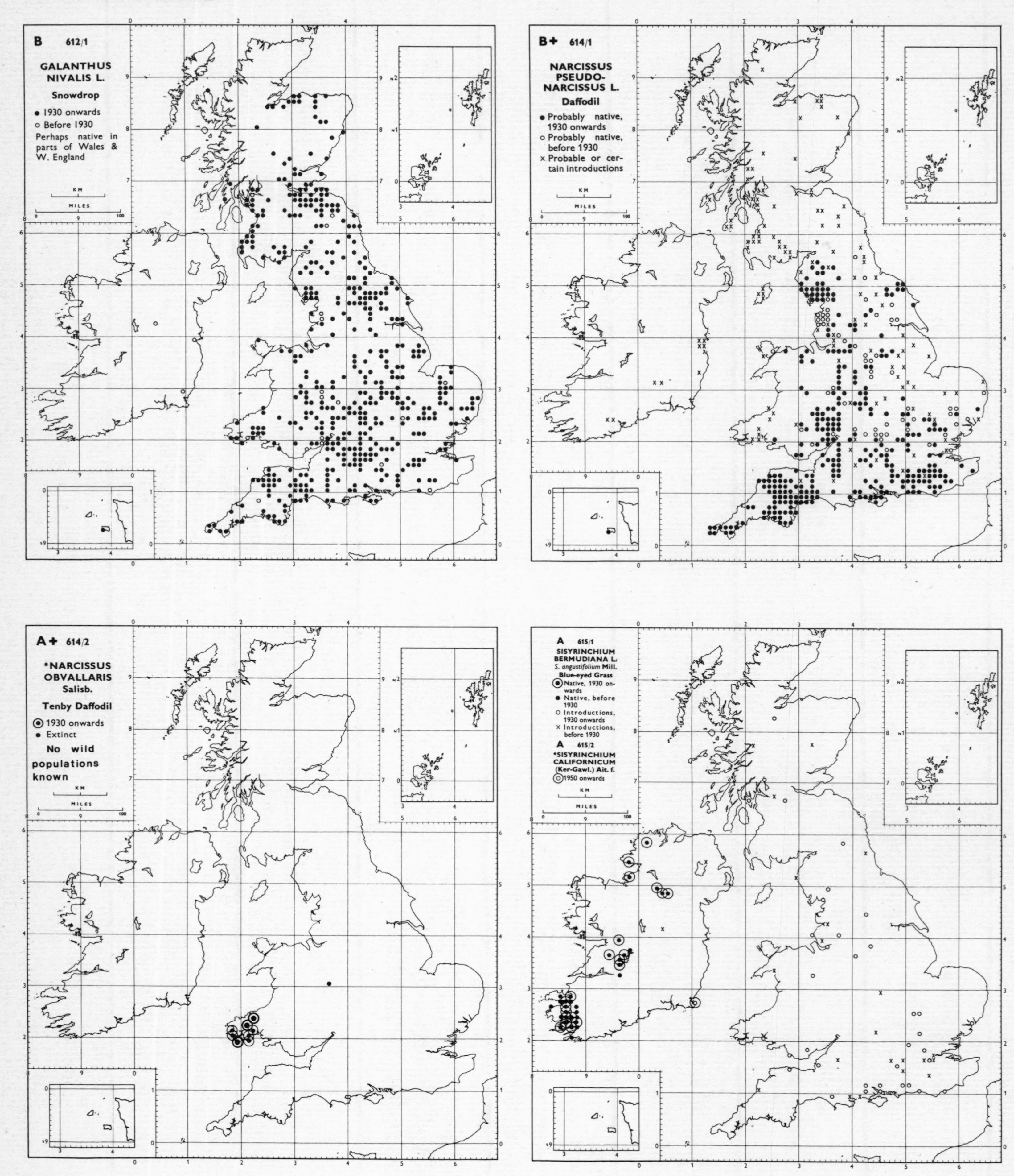

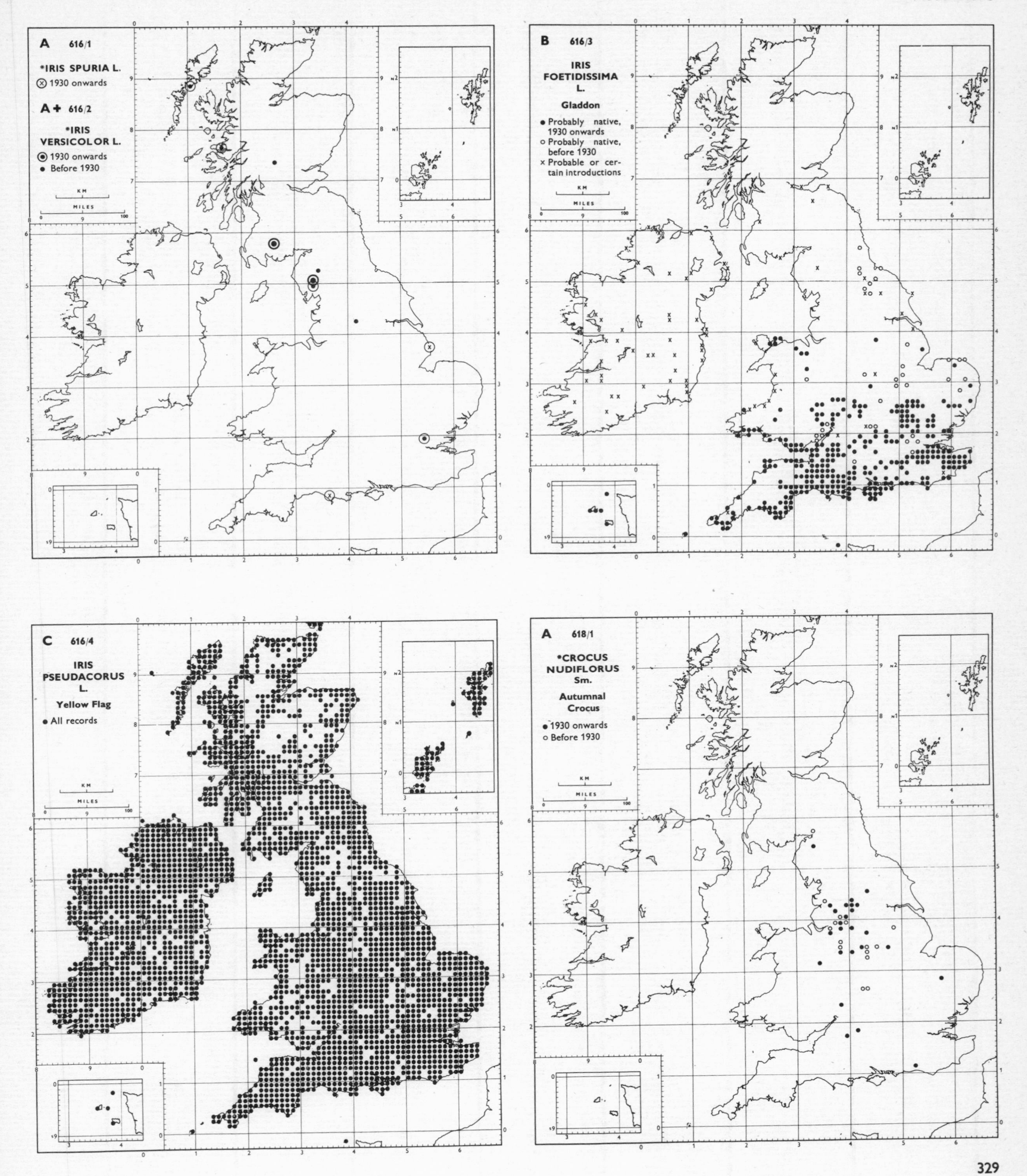
A 616/1
*IRIS SPURIA L.
⊗ 1930 onwards
A+ 616/2
*IRIS VERSICOLOR L.
⊙ 1930 onwards
● Before 1930
B 616/3
IRIS FOETIDISSIMA L.
Gladdon
● Probably native, 1930 onwards
○ Probably native, before 1930
× Probable or certain introductions
C 616/4
IRIS PSEUDACORUS L.
Yellow Flag
● All records
A 618/1
*CROCUS NUDIFLORUS Sm.
Autumnal Crocus
● 1930 onwards
○ Before 1930
KM
MILES

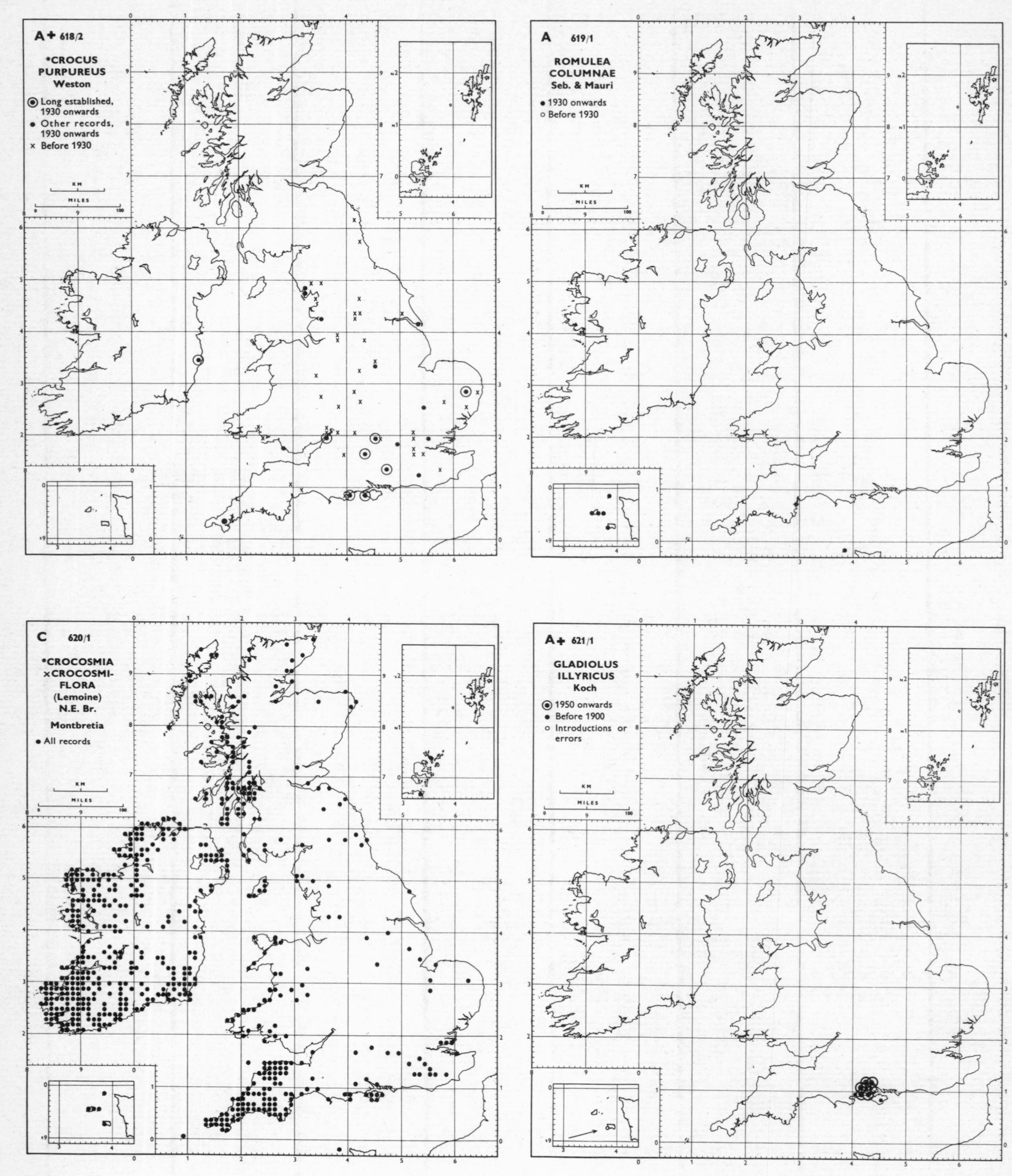
A+ 618/2
*CROCUS PURPUREUS Weston
Long established, 1930 onwards
Other records, 1930 onwards
Before 1930
A 619/1
ROMULEA COLUMNAE Seb. & Mauri
1930 onwards
Before 1930
C 620/1
*CROCOSMIA ×CROCOSMIFLORA (Lemoine) N.E. Br.
Montbretia
All records
A+ 621/1
GLADIOLUS ILLYRICUS Koch
1950 onwards
Before 1900
Introductions or errors
KM
MILES

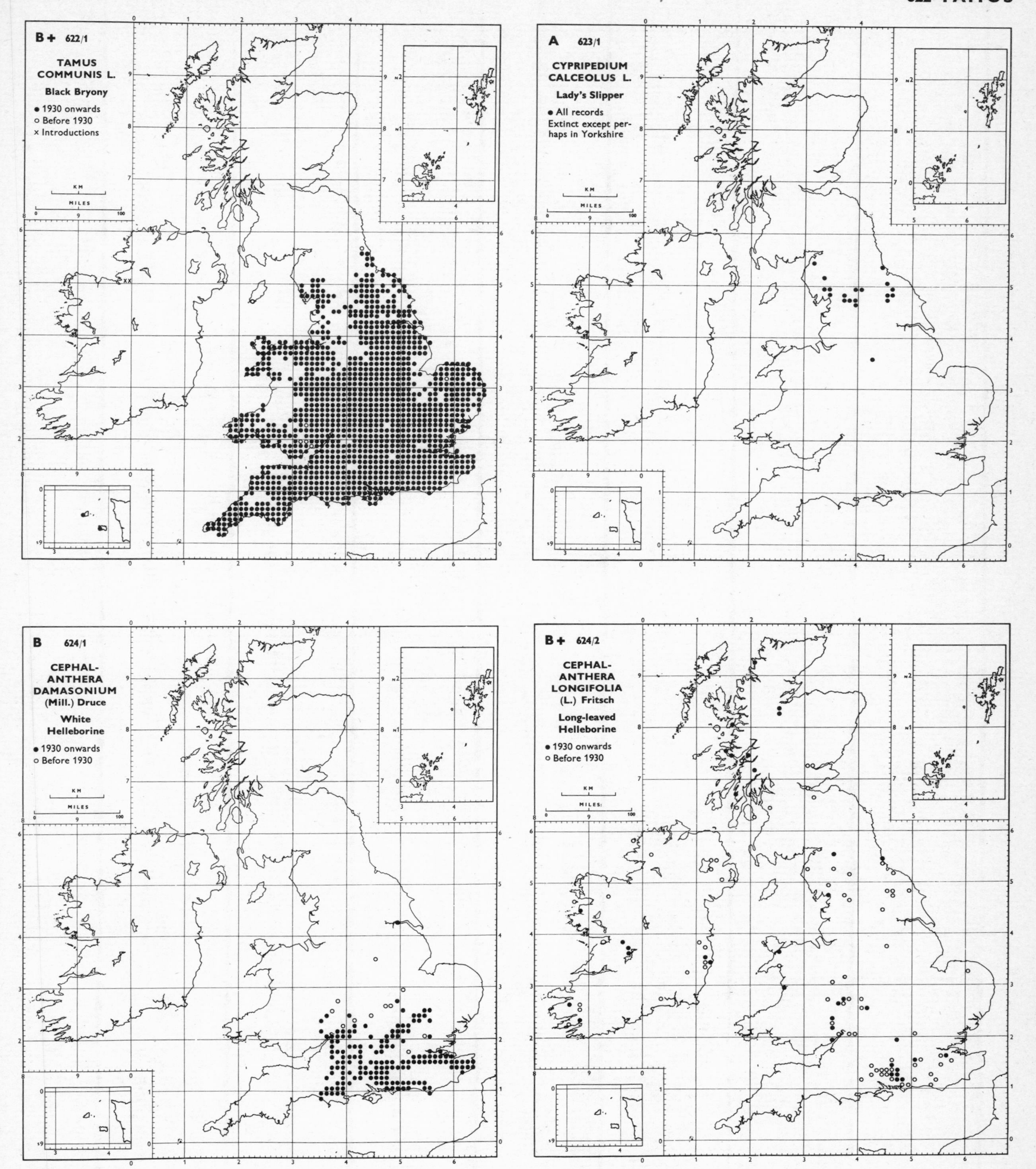
B+ 622/1
TAMUS COMMUNIS L.
Black Bryony
• 1930 onwards
○ Before 1930
× Introductions
A 623/1
CYPRIPEDIUM CALCEOLUS L.
Lady's Slipper
• All records
Extinct except perhaps in Yorkshire
B 624/1
CEPHALANTHERA DAMASONIUM (Mill.) Druce
White Helleborine
• 1930 onwards
○ Before 1930
B+ 624/2
CEPHALANTHERA LONGIFOLIA (L.) Fritsch
Long-leaved Helleborine
• 1930 onwards
○ Before 1930
KM
MILES

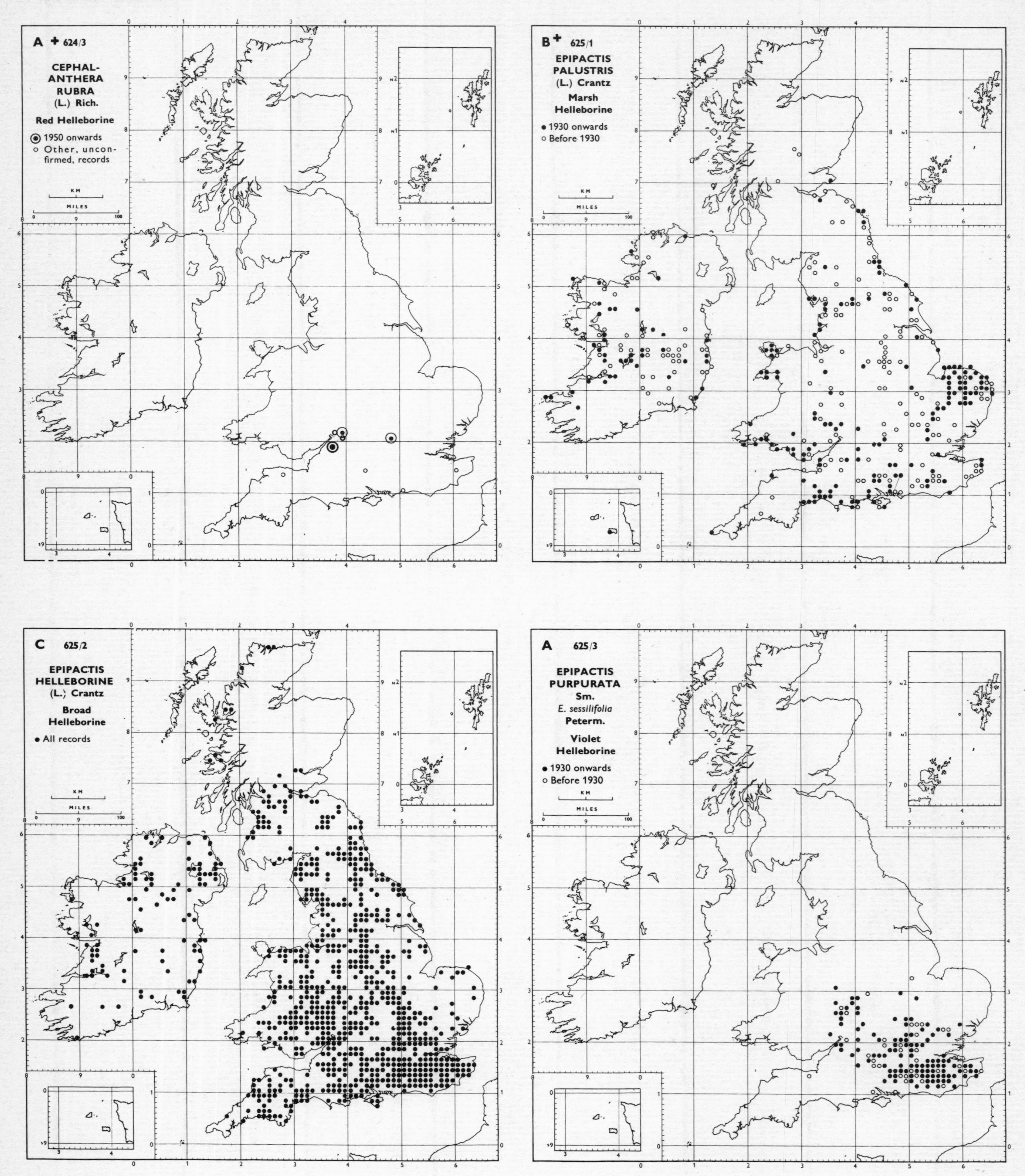

A + 624/3
CEPHALANTHERA RUBRA (L.) Rich.
Red Helleborine
1950 onwards
Other, unconfirmed, records
B + 625/1
EPIPACTIS PALUSTRIS (L.) Crantz
Marsh Helleborine
1930 onwards
Before 1930
C 625/2
EPIPACTIS HELLEBORINE (L.) Crantz
Broad Helleborine
All records
A 625/3
EPIPACTIS PURPURATA Sm.
E. sessilifolia Peterm.
Violet Helleborine
1930 onwards
Before 1930
KM
MILES

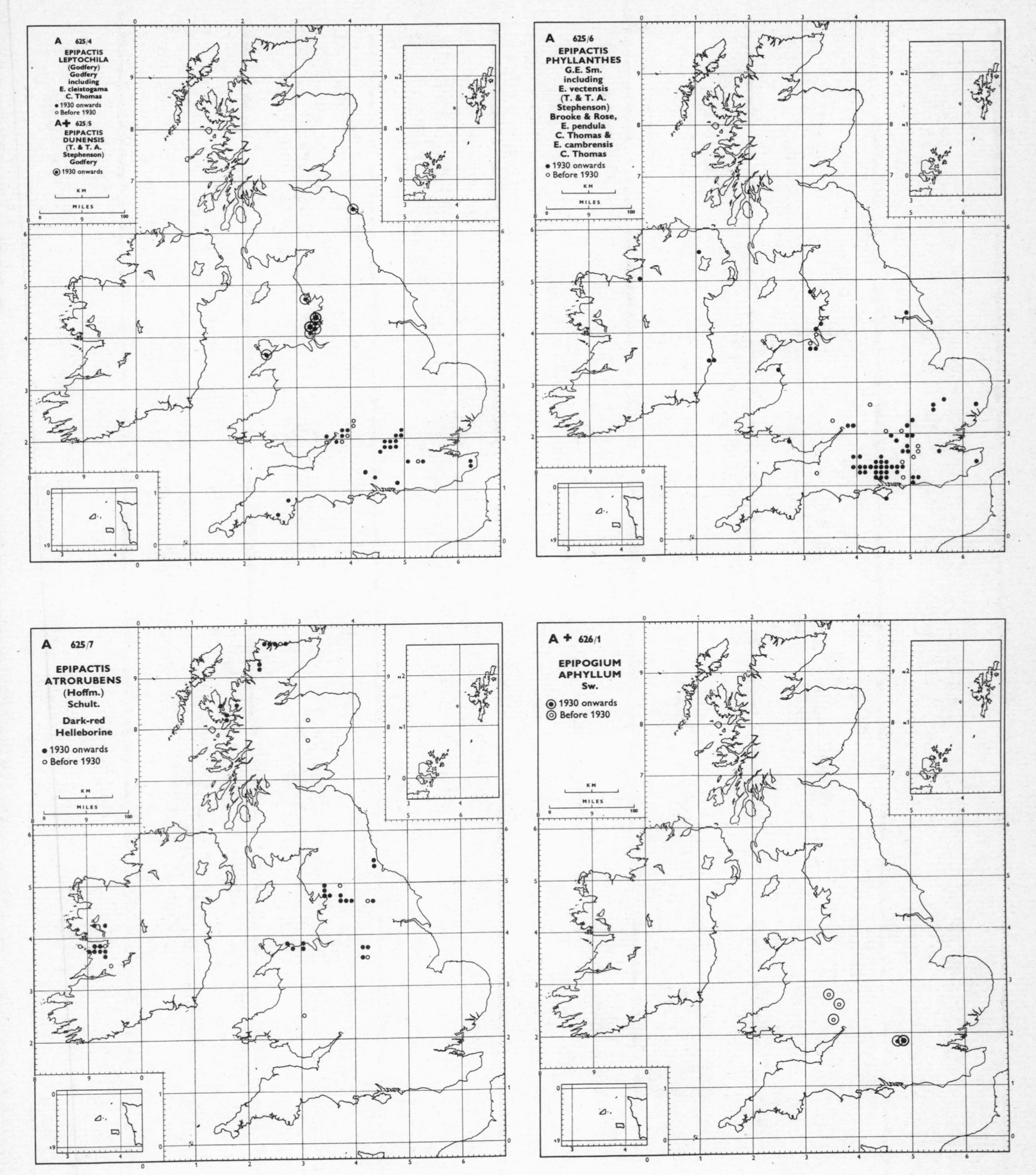

A 625/4
EPIPACTIS LEPTOCHILA (Godfery) Godfery including E. cleistogama C. Thomas
• 1930 onwards
○ Before 1930
A+ 625/5
EPIPACTIS DUNENSIS (T. & T. A. Stephenson) Godfery
◉ 1930 onwards
A 625/6
EPIPACTIS PHYLLANTHES G.E. Sm. including E. vectensis (T. & T. A. Stephenson) Brooke & Rose, E. pendula C. Thomas & E. cambrensis C. Thomas
• 1930 onwards
○ Before 1930
A 625/7
EPIPACTIS ATRORUBENS (Hoffm.) Schult.
Dark-red Helleborine
• 1930 onwards
○ Before 1930
A + 626/1
EPIPOGIUM APHYLLUM Sw.
◉ 1930 onwards
◎ Before 1930
KM
MILES

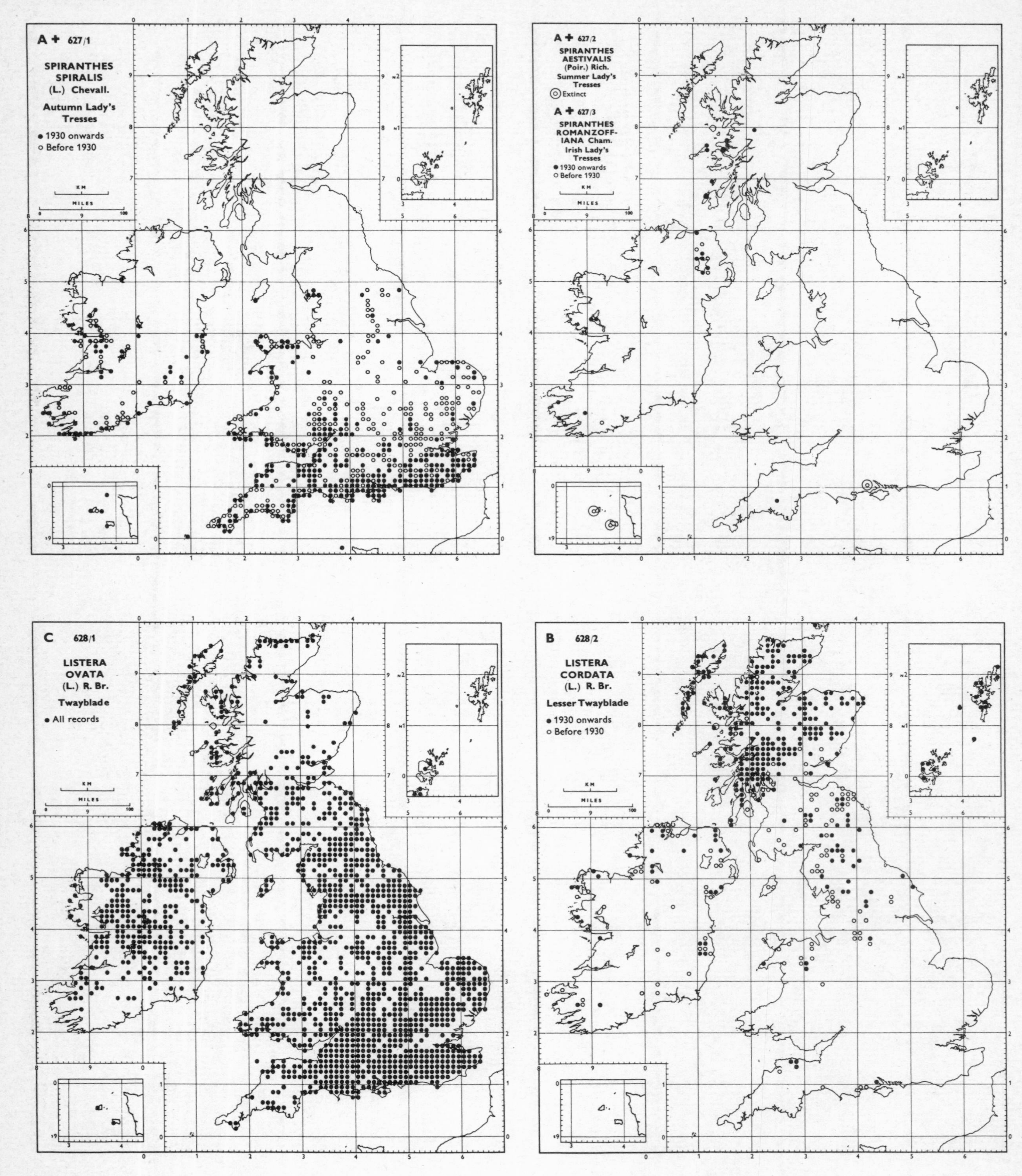
A + 627/1
SPIRANTHES SPIRALIS (L.) Chevall.
Autumn Lady's Tresses
● 1930 onwards
○ Before 1930
A + 627/2
SPIRANTHES AESTIVALIS (Poir.) Rich.
Summer Lady's Tresses
◎ Extinct
A + 627/3
SPIRANTHES ROMANZOFFIANA Cham.
Irish Lady's Tresses
● 1930 onwards
○ Before 1930
C 628/1
LISTERA OVATA (L.) R. Br.
Twayblade
● All records
B 628/2
LISTERA CORDATA (L.) R. Br.
Lesser Twayblade
● 1930 onwards
○ Before 1930
KM
MILES

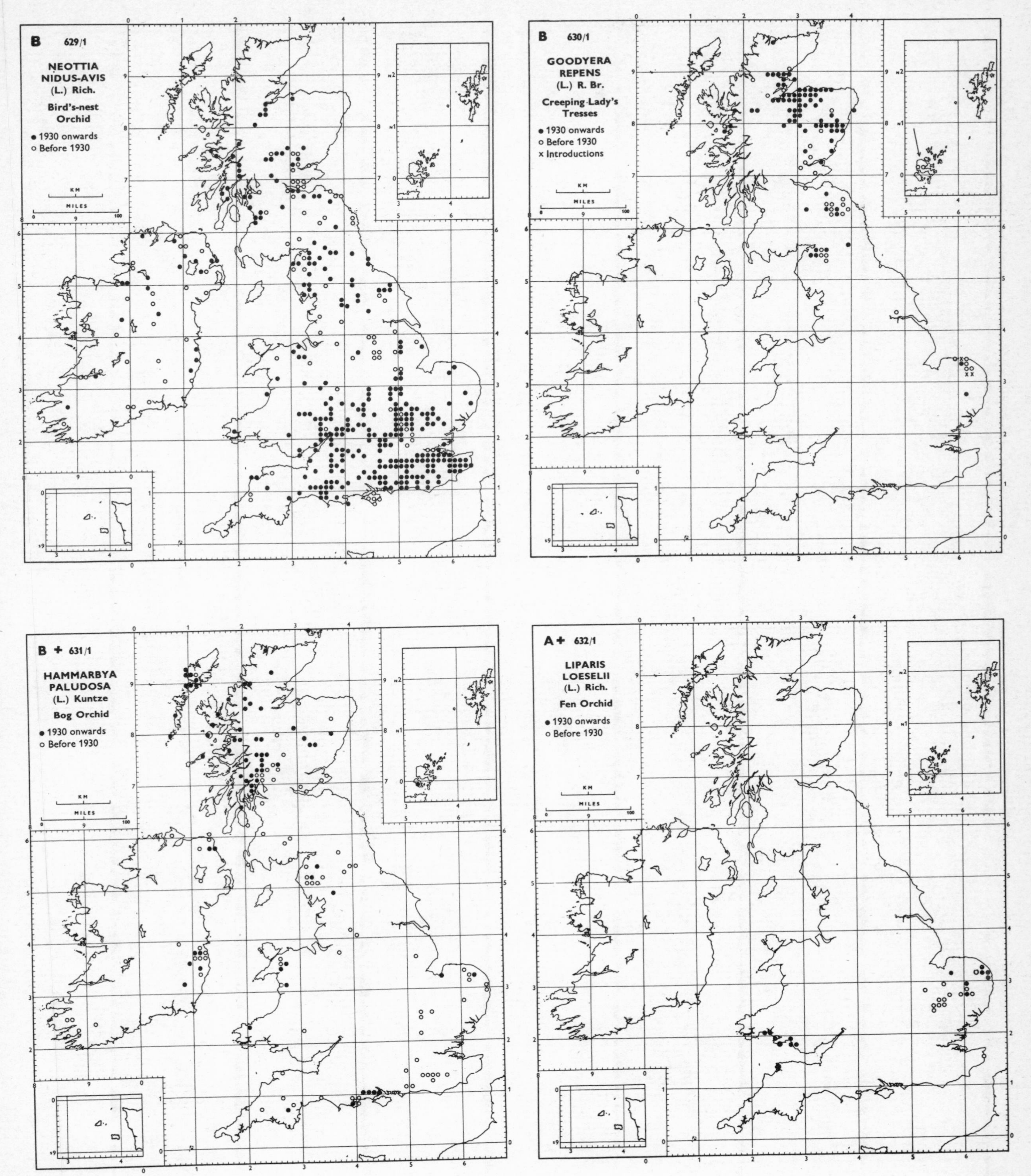
B 629/1
NEOTTIA NIDUS-AVIS (L.) Rich.
Bird's-nest Orchid
● 1930 onwards
○ Before 1930
B 630/1
GOODYERA REPENS (L.) R. Br.
Creeping Lady's Tresses
● 1930 onwards
○ Before 1930
x Introductions
B + 631/1
HAMMARBYA PALUDOSA (L.) Kuntze
Bog Orchid
● 1930 onwards
○ Before 1930
A + 632/1
LIPARIS LOESELII (L.) Rich.
Fen Orchid
● 1930 onwards
○ Before 1930
KM
MILES

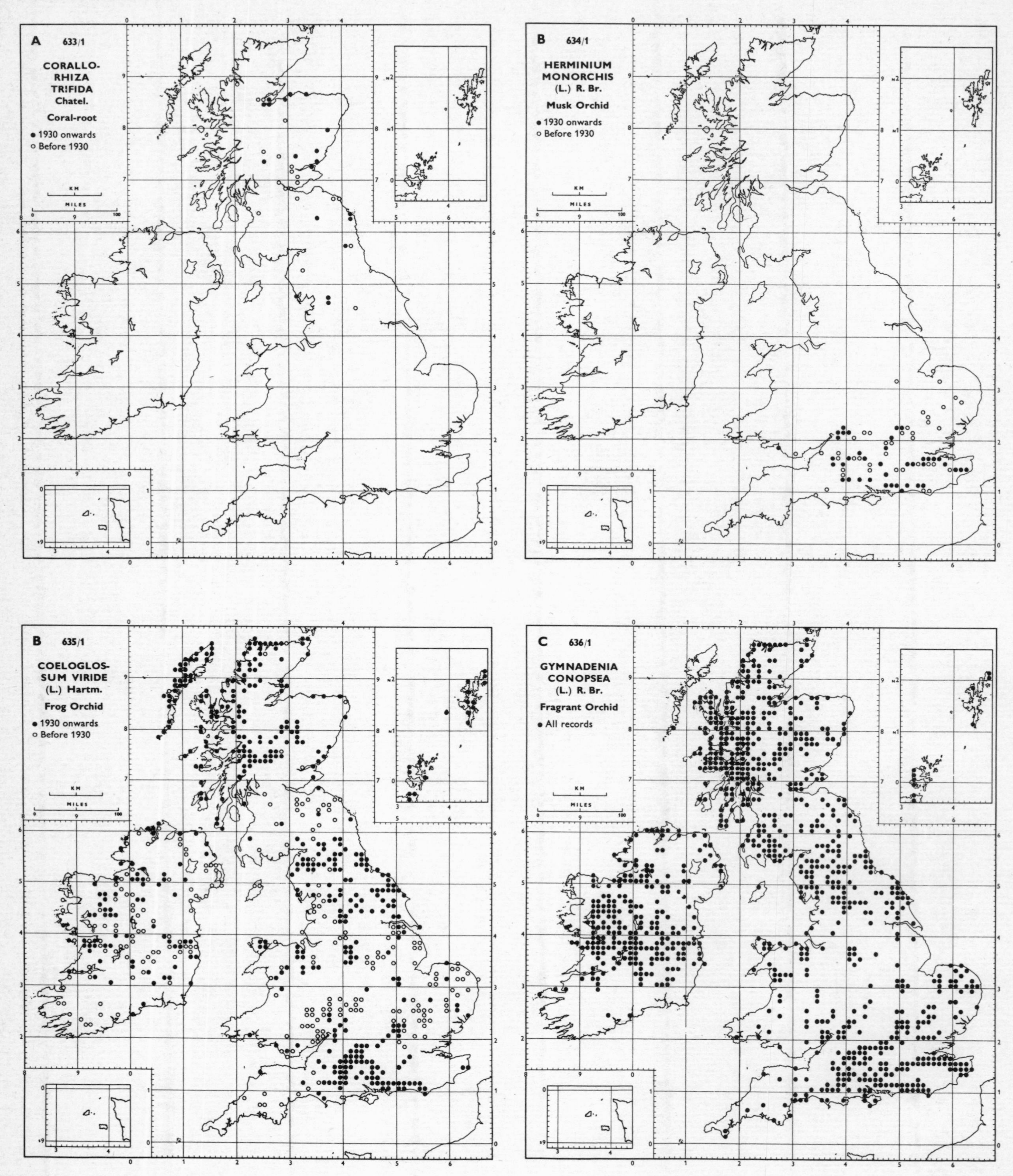
A 633/1
CORALLORHIZA TRIFIDA Chatel.
Coral-root
● 1930 onwards
○ Before 1930
B 634/1
HERMINIUM MONORCHIS (L.) R. Br.
Musk Orchid
● 1930 onwards
○ Before 1930
B 635/1
COELOGLOSSUM VIRIDE (L.) Hartm.
Frog Orchid
● 1930 onwards
○ Before 1930
C 636/1
GYMNADENIA CONOPSEA (L.) R. Br.
Fragrant Orchid
● All records

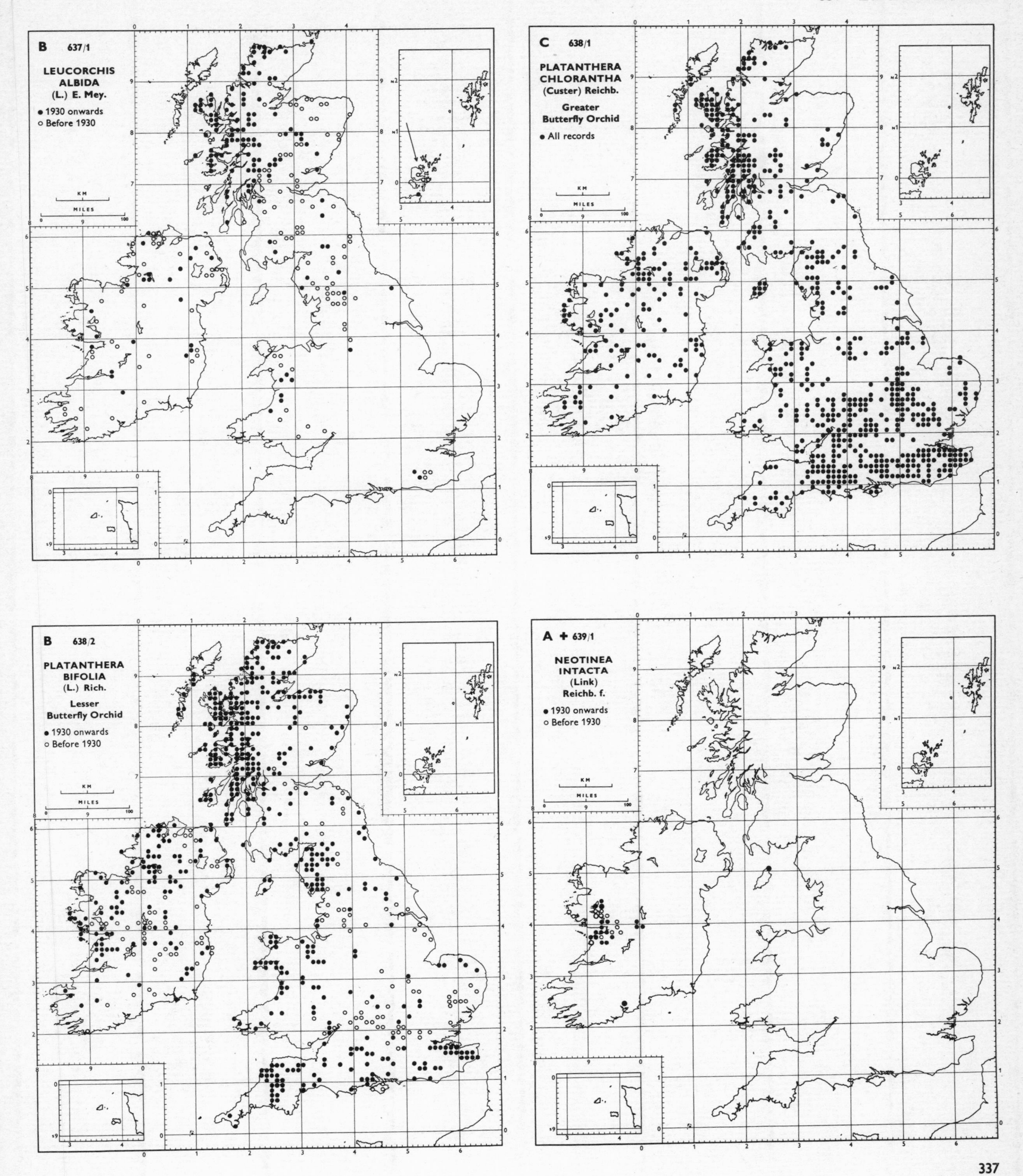
B 637/1
LEUCORCHIS ALBIDA (L.) E. Mey.
● 1930 onwards
○ Before 1930
C 638/1
PLATANTHERA CHLORANTHA (Custer) Reichb.
Greater Butterfly Orchid
● All records
B 638/2
PLATANTHERA BIFOLIA (L.) Rich.
Lesser Butterfly Orchid
● 1930 onwards
○ Before 1930
A + 639/1
NEOTINEA INTACTA (Link) Reichb. f.
● 1930 onwards
○ Before 1930
KM
MILES

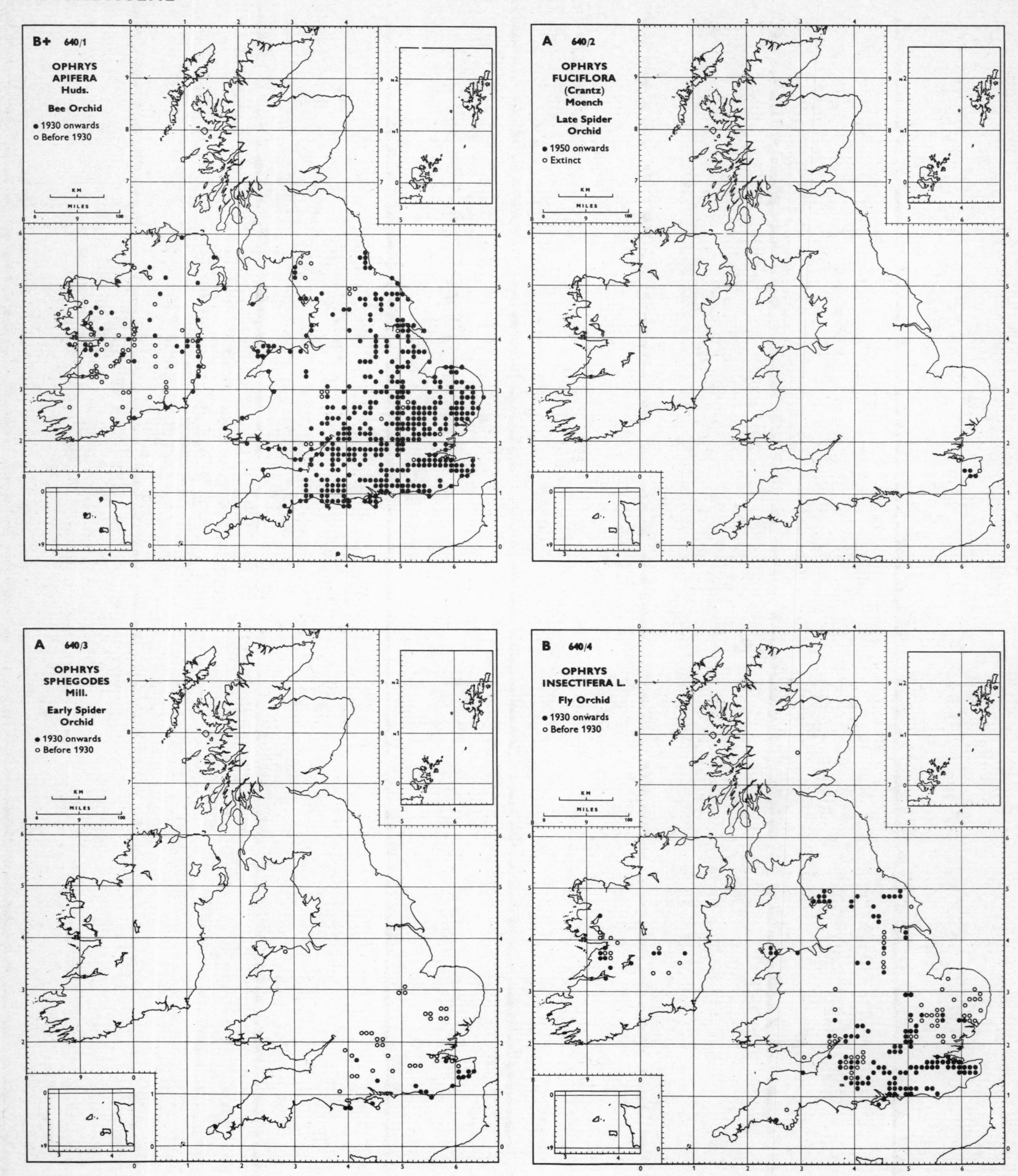
B+ 640/1
OPHRYS APIFERA Huds.
Bee Orchid
● 1930 onwards
○ Before 1930
A 640/2
OPHRYS FUCIFLORA (Crantz) Moench
Late Spider Orchid
● 1950 onwards
○ Extinct
A 640/3
OPHRYS SPHEGODES Mill.
Early Spider Orchid
● 1930 onwards
○ Before 1930
B 640/4
OPHRYS INSECTIFERA L.
Fly Orchid
● 1930 onwards
○ Before 1930

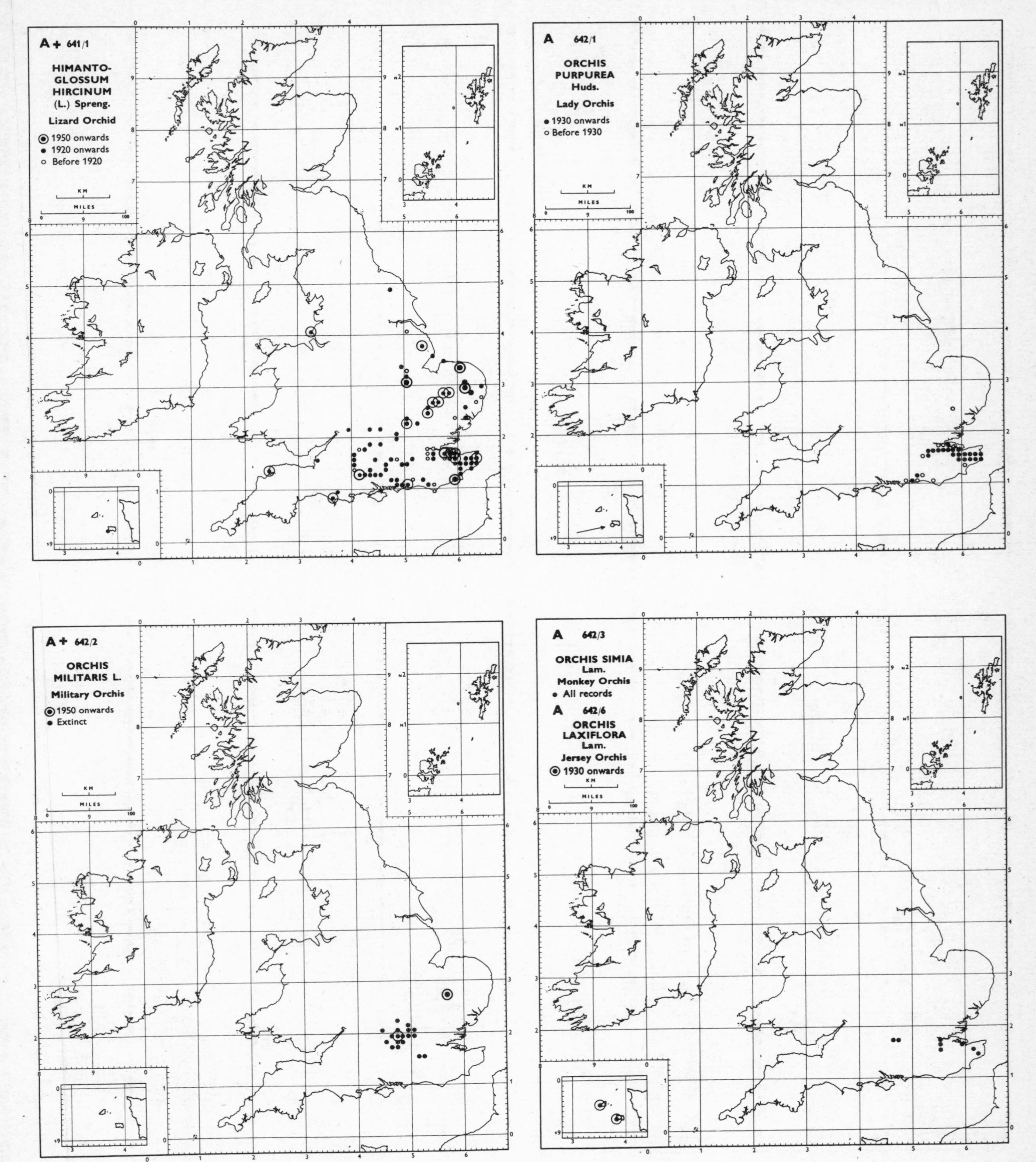
A + 641/1
HIMANTO-GLOSSUM HIRCINUM (L.) Spreng.
Lizard Orchid
1950 onwards
1920 onwards
Before 1920
KM
MILES
A 642/1
ORCHIS PURPUREA Huds.
Lady Orchis
1930 onwards
Before 1930
A + 642/2
ORCHIS MILITARIS L.
Military Orchis
1950 onwards
Extinct
A 642/3
ORCHIS SIMIA Lam.
Monkey Orchis
All records
A 642/6
ORCHIS LAXIFLORA Lam.
Jersey Orchis
1930 onwards

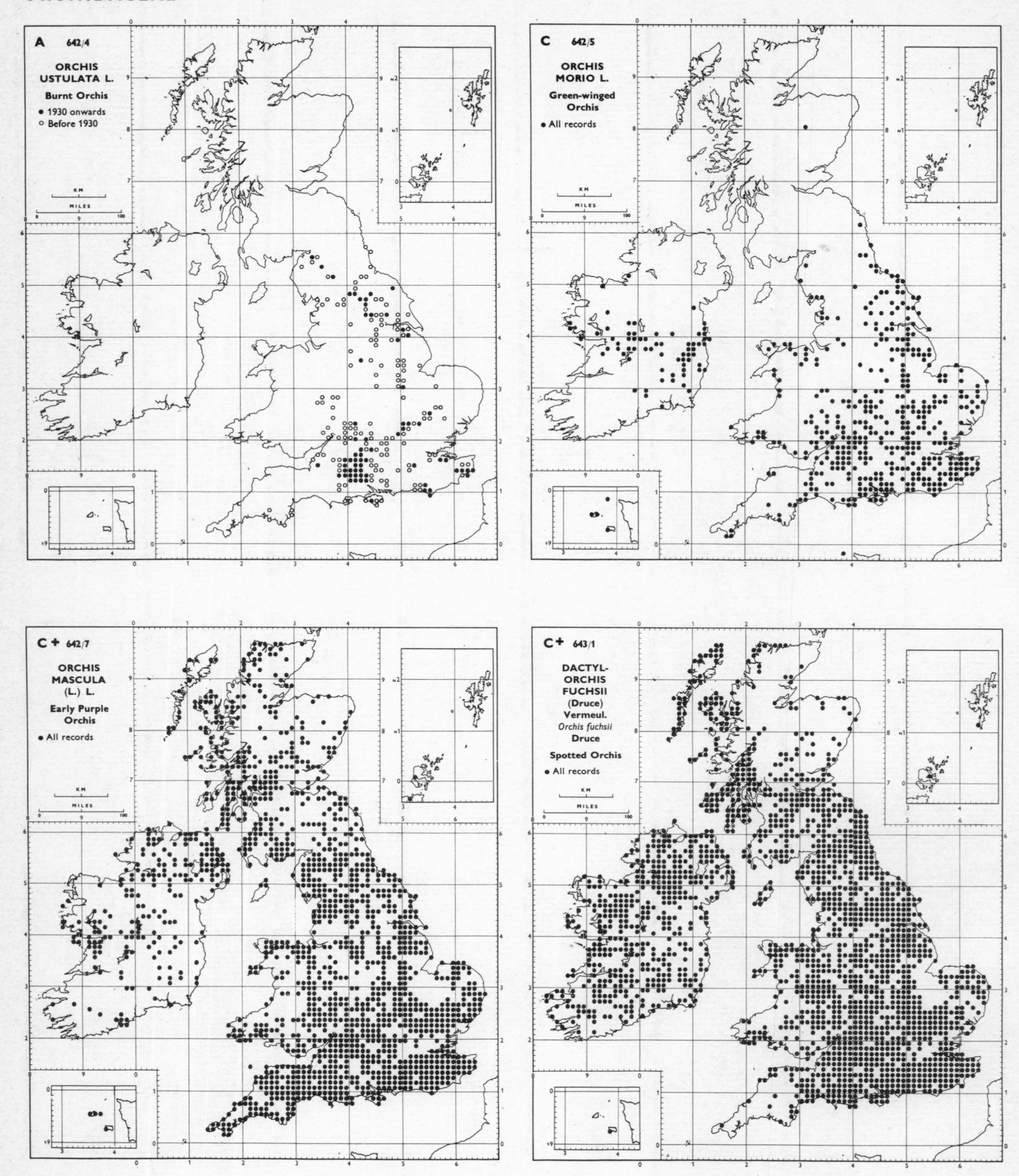
A 642/4
ORCHIS USTULATA L.
Burnt Orchis
1930 onwards
Before 1930
C 642/5
ORCHIS MORIO L.
Green-winged Orchis
All records
C + 642/7
ORCHIS MASCULA (L.) L.
Early Purple Orchis
All records
C+ 643/1
DACTYLORCHIS FUCHSII (Druce) Vermeul.
Orchis fuchsii Druce
Spotted Orchis
All records
KM
MILES

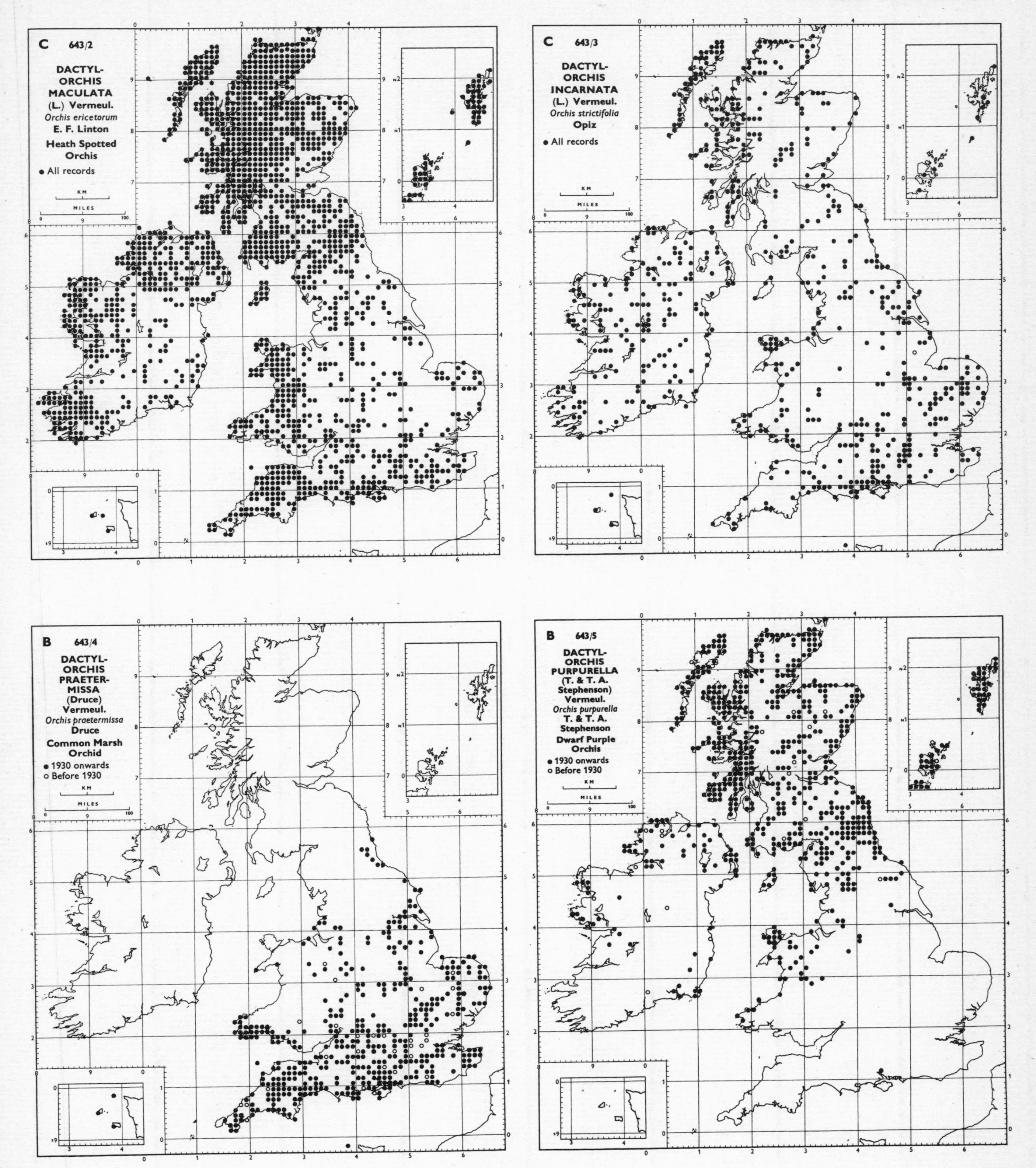
C
643/2
DACTYL-ORCHIS MACULATA
(L.) Vermeul.
Orchis ericetorum
E. F. Linton
Heath Spotted Orchis
● All records
KM
MILES
C
643/3
DACTYL-ORCHIS INCARNATA
(L.) Vermeul.
Orchis strictifolia
Opiz
● All records
KM
MILES
B
643/4
DACTYL-ORCHIS PRAETER-MISSA
(Druce)
Vermeul.
Orchis praetermissa
Druce
Common Marsh Orchid
● 1930 onwards
○ Before 1930
KM
MILES
B
643/5
DACTYL-ORCHIS PURPURELLA
(T. & T. A. Stephenson)
Vermeul.
Orchis purpurella
T. & T. A. Stephenson
Dwarf Purple Orchis
● 1930 onwards
○ Before 1930
KM
MILES

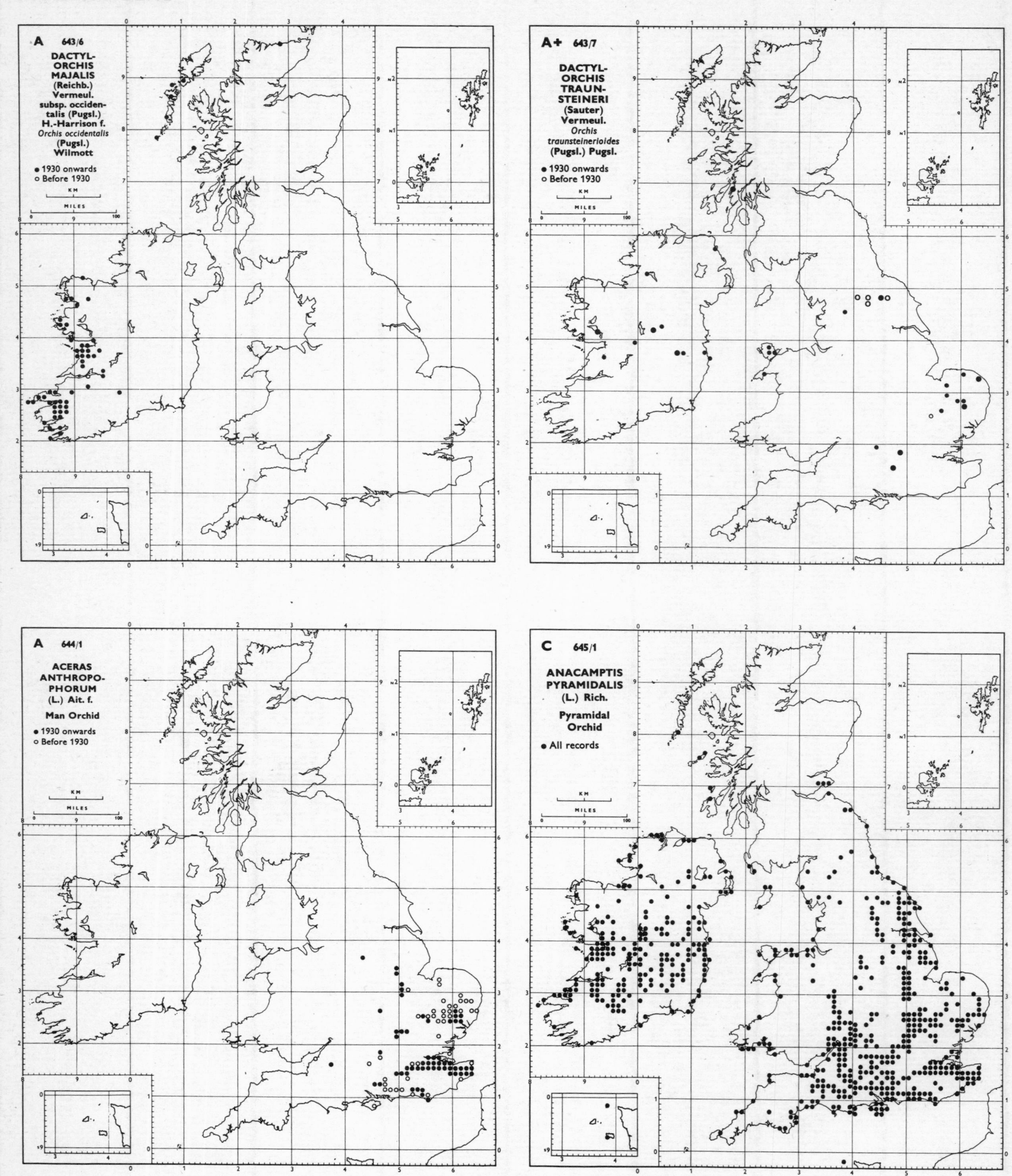
A 643/6
DACTYL-
ORCHIS
MAJALIS
(Reichb.)
Vermeul.
subsp. occiden-
talis (Pugsl.)
H.-Harrison f.
Orchis occidentalis
(Pugsl.)
Wilmott
• 1930 onwards
○ Before 1930
KM
MILES
A+ 643/7
DACTYL-
ORCHIS
TRAUN-
STEINERI
(Sauter)
Vermeul.
Orchis
traunsteinerioides
(Pugsl.) Pugsl.
• 1930 onwards
○ Before 1930
KM
MILES
A 644/1
ACERAS
ANTHROPO-
PHORUM
(L.) Ait. f.
Man Orchid
• 1930 onwards
○ Before 1930
KM
MILES
C 645/1
ANACAMPTIS
PYRAMIDALIS
(L.) Rich.
Pyramidal
Orchid
• All records
KM
MILES

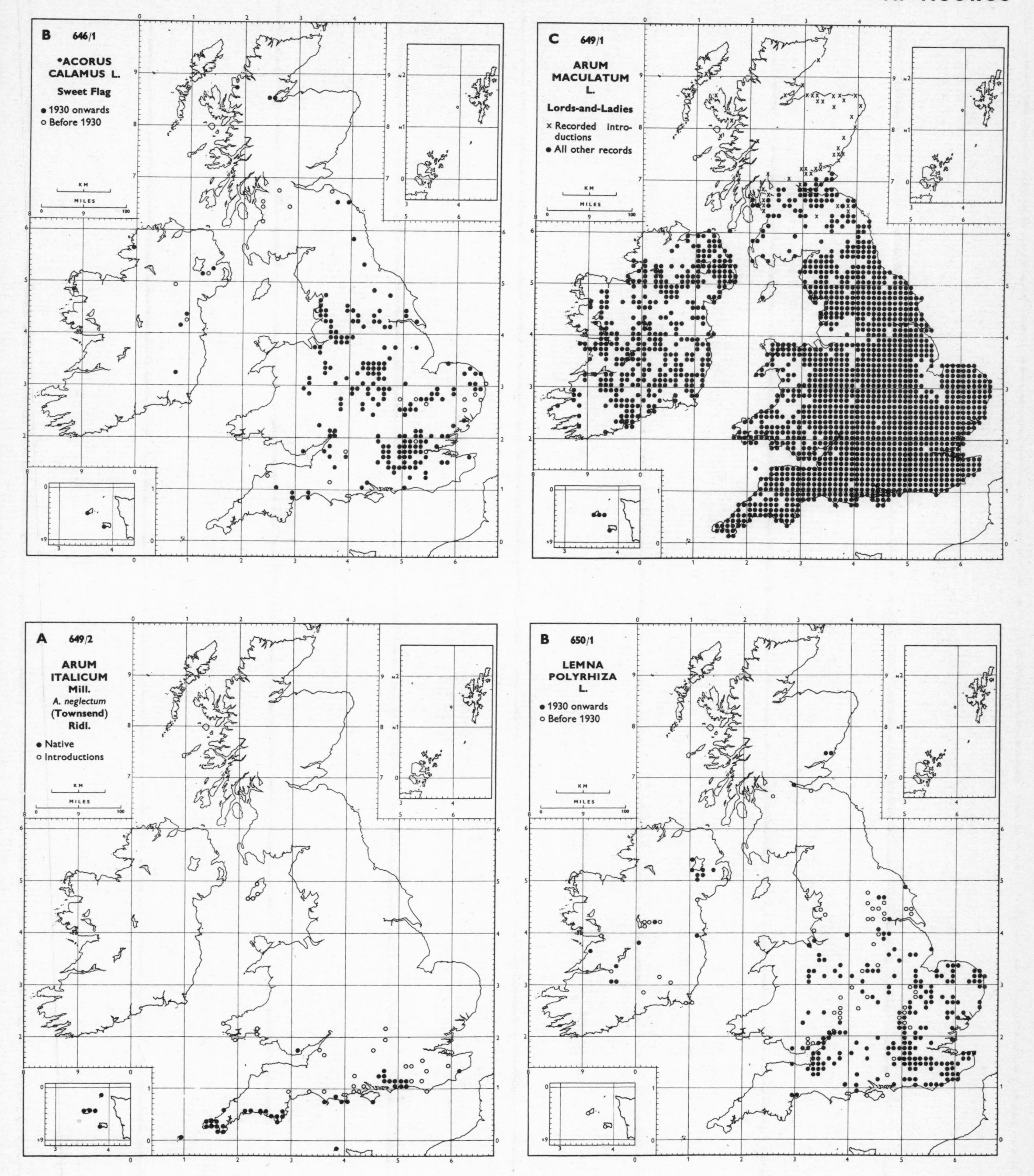
B 646/1
*ACORUS CALAMUS L.
Sweet Flag
● 1930 onwards
○ Before 1930
C 649/1
ARUM MACULATUM L.
Lords-and-Ladies
× Recorded introductions
● All other records
A 649/2
ARUM ITALICUM Mill.
A. neglectum (Townsend) Ridl.
● Native
○ Introductions
B 650/1
LEMNA POLYRHIZA L.
● 1930 onwards
○ Before 1930
KM
MILES

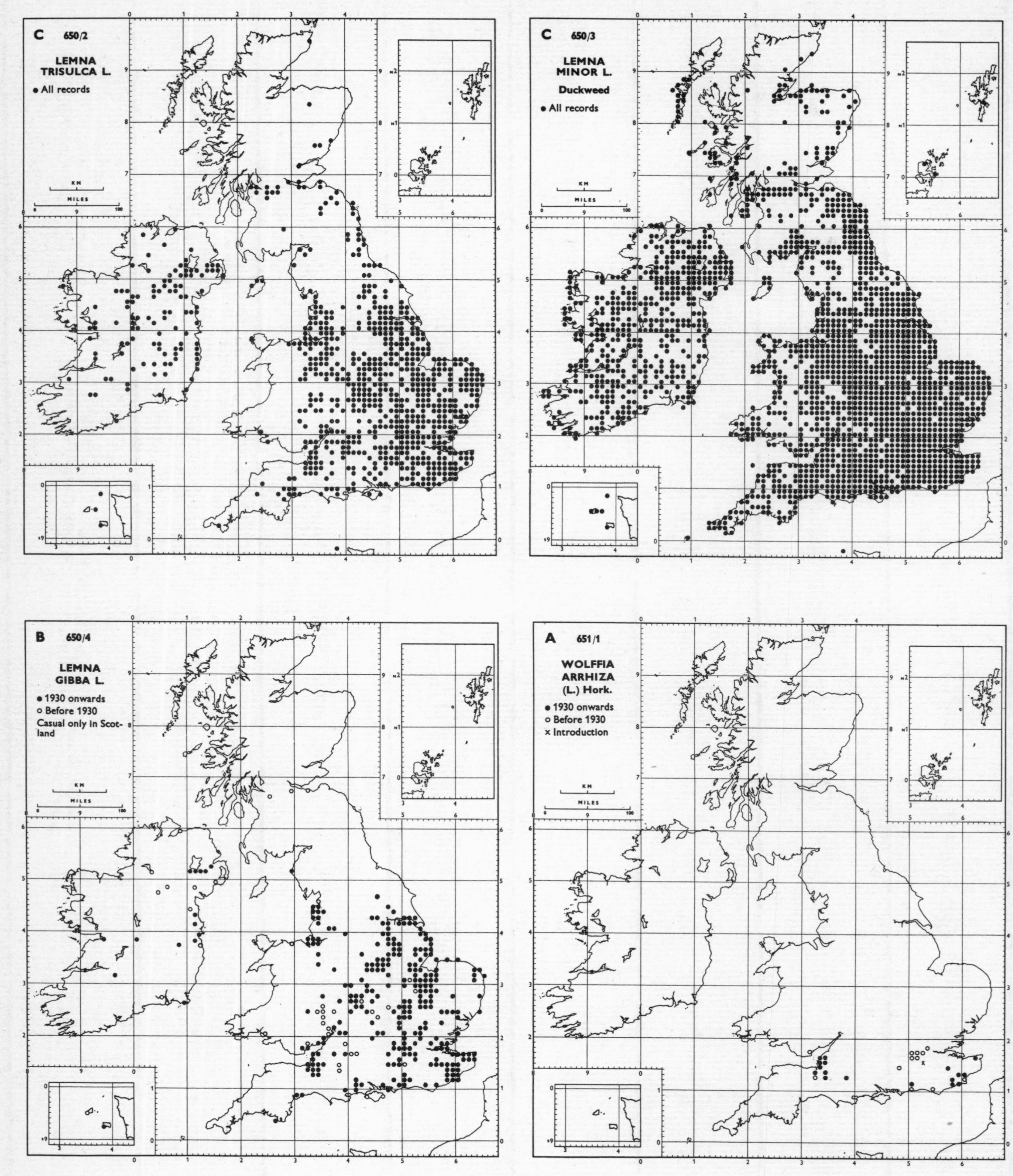
C 650/2
LEMNA TRISULCA L.
● All records
C 650/3
LEMNA MINOR L.
Duckweed
● All records
B 650/4
LEMNA GIBBA L.
● 1930 onwards
○ Before 1930
Casual only in Scotland
A 651/1
WOLFFIA ARRHIZA (L.) Hork.
● 1930 onwards
○ Before 1930
x Introduction
KM
MILES

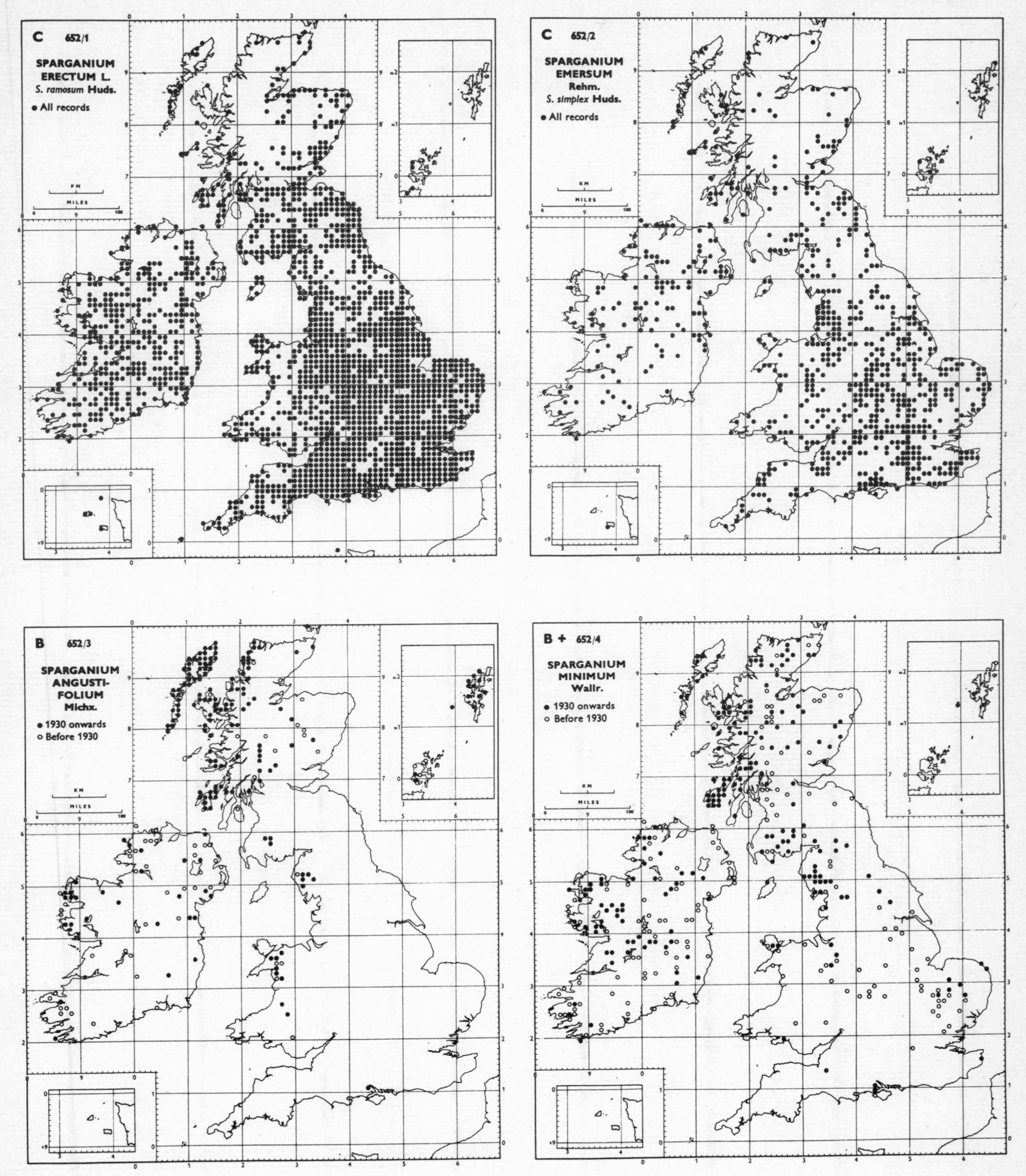
C 652/1
SPARGANIUM ERECTUM L.
S. ramosum Huds.
● All records
C 652/2
SPARGANIUM EMERSUM Rehm.
S. simplex Huds.
● All records
B 652/3
SPARGANIUM ANGUSTIFOLIUM Michx.
● 1930 onwards
○ Before 1930
B + 652/4
SPARGANIUM MINIMUM Wallr.
● 1930 onwards
○ Before 1930
KM
MILES

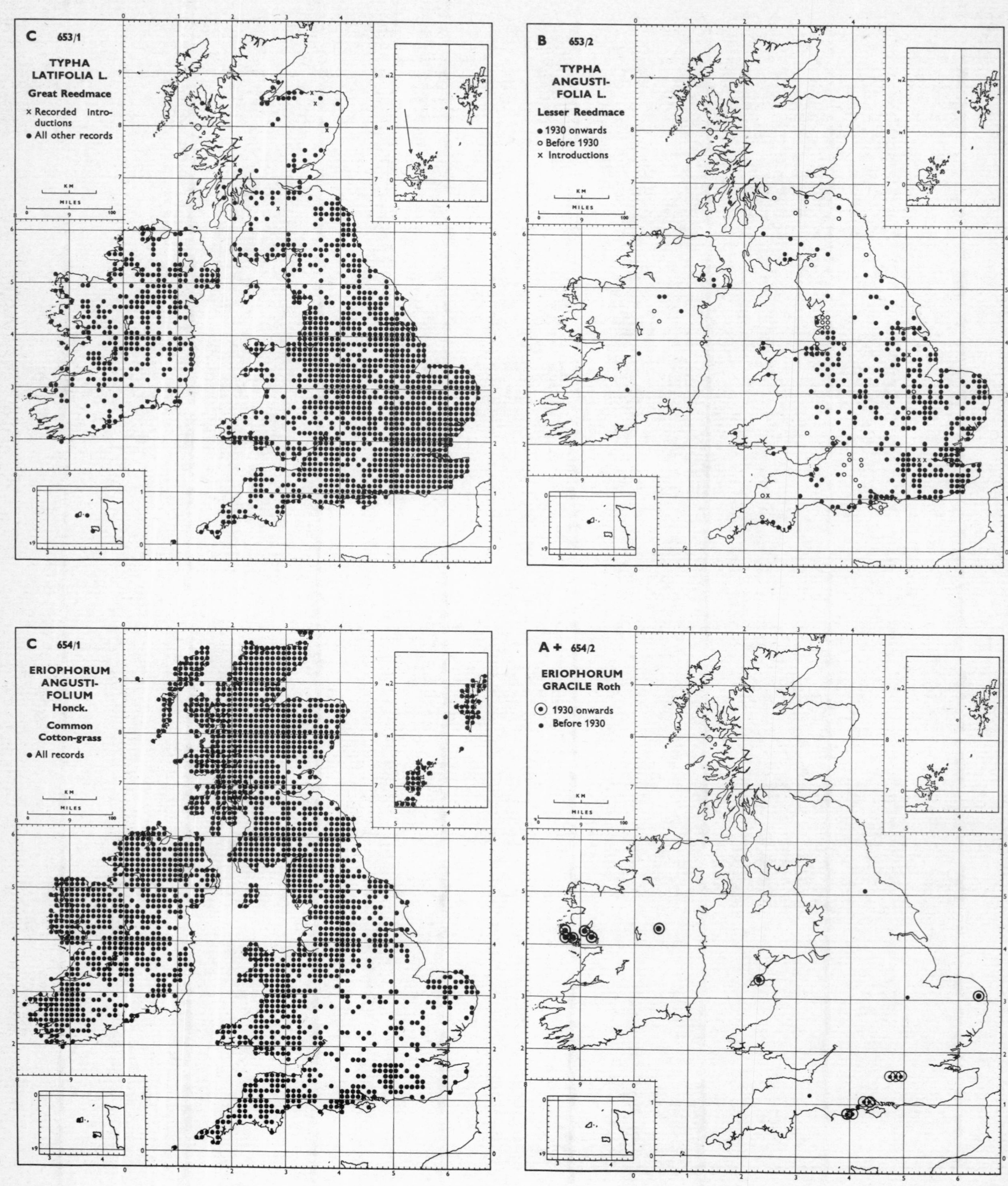
C 653/1
TYPHA LATIFOLIA L.
Great Reedmace
× Recorded introductions
● All other records
B 653/2
TYPHA ANGUSTIFOLIA L.
Lesser Reedmace
● 1930 onwards
○ Before 1930
× Introductions
C 654/1
ERIOPHORUM ANGUSTIFOLIUM Honck.
Common Cotton-grass
● All records
A+ 654/2
ERIOPHORUM GRACILE Roth
⊙ 1930 onwards
● Before 1930
KM
MILES

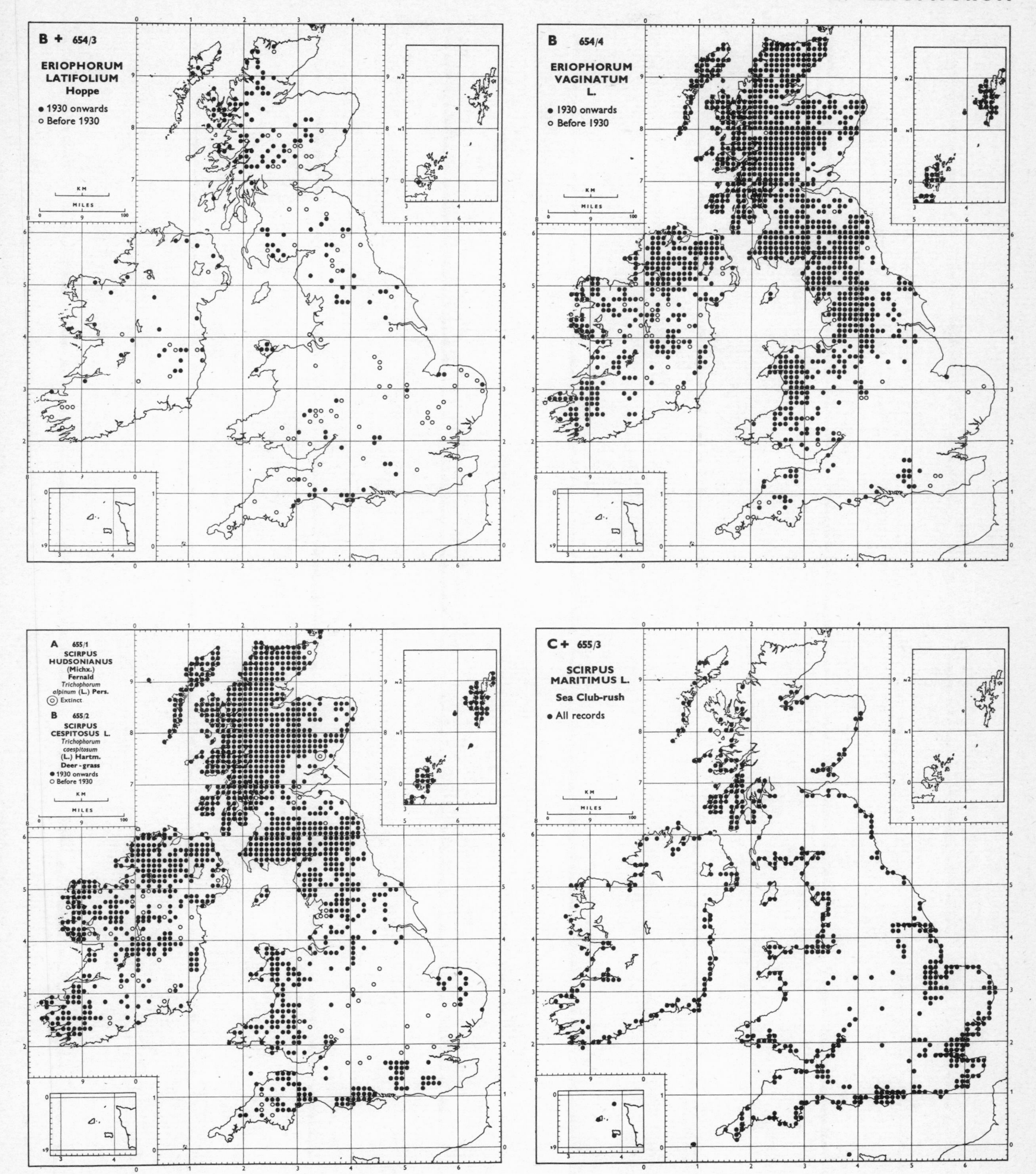
B + 654/3
ERIOPHORUM LATIFOLIUM Hoppe
● 1930 onwards
○ Before 1930
KM
MILES
B 654/4
ERIOPHORUM VAGINATUM L.
● 1930 onwards
○ Before 1930
A 655/1
SCIRPUS HUDSONIANUS (Michx.) Fernald
Trichophorum alpinum (L.) Pers.
◎ Extinct
B 655/2
SCIRPUS CESPITOSUS L.
Trichophorum caespitosum (L.) Hartm.
Deer-grass
● 1930 onwards
○ Before 1930
C + 655/3
SCIRPUS MARITIMUS L.
Sea Club-rush
● All records

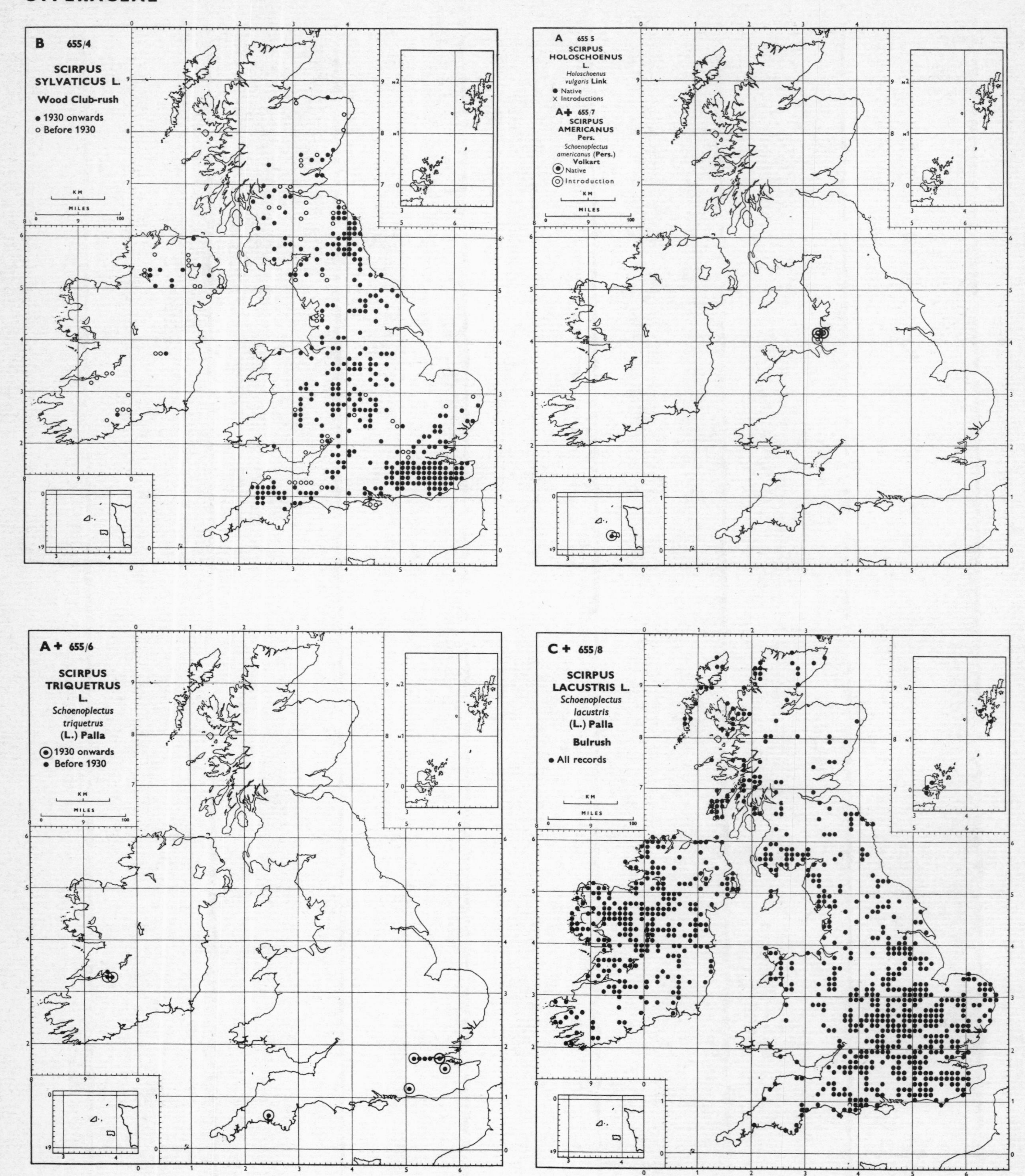
B 655/4
SCIRPUS SYLVATICUS L.
Wood Club-rush
● 1930 onwards
○ Before 1930
A 655/5
SCIRPUS HOLOSCHOENUS L.
Holoschoenus vulgaris Link
● Native
X Introductions
A+ 655/7
SCIRPUS AMERICANUS Pers.
Schoenoplectus americanus (Pers.) Volkart
Native
Introduction
A+ 655/6
SCIRPUS TRIQUETRUS L.
Schoenoplectus triquetrus (L.) Palla
1930 onwards
● Before 1930
C+ 655/8
SCIRPUS LACUSTRIS L.
Schoenoplectus lacustris (L.) Palla
Bulrush
● All records
KM
MILES

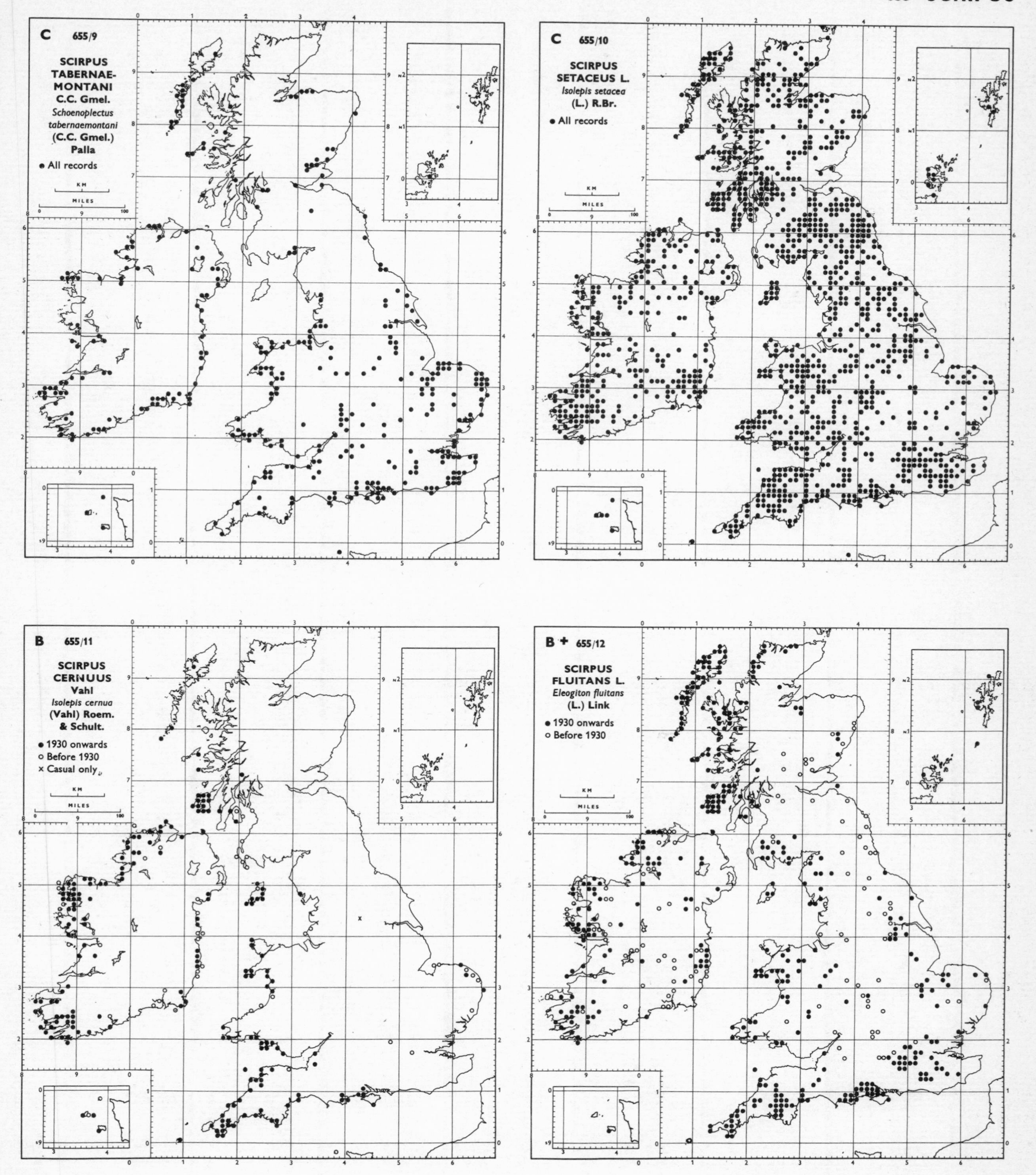
C 655/9
SCIRPUS TABERNAE-MONTANI C.C. Gmel.
Schoenoplectus tabernaemontani (C.C. Gmel.) Palla
• All records
C 655/10
SCIRPUS SETACEUS L.
Isolepis setacea (L.) R.Br.
• All records
B 655/11
SCIRPUS CERNUUS Vahl
Isolepis cernua (Vahl) Roem. & Schult.
• 1930 onwards
○ Before 1930
× Casual only
B + 655/12
SCIRPUS FLUITANS L.
Eleogiton fluitans (L.) Link
• 1930 onwards
○ Before 1930
KM
MILES

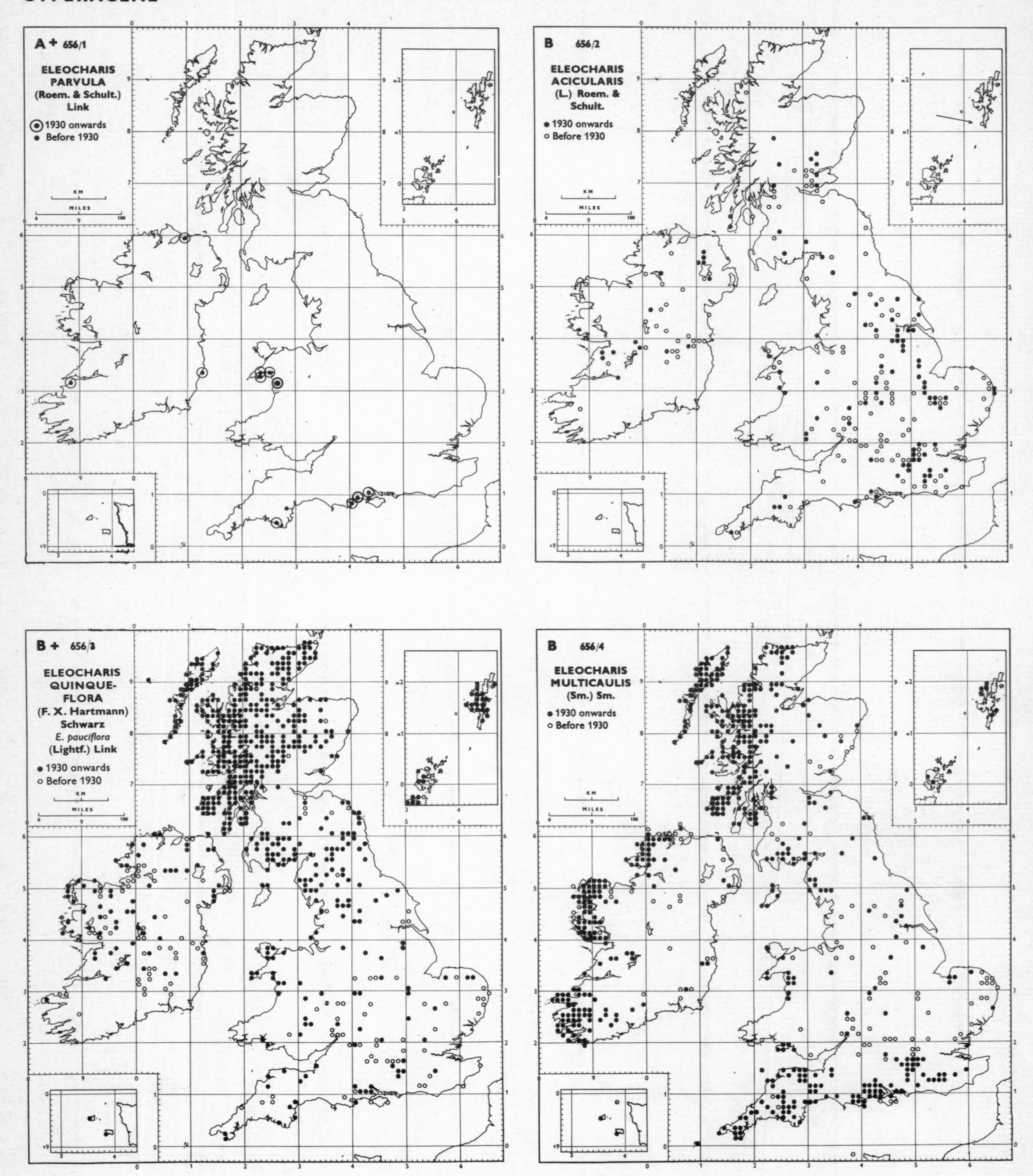

A + 656/1
ELEOCHARIS PARVULA (Roem. & Schult.) Link
1930 onwards
Before 1930
B 656/2
ELEOCHARIS ACICULARIS (L.) Roem. & Schult.
1930 onwards
Before 1930
B + 656/3
ELEOCHARIS QUINQUE-FLORA (F. X. Hartmann) Schwarz
E. pauciflora (Lightf.) Link
1930 onwards
Before 1930
B 656/4
ELEOCHARIS MULTICAULIS (Sm.) Sm.
1930 onwards
Before 1930

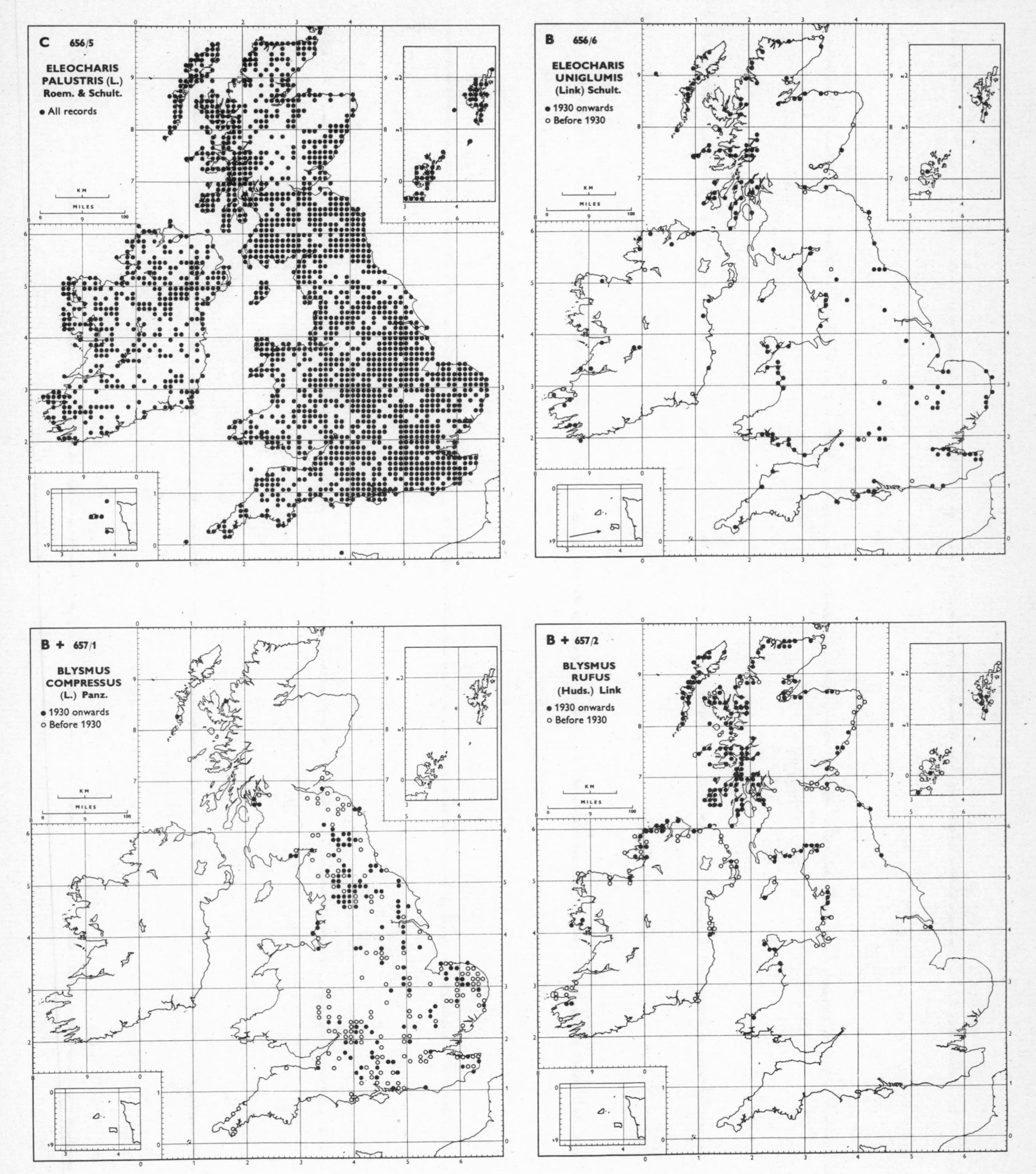
C 656/5
ELEOCHARIS PALUSTRIS (L.) Roem. & Schult.
● All records
B 656/6
ELEOCHARIS UNIGLUMIS (Link) Schult.
● 1930 onwards
○ Before 1930
B + 657/1
BLYSMUS COMPRESSUS (L.) Panz.
● 1930 onwards
○ Before 1930
B + 657/2
BLYSMUS RUFUS (Huds.) Link
● 1930 onwards
○ Before 1930
KM
MILES

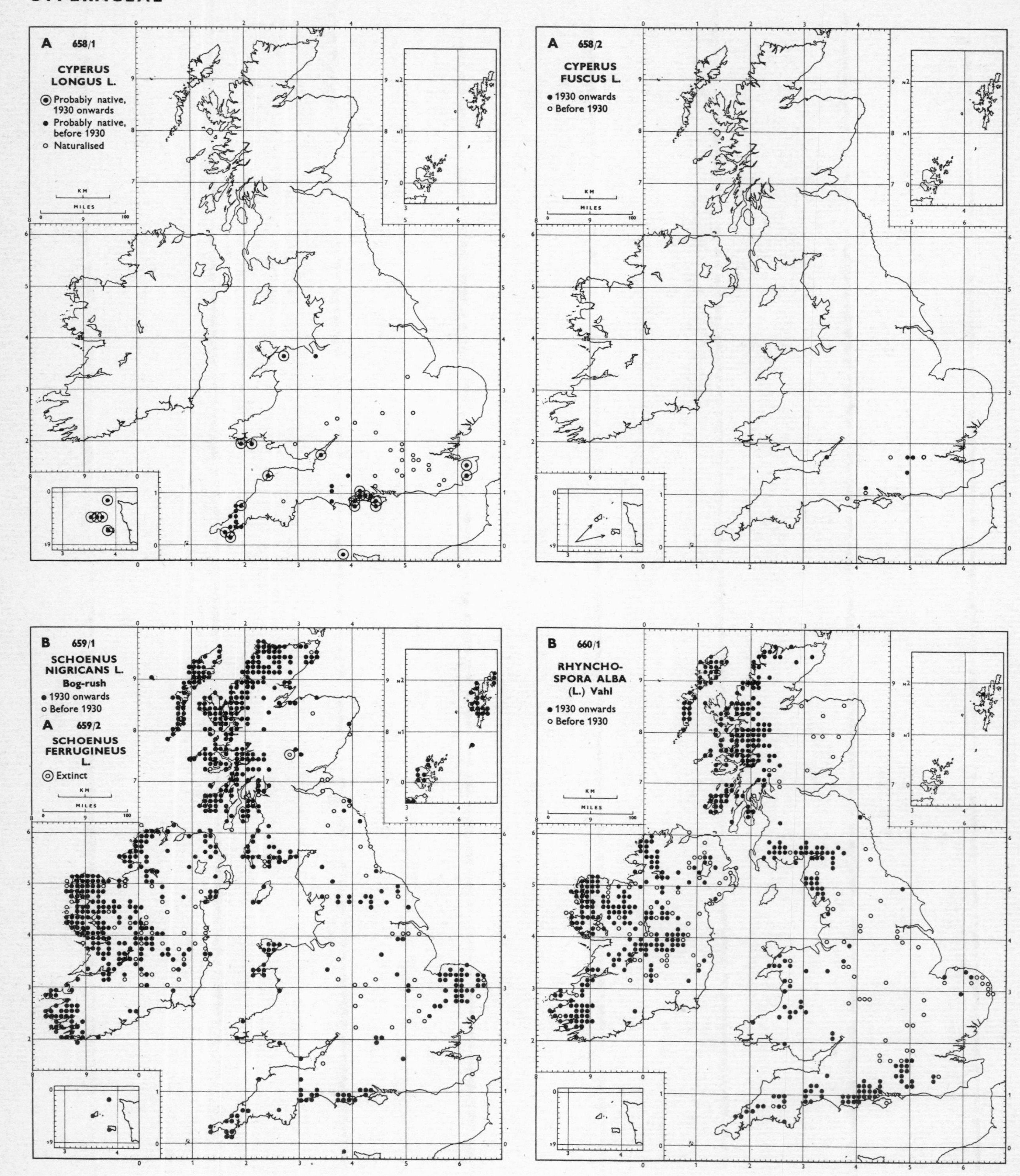
A 658/1
CYPERUS LONGUS L.
Probably native, 1930 onwards
Probably native, before 1930
Naturalised
A 658/2
CYPERUS FUSCUS L.
1930 onwards
Before 1930
B 659/1
SCHOENUS NIGRICANS L.
Bog-rush
1930 onwards
Before 1930
A 659/2
SCHOENUS FERRUGINEUS L.
Extinct
B 660/1
RHYNCHOSPORA ALBA (L.) Vahl
1930 onwards
Before 1930
KM
MILES

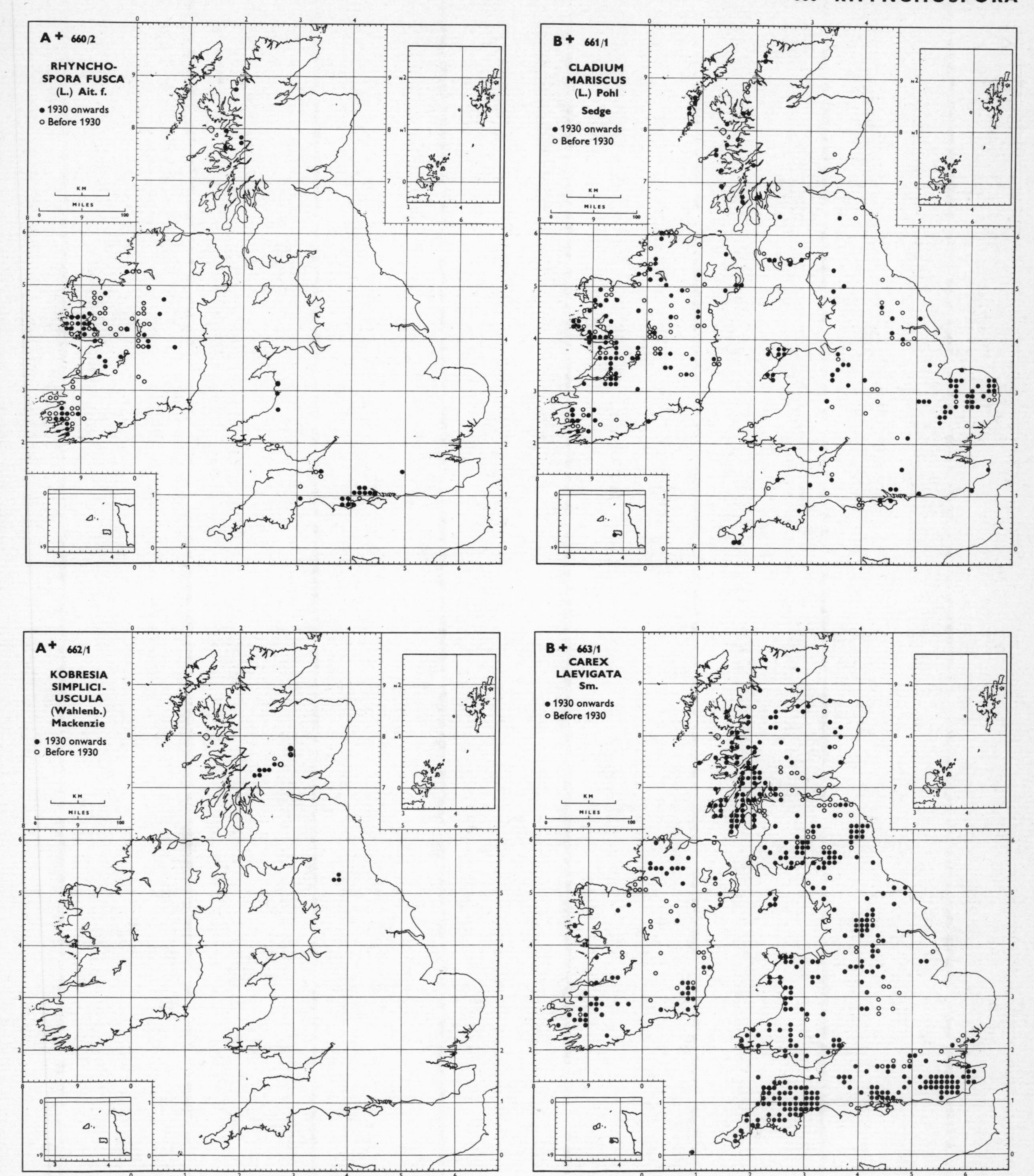
A+ 660/2
RHYNCHO-SPORA FUSCA (L.) Ait. f.
● 1930 onwards
○ Before 1930
KM
MILES
B+ 661/1
CLADIUM MARISCUS (L.) Pohl
Sedge
● 1930 onwards
○ Before 1930
A+ 662/1
KOBRESIA SIMPLICI-USCULA (Wahlenb.) Mackenzie
● 1930 onwards
○ Before 1930
B+ 663/1
CAREX LAEVIGATA Sm.
● 1930 onwards
○ Before 1930

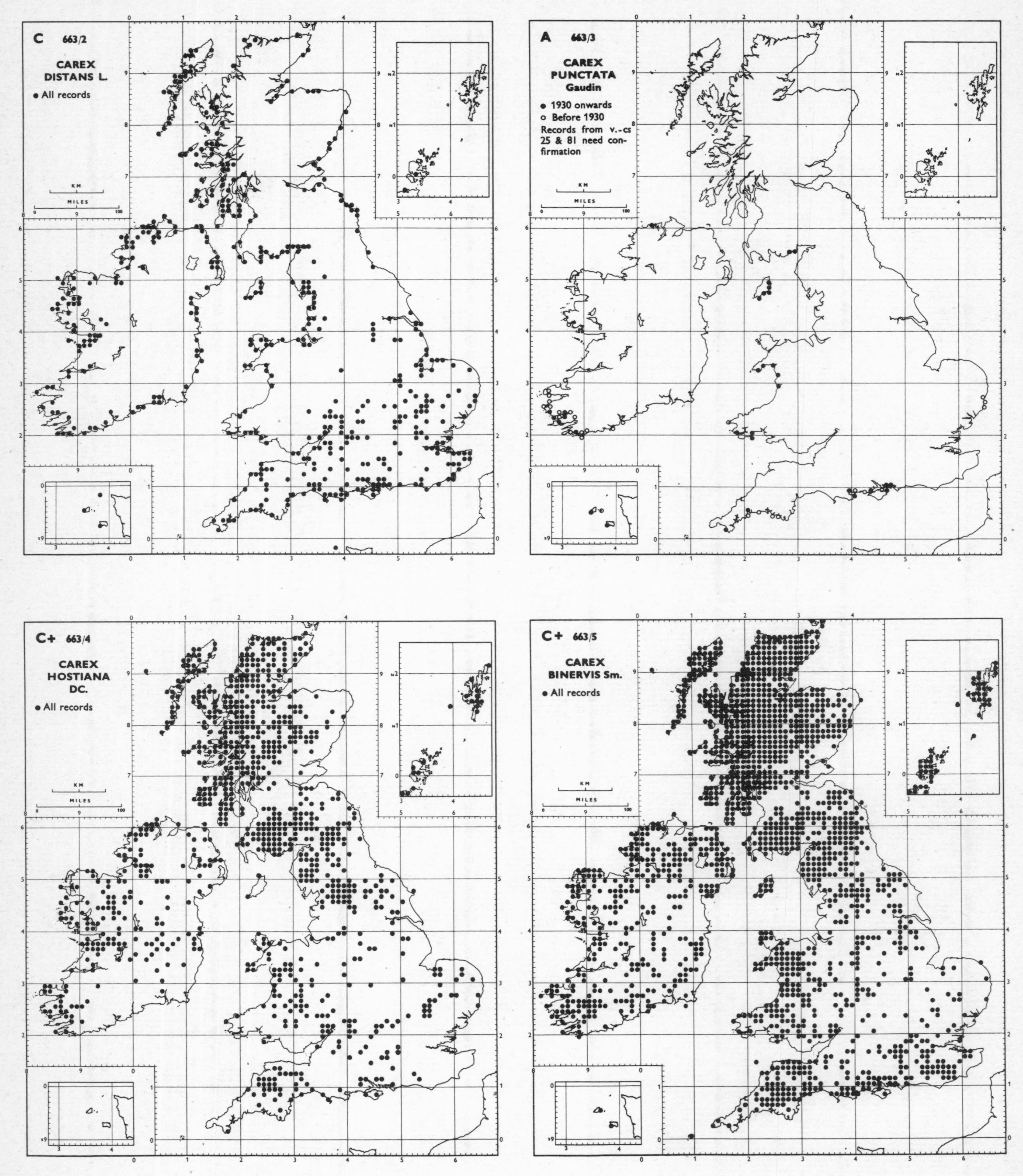
C 663/2
CAREX DISTANS L.
● All records
A 663/3
CAREX PUNCTATA Gaudin
● 1930 onwards
○ Before 1930
Records from v.-cs 25 & 81 need confirmation
C+ 663/4
CAREX HOSTIANA DC.
● All records
C+ 663/5
CAREX BINERVIS Sm.
● All records
KM
MILES

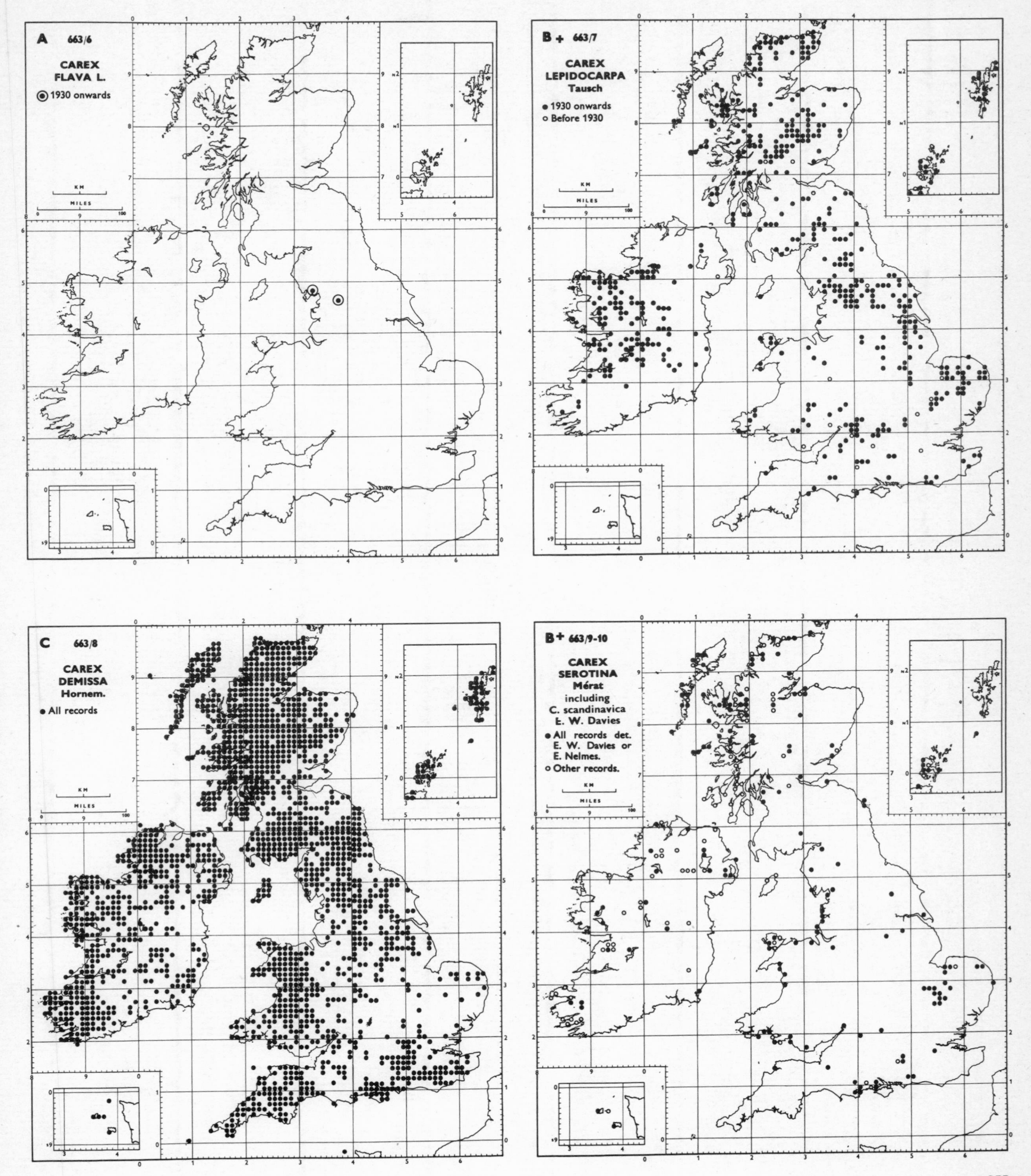
A 663/6
CAREX
FLAVA L.
1930 onwards
B + 663/7
CAREX
LEPIDOCARPA
Tausch
• 1930 onwards
○ Before 1930
C 663/8
CAREX
DEMISSA
Hornem.
• All records
B + 663/9-10
CAREX
SEROTINA
Mérat
including
C. scandinavica
E. W. Davies
• All records det. E. W. Davies or E. Nelmes.
○ Other records.
KM
MILES

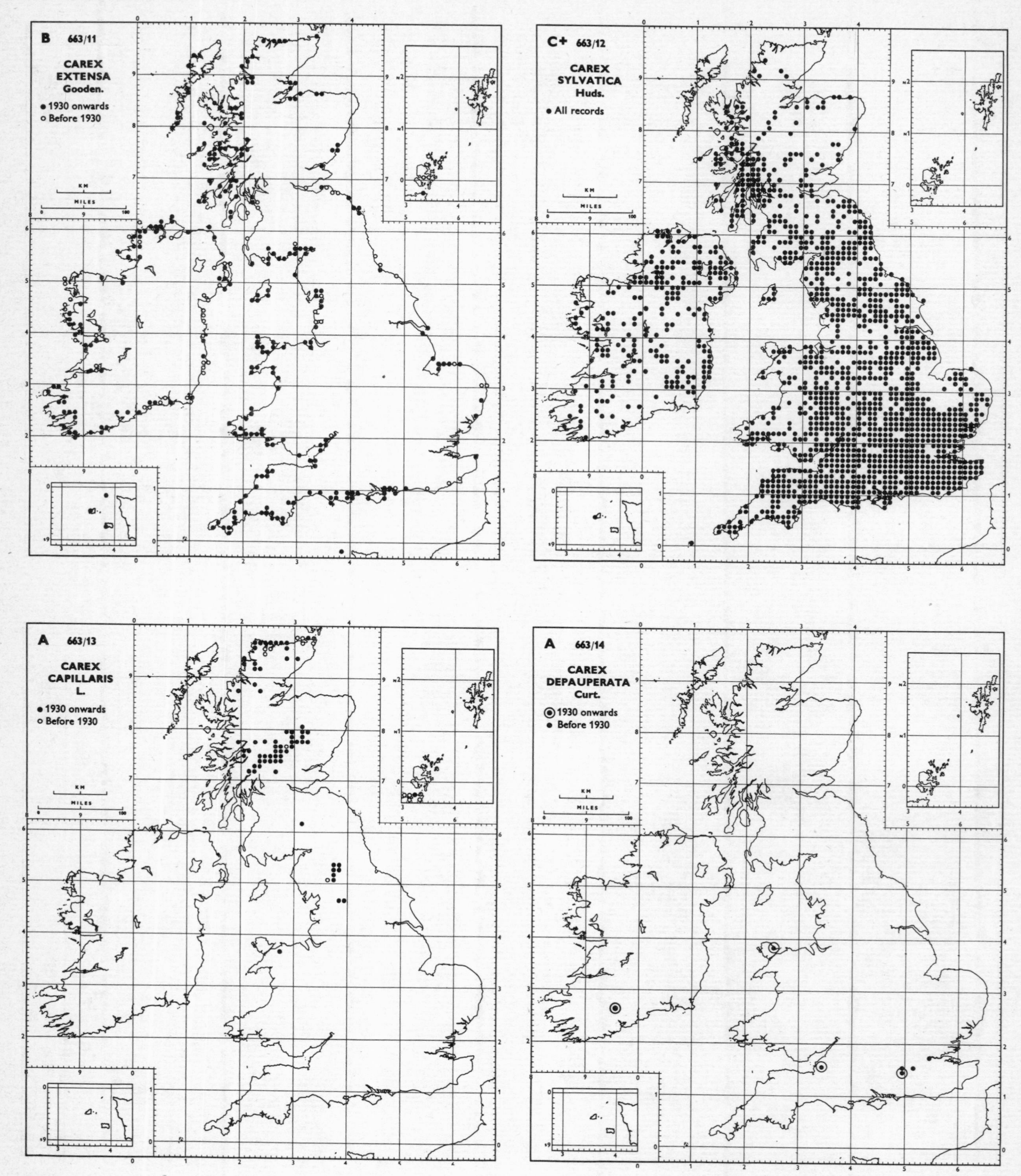
B 663/11
CAREX
EXTENSA
Gooden.
• 1930 onwards
○ Before 1930
C+ 663/12
CAREX
SYLVATICA
Huds.
• All records
A 663/13
CAREX
CAPILLARIS
L.
• 1930 onwards
○ Before 1930
A 663/14
CAREX
DEPAUPERATA
Curt.
⊙ 1930 onwards
• Before 1930
KM
MILES

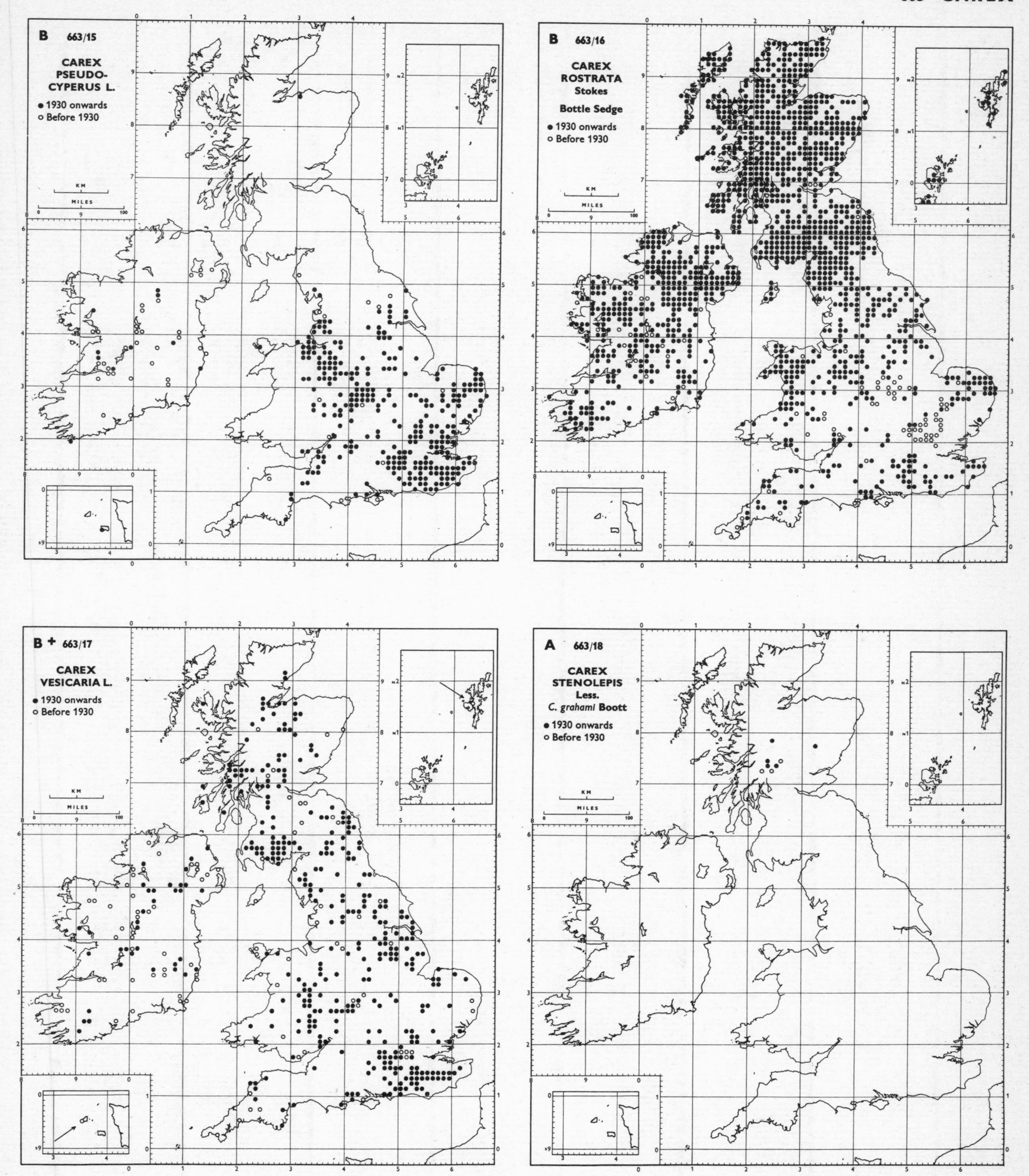
B 663/15
CAREX PSEUDO-CYPERUS L.
● 1930 onwards
○ Before 1930
B 663/16
CAREX ROSTRATA Stokes
Bottle Sedge
● 1930 onwards
○ Before 1930
B + 663/17
CAREX VESICARIA L.
● 1930 onwards
○ Before 1930
A 663/18
CAREX STENOLEPIS Less.
C. grahami Boott
● 1930 onwards
○ Before 1930

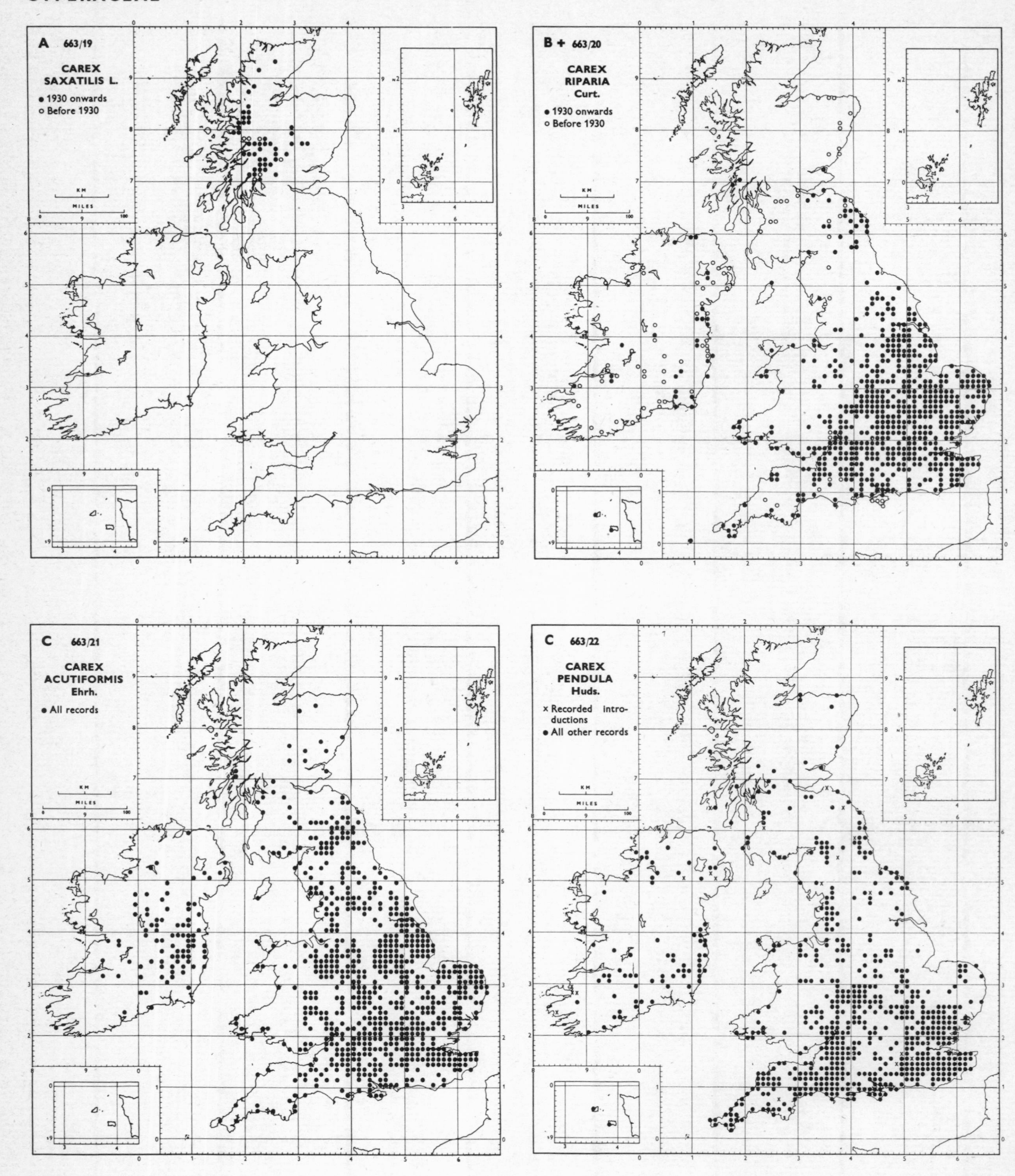
A 663/19
CAREX
SAXATILIS L.
● 1930 onwards
○ Before 1930
B + 663/20
CAREX
RIPARIA
Curt.
● 1930 onwards
○ Before 1930
C 663/21
CAREX
ACUTIFORMIS
Ehrh.
● All records
C 663/22
CAREX
PENDULA
Huds.
× Recorded introductions
● All other records

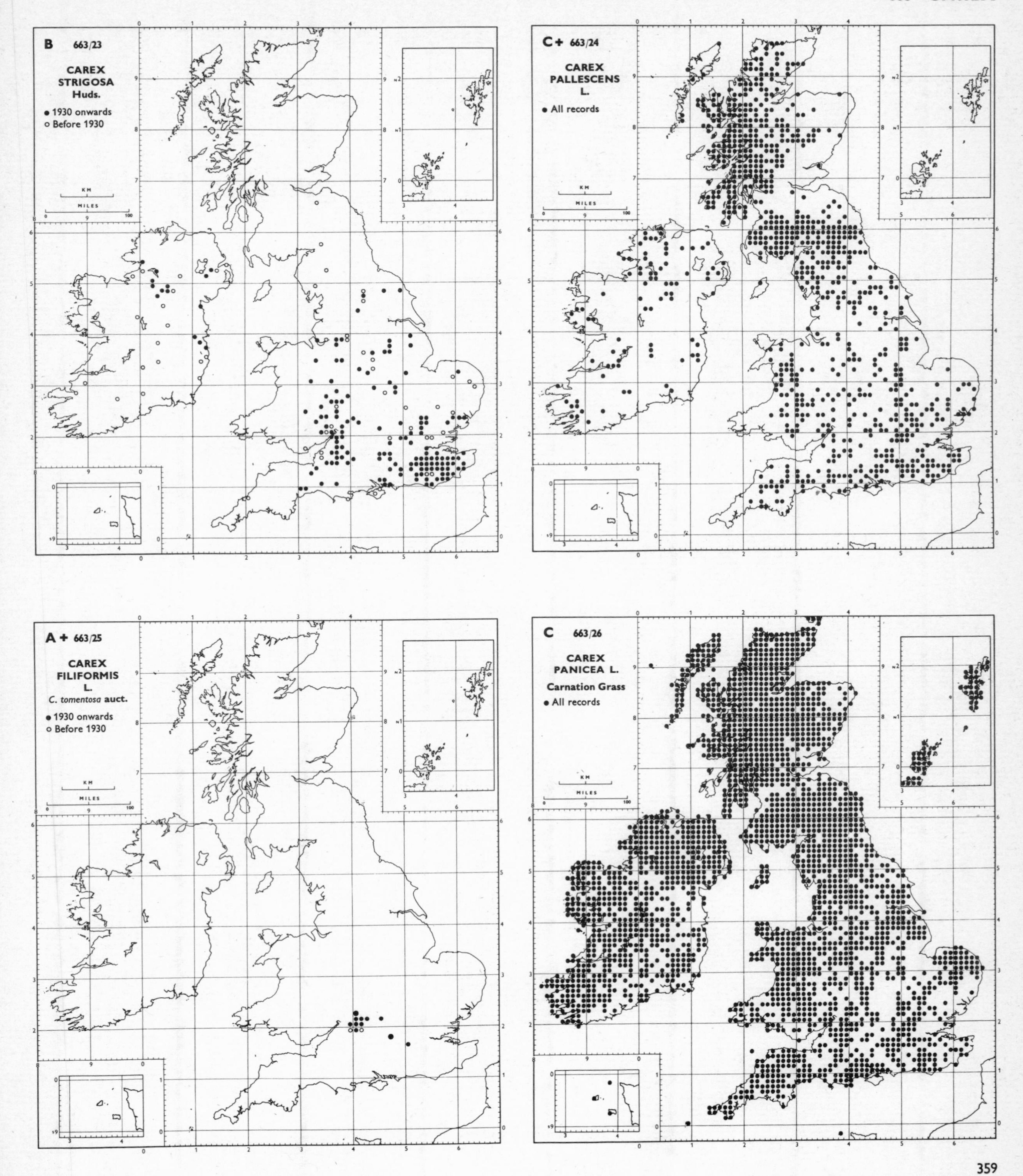
B 663/23
CAREX
STRIGOSA
Huds.
1930 onwards
Before 1930
C+ 663/24
CAREX
PALLESCENS
L.
All records
A+ 663/25
CAREX
FILIFORMIS
L.
C. tomentosa auct.
1930 onwards
Before 1930
C 663/26
CAREX
PANICEA L.
Carnation Grass
All records
KM
MILES

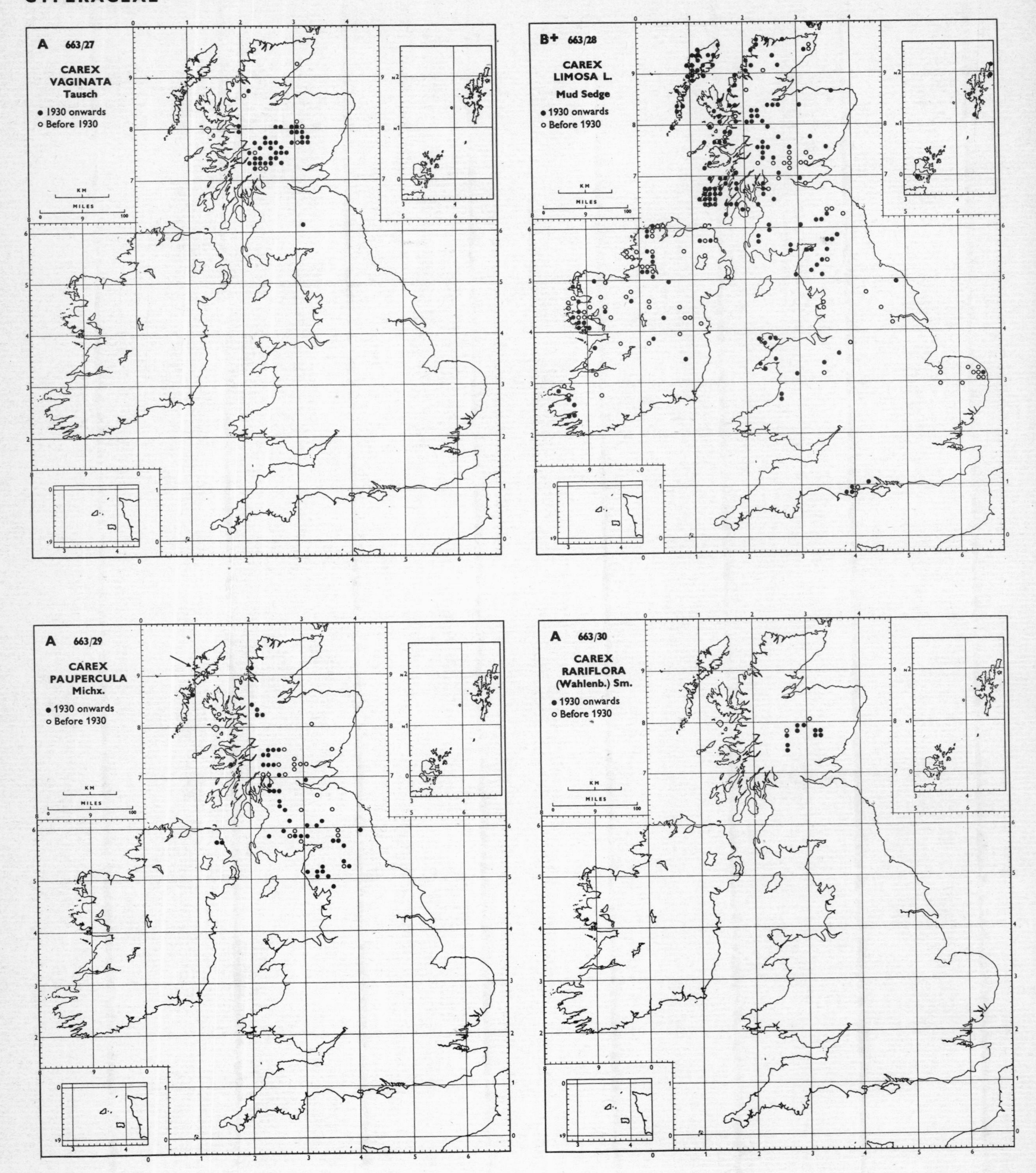
A 663/27
CAREX
VAGINATA
Tausch
● 1930 onwards
○ Before 1930
B+ 663/28
CAREX
LIMOSA L.
Mud Sedge
● 1930 onwards
○ Before 1930
A 663/29
CAREX
PAUPERCULA
Michx.
● 1930 onwards
○ Before 1930
A 663/30
CAREX
RARIFLORA
(Wahlenb.) Sm.
● 1930 onwards
○ Before 1930
KM
MILES

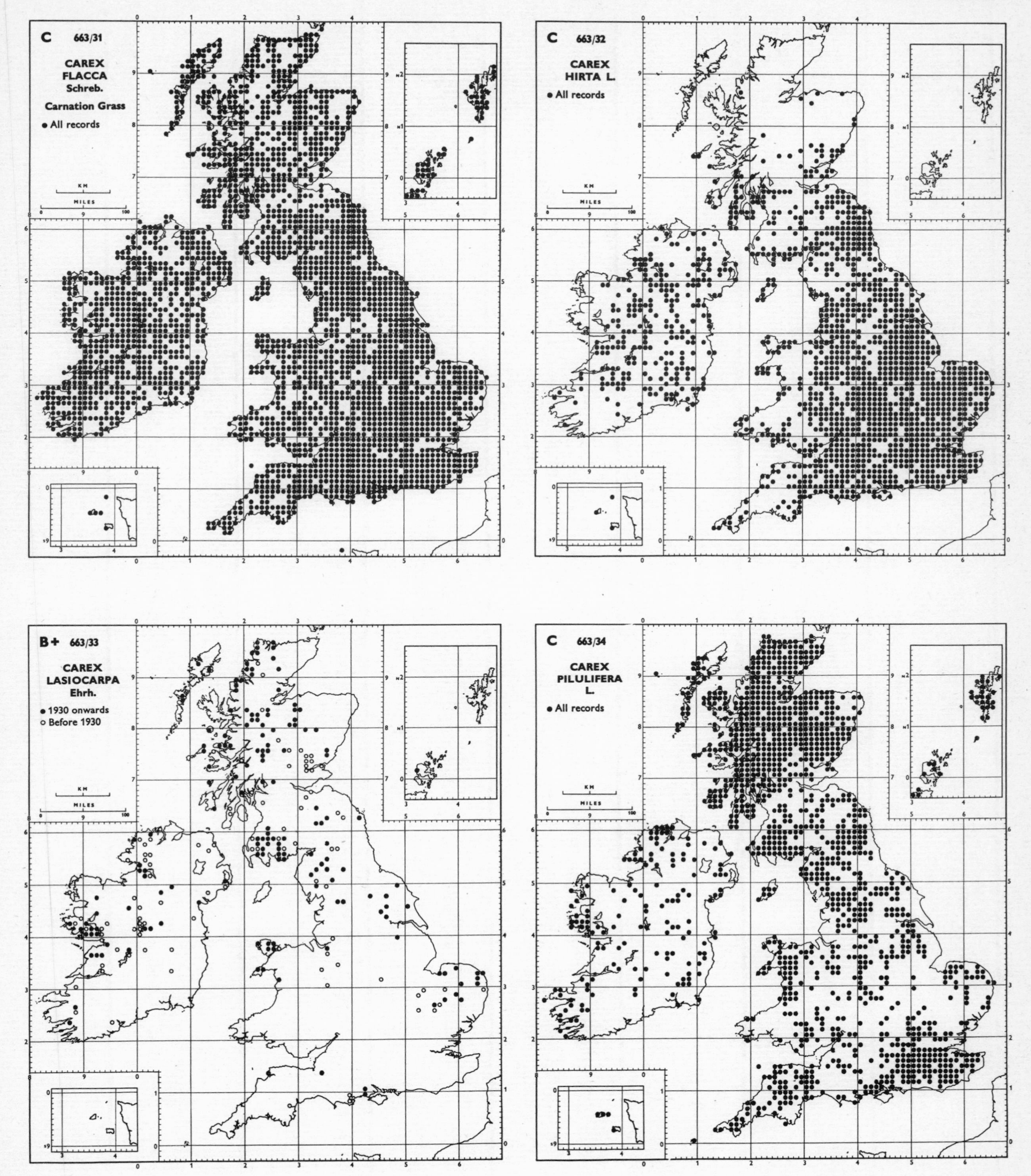
C 663/31
CAREX
FLACCA
Schreb.
Carnation Grass
● All records
C 663/32
CAREX
HIRTA L.
● All records
B+ 663/33
CAREX
LASIOCARPA
Ehrh.
● 1930 onwards
○ Before 1930
C 663/34
CAREX
PILULIFERA
L.
● All records
KM
MILES

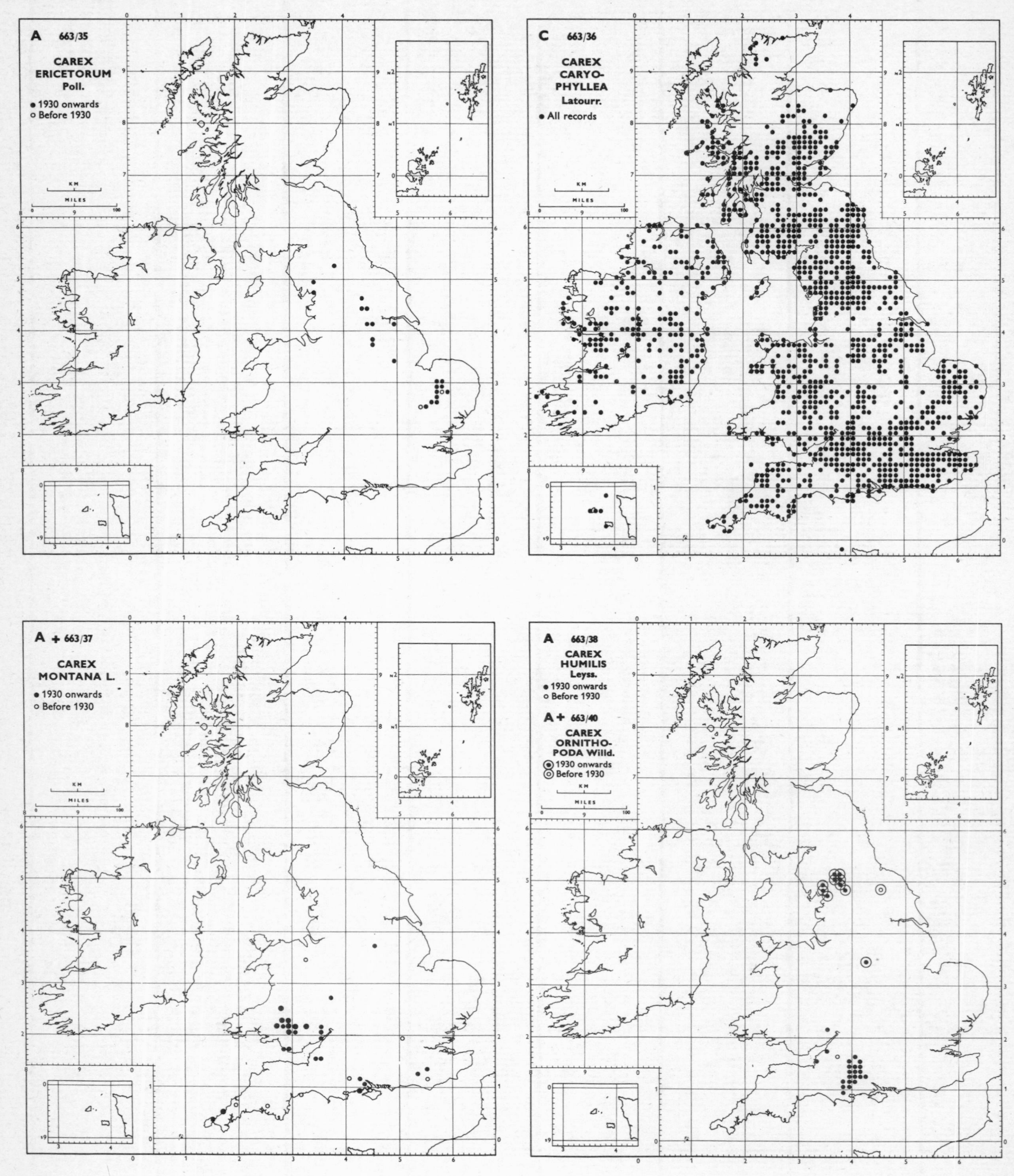
A 663/35
CAREX ERICETORUM Poll.
● 1930 onwards
○ Before 1930
C 663/36
CAREX CARYOPHYLLEA Latourr.
● All records
A + 663/37
CAREX MONTANA L.
● 1930 onwards
○ Before 1930
A 663/38
CAREX HUMILIS Leyss.
● 1930 onwards
○ Before 1930
A + 663/40
CAREX ORNITHOPODA Willd.
◉ 1930 onwards
◎ Before 1930
KM
MILES

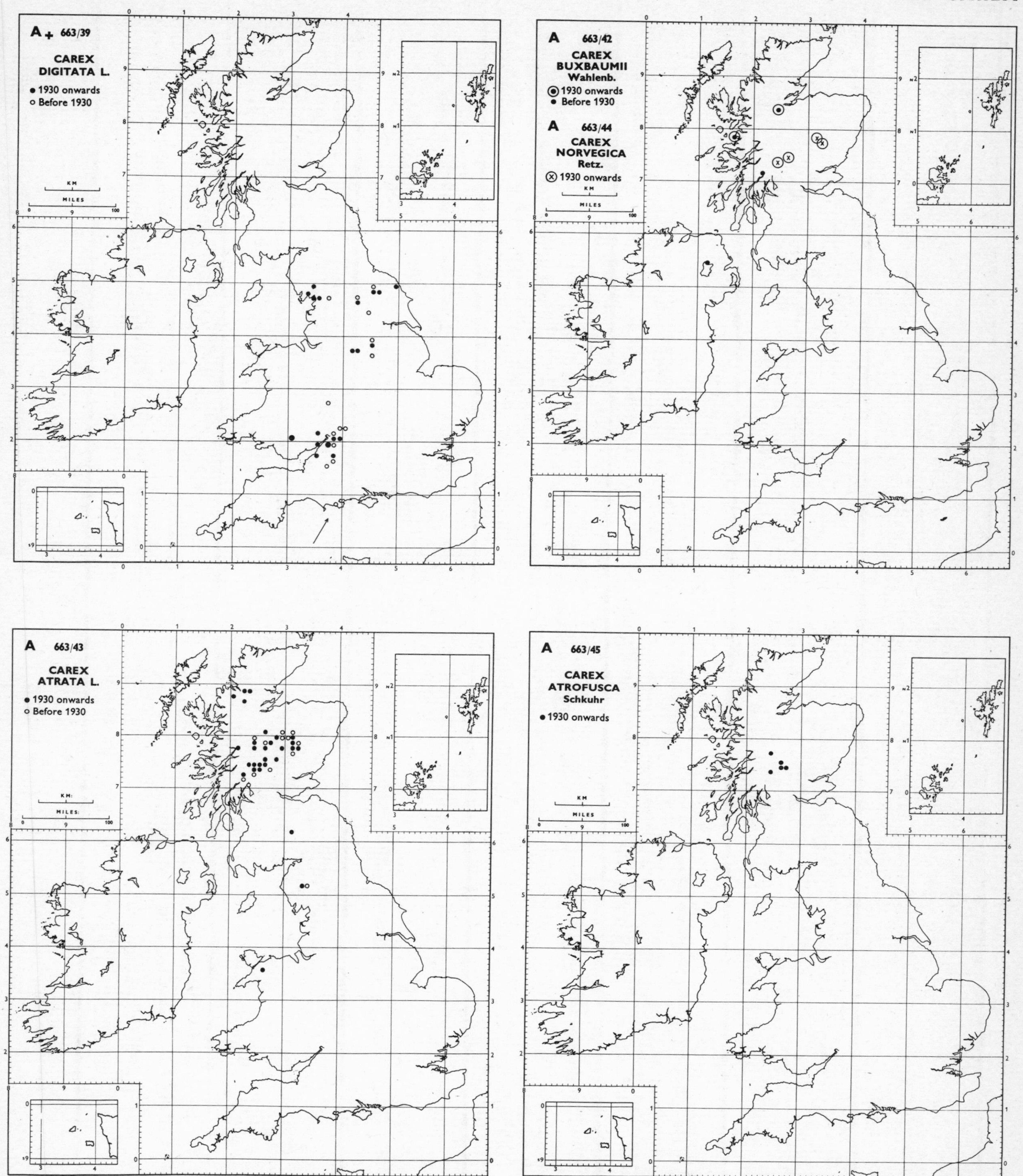
A + 663/39
CAREX DIGITATA L.
● 1930 onwards
○ Before 1930
A 663/42
CAREX BUXBAUMII Wahlenb.
◉ 1930 onwards
● Before 1930
A 663/44
CAREX NORVEGICA Retz.
⊗ 1930 onwards
A 663/43
CAREX ATRATA L.
● 1930 onwards
○ Before 1930
A 663/45
CAREX ATROFUSCA Schkuhr
● 1930 onwards
KM
MILES

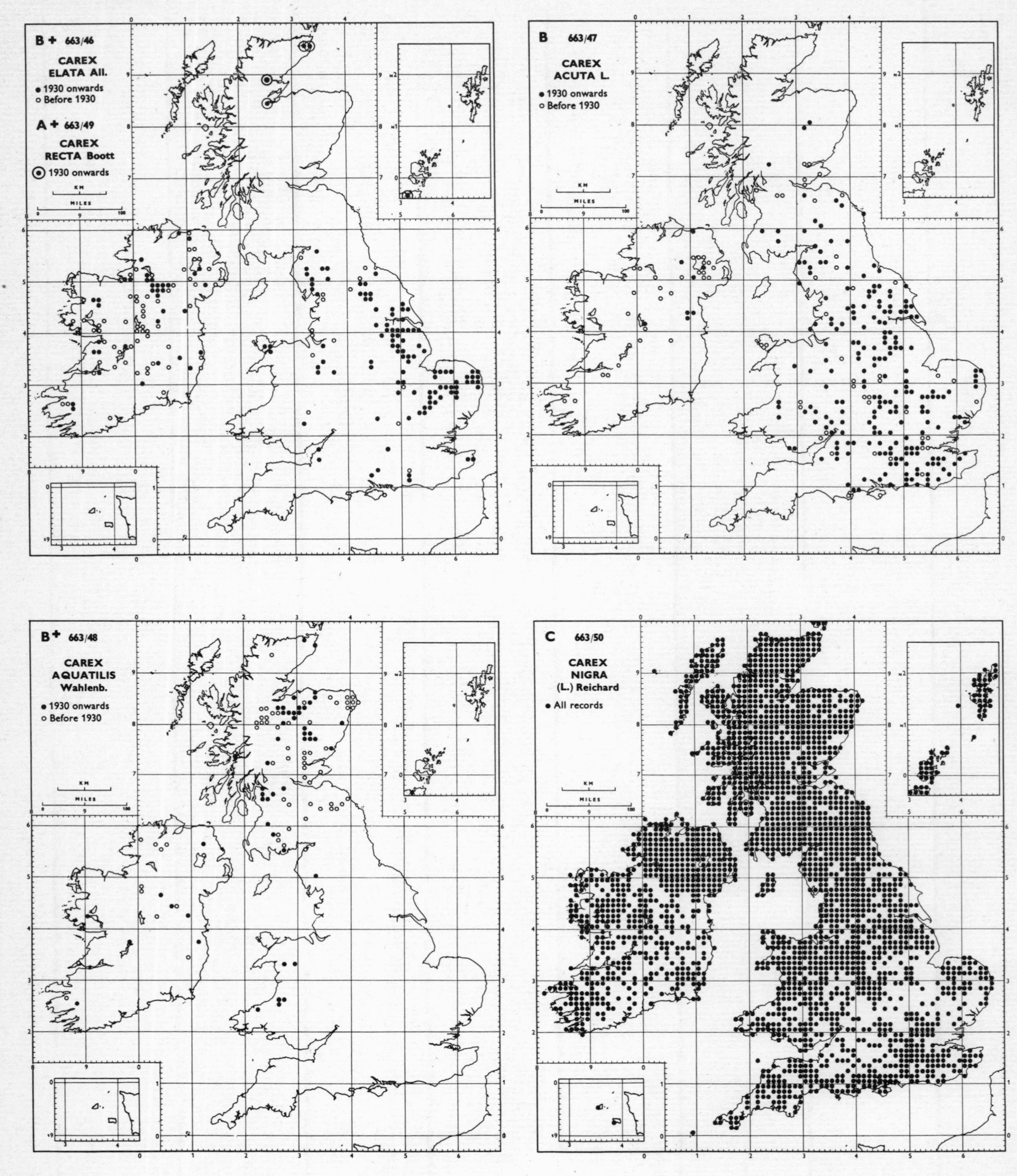

B + 663/46
CAREX
ELATA All.
• 1930 onwards
○ Before 1930
A + 663/49
CAREX
RECTA Boott
⊙ 1930 onwards
KM
MILES
B 663/47
CAREX
ACUTA L.
• 1930 onwards
○ Before 1930
B + 663/48
CAREX
AQUATILIS
Wahlenb.
• 1930 onwards
○ Before 1930
C 663/50
CAREX
NIGRA
(L.) Reichard
• All records

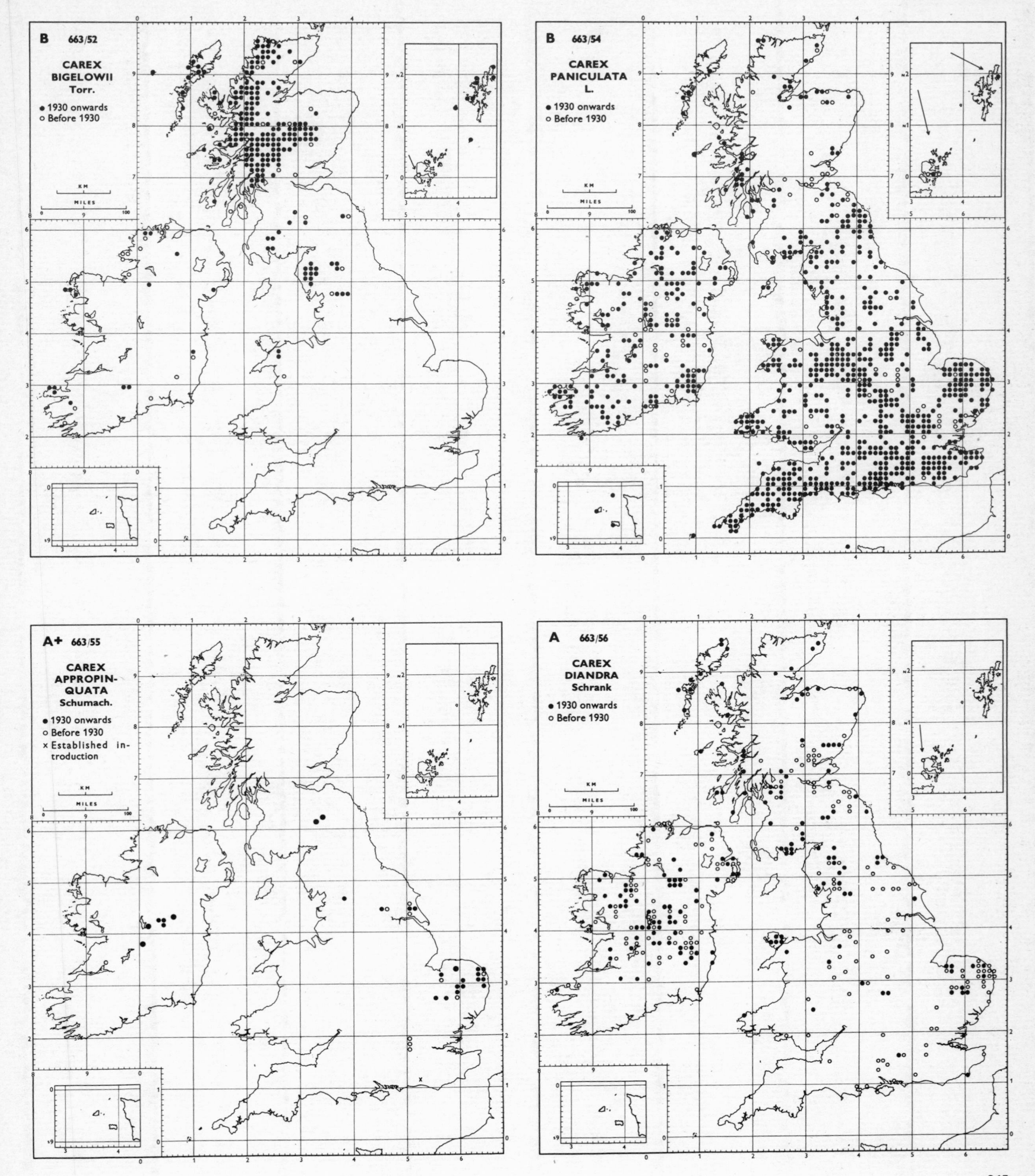
B 663/52
CAREX BIGELOWII Torr.
● 1930 onwards
○ Before 1930
KM
MILES
B 663/54
CAREX PANICULATA L.
● 1930 onwards
○ Before 1930
A+ 663/55
CAREX APPROPIN-QUATA Schumach.
● 1930 onwards
○ Before 1930
× Established introduction
A 663/56
CAREX DIANDRA Schrank
● 1930 onwards
○ Before 1930

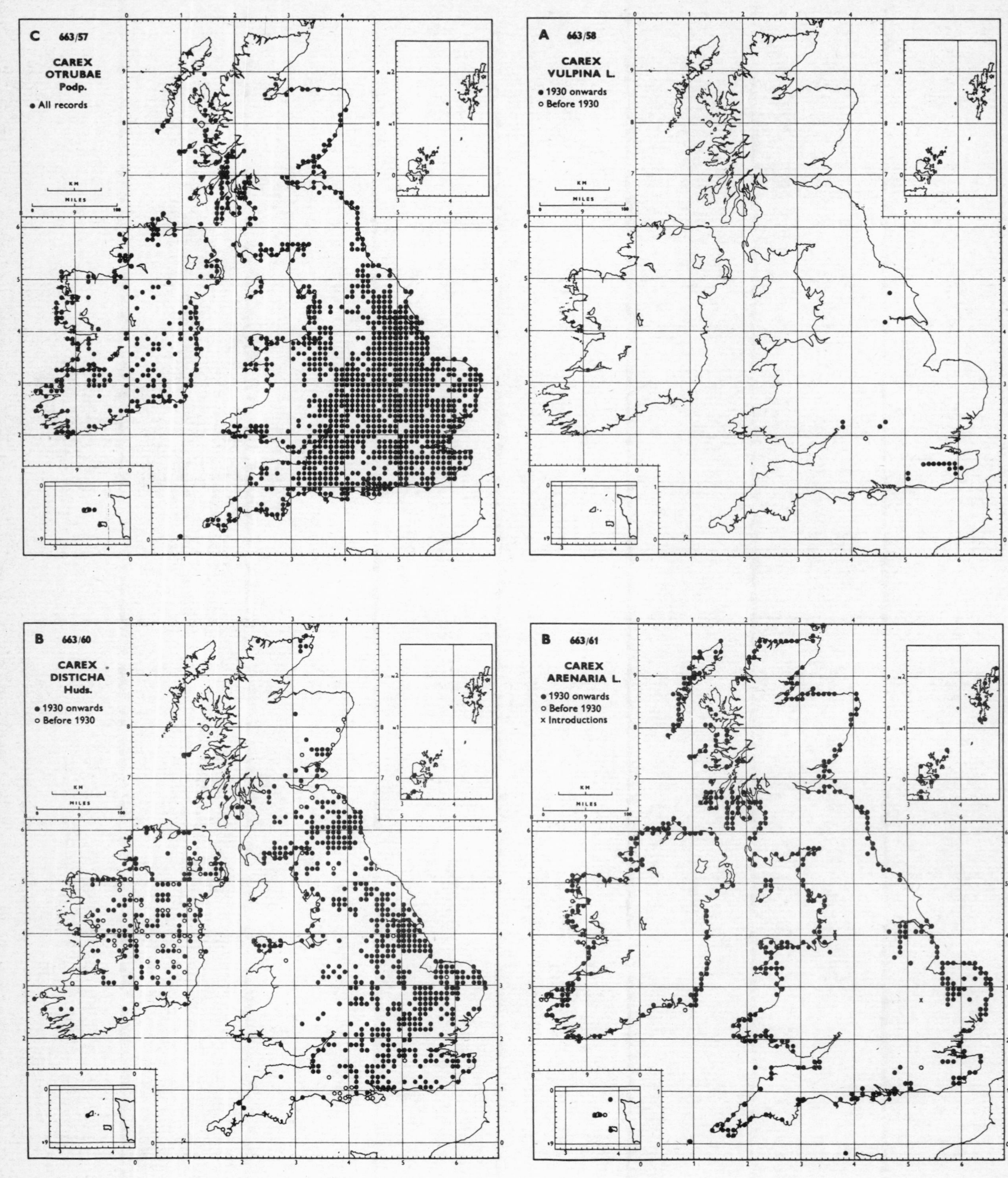
C 663/57
CAREX OTRUBAE Podp.
• All records
A 663/58
CAREX VULPINA L.
• 1930 onwards
○ Before 1930
B 663/60
CAREX DISTICHA Huds.
• 1930 onwards
○ Before 1930
B 663/61
CAREX ARENARIA L.
• 1930 onwards
○ Before 1930
x Introductions

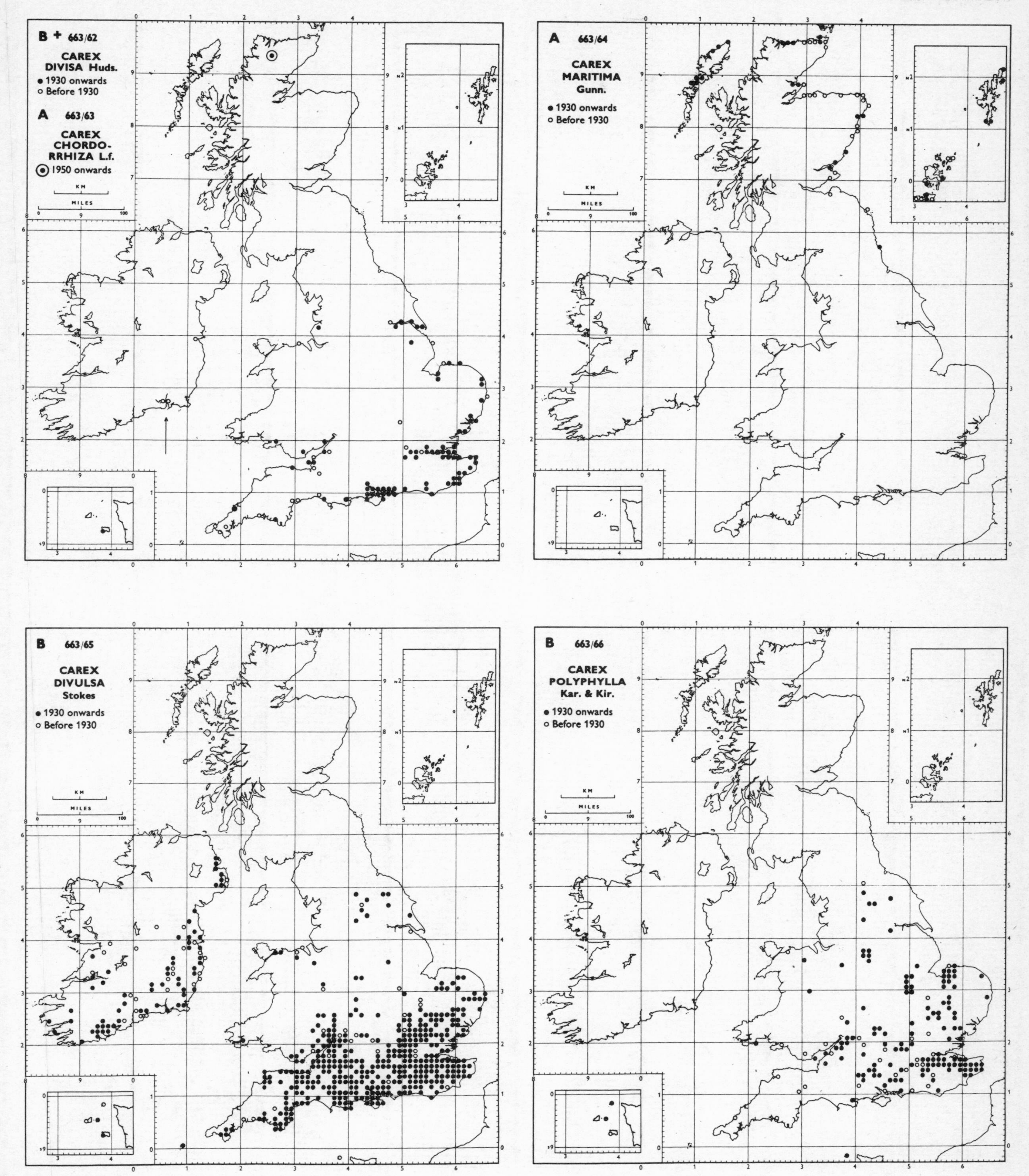
B + 663/62
CAREX DIVISA Huds.
● 1930 onwards
○ Before 1930
A 663/63
CAREX CHORDO-RRHIZA L.f.
⊙ 1950 onwards
A 663/64
CAREX MARITIMA Gunn.
● 1930 onwards
○ Before 1930
B 663/65
CAREX DIVULSA Stokes
● 1930 onwards
○ Before 1930
B 663/66
CAREX POLYPHYLLA Kar. & Kir.
● 1930 onwards
○ Before 1930
KM
MILES

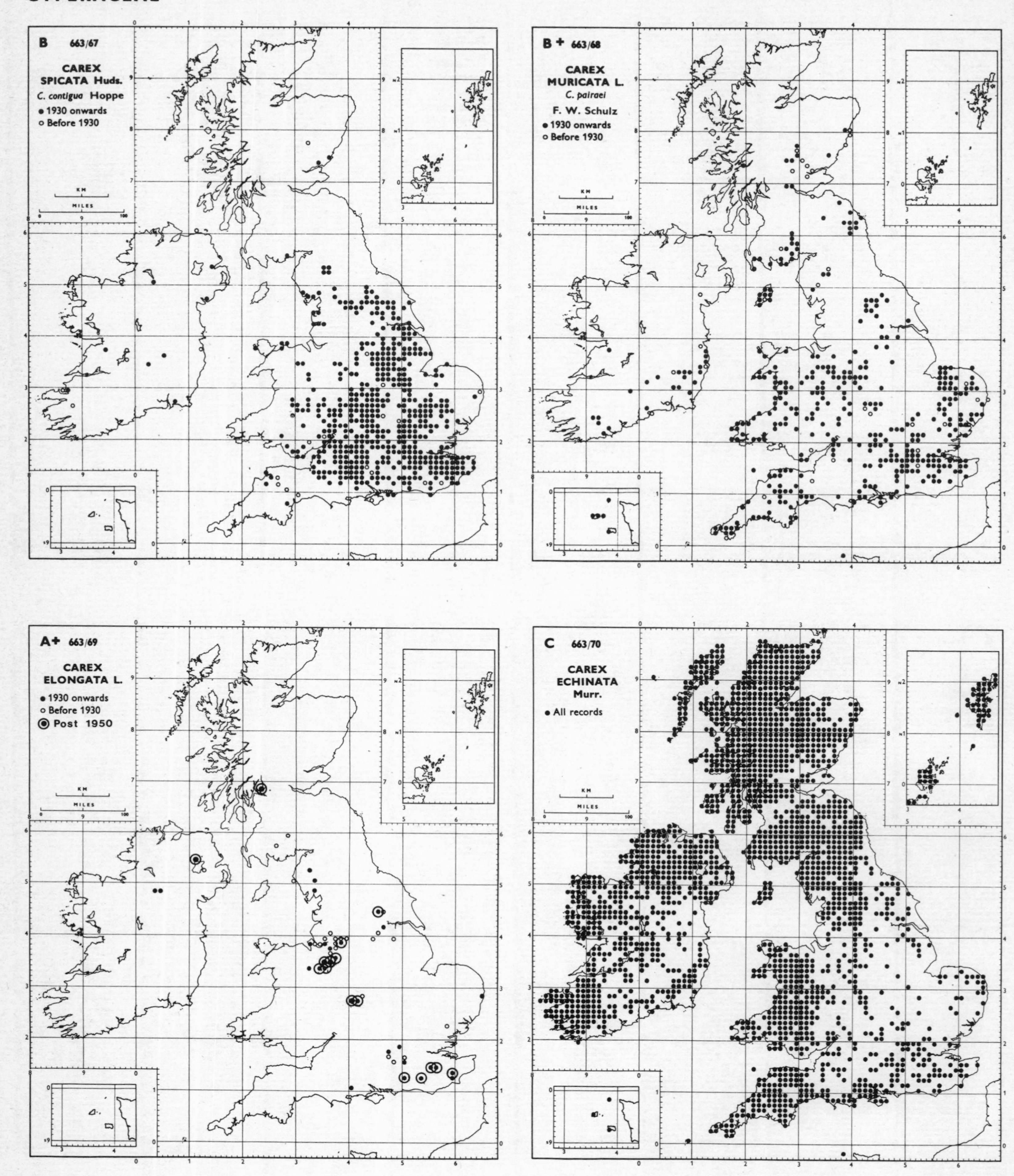
B 663/67
CAREX
SPICATA Huds.
C. contigua Hoppe
• 1930 onwards
○ Before 1930
B+ 663/68
CAREX
MURICATA L.
C. pairaei
F. W. Schulz
• 1930 onwards
○ Before 1930
A+ 663/69
CAREX
ELONGATA L.
• 1930 onwards
○ Before 1930
⊙ Post 1950
C 663/70
CAREX
ECHINATA
Murr.
• All records

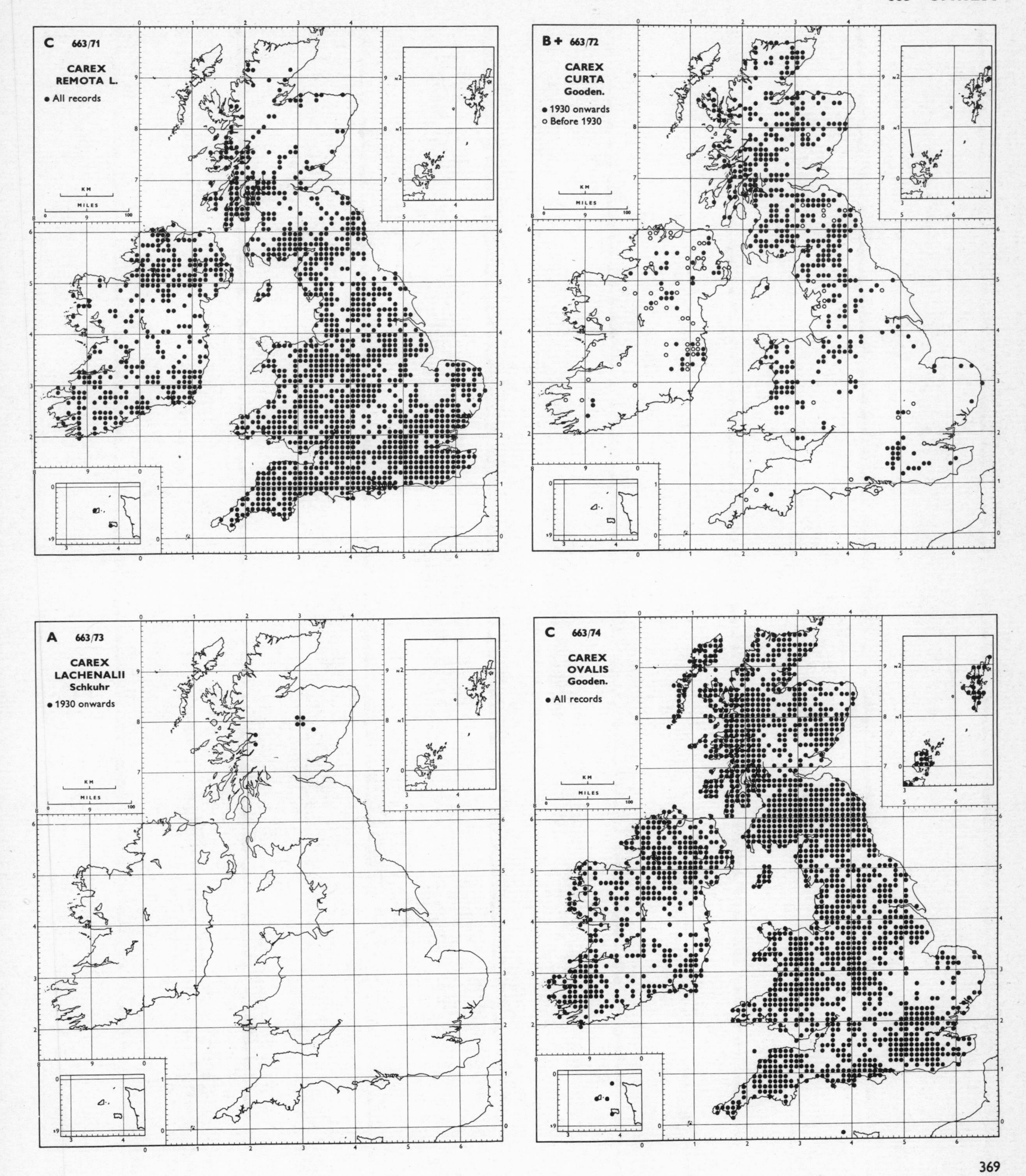
C 663/71
CAREX
REMOTA L.
• All records
B + 663/72
CAREX
CURTA
Gooden.
• 1930 onwards
○ Before 1930
A 663/73
CAREX
LACHENALII
Schkuhr
• 1930 onwards
C 663/74
CAREX
OVALIS
Gooden.
• All records

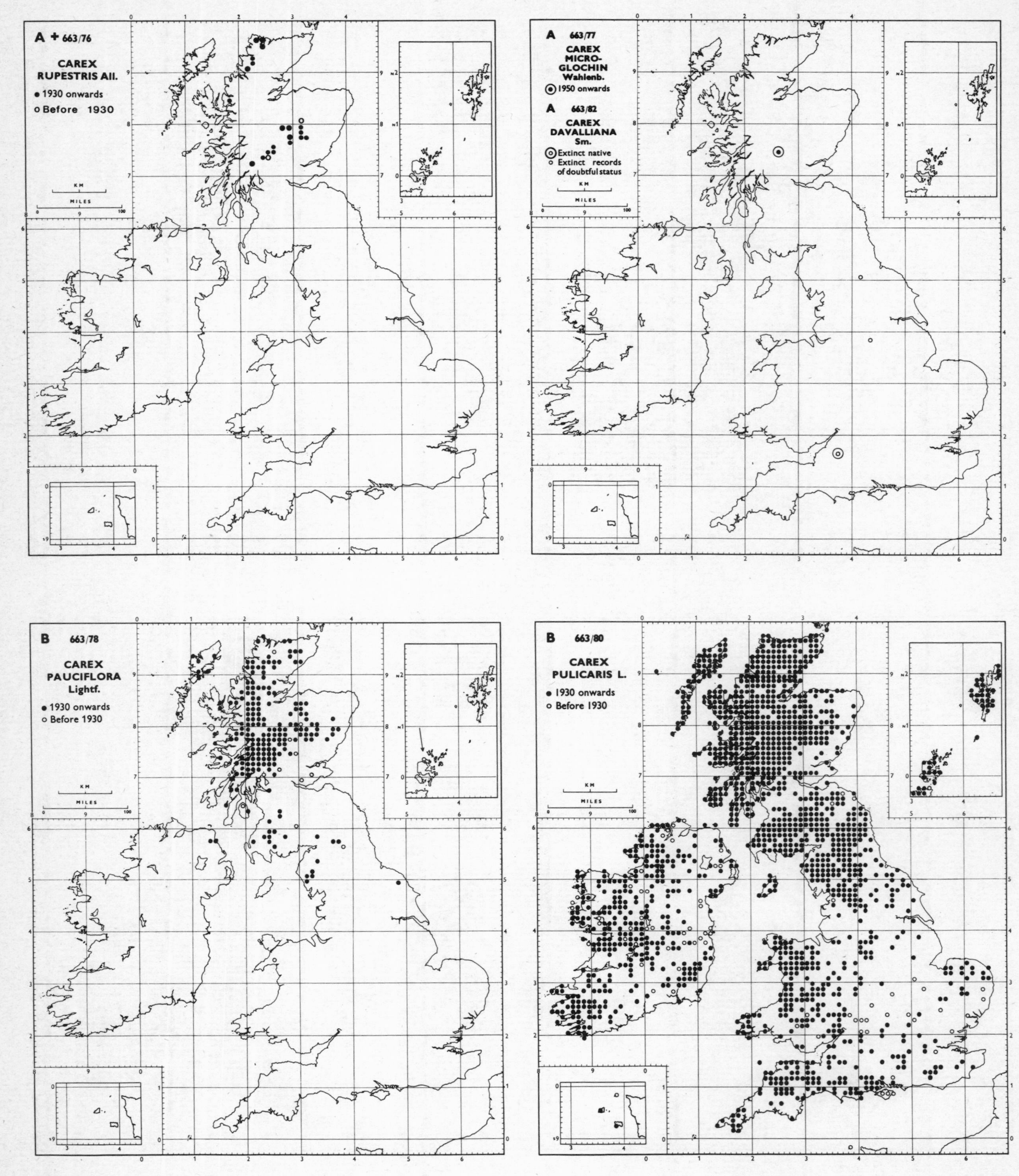
A + 663/76
CAREX
RUPESTRIS All.
● 1930 onwards
○ Before 1930
A 663/77
CAREX
MICRO-
GLOCHIN
Wahlenb.
⊙ 1950 onwards
A 663/82
CAREX
DAVALLIANA
Sm.
◎ Extinct native
○ Extinct records of doubtful status
B 663/78
CAREX
PAUCIFLORA
Lightf.
● 1930 onwards
○ Before 1930
B 663/80
CAREX
PULICARIS L.
● 1930 onwards
○ Before 1930
KM
MILES

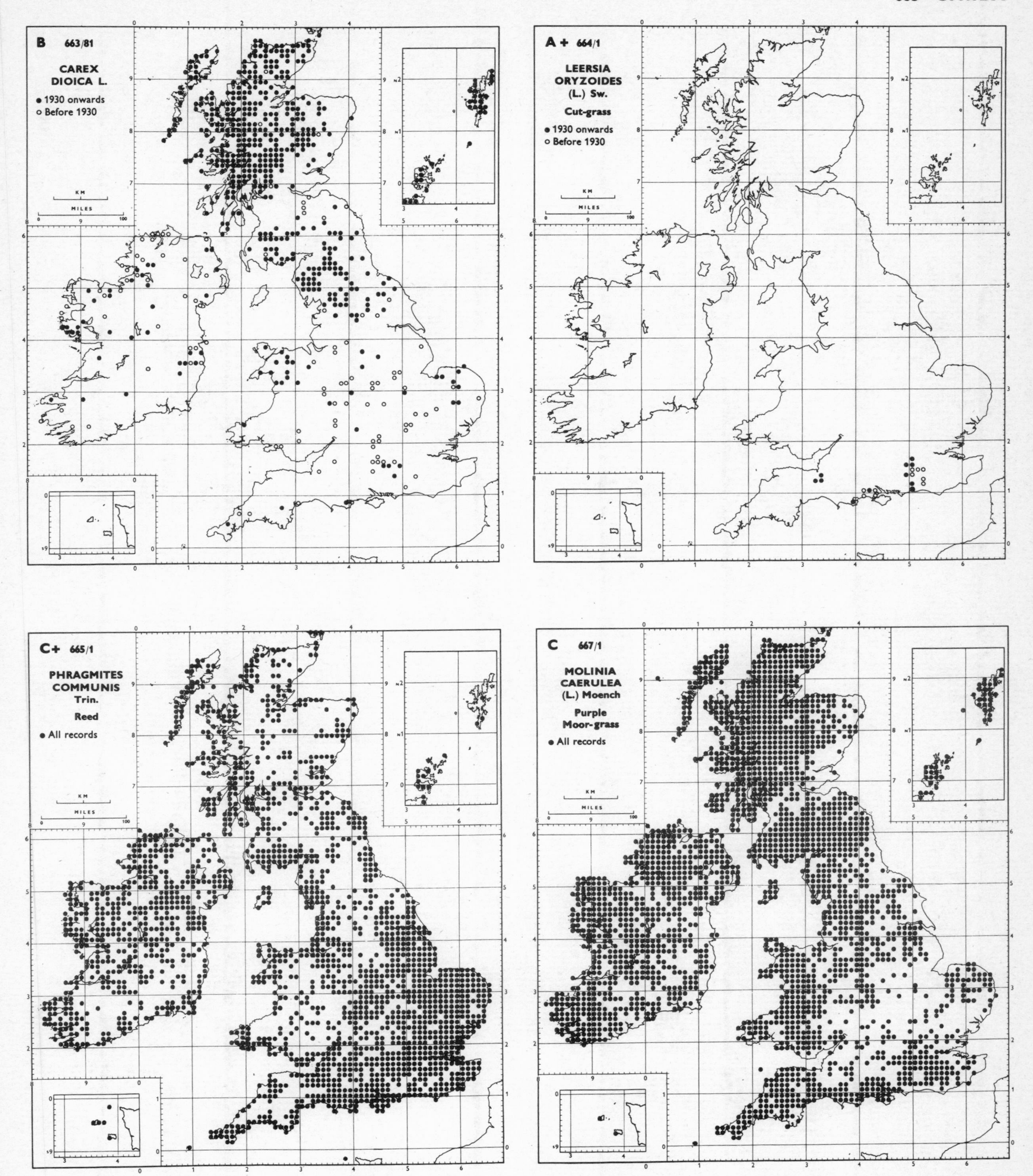
B 663/81
CAREX DIOICA L.
● 1930 onwards
○ Before 1930
A + 664/1
LEERSIA ORYZOIDES (L.) Sw.
Cut-grass
● 1930 onwards
○ Before 1930
C+ 665/1
PHRAGMITES COMMUNIS Trin.
Reed
● All records
C 667/1
MOLINIA CAERULEA (L.) Moench
Purple Moor-grass
● All records

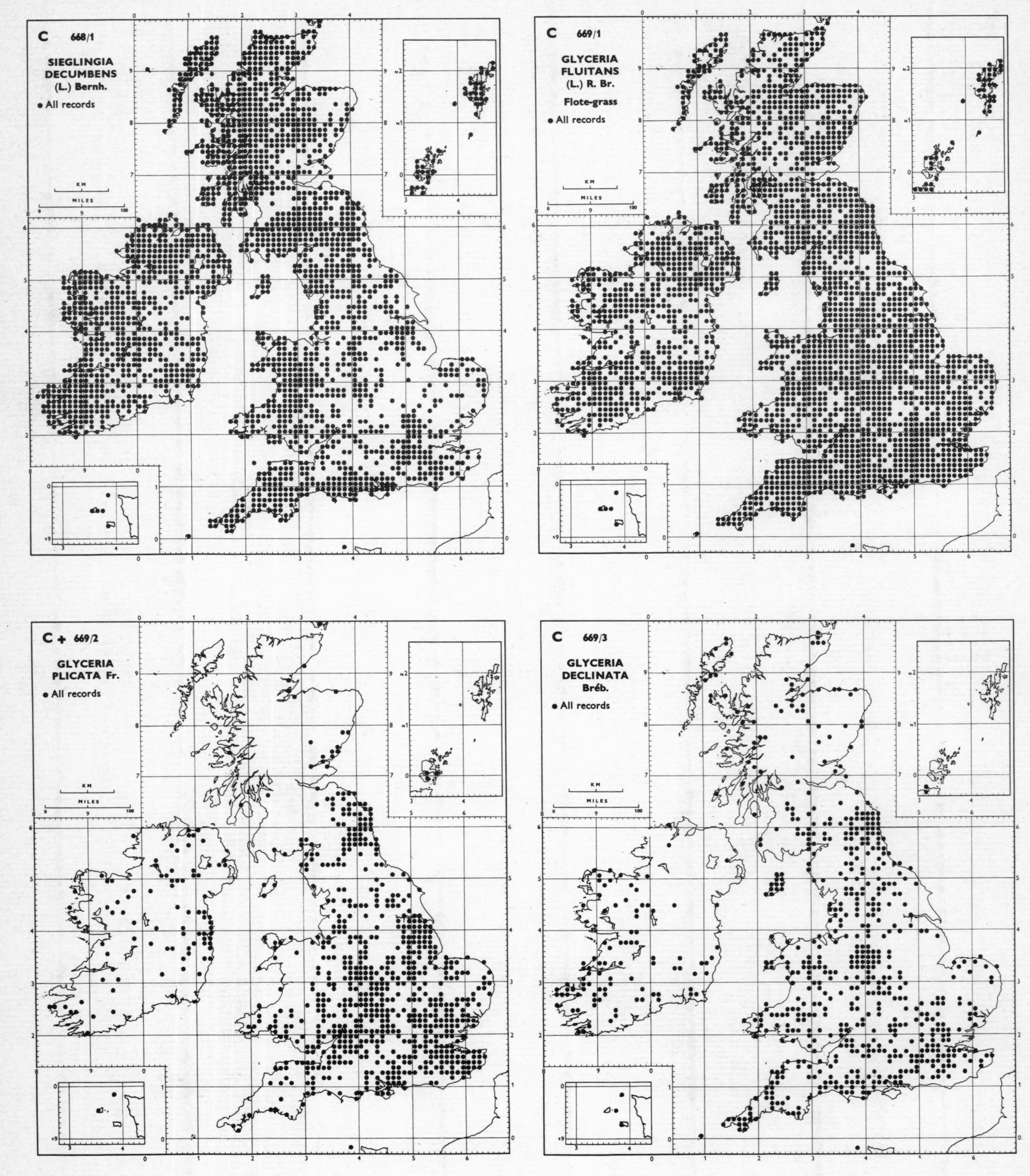
C 668/1
SIEGLINGIA DECUMBENS (L.) Bernh.
• All records
C 669/1
GLYCERIA FLUITANS (L.) R. Br.
Flote-grass
• All records
C + 669/2
GLYCERIA PLICATA Fr.
• All records
C 669/3
GLYCERIA DECLINATA Bréb.
• All records
KM
MILES

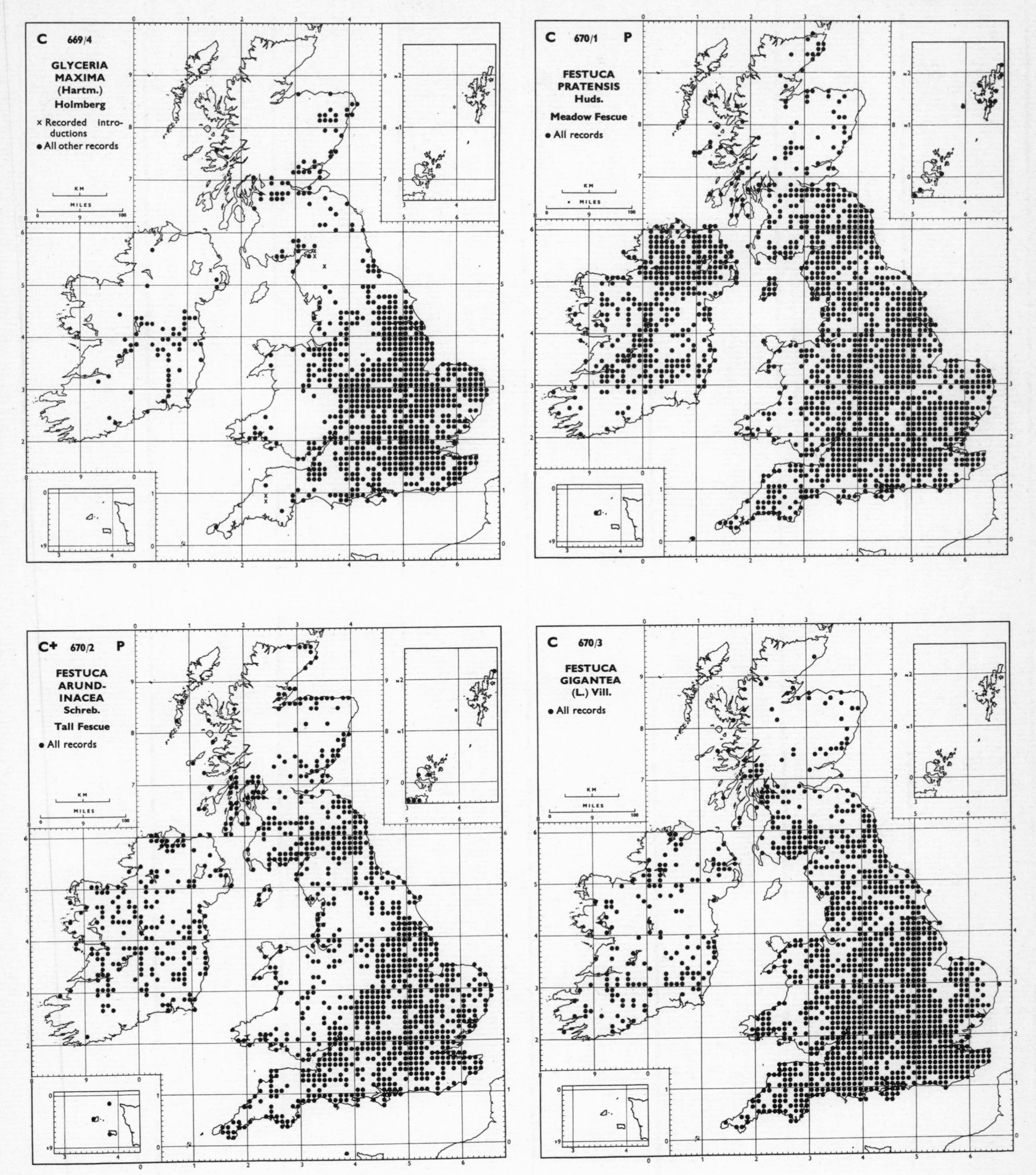
C 669/4
GLYCERIA MAXIMA (Hartm.) Holmberg
× Recorded introductions
● All other records
C 670/1 P
FESTUCA PRATENSIS Huds.
Meadow Fescue
● All records
C+ 670/2 P
FESTUCA ARUNDINACEA Schreb.
Tall Fescue
● All records
C 670/3
FESTUCA GIGANTEA (L.) Vill.
● All records

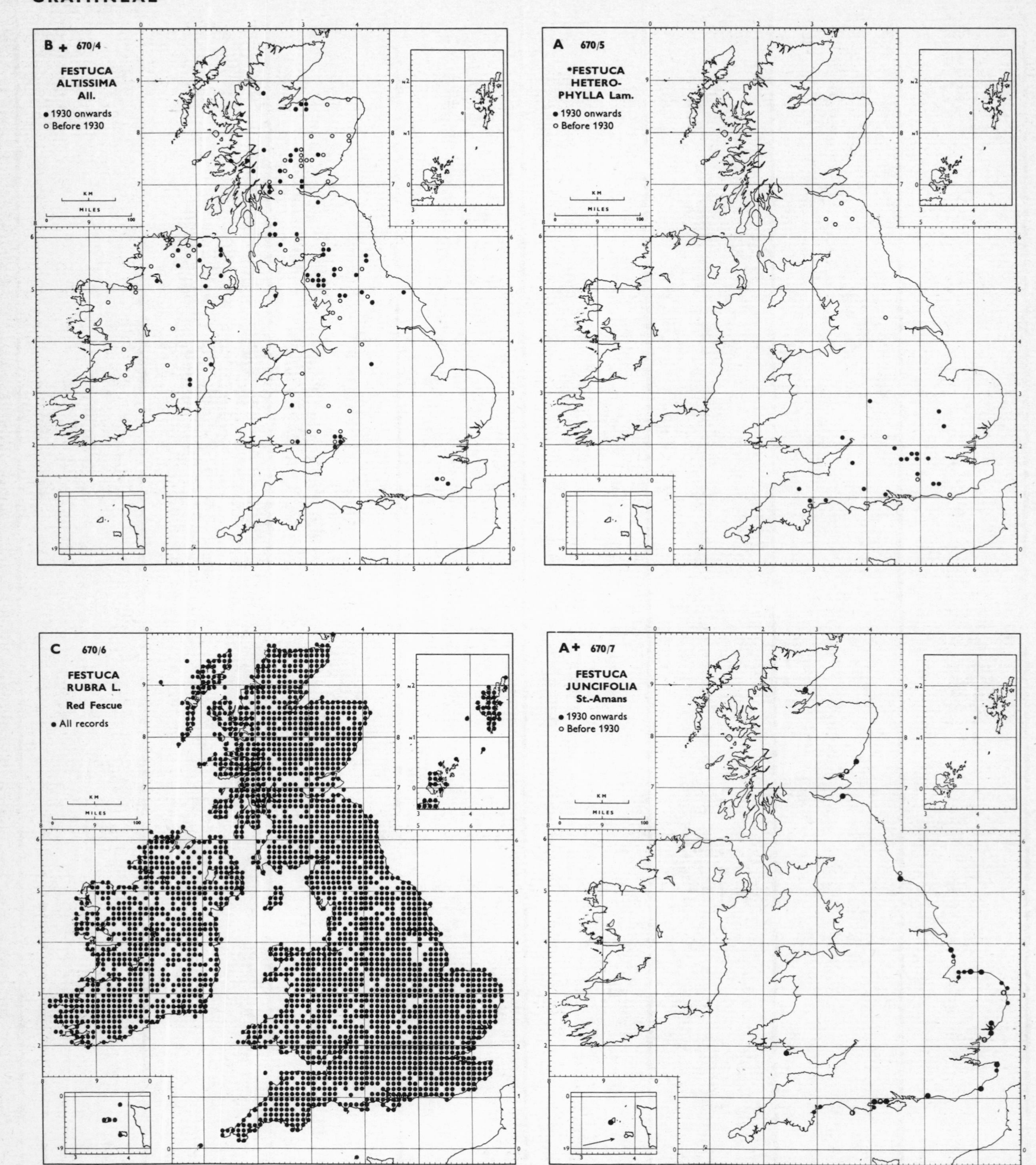
B + 670/4
FESTUCA
ALTISSIMA
All.
• 1930 onwards
○ Before 1930
A 670/5
*FESTUCA
HETERO-
PHYLLA Lam.
• 1930 onwards
○ Before 1930
C 670/6
FESTUCA
RUBRA L.
Red Fescue
• All records
A+ 670/7
FESTUCA
JUNCIFOLIA
St.-Amans
• 1930 onwards
○ Before 1930
KM
MILES

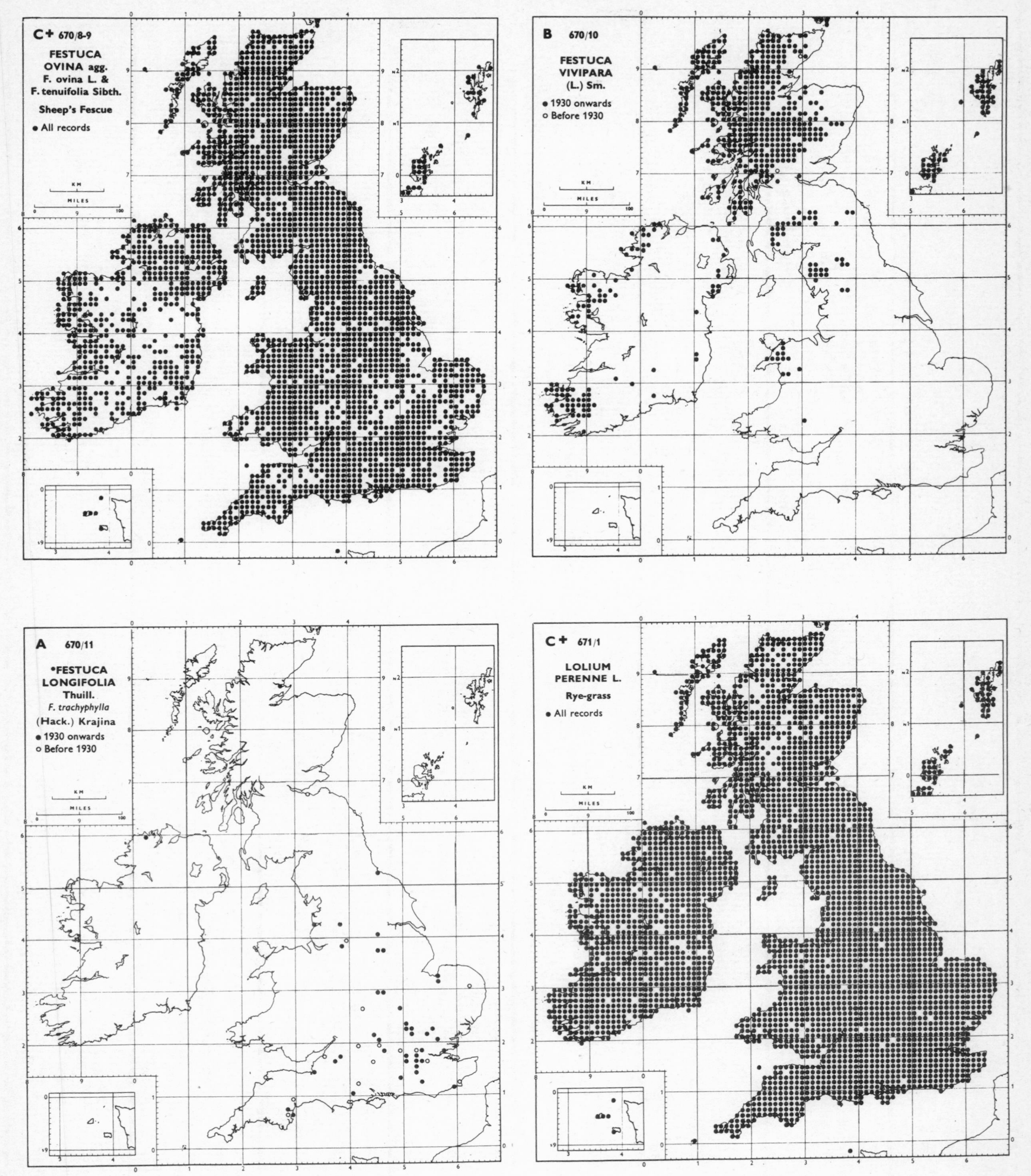
C+ 670/8-9
FESTUCA
OVINA agg.
F. ovina L. &
F. tenuifolia Sibth.
Sheep's Fescue
• All records
B 670/10
FESTUCA
VIVIPARA
(L.) Sm.
• 1930 onwards
○ Before 1930
A 670/11
•FESTUCA
LONGIFOLIA
Thuill.
F. trachyphylla
(Hack.) Krajina
• 1930 onwards
○ Before 1930
C+ 671/1
LOLIUM
PERENNE L.
Rye-grass
• All records

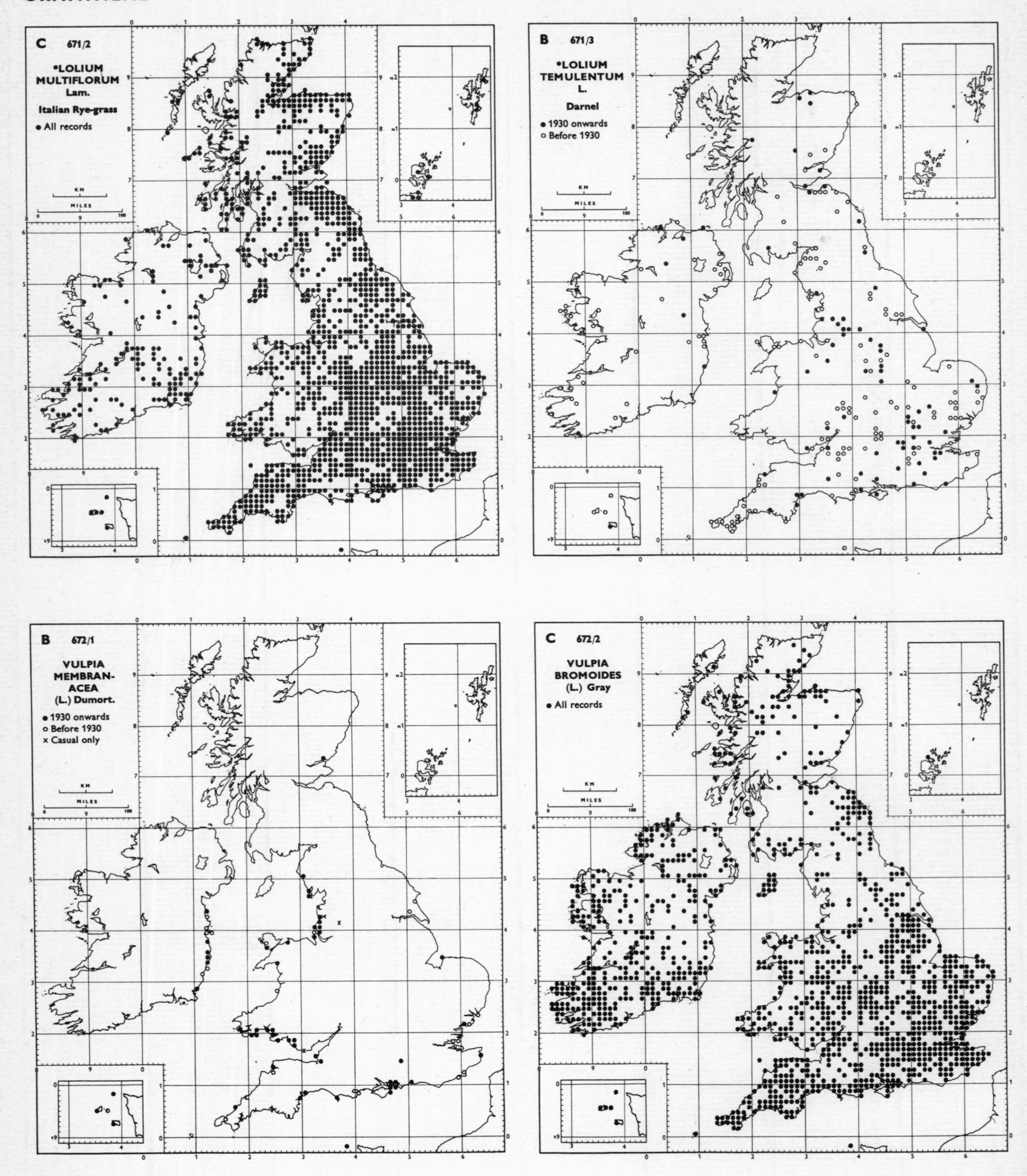
C 671/2
*LOLIUM MULTIFLORUM Lam.
Italian Rye-grass
● All records
B 671/3
*LOLIUM TEMULENTUM L.
Darnel
● 1930 onwards
○ Before 1930
B 672/1
VULPIA MEMBRANACEA (L.) Dumort.
● 1930 onwards
○ Before 1930
× Casual only
C 672/2
VULPIA BROMOIDES (L.) Gray
● All records
KM
MILES

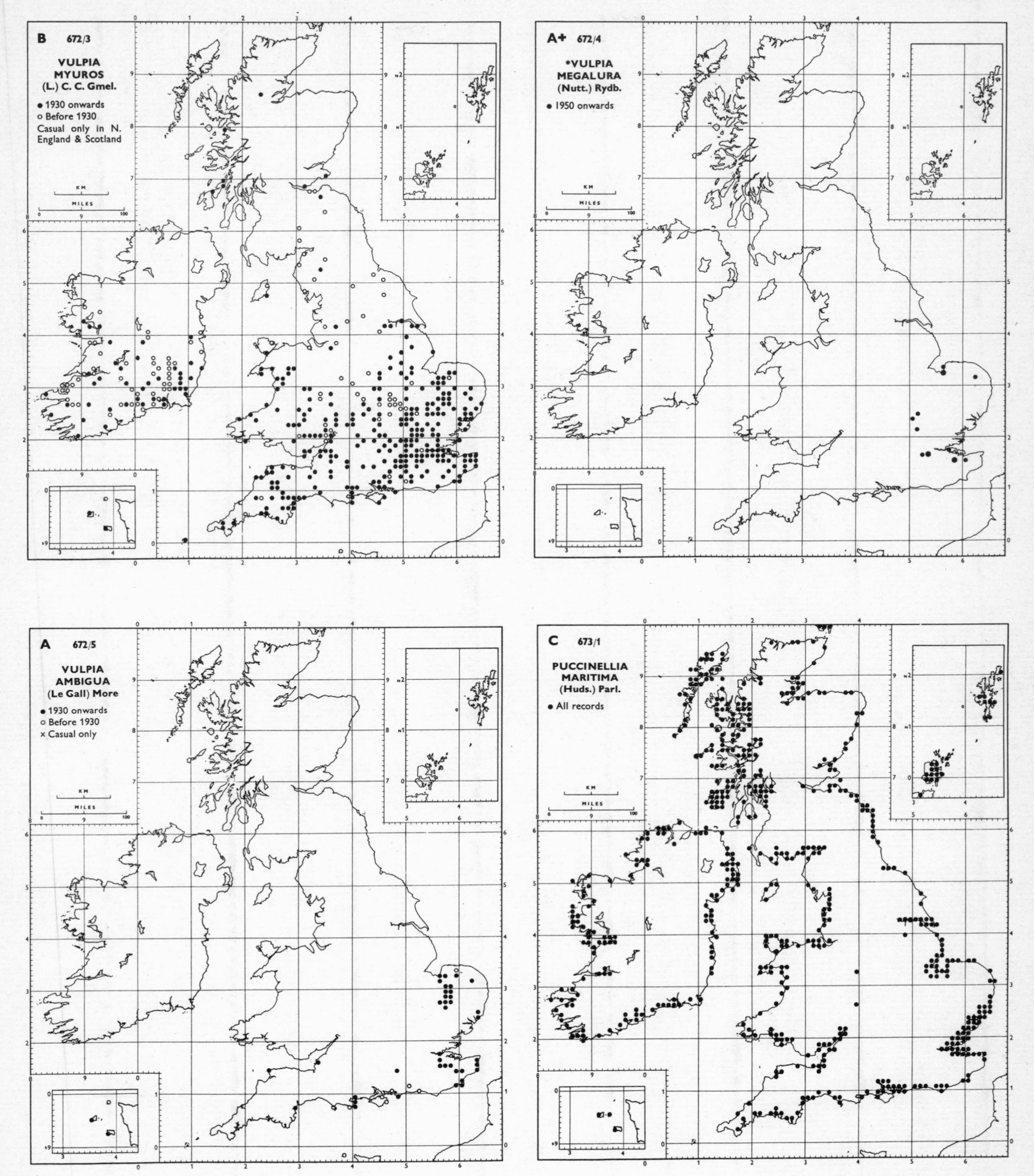
B 672/3
VULPIA MYUROS (L.) C. C. Gmel.
● 1930 onwards
○ Before 1930
Casual only in N. England & Scotland
A+ 672/4
*VULPIA MEGALURA (Nutt.) Rydb.
● 1950 onwards
A 672/5
VULPIA AMBIGUA (Le Gall) More
● 1930 onwards
○ Before 1930
× Casual only
C 673/1
PUCCINELLIA MARITIMA (Huds.) Parl.
● All records

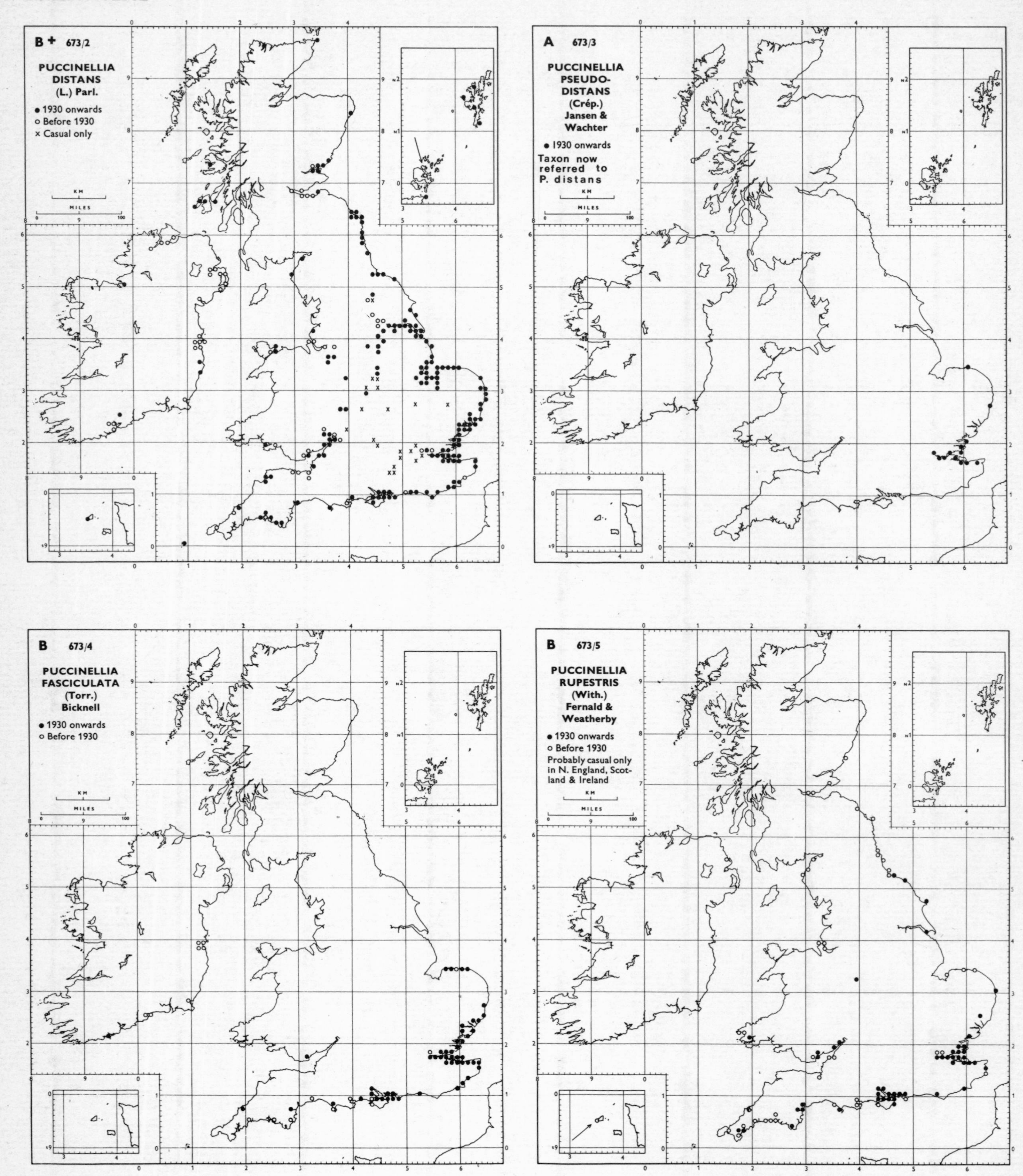
B + 673/2
PUCCINELLIA DISTANS (L.) Parl.
● 1930 onwards
○ Before 1930
× Casual only
A 673/3
PUCCINELLIA PSEUDO-DISTANS (Crép.) Jansen & Wachter
● 1930 onwards
Taxon now referred to P. distans
B 673/4
PUCCINELLIA FASCICULATA (Torr.) Bicknell
● 1930 onwards
○ Before 1930
B 673/5
PUCCINELLIA RUPESTRIS (With.) Fernald & Weatherby
● 1930 onwards
○ Before 1930
Probably casual only in N. England, Scotland & Ireland
KM
MILES

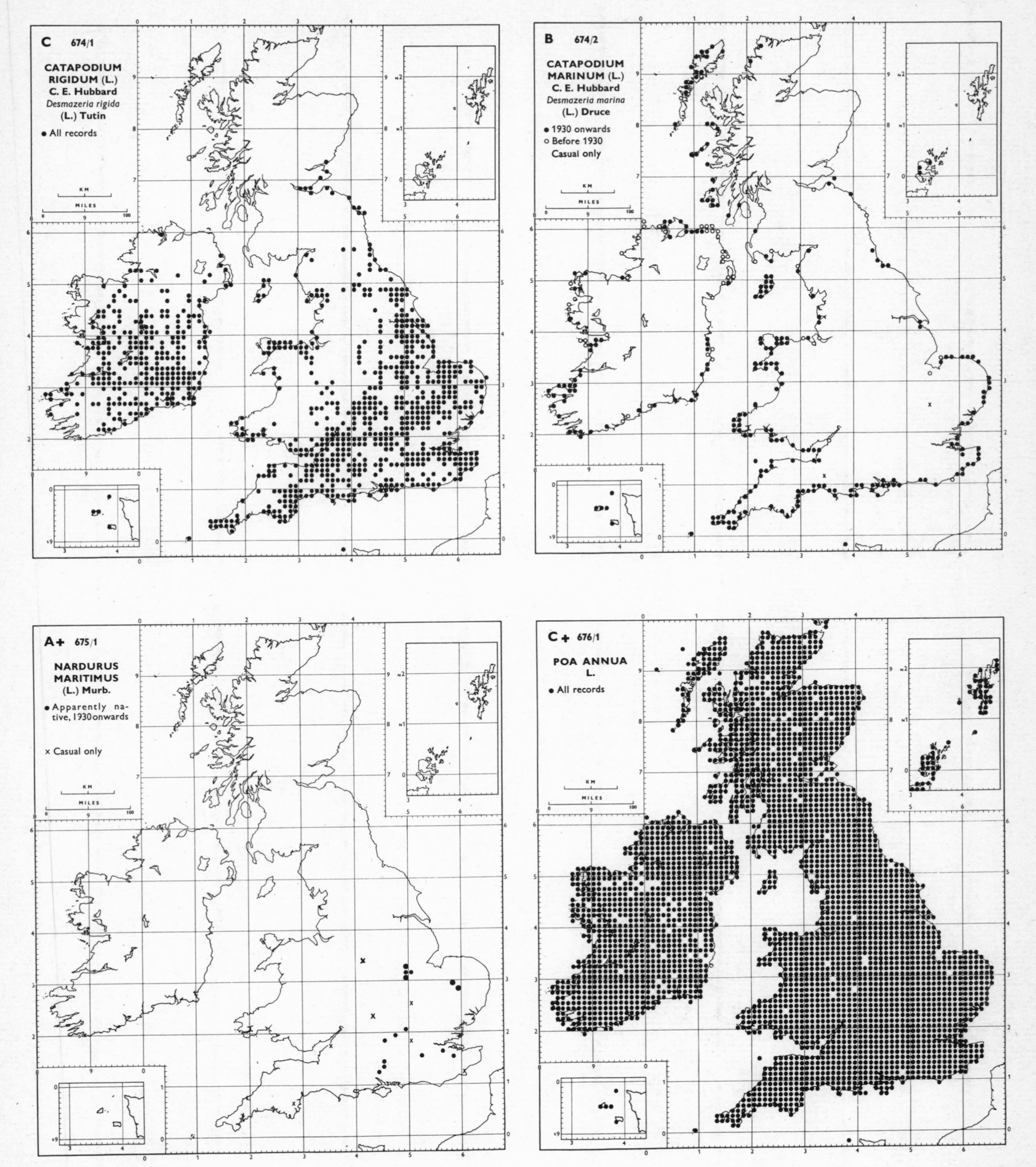
C 674/1
CATAPODIUM RIGIDUM (L.) C. E. Hubbard
Desmazeria rigida (L.) Tutin
● All records
B 674/2
CATAPODIUM MARINUM (L.) C. E. Hubbard
Desmazeria marina (L.) Druce
● 1930 onwards
○ Before 1930
Casual only
A+ 675/1
NARDURUS MARITIMUS (L.) Murb.
● Apparently native, 1930 onwards
x Casual only
C+ 676/1
POA ANNUA L.
● All records
KM
MILES

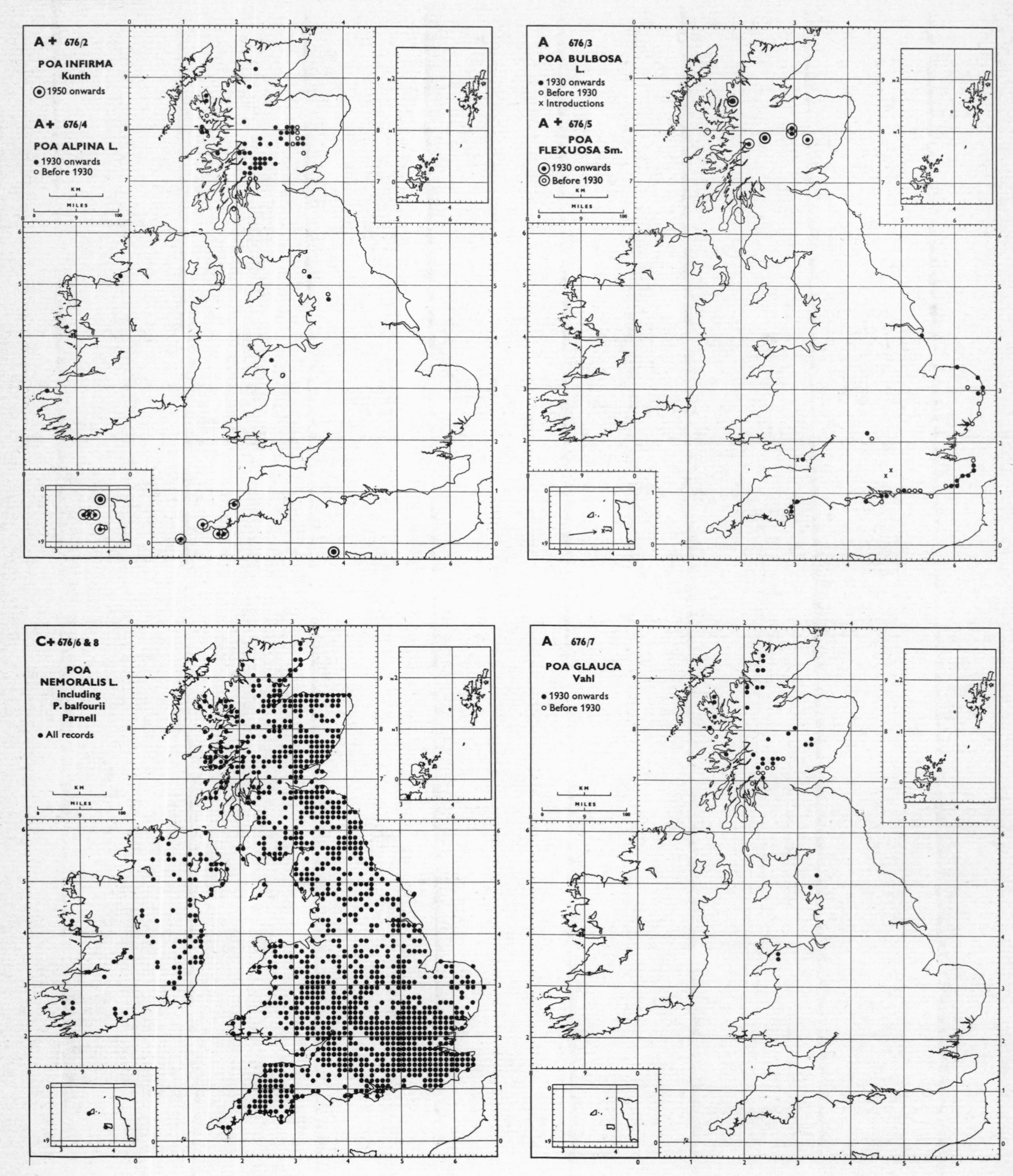
A + 676/2
POA INFIRMA
Kunth
1950 onwards
A + 676/4
POA ALPINA L.
• 1930 onwards
○ Before 1930
A 676/3
POA BULBOSA
L.
• 1930 onwards
○ Before 1930
× Introductions
A + 676/5
POA
FLEXUOSA Sm.
1930 onwards
Before 1930
C + 676/6 & 8
POA
NEMORALIS L.
including
P. balfourii
Parnell
• All records
A 676/7
POA GLAUCA
Vahl
• 1930 onwards
○ Before 1930
KM
MILES

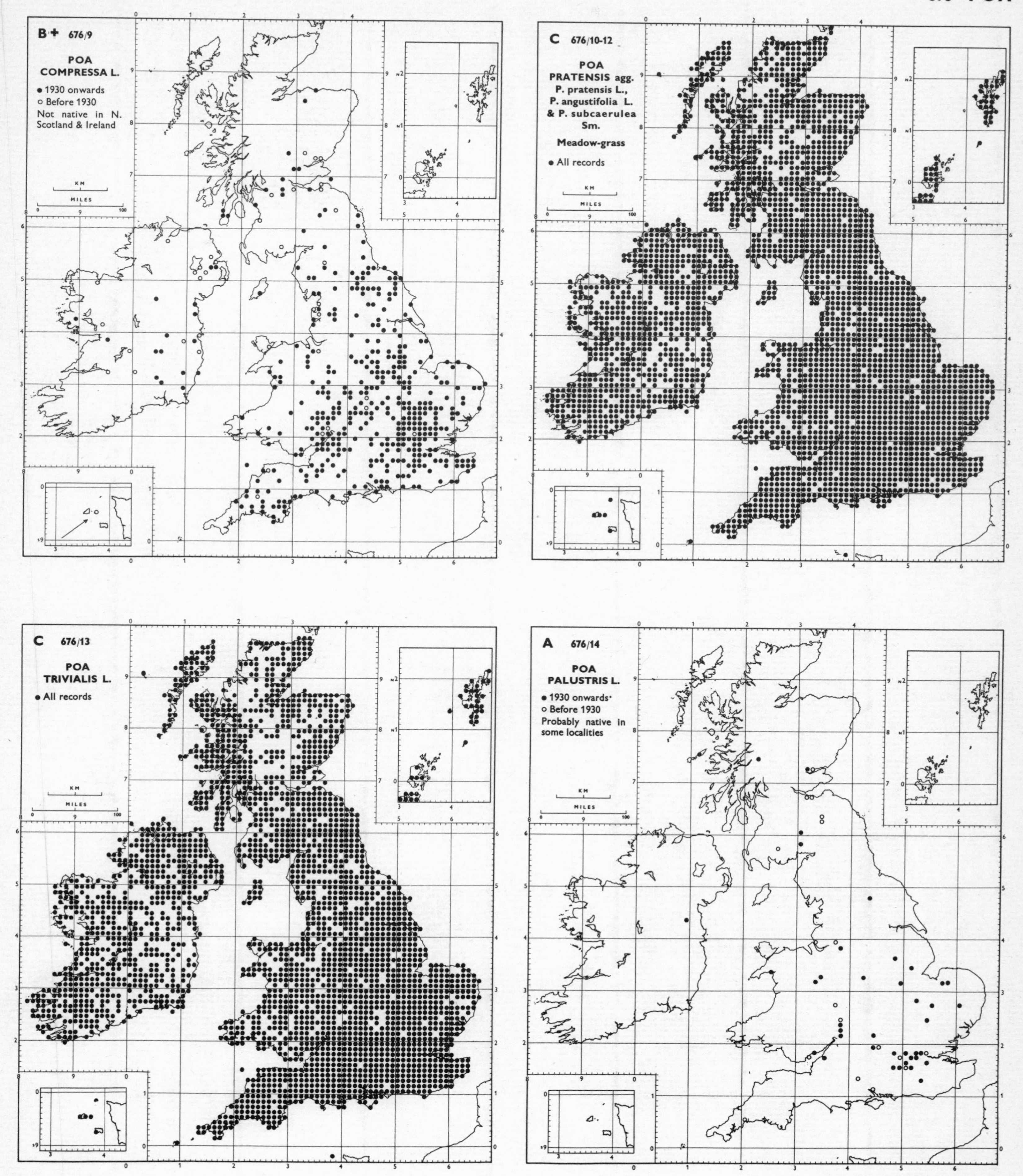
B+ 676/9
POA
COMPRESSA L.
● 1930 onwards
○ Before 1930
Not native in N. Scotland & Ireland
C 676/10-12
POA
PRATENSIS agg.
P. pratensis L.,
P. angustifolia L.
& P. subcaerulea Sm.
Meadow-grass
● All records
C 676/13
POA
TRIVIALIS L.
● All records
A 676/14
POA
PALUSTRIS L.
● 1930 onwards
○ Before 1930
Probably native in some localities
KM
MILES

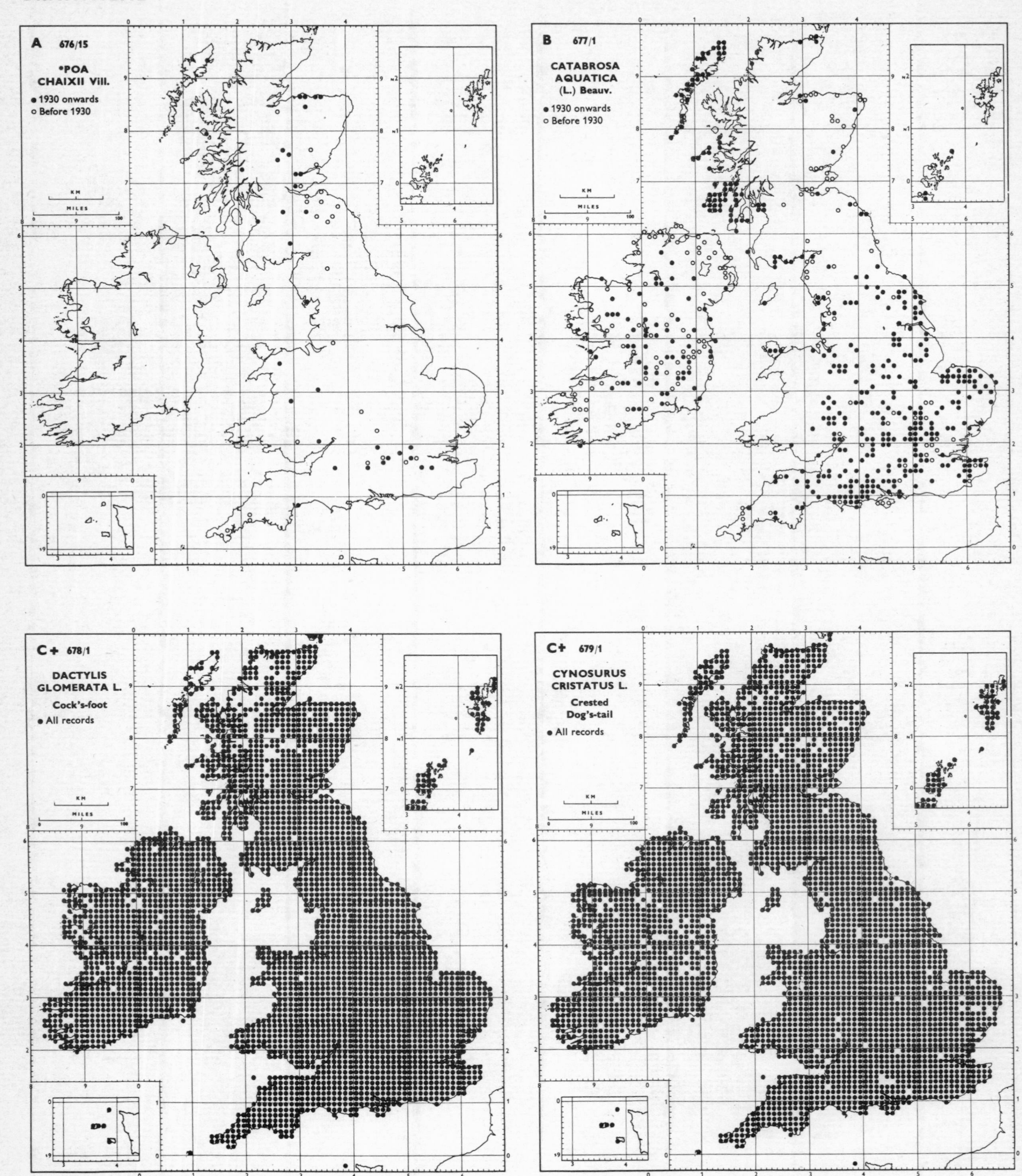
A 676/15
*POA CHAIXII Vill.
● 1930 onwards
○ Before 1930
B 677/1
CATABROSA AQUATICA (L.) Beauv.
● 1930 onwards
○ Before 1930
C+ 678/1
DACTYLIS GLOMERATA L.
Cock's-foot
● All records
C+ 679/1
CYNOSURUS CRISTATUS L.
Crested Dog's-tail
● All records
KM
MILES

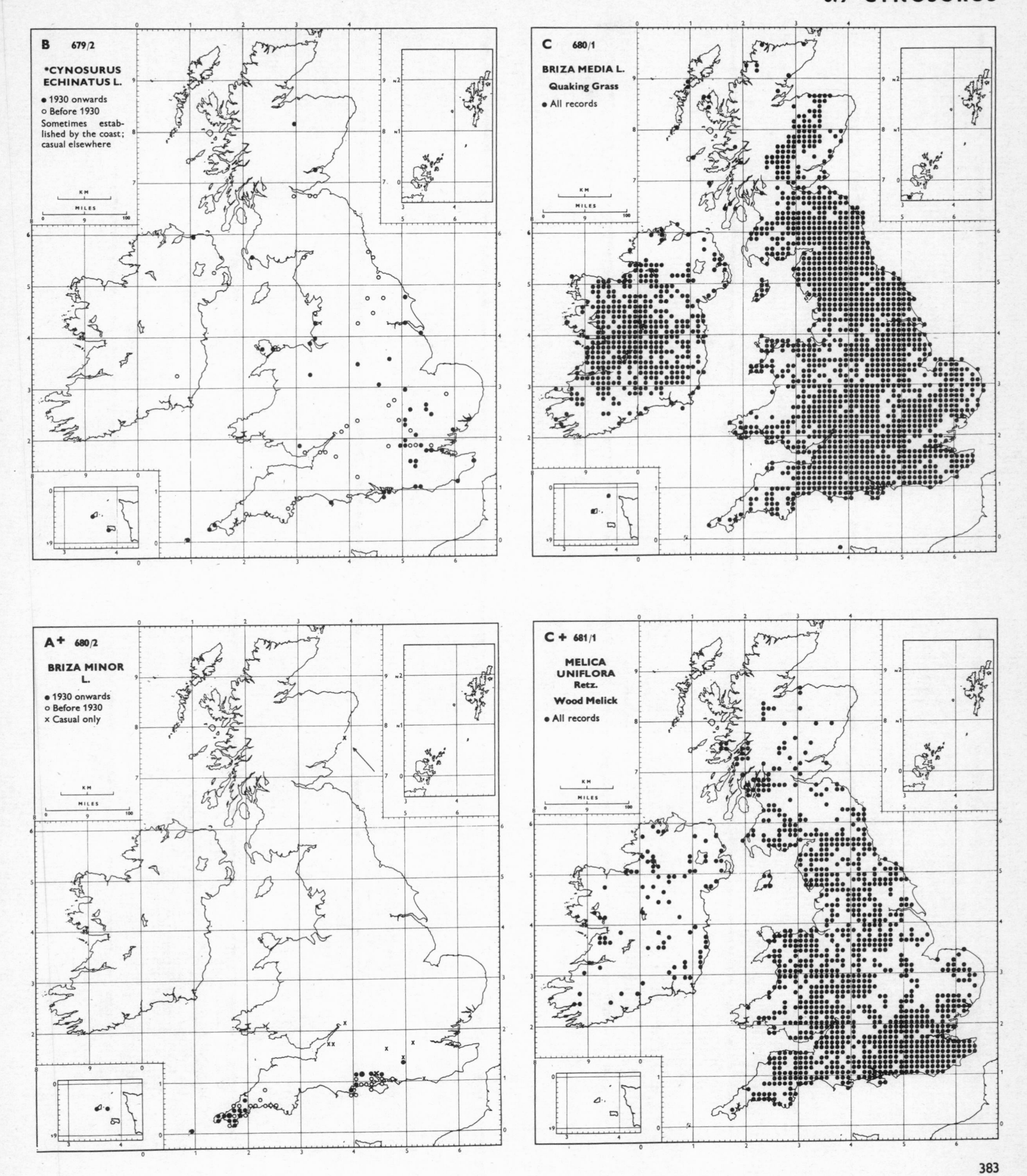
B 679/2
*CYNOSURUS ECHINATUS L.
● 1930 onwards
○ Before 1930
Sometimes established by the coast; casual elsewhere
C 680/1
BRIZA MEDIA L.
Quaking Grass
● All records
A+ 680/2
BRIZA MINOR L.
● 1930 onwards
○ Before 1930
x Casual only
C + 681/1
MELICA UNIFLORA Retz.
Wood Melick
● All records

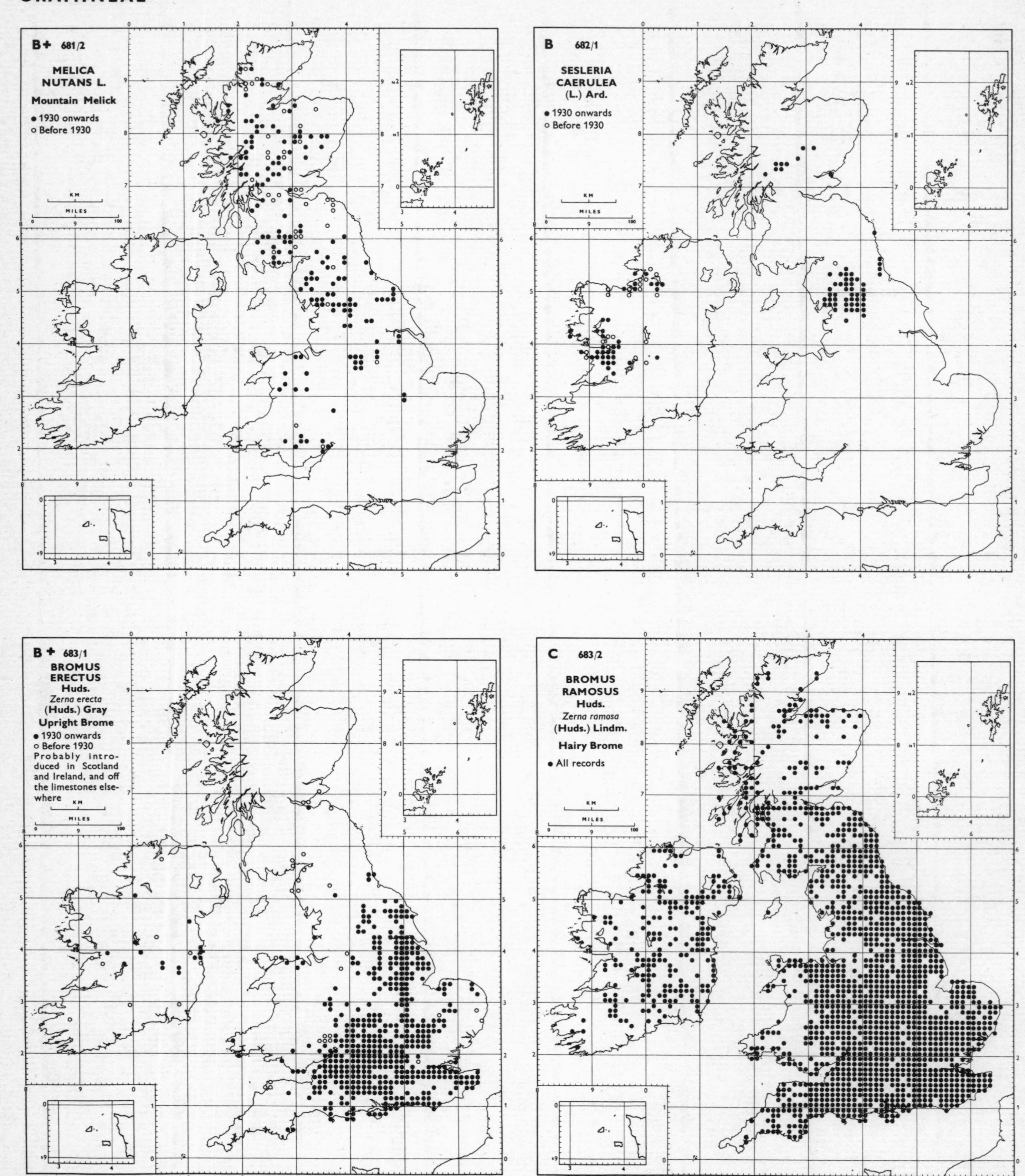
B + 681/2
MELICA NUTANS L.
Mountain Melick
● 1930 onwards
○ Before 1930
B 682/1
SESLERIA CAERULEA (L.) Ard.
● 1930 onwards
○ Before 1930
B + 683/1
BROMUS ERECTUS Huds.
Zerna erecta (Huds.) Gray
Upright Brome
● 1930 onwards
○ Before 1930
Probably introduced in Scotland and Ireland, and off the limestones elsewhere
C 683/2
BROMUS RAMOSUS Huds.
Zerna ramosa (Huds.) Lindm.
Hairy Brome
● All records
KM
MILES

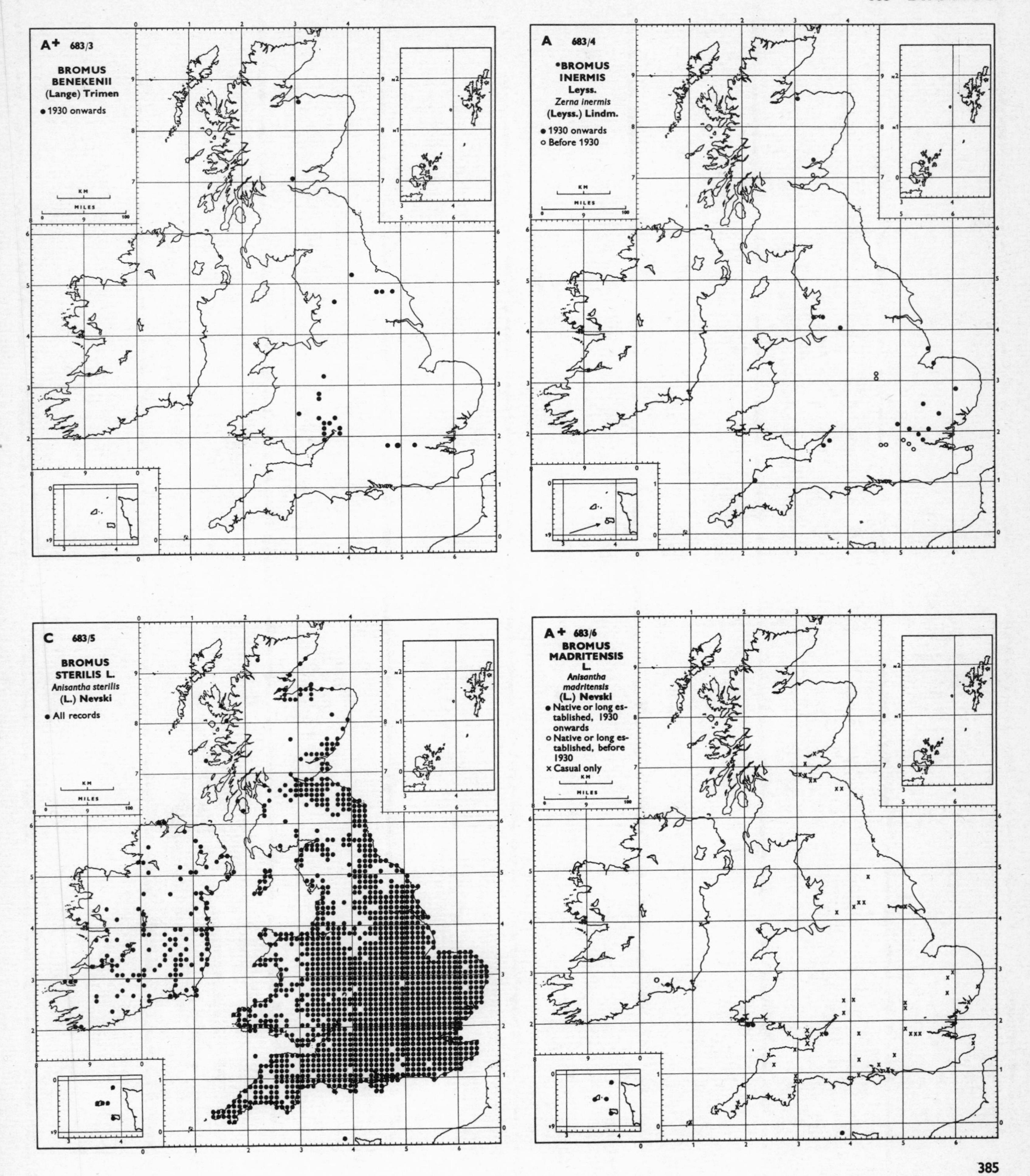
A+ 683/3
BROMUS BENEKENII (Lange) Trimen
● 1930 onwards
A 683/4
*BROMUS INERMIS Leyss.
Zerna inermis (Leyss.) Lindm.
● 1930 onwards
○ Before 1930
C 683/5
BROMUS STERILIS L.
Anisantha sterilis (L.) Nevski
● All records
A+ 683/6
BROMUS MADRITENSIS L.
Anisantha madritensis (L.) Nevski
● Native or long established, 1930 onwards
○ Native or long established, before 1930
× Casual only
KM
MILES

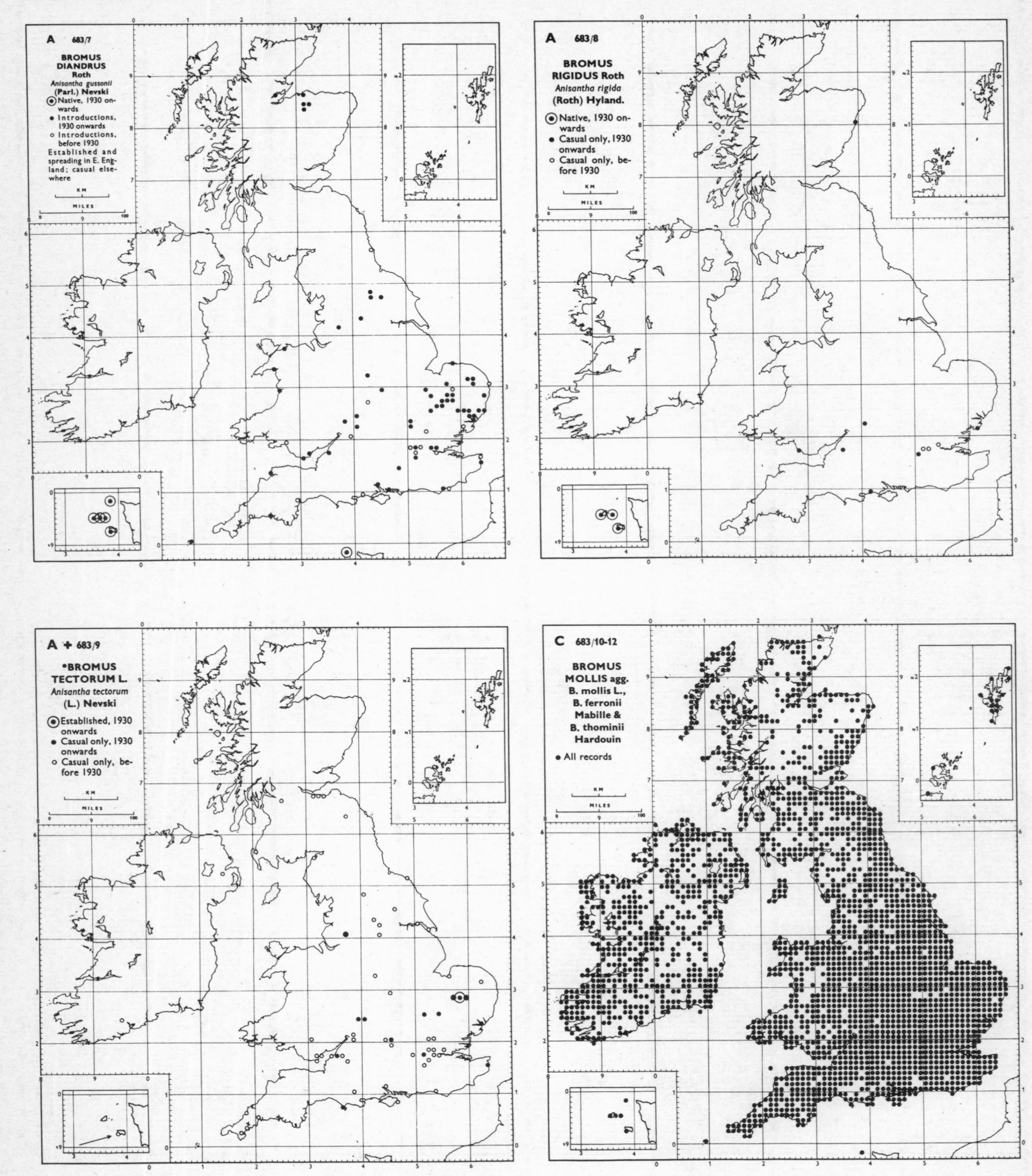
A 683/7
BROMUS DIANDRUS Roth
Anisantha gussonii (Parl.) Nevski
Native, 1930 onwards
Introductions, 1930 onwards
Introductions, before 1930
Established and spreading in E. England; casual elsewhere
KM
MILES
A 683/8
BROMUS RIGIDUS Roth
Anisantha rigida (Roth) Hyland.
Native, 1930 onwards
Casual only, 1930 onwards
Casual only, before 1930
KM
MILES
A + 683/9
*BROMUS TECTORUM L.
Anisantha tectorum (L.) Nevski
Established, 1930 onwards
Casual only, 1930 onwards
Casual only, before 1930
KM
MILES
C 683/10-12
BROMUS MOLLIS agg.
B. mollis L., B. ferronii Mabille & B. thominii Hardouin
All records
KM
MILES

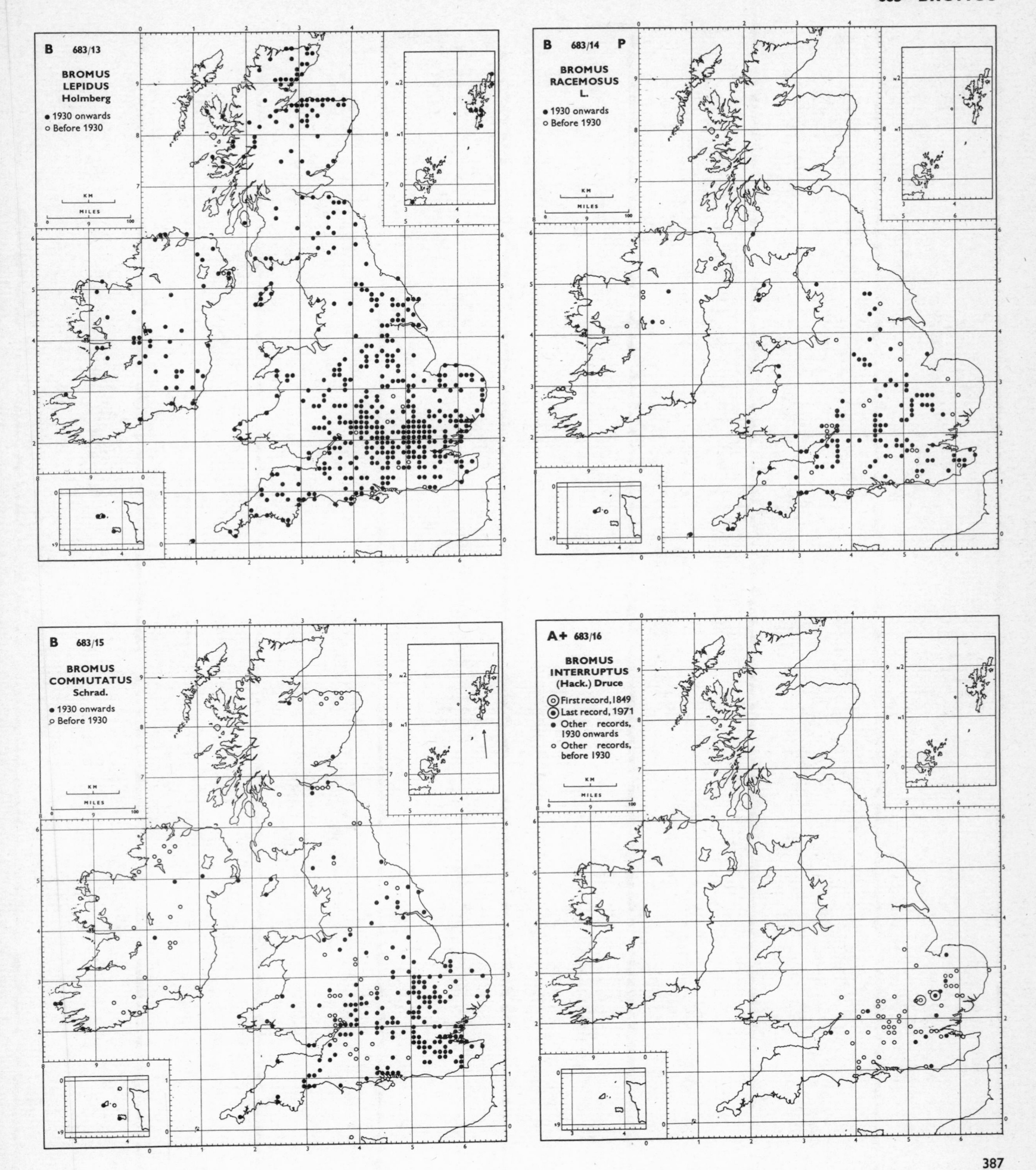
B 683/13
BROMUS LEPIDUS Holmberg
• 1930 onwards
○ Before 1930
KM
MILES
B 683/14 P
BROMUS RACEMOSUS L.
• 1930 onwards
○ Before 1930
KM
MILES
B 683/15
BROMUS COMMUTATUS Schrad.
• 1930 onwards
○ Before 1930
KM
MILES
A+ 683/16
BROMUS INTERRUPTUS (Hack.) Druce
First record, 1849
Last record, 1971
Other records, 1930 onwards
Other records, before 1930
KM
MILES

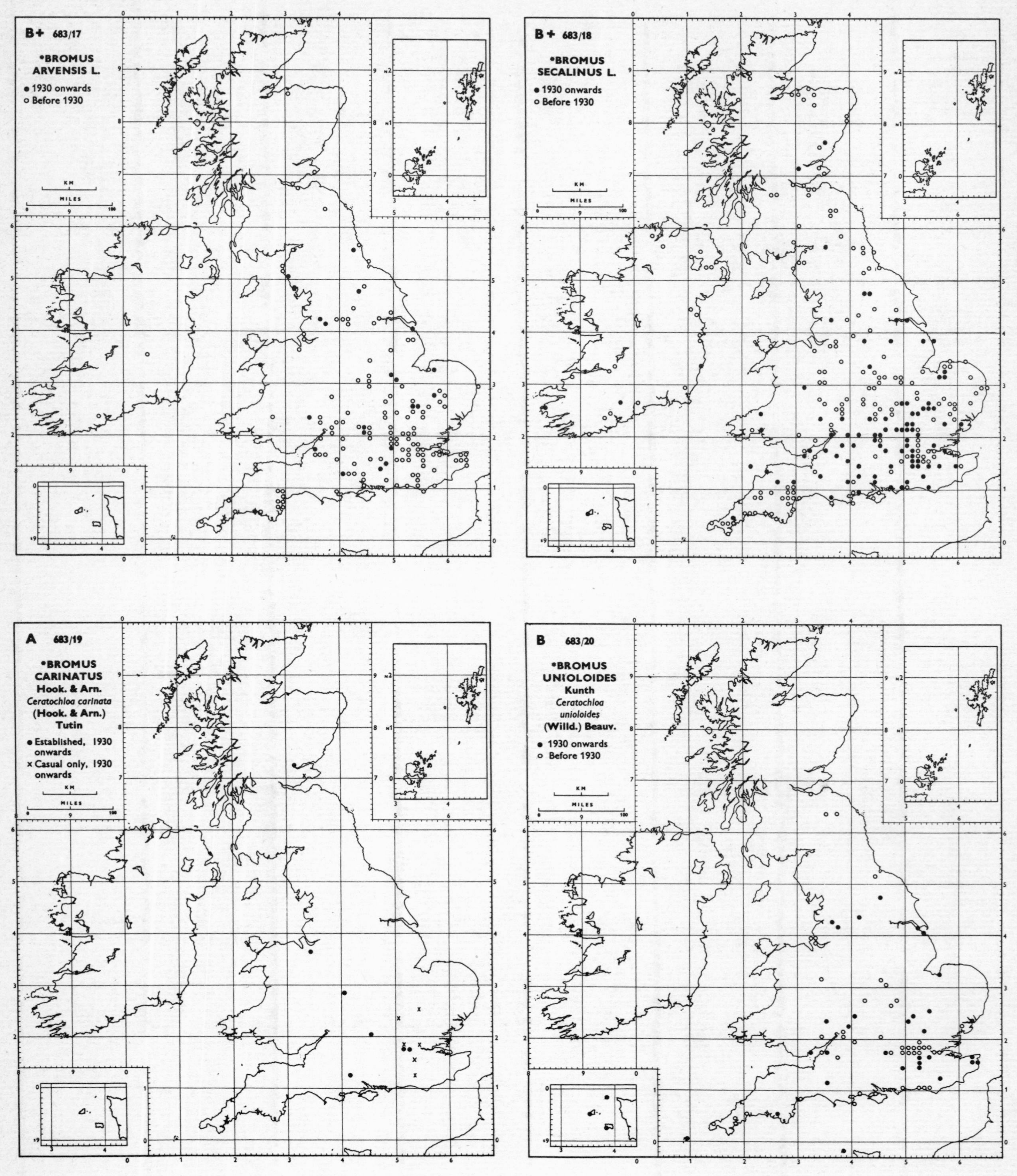
B+ 683/17
*BROMUS ARVENSIS L.
● 1930 onwards
○ Before 1930
KM
MILES
B+ 683/18
*BROMUS SECALINUS L.
● 1930 onwards
○ Before 1930
A 683/19
*BROMUS CARINATUS Hook. & Arn.
Ceratochloa carinata (Hook. & Arn.) Tutin
● Established, 1930 onwards
× Casual only, 1930 onwards
B 683/20
*BROMUS UNIOLOIDES Kunth
Ceratochloa unioloides (Willd.) Beauv.
● 1930 onwards
○ Before 1930

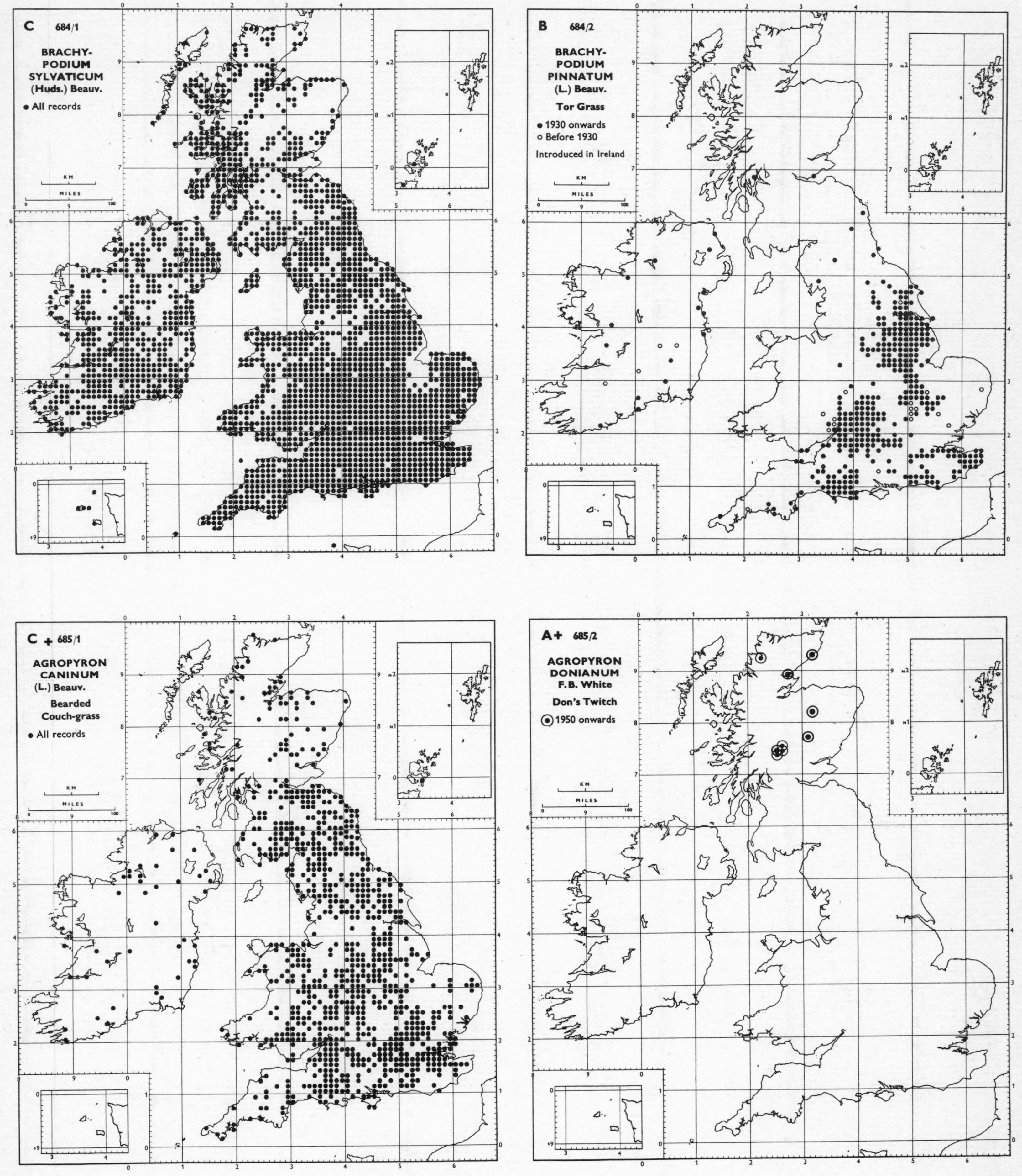
C 684/1
BRACHY-
PODIUM
SYLVATICUM
(Huds.) Beauv.
All records
B 684/2
BRACHY-
PODIUM
PINNATUM
(L.) Beauv.
Tor Grass
1930 onwards
Before 1930
Introduced in Ireland
C + 685/1
AGROPYRON
CANINUM
(L.) Beauv.
Bearded
Couch-grass
All records
A+ 685/2
AGROPYRON
DONIANUM
F.B. White
Don's Twitch
1950 onwards

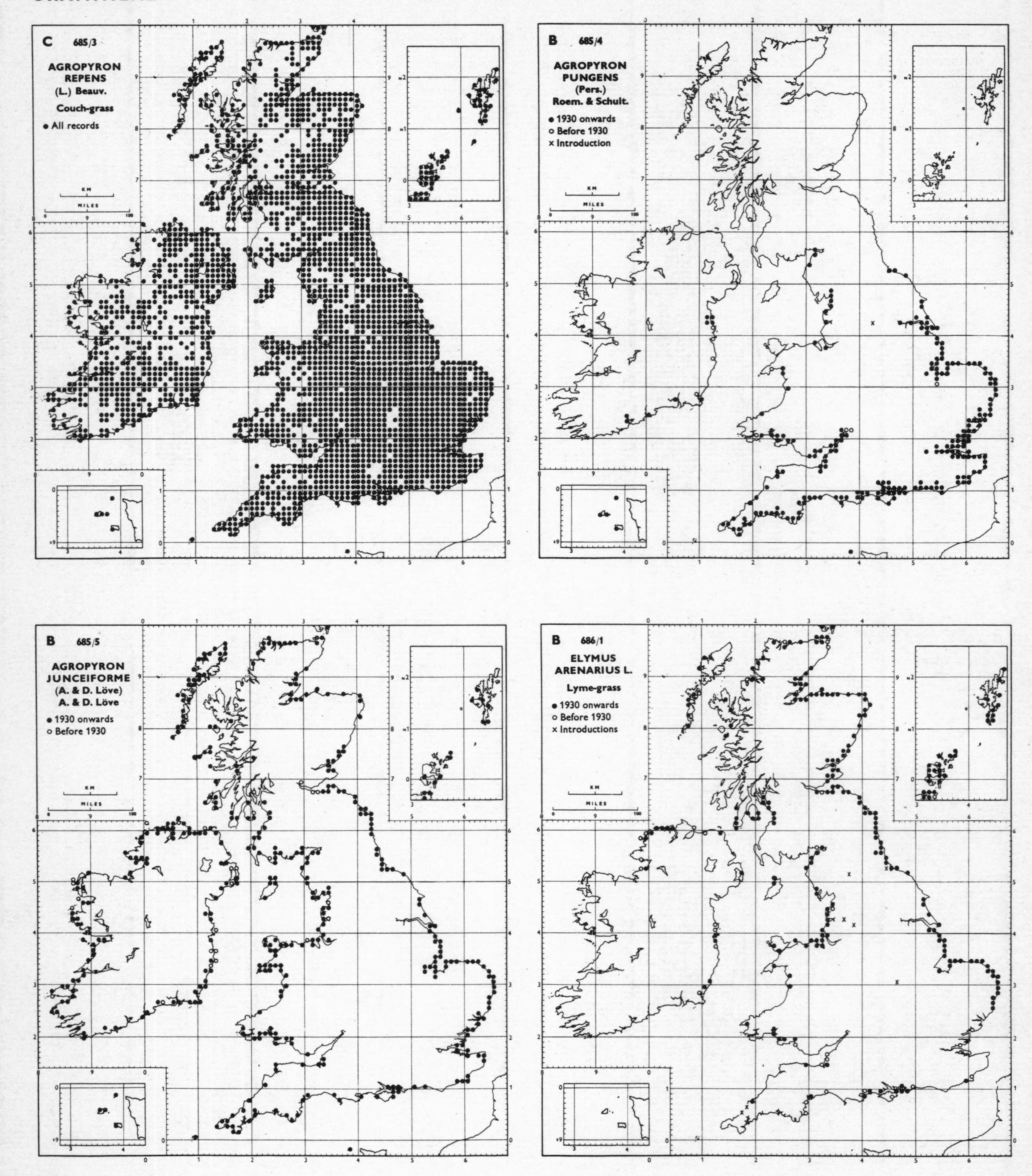
C 685/3
AGROPYRON REPENS (L.) Beauv.
Couch-grass
• All records
B 685/4
AGROPYRON PUNGENS (Pers.) Roem. & Schult.
• 1930 onwards
○ Before 1930
x Introduction
B 685/5
AGROPYRON JUNCEIFORME (A. & D. Löve) A. & D. Löve
• 1930 onwards
○ Before 1930
B 686/1
ELYMUS ARENARIUS L.
Lyme-grass
• 1930 onwards
○ Before 1930
x Introductions
KM
MILES

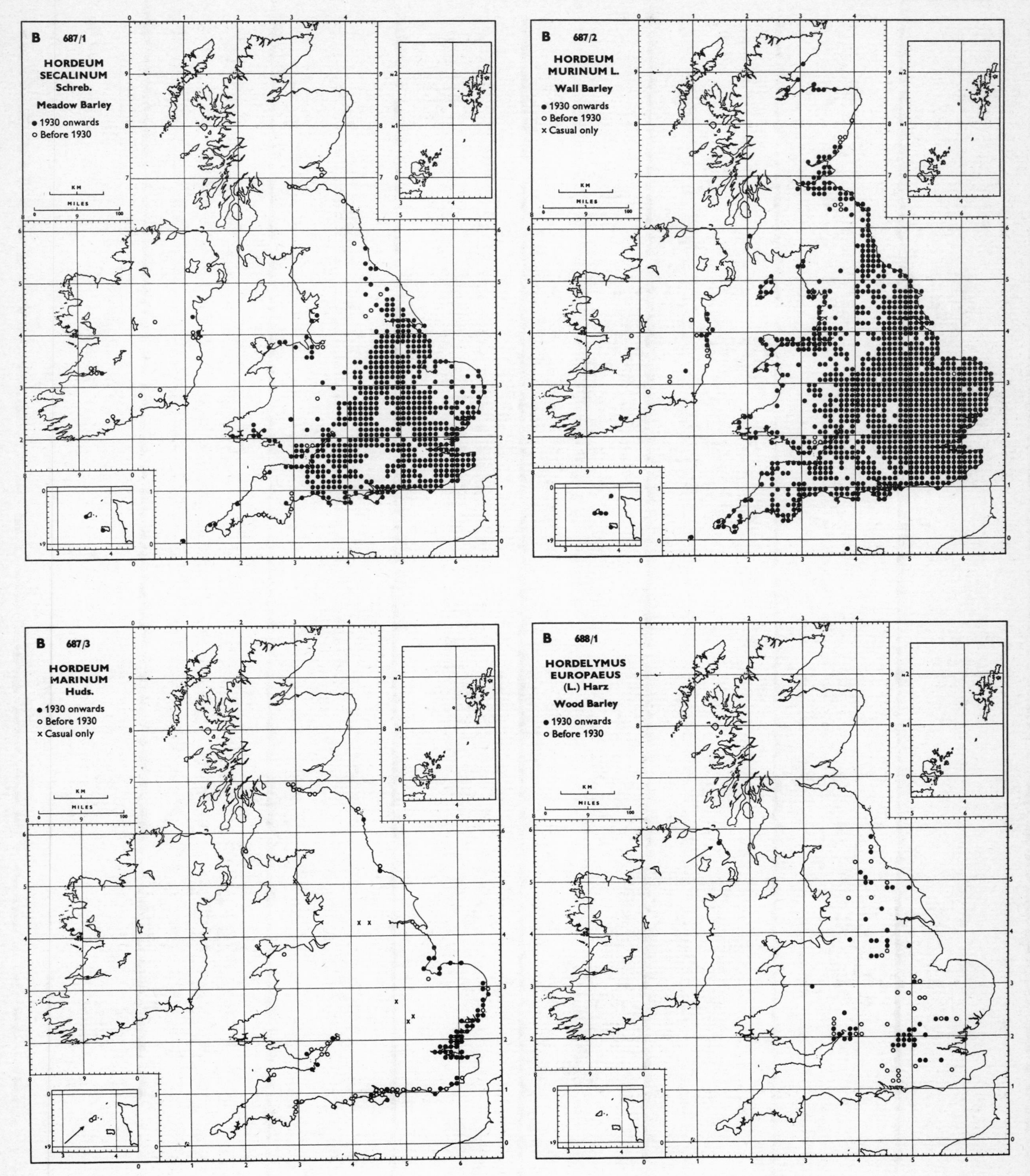
B 687/1
HORDEUM SECALINUM Schreb.
Meadow Barley
● 1930 onwards
○ Before 1930
B 687/2
HORDEUM MURINUM L.
Wall Barley
● 1930 onwards
○ Before 1930
× Casual only
B 687/3
HORDEUM MARINUM Huds.
● 1930 onwards
○ Before 1930
× Casual only
B 688/1
HORDELYMUS EUROPAEUS (L.) Harz
Wood Barley
● 1930 onwards
○ Before 1930
KM
MILES

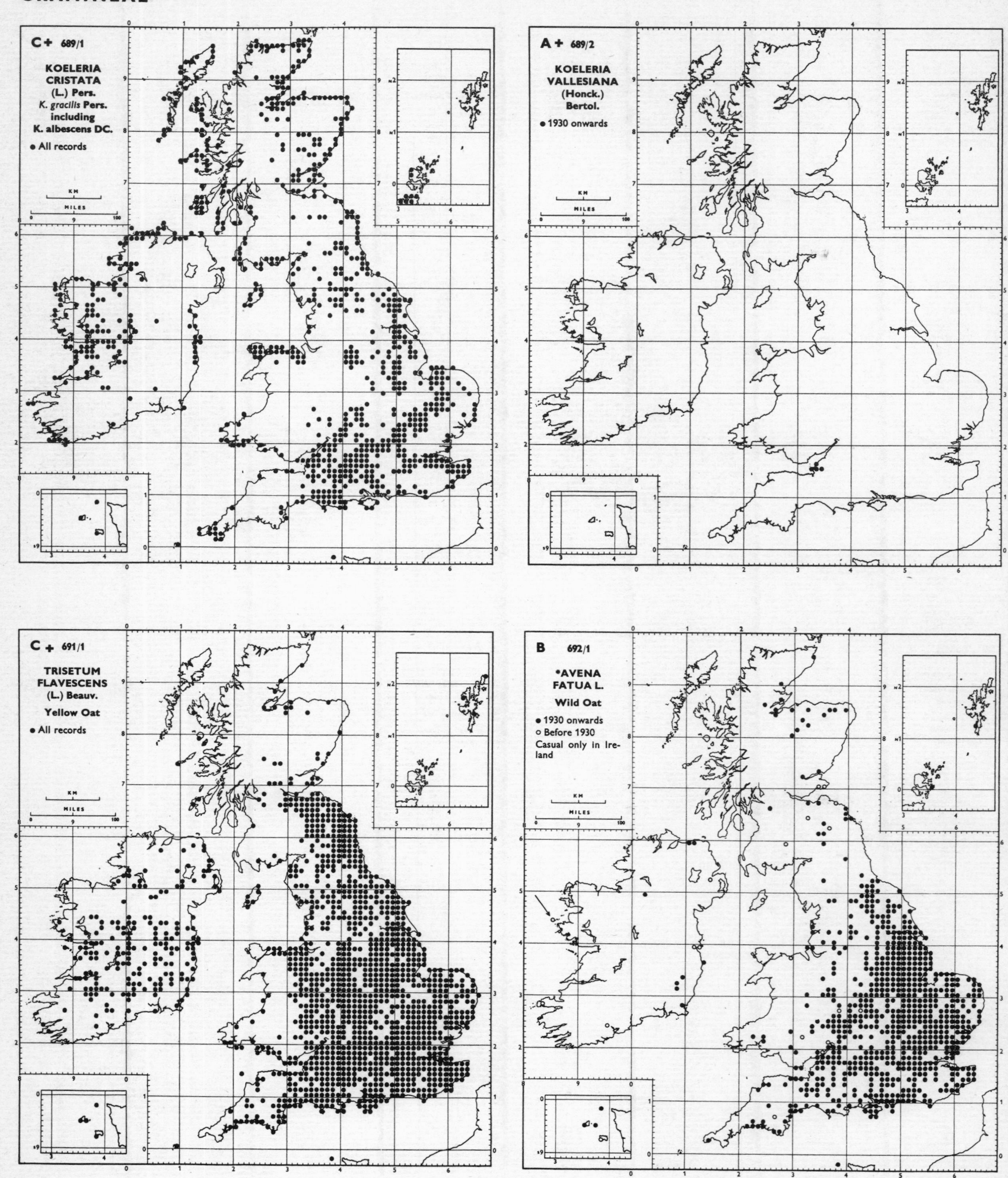
C+ 689/1
KOELERIA CRISTATA (L.) Pers.
K. gracilis Pers. including K. albescens DC.
● All records
A+ 689/2
KOELERIA VALLESIANA (Honck.) Bertol.
● 1930 onwards
C+ 691/1
TRISETUM FLAVESCENS (L.) Beauv.
Yellow Oat
● All records
B 692/1
•AVENA FATUA L.
Wild Oat
● 1930 onwards
○ Before 1930
Casual only in Ireland
KM
MILES

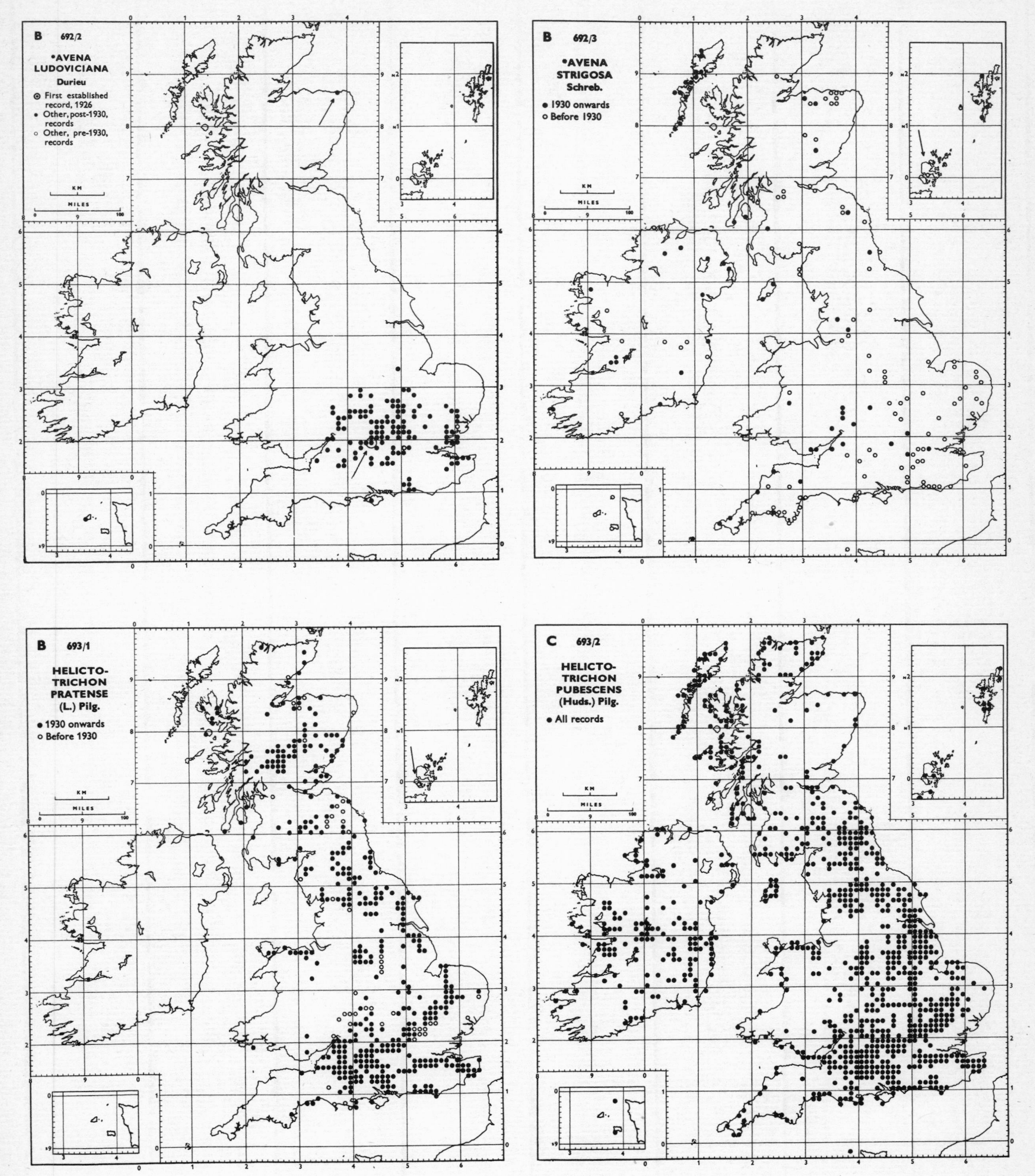
B 692/2
*AVENA LUDOVICIANA Durieu
First established record, 1926
Other, post-1930, records
Other, pre-1930, records
B 692/3
*AVENA STRIGOSA Schreb.
1930 onwards
Before 1930
B 693/1
HELICTOTRICHON PRATENSE (L.) Pilg.
1930 onwards
Before 1930
C 693/2
HELICTOTRICHON PUBESCENS (Huds.) Pilg.
All records

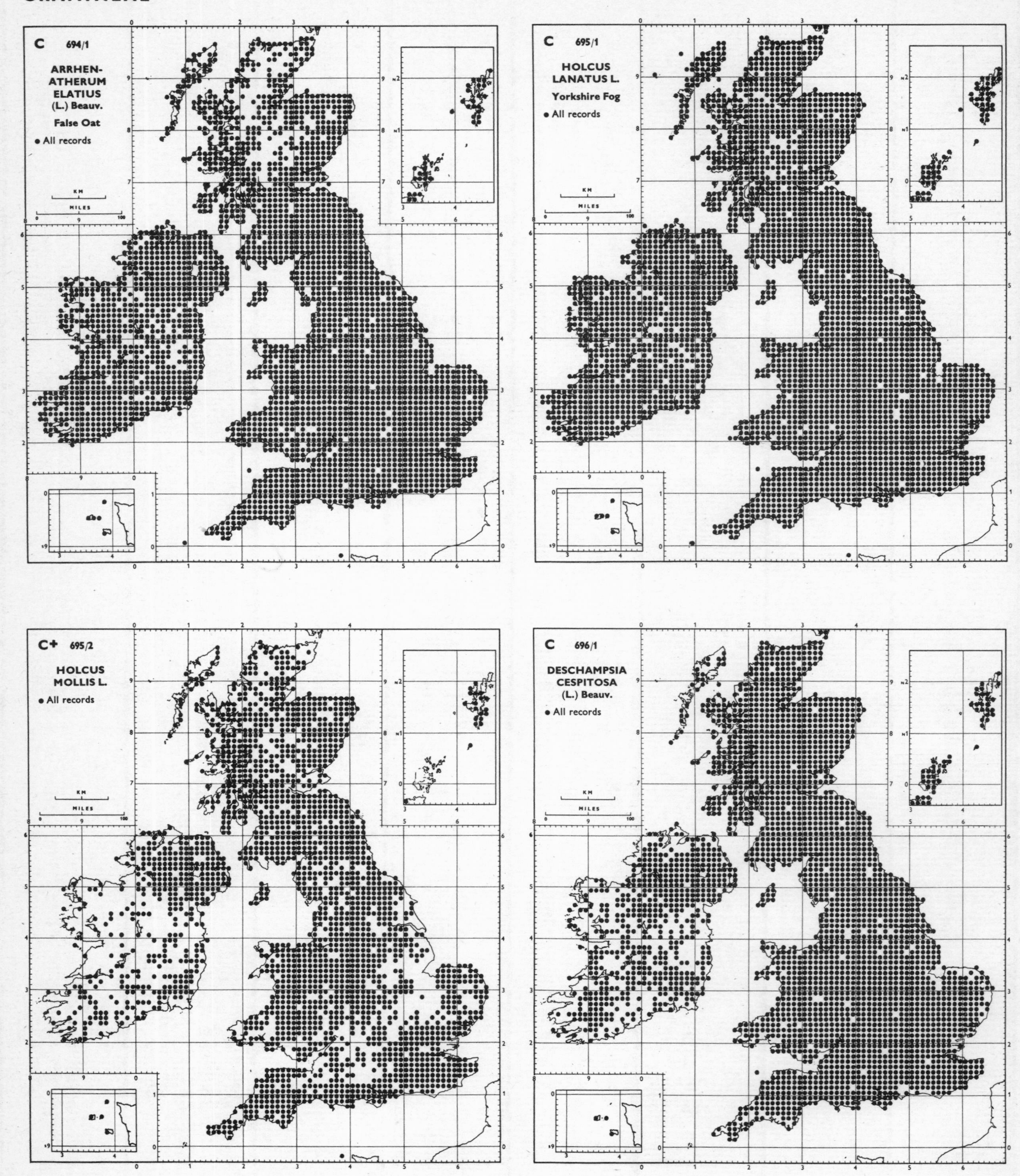
C 694/1
ARRHEN-ATHERUM ELATIUS (L.) Beauv.
False Oat
• All records
C 695/1
HOLCUS LANATUS L.
Yorkshire Fog
• All records
C+ 695/2
HOLCUS MOLLIS L.
• All records
C 696/1
DESCHAMPSIA CESPITOSA (L.) Beauv.
• All records

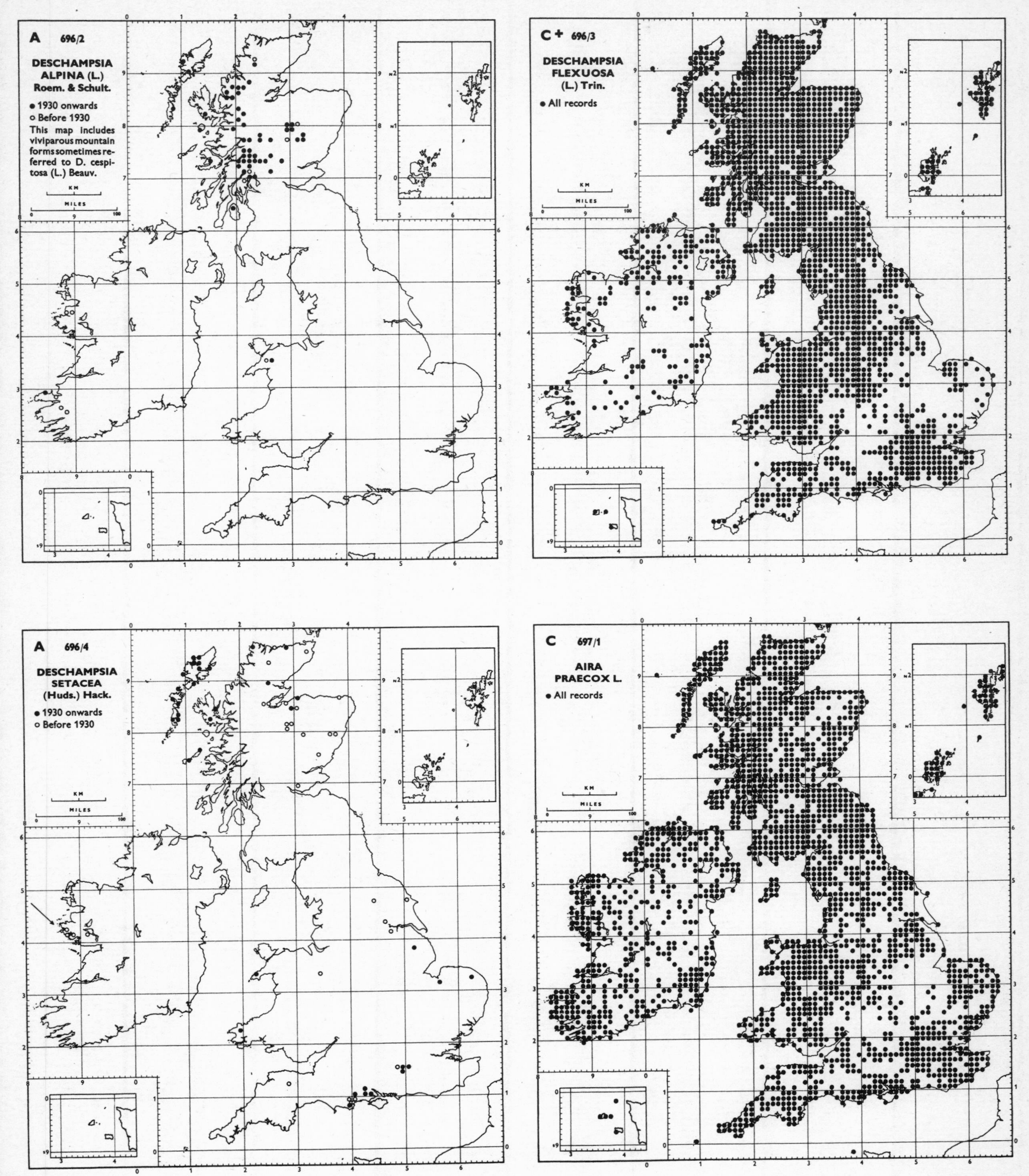
A 696/2
DESCHAMPSIA ALPINA (L.) Roem. & Schult.
● 1930 onwards
○ Before 1930
This map includes viviparous mountain forms sometimes referred to D. cespitosa (L.) Beauv.
C+ 696/3
DESCHAMPSIA FLEXUOSA (L.) Trin.
● All records
A 696/4
DESCHAMPSIA SETACEA (Huds.) Hack.
● 1930 onwards
○ Before 1930
C 697/1
AIRA PRAECOX L.
● All records
KM
MILES

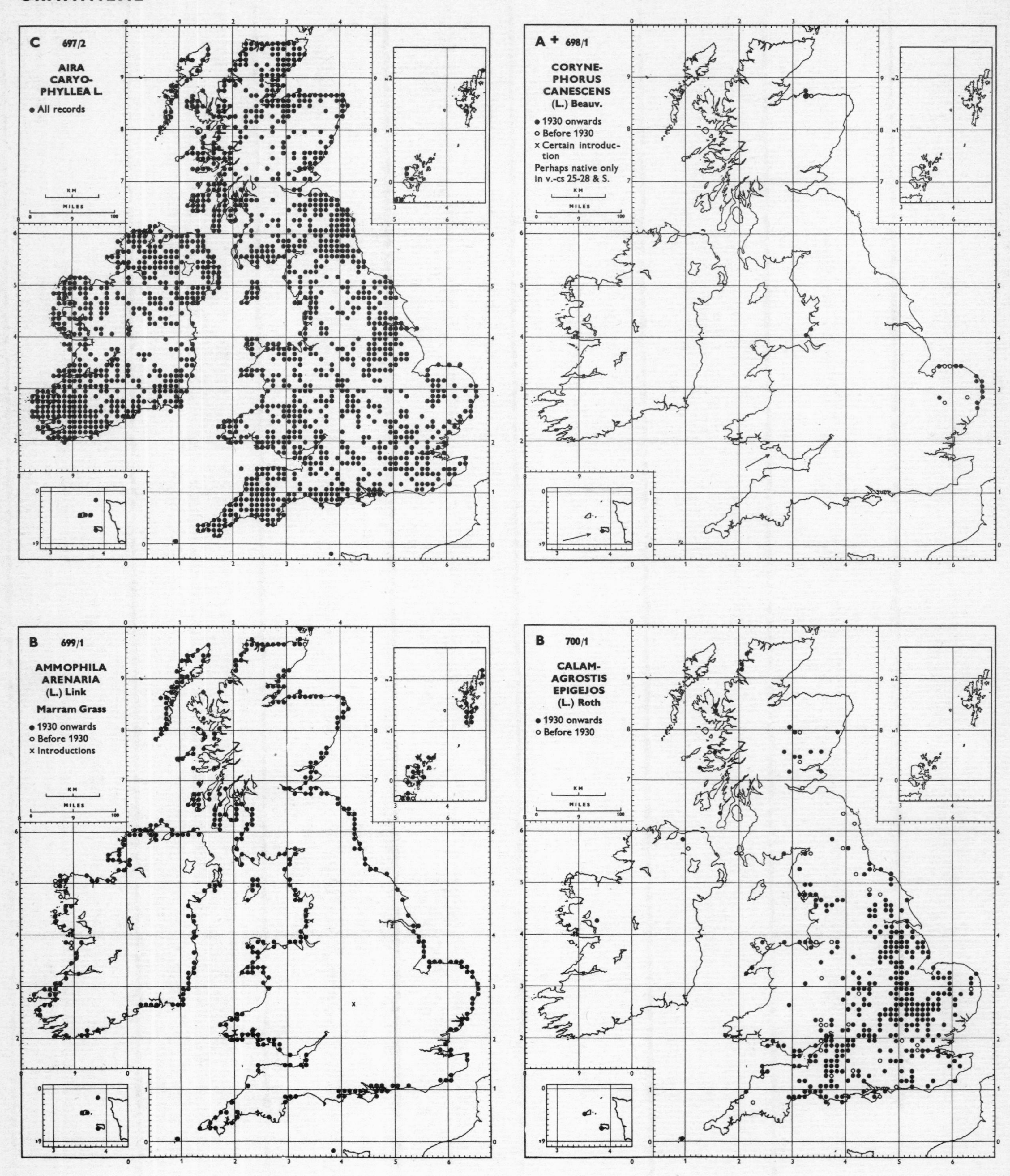
C 697/2
AIRA CARYO-PHYLLEA L.
• All records
KM
MILES
A + 698/1
CORYNE-PHORUS CANESCENS (L.) Beauv.
• 1930 onwards
○ Before 1930
× Certain introduction
Perhaps native only in v.-cs 25-28 & S.
KM
MILES
B 699/1
AMMOPHILA ARENARIA (L.) Link
Marram Grass
• 1930 onwards
○ Before 1930
× Introductions
KM
MILES
B 700/1
CALAM-AGROSTIS EPIGEJOS (L.) Roth
• 1930 onwards
○ Before 1930
KM
MILES

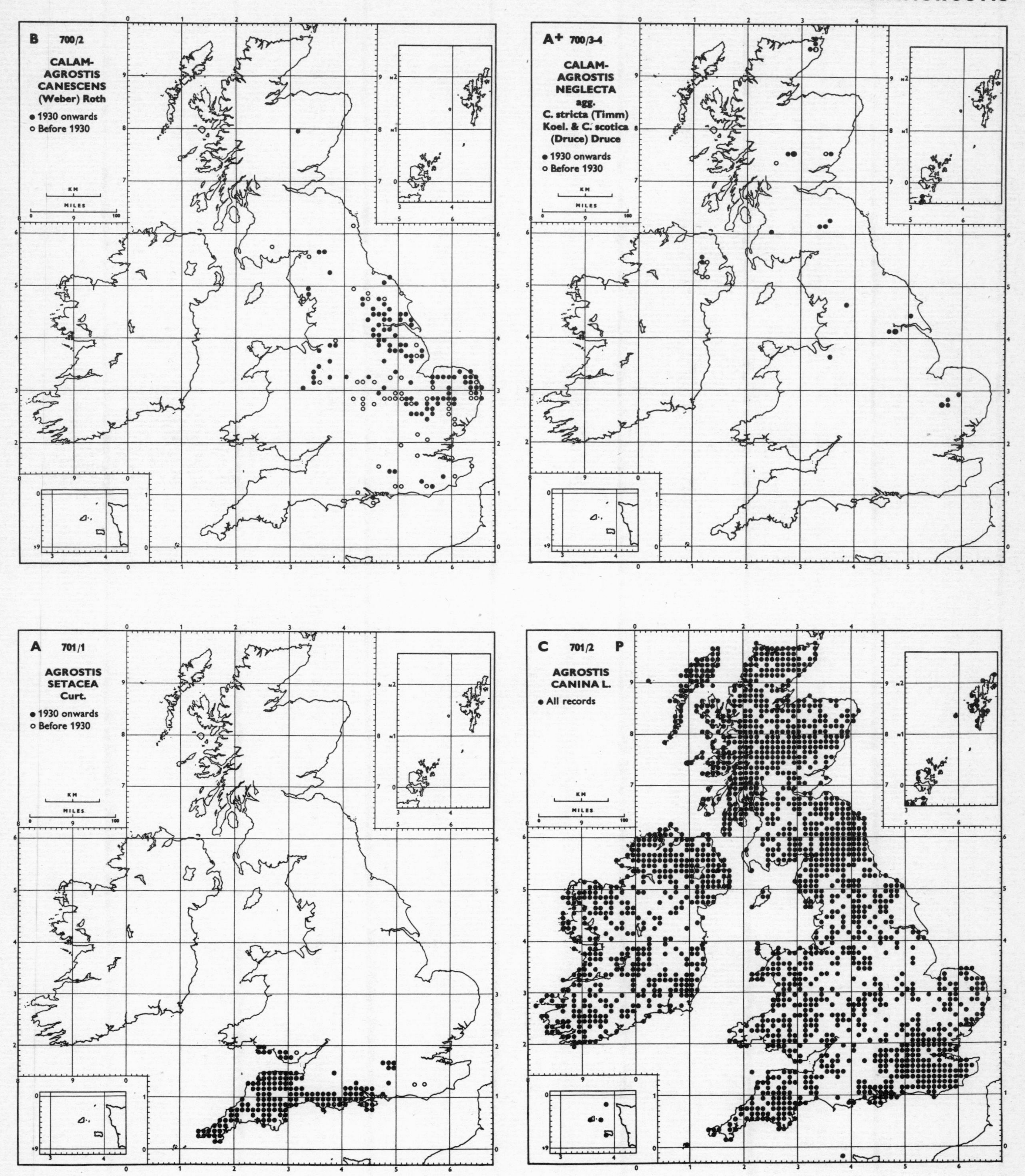
B 700/2
CALAMAGROSTIS CANESCENS (Weber) Roth
● 1930 onwards
○ Before 1930
A+ 700/3-4
CALAMAGROSTIS NEGLECTA agg.
C. stricta (Timm) Koel. & C. scotica (Druce) Druce
● 1930 onwards
○ Before 1930
A 701/1
AGROSTIS SETACEA Curt.
● 1930 onwards
○ Before 1930
C 701/2 P
AGROSTIS CANINA L.
● All records
KM
MILES

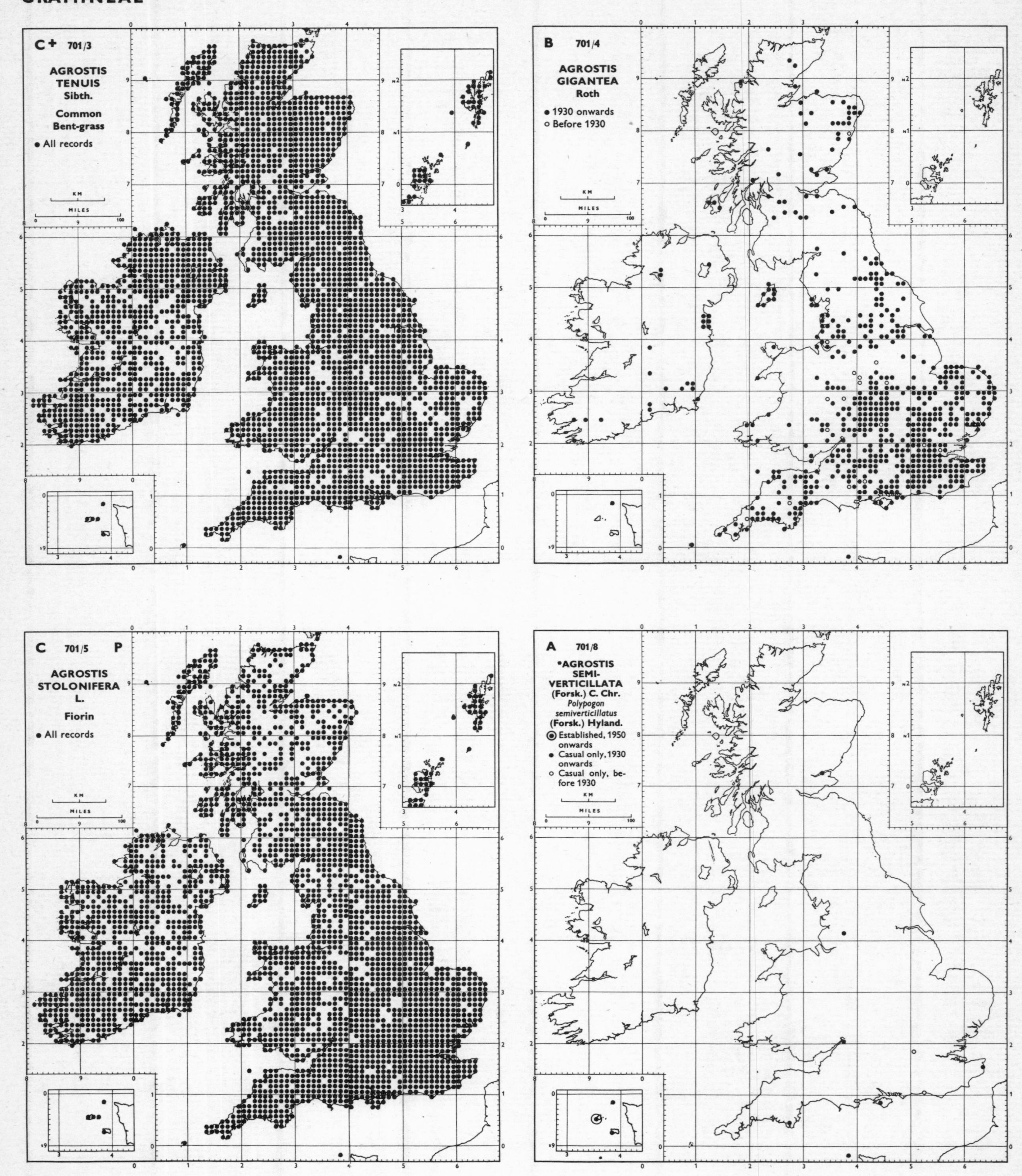
C+ 701/3
AGROSTIS
TENUIS
Sibth.
Common
Bent-grass
• All records
KM
MILES
B 701/4
AGROSTIS
GIGANTEA
Roth
• 1930 onwards
○ Before 1930
KM
MILES
C 701/5 P
AGROSTIS
STOLONIFERA
L.
Fiorin
• All records
KM
MILES
A 701/8
*AGROSTIS
SEMI-
VERTICILLATA
(Forsk.) C. Chr.
Polypogon
semiverticillatus
(Forsk.) Hyland.
Established, 1950 onwards
• Casual only, 1930 onwards
○ Casual only, before 1930
KM
MILES

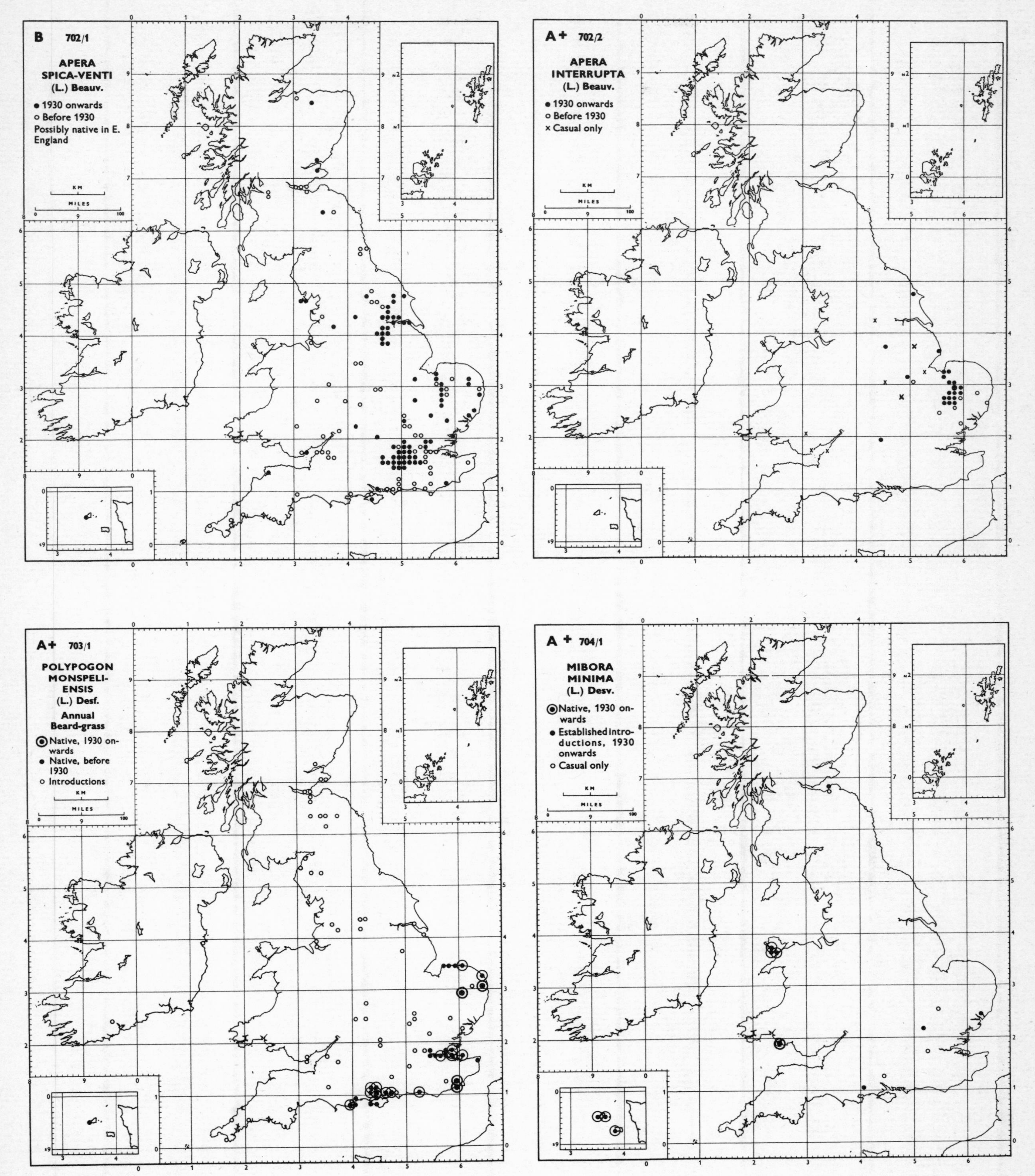
B 702/1
APERA SPICA-VENTI (L.) Beauv.
• 1930 onwards
○ Before 1930
Possibly native in E. England
A+ 702/2
APERA INTERRUPTA (L.) Beauv.
• 1930 onwards
○ Before 1930
× Casual only
A+ 703/1
POLYPOGON MONSPELIENSIS (L.) Desf.
Annual Beard-grass
◉ Native, 1930 onwards
• Native, before 1930
○ Introductions
A + 704/1
MIBORA MINIMA (L.) Desv.
◉ Native, 1930 onwards
• Established introductions, 1930 onwards
○ Casual only
KM
MILES

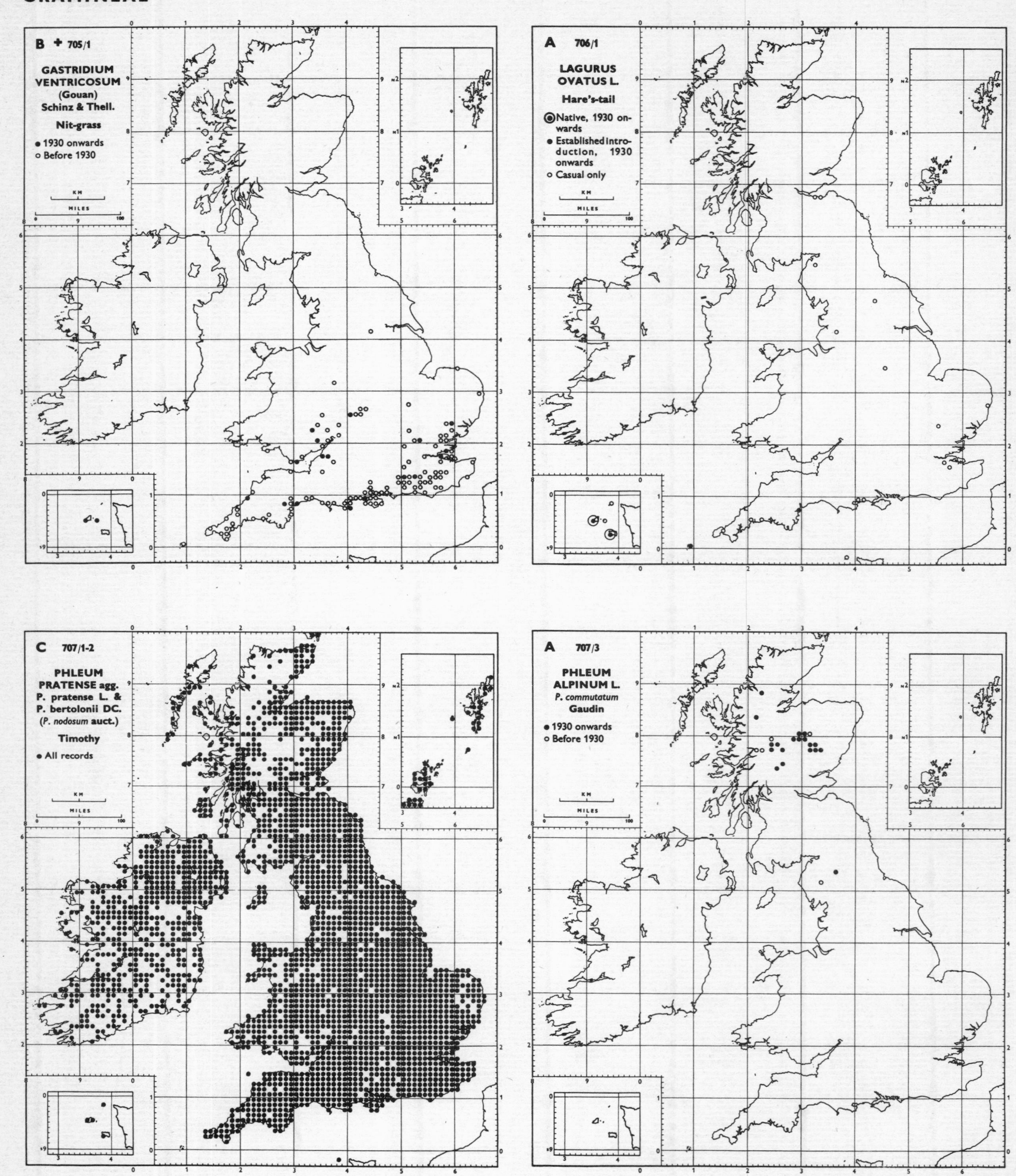
B ✚ 705/1
GASTRIDIUM
VENTRICOSUM
(Gouan)
Schinz & Thell.
Nit-grass
● 1930 onwards
○ Before 1930
A 706/1
LAGURUS
OVATUS L.
Hare's-tail
Native, 1930 onwards
● Established introduction, 1930 onwards
○ Casual only
C 707/1-2
PHLEUM
PRATENSE agg.
P. pratense L. &
P. bertolonii DC.
(P. nodosum auct.)
Timothy
● All records
A 707/3
PHLEUM
ALPINUM L.
P. commutatum
Gaudin
● 1930 onwards
○ Before 1930

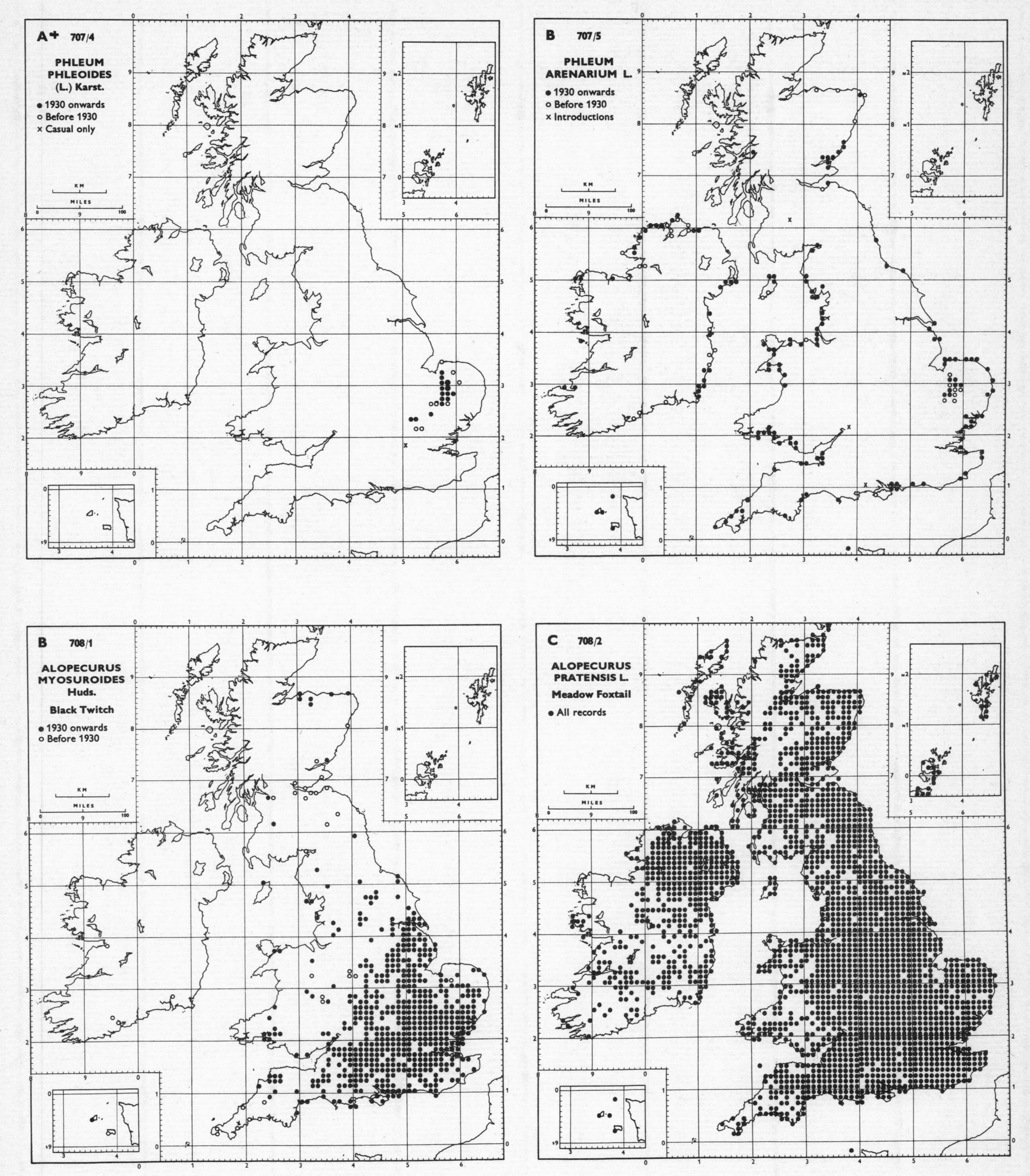
A+ 707/4
PHLEUM PHLEOIDES (L.) Karst.
● 1930 onwards
○ Before 1930
× Casual only
B 707/5
PHLEUM ARENARIUM L.
● 1930 onwards
○ Before 1930
× Introductions
B 708/1
ALOPECURUS MYOSUROIDES Huds.
Black Twitch
● 1930 onwards
○ Before 1930
C 708/2
ALOPECURUS PRATENSIS L.
Meadow Foxtail
● All records
KM
MILES

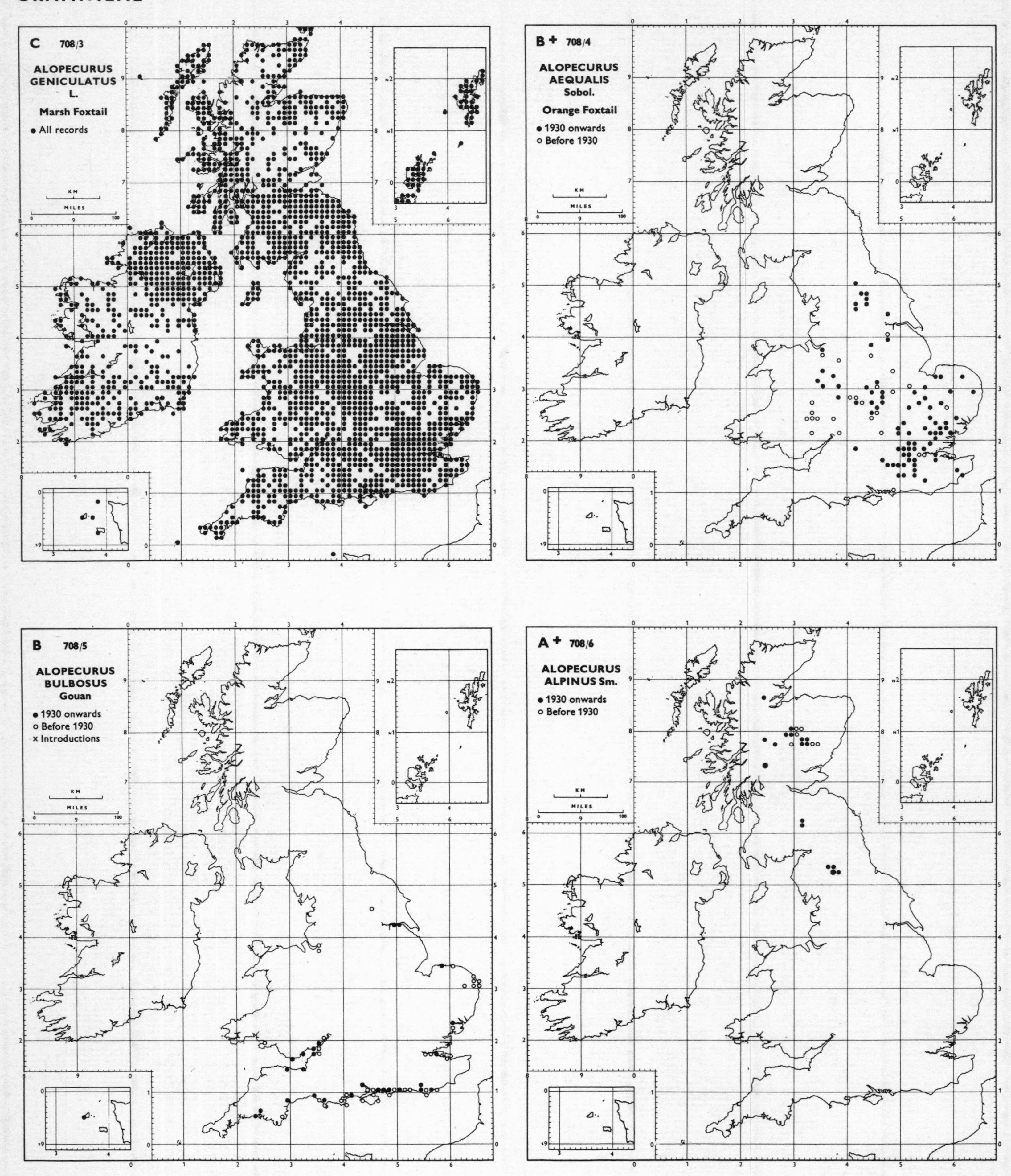
C 708/3
ALOPECURUS GENICULATUS L.
Marsh Foxtail
• All records
B + 708/4
ALOPECURUS AEQUALIS Sobol.
Orange Foxtail
• 1930 onwards
○ Before 1930
B 708/5
ALOPECURUS BULBOSUS Gouan
• 1930 onwards
○ Before 1930
× Introductions
A + 708/6
ALOPECURUS ALPINUS Sm.
• 1930 onwards
○ Before 1930
KM
MILES

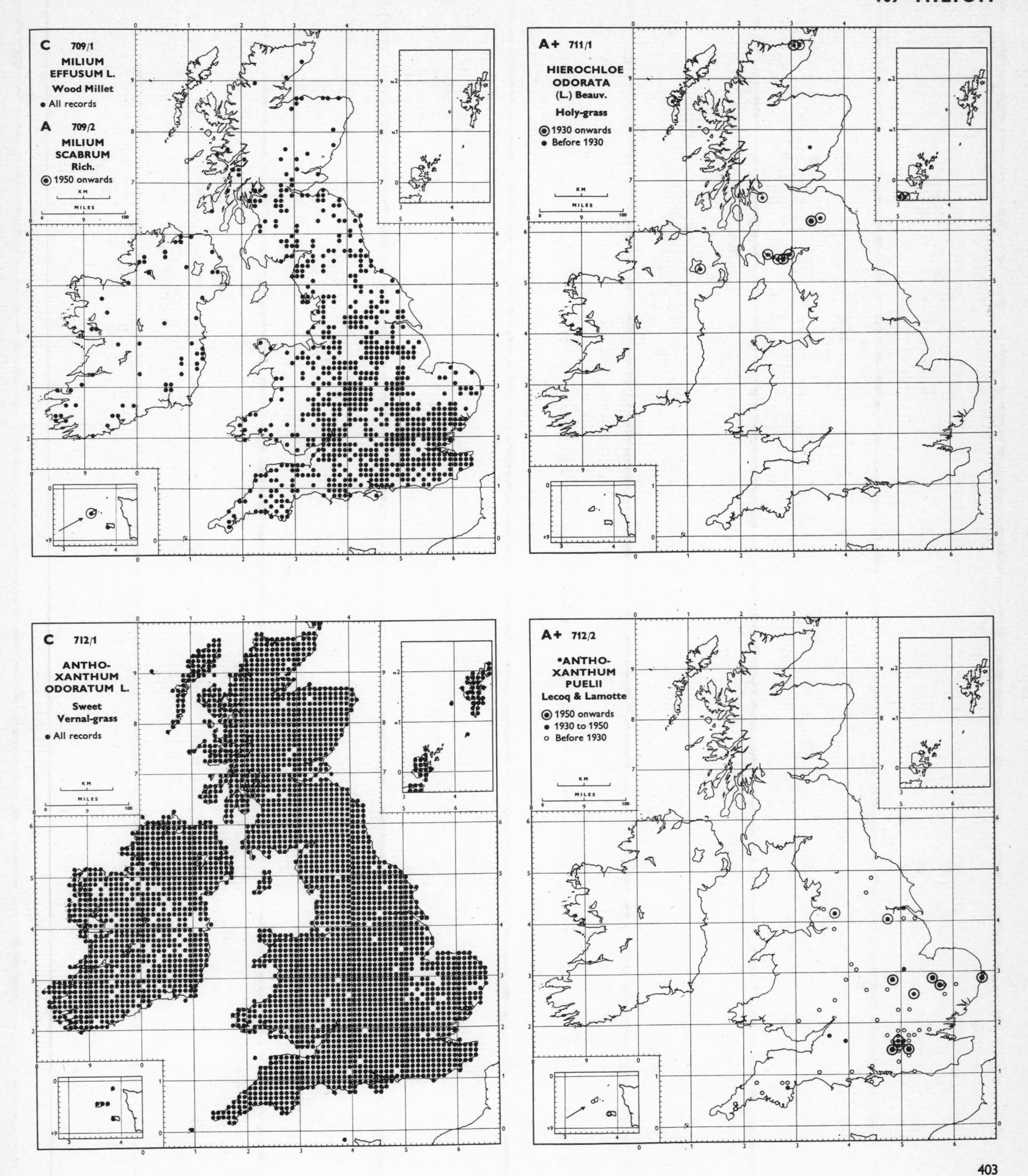
C 709/1
MILIUM EFFUSUM L.
Wood Millet
• All records
A 709/2
MILIUM SCABRUM Rich.
◉ 1950 onwards
A+ 711/1
HIEROCHLOE ODORATA (L.) Beauv.
Holy-grass
◉ 1930 onwards
• Before 1930
C 712/1
ANTHOXANTHUM ODORATUM L.
Sweet Vernal-grass
• All records
A+ 712/2
*ANTHOXANTHUM PUELII Lecoq & Lamotte
◉ 1950 onwards
• 1930 to 1950
○ Before 1930

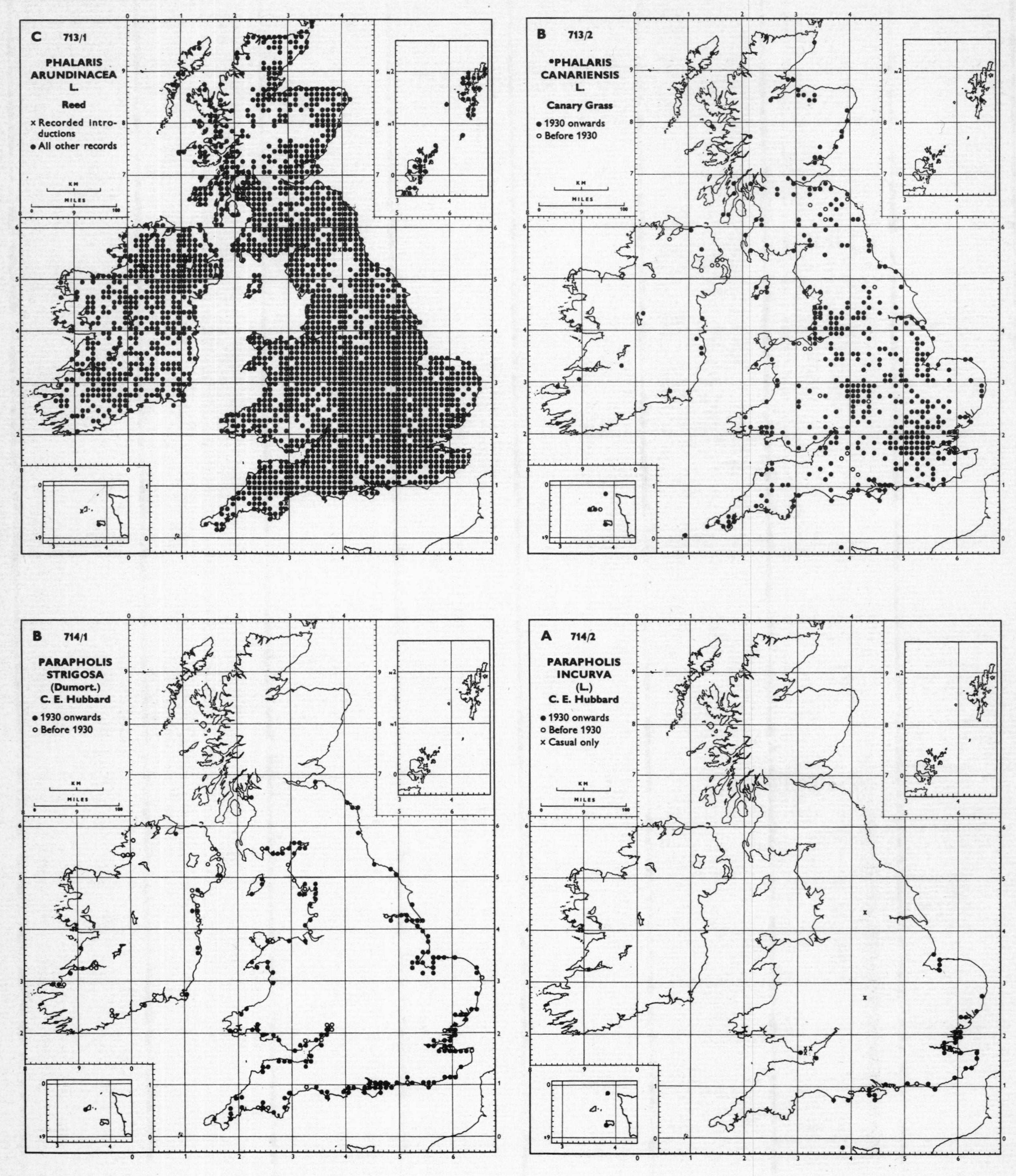

C 713/1
PHALARIS ARUNDINACEA L.
Reed
× Recorded introductions
● All other records
B 713/2
*PHALARIS CANARIENSIS L.
Canary Grass
● 1930 onwards
○ Before 1930
B 714/1
PARAPHOLIS STRIGOSA (Dumort.) C. E. Hubbard
● 1930 onwards
○ Before 1930
A 714/2
PARAPHOLIS INCURVA (L.) C. E. Hubbard
● 1930 onwards
○ Before 1930
× Casual only

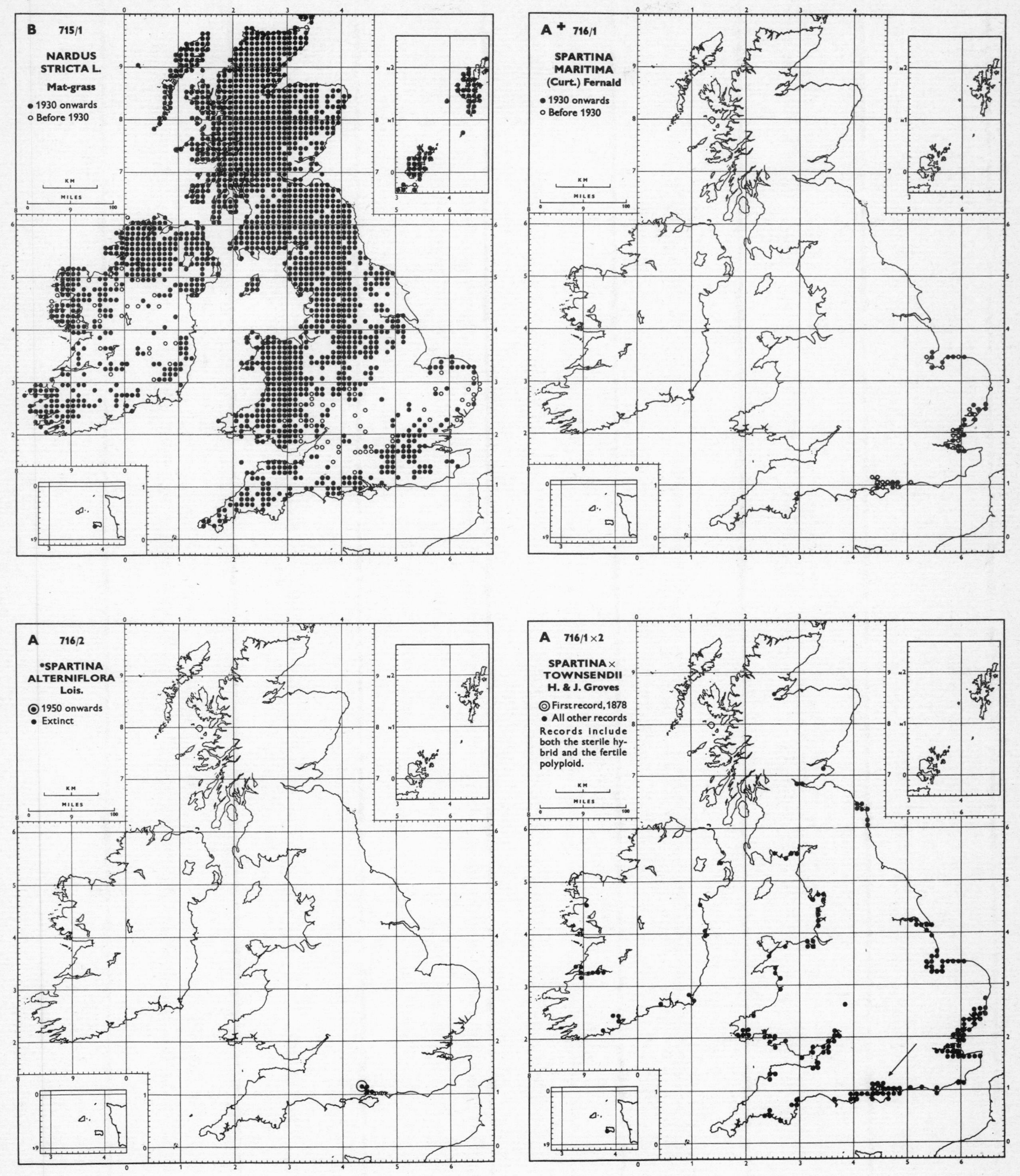
B 715/1
NARDUS
STRICTA L.
Mat-grass
● 1930 onwards
○ Before 1930
A + 716/1
SPARTINA
MARITIMA
(Curt.) Fernald
● 1930 onwards
○ Before 1930
A 716/2
•SPARTINA
ALTERNIFLORA
Lois.
◎ 1950 onwards
● Extinct
A 716/1×2
SPARTINA×
TOWNSENDII
H. & J. Groves
◎ First record, 1878
● All other records
Records include
both the sterile hybrid and the fertile
polyploid.
KM
MILES

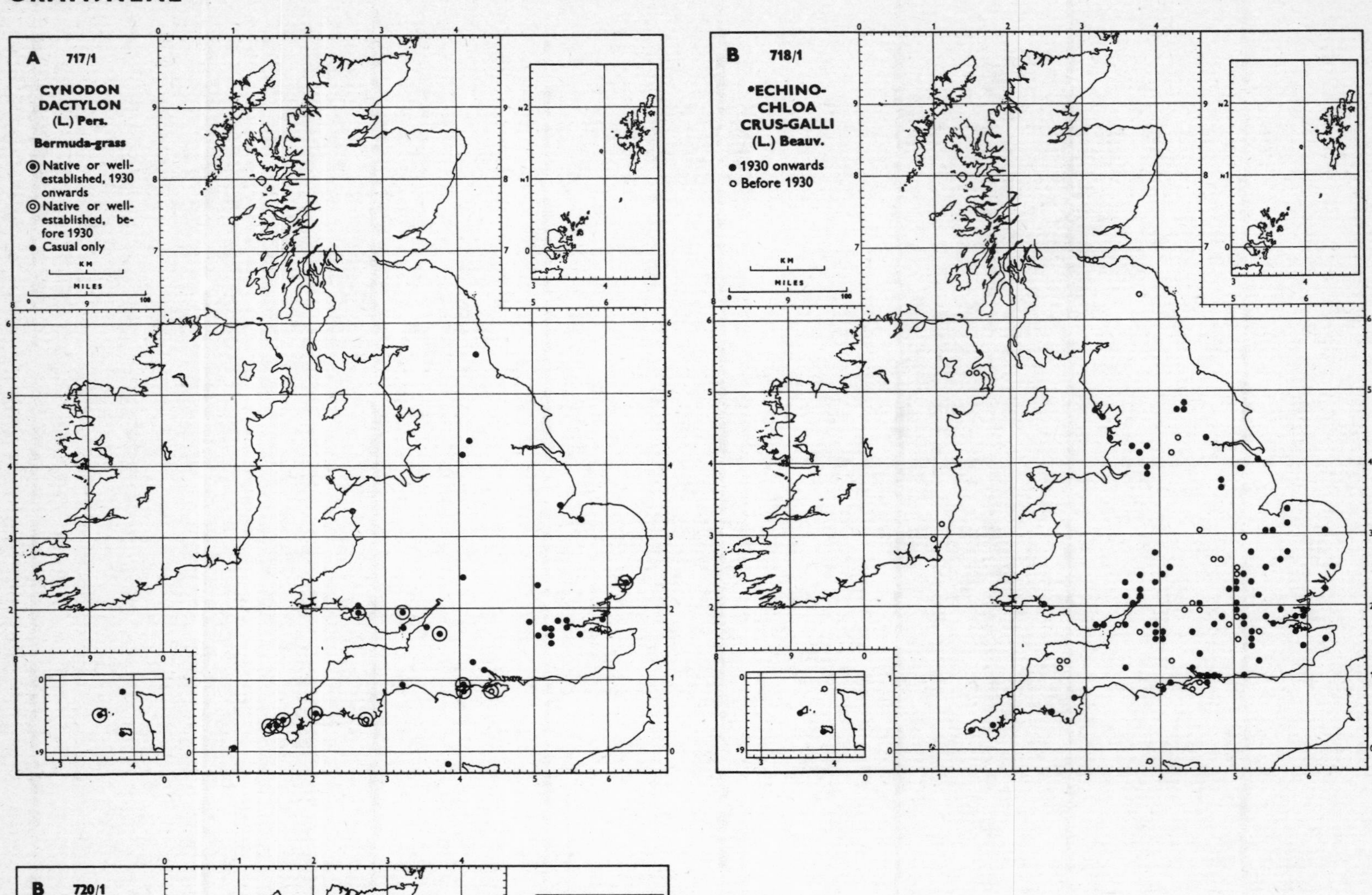
A 717/1
CYNODON
DACTYLON
(L.) Pers.
Bermuda-grass
Native or well-established, 1930 onwards
Native or well-established, before 1930
Casual only
KM
MILES
B 718/1
•ECHINO-
CHLOA
CRUS-GALLI
(L.) Beauv.
• 1930 onwards
○ Before 1930
KM
MILES

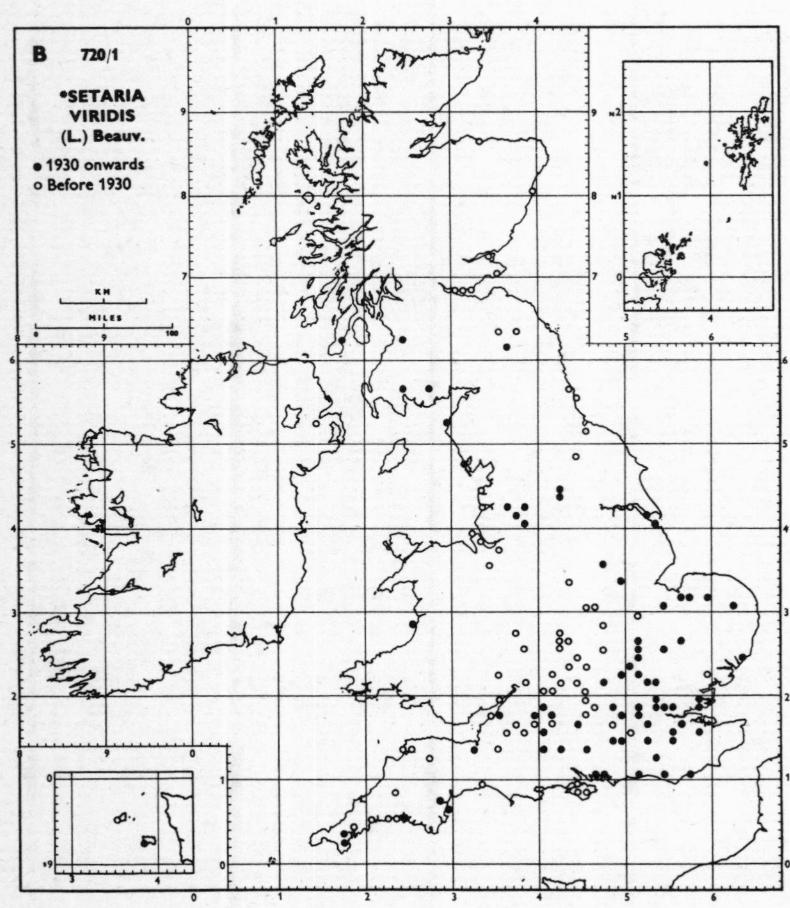
B 720/1
•SETARIA
VIRIDIS
(L.) Beauv.
• 1930 onwards
○ Before 1930
KM
MILES

Appendix I

LIST OF AGGREGATE AND PROVISIONAL MAPS

AGGREGATE MAPS

21/1–2 Dryopteris filix-mas (+D. borreri). *D. borreri* Newm. is also mapped separately.
58/2–3 Papaver dubium (+P. lecoqii).
88/1–3 Cochlearia officinalis (+C. alpina & C. micacea).
95/1–3 Erophila verna (+E. spathulata & E. praecox).
102/1–2 Rorippa nasturtium-aquaticum (+R. microphylla).
120/1–2 Tamarix gallica (+T. anglica).
141/1–2 Arenaria serpyllifolia (+A. leptoclados).
160/2–4 Salicornia (S. dolichostachya, S. europaea & S. ramosissima).
222/1–2 Aphanes arvensis (+A. microcarpa).
225/8–10 Rosa canina (+R. dumalis & R. obtusifolia).
225/11–13 Rosa villosa (+R. tomentosa & R. sherardii).
225/14–17 Rosa rubiginosa (+R. micrantha, R. elliptica & R. agrestis).
258/2–3 Circaea alpina (+C. intermedia).
262/1–2 Callitriche stagnalis (+C. platycarpa).
319/14–15 Euphorbia esula (+E. uralensis).
330/3–6 Ulmus carpinifolia (+U. angustifolia, U. coritana & U. plotii).
342/4–5 Populus nigra (+P.×canadensis).
344/1–2 Ledum palustre (+L. groenlandicum).
409/1–2 Lycium chinense (+L. halimifolium).
442/1–2 Utricularia vulgaris (+U. neglecta).
538/2–4 Arctium minus (+A. nemorosum & A. pubens).
544/6–7 Centaurea nigra (+C. nemoralis).
663/9–10 Carex serotina (+C. scandinavica).
670/8–9 Festuca ovina (+F. tenuifolia).
676/10–12 Poa pratensis (+P. angustifolia & P. subcaerulea).
683/10–12 Bromus mollis (+B. ferronii & B. thominii).
700/3–4 Calamagrostis neglecta (C. stricta & C. scotica).
707/1–2 Phleum pratense (+P. bertolonii).

PROVISIONAL MAPS

21/6 Dryopteris lanceolatocristata.
Some forms of *D. dilatata* have more or less concolorous scales on the rachis, and such plants are often recorded as *D. lanceolatocristata*.

43/1 Anemone nemorosa &
46/24 Ranunculus ficaria.
These are examples of a group of species probably under-recorded in south-west Ireland (see Introduction, page xv).

67/2–3 Brassica napus & B. rapa.
Many recorders have ignored plants referable to one or other of these two species because of their frequent occurrence as relics of cultivated crops. In addition these species are not always readily distinguishable.

88/4 Cochlearia scotica.
Some plants referred to this species differ considerably from the type, and the limits of the species are very uncertain.

136/2 Sagina ciliata.
The taxonomy of this species and *S. apetala* is not properly worked out. Most competent recorders have distinguished two species but plants occur with various combinations of the diagnostic characters.

156/3–4 Atriplex hastata & A. glabriuscula.
The distinction between these two species in maritime habitats is very unsatisfactory.

175/1 Aesculus hippocastanum &
184/1 Laburnum anagyroides.
These are examples of alien trees which have been omitted from lists sent in by some recorders (see Introduction, page xix).

211/9 Rubus caesius.
It is probable that this species has been over-recorded in error for taxa of *Rubus fruticosus* agg. particularly in north and west Britain.

254/5 Epilobium roseum.
This species is perhaps over-recorded in error for *E. adenocaulon* and other species and hybrids.

254/7–8 Epilobium adnatum & E. lamyi.
These two species differ in leaf shape and in size of floral parts but lack the precise qualitative distinction present in most other species of *Epilobium*. Continental authorities often treat them as sub-species. Our records reflect this taxonomic difficulty.

254/9 Epilobium obscurum.
One of the commonest species of *Epilobium* in the country except in parts of eastern England, but apparently overlooked and under-recorded.

256/1–2 Oenothera biennis & O. erythrosepala.
There is acknowledged taxonomic confusion here and the records should be treated with caution.

262/4 Callitriche intermedia.
This species is often recorded in error for submerged forms of *C. stagnalis* agg.

319/14–16 Euphorbia esula agg. & E. cyparissias.
The treatment of this group in Clapham, Tutin & Warburg (1952) is acknowledged to be unsatisfactory, and the boundaries of the taxa are not clear. Even *E. cyparissias* is not always clearly distinguishable from plants of the *E. esula* agg.

320/11 Polygonum nodosum.
The group of taxa to which *P. nodosum* and *P. lapathifolium* belong is difficult in Europe and the British material is not yet clearly understood.

343/14 Salix nigricans.
There is no unanimity amongst the experts as to the limits of this species.

378/1 Ligustrum vulgare.
This is a widely planted species, therefore the status of the records is often uncertain. In addition there is confusion between the native *L. vulgare* and garden species.

392/1 Symphytum officinale.
The limits of this species are disputed. The cream-flowered plant of fens and marshes in eastern England is generally regarded as native, but many, if not all, purple-flowered variants referred to this species are doubtfully native and possibly of hybrid origin. It has therefore not been possible to separate records of *S. officinale* from the hybrid (*S.×uplandicum* Nyman).

400/1 & 4 Myosotis scorpioides (*M. palustris*) & M. caespitosa.
It is probable that *M. scorpioides* has been over-recorded in error for *M. caespitosa*, particularly in north and west Britain.

406/1 & 3 Calystegia sepium & C. silvatica.
Although there is little difficulty in distinguishing good material of these two species, three factors may have confused the records.
(i) The occasional occurrence of intermediate populations.
(ii) The existence of a third and, until recently, overlooked taxon, *C. pulchra*.
(iii) The continued use of *C. sepium* as an aggregate name.

440/9 Orobanche picridis.
The taxonomic distinction between this species and varieties of *O. minor* is not understood, and old herbarium material cannot be used satisfactorily.

538/1 Arctium lappa.
In spite of careful checking of the records the species is perhaps still over-recorded because of the continued use of the name in the aggregate sense. However, the map probably indicates correctly the limits of distribution.

550/2–3 Leontodon hispidus & L. taraxacoides (*L. leysseri*).
Although these two species are not in any sense critical the number of erroneous records received suggests that both may have been over-recorded in error for other yellow composites.

559/2 Crepis vesicaria (*C. taraxacifolia*).
Large-flowered plants of *C. capillaris*, which occur in north and west Britain, have been recorded as this species in error.

559/5 Crepis biennis.
A distinct, but rather local, species which has probably been over-recorded in error for *C. vesicaria.*

605/4 Juncus compressus.
There are two factors which may have confused the records for this species.
(i) Misidentification of specimens of *J. gerardii* near the coast. Northern records are particularly suspect.
(ii) "Juncu com" on the Regional Record Card has been interpreted as *Juncus communis* by some recorders, an aggregate name for *J. effusus* and *J. conglomeratus.* Most errors of this kind have been eliminated.

670/1–2 Festuca pratensis & F. arundinacea.
Although typical specimens of these two species are easily identified some material is difficult: the determination of such material has probably been inconsistent. However, the general features of the two maps seem reasonable.

683/14 Bromus racemosus.
A rare and somewhat critical species which has probably been over-recorded in error for *B. commutatus.*

701/2 Agrostis canina &
701/5 Agrostis stolonifera.
Probably rather seriously under-recorded because of extra taxonomic difficulty, but a special case of the general tendency for the common sedges and grasses to be overlooked (see Introduction, page xvii).

UNDER-RECORDED SQUARES

Great Britain

08/77	33/49	42/68
08/96	33/58	42/72
08/97	34/40	42/73
16/96	34/41	42/74
18/78	34/79	42/75
18/86	34/80	42/77
19/01	34/81	42/78
19/02	34/91	43/47
20/05	35/51	43/49
21/89	35/81	44/07
21/90	35/93	44/32
21/99	35/94	44/59
26/35	37/50	44/69
26/58	38/22	44/96
26/68	38/31	45/03
26/78	38/62	45/14
26/79	38/63	45/61
26/88	39/36	54/07
26/96	39/39	57/53
27/56	39/48	57/62
27/88	41/81	57/63
28/16	41/86	57/64
28/26	41/89	57/74
28/40	42/00	57/75
29/27	42/10	62/02
29/95	42/11	62/11
31/19	42/30	62/12
31/31	42/62	68/48
31/41	42/63	68/49
31/52	42/64	68/58
31/63	42/65	68/59
31/72	42/66	69/50
32/95	42/67	

Ireland

84/83	03/00	13/06
84/94	03/01	13/16
84/98	03/34	13/19
85/81	03/39	14/00
92/10	03/49	14/01
92/95	03/56	14/02
93/25	03/58	14/10
93/45	03/67	14/11
93/66	03/74	
93/82	03/76	
94/01	03/79	
94/08	03/86	
94/10	03/87	
94/15	03/94	
94/16	03/96	
94/17	04/07	
94/40	04/08	
94/42	04/18	
94/43	04/23	
94/46	04/27	
94/52	04/28	
94/61	04/31	
94/69	04/33	
94/82	04/45	
94/83	04/53	
94/88	04/55	
94/93	04/74	
94/99	04/75	
95/01	04/81	
95/10	04/84	
95/90	04/90	
02/29	13/04	
02/38	13/05	

Appendix II

VICE-COUNTY RECORDS OMITTED FROM THE MAPS

THE VICE-COUNTY NUMBERS AND CORRESPONDING VICE-COUNTIES

The vice-county numbers follow Watson (1873) and Praeger (1901). The names have been altered to agree with current usage, but the old names have been added in brackets where they differ significantly from modern ones.

ENGLAND AND WALES

1. West Cornwall (with Scilly)
2. East Cornwall
3. South Devon
4. North Devon
5. South Somerset
6. North Somerset
7. North Wiltshire
8. South Wiltshire
9. Dorset
10. Isle of Wight
11. South Hampshire
12. North Hampshire
13. West Sussex
14. East Sussex
15. East Kent
16. West Kent
17. Surrey
18. South Essex
19. North Essex
20. Hertfordshire
21. Middlesex
22. Berkshire
23. Oxfordshire
24. Buckinghamshire
25. East Suffolk
26. West Suffolk
27. East Norfolk
28. West Norfolk
29. Cambridgeshire
30. Bedfordshire
31. Huntingdonshire
32. Northamptonshire
33. East Gloucestershire
34. West Gloucestershire
35. Monmouthshire
36. Herefordshire
37. Worcestershire
38. Warwickshire
39. Staffordshire
40. Shropshire (Salop)
41. Glamorgan
42. Breconshire
43. Radnorshire
44. Carmarthenshire
45. Pembrokeshire
46. Cardiganshire
47. Montgomeryshire
48. Merionethshire
49. Caernarvonshire
50. Denbighshire
51. Flintshire
52. Anglesey
53. South Lincolnshire
54. North Lincolnshire
55. Leicestershire (with Rutland)
56. Nottinghamshire
57. Derbyshire
58. Cheshire
59. South Lancashire
60. West Lancashire
61. South-east Yorkshire
62. North-east Yorkshire
63. South-west Yorkshire
64. Mid-west Yorkshire
65. North-west Yorkshire
66. Durham
67. South Northumberland
68. North Northumberland (Cheviot)
69. Westmorland with North Lancashire
70. Cumberland
71. Isle of Man

S Channel Isles

SCOTLAND

72. Dumfriesshire
73. Kirkcudbrightshire
74. Wigtownshire
75. Ayrshire
76. Renfrewshire
77. Lanarkshire
78. Peeblesshire
79. Selkirkshire
80. Roxburghshire
81. Berwickshire
82. East Lothian (Haddington)
83. Midlothian (Edinburgh)
84. West Lothian (Linlithgow)
85. Fifeshire (with Kinross)
86. Stirlingshire
87. West Perthshire (with Clackmannan)
88. Mid Perthshire
89. East Perthshire
90. Angus (Forfar)
91. Kincardineshire
92. South Aberdeenshire
93. North Aberdeenshire
94. Banffshire
95. Moray (Elgin)
96. East Inverness-shire (with Nairn)
97. West Inverness-shire
98. Argyll Main
99. Dunbartonshire
100. Clyde Isles
101. Kintyre
102. South Ebudes
103. Mid Ebudes
104. North Ebudes
105. West Ross
106. East Ross
107. East Sutherland
108. West Sutherland
109. Caithness
110. Outer Hebrides
111. Orkney Islands
112. Shetland Islands (Zetland)

IRELAND

H.1 South Kerry
H.2 North Kerry
H.3 West Cork
H.4 Mid Cork
H.5 East Cork
H.6 Waterford
H.7 South Tipperary
H.8 Limerick
H.9 Clare
H.10 North Tipperary
H.11 Kilkenny
H.12 Wexford
H.13 Carlow
H.14 Leix (Queen's County)
H.15 South-east Galway
H.16 West Galway
H.17 North-east Galway
H.18 Offaly (King's County)
H.19 Kildare
H.20 Wicklow
H.21 Dublin
H.22 Meath
H.23 West Meath
H.24 Longford
H.25 Roscommon
H.26 East Mayo
H.27 West Mayo
H.28 Sligo
H.29 Leitrim
H.30 Cavan
H.31 Louth
H.32 Monaghan
H.33 Fermanagh
H.34 East Donegal
H.35 West Donegal
H.36 Tyrone
H.37 Armagh
H.38 Down
H.39 Antrim
H.40 Londonderry

THE OMITTED RECORDS

The species numbers and nomenclature follow Dandy (1958). Numbers outside brackets represent vice-counties from which species have been reliably recorded in the past but for which no localizable record has been traced. Numbers in round brackets represent vice-counties from which native species have been reliably recorded as an introduction but for which no localized record has been traced. Non-native species are marked with an asterisk before the species number. Numbers in square brackets represent vice-counties from which species are known or suspected to have been recorded in error.

The list of numbers is mainly based upon a comparison between the records received by the Maps Scheme and the *Comital Flora* (Druce, 1932). Certain critical genera have been omitted from the list because experience has shown that vice-county records previously published are so liable to error as to be useless.

1/1 Lycopodium selago 45, 74, 79, 82, [7, 13]
1/2 inundatum 5, 31, 61, 108, [7, 49]
1/3 annotinum 78, 100, 103, 109, [55, 59, 64, 66]
1/4 clavatum 52, 66, 79, 86, [7]
1/5 alpinum 51, 59, 102, 103, 110, [58, H.28]
2/1 Selaginella selaginoides 51, 74
3/1 Isoetes lacustris 42, 50, 52, 85, 94
3/2 echinospora 87, 97, 107, [H.40]
4/1 Equisetum hyemale 15, 40, 50, 59, [4, 7, 8, 21, 51, 87]
4/4 variegatum 100, [20, 28, 42, H 16, 26]
4/7 sylvaticum 51, 78, [1, 25, 26]
4/8 pratense 78, [59]
4/10 telmateia H.3, 26, [90–2, 111]
5/1 Osmunda regalis 51, 84, 93, [7, 23, 29, 33, 94]
6/1 Trichomanes speciosum 70, [101]
7/1 Hymenophyllum tunbrigense 78, [13, 69]
7/2 wilsonii 40, 52, 109, [6, 9]
9/1 Cryptogramma crispa 93, 95, 109, [4, 36, 55]
11/1 Adiantum capillus-veneris [48]
14/1 Phyllitis scolopendrium 96
15/2 Asplenium obovatum [15, 40, 42, 71, H.1]
15/4 marinum 99, [15, 50, 72]
15/6 viride 106, [3, 13, 15, 17, 38, 55, 58, 63, 83, 84, 103]
15/8 septentrionale 78
16/1 Ceterach officinarum 77, [58]
18/2 Athyrium alpestre 86, 89, 107
19/1 Cystopteris fragilis 12, 52, 74, H.32, [21, 26]
19/3 montana 94, 104, [49]
20/1 Woodsia ilvensis 103, [88]
20/2 alpina 104
21/4 Dryopteris villarii (H.31), [59]
21/5 cristata [8, 9, 30, 37, 74, 84, 92]
21/6 lanceolatocristata 76, 79, 85, 94–6, 106 [111]
21/8 aemula 51, 89, 99, 109, [11, 52, 64, 65, 80, 81, 90, 106]
22/1 Polystichum setiferum 50, 51, 67, 74, 77, 82, 83, [57]
22/3 lonchitis [23, 36, 48, 68, 95, 103]
24/2 Thelypteris palustris 19, 51, [71]
24/3 phegopteris 84, [33, 61, H.15]
24/4 dryopteris 38, 45, 51, 52, 102, [11, 61]
24/5 robertiana 47, [37, 45]
26/1 Pilularia globulifera 26, 47, 64, 65, 76, [6, 24, 33]
28/1 Botrychium lunaria 20, 43, 79, 84, H.32
29/1 Ophioglossum vulgatum 74, 79
34/1 Juniperus communis H.8, [50, 52]
35/1 Taxus baccata 75, 76, 87
37/1 Trollius europaeus 79, 102, [H.32]
38/1 Helleborus foetidus (81, 82, 87, 89)
38/2 viridis (51, 102)
*39/1 Eranthis hyemalis 37, 42, Perthshire
43/1 Anemone nemorosa H.4, 7, 10, 11, 22
46/3 Ranunculus bulbosus H.4
46/5 arvensis 44, 48, 51, 68, 73, 75, 86, 109, [111]
46/7 sardous 47, 48, 49, 51, 56, 73, 79, 86, 87, 91–3, 95, 102, 104, [23, 32, 33, 82]
46/9 parviflorus 48, 65, 69, (H.39)
46/10 auricomus 74, 79, 85, 93, 98, H.9
46/11 lingua 35, 96, 109, [3, 4, 74, 105, 111]
46/16 hederaceus 86, 97, 99, H.22
46/17 lenormandii 74, 102, [7]
46/18 tripartitus [7, 15]
46/19 fluitans 90, 91, 103, [41, 44, 46, 48, 50, 63]
46/20 circinatus 5, 81, 86, 90, 92, [2, 10, 51, 65, 105]
46/21 trichophyllus 48, 50, 51, 74, 76, 79, 86, 97, 101, 102, 104–6, H.8, 13
46/23 baudotii 50, 51, 74, 81, 93, 96, 109, [30, 32, 38, 55, 57, 71, 80, 84, H.39]
*47/1 Adonis annua 5, 37, 50, 51, 54, 68, 75, 83, [1, 4]
48/1 Myosurus minimus 49, 51
49/1 Aquilegia vulgaris (79, 87, H.6, 33)
50/1 Thalictrum flavum 44, 51, 97, [45, 75, H.16]
50/3 minus 79, 96, [12, 30, 37, 61]
53/1 Berberis vulgaris 79, 92, 93, 101, 102, 110, [74]

*54/1 Mahonia aquifolium 9
55/1 Nymphaea alba 79, 86, 106, 111, H.4–7, 18, (35, H.12, 13)
56/1 Nuphar lutea 51, 79, 99, 102, H.4, 6, [1, H.20]
56/2 pumila [56, 58, 67, 73, 77, 91, 95]
57/1 Ceratophyllum demersum [44, 51, 84, 111]
57/2 submersum [29, 83]
58/1 Papaver rhoeas 44, 76, 79, 86, 91, 93, 103, 109
58/2–3 dubium agg. 79, 87, 99–101, H.39
58/4 hybridum 59, (79), [H.4]
58/5 argemone 51, 78, 86–8, 101, 103, 106, 110
*58/6 somniferum 46
59/1 Meconopsis cambrica (9, 39, 86, 92)
61/1 Glaucium flavum 99
62/1 Chelidonium majus 78, 79, 91, 93, H.3–7, 15, 25, 26
65/3 Corydalis claviculata 79, 109, [71]
*67/2 Brassica napus 10, 28, 52
67/3 rapa 30, 51, 73, 78, 79, 86, 87, 89, 91, 93–7, 99, 100, 102, 103, 105–7, 109
67/4 nigra 65, 79, (81)
*68/1 Erucastrum gallicum 5, 45
69/1 Rhynchosinapis monensis 74, (82, 104), [4, 7, 11, 17, 21, 64]
*69/3 cheiranthos 56, 82
*70/2 Sinapis alba 78, 79, 86, 87, 89, 93, 96, 101, 106, 107, 110
*72/1 Diplotaxis muralis 40, 86, 109
72/2 tenuifolia 81, 86, [36]
74/1 Raphanus raphanistrum H.37
74/2 maritimus H.28, [46, 50, 52]
75/1 Crambe maritima 75, 107, [16, 60, H.10]
79/2 Lepidium campestre 42, 52, 67, 74, 79, 93, H.9
79/3 heterophyllum 65, 79
*79/4 Lepidium ruderale 26, 40, 42, 44, 45, 48, 73, 74, 80
79/6 latifolium (39, 44, 50, 51, 80, 107), [5, 60, H.1]
80/1 Coronopus squamatus 44, 86, 88–90, 92, 109, 110
*80/2 didymus 31, 79, 86, 106
*81/1 Cardaria draba 40, 76, 77
82/1 Isatis tinctoria (H.14), [30, 39]
84/1 Thlaspi arvense 109
84/3 perfoliatum [4, 30, 38, 53, 58, 64, 70]
84/4 alpestre 43, [3, 41, 50]
85/1 Teesdalia nudicaulis 79, 81, [6, 33, 34, 41, 42]
87/1 Hornungia petraea [3, 4]
88/1–3 Cochlearia officinalis agg. H.4, 37, [15, 16, 38]
88/4 scotica 93
88/5 danica 35, 109, [90, 93]
88/6 anglica 86, [48, 50, 57, 66, 77, 78, 83, 103, H.7]
89/1 Subularia aquatica 76, 94, 101, [71]
94/2 Draba norvegica 89, 104
94/3 incana [62, H.36]
94/4 muralis 62, 73
95/1–3 Erophila verna agg. H.3, 26
*96/1 Armoracia rusticana 73, 78–80, 84, 92, 96, H.40
97/2 Cardamine amara 35, 49, 79, [1, 2, 7, 8]
97/3 impatiens 49, [2, 4, 51, 53]
97/8 bulbifera (75)
98/1 Barbarea vulgaris 79, 102, 103, H.3
98/2 stricta [S, Ireland]
*98/3 intermedia 43, [9]
*98/4 verna 73, 74, 89, 101, [42]
99/1 Cardaminopsis petraea 100, 107
100/4 Arabis hirsuta 79, 87
100/5 brownii H.3
101/1 Turritis glabra 87, [14, 61]
102/3 Rorippa sylvestris 79, 104, 110
102/4 islandica 79, 87, 89, 92
102/5 amphibia 47, 60, 65, [3, 9, 69, 81, 83, 87, 92, 94, H.3]
103/1 Matthiola incana [45]
103/2 sinuata [14, 15, 51]
*104/1 Hesperis matronalis 46, 86, 100, H.30
105/1 Erysimum cheiranthoides 45, 60, 73, 81, 84, 96, 100, 109

*106/1 Cheiranthus cheiri 79, 86–9, 96, 101, 109, H.14, [H.18]
107/1 Alliaria petiolata 86, 100, 105, H.17, [H.3, 15]
108/1 Sisymbrium officinale 109
*108/2 irio [1, 2, 10, 14, 20, 22, 24, 26, 27, 53]
*108/4 orientale H.30, 34
*108/5 altissimum 12, 42–3, 45, 47–9, 68, 72–5, 77–81, 87, 89, 92–4, 111, H.3, 16, 30, 40
109/1 Arabidopsis thaliana 102
111/1 Descurainia sophia 44, 86, 88, 92, 107, 109
112/1 Reseda luteola 73, 79, 86, 93, 96, 107, H.1, 3, 4, 30
112/2 lutea 42, 48, 50, 84, 86, 92, 96, 106
113/1 Viola odorata H.20, (74, 86, 89, H.32)
113/2 hirta 59, 72
113/5 reichenbachiana 79, 102–4, 109, 112
113/6 canina 42, 76, 86, 107
113/7 lactea [6, 27, 29, 37, 48, 50, 63]
113/8 stagnina [11, 41, 53]
113/9 palustris 84, H.22, 23, 31, [26]
113/11 lutea [1, 4, 5, 26, 52, 60, 100, 102, 104, 110–12]
114/3 Polygala calcarea [1, 29, 108]
114/4 amara [63, S]
115/1 Hypericum androsaemum 106 (81)
*115/3 hircinum 51, Kent
115/5 perforatum 79, 106, H.27, 29
115/6 maculatum 77, 79, 82, 91, 94, 99, 102, 103, [21, 61, 85]
115/8 tetrapterum (111)
115/9 humifusum 109, 112
115/10 linarifolium (40)
115/12 hirsutum 45, 49, 74, 84, 86, 111, [H.9]
115/13 montanum 44, [7, 8, 14, 20, 30, 75, 106]
115/14 elodes 51, H.32, [7, 82]
118/1 Helianthemum chamaecistus 45, [46]
118/3 canum [42]
*120/1–2 Tamarix gallica agg. 16, [12]
121/1 Frankenia laevis (4, 66)
122/1 Elatine hexandra 102, 103, [37, 55, H.39]
122/2 hydropiper 50
123/1 Silene vulgaris 87, H.40
123/2 maritima 35, 66, 77
123/3 conica (32, 46, 66), [7]
123/6 gallica 75, 86, 93, 96
123/10 nutans [27, 32, 33]
123/11 italica 91, [14]
123/12 noctiflora (75, 76, 86, 92), [35]
123/14 alba 76, 79, H.23
124/2 Lychnis viscaria [48, 72, 106]
125/1 Agrostemma githago 37, 44, 63, 84, 86–93, 102, 104, 106, 109, H.37
127/1 Dianthus armeria 52, (65, 73, 74), [35, 87]
127/8 deltoides 44, 50, 75, (23, Cork), [3, 87, 101, 106, H.39]
*129/1 Saponaria officinalis 74, 93, H.16, 23, 27, 30, 36
130/1 Kohlrauschia prolifera [Cornwall, Perthshire]
131/1 Cerastium cerastoides 70, 107
131/2 arvense 49, 52, 79, 86, 111
131/4 alpinum [94, 99]
131/5 arcticum [87, 90, 108, 110]
131/8 glomeratum 111
131/10 atrovirens 22, 76, 99, H.37, [55]
131/11 pumilum [1, 37, 41, 46]
131/12 semidecandrum 42, 43, 80, 86, 97, 107, [72]
132/1 Myosoton aquaticum 49, [48, 77, 82, 86, 90, 111]
133/1 Stellaria nemorum 42, 74, 79, 92, 109, [41]
133/3 pallida 40, 63, 102, H.14, 15, Cork
133/4 neglecta 28, 67, 89, 108, 109
133/5 holostea 110, 111, H.28
133/6 palustris 19, 58, 74, 79, [41, 49, 50]
135/1 Moenchia erecta 70, [35, 41]
136/1 Sagina apetala 102, [112]
136/2 ciliata 42, 44, 51, 52, 77–9, 87, 91
136/3 maritima 44, 50, 86

136/6 Sagina saginoides [36, 75, 111, 112]
136/7 normaniana 87, [104]
136/8 intermedia [90, 98, 99, 104]
136/9 subulata 51, 99, 106, [32, 36, 37, 61, 70]
136/10 nodosa H.4, 5, [16, 31, 35]
137/1 Minuartia verna 50, 63, 72, 89, 92, [5, 36, 71, 80, 102, 112]
137/2 rubella 107
137/4 hybrida 15, [49, 51]
138/1 Cherleria sedoides 110
139/1 Honkenya peploides 89, 92, 99, H.4, 8, 17, [77–80]
140/1 Moehringia trinervia 100
141/1–2 Arenaria serpyllifolia agg. 111, H.32
143/1 Spergularia rubra 44, 51, 86, 87, 91, 103, [98, 99]
143/2 bocconii [19]
143/3 rupicola 93, (36), [27, 35, 61, 82, 83]
143/4 media H.4, [50, 65]
143/5 marina 67, 92, 96, 107, H.40, [38, 63]
144/1 Polycarpon tetraphyllum [4, 21, 86, 104]
146/1 Herniaria glabra (31), [3, 4, 6, 16, 25, 27]
146/2 ciliolata [21, 110, H.2]
147/1 Illecebrum verticillatum [3, 4]
148/1 Scleranthus annuus 51, 65, 78, 79, 97, 99, 102, 107–9
149/1 Montia fontana H.31
*149/2 perfoliata 88, 92, 101 [2, 35]
*149/3 sibirica 49–51, 79, Donegal
*154/1 Chenopodium bonus-henricus 73, 87, H.16, 17
154/2 polyspermum 49, (90, 92, H.21)
154/3 vulvaria 26, (30, 36, 60, 79, Cork)
154/9 ficifolium (1, 37, 40, 57, 79, 92), [8, H.3, 21, 39]
154/11 murale 12, 45, 74, 77, 79, 83, [35, 48, 50]
*154/12 urbicum 50, [32, 48]
*154/13 hybridum [111]
154/14 rubrum 67, 74, 80, 86, 92, 108, [H.4]
154/15 botryodes [2, 21, 28, 49, 54, 67, 81, 85]
*154/16 glaucum [31, 68, 77]
155/1 Beta vulgaris [112]
156/1 Atriplex littoralis 73, 74, [81, 97, 110],111
156/2 patula 99, H.36
156/3 hastata 65, 76, 79, 80, 86, 88, 89, 92, 93, 97, 99, 109, H.11, 15, 17, 18, 29, 33
156/4 glabriuscula 76, 84, 87, 92, H.29, 37, [63–5]
156/5 laciniata 76, 92, 93, 96, 98, 104, 107, 108, [99, 111]
157/1 Halimione portulacoides 75
157/2 pedunculata [44]
158/1 Suaeda maritima 76, 83, 86, H.8, 37, 39, 40
158/2 fruticosa [3, 6, 10, 50, 62]
159/1 Salsola kali 67, H.4, 5, 39, 40
160/1 Salicornia perennis 53, [4, 41, 46]
160/2–4 agg. 92, 109, H.28, 37
160/5 pusilla 66, [17]
162/2 Tilia cordata 51, (75)
163/1 Malva moschata 86, 100, H.14, 30, (108)
163/2 sylvestris 76, 79, 89, 101, H.24, 30, 37, 40
163/4 neglecta 78, 80, 111, H.33, [73]
165/1 Althaea officinalis 5, [1, 2, 4, 30, 58, 64, 67, 69, 86, 99]
166/1 Linum bienne 51, [22, 32, 38, 59, 77]
166/3 anglicum [17, 63]
167/1 Radiola linoides 30, 47, 66, 67, 98, 99
168/1 Geranium pratense (107, H.15, 20, 37)
168/2 sylvaticum 59, 63, 74, 100, 102, 103, [1, 41, 57, 111]
*168/3 endressii 2, 65, 81
*168/4 versicolor 49, 92, [69, H.2]
*168/6 phaeum 48, 92, 109, H.4
168/7 sanguineum 39, 44, 63, 102
168/9 pyrenaicum 75, 91, 100, H.16, [H.37]
168/10 columbinum 102, H.17, [77, 99]
168/12 rotundifolium 44, [36, 40, 50]
168/14 pusillum 100
168/15 lucidum 59, 74, 76, 93, 106, 109, (H.12), [111]

169/1 Erodium maritimum 59, 69, [47, 50, 58, 70, H.16, 39, 40]
169/2 moschatum (92)
169/3 cicutarium 76, 79, 86, 99, H.31
171/1 Impatiens noli-tangere [13, 29, 38, 39, 41, 45, 49–51, 53–5, 57, 67, 72–4, 83, 86, 95]
*171/2 capensis 73
*171/3 parviflora [3, 4]
*171/4 glandulifera 92
173/3 Acer campestre (71, 102, 109)
*175/1 Aesculus hippocastanum 86
177/1 Euonymus europaeus 79, H.26, 40, (74, 109)
179/1 Rhamnus catharticus (79, 92)
180/1 Frangula alnus 50, 51, 75, 78, 91–3
*183/1 Lupinus nootkatensis 81, 93, [41]
*184/1 Laburnum anagyroides Gloucestershire
185/2 Genista anglica 74, 79, 87, [33]
185/3 pilosa [37]
187/2 Ulex gallii (20), [28, 29, 83, 86, 87, 90, 92, 93, H.40]
187/3 minor [1, 5, 6, 34, 37, 39, 40, 48, 49, 52, 58, 63, 64, 66, 71–5, 83, 85, 90, 95, 98]
189/1 Ononis repens 78, 86, 98, 111
189/2 spinosa 45, 67, 74, 81, [4, 10, 71, 85, 87, 90, 99, S, H.4, 5]
189/3 reclinata [11, 110]
190/1 Medicago falcata (5, 8, 12, 48, 52, 57, 60, 64, 76, 83, 86, 92, 101, 111), [4, 62]
*190/2 sativa 74, 76, Perthshire
190/3 lupulina 76, 101
190/4 minima (78), [2, 6, 17, Devonshire]
190/5 polymorpha (42, 73, 74, 76, 78, 79, 83, 85–92)
190/6 arabica 44, 48, (77–9, 99)
*191/1 Melilotus altissima 40, 50, 51, 67, 74, 79, 81, 91, 101, 109, [42, 43]
*191/2 officinalis 45, 73, 76
*191/3 alba 5, 71, 74, 87, 99
*191/4 indica 26, 72, 74, 86, H.16, [48]
192/1 Trifolium ornithopodioides 42, [7, 23, 55, 62, 64, 70, 75, 76, 90]
192/3 ochroleucon [1, 3, 38, 39, 49, 51]
192/4 medium [H.3]
192/5 squamosum (89), [H.9]
*192/6 stellatum [2, 3]
192/9 arvense 79, 86, 93, 104, 109, (111)
192/10 striatum 72, 74, 79, 101
192/11 scabrum 63, [30, 51, 58]
192/12 bocconei [102]
192/13 subterraneum 51
192/14 strictum [2]
192/15 glomeratum 44, [49, 50, 55]
192/16 suffocatum 18, [2, 51, 60]
*192/17 hybridum 89, H.4, 16, 25, 34, 40
192/19 fragiferum 65, 75, 83, [92, 103]
192/21 campestre 79, H.29, 36, 40, (111)
192/24 micranthum 53, H.9, [77, 82–5, 90–4, 106, H.13]
193/1 Anthyllis vulneraria H.37
195/2 Lotus tenuis [40, 47, 48, 52, 58, 62, 65–8, 81, 87, 90, 92, 107]
195/5 angustissimus [6, 13, 25]
200/1 Astragalus danicus 36, 40, 75, 77, 87, 108
200/3 glycyphyllos 51, 70, 93, [77]
202/1 Ornithopus perpusillus 87, 89, 92, 93, [31, H.9, 31]
*203/1 Coronilla varia 12, 20, 22, 27, 40, 48, 51, 77, 93
204/1 Hippocrepis comosa 63, 91, [5, 40]
206/1 Vicia hirsuta 79, 108, 109
206/2 tetrasperma 44, 51, 74, 81, 84, 92, 107, [68, 72, 86, 103, 112]
206/3 tenuissima [41, 55]
206/9 orobus 37, 68, 71, 74, [38, 65, 66, 92, H.34]
206/10 sylvatica 79, H.37, [1, 13, 28, 45, 48]
206/12 lutea (4, 92, 94), [26, 88]
206/15 angustifolia 86, (112)
206/16 lathyroides 47, 72, (73), [1, 2, 7, 8, 18, 33, 34, 40, 55, 63]
207/1 Lathyrus aphaca (74, 77, 79, 92, 93), [52]
207/2 nissolia 40, [4, 42, 70, 75, 90]

*207/5 Lathyrus tuberosus [35]
207/6 sylvestris 74, 91, 103, [68]
207/9 palustris [22, 41, 57, 59, 62, 69, 74]
207/11 montanus H.25, [26, 28]
*209/1 Spiraea salicifolia 21, 22, 73, 79
210/1 Filipendula vulgaris [72, 77, 91, 92, 109]
211/1 Rubus chamaemorus 45, 49, 73
211/2 saxatilis 59, 79, [1, 2, 51]
211/3 idaeus 111
211/9 caesius 86, 88, 91, [75, 77]
212/1 Potentilla fruticosa [77, H.4]
212/2 palustris (23), [7, 34]
212/3 sterilis [111, 112]
212/4 rupestris [48]
212/6 argentea 50, 81, [5, 7, 35, 49, 52]
*212/8 norvegica 39, 58, 70, 88, 92, 96, [7, 28]
212/11 tabernaemontani [3, 25, 39, 65, 68, 70, 81]
212/12 crantzii [46]
212/14 anglica 68, 78, 79, 81, 86, 87, 98–100, [90–6]
212/15 reptans 84, 86, 107, [105, 111, 112], 101
213/1 Sibbaldia procumbens 70, 73, 78, 91, 93
216/1 Geum urbanum 112
217/1 Dryas octopetala 87, [39, 57]
218/1 Agrimonia eupatoria 99, 109, H.40
218/2 odorata 79, 82
220/1 Alchemilla alpina [57, 62–4]
220/3 vulgaris H.12
222/1 Sanguisorba officinalis 6, 13, 45, 68, 79, [H.40]
223/1 Poterium sanguisorba 90, H.24
*223/2 polygamum 42–4, 46, 72, 77, 83, 90
225/1 Rosa arvensis (81, 84, 88, 89, 91, 106)
225/4 pimpinellifolia 39, 53, 86, 87, H.4, 5, 7, 37, [H.32]
225/14–17 rubiginosa agg. 76, 79, 107, H.16
*226/2 Prunus domestica 74, 78, 79, 89, 91, 98, 100, 107
*226/3 cerasifera 46
226/5 cerasus (59, 83, 90, 91, 94), [6, 30]
226/6 padus 74
*227/2 Cotoneaster simonsii 50, 79, 80
*227/4 microphyllus 39, 56, 65, 67, 70, 80, 88, 89
229/1 Crataegus oxyanthoides [62, 63, 66, 78, 90, 104, S, Ireland]
232/5 Sorbus aria 81, 96
232/7 torminalis 26, 44, 45
*233/1 Pyrus communis 5, 43, 88
233/2 cordata [34–6, 54]
234/1 Malus sylvestris 93, 106
235/1 Sedum rosea 60
235/2 telephium 51, 84, (105, H.13), [H.29]
*235/3 spurium 33
*235/4 dasyphyllum H.37, [H.39]
235/5 anglicum 65, 106, H.25, [19, 28, 38, 39, 92, 112]
235/6 album (79, H.37)
235/8 acre 79
235/10 forsteranum (37, North Yorkshire), [23]
*235/11 reflexum 67, 92, H.12, 32, [105]
235/12 villosum 98
238/1 Umbilicus rupestris 74
239/1 Saxifraga nivalis [83, 103, 111]
239/2 stellaris [42]
239/5 spathularis (H.34)
239/8 tridactylites 67, 68, 71, 79, 96, 104
239/9 granulata 86, 87, (H.37), [105, 111]
239/12 cespitosa 96
239/15 hypnoides 77, 81, 91, [111, H.17]
239/16 aizoides [41, 49, 71, 72]
239/17 oppositifolia 91, 109, [65]
242/2 Chrysosplenium alternifolium [24, 31, 32]
243/1 Parnassia palustris 50, 84, H.37, [46, 48, 110, H.38]
246/1 Ribes sylvestre 73, 79, 101, 109, Cork, [H.2]
246/2 spicatum 61
246/3 Ribes nigrum 46, 51, 52, 86, 99, [H.40]
246/5 alpinum (44, 76, 91, 99, 106), [41]
246/6 uva-crispa 109
247/1 Drosera rotundifolia 56
247/2 anglica 77, [13, 51, 65]
247/3 intermedia 5, 50, 51, 107, [30, 65]
249/1 Lythrum salicaria 77, 79
249/2 hyssopifolia [2, 70, H.12]
250/1 Peplis portula 51, 79, 86, 99, 108, 109
251/1 Daphne mezereum (63, 74, 75), [3, 5, 41, 59, 82]
251/2 laureola (86, 88, 91)
253/1 Ludwigia palustris [13]
254/1 Epilobium hirsutum 79, 87
254/4 lanceolatum [46, 48, 69–79]
254/5 roseum 52, 79, 89, [68, 70]
254/7 adnatum 72, 78, 95, 97, [88, 89, 111, H.21, 40]
254/8 lamyi Cork, [100]
254/9 obscurum 82, H.31
254/11 anagallidifolium 99
254/12 alsinifolium 91
*254/13 nerterioides 23
255/1 Chamaenerion angustifolium H.8, 31
*256/1 Oenothera biennis 77, 90, [35, 50, 51]
*256/3 stricta 18, 79, [32, 63]
258/2–3 Circaea alpina agg. 40, 50, 76, 93, 109, 110, [6, 11, 25, 45, 55, 67, 68, H.16]
259/1 Myriophyllum verticillatum 47, H.4, 9, [1, 41, 45, 50, 51, 95, 111]
259/2 spicatum 78, 79, 87, 93, 97, 99, 100, 107, 109
259/4 alterniflorum 66, 91, 99, [28]
261/1 Hippuris vulgaris 74, H.6, [47]
262/3 Callitriche obtusangula 21, 37, 51, [42]
262/4 intermedia H.5, 30
262/5 hermaphroditica 84, 87, 88, 94, [3, 9, 12, 15, 17, 19, 21, 25, 30, 36–8, 41, 46, 55, 56, 62, 65, 70
262/6 truncata 33,
263/1 Viscum album (67, 68, 70, 72, 74, 89)
264/1 Thesium humifusum [3, 30, 40]
265/1 Thelycrania sanguinea 1, (86, 92, 101, H.3, 32)
270/1 Sanicula europaea 79, 110
272/1 Eryngium maritimum 50, 77
272/2 campestre 1, [71]
273/1 Chaerophyllum temulentum 77, 99, 101
274/1 Anthriscus caucalis 74, 78, 79, 81, 86, 90, 93, [46, 111]
275/1 Scandix pecten-veneris 42, 43, 68, 74, 76, 78, 80, 86–8, 96, 98, 100–2, 106, 108, 109, 112
*276/1 Myrrhis odorata [1, 6, 19]
277/1 Torilis japonica 109
277/2 arvensis 42–4, 47–52, 59, 69
277/3 nodosa 43, 66, 67, 73, 92, 94, 102, 106, [H.22]
*278/1 Caucalis platycarpos 58, 59
*278/2 latifolia 44, 59, 92, 94, [14, 38]
*280/1 Smyrnium olusatrum 86, H.26
281/1 Physospermum cornubiense 1
282/1 Conium maculatum H.40
283/2 Bupleurum rotundifolium (48, 49, 74, 83, 92)
283/4 tenuissimum 52, [2, 3, 60]
284/1 Trinia glauca [7]
285/1 Apium graveolens 67, 89, 104
285/2 nodiflorum 77, 97, [72, 92–5, 112]
285/4 inundatum 88, 99, 107, H.12, 26
*286/1 Petroselinum crispum 12, 17, 40, 58, 75, [51]
286/2 segetum 50, [36, 49, 103]
287/1 Sison amomum 43, 44,[111]
288/1 Cicuta virosa 51, 66, 96–8, 101, [8, 9, 13–16, 35–7, 48, 49, 63, 64, 67, 70, 81, 92, 105, 111
291/1 Carum verticillatum 77, 96, 110
*291/2 carvi 4, 16, 33, 73, 74, 77, 91, H.3, 6, 26, 29
292/1 Bunium bulbocastanum (12, 33), [11]
294/1 Pimpinella saxifraga 97, 98, 109, H.37, 40, [112]

294/2 Pimpinella major H.17, [28, 42, 44, 48, 49, 83]
296/1 Sium latifolium H.32, [37–9, 41, 48, 49, 52, 64, 69, 70, 74, 75, 86, H.31, 34, 39, 40]
297/1 Berula erecta 48, 67, 74, 87, 91, 93
298/1 Crithmum maritimum 19, 50, [27, 59, 83, H.29]
300/1 Oenanthe fistulosa 74, 81, [4, 75, 85]
300/2 pimpinelloides [25, 27, 28, 41, 52, 70, 72–4, 87, 88]
300/3 silaifolia (40), [26–8, 35]
300/4 lachenalii 86, 96, [H.13]
300/5 crocata 53, 54, H.17, [26]
300/6 aquatica 60, H.4, 11, 13, 14, 22, 27, 30, 31, [3, H.6, 26]
300/7 fluviatilis 32, H.10, 14, 26–8, 31, 32, 37, [49, 52, H.18]
301/1 Aethusa cynapium 79, 91, 93, 98, 102–4
*302/1 Foeniculum vulgare 82, 93, 101, [111]
303/1 Silaum silaus 44, 49, 52, 59, 74, 102, [46]
304/1 Meum athamanticum [64]
306/1 Ligusticum scoticum 99, 106
*307/2 Angelica archangelica 39, [20]
309/1 Peucedanum officinale 16
309/2 palustre [53, 56, 59, 69, 76, 83, 99]
*309/3 ostruthium 73, 84, 89, 99, [61]
310/1 Pastinaca sativa 51, (75, 86, 92, 99, 100, H.33)
*311/2 Heracleum mantegazzianum 9
314/1 Daucus carota 78, 79, 84, 86, 87, 99
315/1 Bryonia dioica 48, [4]
*316/1 Asarum europaeum 50, 59, 68, 76, 93
*317/1 Aristolochia clematitis [18]
318/1 Mercurialis perennis 110
318/2 annua 59, 62, 90, H.8, 9
319/1 Euphorbia peplis 14, [41]
319/2 lathyrus (39, 75, 77), [46, H.40]
319/5 hyberna (6), [H.25]
*319/6 dulcis 86
319/7 platyphyllos [68]
319/8 stricta [17]
319/9 helioscopia 79, 99
319/10 peplus 79, 98, 109
319/11 exigua 44, 67, 68, 74, 109, [87]
319/12 portlandica 50
319/13 paralias [12, H.4]
*319/14–15 esula agg. 42, 51, 58, 75
*319/16 cyparissias 9, 20, 75, 92, 97, [35]
319/17 amygdaloides 28, 49, (95)
320/2 Polygonum raii 76, 82, 83, 97, 102, [18]
320/3 maritimum [9, 13, 14, 28, 46, 49, 75, H.5, 6, 38]
320/5 viviparum 75
320/6 bistorta 74, 79, H.37, [H.2]
320/10 lapathifolium 87, 97, 99, 106, H.4, 7, 23, 30
320/11 nodosum 59, 74, 79, 83, 87, 97
320/12 hydropiper 109
320/13 mite 2, 20, 39, 60, 70, 86, [47]
320/14 minus 60, 73, 91–3, [30, 31, 46, 61, 63]
320/16 dumetorum 19, 35, [23]
*321/1 Fagopyrum esculentum 33, 34, 46, H.21
324/1 Oxyria digyna 93, [60]
325/4 Rumex hydrolapathum 74, 76, 81, 86, 95, [H.7]
325/8 longifolius 57, 74, 79, 84, 87, 108
*325/10 patientia [61]
325/13 pulcher (44, 49, 67, 68, 70, 75, H.4), [65]
325/14 sanguineus 76, 86, 91–3, 96, 109
325/15 conglomeratus 84, 91, 92, 99, 103, 109
325/16 rupestris [11, 13]
325/17 palustris [1, 9, 38, 39, 50, 52, H.5, 21]
325/18 maritimus [2, 4, 36, 42, 57, 58, 69, 70, 72, 85–7, 90, 92, 94, 104, 111, H.5]
326/1 Parietaria diffusa 43, 74, 78, 79, 86, 92, 96, 106
328/1 Urtica urens 87
329/1 Humulus lupulus 76, 78, 79, 89, 91, 92, 96, 98, 109
330/3–6 Ulmus carpinifolia agg. 22, [42]
333/1 Myrica gale 79, H.19, [38, 42, 84]
335/1 Betula pendula [111]
335/3 nana 68, 78, 81, 95, 109
336/1 Alnus glutinosa [111]
337/1 Carpinus betulus (87, 100, 101, 103, 105)
*340/1 Castanea sativa 52
*341/2 Quercus ilex 24, 40
342/2 Populus canescens (52, 68, 69, 73, 77–9, 81, 97)
342/3 tremula 52
342/4–5 nigra agg. 49, 52, 79, 80, 87, 97, 100, 104, 108–11, H.16, 40, [112]
343/1 Salix pentandra 49, 78, 82, 84, 96, 99, H.10
343/2 alba (111)
343/4 fragilis 99
343/5 triandra 50, 65, 91, [71]
343/6 purpurea 97, 99, 108–10, [1, 48]
343/11 caprea [112]
343/14 nigricans 74, 78, 79, 102, 106, 107, [27, 36, 41, 56, H.34]
343/15 phylicifolia 100, [35, 62, 68, 85]
343/16 repens 99, H.5, 22, [37, 43, 47, 79]
343/17 lapponum 99, [83, 111]
343/19 arbuscula [111]
343/20 myrsinites 99, [H.28]
343/21 herbacea 75
343/22 reticulata [87, 111]
*344/1–2 Ledum palustre agg. [H.27]
346/1 Loiseleuria procumbens 110
350/1 Andromeda polifolia 44, 49, 102
355/1 Arctous alpinus [92]
357/2 Erica mackaiana [H.3]
357/3 ciliaris [12, 17, 41, 52, H.3, 16]
357/4 Erica cinerea H.24
357/7 mediterranea (2, 41, H.28)
358/1 Vaccinium vitis-idaea 79, [6, 15]
358/4 oxycoccos 52, 53, 100, 102–4, 106, H.8
359/1 Pyrola minor 68, 86, 101, [7, 38, 49, 51, 52]
359/2 media 40, 102, 108, [8, 15, 20, 23, 26, 35, 38, 51, 60, 78]
359/3 rotundifolia 77, 91, [13, 36, 40, 67, 72, 73, 81, 107, 109, H.18]
360/1 Orthilia secunda 73, 102, [80, 82–4]
361/1 Moneses uniflora 72, 90, 91, 97
362/1 Monotropa hypopitys 58, [H.38, 39]
365/1 Limonium vulgare [1, 2, 51]
365/2 humile 72, [16, 27, 44, 50, 68]
365/3 bellidifolium [25, 53, S]
365/5 binervosum 18, 50, 51
366/1 Armeria maritima 77, 86, 87, [63]
367/1 Primula farinosa 68, [57, 63]
367/3 veris 74, 86, 102, 103
367/4 elatior [53]
368/1 Hottonia palustris 36, 44, 48, 50, 51, 96, [H.30]
369/1 Cyclamen hederifolium (10, 12, 18, 56, Donegal)
370/2 Lysimachia nummularia 74, 82, 100
370/3 vulgaris 79, 103
*370/5 punctata 9
370/7 thyrsiflora (88), [16]
372/1 Anagallis tenella 77, 92, H.32, 37
372/2 arvensis 86, 99, 105, 108, 109, 112
372/4 minima 50, 65, 67, 86, 92, 96, H.32, [72, 76, 77]
373/1 Glaux maritima 86, 88, 92, [42, 43, 77]
374/1 Samolus valerandi 76, 86, H.32, 36, [50]
376/1 Fraxinus excelsior 111
*377/1 Syringa vulgaris 9, 40
378/1 Ligustrum vulgare (92)
379/1 Vinca minor 86, 91, 102, 103, 109
*379/2 major 59, 94
382/1 Centaurium pulchellum 51, 102, [70]

382/4 erythraea 86, 93, 96, 106
382/5 capitatum 44, 49, [11, 15]
382/6 littorale 71, [9, 50]

383/1 Blackstonia perfoliata 73, [67]
384/1 Gentiana pneumonanthe 15, [8, 44]
384/2 verna [62]
384/3 nivalis 108
385/1 Gentianella campestris 43, 44, 47, 56, 79, 84, [5, 6, 24, 30]
385/2 germanica [5, 7, 45, 57, 101]
385/3 amarella 47, 86, [H.1, 2, 6, 29, 32, 36–40]
385/4 anglica [14]
385/5 uliginosa [57, 65, 90, 96, 107]
387/1 Nymphoides peltata [54]
389/1 Cynoglossum officinale 66, 88
389/2 germanicum [90]
392/1 Symphytum officinale 79, (97, H.6, 18, 19, 24, 26, 29, 31)
392/6 tuberosum 79, (6, 25, 57, 61, Cork), [24, 29, 35]
*395/1 Pentaglottis sempervirens [42]
397/1 Lycopsis arvensis 79, 86, 87, 99
399/1 Pulmonaria longifolia [17, 51, 61, 63]
399/2 officinalis (66, 71, 73, 74)
400/2 Myosotis secunda 56, [21, 25, 26, H.14]
400/7 sylvatica 74, 86
400/10 ramosissima 73, 109, [75–7, 102]
401/2 Lithospermum officinale (76, 78, 82, 87, 91, 101, 111), [H.38]
401/3 arvense 44, 45, 48, 50, 51, 65, (73, 78, 80, 81, 86, 88, 96, 100, 101, 106)
402/1 Mertensia maritima [1, 4, 11, 54]
403/1 Echium vulgare 43, 73, 96, 98, 102, 109
403/2 lycopsis (10–12, 37)
405/1 Convolvulus arvensis 79, 87, 93, 107, H.18, 29
406/4 Calystegia soldanella [H.34]
407/1 Cuscuta europaea (39, 57, 69, 72, 77, 83, 92), [41, 53, 90]
407/2 epithymum 43, (72, 75–7, 85–7, 90, 92, 93), [42]
410/1 Atropa bella-donna 47, (42, 45, 76, 81, 86, 87, 98, 99)
411/1 Hyoscyamus niger (66, 72, 77, 101, 108)
413/1 Solanum dulcamara 97, [H.10]
413/3 nigrum (76, 86–9, 92)
*415/1 Datura stramonium H.18
416/1 Verbascum thapsus 102, 111, H.16
416/4 lychnitis (62, 64, 89, 99), [10, 14, 29]
416/5 pulverulentum (36)
416/7 nigrum (87–9, 95)
*416/9 blattaria 22, 48, [51]
416/10 virgatum (10, 39, 83), [22, 53]
417/1 Misopates orontium (65)
*418/1 Antirrhinum majus 33
*420/2 Linaria purpurea 74
420/3 repens (87, 93, 100, H.22), [H.40]
421/1 Chaenorhinum minus 49, [S]
422/1 Kickxia spuria [46]
422/2 elatine (75, 79, 86, 90)
424/2 Scrophularia aquatica 68, 78, [75–7, 87]
424/3 umbrosa 39, [1, 2, 13, 24]
424/4 scorodonia [20, H.1, 40]
*424/5 vernalis 18, 21, 73, 86, 93, 99, [27, 48, 55, 61, 75]
*425/1 Mimulus guttatus 32, H.19
*425/3 moschatus [H.37]
426/1 Limosella aquatica 44, [8, 49, 62, 90, 48]
430/2 Veronica anagallis-aquatica H.30
430/3 catenata 10, 90, 112, H.5, 12, 21, 24, 28, 33, 34, 36
430/4 scutellata H.31
430/6 montana 93, [31]
430/10 fruticans [107, 108]
430/12 alpina 72, 87
*430/14 peregrina Yorkshire
430/19 triphyllos [38, H.21]
430/20 hederifolia 74, 76, 86, 97, 99, 102, H.2–4, 16, 24, 29, 33, 40
430/22 polita 72, 76, 78, 79, 86, 91–3, 98, 102, 109, H.31, 33, 37
430/23 agrestis 52
432/1 Pedicularis palustris 51
432/2 sylvatica H.22
433/1 Rhinanthus serotinus 60, 65, 69, 91, 96, 102, 106, [6, 20, 24, 30, 33 H.8]
434/1 Melampyrum cristatum [7, 8, 37, 40, 66, 70]
434/2 arvense [32, 33, 38, 58]
434/3 pratense 79, H.30, 31
434/4 sylvaticum [67, 68, 73, 75, 80, 94, 106, H.35, 38]
435/2 Euphrasia salisburgensis [3, H.3, 40]
437/1 Parentucellia viscosa 101, (49, 56) , [85]
438/1 Bartsia alpina 97, 105
439/1 Lathraea squamaria 79, 97, H.25
440/2 Orobanche purpurea [1, 12, 13, 22, 41, 48, 56]
440/3 rapum-genistae 44, 45, [23, 53, 60, 85, 88, 110]
440/4 alba 62, H.29, [11, 41, 56]
440/5 caryophyllacea [26, 98, Devonshire]
440/6 elatior 10, 38, 39, 40, 42, 49
440/7 reticulata 27, [42]
440/8 minor 40, [48, 67]
440/9 picridis 27
440/10 hederae 11, 42
440/11 maritima [10, 14]
441/1 Pinguicula lusitanica 86, 96, H.30, [41, H.22]
441/3 vulgaris 84, H.37, [5, 8]
442/1–2 Utricularia vulgaris agg. 74, 106, 109, 111, [51, 77]
442/3 intermedia 26, 60, 85, 107, 111, H.4, [H.40]
442/4 minor 74, [8, 23, 99]
444/1 Verbena officinalis 65
445/2 Mentha pulegium 6, 51, [64, 65, 83, 85, 95, 109, H.38]
445/3 arvensis [111, 112]
*445/5 spicata 73, 84, 87, H.4, [74]
*445/6 longifolia 47, 58, 68, 70, 79, 82, 86, 96, 99, 107, 111, H.2, 5, 11, 20, 36, [10, 105, S]
445/7 rotundifolia (100, H.37, 40)
447/1 Origanum vulgare 74, 79, 93, 109, (102)
451/2 Calamintha ascendens [68, 85, 87]
451/3 nepeta 33, 36, 38, [35, 43, 45, 49, 51, 54, H.2, 3]
452/1 Acinos arvensis 96, 107, (59, 74, 75, 77)
453/1 Clinopodium vulgare 74, 96, 98
*454/1 Melissa officinalis 25, 26, 28, 51, 63, 70, 80, 91, 93
455/2 Salvia pratensis [21, 25, 26, 34, 41, 52, 56, 62]
455/4 horminoides 44, 50, 51, 75, 86, 106, (48)
456/1 Melittis melissophyllum [34]
458/1 Betonica officinalis 76, [84, H.9, 10]
459/3 Stachys arvensis 84, 99, 107
459/1 germanica [3, 10, 11, 22, 24, 27, 57, 62]
459/5 alpina [14]
460/1 Ballota nigra 86, H.3, 6, 32, [76]
461/1 Galeobdolon luteum (72, 79, 86, 97)
462/1 Lamium amplexicaule 51, 78, 87
462/2 moluccellifolium 78, 92, [31, 50, 63, 84]
462/3 hybridum 67, 78, 80, 84, 86, 92–4, 99, 100, 106
462/5 album 109
465/1 Galeopsis angustifolia 56, 59, 65, 75, 86, 92, 99, [104, 111, H.5]
465/3 segetum 61, 62, [38, 59, 64, 71]
465/5 speciosa H.24, 30, (6, 10), [H.16]
466/1 Nepeta cataria 45, (74, 86), [44, H.3, 19]
468/1 Marrubium vulgare (73)
469/1 Scutellaria galericulata 51, 79, H.7, 10, 11
469/2 minor 19, 72, 74, 75, 77, [33]
*470/1 Teucrium chamaedrys 11, 44, 56, [28]
470/2 scordium [3, 26, H.5, 6]
470/4 scorodonia H.30, [112]
471/1 Ajuga chamaepitys [7, 10, 22, 32, 45, 47]
471/2 reptans [112]
472/2 Plantago media 79, 84, 86–8, 97, 99, (71)
472/4 maritima H.33, [H.13, 14]
472/5 coronopus 76, 92
473/1 Littorella uniflora 84, H.4, 8, 23, 34, [8]
474/1 Wahlenbergia hederacea [33, 37, 66]
*475/3 Campanula rapunculoides 91
475/5 persicifolia (1, 84, 99)
475/7 rotundifolia H.37, [H.5]

475/8 Campanula patula [9, 24, 32, 41, 54, 62, 64, 66]
*475/9 rapunculus 40, [32, 35, 56, 83]
476/1 Legousia hybrida 67, 68
478/1 Phyteuma tenerum [9]
478/2 spicatum (73)
479/1 Jasione montana 79, H.22, 30
480/2 Lobelia dortmanna 43, 77
481/1 Sherardia arvensis 76, 78, 79, H.24
483/2 Asperula cynanchica 50, [46, Cornwall, H.5, 6]
484/1 Cruciata chersonensis 97, 103, [1, 93]
485/1 Galium odoratum 102, H.4, 5, 7, 34
485/2 boreale H.8, (36, 111, 112), [41, 50, 63]
485/3 mollugo 87, 93
485/9 debile [1, 67, 90]
485/10 uliginosum 109, [45]
485/11 tricornutum (67, 92, 93, 95), [48]
485/13 spurium 66, [18, 37]
485/14 parisiense (90), [6, 23, 38, 39, 58, 80]
486/1 Rubia peregrina [14, 57]
487/1 Sambucus ebulus 45, H.16, [83]
488/1 Viburnum lantana 45, (72, 86, 96)
490/1 Linnaea borealis [11, 68, 105]
491/1 Lonicera xylosteum Norfolk, [3, 25, 54, H.11]
*491/4 caprifolium 10
494/1 Valerianella locusta 67, 87, 105, 107, 111
494/2 carinata [53]
494/3 rimosa 18, 37, 44, [35, 50]
*494/4 eriocarpa 45, [29, 37, 41]
494/5 dentata 42–4, 48, 66, 67, 74, 75, 87, 110, H.30, [77, 83]
495/1 Valeriana officinalis [112, S]
495/3 dioica 75, 86, 102, 108, 109, [105]
497/1 Dipsacus fullonum (86, 92, 95, 99, 100)
497/2 pilosus [49]
498/1 Knautia arvensis 74, 96, 98, 99, 102, 107, 109
499/1 Scabiosa columbaria 86, (59, 87, 92)
502/1 Bidens cernua 80, 84, 98, 101
502/2 tripartita 51, 73, 74, 79, 80, 86, 92, 101
*503/1 Galinsoga parviflora 31, 80, [6, 14, 29, 38, 49, 51, 66]
506/3 Senecio erucifolius [48, H.31]
*506/4 squalidus 78, H.7
506/6 sylvaticus 99, H.4, 6, 15, 17, 28, 37
*506/13 fluviatilis 74, 91, 92, 106, H.14, [H.15]
506/16 palustris [53]
506/17 integrifolius 54
*506/18 cineraria [18]
*507/1 Doronicum pardalianches 76, 89, 99, 102, H.7, 10, 12
*507/2 plantagineum 86, 89, 91, [35, 37, 63]
509/1 Petasites hybridus 107, H.16

*509/4 fragrans 67, 80, 88
*512/1 Inula helenium 54, 65, 74
512/4 conyza [67]
512/5 crithmoides [25, 27, 44, 53, 54]
513/1 Pulicaria dysenterica 72, 77, 86, 103, 104, 106, H.32, 36
513/2 vulgaris [1, 41, 47, 48, 63, Somerset, H.1]
514/1 Filago germanica 63, 79, 80, 86, 87, 97, 101
514/2 apiculata [14]
514/3 spathulata 25, (83), [S]
*514/4 gallica (48), [83]
514/5 minima 31, 45, 76, 79, 84, 86, 87, [H.28–33]
515/1 Gnaphalium sylvaticum 1, 2, 79, 99
515/2 norvegicum [88, 89, 109]
515/4 uliginosum 109
*516/1 Anaphalis margaritacea 79, 83, 91, [18, 52]
517/1 Antennaria dioica 43, 45, 51, 56, 71, 79, 99, H.3, 5, 6, 32
518/1 Solidago virgaurea H.24, 30, 37
519/1 Aster tripolium 83, 93, [42, 43, 77, 81]
520/1 Crinitaria linosyris (32), [50, 60]
*522/1 Conyza canadensis 49, 69, 86

525/1 Eupatorium cannabinum 80, 86, 92, 108, H.32, 37
526/2 Anthemis cotula 44, 90, H.6, 40, [112]
526/3 arvensis 51, 66, 67, 74, 76, 78, 87, 99, 109
527/1 Chamaemelum nobile 44, 71, 104, (51, 79, 81, 109, 112), [30, 35, 47, H.23]
528/3 Achillea ptarmica H.6, 11, 13
529/1 Otanthus maritimus 49, [13, 41, 48]
531/1 Tripleurospermum maritimum H.37
532/1 Matricaria recutita (86, 92, 99, 100, 107, H.37, Galway)
533/1 Chrysanthemum segetum 78
533/5 vulgare 79, H.19
535/1 Artemisia vulgaris 78
535/6 absinthium [111]
535/7 maritima 50, 51, 67, 93, 99, (2, 46, 75), [33]
535/8 campestris (17, 38)
537/1 Carlina vulgaris 79
539/1 Carduus tenuiflorus 12, 73, 79, 80, 84, 86, 93, 99, H.26
539/3 nutans 67, 79, 86, 90, 91, (75, 104), [105, H.1, 3, 19, 26]
539/4 acanthoides 79, 87, 104, [H.5]
540/1 Cirsium eriophorum 18, 44, 45, 47, [1, 2, 13, 14, 48, 60, 69, 70, 77, 86, 87, 90, 92, 98, 99]
540/7 heterophyllum 52, [41]
540/8 dissectum 48, H.6, 37, [50]
540/9 tuberosum (33)
*541/1 Silybum marianum 48, 60, 93, 99, Perthshire, H.11, [74]
542/1 Onopordum acanthium 44–7, 60, 79, 84, 86, 106
543/1 Saussurea alpina 77, 102, [H.10]
544/1 Centaurea scabiosa (76, 92, 98–100), [71]
544/3 cyanus 65, 73, 74, 78, 79, 84, 86, 90, 98, 99, 108, 109, H.3, 4, 37, 40
*544/5 jacea 3, 7, 8, 11, 16, 20, 24, 38, 69, Perthshire
544/8 aspera (38)
544/9 calcitrapa 10, 22, 61, 72, 79
*544/10 solstitialis 12, 61
545/1 Serratula tinctoria [111]
546/1 Cichorium intybus 43, 76, 78, 86, 87, 109, [H.33]
548/1 Arnoseris minima 66, 93, 94, [23]
549/2 Hypochoeris glabra 36, 75, 79, 87, 92, 100, [39, H.2, 4, 5, 11, 21]
549/3 maculata 60, 66
550/2 Leontodon hispidus 84, 88, 90, 91, 97, 99, 100, 108, [111]
550/3 taraxacoides 92, 99, [88]
551/1 Picris echioides (74, 79, 85, 92)
551/2 hieracioides 48, 60, 74, [67, H.9, 13]
552/1 Tragopogon pratensis 79, 86, 87, [H.6, 38]
*552/2 porrifolius 40
554/3 Lactuca saligna (6, 36)
555/1 Mycelis muralis 1, 91, 93
559/1 Crepis foetida [11, 17, 23–5, 28, 29, 32]
*559/2 vesicaria 39, [71, 88, 94–6, 98]
559/4 mollis 50, 73, 78, 99, [11, 49]
559/5 biennis [83, 84, 88, 92]
561/1 Baldellia ranunculoides 50, 80, 92, 104, H.4, 5, 12, 34, [8, 89]
562/1 Luronium natans [36, 51, 54, 61, 74, 75, H.2, 9, 29]
563/1 Alisma plantago-aquatica 79, 93
564/1 Damasonium alisma 54, [H.16, 21]
565/1 Sagittaria sagittifolia [1, 48, 49, 85]
566/1 Butomus umbellatus 45, (79, 93), [2, 4, 46]
567/1 Hydrocharis morsus-ranae 49, 51
568/1 Stratiotes aloides (18, 62, 89)
573/1 Scheuchzeria palustris [54, 56, 61, 64, 65, 67, 68]
574/2 Triglochin maritima 76, 92
576/1 Zostera marina 44, 66, 67, 74, 84, 91, 96, 107, 109, H.8, 22, [29, 46]
576/2 angustifolia 104, H.39, 40
576/3 noltii 103, [6, 41, 44, 111]
579/1 Ruppia spiralis 102, [46, 48, 62]
579/2 maritima 108, [50]
580/1 Zannichellia palustris 72, 74, 92, 93, 97, 101, 106
581/1 Najas flexilis (85), [9, 104]
583/1 Tofieldia pusilla 99, 109, [95]
584/1 Narthecium ossifragum [33, 53]

588/1 Convallaria majalis 51, (68, 82, 86, 94, 109)
589/1 Polygonatum verticillatum 72, [21]
589/2 odoratum (9, 20, 25, 27, 28, 40, 47, 70, 92, 102), [8, 12, 16]
589/3 multiflorum 30, 50, (68), [1, 13]
591/1 Asparagus officinalis 51, (32), [46]
592/1 Ruscus aculeatus (73, 76, 100), [88, 109]
593/1 Lilium martagon (41)
594/1 Fritillaria meleagris 28, (53, 54, 75), [2, 62]
*595/1 Tulipa sylvestris 82
597/1 Gagea lutea 13, 92, 96, (58), [14, 41]
598/1 Ornithogalum umbellatum (42, 50, 92, 93, 102)
*598/2 nutans 50, Somerset, [9, 70]
598/3 pyrenaicum [11, 37]
599/1 Scilla verna [54]
600/1 Endymion non-scriptus H.22
602/1 Colchicum autumnale 24, 60, [10, 20, 51, 83, 85, 87, 94, 107, H.8, 21, 37]
603/1 Paris quadrifolia [H.2]
605/1 Juncus squarrosus H.23, [7]
*605/2 tenuis 51, 79, 83, 87
605/4 compressus 75, 100, 103, [41, 48, 72, 73, 80, 89, 102, 105, S]
605/5 gerardii 56, 86, 88, [7, 8, 40, 77]
605/6 trifidus 111
605/8 inflexus 75, 79, 86, 108, 109, [71]
605/12 filiformis 99, [95]
605/13 balticus [73]
605/14 maritimus 50, 76, 86, 96, H.4, [77]
605/15 acutus [1, 5, 6, 9, 18, 25, 33, 51, 54, 69, 73]
605/17 subnodulosus 43, 51, 75, 76, 102, [77, 83, 84, H.5]
605/18 acutiflorus 111
605/20 alpinoarticulatus [27, 41, 87, 90, 102]
605/24 castaneus 96, [99]
605/25 biglumis 87
605/26 triglumis 101
606/2 Luzula forsteri [6, 39, 41, 46, 70, 75, 109]
606/3 sylvatica 52, H.15
*606/4 luzuloides 38, 40, 60, 83
606/6 spicata 109, [49]
606/7 arcuata 89, 90
606/9 multiflora H.31
606/10 pallescens [85, H.2]
607/3 Allium scorodoprasum [1, 9, 37, 44, 48, 60, H.35]
607/5 vineale 79, 93
607/6 oleraceum 44, 45, 74, 81, 84, [30, 85, H.38]
*607/7 carinatum [H.21]
*607/10 triquetrum [13, 21, 27, 70]
611/2 Leucojum aestivum H.7, [29]
612/1 Galanthus nivalis (86, 89, 93)
614/1 Narcissus pseudonarcissus 43, (74, 76, 86, 87, 89, 92, Kerry)
615/1 Sisyrinchium bermudiana 1
616/3 Iris foetidissima (74, 100)
*618/1 Crocus nudiflorus [14, 34, 65]
*618/2 purpureus 5, 53, 54, [11]
619/1 Romulea columnae [41]
624/1 Cephalanthera damasonium [48, 55, 57, 62, 69, 70]
624/2 longifolia 15, 57, 67
625/1 Epipactis palustris 43, [21, 38, 46, 73, 104]
625/2 helleborine 74, 90, 100
625/3 purpurata [3, 4, 9, 10, 18, 27–9, 39, 46, 49, 54, 57, 61–5, 69, 70, 84, 89]
625/7 atrorubens 65, [3, 4, 6, 13, 17, 28, 34–6, 50, 102]
628/1 Listera ovata 106, H.3
628/2 cordata [3, H.17]
629/1 Neottia nidus-avis 1, 44, 86, 91, 92, 94
631/1 Hammarbya paludosa 39, 57, 60, 89, 102
632/1 Liparis loeselii [54, H.11]
634/1 Herminium monorchis [21, 31, 32, 35]
635/1 Coeloglossum viride 18, 42, 79, 84, 87, 99, H.22, 30
636/1 Gymnadenia conopsea 31, 43, 79, 84, H.6
637/1 Leucorchis albida 51, 79, 99, [11, 15, 68, 112]
638/1 Platanthera chlorantha 79, 82, 92, 109
638/2 bifolia 43, [29, 57]
639/1 Neotinea intacta [H.27]
640/1 Ophrys apifera [72–7]
640/2 fuciflora [9, 16, 17, 26]
640/3 sphegodes [7, 30, 40]
640/4 insectifera [105]
641/1 Himantoglossum hircinum [34, 56, 57, 70, H.9]
642/1 Orchis purpurea [21, 24, 53, 54, Cork]
642/2 militaris [15, 16, 19]
642/4 ustulata [H.5, 21]
642/5 morio 43, [2]
643/2 Dactylorchis maculata H.19, 22, 24, 29
643/3 incarnata 38, [37]
643/4 praetermissa [47, 69, 70, 74, 79, 81, 89–92, 96, 104, 105, 110–12, H.3, 9, 16, 21, 34]
643/5 purpurella 44, H.5, 6, 17, 22, 29, 31, 32, 34, [41]

644/1 Aceras anthropophorum [9, 22, 24, 33, 54, 63]
645/1 Anacamptis pyramidalis 73
*646/1 Acorus calamus 65, 72, 86, 93, [8, 84]
649/1 Arum maculatum 79, 102
650/1 Lemna polyrhiza 44, 45, 49, 51, 52, 84
650/2 trisulca 45, 49, 75
650/4 gibba 75, 86, 109, [49, 52, 85, H.9]
652/2 Sparganium emersum 79, 99, H.6, 10, 12, 13, 15, 29
652/3 angustifolium [17, 27, 28, 34, 44, 45, 52, 54, 67, 71, 72, 74, 75, 79–86, 91, 93, 94]
652/4 minimum 79, 111, H.4, [33–5, H.40]
653/1 Typha latifolia 79, 107
653/2 angustifolia 68, [48]
654/2 Eriophorum gracile [37]
654/3 latifolium 51, 60, 81, 99, [7, 30, 53]
654/4 vaginatum 52, [7, 10, 15, 22, 23, 33]
655/2 Scirpus cespitosus [15]
655/3 maritimus 86
655/4 sylvaticus 44, 45, 79, 98
655/5 holoschoenus 11, S, [37]
655/6 triquetrus [14, 27, 55]
655/8 lacustris 79, 88, 99, H.4, [1]
655/9 tabernaemontani 70, 72, 74, 104, 106, H.5
655/10 setaceus 51. H.23
655/11 cernuus 51, 97, 106
655/12 fluitans 51, 99, 107, 109, H.29, [H.23]
656/1 Eleocharis parvula [8, H.29]
656/2 acicularis 26, 43, 51, 66, 74, 77, 78, 87, 96, [71, 108, S]
656/3 quinqueflora 45, 50, 79, 82, 87, H.37
656/4 multicaulis 44, 51, 65, 78, 79, 81, 86, 93, 109, H.6, [30, 31, 35, 47, 87]
656/6 uniglumis 45, 83, 88, 91, H.8, 21
657/1 Blysmus compressus [H.3]
657/2 rufus 50, 76, 77, 83, [25, 27, 46]
658/1 Cyperus longus [27, 39]
658/2 fuscus [62]
659/1 Schoenus nigricans 50, 57, 90, H.4, 6, 12, 29, 37, [63]
660/1 Rhynchospora alba 50–2, 77, 86, 106, H.4, 5, [23, 26, 61]
660/2 fusca [2, 4, 40, 42, 62, 98, H.38]
661/1 Cladium mariscus 46, 72, [51]
663/1 Carex laevigata 76, 78, [23, 33]
663/2 distans 86, [112]
663/3 punctata 52, [70]
663/4 hostiana 15, H.5, 6, 29, 32
663/5 binervis H.22
663/7 lepidocarpa 1, 2, 10, 18, 24, 31, 38, 39, 44, 46, 58, 74, 84–6, H.21, 32, 37
663/11 extensa 50, 99, [53, 83, 92]
663/12 sylvatica H.7
663/13 capillaris 104, 112, [87]
663/15 pseudocyperus 65, 66, Cornwall, [49, 102, H.21]

663/16 Carex rostrata 53
663/17 vesicaria 79, 85, 100, 105, 109, [30]
663/18 stenolepis [34]
663/19 saxatilis 72, 104
663/20 riparia 73, 78, 96, 105, 106, 109, [48]
663/21 acutiformis 78, 82, 99, 100, 106, 109, [93]
663/22 pendula 84
663/23 strigosa H.4, [1, H.27]
663/24 pallescens 52, 78
663/25 filiformis Sussex
663/27 vaginata 86, 95, 99
663/28 limosa [37, 84]
663/29 paupercula 66, [9]
663/32 hirta 87, 96, H.16
663/33 lasiocarpa 60, 81, H.36
663/34 pilulifera H.4–6, 29
663/36 caryophyllea H.5, [111]
663/37 montana [15, 17]
663/39 digitata [3]
663/46 elata H.12, [1–5, 7–9, 14, 17–19, 24, 33, 34, 37, 38, 41, 67, 73, 74, 76, 77, 83–5, 95, 98, 99, H.21]
663/47 acuta 49, 50, 74, 98, 99, 101, 109, [1–4, 102, H.35]
663/48 aquatilis 84
663/52 bigelowii 60, 74, [80]
663/54 paniculata 84, 86, 91, 92, 96
663/55 appropinquata 78, [57]
663/56 diandra 94, 97, 108, [3]
663/57 otrubae 77, 78, H.25, 26, 33, 37, [76, 83]
663/58 vulpina 7, 14
663/60 disticha 48, 84, 93, 107, H.4, [H 3]
663/61 arenaria 76, H.4
663/62 divisa 44, 48, 50, 67, 68, 76, 90, [62]
663/64 maritima [70, 71]
663/65 divulsa 42, 43, 47, 48, 52, 53, 57, 58, 66, 67, 69
663/66 polyphylla 52
663/67 spicata 46
663/68 muricata 32, [33]
663/69 elongata [27, 29, 37, 62, H.40]
663/71 remota H.24, 31
663/72 curta 51, 82, [4, 26, 34, 55]
663/80 pulicaris 84, H.24
663/81 dioica 51, 84
669/2 Glyceria plicata 82, 84, 86, 108, H.5, 10, [98, 100]
669/4 maxima 48, 52, 74, 101, 103, [2]
670/2 Festuca arundinacea 50, 87, 97, 107, H.32
670/3 gigantea 87, [S]
670/4 altissima [2, 3, 7, 16, 39, 101]
*670/5 heterophylla 12, 67, [7]
670/7 juncifolia 90
*670/11 longifolia 110
*671/3 Lolium temulentum 31, 74, 75, 92, 101, H.11–15, 17, 18, 23–5, 28, 29, 31–3
672/1 Vulpia membranacea 5, (42)
672/2 bromoides 78, 79, 87, 98, 99, 112
672/3 myuros 44, 50, (67, 73, 77, 78, 81, 90, 101)
672/5 ambigua [30, 41]
673/1 Puccinellia maritima 86, [H.36]
673/2 distans 50, 51, 73, 75, 81, 86, 87, 92, 106, 109, H.3, (79), [7, 20, 30]
673/4 fasciculata 85, 90, [1, 35]
673/5 rupestris 50, (75, 83, 86, 91), [8, H.5, 21, 38]
674/1 Catapodium rigidum 87, 91, 92, 106, H.37
674/2 marinum 72, 76, 84, 86, [67, 90]
676/3 Poa bulbosa [2, 16, 21]
676/6 & 8 nemoralis [111]
676/7 glauca [64, 94]
676/9 compressa 106, H.4, [111]
676/14 Poa palustris 38, 39, 92, 110, [66, H.32]
*676/15 chaixii 39, 92, [97]
677/1 Catabrosa aquatica 44, 45, 48, 50, 78–80, 86, 87, 96, 99, 105–7
*679/2 Cynosurus echinatus 22, 26, 27, 36, 38, 53, 68, 69, 89, [4, 5, 65 Cork]
680/1 Briza media 76, 96
680/2 minor (39, 51, 59), [6, 37, 58]
681/1 Melica uniflora 78–80, 107, H.3, 23, 25, 26, 31
681/2 nutans [3, 58]
682/1 Sesleria caerulea 42, 96, [1, 40, 105]
683/2 Bromus ramosus 1, H.40, [S]
*683/4 inermis Yorkshire, Lothians
683/5 sterilis 87, 93, 98, [111]
683/6 madritensis (35, 78), [14, 19, 23]
*683/9 tectorum [3]
683/14 racemosus 18, 25, 40, 50, H.8, 21, [43, 44, 47, 72–4, 76–8, 81, 82, 87, 91–3, 96, 97, 105, 107, 109, 111, H.20, 40]
683/15 commutatus 43, 47, 48, 65, 67, 72, 76, 78, 79, 81, 85–8, 91–3, 97–101, 106–8, 111
683/16 interruptus [30, 61]
*683/17 arvensis 74, 92, 109, [10, 41, 50, 57, H.21]
*683/18 secalinus 50, 51, 60, 68, 74–6, 86, 91, 96–102, 106–8, 111, [42, 43, 45, 112]
684/2 Brachypodium pinnatum 26
685/4 Agropyron pungens 59, [82, 85, 91, 93, 103]
685/5 junceiforme 92, 97, H.31, [34]
686/1 Elymus arenarius 97, [12, S, H.3, 5]
687/1 Hordeum secalinum 50, 65, 86, [75, 83, H.27]
687/2 murinum 74, 79, 86, 109, [H.4]
687/3 marinum (1, 2), [17, 48, 58, 90, 94, H.5, 21]
688/1 Hordelymus europaeus [7, 8, H.21]
689/1 Koeleria cristata 79, 99
689/2 vallesiana (83)
691/1 Trisetum flavescens 74, 97, 100, 101, 108, 111, H.27, 29, 30, [110]
*692/1 Avena fatua 45, 46, 49, 75, 79, 87, 89, 91, 92, 98
*692/3 strigosa 52, 102, Perthshire, [29]
693/1 Helictotrichon pratense 35, [4, 71]
693/2 pubescens 87, 99, H.3, 28
696/2 Deschampsia alpina 102, 108
696/3 flexuosa H.15, 23
696/4 setacea 58, 95, 105, [97]
699/1 Ammophila arenaria 86, H.8, [26, 29, 42, 77]
700/1 Calamagrostis epigejos 44, 86, 110
700/2 canescens 37, [3, 6, 9, 30, 33–5, 41, H.38]
700/3–4 neglecta agg. Aberdeenshire, [62]
702/1 Apera spica-venti 39, 50, 70
703/1 Polypogon monspeliensis (20, 24, 39, 82, 86, 91)
704/1 Mibora minima [18, H.16]
705/1 Gastridium ventricosum (Lothians), [25, 50, 51]
707/3 Phleum alpinum (11, 22), [99, H.16]
707/4 phleoides [33]
707/5 arenarium 68, 73, 86, [34, 83, H.4]
708/1 Alopecurus myosuroides 36, 49, 66, 68, 74, 78, 86, 91, 96–8, 108, 109, H.21
708/2 pratensis H.26, 27
708/4 aequalis 50, 59, [3, 5, 6, 8, 34, 41]
708/5 bulbosus 44, 45, 108, 109
708/6 alpinus [88]
709/1 Milium effusum 74, 96, [100, H.22]
711/1 Hierochloe odorata [72, 89]
*712/2 Anthoxanthum puelii 1, 76, 80, [14, 92]
*713/2 Phalaris canariensis 99, 100
714/1 Parapholis strigosa 103, H.3, [78, 79, 83, 84, 102]
717/1 Cynodon dactylon (83)
*718/1 Echinochloa crus-galli 39, 66, 79, 83, 86, 92
*720/1 Setaria viridis 41, 49, 51, 52, 78, 83, 86, 90
695/2 Holcus mollis [111]

Appendix III

ACKNOWLEDGEMENTS

COUNTY LISTS

The following supplied more or less complete vice-county lists:

7–8	Wiltshire	J. D. Grose
9	Dorset	R. D'O. Good
15–16	Kent	F. Rose
17	Surrey	Flora Committee
20	Hertfordshire	J. G. Dony
21	Middlesex	D. H. Kent
25–26	Suffolk	Flora Committee
30	Bedfordshire	J. G. Dony
31	Huntingdonshire	J. L. Gilbert
36	Herefordshire	Flora Committee
38	Warwickshire	Flora Committee
39	Staffordshire	E. S. Edees
52	Anglesey	A. D. Q. Agnew and S. W. Greene
53–54	Lincolnshire	Miss J. Gibbons
56	Nottinghamshire	Mr and Mrs R. C. L. Howitt
57	Derbyshire	Flora Committee
61	East Yorkshire	R. D'O. Good and Miss E. Crackles
62 & 65	North Yorkshire	Miss C. M. Rob
71	Isle of Man	D. E. Allen
90	Angus	Miss U. K. Duncan
93	North Aberdeenshire	Flora Committee
95	Moray	Miss M. McCallum Webster
98	Argyll Main	K. N. G. MacLeay
102	South Ebudes	J. K. Morton
H.33	Fermanagh	R. D. Meikle

HELP WITH PARTICULAR SPECIES

The following provided additional data for particular species. Maps of those species marked with an asterisk have been published in the "Biological Flora" series in the *Journal of Ecology* and the year of publication is given in each case. Those marked with a dagger have been published elsewhere and the reference will be found in the Bibliography. A cross indicates species for which there are published papers giving localised distribution data. References to these will also be found in the Bibliography.

15/6	Asplenium viride	C. D. Pigott
21/2	†Dryopteris borreri	Miss P. J. Pugh (1953)
33/1	†Pinus sylvestris	H. M. Steven and A. Carlisle (1959)
45/1	Clematis vitalba	E. Milne-Redhead
46/16	Ranunculus hederaceus	C. D. K. Cook
46/23	baudotii	C. D. K. Cook
56/2	*Nuphar pumila	Mrs Y. Heslop Harrison (1955)
66.	Fumaria species	N. Y. Sandwith
87/1	*Hornungia petraea	Denis Ratcliffe (1959)
90/2	Bunias orientalis	B. M. G. Jones
94/4	*Draba muralis	Denis Ratcliffe (1960)
100/4	Arabis hirsuta	B. M. G. Jones
100/5	brownii	B. M. G. Jones
113/7	*Viola lactea	D. M. Moore (1958)
114.	Polygala species	D. R. Glendinning
117/1	*Tuberaria guttata	M. C. F. Proctor (1960)
118.	*Helianthemum species	M. C. F. Proctor (1956)
123/10	*Silene nutans	F. N. Hepper (1956)
127/1	Dianthus armeria	Miss S. S. Hooper
130/1	Kohlrauschia prolifera	Miss S. S. Hooper
131/11	Cerastium pumilum	E. Milne-Redhead
137.	Minuartia species	G. Halliday
142/1	*Spergula arvensis	J. K. New (1961)
146.	Herniaria species	D. E. Coombe and L. C. Frost
154.	Chenopodium species	J. P. M. Brenan
160.	Salicornia species	P. W. Ball
162/1	Tilia platyphyllos	C. D. Pigott
168/17	†Geranium purpureum	H. G. Baker (1955)
170/2	×Oxalis corniculata	D. P. Young (1958)
170/4	× europaea	D. P. Young (1958)
171/1	Impatiens noli-tangere	D. E. Coombe
171/2	capensis	D. E. Coombe
171/3	* parviflora	D. E. Coombe (1956)
180/1	Frangula alnus	P. W. Richards and others
187/2	Ulex gallii	M. C. F. Proctor
187/3	minor	M. C. F. Proctor
192/8	Trifolium molinerii	D. E. Coombe
192/12	bocconei	D. E. Coombe
192/14	strictum	D. E. Coombe
200/1	†Astragalus danicus	C. D. Pigott (1951)
217/1	†Dryas octopetala	C. D. Pigott (1956a)
229/1	Crataegus oxyacanthoides	A. D. Bradshaw
242/2	Chrysosplenium alternifolium	F. Rose
252/1	†Hippophae rhamnoides	E. W. Groves (1958)
254/13	*Epilobium nerterioides	Miss A. J. Davey (1961)
262.	Callitriche species	J. P. Savidge
285/3	Apium repens	R. D. Meikle
320/2	Polygonum raii	B. T. Styles
325.	Rumex species (rare)	J. E. Lousley
330.	Ulmus species	R. H. Richens
335/3	†Betula nana	Miss A. Conolly (1950)
343/21	†Salix herbacea	Miss A. Conolly (1950)
349/1	*Daboecia cantabrica	S. R. J. Woodell (1958)
353/1	*Arbutus unedo	J. R. Sealy and D. A. Webb (1950)
367/2	*Primula scotica	J. C. Ritchie (1954)
385.	Gentianella species	N. M. Pritchard
388/1	*Polemonium caeruleum	C. D. Pigott (1958)
392/2	Symphytum asperum	A. E. Wade
407/1	*Cuscuta europaea	B. Verdcourt (1948)
430/24	†Veronica filiformis	E. B. Bangerter and D. H. Kent (1957)
434/3	Melampyrum pratense	A. J. E. Smith
434/4	sylvaticum	A. J. E. Smith
445.	Mentha species	R. A. Graham and R. M. Harley
448.	*Thymus species	C. D. Pigott (1955)
462/5	Lamium album	Miss A. Conolly and Mrs D. Walker
485/6	Galium pumilum	K. M. Goodway
485/7	sterneri	K. M. Goodway
503.	†Galinsoga species	W. S. Lacey (1957)
506/4	×Senecio squalidus	D. H. Kent (1956, 1960)
513/2	Pulicaria vulgaris	C. D. Pigott
514/2	Filago apiculata	J. E. Lousley
514/3	spathulata	J. E. Lousley
540/6	Cirsium acaulon	C. D. Pigott (1956b)
544/5	×Centaurea jacea	E. M. Marsden-Jones and W. B. Turrill (1954)
559.	Crepis species	J. B. Marshall
562/1	Luronium natans	A. C. Jermy
564/1	Damasonium alisma	C. D. Pigott
573/1	†Scheuchzeria palustris	W. A. Sledge (1949)
577.	×Potamogeton species	J. E. Dandy and G. Taylor (1938–42)
602/1	*Colchicum autumnale	R. W. Butcher (1954)
625.	†Epipactis species	D. P. Young (1952)

641/1	†Himantoglossum hircinum	R. D'O. Good (1936)
643/6	×Dactylorchis majalis	P. M. Hall (1937)
643/7	† traunsteineri	J. Heslop Harrison (1953)
643.	Other Dactylorchis species	E. Milne-Redhead and V. S. Summerhayes
649/2	*Arum italicum	C. T. Prime (1954)
652.	Sparganium species	C. D. K. Cook
657.	Blysmus species	Mrs G. Crompton
663/10	†Carex serotina	Miss E. W. Davies (1953)
663/38	humilis	D. E. Coombe
663/56	diandra	G. Halliday
675/1	Nardurus maritimus	C. A. Stace
683/3	Bromus benekenii	A. Melderis
692/2	†Avena ludoviciana	Miss J. M. Thurston (1954)
701/1	*Agrostis setacea	R. B. Ivimey-Cook (1959)
715/1	*Nardus stricta	M. J. Chadwick (1960)
716.	†Spartina species	Miss J. M. Lambert (1959)

COUNTY REFEREES

The following checked the field cards for the vice-counties listed:

1 O. V. Polunin
1a (Scillies) J. E. Lousley
2 R. W. David
3 & 4 W. Keble Martin
5 C. C. Townsend
6 N. Y. Sandwith
7 & 8 J. D. Grose
9 R. D'O. Good
10 A. W. Westrup
11 N. D. Simpson
12 E. C. Wallace
13 O. Buckle
14 E. C. Wallace
15 & 16 F. Rose
17 D. P. Young
18 & 19 S. Jermyn and B. T. Ward
20 J. G. Dony
21 D. H. Kent
22 & 23 E. F. Warburg
24 R. A. Graham and R. M. Harley
25 & 26 F. W. Simpson
27 & 28 E. L. Swann
30 J. G. Dony
31 J. Gilbert
32 I. Hepburn
33 C. C. Townsend
34 C. C. Townsend and N. Y. Sandwith
35 A. E. Wade
36 F. M. Day
37 R. C. L. Burges
38 J. G. Hawkes
39 E. S. Edees
40 R. C. L. Burges
41–43 A. E. Wade
44 Mrs I. M. Vaughan
45–47 A. E. Wade
48 P. Benoit
49–52 A. E. Wade
53 & 54 Miss J. Gibbons
55 T. G. Tutin
56 R. C. L. Howitt
57 A. R. Clapham and C. D. Pigott
58 W. D. Graddon
59 & 60 Miss V. Gordon
61 R. D'O. Good
62 Miss C. M. Rob
63 & 64 W. A. Sledge
65 Miss C. M. Rob
66–68 D. H. Valentine
69 G. Wilson
70 Derek Ratcliffe
71 D. E. Allen
72–74 H. Milne-Redhead
75–77 R. Mackechnie
78 P. S. Green
79 & 80 B. L. Burtt
81 P. S. Green
82–84 B. L. Burtt
85 Miss C. W. Muirhead
86 & 87 B. W. Ribbons
88 & 89 J. Grant Roger
90 Miss U. K. Duncan
91 J. Grant Roger
92–95 Miss M. McCallum Webster
96 A. Slack
96b (Nairn) Miss M. McCallum Webster
97 E. C. Wallace
98 K. N. G. MacLeay
99 A. McG. Stirling
100 D. Patton
101 Miss M. H. Cunningham
102 J. K. Morton
103 Miss C. W. Muirhead
104 G. Halliday
105–108 E. C. Wallace
109 Miss M. McCallum Webster
110 J. W. Heslop Harrison and Miss M. S. Campbell
111 I. Hedge
112 D. H. N. Spence
S D. McClintock
H.1–31 D. A. Webb
H.32 J. Heslop Harrison and Miss M. P. H. Kertland
H.33 R. D. Meikle
H.34 J. Heslop Harrison and Miss M. P. H. Kertland
H.35 D. A. Webb
H.36–40 J. Heslop Harrison and Miss M. P. H. Kertland

GENERAL ACKNOWLEDGEMENTS

The following organisations contributed records:

Natural History Societies

Aberdeen Natural History Society
Altrincham Natural History Society
Andersonian Naturalists of Glasgow
Barnsley Naturalist and Scientific Society
Belfast Naturalists' Field Club
Birmingham Natural History and Philosophical Society
Bishop's Stortford and District Natural History Society
Blackburn Naturalists' Field Club
Bournemouth Natural Science Society
Bradford Natural History Society
Bristol Naturalists' Society
Bury Field Naturalists' Society
Cambridge Natural History Society
Caradoc and Severn Valley Field Club
Carlisle Natural History Society
Castleford and District Naturalists' Society
Chepstow Society
Cotteswold Naturalists' Field Club
Craven Naturalists' and Scientific Association
Croydon Natural History and Scientific Society
The Devonshire Association for the Advancement of Science, Literature and Art
Dublin Naturalists' Field Club
Edinburgh Botanical Society
Essex Field Club
Exeter Natural History Society
Hampshire Field Club
Harrogate and District Naturalist and Scientific Society
Haslemere Natural History Society
Herefordshire Botanical Society
Horsham Natural History Society
Isle of Wight Natural History and Archaeological Society
Kettering and District Naturalists' Society and Field Club
Leicester Literary and Philosophical Society
Liverpool Botanical Society
Liverpool Naturalists' Field Club
London Natural History Society
London Region Youth Hostels Association (Field Group)
Loughborough Naturalists' Club
Mid-Somerset Naturalist Society
Montgomeryshire Field Society
Norfolk and Norwich Naturalists' Society
Northamptonshire Natural History Society and Field Club
North Gloucestershire Naturalists' Society
Oswestry Branch, B.E.N.A.
Otley Naturalists' Society
Perthshire Society of Natural Science
Reading and District Natural History Society
Rye Natural History Society
Société Jersiaise
Somerset Archaeological and Natural History Society (Natural History Section)
South Essex Natural History Society
Southampton Natural History Society
South-west Ross Field Study Group
Suffolk Naturalists' Society
Torridge and District Branch, B.E.N.A.
Wakefield Naturalists' Society
West Wales Field Society
Weybridge Natural History Society
Wharfedale Naturalists' Society
Whitby Naturalists' Club, Botanical Section
Wigan and District Field Club
Wild Flower Society
Worcestershire Naturalists' Club
Yorkshire Naturalists' Union

Schools and Colleges

Ackworth School, Yorkshire
Askrigg C. E. School, N. Yorkshire
Babington House School, London, S.E.9
Bartley County Secondary School, Southampton
Batley Primary School, W. Yorkshire
Bedales School, Petersfield, Hampshire
Bede Grammar School, Sunderland
Bedstone School, Shropshire
Beverley Park Camp School, Pateley Bridge, Yorkshire
Bishop's Stortford College, Hertfordshire
Blundell's School, Tiverton, Devon
Blyth School, Norwich
Bootham School, York
Brighton College, Sussex
Bridlington High School for Girls
Brinkley School, Cambridgeshire
Brockhill School, Hythe, Kent
Buckhaven High School, Fife
Cedars School, Leighton Buzzard, Bedford
Clunbury C.E. School, Shropshire
Collyer's School, Horsham, Sussex
County Secondary Grammar School, Newport, Isle of Wight
Dauntsey's School, Devizes, Wiltshire
Doncaster Grammar School, Yorkshire
Dulwich College, London, S.E.21
East End Boys' School, Pembroke
East Suffolk Primary Schools
Essex Primary Schools
Eton College, Buckinghamshire
Fakenham Grammar School, Norfolk
Felsted School, Essex
Friends' School, Saffron Waldon, Essex
Furzedown College, London, S.W.17
Glendale County Secondary School, Wooler, Northumberland
Greek Street High School, Stockport, Cheshire
Gresham's School, Holt, Norfolk
Halstead Grammar School, Essex
Hammond's Grammar School, Swaffham, Norfolk
Hanbury Secondary Modern School, Church Langton, Leicestershire
Harrogate Grammar School, Yorkshire
Harrow School, Middlesex
Hill Brow School, Brent Knoll, Somerset
Hull Grammar School, E. Yorkshire
Hutton Grammar School, Preston, Lancashire
Ilfracombe Grammar School, N. Devon
Ipswich High School, Suffolk
Kennington County Secondary School, London, S.W.9
Kichen Grammar School, Bitterne, Hampshire
Kimbolton School, Huntingdonshire
King Edward's High School for Girls, Birmingham, 15
Kingham Hill School, Kingham, Oxfordshire
King's School, Worcester
King's Warren School, London, S.E.18
Kirkby Lonsdale National School, Westmorland
Lady Eleanor Holles School, Hampton, Middlesex
Lancing College, Sussex
Leighton Park School, Reading, Berkshire
Liskeard Grammar School, Cornwall
Llandysul Grammar School, Cardiganshire
Lord Wandsworth College, Long Sutton, Hampshire
Magdalen College School, Oxford
Manchester High School, Manchester, 14
Marlborough College, Wiltshire
Merthyr Tydfil County Grammar School, Glamorgan
Midhurst Grammar School, Sussex
Minehead Grammar School, Somerset
Neath Grammar School, Glamorgan
Newquay County School, Cornwall
Norfolk Primary Schools
Otterburn County Primary School, Northumberland
Oundle School, Northamptonshire
Palmer's School, Grays, Essex
Penistone Grammar School, W. Yorkshire
Pontygof Girls' School, Ebbw Vale, Monmouthshire
Queen's University Natural History Society, Belfast
Richmond High School, Yorkshire
St Hilda's School, Whitby, N. Yorkshire
St Margaret's P.N.E.U. School, Ludlow, Shropshire
Scalby Modern School, Newby, Scarborough, N. Yorkshire
Scarborough Girls' High School, N. Yorkshire
Sexey's School, Bruton, Somerset
Southport High School for Girls, Lancashire
Stoke Row Primary School, Henley, Oxford
Stranmillis Training College Field Study Society
Thornbury County Secondary School, Gloucestershire
Tiffin Boys' School, Kingston-on-Thames, Surrey
Uppingham School, Rutland
Upton County School, Wiveliscombe, Somerset
Ursuline Convent School, Co. Sligo
Wellingborough School, Northamptonshire
Wellingborough Grammar School, Northamptonshire
Wellingborough High School, Northamptonshire
Whittingham C.E. School, Alnwick, Northumberland
Winchester College, Hampshire
Withycombe C.E. School, Somerset

Other organisations

Field Study Centres
    Dale Fort, Pembrokeshire
    Flatford Mill, Suffolk
    Juniper Hall, Surrey
    Malham Tarn, Yorkshire
    Preston Montford, Shropshire
    Slapton, Devon
Officers of the National Agricultural Advisory Service

INDIVIDUAL CONTRIBUTORS

W. L. Abbott, Mr F. W. Adams, The Revd Canon J. H. Adams, Mr. K. J. Adams, Dr E. M. Adcock, Mr A. D. Q. Agnew, Dr J. K. Aiken, Mr J. Ainsworth, Mr J. R. Aitken, Miss P. Alexander, Mr D. E. Allen, Miss G. E. Allen, Mr K. G. Allenby, Miss R. Allinson, Mrs E. L. Almond, Mr A. H. G. Alston, Dr K. L. Alvin, Mr F. Ambrose, Miss S. R. Amner, Mr D. J. Anderson, Mrs E. Anderson, Miss M. C. Anderson, Mr C. E. A. Andrews, Mr M. V. Angel, Dr T. H. Angel, Mrs A. B. Angus, Mr J. Anthony, Mrs J. Appleyard, Miss K. M. Archard, Miss M. F. Archer, Mrs B. Archibald, Miss P. R. K. Armitage, Mr M. A. Arnold, Mr G. M. Ash, Miss M. Ash, Mrs G. Ashton, Mr D. L. Ashworth, Mr J. Ashworth, Mrs D. Aston, Mr R. S. Atkinson, Mrs B. Auld, Mr R. A. Avery, Mr P. G. Awcock.

Capt. D. H. L. Back, Miss H. Baelz, Mrs G. D. Baigent, The Revd Dr D. S. Bailey, Mr G. S. Bailey, Miss E. C. Bain, Miss M. Baker, Mr A. Ball, Mrs M. D. Ball, Dr P. W. Ball, Miss R. M. Ball, Mr J. O. Ballard, Mr E. B. Bangerter, Miss S. M. Banks, Mr S. D. Bannister, Mr P. J. T. Barbary, Mrs Barker, Miss J. Barker, Mrs J. E. Barker, Miss L. Barker, Mr P. A. Barker, Mr R. W. Barker, Mr W. G. Barker, Mr F. C. Barnes, Miss R. M. Barnes, Miss M. E. Barnsdale, Mr W. M. M. Baron, Miss E. M. Barraud, Mr C. P. Barrett, Mrs J. H. M. Bartlett, Mr D. D. Bartley, Miss F. M. Barton, Col. H. R. Barton, Mr T. H. C. Bartrop, Mr E. B. Basden, Miss M. E. Bastow, Lady Olivia Bates, Mr S. Batey, Mr D. J. Bauer, Miss D. Baylis, Miss J. M. Beadon, Mrs. L. W. Beasley, Miss E. P. Beattie, Mr K. A. Beckett, The Revd A. J. C. Beddow, Miss G. Bell, Mr T. M. Bell, Mr M. Bendix, Mr H. Bendorffe, Mrs M. W. Benn, Mr P. M. Benoit, Lieut-Col. C. J. F. Bensley, Mr B. Bentley, Mr F. Bentley, Miss P. M. Bevan, Dr Biggar, Miss E. I. Biggar, Miss M. Biller, Miss M. B. Bing, Mr F. J. Bingley, Mr J. D. Birchall, Mr A. J. Bird,

Mr E. L. Birse, Mr O. N. Bishop, Miss D. B. Blackburn, Dr K. B. Blackburn, Miss P. Blackhall, Miss F. M. Blackhurst, Dr H. Blackler, Miss E. M. Blackwell, Mrs K. M. Blades, Miss N. M. Blaikley, Mrs P. Bland, Mr D. W. Bloodworth, Mrs M. L. Bolitho, Mr W. A. Bollard, Mrs A. V. Bolton, Mrs G. F. Bolton, Miss K. Bolton, Mr C. J. Bond, Mr R. A. Boniface, Mr H. W. Boon, Miss B. Booth, Miss E. M. Booth, Mr B. N. Boothby, Mr S. T. Bormond, Mr T. A. Bowbeer, Dr H. J. M. Bowen, Miss S. Bower, Miss J. Bowman, Mr R. P. Bowman, Miss M. P. Boycott, Mrs A. K. Boyd, Mr H. J. D. Boyd, Mr J. Boyd, Mr R. K. Braddon, Miss V. Bradley, Mr A. D. Bradshaw, Mr E. Bradshaw, Dr M. E. Bradshaw, Mr J. P. M. Brenan, Mrs M. A. Brewins, Lady Anne Brewis, Mr J. N. Brierly, Dr D. Briggs, Mrs M. Briggs, Miss M. A. Briggs, Mr D. Brightmore, Mr J. Bromwich, Mr B. S. Brookes, Mr H. A. Brookman, Miss B. Brown, Mrs E. D. Brown, Mr G. M. Brown, Mr J. J. Brown, Miss M. I. Brown, Dr N. Brownbridge, Mr R. K. Brummitt, Mr J. P. Brunker, Mr C. J. Bruxner, Miss J. Buchanan, Mr C. Bucke, Mr O. Buckle, Mr K. Budd, Mr A. L. Bull, Mr A. T. Bull, Miss F. M. Bull, Mr J. G. Bull, Mr K. E. Bull, Miss E. R. Bullard, Mrs D. E. Bunce, Mr T. F. Buntin, Mr B. T. Bunting, Mr W. Bunting, Mr D. G. Burch, Miss C. Burchardt, Miss P. H. Burford, Dr R. C. L. Burges, Mr I. H. Burkill, Mr K. F. P. Burkitt, Sir David Burnett, Mr J. Burns, Jnr., Mr J. S. Burns, Miss B. A. Burrough, Mr E. A. Burrows, Mr J. S. Burton, Mr R. F. Burton, Mr B. L. Burtt, Miss E. A. Bush, Dr R. W. Butcher, Mrs E. Butler, Miss G. Butler, Miss J. D. Butler, Mrs Y. M. Butler, Mr A. Butterfield.

Miss D. A. Cadbury, Mr J. Cadbury, Mrs C. M. A. Cadell, Mr J. R. Cadman, Mr A. G. Cadogan, Dr H. L. Caldwell, Miss M. W. Caldwell, Dr E. O. Callen, Miss K. Calverley, Mr W. D. Calvert, Mr W. V. Calvert, Dr A. J. Campbell, Mrs G. Campbell, Miss M. Campbell, Miss M. S. Campbell, Mr J. F. M. Cannon, Mr K. Scott Cansdale, Mrs E. N. Cardo, Mr J. Carlyle, Miss D. Carr, Mr J. W. Carr, Mr E. N. Carrothers, Miss V. M. Caswall, Mr R. G. Cave, Mr M. J. Chadwick, Mr T. Chadwick, Mr D. Chamberlain, Mr E. Chambers, Miss M. Chambers, Mrs A. H. Chandler, Mr J. H. Chandler, Mr G. M. Chapman, Professor V. J. Chapman, Mr S. G. Charles, Mr A. O. Chater, Dr E. H. Chater, Mr D. M. Cheason, Mr E. Chicken, Miss L. M. Child, Mr D. M. Chillingworth, Mr L. J. Churchill, Professor A. R. Clapham, Mr C. Clapham, Mr M. C. Clark, Dr W. A. Clark, Mrs D. A. Clarke, Mr J. H. Clarke, Mr J. W. Clarke, Mr R. Clarke, Mr R. A. R. Clarke, Mrs R. R. Clarke, Mr S. J. Clarke, Mr D. V. Clish, Dr R. S. Clymo, Miss L. E. Cobb, Miss A. W. Cochrane, Mrs O. Cochrane, Mr M. H. Cocke, Lieut-Col. J. Codrington, Miss C. I. Coe, Mrs M. C. Cohen, Mr J. A. Cole, Dr M. J. Cole, Miss I. Coles, Mr M. G. Collett, Mr T. G. Collett, Miss P. Collier, Mr R. E. Collins, Miss E. Collyer, Miss E. R. T. Conacher, Mr W. Condry, Miss A. Conolly, Dr C. D. K. Cook, Mr F. Cooke, Dr D. E. Coombe, Mrs J. I. Coomber, Mrs H. C. Copestake, Dr R. E. C. Copithorne, Mr W. O. Copland, Mrs M. Cordiner, Mrs E. A. Cormack, Mr R. Corner, Mrs W. Corry, Mr V. Cory, Mr C. P. J. Coulcher, Mr D. A. Coult, Mrs J. M. Courtenay, Mr J. B. Coutts, Mr L. A. Cowcill, Mrs K. Coxhead, Mr J. A. Crabbe, Miss E. Crackles, Mr E. Crapper, Mr W. S. Craster, Mr S. A. Craven, Dr A. Crawford, Mr G. I. Crawford, Miss M. Crocker, Mrs H. E. Crockett, Mrs G. Crompton, Mrs M. Crookston, Mr A. C. Crundwell, Mr E. H. Crute, Mr J. Cullen, Mrs M. Cullen, The Revd J. C. Culshaw, Mr B. M. Cunliffe, Miss M. H. Cunningham, Miss C. Curle, Mr A. Currie, Mrs G. C. Curtis, Mrs R. Cusack, Mr J. Cusden.

Miss A. Daisley, Dr D. H. Dalby, Mrs H. Dales, Miss D. G. D'Alton, Mr M. D'Alton, Miss E. Dampier-Child, Mr J. E. Dandy, Miss D. Darlow, Mrs A. Daughty, Dr A. J. Davey, Mrs M. Davey, Mr F. David, Mr R. W. David, Dr J. H. Davie, Mr H. Q. Davies, Mr R. G. Davies, Mr S. J. J. Davies, Mr W. J. Davies, Dr P. H. Davis, Mr T. A. W. Davis, Mr R. Davison, Miss V. Davison, Miss N. Dawson, Mr F. M. Day, Mr G. Day, Mr R. O. J. Day, Mr B. de Graff, Mrs E. Deighton, Miss M. P. Denne, Mr T. E. Dennis, Mrs B. Denny, Miss K. Dent, Mr B. J. Deverall, Miss D. E. de Vesian, Dr A. F. Devonshire, Mrs G. E. Dickinson, Mr J. Dickson, Mr R. Dix, Mr A. J. Dodd, Mr J. D. Dodge, Mr C. J. Dolloway, Mr H. P. Donald, Mrs W. E. Donaldson, Dr J. G. Dony, Mrs E. F. Doran, Miss N. J. Drake, Mr R. Drane, Mr M. Dransfield, Miss G. Drennan, Miss J. M. Drewitt, Mr B. F. T. Ducker, Miss B. Duckers, Mrs D. L. Duckworth, Dr E. Duffey, Mrs J. E. Duncan, Mrs M. J. Duncan, Miss U. K. Duncan, Mrs E. Dunn, Dr M. D. Dunn, Mrs K. Dunstan, Mr T. W. J. D. Dupree, Mr R. C. Dyason, Mr W. J. P. Dyce.

Dr N. B. Eales, Mr L. H. Easson, Miss J. G. Eastwood, Dr N. B. Eastwood, Mrs G. M. Ede, Mr E. S. Edees, Mr G. A. Edgill, Mr M. Edmunds, Mr R. Edwards, Mrs R. S. Edwards, Dr W. J. Eggeling, Mrs F. M. Elder, Lady Alethea Eliot, Mr T. T. Elkington, Mrs E. I. A. Ellershaw, The Revd E. A. Elliott, Mrs G. Elliott, Dr R. J. Elliott, Mr E. A. Ellis, Miss G. Elwell, Mr R. E. Emms, Miss M. Etherington, Mrs S. W. Eunson, Mr A. S. Evans, Dr E. M. Evans, Mrs F. Evans, Mr G. Evans, Mr G. L. Evans, Mr I. M. Evans, Mr J. B. Evans, Mr M. Evans, Mr T. Evans, Mrs B. Everard, Mr G. H. Ewing, Mr A. W. Exell.

Mr J. Fairweather, Miss M. W. Fallows, Mr D. B. Fanshawe, Mrs E. Farnol, Mr D. Farren, Mr J. R. Faulkner, Dr A. F. Fenton, Mr R. E. C. Ferreira, Miss J. M. Ferrier, Mr W. E. H. Fiddian, Mr J. A. Field, Mr J. L. Fielding, Mr F. Fincher, Mr I. C. Fitch, Mr R. S. R. Fitter, Mrs V. H. Fitzgerald, Mrs M. A. Fixsen, Mr B. Flannigan, Mr G. J. Fleming, Mr J. R. Flenley, Mr W. W. Fletcher, Miss D. Fonge, Miss C. Forrest, Mr J. D. Forrest, Mr J. L. Forrest, Mr J. T. Forrest, Miss C. Forsyth, Mrs E. Foster, The Revd A. F. Fountain, Dr B. W. Fox, Mrs W. Frame, Mrs S. Francis, Mr K. A. Franey, Mr B. Frankland, Mr J. N. Frankland, Mr J. Frankton, Miss A. D. French, Lieut-Col. F. H. R. French, Dr L. C. Frost, Miss L. W. Frost, Miss D. M. Frowde, Mr F. G. Fuller, Mr L. Fullerton, Mr J. Fulton, Mrs S. Furse.

Mr J. C. Gardiner, Mr R. J. Garland, Mr G. W. Garlick, Mrs B. E. M. Garratt, Mr D. Garrett, Miss M. Garvie, Mrs E. Gass, Mr A. Gaunt, Dr J. Gay, Dr P. A. Gay, Mr C. J. Gent, Miss M. B. Gerrans, Miss J. Gibbons, Miss A. R. Gibbs, Mrs A. N. Gibby, Mrs G. M. Gibson, Mr J. L. Gilbert, Mr O. L. Gilbert, Mr H. Gilbert-Carter, Mr B. W. Gill, Mr J. A. Gilleghan, Mrs S. Gillett, Mr J. B. Gillies, The Revd R. A. Gilman, Mr J. S. L. Gilmour, Mrs M. R. Gilson, Dr C. H. Gimingham, Mr D. R. Glassford, Mr D. R. Glendinning, Mrs J. Glennie, Miss G. Glover, Professor H. Godwin, Professor R. D'O. Good, Mr A. A. Goodhead, Mr F. D. Goodliffe, Miss C. M. Goodman, Mr G. T. Goodman, Dr K. M. Goodway, Dr A. K. Gordon, Mrs S. Gordon, Miss V. Gordon, Mr M. A. T. Gove, Mr W. D. Graddon, Mrs E. Graham, Mrs E. I. Graham, The Revd G. G. Graham, Mr R. A. Graham, Mr G. G. Grahame, Mr D. R.Grant, Mr T. E. C. Graty, Miss I. F. Gravestock, Mrs D. H. Grayson, Mrs M. Green, Mrs M. R. Green, Mr P. S. Green, Mrs B. M. Greene, Mr S. W. Greene, Miss A. Greenwood, Mrs M. Greenwood, Dr J. Greer, Mr J. B. Gregory, Miss M. Gregory, Capt. O. Greig, Mrs J. Grieve, Miss G. A. Griffiths, Mr J. D. Grose, Mrs R. G. B. Groundwater, Mr E. W. Groves, Miss P. M. Guest, Mrs A. H. Gurney, Miss C. Gurney.

Mr E. C. M. Haes, Mr F. T. Hall, Mr & Mrs P. C. Hall, Mr R. H. Hall, Mr W. A. Hall, Mr & Mrs A. D. Hallam, Dr G. Halliday, Mrs J. Halliday, Mr M. P. Halstead, Mrs A. P. S. Hamilton, Miss M. N. Hamilton, Mr T. S. Hamilton, Miss S. Hamilton-Meikle, Miss M. F. Hancock, Miss M. Handley, Mr W. Handyside, Mr M. K. Hanson, Mrs M. Harber, Mr W. H. Hardaker, Mr R. E. Hardy, Mrs E. N. Harford, Mrs A. Hargreaves, Mr R. M. Harley, Mr C. E. J. Harper, Mr J. Harper, Miss A. S. Harris, Mrs J. A. Harris, Mrs S. Harris, Dr B. D. Harrison, Professor & Mrs J. Heslop Harrison, Professor J. W. Heslop Harrison, Mrs R. N. Harrison, Mrs H. G. Harvey, Mr M. J. Harvey, Dr J. K. Hasler, Mrs A. W. Haslett, Mr J. P. Hatton, Professor J. G. Hawkes, Mr J. H. Hawkins, Mrs N. M. Hawkins, Mr J. A. Hay, Miss M. G. Hay, Mrs D. E. Haythornthwaite, Dr L. A. F. Heath, Mr I. C. Hedge, Mr G. E. C. Hemingway, Miss C. Hemsley, Mr J. H. Hemsley, Mr D. M. Henderson, Mrs P. Henderson, Mrs R. Henning, Miss P. Hensman, Mr I. Hepburn, Mrs M. I. Hepburn, Mr F. N. Hepper, Mr T. F. Hering, The Revd Canon G. A. K. Hervey, Miss R. T. Heward, Mr D. G. Hewett, Dr V. H. Heywood, Mr R. H. Higgins, Mrs R. Hild, Mrs E. M. Hill, Miss P. M. Hill, Mr S. T. Hill,

Mrs M. P. Hill-Cottingham, Miss E. M. Hillman, Mrs J. C. Hilton, Mr G. Hind, Mr D. J. Hinson, Mrs M. C. Hockaday, Miss B. Hodgson, Mr J. Hodgson, Mrs L. Hoggett, Miss S. Holbourn, Miss K. M. Hollick, Mr H. C. Holme, Mr P. H. Holway, Miss B. Hooper, Miss S. S. Hooper, Dr J. F. Hope-Simpson, Mr B. Hopkins, Miss B. A. Hopkins, Mr W. J. Hopkins, Mr J. M. Hopkinson, Mrs S. E. Hopton, Dr D. A. Hopwood, Mrs M. H. Hornby, Mr W. B. Hornby, Mr J. Horsman, Mrs L. M. Hort, Mr E. K. Horwood, Mrs G. I. Hoskins, Mrs F. Houseman, Miss M. A. Howard, Mr C. A. Howe, Miss J. S. Howes, Mr & Mrs R. C. L. Howitt, Dr C. E. Hubbard, Miss J. M. Hubbard, Miss G. M. Hughes, Dr M. G. Hughes, Mr R. Hull, Mr D. C. Hulme, Miss S. A. Humphreys, Mr D. A. J. Hunford, Mrs A. K. Hunt, Mrs D. E. Hunt, Mr J. B. Hunt, Mr P. F. Hunt, Mrs S. Hurd, Miss B. Hurst, Mrs C. C. Hurst, Mr G. R. Hutson, Mrs L. T. Hyde.

Mr B. Ing, Capt. E. R. Ingles, Mrs W. Inglesfield, Miss H. Inglis, Mr H. A. P. Ingram, Miss S. M. Ingram, Mr W. F. Irvine, Miss M. Irwin, Miss E. M. C. Isherwood, Dr R. B. Ivimey-Cook.

Mr A. L. Jackson, Mr K. Jackson, Mr P. G. Jackson, Mrs J. T. Jacobs, Miss P. Jagoe, Mr H. I. James, Miss R. D. Jaques, Miss S. Jaques, Mr N. Jardine, Dr F. M. Jarrett, Mr N. Jee, Mr R. L. Jefferies, Mr C. Jeffrey, Miss R. Jellis, Mr A. C. Jermy, Mr S. T. Jermyn, Miss H. Jerrold, Mrs S. Jewell, Mrs R. V. Johns, Miss M. S. Johnson, Miss F. K. Johnston, Mr T. O. Johnston, Mr M. F. Johnstone, Mr A. S. Jones, Mr A. Vaughan Jones, Mr B. D. Jones, Mr B. M. G. Jones, Dr D. G. Jones, Dr E. W. Jones, Mr H. S. Jones, Mrs K. Jones, Dr M. D. G. Jones, Mr T. W. W. Jones, Mr K. Joy, Mr B. E. Juniper.

Mr J. M. Kahn, Mr S. L. M. Karley, Mr R. Kaye, Mrs G. D. Keech, Mrs A. Kelham, Mrs C. Kendon, Mr F. M. Kendrick, Miss A. Kennedy, Mr A. Kenneth, Mr D. H. Kent, Miss M. B. Kent, Miss M. P. H. Kertland, Mr D. Keys, Mr L. N. Kidd, Mr J. A. Kiernan, Dr B. A. Kilby, Miss R. Kilby, Mr & Mrs D. E. Kimmins, Mrs E. L. King, Miss J. Kirk, Mr F. W. Knaggs, Miss B. A. Kneller, Dr J. T. H. Knight, Mr M. E. Knight, Mr R. Knowles, Mr R. B. Knox.

Dr W. S. Lacey, Miss J. Lamb, Dr J. M. Lambert, Mr R. A. Langdale-Smith, Miss N. E. Langley, Mr C. Langridge, Mr D. A. Lannon, Mr L. Larsen, Mr J. R. Laundon, Mr G. F. Lawrence, Mr I. C. Lawrence, Mr G. E. Laycock, Mr F. Leach, Miss V. M. Leather, Mr C. H. Lee, Mr A. W. Leftwich, Lady Edith Legard, Miss E. Lemon, Lady Lennard, Mrs F. Le Sueur, Dr B. J. Levy, Mr H. R. Lewis, Miss I. D. Lewis, Mrs L. Lewis, Mr R. Lewis, Mr W. D. Linton, Mrs M. Little, Mr K. G. Littlejohn, Mrs E. R. Littlewood, Miss A. Livesey, Miss S. Livingstone, Mr P. S. Lloyd, Miss S. F. Lloyd, Mrs N. Lockhart-Mure, Mrs Long, Mr A. G. Long, Miss D. A. C. Long, Mr R. A. Long, Miss C. E. Longfield, Mr R. E. Longton, Mr J. W. Longworth, Mr J. E. Lousley, Mr P. F. Lumley, Miss F. E. Lutener, Dr A. G. Lyon, Dr A. M. Lysaght.

Miss C. McAlister, Mr D. McClintock, Miss J. McClure, Miss A. M. McCosh, Mr D. J. McCosh, Miss M. M. Macdonald, Mr W. A. B. Macdonald, Miss B. L. McFarlane, Mr J. McGrath, Mrs M. MacInnes, Mrs F. M. McIver, Mr J. R. Mackay, Miss M. McKechnie, Mr R. Mackechnie, Mr G. C. Mackenzie, Mrs G. Mackie, Mrs M. P. Mackie, Professor K. N. G. MacLeay, Mr G. MacLennan, Mr A. MacLeod, Dr A. M. Macleod, Mr R. MacLeod, Mr R. D. Macleod, Miss C. Macmahon, Mrs N. McMillan, Miss V. J. Macnair, Mr I. H. McNaughton, Dr J. McNeill, Dr P. Macpherson, Dr D. N. McVean, Mr I. M'Whan, Mr H. Mace, Mr L. Magee, Mrs M. B. Mallinson, The Revd T. Maloney, Miss A. M. Maltby, Mr S. A. Manning, Miss A. L. Marchant, Mr L. J. Margetts, Mr M. J. Marriott, Miss P. Marsh, Mr J. B. Marshall, Miss A. Martin, Mr J. Martin, The Revd W. Keble Martin, Mr W. R. Masefield, Mrs K. M. Mason, Mr G. A. Matthews, Mr R. F. May, Mr R. Maycock, Mr H. T. Mayo, Mr T. F. Medd, Miss H. D. Megaw, Miss S. Megaw, Mr R. D. Meikle, Professor M. F. M. Meiklejohn, Dr A. Melderis, Dr R. Melville, Miss M. J. Mence, Mr G. Messenger, Mr H. Meyer, Miss B. Miall, Lady Miles, Mr B. Miles, Mr G. G. H. Miller, Mr G. R. Miller, Mr H. Miller, Miss H. M. Miller, Dr J. N. Mills, Mr A. T. Milne, Mr G. G. Milne, Dr J. Milne, Mr E. Milne-Redhead, Dr H. Milne-Redhead, Miss M. E. Milward, Mr R. Minor, Mrs N. Mitchinson, Mr H. Molgaard, Mrs E. J. Montgomery, Mr J. McK. Moon, Mr D. M. Moore, Miss F. A. Moore, Mrs L. Moore, Miss B. M. C. Morgan, Mr G. H. Morgan, Mrs M. J. Morgan, Mr L. E. Morris, Mr M. Morris, Mr C. M. Morrison, Miss E. M. Morrison, Dr M. E. S. Morrison, Mr N. R. Morrison, Mrs R. H. Mortis, Mr J. D. Morton, Dr J. K. Morton, Miss M. Morton, Miss A. Mowat, Miss C. W. Muirhead, Miss V. Muller, Mrs I. S. Mulloy, Mrs D. Munro-Cape, Mrs M. V. Murdoch, Mr F. Murgatroyd, Miss V. W. Mugatroyd, Miss R. J. Murphy, Mrs C. W. Murray, Mrs D. Murray, Mr M. Murray, Mrs M. A. L. Murray, Mrs M. S. Myers.

Miss B. Nash, Mrs C. D. Needham, Miss M. Needham, Mr E. C. Neighbour, Dr G. A. Nelson, Mr P. J. M. Nethercott, Mrs M. Neville, Miss B. Nevinson, Dr J. Newbould, Dr P. J. Newbould, Miss S. Newlands, Miss S. E. Newton, Miss I. G. Nicholson, Mr E. Nimmo, Miss E. R. Noble, Mr C. R. Nodder, Mrs J. M. Nolder, Miss G. Norman, Dr L. G. Norman, Mrs M. Norman, Miss M. M. Norman, Mr P. R. Norman, Mr R. F. Norris, Dr V. Norris, Miss D. E. North, Mr M. Northway, Mr A. P. Norwood.

Mr T. G. & Miss S. Odell, Mrs J. E. Ogston, The Revd Father P. H. O'Kelly, Miss H. Öpik, Mr E. E. Orchard, Mr P. D. Orton, Mr J. T. Osborne, Mr P. H. Oswald, Miss F. E. Other, Mr J. Ounsted, Miss S. E. Outhwaite, Mr G. Owen, Mr R. E. Owens, Mr J. E. C. Oxenham.

Mr J. E. Page, Mr R. C. Palmer, Mr W. H. Palmer, Mr J. S. R. Pankhurst, Mrs C. B. Parker, Mr H. Parker, Mr P. F. Parker, Mr R. E. Parker, Mrs M. Parkinson, Mrs P. A. Parr, Mrs A. Parris, Miss G. I. Parry, Mrs D. Parsons, Miss N. Parsons, Mrs F. Partridge, Dr J. S. Pate, Mr A. Paterson, Mrs E. F. Paterson, Mrs D. Paton, Mrs P. C. Paton, Dr D. Patton, Mrs V. N. Paul, Miss E. M. Payne, Mrs J. Payne, Mr R. M. Payne, Mr B. W. Pearson, Mr E. L. P. Pearson, Dr R. C. Pecket, Miss H. R. Peden, Mrs M. H. Peebles, Mrs J. H. Pennington, Mr T. D. Pennington, Dr M. S. Percival, Mr C. Perraton, Mr S. Perring, Dr C. P. Petch, Mr J. H. G. Peterken, Mrs J. D. Peterson, Mrs M. C. Peterson, Mrs J. Petford, Mr A. Pettet, Miss A. K. Pettit, Mr D. Philcox, The Hon. Gwenllian Philipps, Miss L. G. Phillipps, Miss V. I. Phillips, Mr G. G. Pierce, Dr C. D. Pigott, Mr F. M. Pilkington, Dr M. G. Pitman, Mr P. B. Pitman, Mr H. J. Platt, Mr G. Platten, Miss D. Pocock, Mr R. J. Pollitt, Mr O. V. Polunin, Professor & Mrs M. E. D. Poore, Mrs H. M. Porteous, Mr I. T. Prance, Mr H. M. Pratt, Mr T. F. Preece, Mrs R. Prendergast, Mrs A. E. Pressly, Mrs J. M. R. Price, Mr J. W. Price, The Revd A. L. Primavesi, Dr C. T. Prime, Dr N. M. Pritchard, Mr G. A. Probert, The Revd H. G. Proctor, Dr M. C. F. Proctor, Dr D. C. Prowse, Mr R. D. Pucknell, Mr D. Punter, Miss M. I. Pye, Mrs T. H. Pyke, Mrs J. Pym, Mrs C. Pyrah.

Mr O. Rackham, Miss S. Ramage, Dr D. S. Ranwell, Dr Denis Ratcliffe, Dr Derek Ratcliffe, Mr J. E. Raven, Mrs E. M. Rawson, Miss M. N. Read, Mr R. C. Readett, Mr D. E. Redfearn, Dr A. N. Redfern, Mr E. J. Redshaw, Miss M. Rees, Mr B. G. Reeves, Miss P. Reidy, Mrs M. E. Reis, Mr G. C. Rhodes, Mr B. W. Ribbons, Mr D. J. Rice, Miss A. N. Richards, Miss C. M. Richards, Mrs M. Richards, Professor P. W. Richards, Mr F. D. S. Richardson, Dr J. A. Richardson, Mr W. E. Richardson, Mr R. H. Richens, Miss V. I. Ricketts, Mrs V. J. Ricketts, Mr H. Riley, Miss C. M. Rob, Miss W. Roberts, Miss B. F. Robertson, Miss M. A. Robertson, Mr A. E. Robinson, Mr C. A. Robinson, Mr F. A. Robinson, Mr A. W. Robson, Mrs C. I. Robson, Mr N. K. B. Robson, Dr J. Roche, Capt. & Mrs R. G. B. Roe, Mr J. Grant Roger Mrs B. C. Rogers, Miss C. M. Rogers, Miss S. Rogers, Mr J. Rogerson Miss M. E. Roper, Mr R. Roscoe, Dr F. Rose, Mr R. Ross, Dr E. M. Rosser Mr J. Rossiter, Mr G. L. Rowe, Dr C. H. Fraser Rowell, Mr P. Rowling Mr D. Royle, Mr A. Runnalls, Mrs B. H. S. Russell, Mr A. Russell-Smith Mr E. M. Rutter, Mr M. G. Rutterford, Mr J. S. Ryland.

Dr H. A. Salzen, Mr R. Sandell, Mrs C. Sandeman, Mr N. Y. Sandwith, Mr J. Sankey, Mr B. P. Sargeant, Mr H. B. Sargent, Mrs G. Satow, Miss A. Saunders, Mrs N. Saunders, Mr J. P. Savidge, Mr M. M. Sayer, Miss M. Scannell, Miss C. Schelwald, Miss J. Schofield, Miss S. M. Schofield, Mr T. Schofield, Miss M. A. R. S. Scholey, Miss I. Scholley, Miss D. Schuftan, Mr D. I. G. Scott, Dr E. Scott, Dr G. A. M. Scott, Mr R. Scott, Mr Walter Scott, Mr Wilfred Scott, Mr W. A. Scott, Mr W. K. Scudamore, Mrs M. M. Seabroke, Miss A. Seabrook, Mr A. Searle, Miss A. V. Seaton, Mr M. R. D. Seaward, Mrs P. Seaward, Dr B. Seddon, Mrs J. Sell, Mr P. D. Sell, Mr B. F. C. Sennitt, Dr E. R. Seppings, Mr C. Seymour, Miss C. Shaddick, Mr J. E. M. Shakeshaft, Miss M. R. Shanahan, Mrs D. Sharrock, Mr C. Shaw, The Revd C. E. Shaw, Mr G. A. Shaw, Miss M. S. Shaw, Mr P. G. Sheasby, Mr R. Sheldrick, Mrs I. L. Shelton, Mrs K. Sidaway, Mr & Mrs A. G. Side, Mrs A. M. Simmonds, Mr P. H. Simon, Mr B. Simpson, Mr F. W. Simpson, Mrs M. H. Simpson, Mr N. D. Simpson, Mr M. Sinclair, Mr R. Sinclair, Mr C. Sinker, Mrs A. Skimming, Miss K. M. Skinner, Mr A. Slack, Mrs A. Slansky, Mrs F. M. Slater, Mr R. J. Slatter, Dr W. A. Sledge, Mrs P. Sleigh, Mr P. Sleight-Holme, Mrs L. M. Small, Mr A. Smith, Mr A. E. Smith, Mr A. J. E. Smith, Mr A. Malins Smith, Mr A. R. Smith, Mr D. L. Smith, Mr E. R. Smith, Mr E. S. Smith, Mr G. L. Smith, Mrs G. W. Smith, Miss H. D. Smith, Mrs J. E. Smith, Mrs K. Pickard Smith, Mr L. R. Smith, Mrs M. Smith, Miss M. Smith, Mrs M. M. Smith, Miss N. Smith, Mr P. Bevington Smith, Mr R. Smith, Miss U. K. Smith, Mr R. W. Snaydon, Mrs M. L. Sneyd, Mrs A. H. Sommerville, Miss M. Songhurst, Mr R. Soper, Mr J. E. S. Souster, Mr A. J. Souter, Mrs M. Southwell, Mr W. H. B. Sowerby, Miss I. Spalding, Dr D. H. N. Spence, Mr A. G. Spencer, Mr P. Spicer, Miss M. R. Spiller, Dr T. A. Sprague, Miss A. V. Spratt, Mrs B. Spurgon, Mr B. Spurgon, Mr C. A. Stace, Mr J. Langford Stacey, Dr P. C. Stafford, Miss F. E. A. Stafforth, Dr M. W. Stanier, Miss E. Starr, Dr W. T. Stearn, Mr A. W. Stelfox, Mr J. D. Stephen, Mrs E. K. Stephenson, Mrs G. M. Steuart, Miss A. B. Stevens, Miss E. H. Stevenson, Mrs D. L. Stewart, Miss J. C. Stewart, Mr A. McG. Stirling, Miss M. Stirling, Mr B. Storer, Miss E. E. Storey, Miss A. Straker, Mr J. P. A. Strange, Mr D. T. Streeter, Dr J. M. Stuart, Miss B. M. Sturdy, Mrs M. A. Stutchbury, Dr B. T. Styles, Mr V. S. Summerhayes, Mr J. Summerton, Miss C. A. Swain, Dr G. A. Swan, Mr E. L. Swann, Dr T. D. V. Swinscow, Mrs E. K. Swinton, Miss R. M. Sworder.

Mr J. Taggart, Mr J. M. Taverner, Lady Taylor, Mr F. J. Taylor, Sir George Taylor, Miss I. Taylor, Mr J. Taylor, Mr P. Taylor, Mr H. F. Tebbs, Miss M. I. Tetley, Mr T. F. Teversham, Mr A. Tewnion, Mr W. L. Theobald, Mrs E. Thesigner, Dr E. Thomas, Dr G. E. Thomas, Mr J. D. Thomas, Mr J. F. Thomas, Mrs A. Thompson, Mrs J. Thompson, Mr M. G. V. Thompson, Mr R. Thompson, Mrs A. Thomson, Mr A. L. Thorpe, Miss S. C. Thurman, Miss J. M. Thurston, Mr J. Timson, Mr A. R. Tindall, Mrs J. Tinsley, Miss J. M. Tobias, Mr C. Toft, Mr M. R. Tomlinson, Mr C. C. Townsend, Major E. W. Townsend, Mr P. J. O. Trist, Mr L. C. Tromans, Miss P. A. Tromans, Mrs M. Trost, Mr T. Trought, Mr J. E. Trowell, Miss J. Tucker, Mr W. E. Tucker, Mr W. H. Tucker, Miss E. M. Tully, Miss J. M. Tunstall, Mrs A. R. Turnbull, Mr C. Turner, Mrs M. Turner, Miss M. A. Turner, Professor T. G. Tutin, Dr F. H. Tyrer, Mr P. Tyson.

Mrs E. Underwood, Mr J. G. Urquhart.

Professor D. H. Valentine, Mr F. R. Vane, Mr A. H. Vaughan, Mrs I. M. Vaughan, Miss K. A. Veitch, Mr L. S. V. Venables, Brig. F. E. W. Venning, Mr B. Verdcourt, Mr H. S. Vere-Hodge, Miss N. G. Vernon, Mrs P. G. Vicars-Miles, Miss M. E. Vince, Mr C. J. Vyle.

Mr N. M. Wace, Mr A. E. Wade, Mr G. B. Wakefield, Mr R. W. Wakeley, The Hon. Mrs Waldy, Mr A. D. Walker, Dr & Mrs D. Walker, Miss D. R. Walker, Dr S. Walker, Mr D. W. Wallace, Mr E. C. Wallace, Mr R. Wallace, Mr. T. J. Wallace, Miss L. J. Walley, Mr O. H. Wallis, Mr R. C. Walls, Mr M. L. Walsh, Mr G. Walton, Professor J. R. Walton, Mr T. W. Wanless, Mr P. J. Wanstall, Dr E. F. Warburg, Mr B. T. Ward, Mr C. W. Ward, Mr H. G. Ward, Mr J. F. C. Ward, Professor P. F. Wareing, Mrs S. Warren, Mr W. E. Warren, Mr R. L. Wastie, Mrs D. M. Watson, Dr E. V. Watson, Miss M. Watson, Dr A. S. Watt, Dr G. Watt, Mr J. S. Watt, Miss O. E. Watt, Mr W. Boyd Watt, Mr W. Watts, Dr W. A. Watts, Professor D. A. Webb, Mrs I. Webster, Miss M. McCallum Webster, Mr A. Wegener, Mr J. Weir, Mrs B. Welch, Mr D. Welch, Mr H. Weller, Mr R. V. Wells, Mr E. J. Wenham, Dr C. West, Mr J. B. Weston, Mr A. W. Westrup, Mr J. A. Whellan, Mr D. White, Miss K. White, Mrs K. M. W. White, Mr P. H. F. White, Miss T. White, Dr F. H. Whitehead, Mrs L. E. Whitehead, Dr H. L. K. Whitehouse, Miss M. M. Whiting, Dr T. C. Whitmore, Miss J. M. Whyte, Mr E. D. Wiggins, Miss A. M. Wigham, Mrs C. A. Wilkins, Mr D. A. Wilkins, Mrs B. M. Wilkinson, Miss G. M. Wilkinson, Mr P. R. Wilkinson, Mrs V. M. Wilkinson, Mrs D. Will, Mr & Mrs J. E. Willé, Mr J. T. Williams, Mr J. W. Williams, Mrs L. M. Williams, Mrs M. R. Williams, Dr W. B. Williams, Miss J. D. Williamson, Dr A. J. Willis, Miss J. C. N. Willis, Mr J. H. Willis, Mr P. W. Wilmot-Dear, Mrs D. P. Wilson, Mrs E. Wilson, Mr G. Wilson, Mrs J. Y. Wilson, Miss M. C. E. Wise, Mr J. R. Ironside Wood, Mrs M. Woodall, Dr S. R. J. Woodell, Mr M. S. Woodgates, Mrs F. L. Woodman, Mrs D. V. G. Woods, Lady Woodward, Mr E. V. Wray, Dr F. R. Elliston Wright, Mr J. D. Wright, Miss P. M. Wright, Mr W. S. Wright, Mr T. C. Wrigley, Miss M. M. Wynn.

Mrs A. Yendell, Dr P. F. Yeo, Miss B. E. Young, Miss B. J. Young, Dr D. P. Young, Mrs H. J. Younger, Mr G. Youngs.

Mrs R. Ziman.

Bibliography

Baker, H. G. (1955). *Geranium purpureum* Vill. and *G. robertianum* L. in the British flora. I. *Geranium purpureum. Watsonia*, **3**, 160–7.

Bangerter, E. B. & Kent, D. H. (1957). *Veronica filiformis* Sm. in the British Isles. *Proc. B.S.B.I.*, **2**, 197–217.

Bennett, A., Salmon, C. E. & Matthews, J. R. (1929–30). Second supplement to Watson's Topographical Botany. *J. Bot. Lond.*, **67** & **68** (suppl.).

Bentham, G. & Hooker, J. D. (1924). *Handbook of the British Flora.* Ed. vii revised by A. B. Rendle. Ashford, Kent.

Christy, M. (1884). On the species of the genus *Primula* in Essex. *Trans. Essex Field Club*, **3**, 148–211.

Clapham, A. R., Tutin, T. G. & Warburg, E. F. (1952). *Flora of the British Isles.* Cambridge.

Conolly, A. P. *et al.* (1950). Studies in post-glacial history of British vegetation. XI. Late-glacial deposits in Cornwall. *Phil. Trans. Roy. Soc.* **234**, B, 397–469 [Maps of *Betula nana* L. and *Salix herbacea* L.].

Dandy, J. E. (1958). *List of British Vascular Plants.* London.

Dandy, J. E. & Taylor, G. (1938–42). Studies of British Potamogetons. I–XVIII. *J. Bot. Lond.*, **76–80**.

Davies, E. W. (1953). Notes on *Carex flava* and its allies. I. A Sedge new to the British Isles. *Watsonia*, **3**, 66–9.

Druce, G. C. (1932). *Comital Flora of the British Isles.* Arbroath.

Godwin, H. (1956). *History of the British Flora.* Cambridge.

Good, R. D'O. (1936). On the distribution of the Lizard Orchid (*Himantoglossum hircinum* Koch). *New Phyt.*, **35**, 142–70.

Groves, E. W. (1958). *Hippophae rhamnoides* in the British Isles. *Proc. B.S.B.I.*, **3**, 1–21.

Hall, P. M. (1937). The Irish Marsh Orchids. *Rept Bot. Soc. & Exch. Club*, **11**, 330–52.

Harrison, J. Heslop (1953). Studies in *Orchis* L. II. *Orchis traunsteineri* Saut. in the British Isles. *Watsonia*, **2**, 371–91.

Hoffmann, H. (1860). Vergleichende Studien zur Lehre von der Bodenstetigkeit der Pflanzen. *Ber. Oberhess. Ges. Natur. & Heilk.*, **8**, 1–12.

Hoffmann, H. (1867–9). Pflanzenarealstudien in den Mittelrheingegenden. *Ber. Oberhess. Ges. Natur. & Heilk.*, **12**, 51–60; **13**, 1–63.

Hoffmann, H. (1879). Nachträge zur Flora des Mittelrheingebietes. *Ber. Oberhess. Ges. Natur. & Heilk.*, **18**, 1–48.

Hultén, E. (1950). *Atlas över Kärlväxterna i Norden.* Stockholm.

Ihne, E. (1879). Studien zur Pflanzengeographie. *Ber. Oberhess. Ges. Natur. & Heilk.*, **18**, 49–82.

Kent, D. H. (1956). *Senecio squalidus* L. in the British Isles. I. Early records (to 1877). *Proc. B.S.B.I.*, **2**, 115–18.

Kent, D. H. (1960). *Senecio squalidus* L. in the British Isles. II. The spread from Oxford (1879–1939). *Proc. B.S.B.I.*, **3**, 375–9.

Lacey, W. S. (1957). A comparison of the spread of *Galinsoga parviflora* and *G. ciliata* in Britain. In Lousley (1957), 109.

Lambert, J. M. in Goodman, P. J. *et al.* (1959). Investigations into 'Die-back' in *Spartina townsendii* agg. *J. Ecol.*, **47**, 651–77.

Linton, D. L. (Ed.) (1956). *Sheffield and its Region.* Brit. Assn. Sheffield.

Lousley, J. E. (Ed.) (1951). *The Study of the Distribution of British Plants.* Arbroath.

Lousley, J. E. (Ed.) (1957). *Progress in the Study of the British Flora.* Arbroath.

Marsden-Jones, E. M. & Turrill, W. B. (1954). *British Knapweeds.* London.

Matthews, J. R. (1937). The geographical relationships of the British Flora. *J. Ecol.*, **25**, 1–90.

Matthews, J. R. (1942). The germination of *Trientalis europaea. J. Bot. Lond.*, **80**, 12–16.

Meteorological Office (1938). Air Ministry Publication no. 421.

Meyer, A. (1926). Über einige Zusammenhänge zwischen Klima und Boden in Europa. *Chemie der Erde*, **2**, 209–347.

Moss, C. E. (1914). *The Cambridge British Flora.* Vol. **2**. Cambridge.

Pigott, C. D. (1951). In Lousley (1951), 19. [Map of *Astragalus danicus* Retz.]

Pigott, C. D. (1956a). The vegetation of Upper Teesdale in the North Pennines. *J. Ecol.*, **44**, 545–86. [Map of *Dryas octopetala* L.].

Pigott, C. D. (1956b). In Linton (1956), 86. [Map of *Cirsium acaulon* (L.) Scop.].

Praeger, R. Ll. (1901). *Irish Topographical Botany.* Dublin.

Praeger, R. Ll. (1902). On types of distribution in the Irish flora. *Proc. Roy. Irish Acad.*, **24**, B, 1–60.

Praeger, R. Ll. (1909). *Tourist's Flora of the West of Ireland.* Dublin.

Praeger, R. Ll. (1934). *The Botanist in Ireland.* Dublin.

Pugh, P. J. (1953). The distribution of *Dryopteris borreri* Newm. in the British Isles. *Watsonia*, **3**, 57–65.

Salisbury, E. J. (1932). The East Anglian flora. *Trans. Norf. & Norw. Nat. Soc.*, **13**, 191–263.

Skene, M. (1924). *The Biology of the Flowering Plants.* London.

Sledge, W. A. (1949). The distribution and ecology of *Scheuchzeria palustris* L. *Watsonia*, **1**, 24–35.

Stapf, O. (1917). A cartographic study of the southern element in the British flora. *Proc. Linn. Soc. Lond.*, sess. **129**, 81–92.

Steven, H. M. & Carlisle, A. (1959). *The Native Pinewoods of Scotland.* Edinburgh.

Thurston, J. M. (1954). A survey of wild oats in England and Wales in 1951. *Ann. Appl. Biol.*, **41**, 619–36.

Watson, H. C. [1832]. *Outlines of the Geographical Distribution of British Plants.* Edinburgh.

Watson, H. C. (1835). *Remarks on the Geographical Distribution of British Plants.* London.

Watson, H. C. (1836). Observations on the construction of maps for illustrating the distribution of plants. *Loud. Mag. Nat. Hist.*, **9**, 17–21.

Watson, H. C. (1843). *Geographical Distribution of British Plants.* 'Ed. iii'. London.

Watson, H. C. (1847). *Cybele Britannica.* London.

Watson, H. C. (1873). *Topographical Botany.* London.

Webb, D. A. (1952). Irish plant records. *Watsonia*, **2**, 217–36.

Webb, D. A. (1955). The Distribution Maps Scheme: a provisional extension to Ireland of the British National Grid. *Proc. B.S.B.I.*, **1**, 316–19.

West, G. (1904). A comparative study of the dominant phanerogamic and higher cryptogamic flora of aquatic habit in three lake areas of Scotland [Flora of Scottish Lakes]. *Proc. Roy. Soc. Edin.*, **25**, 967–1023.

West, G. (1910). A further contribution to a comparative study of the dominant phanerogamic and higher cryptogamic flora of aquatic habit in Scottish lakes [Flora of Scottish Lakes]. *Proc. Roy. Soc. Edin.*, **30**, 65–181.

Young, D. P. (1952). Studies in the British *Epipactis. Watsonia*, **2**, 253–76.

Young, D. P. (1958). *Oxalis* in the British Isles. *Watsonia*, **4**, 51–69.

Index

The roman numerals refer to the Introduction; the letters *a* and *b* which follow them refer to the left and right hand column of the page respectively. The arabic numerals give the page on which the map is to be found.